W0106519

The Future of Muon Physics

K. Jungmann V. W. Hughes G. zu Putlitz (Eds.)

The Future of Muon Physics

Proceedings of the International Symposium on
The Future of Muon Physics,
Ruprecht-Karls-Universität Heidelberg,
Heidelberg, Federal Republic of Germany,
7–9 May, 1991

With 266 Figures

Springer-Verlag Berlin
Heidelberg GmbH

Dr. Klaus Jungmann

Physikalisches Institut, Universität Heidelberg
Philosophenweg 12, W-6900 Heidelberg, FRG

Professor Dr. Vernon W. Hughes

Physics Department, Yale University
New Haven, CT 06520, USA

Professor Dr. Gisbert zu Putlitz

Physikalisches Institut, Universität Heidelberg
Philosophenweg 12, W-6900 Heidelberg, FRG

Cover figure shows the Feynman diagram of the muon decay $\mu \to e\nu_\mu + \bar{\nu}_e$
in the standard model of the weak theory

This book originally appeared as a supplement to Vol. 56, 1992, of the journal
Zeitschrift für Physik C – Particles and Fields ISSN 0170-9739

ISBN 978-3-642-77962-6 ISBN 978-3-642-77960-2 (eBook)
DOI 10.1007/978-3-642-77960-2

Originally published by Springer-Verlag Berlin Heidelberg New York in 1992
Softcover reprint of the hardcover 1st edition 1992

57/3140 – 5 4 3 2 1 0 – Printed on acid-free paper

Contents

Preface

This volume comprises a collection of invited papers presented at the international symposium *"The Future of Muon Physics"*, May 7-9 1991, at the Ruprecht-Karls-Universität in Heidelberg. In the inspiring atmosphere of the Internationales Wissenschaftsforum researchers working worldwide at universities and at many international accelerator centers came together to review the present status of the field and to discuss the future directions in muon physics.

The muon, charged lepton of the second generation, was first oberved some sixty years ago. Despite many efforts since, the reason for its existence still remains a secret to the scientific community challenging both theorists and experimentalists. In modern physics the muon plays a key role in many topics of research.

Atomic physics with negative muons provides excellent tests of the theory of quantum electrodynamics and of the electro-weak interaction and probes nuclear properties. The purely leptonic hydrogen-like muonium atom allows tests of fundamental laws in physics and the determination of precise values for fundamental constants. New measurements of the anomalous magnetic moment of the muon will probe the renormalizability of the weak interaction and will be sensitive to physics beyond the standard model. The muon decay is the most carefully studied weak process. Searches for rare decay modes of muons and for the conversion of muonium to antimuonium examine the lepton number conservation laws and new speculative theories. Nuclear muon capture addresses fundamental questions like tests of the CPT theorem. Deep inelastic scattering of muons has enlarged our understanding of the spin structure of the proton. The wide field of neutrino physics is intimately related to muon physics in many repects. Muon catalyzed fusion research is of interest for molecular physics and for future power plants. The growing international μSR community utilizes the muon advantageously as a probe for condensed matter.

At many accelerator centers plans for upgrading existing facilities or for constructing new machines have been developed (KEK, LAMPF, PSI, RAL, TRIUMF). Muon physics will greatly benefit from higher muon fluxes, from intense pulsed sources, from the new techniques for producing slow muons and from many other improvements in the muon beam lines.

In view of the short notice at which the participants could be invited to this meeting and in view of the tight beam time schedules at the accelerators for many of our colleagues we have extended the deadline for this proceedings volume in favour of a more complete overview of the field. The relevance has not suffered in the year since the conference took place, in particular since some authors have included new results and developments in their contributions.

The organization and the local arrangements of the meeting were in the hands of staff members of the IWF and members of the Physics Institute of the university. A. Binding and F. Holzwarth of Springer-Verlag have prepared the TEX macros for the layout of the articles in the style of *Zeitschrift für Physik C*. S. Messner-Eberle and M. Zinser helped many of the authors in bringing their papers into the final format. Financial support was provided by the Deutsche Forschungsgemeinschaft, the State of Baden-Württemberg, the University of Heidelberg, Balzers Hochvakuum GmbH, Coherent GmbH, and IBM Germany. We are very grateful to all of them.

August 1992

K. Jungmann
V.W. Hughes
G. zu Putlitz

Our Favourite – the MUON

Known from a long time,
A lepton of import prime,
Gifted with charge states two,
Its own lepton flavour too,
Born of the charged parent pion,
Its our enigmatic favourite – the busy MUON.

It fleets through its microsecond life,
It streaks through space and time –
Flitting from atom to atom,
Indulging pleasures exotic form,–
Its story in matter when told
Makes history exciting and bold.

An incisive electromagnetic nuclear probe,
It can make a fusion trove.
Changing roles with its neutral ghost,
Weak death is however sadly forced.

Oh that Nature were generous more
With her treasure of muons for us to store –
Borrowing from Grays' Elegy,
The Muon inspires its eulogy :

"Full many a muon is born to die unseen,
And waste its flavour in the barren air –
Yet so many more in their brief lifetime
Do fuse their hosts – so bles't they were".

Lali Chatterjee

Z. Phys. C – Particles and Fields 56, S3–S4 (1992)

Zeitschrift für Physik C Particles and Fields

Opening of the Symposium "The Future of Muon Physics" May 7-9, 1991 Ruprecht-Karls-Universität Heidelberg

Gisbert zu Putlitz

Physikalisches Institut der Universität Heidelberg, Philosophenweg 12, 6900 Heidelberg, Germany

17-August-1992

Dear colleagues, ladies and gentlemen!

Nearly sixty years ago an elementary particle made its first appearance not far from here in the old Hansetown of Rostock at the Baltic Sea. Today, six decades later we have come together here to discuss the nature of this particle still not fully understood, to envisage further steps to unveal its properties, and to enlarge its merit as a unique tool both in fundamental and applied physics. Indeed, the persistance of this particle to resist its complete insight and understanding despite the enduring efforts of generations of most talented physicists is kind of unique in modern science.

Let me remind you at the beginning of this symposium on "The Future of Muon Physics" of some of the most important steps in the discovery of this particle.

Paul Kunze in Rostock took pictures of the socalled "Ultrastrahlung" in a Wilson chamber [1] and observed in one of the pictures taken (Fig.1) an electron and "a considerably stronger ionizing positive particle of smaller curvature". He comments "The nature of this particle is unknown. It ionizes too little for a proton but too much for an electron". Kunze attributes the trace to something originating from a nuclear explosion and misses to claim the discovery of a new elementary particle. - Only three years later Seth H. Neddermeyer and Carl D. Anderson of the Californian Institute of Technology in Pasadena carried out measurements of the energy loss of particles occurring in cosmic ray showers in heavy material like platinum [2]. They observed a group of particles penetrating the absorber much stronger than electrons and concluded from their observations "that there exist particles of unit charge, but with a mass larger than that of a normal free electron and much smaller than that of a proton". Since there is no evidence for the existence of such particles in ordinary matter, they suggested "it seems likely that there must exist some very effective process for removing them". At this point the most basic properties of what became later known as the muon were spelled out: Unit charge, mass between electron and proton, short lifetime.

The different steps and concepts in the discovery of the muon have been described in the literature [3-5]. The hope that the new species was the predicted meson by Yukawa [6] collapsed soon after no strong interaction could bee seen. The uniqueness of this particle was unvealed step by step: No strong interaction, two kinds with opposite charges, a rather long lifetime of μsec, properties like an electron except for the much larger mass, the same coupling constant for μ decay, μ capture, and for the nuclear β decay - all these observations raised questions rather than giving answers to the real nature of its existence. Comments of famous physicists relating to this situation are well known [3-5]: On the discovery: I.I. Rabi "Who ordered that?" and later M. Gell-Mann and Rosenbaum: "The muon was the unwelcome baby on the doorsteps, signifying the end of the days of innocence." Fermi commented at the 1950 Silliman lecture at Yale University with respect to the equality of coupling constants in electronic and muonic reactions, that this "is not an accident but has some deep meaning not understood".

A major step forwards in muon research was the copious production of muons by accelerators after 1957, led by the Nevis Columbia Cyclotron and the CERN-SC. In 1957 the asymmetric decay of muons was discovered by Garwin, Ledermann, Weinrich, Friedman and Telegdi [6-7], the prerequisite of many current applications of muons in nuclear, atomic and solid state physics, in chemistry and metrology. The anomalous g-factor of both μ^+ and μ^- were measured by Picasso, Farley and collaborators [8] with unprecedented precision and muonium, the lighter leptonic brother of the H-atom was synthesized by Hughes and coworkers [9] from Yale. In the high energy domain μ scattering confirmed its electron like nature today referred to as μ-e universality of the muon. But still, after six decades of intense research, after unification of the electromagnetic and weak inter-

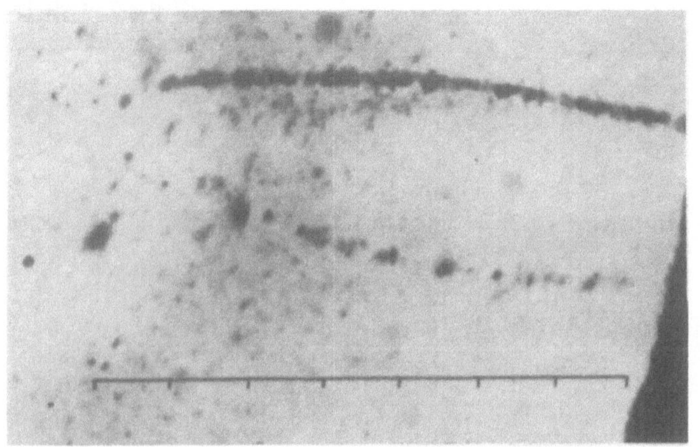

Fig. 1. The first observed muon by Paul Kunze in Rostock [1]: "Double track as a result of a probable nuclear explosion. Lower track: electron of 37 000 000 V. The nature of the upper positive particle is unknown."

action most important for muons, after the discovery of the τ, a third kind of lepton, the question of their existence, their mass difference, their point like nature and their magnetism, their conservation laws for quantum numbers, the different nature of their decay neutrinos and many other properties remain open.

In view of the many open questions summarized above there is no doubt that muon physics has its future. Many prerequisites for more progress in this field have been changed for the better. The standard model provides much more insight in the interaction of particles and their fields. Hence much better questions can be asked with respect to the muon. On the experimental side much more powerful tools and methods are provided in these days: Accelerators are available or under construction with very high fluxes of muons, excellent luminosity, best beam quality, and very short pulses of muons if desired. In addition greatly improved methods for the detection of muons and their decay particles became accessible as a spin-off from the field of particle physics. As a consequence muon physics is a field with many opportunities to be exploited, even today.

Let us discuss in our symposium how we should progress in this field, how we can strengthen our efforts through international collaboration, how we can develop the best instruments, and most important of all, how we can get the incoming young researchers to share our enthusiasm for muon physics.

References

1 Kunze, P., Z. Phys. 83, (1933) 1
2 Neddermeyer, S.H., Anderson, C.D., Phys. Rev. 51, (1937) 884
3 Wu, C.S. and Hughes, V.W., in: Muon Physics, Hughes, V.W and Wu, C.S. (eds.), Academic Press New York (1977)
4 Scheck, F., Phys. Reports 44, (1978) 187
5 Gerber, H.-J., in: Proceedings of the International Europhysics Conference on High Energy Physics, July 1987.
6 Garwin, R.L., Ledermann, L.M., Weinrich, W., Phys. Rev. 105, (1957) 1415
7 Friedman, J., Telegdi, V.L., Phys. Rev. 105, (1957) 1681
8 Bailey, J., et al., Nucl.Phys. B150 (1979) 1
9 Hughes, V.W., et al., Phys.Rev.Lett. 5 (1960) 63

This article was processed using Springer-Verlag TEX Z.Physik C macro package 1991
and the AMS fonts, developed by the American Mathematical Society.

Z. Phys. C - Particles and Fields 56, S5-S12 (1992)

Zeitschrift für Physik C Particles and Fields

Muon physics – Survey

Florian Scheck

Institut für Physik, Johannes Gutenberg-Universität, Postfach 3980, D-(W)6500 Mainz, Germany

1 December 1991

Abstract. The empirical basis of the minimal standard model has been consolidated in an impressive way, over the last seventeen years, by precision experiments at the meson factories. I illustrate this by means of selected examples of muonic weak interaction processes. I then describe an extension of Yang-Mills theory, inspired by noncommutative geometry, that yields precisely the standard model but fixes and explains some of its empirical input. In particular, this new approach yields a simple geometrical interpretation of spontaneous symmetry breaking. The algebraic framework of this approach offers a natural place for the lepton and quark matter fields and for inter-family mixing.

1 Introduction

It is a great honour and an even greater pleasure for me to open this workshop on Muon Physics one of whose purposes is to mark and to celebrate Vernon Hughes' seventieth and Gisbert zu Putlitz' sixtieth birthdays. As is well-known the muon is an important probe for electromagnetic interactions at low energies which are governed by Sommerfeld's fine structure constant. Thus, looking back on the main themes of Vernon's and Gisbert's life-long research, it is gratifying to realize that the sum of their present ages equals α^{-1}, to good approximation,[1]

$$\mathcal{A}(V) + \mathcal{A}(G) \simeq 1/\alpha \quad .$$

Vernon Hughes and Gisbert zu Putlitz both founded or triggered schools of research whose excellence, competence and leadership are well-known and appreciated in the physics community. Thus, before entering the subject matter of my contribution, I take this opportunity to wish them both many more years of happiness, a lot of fun with physics, and – last but no less important – continuing success in science management, to the benefit of all of us.

Over the last sixteen years, ever since the meson factories came into operation, muon physics has become a field of research of great diversity and impressive precision. Besides its role as a clean and precise probe in atomic, nuclear and condensed matter physics, the muon has been a tool of primary importance in searches for new physics, at low energies, that would go beyond the standard model. Great efforts went into this area of elementary particle physics and the precision of experiments was improved dramatically as compared to the state of knowledge in the late sixties. As of today, no deviation from the minimal standard model was found, neither in the physics of muons nor anywhere else in the physics of charged leptons and neutrinos. Although somewhat disappointing because of the lack of new discoveries, this result is truly remarkable. When this period of research started about seventeen years ago, many of us expected to find small but characteristic deviations from the minimal pattern assumed by the standard model which would be – mostly indirect – signals of new physics at scales between the masses of the W and Z^0 and a few TeV. As nothing of this kind was found, it may be time to find better theoretical justification for the fact that the standard model works so well, a model which, so far, was based primarily on phenomenological information and seemed to contain a great amount of arbitrariness. To quote just one example, maximal violation of parity in charged weak interactions was built in, although it seemed more natural to assume that parity violation was a low energy phenomenon, perhaps due to spontaneous symmetry breaking [1] and parity symmetry would be restored at energies well beyond m_W. As a consequence, in extended gauge theories with intrinsic left-right symmetry, specific deviations from the chirality pattern of leptons as demanded by maximal parity violation should have been seen in the recent precision experiments on muon and pion decays. As this was not

[1] Actually, as pointed out by José Bernabéu, the agreement improves even further if we let $\alpha(q^2)$ run to the scale of m_W^2.

the case we should take the minimal standard model for granted – as a working hypothesis – and try to discover a deeper foundation for its specific structure. It seems that nature, through the peculiar chirality pattern in which fermionic matter fields appear in electroweak and strong interactions, wants to tell us something which goes beyond the structure of Yang-Mills theories.

In this survey, after a short historical remark, I start by quoting some examples of such tests involving muons (Sect. 2). In Sect. 3 I remind the reader of the conventional construction of the standard model starting from observed and extrapolated phenomenology. Sect. 4 describes in a somewhat simplified and sketchy manner a new, classical, "noncommutative" extension of Yang-Mills theories which, when applied to the standard model of electroweak interactions, eliminates some of the apparent arbitrariness in its construction and exhibits some interesting pattern in the fermionic multiplets and their interfamily mixing. The concluding section 5 gives an outlook and lists some open questions.

2 Muons within the standard model

2.1 The muon as a probe

As is well-known there are two main aspects of muon physics at low energies which have been fully exploited during the era of the meson factories and both of which are discussed at length at this workshop: On the one hand the muon is a precise and particularly neat probe in a variety of applications:

(i) As advocated already in the classical paper by Garwin, Lederman and Weinrich [2] muon spin precession and muon spin relaxation are probes of internal magnetic fields in solids which are complementary to other more classical methods. This technique has since developed into a respectable branch of condensed matter physics which is covered regularly by an independent series of conferences.

(ii) There are many beautiful applications of muons in atomic and nuclear physics (see e.g. Ref. [3]) which rank among the most beautiful experiments at the meson factories. The specific spatial and temporal pattern of muonic atoms allows for the determination of quantities which otherwise are not directly accessible. For example, atomic cascade times of muonic atoms are of the order of 10^{-15} to 10^{-19} seconds while the lifetime in the 1s-state, before the muon decays or is captured by the nucleus, is on the order of 10^{-6} to $10^{-7}s$. The lifetimes of typical low-lying nuclear excited levels which range somewhere between 10^{-8} and 10^{-12} seconds, fall right between these two rather distant muonic time windows. This opens the possibility of studying nuclear transitions from states which are excited during the muon's cascade, in the presence of the muon in its 1s-state where it strongly penetrates the nucleus. For instance, in the case of nuclei with nonvanishing spin, this allows to measure the spatial distribution of nuclear magnetic densities through the observed hyperfine structure [3].

(iii) The spatial closeness of the muon to the nucleus, as compared to the electron's Compton wave length, is the reason why the study of radiative corrections in muonic atoms is the prime source of information on vacuum polarization (for an introduction see [4], the present state of the art is described in L. Schaller's contribution to the workshop).

(iv) Similarly, the magnetic moment anomaly of the muon has been an interesting testing ground, besides its importance in testing QED and weak corrections, for nonperturbative QCD (through the hadronic contribution to $g-2$) and possible extensions of the standard model.

(v) Finally, there are the muonic "clockworks" muonium μ^+e^- and muonic Helium μ^3He, μ^4He which continue to yield a wealth of precision information on QED, the muon's magnetic moment, the μ/e mass ratio, etc., and which are also discussed at length at this workshop.

2.2 Muons testing for physics beyond the standard model

In tests of the standard model at low energies the main issues may be summarized as follows:

1. Complete determination of the Lorentz structure of charged weak interactions in purely leptonic interactions. 2. Understanding the role of the global lepton family numbers $L_e, L_\mu, L_\tau, \ldots$ and testing to which degree these are conserved individually. 3. Search for direct and indirect manifestations of neutrino masses, including the search for neutrino oscillations. 4. Understanding the structure of the quark mass matrix and of the Cabibbo-Kobayashi-Maskawa mixing matrix. 5. Learning more about PC-violation.

The experimental knowledge of the properties of leptons has improved dramatically over the last seventeen years. Most of what concerns the contributions of muon physics to these topics is covered at this workshop in the contributions by W. Fetscher, R. Frosch, and H. K. Walter. Here I concentrate on a few selected examples from the series of precision experiments on polarized muon decay $\mu^+_\Uparrow \to e^+_\Uparrow \nu_e \overline{\nu_\mu}$ that were carried out at the meson factories. As we wish to determine the Lorentz structure of the charged interaction, i.e. the deviation from the assumed "$v-a$" coupling, it is useful to parametrize the effective interaction by the *chirality projection form* as first proposed in [4] and [5].

For this we define the symbols

$$s_{(1)} \equiv \overline{f_1(x)}\mathbb{1}\nu_1(x) \quad , \quad s^{(0)} \equiv \overline{\nu_0(x)}\mathbb{1}f_0(x) \tag{1}$$

for the scalar terms, and analogous expressions with $\mathbb{1}$ replaced by γ_5 for the p-terms, by γ_α for the v-terms, by $\gamma_\alpha\gamma_5$ for the a-terms, by $\sigma_{\alpha\beta}/\sqrt{2}$ for the t-terms, and by $\sigma_{\alpha\beta}\gamma_5/\sqrt{2}$ for the t'-terms. The most general, derivative-free, four-fermion effective hamiltonian reads

Table 1. Handednesses/helicities in the decay $\mu^- \to e^- \overline{\nu_e} \nu_\mu$

coupling	μ^-	e^-	ν_μ	$\overline{\nu_e}$
h_{11}	R	L	L	L
h_{12}	L	L	R	L
h_{21}	R	R	L	R
h_{22}	L	R	R	R
g_{11}	R	R	R	L
g_{12}	L	R	L	L
g_{21}	R	L	R	R
g_{22}	L	L	L	R
f_{11}	R	L	L	L
f_{22}	L	R	R	R

$$\mathcal{H} = \frac{G_F}{\sqrt{2}} \cdot \Big\{ h_{11}(s+p)_{(1)}(s+p)^{(0)} + h_{12}(s+p)_{(1)}(s-p)^{(0)} + h_{21}(s-p)_{(1)}(s+p)^{(0)} + h_{22}(s-p)_{(1)}(s-p)^{(0)} + g_{11}(v+a)^{\alpha}_{(1)}(v+a)^{(0)}_{\alpha} + g_{12}(v+a)^{\alpha}_{(1)}(v-a)^{(0)}_{\alpha} + g_{21}(v-a)^{\alpha}_{(1)}(v+a)^{(0)}_{\alpha} + g_{22}(v-a)^{\alpha}_{(1)}(v-a)^{(0)}_{\alpha} + f_{11}(t+t')^{\alpha\beta}_{(1)}(t+t')^{(0)}_{\alpha\beta} + f_{22}(t-t')^{\alpha\beta}_{(1)}(t-t')^{(0)}_{\alpha\beta} + \text{h.c.} \Big\} \quad (2)$$

The individual terms in Eq. (2) describe transitions between states of definite chirality. As an example, the handednesses of the charged and neutral particles in negative muon decay, for a given type of covariant in Eq. (2), are as shown in Table 1.

This form of the effective interaction is particularly well adapted to the analysis of the measured quantities in muon decay for several reasons: In the limit where the masses of the particles in the final state can be neglected as compared to the decay energy, any two amplitudes which correspond to the same type of covariant (s,p) or (v,a) or (t,t') but to different chiralities do not interfere. Hence the observables depend only on absolute squares of coupling constants and, possibly, a minimal number of interference terms [4]. Furthermore, in the minimal standard model out of the ten terms of Eq. (2) only one term is present, viz.

$$\frac{G_F}{\sqrt{2}} = \frac{g^2}{8m_W^2}\ , \ g_{22} = 1 \qquad g_{11} = g_{12} = g_{21} = 0\ , \ h_{ik} = 0\ , \ f_{ii} = 0 \quad (3)$$

Thus, any deviation of the effective interaction which may stem from modifications of the standard model or from new physics at higher energy scales is easily identified as a correction to Eq. (3) and to the specific chirality pattern connected with it, cf. Table 1, line g_{22}.

For example, extended models with hidden left-right symmetry contain a gauge boson W_R that couples to right-handed currents [1]. This yields an effective coupling constant g_{11}, as well as mixed terms $g_{12} = g_{21}$, in case the weak eigenstates W_L, W_R do not coincide with the mass eigenstates in the charged gauge boson sector. The data on muon decays [6] exclude such contributions at the following level. Under the assumption that the left-right symmetry is manifest (i.e. that the neutral partners of the right-handed charged leptons are light) the heavier of the two charged boson masses cannot be lighter than about 480 GeV, while the mixing angle is bounded by about 0.05.

The electron (positron) observables in μ^- (μ^+) decay may be grouped into 6 spectrum parameters, 3 spin-momentum and spin-spin correlation parameters, and 6 polarization parameters, [4]. Out of these the spectrum parameters $\{\rho, \eta, \delta\}$, the correlation parameters $\{\xi, \xi', \xi''\}$ and all four polarization parameters $\{\alpha, \beta, \alpha', \beta'\}$ are known [6-8]. In particular, a measurement of the decay asymmetry near the upper end of the spectrum, viz.

$$\left(\frac{d^2\Gamma}{dx\, d\cos\theta}\right)_{x\to 1} = \Gamma \frac{4\rho}{3}\left\{1 - P_\mu \frac{\xi\delta}{\rho}\cos\theta\right\} \quad (4)$$

gave the result [6]

$$P_\mu \frac{\xi\delta}{\rho} = 0.9989 \pm 0.0023 \quad (5)\ .$$

Here P_μ is the longitudinal polarization of the muon from the decay $\pi^- \to \mu^- \overline{\nu_\mu}$. As is evident from Eq. (4) with $P_\mu = \pm 1$, the specific combination of parameters $\xi\delta/\rho$ cannot exceed the value 1. Furthermore, by definition, $|P_\mu|$ alone cannot be larger than 1. The result (5) then implies that both $|P_\mu|$ and $|\xi\delta/\rho|$ are very close to 1. By angular momentum conservation, this means, in turn, that the helicity h of $\overline{\nu_\mu}$ and hence of ν_μ is also known very accurately. The sign of $h(\nu_\mu)$ being known to be negative [9], one concludes [10]

$$h(\nu_\mu) = -1 \pm 0.0032 \quad (\text{at } 90\%\ \text{C.L.})\ . \quad (6)$$

This is a truly remarkable result. It constitutes by far the most accurate determination of a neutrino helicity.

Inspection of Table 1 shows at once that by measuring the charged lepton observables in muon decay only, one cannot discriminate between the couplings h_{12}, second line, and g_{22}, eigth line, [5] (see also [11]). Both convert a left-handed μ^- into a left-handed electron (or a right-handed μ^+ into a right-handed positron), and differ only by the helicity assignments of the neutrinos. Adjusting the effective constant G_F so as to absorb an overall normalization, a more detailed analysis [5] shows that muon decay alone, in the best case, would tell us that the combination $(|g_{22}|^2 + |h_{12}|^2/4)$ equals 1, without further identification of the type of coupling. Clearly, in order to resolve this ambiguity, at least one spin-momentum correlation involving neutrinos must be measured. One such possibility is the reaction $\nu_\mu + e^- \to \mu^- + \nu_e$, the so-called "inverse muon decay". As $h(\nu_\mu) = -1$, from the result (6), the ratio of the measured cross section to its value for the "$v - a$" case, Eq. (3), [12]

$$S = 1.006 \pm 0.048 \quad (7)$$

places a *lower* limit on g_{22} and hence an *upper* limit on h_{12}, $|h_{12}| < 4(1-S)$ which is of the order of 0.2, [13].

An alternative could be to measure the forward-backward asymmetry of the ν_e in μ^+ decay. Indeed, integrating over the energy of the ν_e and denoting the angle between the muon spin and the direction in which the neutrino is emitted by γ, we find [14]

$$d\Gamma\left(\mu^+ \to \nu_e(e^+\overline{\nu_\mu})\right) = \frac{\Gamma}{2}\{1-(1-a)\cos\gamma\} \tag{8}$$

with

$$a = \frac{2|h_{12}|^2}{3(4|g_{22}|^2 + |h_{12}|^2})\,.$$

Although difficult, this might be a challenging experiment at a future intense, pulsed, muon beam.

3 Conventional construction of the standard model

The few examples quoted in Sect. 2.2 above illustrate the quality of the information on electroweak interactions that we have gained from precision experiments with muons. Clearly, there are many more examples from neutrino scattering and Z^0 physics all of which seem to confirm the foundations of the standard model. In developing the standard model, at first sight, there is quite some arbitrariness. It is instructive to recall the empirical input on which it is based and which by now seems so well confirmed, (see e.g. [4] or [15]).

The first ingredients are the structure group $G = SU(2)_L \times U(1)$ for the electroweak sector and the colour group $SU(3)_c$ for the strong interactions, both of which are made into gauge groups by the standard procedures of Yang-Mills theories. As we concentrate here on the electroweak part, we do not write the colour group explicitly in the sequel. As is well known G must be broken spontaneously such that the residual symmetry in the electroweak sector be the gauge group $H = U(1)_{e.m.}$ of electromagnetism.

The next ingredients are the fermionic matter fields: In the case of one lepton family, a doublet of left-chiral fields $((\nu_f)_L, f_L^-))$ and a singlet right-chiral field (f_R^-) with weak hypercharges y_D, y_S, respectively; in the case of quarks a doublet of left-chiral fields (u_L, d_L) with hypercharge y and two singlets (u_R, d_R) with hypercharges y_u, y_d, respectively. Notice that, through this choice, maximal parity violation is built in from the start. As the group G is not simple there are two free coupling constants g, g' whose ratio is the tangent of the Weinberg angle θ_W. The weak hypercharges y_D, y_S are fixed by the phenomenological requirement that the neutrino decouple from the photon, giving $y_D = -1$, and that the f^--photon interaction be parity conserving, giving $y_S = -2$. Likewise the electric charge assignments of quarks require $y_u = y+1, y_d = y-1$.

The final ingredient is the Higgs sector which is adjusted as follows. The Higgs multiplet must have a neutral component along the homogeneous space G/H but, for the rest, could be anything. In other terms, it may carry any weak isospin $t \geq \frac{1}{2}$ provided it has a neutral component H_0 with eigenvalue $t_3(H_0)$ of I_3 and weak hypercharge $y(H_0) = -2t_3(H_0)$. This neutral field, as is well-known, is the agent of spontaneous symmetry breaking, through the potential

$$V(\Theta) = \mu^2\overline{\Theta}\Theta + \lambda(\overline{\Theta}\Theta)^2 + \text{const.} \tag{9}$$

with positive λ and negative μ^2. Again, it is an empirical constraint that fixes t to be $\frac{1}{2}$, viz.

$$\begin{aligned}\rho \equiv \frac{M_W^2}{M_Z^2\cos^2\theta_W} &= \frac{t(t+1)-(t_3(H_0))^2}{2(t_3(H_0))^2} \\ &= 1.003 \pm 0.004 \quad (\text{with } m_{top} = 200\,\text{MeV}).\end{aligned} \tag{10}$$

Finally, there are two further remarks which emphasize the arbitrariness in the conventional construction of the standard model. The first is that finite dimensional unitary representations of Lie groups are fully reducible. That means, in case of several families of leptons and quarks, that the gauge bosons act only *within* one family but cannot relate different families. As a consequence, mixing of different generations such as the Cabibbo-Kobayashi-Maskawa mixing of quarks must invoke new physics that is beyond the framework of the model. The second remark is one of conceptual nature: In the standard construction sketched above the Higgses exhibit some features of gauge bosons but, at the same time, also behave like matter particles. Somehow they do not seem to make up their mind which camp they should belong to.

4 A new approach to the standard model

Another way of summarizing the discussion of the conventional construction, Sect. 3, is to state that the framework of Yang-Mills theories is incomplete: the Higgs sector is introduced and adjusted by hand; the matter fields (leptons and quarks) are classified as prescribed by empirical information. In what follows I wish to describe an alternative approach based on ideas borrowed from noncommutative geometry that eliminates much of this apparent arbitrariness [16-20]. This new construction yields precisely the standard model in its final form. As novel features it provides a geometric interpretation of spontaneous symmetry breaking and offers a natural place to inter-family mixing, without invoking new physics beyond the model. Finally, it provides an example of implementing supersymmetry in an alternative manner.

4.1 The basic idea

The bosonic sector of the standard model contains Higgs fields $\Phi_m(x)$ which classically are complex functions or, in the language of differential geometry, zero-forms. They connect left-chiral (L) with right-chiral (R) fermion fields. The neutral component, the agent of spontaneous symmetry breaking, serves to generate masses of fermions without destroying the initial, but hidden, symmetry. Gauge fields, on the other hand, relate left- to left-, and right- to right-chiral fields, and their vertices define the electric, weak (and colour) charges of fermions. Geometrically speaking they are one-forms, more specifically connections A_μ, out of which one constructs the field strength tensors $F_{\mu\nu}$ (which are two-forms) by differentiation. Thus Yang-Mills theories contain exterior forms of even or odd degree, in other terms antisymmetric tensor fields of even and odd rank, and a differential, the Cartan exterior derivative d_C which converts even into odd and vice versa. There is a natural classification of these fields into objects with *even* and *odd* exterior degree – a so-called $\mathbb{Z}_2$ grading – in this part of the theory, with d_C as the characteristic differential.

While the field strengths $F_{\mu\nu}$ are derived from A_μ, the scalar fields stand somewhat apart. They are forms of exterior degree even but are not related to A_μ or $F_{\mu\nu}$. The first question then is whether there is another grading, called matrix degree in the sequel, accompanied by its own derivative d_M, which is such that the gauge fields A_μ and the Higgs fields Φ belong to the same geometrical object of total degree odd. This would require assigning the matrix degree *even* to A_μ (which has exterior degree *odd*, matrix degree *odd* to Φ (whose exterior degree is *even*) so that the sum of the gradings is odd for both. This idea of grouping the gauge fields and the Higgs fields together in the form of what is called a super connection, or super gauge field $\mathcal{A}$, was proposed previously [21]. What is new is the accompanying derivative d_M which is needed for the construction of the corresponding super field strength $\mathcal{F}$ and , therefore, the lagrangian [17].

Before I describe the bosonic sector of the model that is obtained from this idea, I wish to return to the fermions for a moment. A remarkable feature of the standard model as obtained on the basis of the phenomenological input (Sect. 3) is that its fermionic building blocks are fields of definite chirality, L and R. As emphasized in the introduction it seems as though nature wished to tell us something here which is not yet part of the theory. The second question then is whether the new grading alluded to above also relates L- to R-fields. This is indeed rather natural: As pointed out above, A_μ which is odd in exterior degree but even in matrix degree couples (L $\leftrightarrow$ L) and (R $\leftrightarrow$ R), while Φ which is exterior even but matrix odd couples (L $\leftrightarrow$ R). These two ideas can indeed be implemented in a simple and consistent way and lead to the standard model, however, without the freedom of choices discussed above.

Notice that we are introducing supersymmetry in a different realization than hitherto. The supersymmetry does not relate states with integer spin to states with half-integer spin. Rather it relates antisymmetric tensor fields of even rank to such fields of odd rank, and fermionic L-fields to fermionic R-fields.

4.2 The bosonic sector

The ideas sketched above are implemented through the introduction of a graded matrix algebra which is defined as follows. Let

$$\mathrm{M} = \begin{pmatrix} A_{2\times 2} & C_{2\times 1} \\ D_{1\times 2} & B_{1\times 1} \end{pmatrix} \tag{11}$$

be any antihermitean 3×3 matrix with vanishing supertrace

$$\mathrm{Str}\,\mathrm{M} = \mathrm{tr}\,A - \mathrm{tr}\,B\,. \tag{12}$$

A and B are said to be the even part M_0 of M, C and D its odd part M_1,

$$\mathrm{M} = \begin{pmatrix} A & 0 \\ 0 & B \end{pmatrix} + \begin{pmatrix} 0 & C \\ D & 0 \end{pmatrix} \equiv \mathrm{M}_0 + \mathrm{M}_1\,.$$

The set of all such matrices defines the super (or graded) Lie algebra $SU(2|1)$,

$$SU(2|1) \equiv \{\mathrm{M}_{3\times 3}\,|\,\mathrm{M}^\dagger = -\mathrm{M}\,,\,\mathrm{Str}\,\mathrm{M} = 0\}\,. \tag{13}$$

This algebra contains the Lie algebras of $SU(2)$ and $U(1)$ as subalgebras and it may be generated by the 4 generators of the latter,

$$I_i = \begin{pmatrix} \tau_i/2 & 0 \\ 0 & 0 \end{pmatrix}\,,\; Y = \begin{pmatrix} \mathbb{1} & 0 \\ 0 & 2 \end{pmatrix}$$

and 4 more generators for its odd elements, viz.

$$\Omega_+ = \begin{pmatrix} 0 & 0 & 1 \\ 0 & 0 & 0 \\ 0 & 0 & 0 \end{pmatrix}\,,\; \Omega_- = \begin{pmatrix} 0 & 0 & 0 \\ 0 & 0 & 1 \\ 0 & 0 & 0 \end{pmatrix}$$

$$\Omega'_- = \Omega_+^\dagger\,,\; \Omega'_+ = \Omega_-^\dagger\,.$$

Consider now the following element of $SU(2|1)$

$$\eta = i\,(\Omega_+ + \Omega'_-) \in SU(2|1)\,. \tag{14}$$

The matrix derivative is defined by means of this matrix as follows

$$\mathrm{d}_M\mathrm{M} = [\eta, \mathrm{M}_0] + i\{\eta, \mathrm{M}_1\}\,. \tag{15}$$

It is given by the commutator of η with the even part and by the anticommutator of η with the odd part. It is easy to convince oneself that the result of the former is odd, the result of the latter is even. Thus d_M does indeed relate (*even* $\leftrightarrow$ *odd*) as required.

The next step is to replace the constant matrices (11) by matrices whose entries are exterior forms, and to combine the two exterior derivatives in a mathematically consistent way. The algebra (13) is now infinite dimensional and contains two gradings: the grading of forms and the matrix grading. This doubly graded structure becomes clearer if we write down at once the super gauge field (super connection), viz.

$$\mathcal{A} = i\left(\sqrt{2}(W_-I_+ + W_+I_- + W_3I_3) + \frac{1}{\sqrt{6}}W_8Y\right) + \frac{i}{\sigma}\left(\overline{\Phi^{(0)}}\Omega_+ + \overline{\Phi^{(+)}}\Omega_- + \Phi^{(0)}\Omega'_- + \Phi^{(+)}\Omega'_+\right) . \tag{16}$$

Here $(W_\pm, W_3)$ and W_8 denote the usual gauge fields (one-forms) of $SU(2)_L$ and $U(1)$, respectively, $(\Phi^{(0)}, \Phi^{(+)})$ are the neutral and charged Higgs fields (zero-forms), σ is a universal mass scale. As announced above the super gauge field has indeed total grade 1: the gauge fields are multiplied with generators with matrix grade 0, the Higgs fields are multiplied by generators with matrix grade 1. The super field strength (curvature) $\mathcal{F}$ is obtained from $\mathcal{A}$, Eq. (16), in a standard manner [18], [19], while the lagrangian is calculated from the trace of $\mathcal{F}\cdot\mathcal{F}^\dagger$. The details of this calculation are described in [17] and [20] and I do not repeat them here. The main results may summarized as follows. The lagrangian is exactly the one of the standard model and it contains precisely the physical degrees of freedom that are needed, no more and no less [2]. However, all quantum number assignments and all parameters are fixed. For instance, the Weinberg angle is fixed by the relative normalization of I_3 and $Y/2$,

$$\tan^2\theta_W = \frac{a_3}{a_8}\ , \text{ where} \qquad a_3 = \mathrm{tr}\,(I_3)^2\ ,\ a_8 = \mathrm{tr}\left(\frac{Y}{2}\right)^2 . \tag{17}$$

In the defining representation above that we used in Eq. (16), we find $\tan^2\theta_W = \frac{1}{3}$, i.e. $\sin^2\theta_W = 0.25$. The Higgs potential has the desired form (9) with the parameters being given by $\lambda = \frac{1}{2}$, $\mu^2 = -2\sigma^2$. Notice that by the construction of the superconnection $\mathcal{A}$,Eq. (16), the Higgs field must be a doublet with respect to weak isospin and that there is no more the previous freedom of choice discussed in relation with Eq. (10). Furthermore, the field $\Phi^{(0)}$ of Eq. (16) turns out to be the physical neutral Higgs field $\Phi^{(0)} = \Theta^{(0)} - <\Theta^{(0)}>$ and we do not have to make the shift to the minimum of the potential (9) – unlike in the conventional construction.

Perhaps the most interesting feature of this extension of Yang-Mills theory is that it yields spontaneous symmetry breaking automatically and in a geometrical fashion. This can be seen by studying generalized gauge transformations acting on $\mathcal{A}$,

$$\mathcal{A}' = \mathcal{A} + \mathrm{d}\mathcal{E} + [\mathcal{A}, \mathcal{E}] \tag{18}$$

where d is the combined derivative and $\mathcal{E}$ is an element of total grade zero which we take here to be, for simplicity,

$$\mathcal{E} = i\left(\sqrt{2}\,\mathbf{I}\cdot\mathbf{e} + \frac{1}{\sqrt{6}}Ye_8\right) .$$

[2] This is in contrast to earlier work which attempted at gauging the full graded Lie algebra $SU(2|1)$ interpreted as an internal symmetry of the theory [21-23]. For a discussion see [1].

Thus, $\mathbf{e}$ and e_8 are functions (zero-forms). It is not difficult to see that the constant super gauge field

$$\mathcal{A}_0 = -\eta = -i\,(\Omega_+ + \Omega'_-) \tag{19}$$

as well as the corresponding super field strength $\widehat{\mathcal{F}_0}$ are *invariant* under all constant gauge transformations (18). The field $\widehat{\mathcal{F}_0}$ is calculated to be [19][20]

$$\widehat{\mathcal{F}_0} = -i\eta^2 = i\left(I_3 + \frac{Y}{2}\right) . \tag{20}$$

If one subtracts the constant gauge field $\mathcal{A}_0$ from $\mathcal{A}$, and, analogously, the constant field $\widehat{\mathcal{F}_0}$ from $\mathcal{F}$,

$$\mathcal{A}_\odot \equiv \mathcal{A} - \mathcal{A}_0\ ,\ \mathcal{F}_\odot \equiv \mathcal{F} - \widehat{\mathcal{F}_0}\ , \tag{21}$$

then $\mathcal{A}_\odot$ and $\mathcal{F}_\odot$ are seen to coincide with the well-known gauge fields and field strength tensors, respectively, of the usual standard model. Thus, by subtracting the constant, invariant, background field $\widehat{\mathcal{F}_0}$, one recovers the standard model with its original $SU(2)_L \times U(1)$ symmetry, before spontaneous symmetry breaking. On the other hand, the existence of $\mathcal{A}_0$ and $\widehat{\mathcal{F}_0}$ is a direct consequence of the noncommutative structure that we introduced by the derivative (15). As a result,this extension of Yang-Mills theory is characterized by the presence of a constant background field which is proportional to the electric charge, cf. Eq. (20). The full lagrangian has no more than the residual symmetry $U(1)_{e.m.}$ generated by this constant field which is element of the algebra. If one wishes to recover the full symmetry $SU(2)_L \times U(1)$ one must take out this field "by hand". This is somewhat analogous to studying atomic spectra in the presence of a universal constant magnetic field $\mathbf{B}_0$: the spectral lines are split by the Zeeman effect. The residual symmetry is the axial symmetry about the direction of the field. The full rotational symmetry is recovered only after switching off $\mathbf{B}_0$.

4.3 Fermions in the $SU(2|1)$ construction of the standard model

The discussion of the preceding section shows that the noncommutative structure introduced through the definition (15) is embedded in the Lie super algebra $SU(2|1)$. It was noted earlier and independently by Fairlie and Ne'eman that this algebra was relevant for the classification of L- and R-components of leptons fields and quark fields [22],[23]. For example, the simplest (non-typical) representation $[I = \frac{1}{2}]_-$ of $SU(2|1)$ (here I use the notation of ref. [18]) which decomposes into $(i = \frac{1}{2})_{y=-1}$ and $(i = 0)_{y=-2}$ with respect to $SU(2)_L \times U(1)$, accommodates a left-handed lepton doublet and a right-handed singlet, with precisely the eigenvalues of weak hypercharge that are needed to describe one lepton family, cf. Sect. 3. Similarly, the simplest typical representation of $SU(2|1)$, $[y \neq \pm 1, I = \frac{1}{2}]$ which decomposes into $(i = \frac{1}{2})_y, (i = 0)_{y+1}, (i = 0)_{y-1}$ with respect to

$SU(2)_L \times U(1)$, describes a doublet of left-handed quarks and two singlets of right-handed quarks. Furthermore, the absence of anomalies requires quarks to have three colour states, $N = 3$, and y to be $\frac{1}{3}$, [18] – in accordance with what we need to describe one quark generation.

Indeed, there are four local anomaly conditions all of which constrain the weak hypercharges of leptons and quarks [24],[25]. In the nomenclature of ref. [24] and using the notations of Sect. 3 the $U(1)_Y[gravity]^2$ anomaly gives the condition

$$N(2y - y_u - y_d) + 2y_D - y_S = 0 \; ; \tag{22}$$

The $U(1)_Y[SU(3)_c]^2$ anomaly yields the condition

$$2y - y_u - y_d = 0 \; , \tag{23}$$

which when combined with Eq. (22), yields $2y_D - y_S = 0$. Notice that these are nothing but the supertrace condition $\mathrm{Str}\, Y = 0$, Eq. (12), which must hold for quarks and leptons separately. This is in accordance with their classification in terms of representations of $SU(2|1)$. The anomaly condition $U(1)_Y[SU(2)_L]^2$ yields

$$Ny + y_D = 0 \; , \tag{24}$$

while the $(U(1)_Y)^3$ anomaly condition is

$$N\left(2y^3 - y_u^3 - y_d^3\right) + 2y_D^3 - y_S^3 = 0 \; . \tag{25}$$

Combining Eqs. (23), (24), and (25), one finds $y_u = y \mp y_D$, $y_d = y \pm y_D$. The weak hypercharge of the lepton doublet can always be normalized such that $y_D = -1$, giving charge -1 for the electron, so that Eq. (24) becomes $Ny = 1$. As $y = 1$ is excluded, see above, the integer N must be greater than 1. Finally, the absence of the global $SU(2)$ anomaly [26] requires the total number of doublets per generation (1 leptonic and N quark doublets) to be *even*. This implies N to be odd, $N = 3, 5, \ldots$. It is gratifying to see that $N = 3$ is the simplest possibility.

In ref. [18] we point out that, beyond the more obvious classification discussed above, the algebra $SU(2|1)$ also possesses reducible but indecomposable representations which not only classify the fermions correctly but also provide a natural framework for mixing between different families. I mention here only the case of quarks and mixing between two generations, and refer to ref. [18] for a more complete discussion. This new type of multiplet (which does not exist in the case of an ordinary Lie algebra), is constructed by taking the semi-direct sum of two representations $[y, I]$. This means that one of these is an invariant subspace, the other is not. Reducible but indecomposable representations have the triangular form

$$\begin{pmatrix} A_1 & A_{12} \\ 0 & A_2 \end{pmatrix} \tag{26}$$

where the entries are 4×4 block matrices, and where A_{12} describes the transition matrix elements between family 1 and family 2. In particular, the mass matrix of two quark generations will have the form (26). Now, assume that mass differences are due to electroweak interactions only, i.e. that the diagonal blocks in (26) are equal. Diagonalization of the mass matrix then yields a remarkable formula for the Cabibbo angle, viz.

$$|\theta_c| \approx \sqrt{\frac{m_d}{m_s}} - \sqrt{\frac{m_u}{m_c}} \; , \tag{27}$$

in reasonable agreement with experiment.

The case of the full Cabibbo-Kobayashi-Maskawa mixing matrix as well as mixing between leptonic families is being studied presently [27][28].

5 Concluding remarks and some open questions

The studies of muonic weak interactions that were carried out over the last fifteen years, during the era of the meson factories, have contributed much to the consolidation of the assumptions on which the minimal standard model is based. When we started working in this field many of us expected to find specific deviations from the minimal model, possibly signalling new physics at scales beyond the weak interaction scale or indicating the relevance of a larger gauge group than $SU(2)_L \times U(1)$. Nothing of this kind was found. This is illustrated here by some characteristic examples from charged weak interactions involving muons. With this state of affairs the standard model must be taken for granted even though its conventional construction makes use of theoretical assumptions and empirical inputs for which other choices would have been possible. This is to say that the framework of Yang-Mills theory is still too wide, as it provides little constraint on the Higgs sector and practically no constraint on the fermionic matter fields.

An extension of this framework, based on ideas of noncommutative geometry and using the graded Lie algebra $SU(2|1)$ resolves at once many of these puzzles: the classification of the Higgs fields becomes unique. Gauge fields and Higgs fields are combined in one geometric object, a generalized super gauge field. An immediate consequence of the noncommutative structure is the appearance of a universal, constant, background field (in the internal symmetry space) which is proportional to the electric charge operator. This feature provides a beautiful geometrical interpretation of spontaneous symmetry breaking to the $U(1)_{e.m.}$ of electromagnetism.

Fermion matter fields fit naturally into representations of the algebra $SU(2|1)$, with the correct assignment of weak hypercharges. No assignment has to be made by hand. The Weinberg angle becomes computable as soon as a specific representation of the algebra is chosen. For instance, in the defining representation (13) one obtains $\sin^2\theta_W = 0.25$. If, instead, one starts from a (reducible) representation that combines three colours of quarks and one lepton family, Eq. (17) yields $\sin^2\theta_W = 3/8$, a value well-known from grand unified theories.

Another remarkable feature of this approach to the standard model is the appearance of reducible but indecomposable representations which offer a natural place for generation mixing. Our first calculation of the mixing matrix for two generations is encouraging. It yields the correct Cabibbo angle. The detailed study of the full Cabibbo-Kobayashi-Makawa mixing matrix and comparison with experiment is in progess [28]. A similar analysis of lepton family mixing which proceeds along a quite different pattern is also being studied [27].

This new approach yields precisely the standard model and, therefore, has all the virtues of the latter, after quantization. The algebraic structure encoded in Eq. (13) and the matrix derivative (15) provides a natural and, in fact, quite restrictive framework which eliminates much of the apparent arbitrariness in the conventional construction of the standard model. In addition, it provides a formal framework which allows to predict some of the parameters of the model and hence a first step in a long-standing problem of particle physics. Finally, this approach offers a new interpretation of supersymmetry as applied to particle physics that may open up a new line of thought.

Acknowledgement. I wish to thank the organizer of the workshop Klaus Jungmann for having provided a stimulating atmosphere during the meeting and for his patience in preparing the proceedings. I am grateful to my collaborators R. Coquereaux, G. Esposito-Farèse, R. Häußling and N. Papadopoulos for enligthening discussions.

References

1. J.C. Pati, A. Salam: Phys. Rev. Lett. 31 (1973) 663; Phys. Rev. D10 (1974) 275. R.N. Mohapatra, J.C. Pati: Phys. Rev. D11 (1975) 566, 2588. M.A.B. Bég et al.: Phys. Rev. Lett. 38 (1977) 1252
2. R.L. Garwin, L.M. Lederman, M. Weinrich: Phys. Rev. 105 (1957) 1415
3. J. Hüfner, F. Scheck, C.S. Wu: Muonic atoms. In: Muon Physics . V.W. Hughes, C.S. Wu (eds.), pp.202-304. New York, San Francisco, London: Academic Press 1977
4. F. Scheck: Leptons, Hadrons and Nuclei. Amsterdam: North Holland 1983
5. K. Mursula, F. Scheck: Nucl. Phys. B253 (1985) 189
6. J. Carr et al.: Phys. Rev. Lett. 51 (1983) 627. Erratum: 51 (1983) 1222. A. Jodidio et al.: Phys. Rev. D34 (1986) 1967. D.P. Stoker et al.: Phys. Rev. Lett. 54 (1985) 1887
7. ETH Zurich-Mainz-SIN collaboration: H. Burkhard et al.: Phys. Lett. B160 (1985) 343. I. Beltrami et al.: Phys. Lett. B194 (1987) 326. H. Burkhard et al.: Phys. Lett. B150 (1985) 242.
8. B. Balke et al.: Phys. Rev. D37 (1988) 587
9. L.Ph. Roesch et al.: Helv. Phys. Acta 55 (1982) 74
10. W. Fetscher: Phys. Lett. B140 (1984) 117
11. C. Jarlskog: Nucl. Phys. 75 (1966) 659
12. CHARM II collaboration: J. Dorenbosch et al.: Z. Physik C41 (1989) 589. D. Geiregat et al.: Phys. Lett. B247 (1990) 131
13. W. Fetscher, H.J. Gerber, K.F. Johnson: Phys. Lett. B173 (1986) 102
14. K. Mursula, F. Scheck: Leptonic charged weak interactions: Status and prospects. In: Essays in Honour of Matts Roos, Research Institute for High-Energy Physics report series HU-SEFT-1991-19: M. Chaichian, J. Maalampi (eds.), pp. 97-111. Helsinki 1991.
15. L. O'Raifeartaigh: Group Structure of Gauge Theories. Cambridge: Cambridge University Press 1986
16. A. Connes, J. Lott: Nucl. Phys. (Proc. Suppl.) B18 (1990) 29
17. R. Coquereaux, G. Esposito-Farèse, G. Vaillant: Nucl. Phys. B353 (1991) 689
18. R. Coquereaux, G. Esposito-Farèse, F. Scheck: Nucl. Phys. (submitted)
19. R. Häußling, N.A. Papadopoulos, F. Scheck: Phys. Lett. B260 (1991) 125
20. R. Coquereaux, R. Häußling, N.A. Papadopoulos, F. Scheck: Int. Journ. of Mod. Phys. A (in print)
21. Y. Ne'eman, S. Sternberg: Internal Supersymmetry and Superconnections. In: Symplectic Geometry and Mathematical Physics: P. Donato, C. Duval, J. Elhadad, G.M. Tynman (eds.). Basel, Boston, Berlin: Birkhäuser 1991
22. D.B. Fairlie: Phys. Lett. B82 (1979) 97. Y. Ne'eman: Phys. Lett. B81 (1979) 190
23. P.H. Dondi, P.D. Jarvis: Phys. Lett. B84 (1979) 75
24. K.S. Babu, R.N. Mohapatra: Phys. Rev. D42 (1990) 3866, and literature quoted therein
25. F. Scheck: in preparation
26. E. Witten: Phys. Lett. B117 (1982) 324
27. R. Coquereaux, F. Scheck: in preparation
28. R. Coquereaux et al.: in preparation

This article was processed using Springer-Verlag TEX Z.Physik C macro package 1991
and the AMS fonts, developed by the American Mathematical Society.

Z. Phys. C - Particles and Fields 56, S13-S23 (1992)

Two-body QED bound states

Donald R. Yennie

Laboratory of Nuclear Studies, Cornell University, Ithaca, NY 14853

15-September-1991

Abstract. The status of theory and experiment is reviewed for two-body systems in quantum electrodynamics, and places where further work would be useful are pointed out.

1 Introduction

This paper draws heavily upon a review article prepared with Jonathan Sapirstein [1]. Because more complete references can be found in that article, we do not attempt to present a historical survey here. However, some of the information is updated by recent developments.

High precision bound state calculations are different in character from other precision calculations, such as the electron anomaly. The latter requires the highest refinement in perturbative analysis of renormalization, including intricate subtractions of overlapping and nested divergences. On the other hand, precision bound state calculations are essentially nonperturbative in the binding potential. Though they involve kernels which can be written in terms of Feynman graphs, the dimensionless parameters, particularly the fine structure constant α, enter the wave functions and the particle propagators (through the energy) in a nonperturbative way. Integrations over the internal variables of a kernel can produce inverse powers of α. Thus one cannot simply count powers of α in a given kernel in order to determine its ultimate contribution to an energy shift. Typically, a given kernel includes some leading order plus an infinite series of smaller terms. Further, kernels can often be grouped together in such a way that their leading order cancels; it is important to take this into account so as to avoid calculating terms of spurious lower order. As a consequence, the formal development of bound state theory is not at all mechanical. Most calculations are at the two-loop level, rather than the four-loops which presently characterize the anomaly calculation. Although bound state theory is nonperturbative, it is possible to make use of small parameters such as α and m_e/m_N (where m_N is the mass of the nucleus) to develop expressions in increasing orders of smallness. However, the nonperturbative nature of the expansion shows up in nonanalytic dependence on these parameters (such as logarithms).

There are many different types of contributions to bound state energies. In principle, one needs a unified treatment of the problem so that these different contributions need not be combined in a patchwork way. However, it is fortunate that different physics, such as nuclear motion and radiative corrections, can largely be compartmentalized. For example, we can largely ignore dynamical recoil effects (while keeping reduced mass dependence) while treating the radiative corrections, which refer to corrections due to the emission and absorption of photons by the electron and also to vacuum polarization corrections to the Coulomb interaction. The treatment of renormalization is similar to that for free electrons, except that electron propagators in an external Coulomb field are used. Similarly, in treating *recoil corrections*, in which truly dynamical effects due to nuclear recoil are incorporated, radiative corrections may be ignored, except to the extent that it may be legitimate to incorporate a particle's anomalous magnetic moment phenomenologically. Until recently, it was quite possible to keep these two categories separate to the order of interest. Now, the level of accuracy has reached the point where both complications must be treated together as *radiative-recoil* corrections. [At some level it is necessary to include a fourth category consisting of other small effects such as weak interactions, finite nuclear size, nuclear polarizability, etc.]

In Chapter II, we describe the formulation of the relativistic two-body problem. When one or both particles are to be treated relativistically, there is no simple procedure to bring out the reduced mass dependence. In fact, if one starts with a two-body system and lets one of the particle masses become infinite, it requires a nontrivial analysis to demonstrate that the result can be expressed in terms of a relativistic equation for the other particle in a central potential. The physics is clear; it is simply the formalism which is awkward. Chapter III contains a summary of the most important theoretical results in the field and a comparison of them with experimental results. At that point, there is some attempt to describe their meaning, but not their derivation.

An "obvious" approach to the two-body problem is to add the Dirac Hamiltonians of the two particles and introduce an interaction between them, as was done by Breit [2]. His equation has an instantaneous interaction between the particles, but it does incorporate retardation effects at the v^2/c^2 level. As was pointed out by Salpeter [3], the Breit equation corresponds to a single electron theory, rather than to hole theory; and it becomes inaccurate at a certain level. As experimental precision increased, it became necessary to develop a more accurate formalism based on a complete field theoretic treatment of the two-body problem.

A field theoretical treatment was developed by Schwinger [4] and by Salpeter and Bethe [5]; the result is usually referred to as the Bethe-Salpeter equation. It is now generally recognized that this equation, which is a homogeneous four-dimensional integral equation, is not in a practical form for high precision analysis of energy levels. The first systematic approximation procedure for solving this equation was developed by Salpeter [3] who showed how to develop a perturbation expansion in which the starting point is a solvable three-dimensional equation.

In evaluating energy levels, the 1986 adjustment of physical constants [6] are used here. Some relevant ones are:

$$\begin{aligned} c &= 2.99792458 \times 10^{10} cm/sec \\ \alpha^{-1} &= 137.0359895(61) \\ R_\infty &= 109737.315709(18) cm^{-1} \end{aligned} \tag{1}$$

We have in the above presented a more recent determination of R_∞ [7] which disagrees slightly with the 1986 adjustment but is better for consideration of $n = 2$ to $n = 1$ transitions.

2 Formulation of a two-body equation for bound state QED

Consistency requires a field-theoretical approach rather than a two particle Hamiltonian approach. One desires an unperturbed problem that includes the basic nonrelativistic physics of the Schrödinger equation and as much of the relativistic physics as is feasible. The choice described here yields a fairly simple unperturbed problem and also limits the number and complexity of the perturbation kernels required. It also includes most of the reduced mass effects at the unperturbed level. It is the one favored by the author; but there are other, equally viable, approaches.

Inspired by the Bethe-Salpeter method, we arrive at the structure of our unperturbed problem by studying the two-particle four-point function G which describes the scattering of an electron and a proton (or muon). It satisfies a four-dimensional inhomogeneous integral equation. The bound state energies are given by the position of poles of this function as a function of the total energy. At these poles, the integral equation reduces to a homogeneous one for the Bethe-Salpeter wave function. This equation is awkward to solve, but was reduced to a three-dimensional one by Salpeter [3]. The result is still awkward because of a dependence on the relativistic energies (like $\sqrt{p^2 + m_e^2}$). Also, in some of the refined corrections, it artificially separates effects which ought to be kept together.

Between any two two-particle irreducible kernels there occurs the product of two free propagators, S, which in momentum space is given by

$$S = \frac{1}{\not{P}_e - m_e} \times \frac{1}{\not{P}_p - m_p} \tag{2}$$

where, in the center-of-mass frame,

$$(P_e + P_p)_\mu = E g_{\mu 0} \ . \tag{3}$$

Here E is the total energy of the system. It is convenient to split off a fixed energy piece of each four-momentum in some appropriate way and write

$$P_{e\mu} = E' g_{\mu 0} + p_\mu \,, \quad P_{p\mu} = E'' g_{\mu 0} - p_\mu \,. \tag{4}$$

The first step is to find a simple approximation to the four-point function. To do this, it is useful to study the pole structure of the two propagators in (2) as a function of p_0. For nonrelativistic spatial components of momentum, the positions of various poles are illustrated in Fig.1. The dominant feature of the pole structure is that the positive energy parts of the electron and proton propagators have poles close to the origin and on opposite sides of the axis, nearly pinching the p_0-contour. Associated with this pinching region, one finds that the leading term in the numerator is the large component projector for each particle. Consequently, the behavior of the dominant term from this region does agree with that of the Schrödinger equation for the large components. Any treatment must incorporate this dominant feature into the approximation to S, and various treatments differ primarily in how this is done and how the other poles are treated. We call the chosen approximation $\bar{S}$; and to emphasize that p_0 is fixed at a small value, we write

$$S \approx \bar{S} = -2\pi i \delta(p_0) \bar{s} \ , \tag{5}$$

where $\bar{s}$ is now a three-dimensional propagator. It should be emphasized that we do not obtain $\bar{s}$ directly from S by setting $p_0 = 0$; instead, $\bar{s}$ corresponds to an operator residue associated with the dominant pole structure. In general, we represent the separation of S into $\bar{S}$ and other

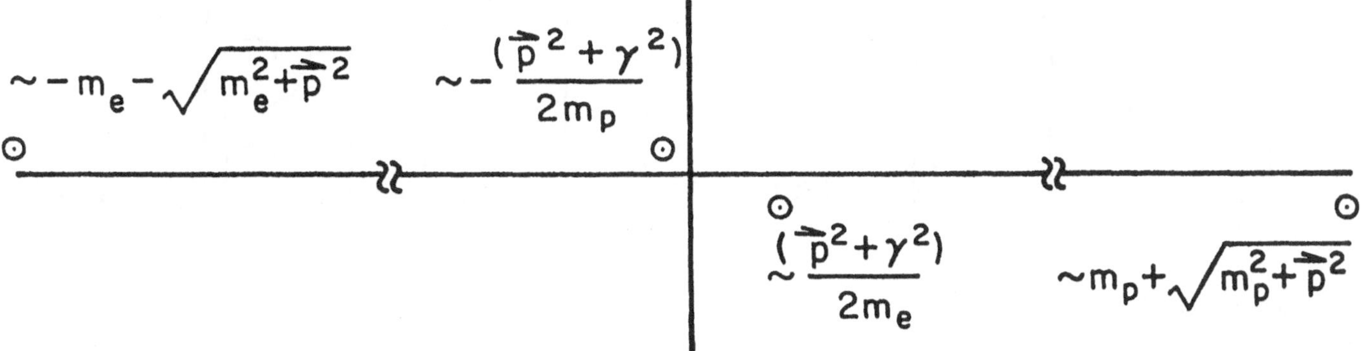

Fig. 1. The location of the poles as a function of p_0 in the ladder configuration.

$$S = N\bar{S} + R$$

Fig. 2. Graphical representation of the decomposition of the propagator product into a three-dimensional piece and a remainder.

terms diagrammatically as in Fig.2. The heavy lines indicate the resulting three-dimensional propagator, and R represents the remainder, which is to be treated along with other perturbations.

The perturbation scheme is straightforwad, but will not be described in detail here. Examples of some perturbation kernels K giving recoil corrections are illustrated in Fig.3. It is interesting to note that the first two kernels independently have non-recoil pieces which cancel when they are added. That is, if one lets the nuclear mass become infinite, only the combination vanishes. Radiative correction kernels are not illustrated. They have additional photon interactions with the electron or vacuum polarization insertions in the exchanged photons. The loop energy in the exchanged photons is fixed by the same type of contour pinching as in the preceding discussion.

Now one organizes the calculation by first considering the set of all diagrams with Coulomb exchanges in a ladder configuration. If one takes the part of the propagator product in which p_0 is fixed between all Coulomb interactions, one finds a propagator $\bar{G}$ with an effective potential interaction between the particles. It satisfies an integral equation of the form

$$\bar{G} = \bar{S} + \bar{S}\mathcal{V}\bar{G} \ , \qquad (6)$$

Various kernels are inserted between factors of $\bar{G}$ and modify this into the complete propagator. The problem defined by (6) gives unperturbed energies and wave functions; and a perturbation formalism can be easily developed to give the energy shifts produced by the kernels. This formalism is very similar to the usual nonrelativistic perturbation theory in quantum mechanics. For example, the leading (and often the only important) term is the expectation value of the kernels for the state of interest. The main difference from nonrelativistic perturbation theory is that the wave functions have three-dimensional arguments while the internal integrations in the kernels are four-dimensional. The kernels contain the field-theoretical content of the theory.

It is plausible that the most important term in K is the one photon exchange contribution (because of the small value of α). It turns out that it is sometimes possible to pack a little extra physics into the choice of the effective potential $\mathcal{V}$; in particular, some effects from the one transverse photon exchange can be included in such a modification.

Let me describe briefly the results of the "Dirac model" treatment of the unperturbed model. This treatment is defined by using the large component projector for the nucleus and the full Dirac numerator for the electron, together with a particularly felicitous choice of the

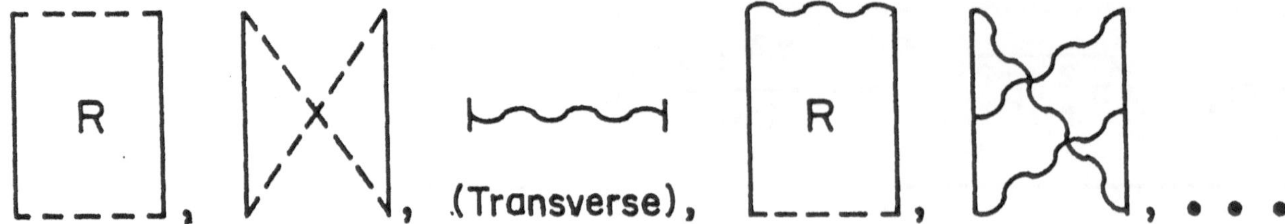

Fig. 3. Examples of perturbation kernels.

effective potential. As described in detail in [1], it turns out that the wave functions and energies of this problem can be defined in terms of the well-known ones for the usual Dirac equation in an external Coulomb potential. However, the fine structure constant, electron mass and energy eigenvalues have values which are modified by effects depending on m_e/m_p. It turns out that the Dirac degeneracy remains valid independently of the mass ratio. Using the Dirac model together with a few simple corrections, one can easily work out the fine structure of hydrogen to order α^4, without approximation in the mass ratio. The result is incorporated in (8) below; however, that equation also contains some terms of higher order in α.

3 Present status of theory and comparison with experiment

It is our purpose in this chapter to present a review of the experimental situation and theoretical status of fine structure and hyperfine structure in one-electron atoms, specifically for muonium and hydrogen, and to some extent for positronium. We limit ourselves to discussion of ground state hyperfine splitting, and to fine structure of only the $n = 1$ and $n = 2$ levels. The chapter is a synopsis of material in [1], to which the reader is referred for details.

The Dirac model described above provides a good starting point since it incorporates reduced mass dependence through order α^4. Other effects of the nuclear motion are regarded as bonafide dynamical recoil effects and are included through perturbation kernels. To make this concrete and to describe what has been done so far, let us first consider the energy levels for the Dirac-Coulomb problem without recoil. These are given by

$$E_{nj} = m_e f(n, j, Z\alpha) \; , \tag{7}$$

where the f's are well known. It is important to recall that the energy levels are independent of the orbital angular momentum given by l, which is related to j by $l = j \pm \frac{1}{2}$. Using the approach outlined in Chapter II, an expression for the total energy in which terms of order $\alpha^6 m_e^3/m_N^2$ have been neglected (totally negligible for hydrogen and muonium) and perturbation kernels through order α^4 have been incorporated is

$$E = M + m_r[f-1] - \frac{m_r^2}{2M}[f-1]^2 + \frac{(Z\alpha)^4 m_r^3}{2n^3 m_N^2}\left[\frac{1}{j+\frac{1}{2}} - \frac{1}{l+\frac{1}{2}}\right](1-\delta_{l0}) \tag{8}$$

where $M = m_e + m_N$. Note that the terms depending on $f(n, j, Z\alpha)$ retain the degeneracies of the original Dirac-Coulomb result. While the second of these does not contribute to the splitting, it does contribute about 22 MHz to the $n = 2$ to $n = 1$ energy difference in hydrogen. The last term contributes a tiny amount, -2 kHz, to the $2S_{1/2} - 2P_{1/2}$ splitting (i.e., to the Lamb shift). The complete correct mass dependence through order α^4, including the small term which breaks the degeneracy, was obtained originally by Barker and Glover [8]. Several authors also derived the result for positronium energy levels to the same order in α [9].

3.1 Lamb Shift

In the following we use a general terminology for the Lamb shift to include any deviation from the energy levels predicted by the Dirac equation with the binding scaled by the reduced mass of the electron, namely the deviation from $M + m_r[f-1]$. Where appropriate, we refer to the $2S_{1/2} - 2P_{1/2}$ splitting as the 'classic' Lamb shift. At first ignoring nuclear motion, the bulk of the Lamb shift due to one photon radiative corrections to the electron and vacuum polarization corrections to the potential can be expressed by the formula

$$\Delta E_n(\text{one}-\text{loop}) \equiv \frac{m_e\alpha(Z\alpha)^4}{\pi n^3} F_n(Z\alpha) \; . \tag{9}$$

For large values of Z the function $F_n(Z\alpha)$ must be determined numerically. For smaller values, one can expand it in powers of $Z\alpha$, but non-analytic factors may occur.

In the non-recoil limit, this expansion involves constants and the logarithmic factor $\ln(Z\alpha)^{-2}$. Certain consequences of nuclear motion can be incorporated through reduced mass dependence. For example, most terms acquire an overall factor of $(m_r/m_e)^2$ or $(m_r/m_e)^3$, which arises simply from the scaling of the wave functions. The argument of the logarithm which comes from radiative corrections to the electron is really the ratio of the electron's rest energy and a characteristic binding energy ($\propto m_r(Z\alpha)^2$). Thus the logarithm takes on an argument $(m_e/m_r)(Z\alpha)^{-2} \equiv \sigma(Z\alpha)^{-2}$. Other nuclear motion corrections which cannot be incorporated with these simple substitutions are recoil corrections and are described later. We then have the expansion

$$
\begin{aligned}
F_n(Z\alpha) = A_{40}(n) + A_{41}(n)\ln[\sigma(Z\alpha)^{-2}] \\
+ (Z\alpha)A_{50}(n) + (Z\alpha)^2\{G(n, Z\alpha) \\
+ A_{61}(n)\ln[\sigma(Z\alpha)^{-2}] \\
+ A_{62}(n)\ln^2[\sigma(Z\alpha)^{-2}]\} \ .
\end{aligned}
\tag{10}
$$

Values of the constants are given in [1]. The functions $G(n, Z\alpha)$ represent non-logarithmic terms of order $m_e\alpha(Z\alpha)^6$ plus all higher order corrections. The work involved in calculating these constants and functions has involved many physicists and has extended over almost four decades. The constants A_{ij} are either known analytically or to high numerical precision. The part of the functions G associated with the electron self-energy is calculated numerically by extrapolating the Z=10, 20, and 30 values calculated by Mohr [10] down to $Z = 1$.

There is one last set of single-loop non-recoil terms to be mentioned. These are vacuum polarization corrections from heavier mass particles. Generally such corrections are reduced because they are proportional to $1/\text{mass}^2$. For example, the muon contributes $(m_e/m_\mu)^2$ as much as the electron, or about 0.6 kHz to the classic Lamb shift in hydrogen. The hadronic contribution is more complicated, but it can be worked out in principle using the Källen-Lehman representation, whose weight function can be determined from the total cross section for $e^+e^- \rightarrow$ hadrons.

This completes the description of the one-loop non-recoil Lamb shift, including reduced mass effects. Because we have not assumed a small mass ratio, the expression is correct, though incomplete, for positronium. However, in that case the effect of self-energy on the positive particle is just as important, and must be added in.

Dynamical recoil corrections of order $(Z\alpha)^5 m_r^3/(m_e m_N)$ were first derived by Salpeter [3]. These require the use of a form of the Bethe-Salpeter equation since the previously used Breit equation had missed terms of this order entirely. Different contributions in this order have parts that involve very different physics. The exchange of a transverse photon between the electron and the nucleus with any number of Coulomb interactions occurring between emission and absorption involves highly nonrelativistic momenta and gives rise to an expression closely related to the lowest order self-energy calculation, but reduced by a factor m_e/m_N. However, there is also a term of the same order associated with various combinations of two-photon exchange kernels, which involves both nonrelativistic and relativistic electron momenta. Recoil terms of order $m_e^2(Z\alpha)^6/m_N$ have recently been completed [11,12]. They yield a contribution of +3.15kHz to the Lamb shift in hydrogen. Radiative-recoil contributions of the order $\alpha(Z\alpha)^5 m_e^2/m_N$ have recently been calculated by Bhatt and Grotch [13]. The overall contribution to the Lamb shift in hydrogen is small, approximately -2.5 kHz.

We now briefly describe higher order radiative corrections. The function $F_n(Z\alpha)$ is nonperturbative inasmuch as its evaluation requires consideration of an infinite set of Coulomb exchanges. At the present level of accuracy, it is necessary to consider higher order terms in the loop expansion. The term of order α^2 can be written in analogy with the one-loop expression as

$$
\Delta E_n(\text{two} - \text{loop}) = \frac{m_e\alpha^2(Z\alpha)^4}{\pi^2 n^3} H_n(Z\alpha) \tag{11}
$$

The complexity of the evaluation of the lowest order term in $F_n(Z\alpha)$ is connected with a sensitivity to the infrared region in one-loop order. However to lowest order in $Z\alpha$, this sensitivity is not present with two loops; and it suffices to carry out a standard vertex calculation using free propagators. The total result, including terms of the same order arising from the two-loop contribution to the electron anomalous magnetic moment and two-loop contributions to vacuum polarization, gives a contribution of 101 kHz to the classic Lamb shift.

The finite extension of the nuclear charge affects primarily the binding of S states, and for nonrelativistic systems leads to the well-known shift

$$
\Delta E_n(\text{finite} - \text{size}) = \frac{2}{3n^3}(Z\alpha)^4 m_r^3 \langle r^2 \rangle \ , \tag{12}
$$

where $\langle r^2 \rangle$ is the mean square radius of the charge distribution, and there is negligible sensitivity to the details of the distribution. We note that the effect of finite nuclear size is particularly important for muonic atoms, in which the muonic wave function has a significant overlap with the nucleus. Another nuclear effect is nuclear polarizability, which refers to the contribution from intermediate excited states of the nucleus. However, that effect should be greatly suppressed for fine structure because of a cancellation between the two two-photon exchange contributions. It is most likely that the effect of nuclear polarizability cannot be detected in hydrogen because of its smallness and the theoretical and experimental uncertainties.

At this point, we have mentioned all the known contributions that apply to fine structure. Important terms that have not been calculated are, in estimated order of importance, as follows:

i) Binding corrections to the two-loop Lamb shift. These corrections have an additional factor of $Z\alpha$ and may contribute at the few kHz level.

ii) Lowest order three-loop Lamb shift. Probably they give a contribution smaller than one kHz.

It is important to note that the numerical uncertainty in G_{SE} leads to a theoretical error of about 8 kHz in the classic Lamb shift for $Z = 1$. Hence the calculations described above must also be accompanied with a more accurate calculation of that quantity in order to reduce theoretical uncertainties to the 1 kHz level.

We turn now to comparison with experiment.

3.1.1 Classic Lamb Shift in Hydrogen

The application of the above procedure to the classic Lamb shift immediately encounters the problem that there are two measurements of the charge radius of the proton in the literature that are discrepant by more than the quoted error bars. As we are in no position to judge one measurement more reliable than the other, we simply carry out the calculation for each radius, and find

$$\begin{aligned} &\mathcal{S} = 1057.855(11)\ \mathrm{MHz} \\ &\qquad\qquad \text{for } \langle r^2\rangle^{1/2} = .805(11)\ \mathrm{fm}\ [14] \\ &\mathcal{S} = 1057.873(11)\ \mathrm{MHz} \\ &\qquad\qquad \text{for } \langle r^2\rangle^{1/2} = .862(12)\ \mathrm{fm}\ [15]\ , \end{aligned} \tag{13}$$

where $\mathcal{S} \equiv \Delta E_{S_{1/2}} - \Delta E_{P_{1/2}}$. The theoretical error arises mainly from the uncertainties in G_{SE} and the proton radius. The experimental result of Lundeen and Pipkin is 1057.845(9) MHz [16]. References to the extensive experimental literature on this transition can be found in a review by Pipkin [17]. A new measurement of the proton charge distribution is clearly needed before theory can be meaningfully confronted with experiment. Barring that, progress in reducing the theoretical uncertainties discussed above has the potential of providing an atomic physics determination of a fundamental property of the proton.

3.1.2 Classic Lamb Shift in Muonium

Muonium provides a very interesting system in which to study the Lamb shift, both because the muon is pointlike and thus has no finite size uncertainty, and because the mass ratio of the electron to the nucleus is significantly larger, by a factor of 8.9, than in hydrogen. Therefore recoil corrections, which test the basic framework of our understanding of the relativistic two-body problem, are larger and more easily studied. This field is still very new; and the most recent experiments

$$\begin{aligned} &\mathcal{S}(\text{muonium}) = 1042^{+21}_{-23}\ \mathrm{MHz}\ [18] \\ &\mathcal{S}(\text{muonium}) = 1070^{+12}_{-15}\ \mathrm{MHz}\ [19]\ , \end{aligned} \tag{14}$$

while consistent with the theoretical result 1047.52(1) MHz, do not yet provide such a test. We note that the calculation of the higher order recoil correction and the radiative-recoil correction, which contribute basically negligible -2.53 kHz and +3.15 kHz to the classic Lamb shift in hydrogen, contribute much larger -22.3 kHz and +27.9 kHz in this atom. This new physics can then be qualitatively tested in muonium if experimental accuracies can be reduced to the order of 10kHz, whereas it is problematical whether the situation in hydrogen will ever allow information about these effects to be obtained.

3.1.3 n=2 Energy Levels in Positronium

The study of this beautiful system of course provides the most rigorous test of recoil, since the mass ratio is now unity. Unfortunately, theory has advanced relatively slowly for two reasons. The first is that one lacks a small expansion parameter to simplify the kernels. The second is that because of the possibility of annihilation into photons, a large number of extra kernels must be considered. However, the result to order $m\alpha^5$ has been known since the mid-1950's. There has been recent experimental progress [20] in measuring the $n = 2$ levels. The measurements were of the $2\,^3S_1 \to 2\,^3P_0$, $2\,^3S_1 \to 2\,^3P_1$, and $2\,^3S_1 \to 2\,^3P_2$ transitions and uncertainties are at the few MHz level. The theoretical splittings are all in reasonable agreement with them. If the next round of experiments reach the 1 MHz level of accuracy, a considerable challenge will be presented to theory to determine the $m\alpha^6$ corrections.

3.1.4 Hydrogenic Ions

We recall that the theoretical interpretation of the hydrogen Lamb shift was made uncertain by conflicting experimental measurements of the finite size of the proton. An advantage of studying hydrogenic ions is that a precise measurement in a system with more completely understood nuclear properties can bypass this problem. In addition, the occurrence of high powers of Z leads to enhanced radiative effects: in particular, the function $G(Z\alpha)$ introduced above, is enhanced relative to the leading order of the Lamb shift by Z^2. Thus, although the experimental error obtained for He^+ is very large compared to that for hydrogen, the study of that system gives information about G_{SE} competitive with the latter experiments. It turns out that theory and experiment are in excellent agreement.

3.2 n=2 to n=1 transitions

Precise measurement of this transition permits determination of the $1S_{1/2}$ energy shift, which is otherwise inaccessible. Since the bulk of the Lamb shift scales as $1/n^3$, the shift is larger than in the $2S_{1/2}$ state. Alternatively, one might accept the validity of the theory and use the measurements to determine the Rydberg precisely. We use the first point of view here.

3.2.1 Hydrogen:

The hydrogenic measurement recently carried out by Beausoleil et al. [21]gives

$$\Delta\nu = 2\ 466\ 061\ 413.8(1.5)\ \mathrm{MHz}\ . \tag{15}$$

Using the new value of the Rydberg constant given in (1), the 1S Lamb shift is determined to be

$$\Delta\nu = 8173.3(1.7)\ \mathrm{MHz}\ . \tag{16}$$

This is consistent with the theoretical value 8172.96 (7) MHz, though the experimental error is relatively large. Because the natural linewidth of the transition is only 1.3 Hz, however, the prospects for a high accuracy test of QED for this state are clearly very bright.

3.2.2 Muonium:

The $n = 2$ to $n = 1$ splitting between $F = 1$ states in muonium has been measured [22,23] to be 2 455 527 936(120)(140) MHz, where the first error is statistical and the second systematic. Without hyperfine splitting, the theoretical prediction is 2 455 528 935 MHz, with error well under 1 MHz and a further uncertainty of 3.5 MHz arising from the reduced mass. Ignoring the small state dependence in hyperfine splitting, we can account for that effect by subtracting 7/8 of one quarter of the ground state hfs (the $F = 1$ part of the splitting) and find 2 455 527 959 MHz, in very good agreement with experiment.

3.2.3 Positronium:

The experiment measuring the energy difference between the $n = 1$ and $n = 2$ triplet states in positronium, which is reviewed by Chu and Mills [24], has determined the splitting [23,25]

$$\Delta E = 1\ 233\ 607\ 218.9(10.7)\ \mathrm{MHz}\ . \tag{17}$$

Turning to theoretical considerations, we note that the bulk of the transition is of course three eighths of a Rydberg, scaled down by a factor of two from the same transition in hydrogen because $m_r = m_e/2$. Corrections of order $\alpha^2 R_\infty$ include not only the fine structure effects already described, but also the equally important effects of hyperfine splitting, which shifts the triplet states we are concerned with here by $\alpha^2 R_\infty/6n^3$, and the effect of annihilation into a single virtual photon, which shifts these states upward by $\alpha^2 R_\infty/2n^3$. At this point experiment lies 1510.9(10.7) MHz below theory.

The next order $\alpha^3 R_\infty$ receives contributions from many sources. These are compiled, and some errors corrected, in [1]. The contribution to $n = 1$ to $n = 2$ splitting from this order turns out to be -1501.5 MHz, which leaves theory 9.5 MHz above experiment. The complete evaluation of the $\alpha^4 R_\infty$ terms will be a very large scale task involving a large number of Feynman diagrams. The nominal order of these terms is 18.6 MHz, so the next round of experiments, which should achieve an order of magnitude reduction in error, will clearly present a major challenge to theory in a system which is most sensitive to our understanding of the relativistic bound state problem.

3.3 Hyperfine Splitting in Muonium

Muonium hyperfine splitting provides one of the important ingredients for testing QED. This is primarily because muonium is essentially a pure QED system to the present level of accuracy. The effect of strong interactions, which enter through the vacuum polarization, is very small and has an uncertainty of less than one part in 10^8. Weak interactions between the electron and muon also have a very small effect. At the same time, the experimental accuracy is excellent and there is no obstacle to working out the theory to the same level; most of that theoretical work has already been accomplished. The principal uncertainties in comparing theory and experiment are currently smaller than 0.3 ppm and are likely to decrease further in the coming years. At the present time, it provides the most strenuous test of relativistic two-body theory in QED.

Hyperfine splitting in muonium is simpler than in hydrogen because it does not suffer the complications of nucleon structure. On the other hand, recoil corrections are relatively more important. Here the term hyperfine splitting refers to the energy (or frequency) separation between the spin 0 and spin 1 S-states of the atom, usually for the ground state. To match the accuracy of the experiment [26]

$$\Delta\nu(\mathrm{exp}) = 4\ 463\ 302.88(16)\ \mathrm{kHz}, \tag{18}$$

it is necessary to know all theoretical contributions through 1 ppm (e.g., through relative order α^3 and $\alpha^2 m_e/m_N$).

The first formulation of the theory of the hyperfine splitting (hfs) was due to Fermi [27], who obtained his result by making a nonrelativistic reduction of the Dirac equation. Traditionally the leading contribution to the ground state hfs is referred to as the "Fermi splitting":

$$E_F \equiv \frac{16}{3}\alpha^2 \frac{m_r^3}{m_e^2 m_N} hcR_\infty\ . \tag{19}$$

Planck's constant disappears in the comparison with the frequency measurement. The Rydberg is known to great accuracy, so the important uncertainties are in the value of α and the ratio of the electron mass to the muon mass. Another part of the same experiment independently determines the magnetic moment of the muon. When this is combined with the value from SIN [28], and the value of the muon anomaly a_μ, one obtains $m_\mu/m_e = 206.768\,262(62)$.

The reduced mass enters (19) through the square of the nonrelativistic wave function at the (spatial) origin. Breit [29] treated the Dirac wave functions without approximation and derived a correction factor of $(1+\frac{3}{2}\alpha^2+...)$ for the ground state. He could not supply the correct reduced mass factor at that time because of the complications of dealing properly with recoil for the Dirac equation.

We refer to the contributions not involving recoil as the QED contributions. To indicate the separation of binding effects from radiative corrections, we use the convention of including factors of $Z\alpha$ for the former in the following expression:

$$\begin{aligned}\Delta E(\text{hfs; QED}) = E_F(1+a_\mu) \cdot &\\ \Big\{1+\frac{3}{2}(Z\alpha)^2 + a_e + \alpha(Z\alpha)\left(\ln 2 - \frac{5}{2}\right)&\\ +\frac{\alpha(Z\alpha)^2}{\pi}\left[-\frac{8}{3}\ln Z\alpha\left(\ln Z\alpha - \ln 4 + \frac{281}{480}\right)\right.&\\ \left.+(15.38\pm 0.29)\frac{1}{1}\right]& \qquad (20)\\ +\frac{\alpha^2(Z\alpha)}{\pi}\left[-\frac{4}{3}\ln^2\frac{1+\sqrt{5}}{2} - \frac{20\sqrt{5}}{9}\ln\frac{1+\sqrt{5}}{2}\right.&\\ \left.+\frac{608}{45}\ln 2 + \frac{\pi^2}{9} - \frac{38\pi}{15} + \frac{91639}{37800} + D_1{}'\right]\Big\}.&\end{aligned}$$

Here a_μ represents the anomalous magnetic moment of the muon. The explicit factor of $(1+a_\mu)$ appearing here means that these contributions are proportional to the muon's total magnetic moment. This is a consequence of the fact that they arise from very small internal momenta ($\leq \mathcal{O}(m_e)$) and hence are not sensitive to the structure of the current distribution in the muon. The calculation of these terms involved the efforts of many physicists and spanned a nearly forty year period. It is still continuing; see [1] for further information. The most recent contributions are the explicitly given ones of relative order $\alpha^2(Z\alpha)$, which involve two virtual photons [30]. The $D_1{}'$ term represents as yet uncalculated radiative corrections involving two virtual photons, as illustrated in Fig.4.

The presently known recoil corrections sum to

$$\begin{aligned}\Delta E(\text{hfs; rec}) = &E_F\Big\{-\frac{3\alpha}{\pi}\frac{m_e m_\mu}{m_\mu^2 - m_e^2}\ln\frac{m_\mu}{m_e}\\ &+\frac{\gamma^2}{m_e m_\mu}\left[2\ln\frac{m_r}{2\gamma} - 6\ln 2 + 3\frac{11}{18}\right]\Big\}.\end{aligned} \qquad (21)$$

where $\gamma \equiv m_r\alpha$. The muon's anomalous moment does not appear here since the internal momenta are large in the first term, and its effect should be counted as part of the radiative-recoil correction.

The radiative-recoil contributions, which arise from both lepton lines and from vacuum polarization, are given by

$$\begin{aligned}\Delta E(\text{hfs; rad}-\text{rec}) = \left(\frac{\alpha}{\pi}\right)^2\frac{m_e}{m_\mu} \cdot &\\ \left[-2\ln^2\frac{m_\mu}{m_e} + \frac{13}{12}\ln\frac{m_\mu}{m_e}\right.& \qquad (22)\\ \left.+\frac{21}{2}\zeta(3) + \frac{\pi^2}{6} + \frac{35}{9} + (1.91\pm 0.26)\right].&\end{aligned}$$

The present importance of these terms was first pointed out by Caswell and Lepage [31], who evaluated the one proportional to $\ln^2(m_\mu/m_e)$. The most recent development is that the lepton line contributions of the non-logarithmic terms were evaluated analytically by Eides, et al [32]. The hadronic vacuum polarization contribution is represented by the 1.9±0.3 term. See [1] for more details.

The complete theoretical result is

$$\Delta\nu(\text{theory}) = 4\;463\;303.73(.66)(0.21)(1.0)\ \text{kHz}\ . \qquad (23)$$

The first uncertainty reflects the uncertainty from the measurement of m_μ, the second from that of α, and the third is an order of magnitude estimate of the uncalculated $D_1{}'$ contribution. The comparison with the experimental result is very satisfactory. The main job remaining for theorists is the evaluation of the $D_1{}'$ coefficient. After that, uncalculated QED contributions will be at the level of one part in 10^8.

3.4 Hyperfine Splitting in Hydrogen

The hyperfine splitting between the spin 0 and spin 1 levels in the hydrogen ground state is one of the most accurately measured quantities in physics. The most recent experimental determinations of this quantity [33,34] give

$$\nu(\text{exp}) = 1420.405\,751\,766\,7(9)\text{MHz}. \qquad (24)$$

The most important theoretical contributions are those given in (20), but with the explicit factor of $(1+a_\mu)$ omitted there and incorporated instead into the definition of E_F in (19). This definition is the more usual one for the hydrogen hfs.

The known QED contributions to the hfs, as given in (20), yield

$$\nu(\text{QED}) = 1420.451\,99(14)\ \text{MHz}\ . \qquad (25)$$

The difference between QED theory and experiment is then

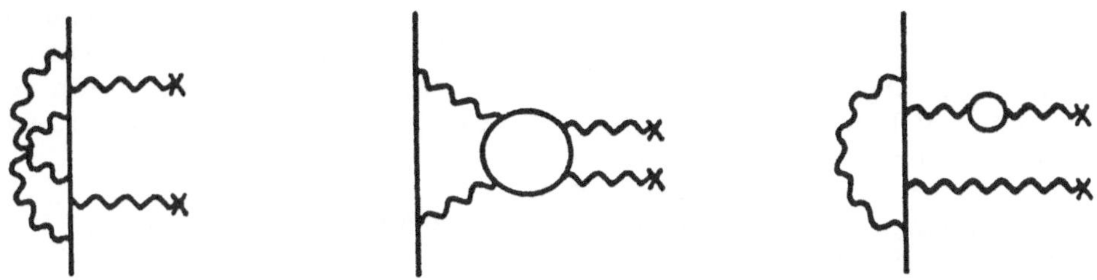

Fig. 4. Graphical examples of contributions to $D_1{}'$. Either of the external interactions is with the nuclear magnetic moment and the other is with its Coulomb potential.

$$\frac{\nu(\text{QED}) - \nu(\text{Exp})}{\nu_F} = 32.55(10)\ \text{ppm}. \tag{26}$$

In order to take full advantage of the refinements described so far, it would be necessary to calculate the recoil and dynamical corrections involving the proton's structure to about 0.1 ppm relative accuracy. Although such a goal has not yet been achieved, substantial progress has been made in working out these structure-dependent corrections. They are usually expressed as

$$\Delta E(\text{hfs; structure}) = E_F\left[\delta_p(\text{rigid}) + \delta_p(\text{polarizability})\right]. \tag{27}$$

The quantity $\delta_p(\text{rigid})$ is computed by using elastic form factors to approximate the electromagnetic interactions of the proton. The term $\delta_p(\text{polarizability})$, the proton polarizability correction, contains all of the effects of the dynamics of the proton that are not included in $\delta_p(\text{rigid})$.

The most important form-factor-dependent correction does not involve recoil; it is known as the "nonrelativistic size correction." In computing this correction, one treats the proton as a nonrecoiling particle with a fixed charge-current distribution of finite extent. It was first analyzed by Zemach [35] and may be written

$$\delta_p(\text{Zemach}) = -2m_r\alpha R_{pr} \equiv \frac{2\alpha m_r}{\pi^2}\int\frac{d^3p}{p^4}\left[\frac{G_E(-p^2)G_M(-p^2)}{1+\kappa} - 1\right], \tag{28}$$

where R_{pr} is a mean radius associated with the proton's charge-current distribution. Accurate calculation of $\delta_p(\text{Zemach})$ requires good knowledge of the elastic electric and magnetic form factors from experiment, as is seen from the second form. When evaluated with the commonly-used dipole parametrization of the elastic form factors, it gives a contribution of -38.72 ppm, which removes most of the difference between theory and experiment. This contribution requires a serious reanalysis. The uncertainty in the total structure-dependent contribution has been estimated to be 0.9 ppm [36].

Now let us turn to the effects of recoil. In hydrogen, the one-loop (relative order $\alpha(m_e/m_p)$) contributions, corresponding to the first term of (21), must be redone. These contributions involve very large characteristic momenta and hence they involve the proton's internal structure in an important way. On the other hand, the second term of (21), which contains no logarithm of the heavy particle mass, involves momenta of order m_e and less. Thus, in the case of the hydrogen hfs, such contributions would not show a sensitivity to the proton's structure. Hence, in analyzing contributions to the hydrogen hfs of this type, one can make use of any results that were derived originally for the muonium hfs. In addition contributions that arise from the anomalous moment of the proton must also be included.

A recent reanalysis of the recoil corrections has been given by Bodwin and Yennie [37], and they find 5.68 ppm. When this is combined with the Zemach correction, the difference between theory and experiment becomes

$$\frac{\nu(\text{theory}) - \nu(\text{Exp})}{\nu_F} = (-0.47 \pm 0.56 \pm \text{unknown})\text{ppm}. \tag{29}$$

The error of 0.56 ppm contains a small contribution from the uncertainty in α, but it arises mainly from the uncertainty in the parameter Λ in the dipole fit to the proton's elastic form factors.

Finally, we turn to the term $\delta_p(\text{polarizability})$. For details, we refer the reader to the review of Hughes and Kuti [38]. Here we merely note that deRafael [39] and Gnädig and Kuti [40] have shown how to use data from inelastic electron scattering with polarized beam and target to put a bound on $\delta_p(\text{polarizability})$. In their review, Hughes and Kuti give $|\delta_p(\text{polarizability})| < 4\,\text{ppm}$. The result (29) leaves little room for a polarizability correction.

The part of the error labeled "unknown" in (29) represents all remaining uncertainties in $\nu(\text{theory})$. The most important sources of these uncertainties are the radiative corrections to the structure-dependent contributions and the systematic errors from the proton's form factors. The radiative corrections to the structure-dependent contributions could potentially contribute at the level of 1 ppm. However, they could probably be calculated to a precision of 0.01 ppm or better. It is difficult to estimate how much the use of more precise expressions for the form factors might shift the central value of ν(theory) or how large the remaining statistical uncertainty in that determination might be. The work of Bodwin and Yennie has shifted the theory by $\approx +1.5$ ppm. Their subjective impression is that a statistical uncertainty of 1 ppm seems to be a reasonable estimate of what might be achieved through a more precise treatment of existing form factor data, together with an evaluation of radiative corrections that are significant at that level. It would not be surprising if such an analysis were to lead to a shift of as much as 1 ppm in the central value of ν(theory). However, it is unlikely that the analysis would reveal any incompatibility with the above bounds on $|\delta_p(\text{polarizability})|$. In fact, by incorporating the refinements in the computation of ν(theory) that we have already mentioned, one could use the hydrogen hfs to determine δ_p(polarizability) with a precision of roughly 1 ppm.

3.5 Hyperfine Structure in Positronium

The theoretical expression for the positronium hfs is given by

$$\begin{aligned}\nu =&\alpha^2 R_\infty \left[\frac{2}{3} + \frac{1}{2} - \frac{\alpha}{\pi}(\ln 2 + \frac{16}{9})\right.\\ &\left. + \frac{5}{12}\alpha^2 \ln \alpha^{-1} + K\alpha^2 + K'(\alpha)\alpha^3\right] .\end{aligned} \tag{30}$$

The α/π term includes radiative corrections as well as a term obtained by taking the equal mass limit of (21). The $\alpha^2 \ln \alpha^{-1}$ term comes partially from (21), but it also has contributions from annihilation kernels. The α^2 term has been only partially evaluated. The principal set of uncalculated terms involves two-loop corrections to the one-photon annihilation graphs. The K' term is uncalculated, but it could be enhanced by factors of $\ln \alpha$. Recently recoil corrections of the type contained in (21) were evaluated by Caswell and Lepage [41] using a novel approach to the bound state problem based on the use of effective Lagrangians. The idea of this technique is to effect a separation of the high-energy part of the calculation, where renormalization is an important issue and Feynman gauge the most convenient gauge, from the low-energy part, which is dominated by the nonrelativistic region and is best calculated in Coulomb gauge.

The most recent experimental result by Ritter et al. [42] is

$$\nu = 203\ 389.10(74) \text{ MHz}. \tag{31}$$

The present theoretical value (Sapirstein and Kinoshita [43]), including all known contributions along with the corrected three-photon answer and the recoil term calculated by Caswell and Lepage, is

$$\nu = 203\ 404.5(0.6)(9.3) \text{ MHz} . \tag{32}$$

The first error is dominated by the numerical uncertainty of the Caswell and Lepage calculation, and the second corresponds to a possible magnitude of unity for uncalculated contributions to K. The difference between theory and experiment is quite compatible with the expected magnitude of the uncalculated terms.

Acknowledgement. Preparation of this paper was supported in part by the National Science Foundation. I wish to thank the organizers of the workshop on The Future of Muon Physics for providing travel support to attend the meeting. I would also like to thank Jonathan Sapirstein for his help in preparing the longer review on which most of this paper is based and Toichiro Kinoshita for compiling an update on the status of muonium hyperfine splitting.

References

1. J. R. Sapirstein and D. R. Yennie, in Quantum Electrodynamics (World Scientific, Singapore, 1990), p. 560.
2. G. Breit, Phys. Rev. **29**, 553 (1929).
3. E. E. Salpeter, Phys. Rev. **87**, 328 (1952).
4. J. Schwinger, Proc. Nat. Acad. Sci. USA **37**, 452, 455 (1951).
5. E. E. Salpeter, and H. A. Bethe, Phys. Rev. **84**, 1232 (1951).
6. E. Richard Cohen and Barry N. Taylor, Rev. Mod. Phys. **59**, 1121 (1987).
7. F. Biraben, J. C. Garreau, L. Julien, and M. Allegrini, Phys. Rev. Lett. **62**, 621 (1989).
8. W. A. Barker and F. N. Glover, Phys. Rev. **99**, 317 (1955).
9. J. Pirenne, Arch. Sci. Phys. Nat. **29**, 121, 207, 265 (1947); V. Berestetski and L. Landau, J. Exp. Theor. Phys. USSR **19**, 673, 1130 (1949); R. A. Ferrell, Phys. Rev. **84**, 858 (1951).
10. P. J. Mohr, Phys. Rev. **A23**, 2338 (1982).
11. G. W. Erickson and H. Grotch, Phys. Rev. Lett. **60**, 2611 (1988); erratum: Phys. Rev. Lett. **63**, 1326 (1989).
12. M. Doncheski, H. Grotch, and G. W. Erickson, Phys. Rev. **A45**, 2152 (1991).
13. G. Bhatt and H. Grotch, Phys. Rev. **A31**, 2794 (1985); Phys. Rev. Lett. **58**, 471 (1987); Ann. Phys. (NY) **178**, 1 (1987).
14. D. J. Drickey and L. N. Hand, Phys. Rev. Lett. **9**, 521 (1962); L. N. Hand, D. J. Miller, and R. Wilson, Rev. Mod. Phys. **35**, 335 (1963).
15. G. G. Simon, Ch. Schmidt, F. Borkowski, and V. H. Walther, Nucl. Phys. **A333**, 381 (1980).
16. S. R. Lundeen and F. M. Pipkin, Phys. Rev. Lett. **46**, 232 (1981); S. R. Lundeen and F. M. Pipkin, Metrologia **22**, 9 (1986).
17. Francis M. Pipkin, in Quantum Electrodynamics (World Scientific, Singapore, 1990), p. 696.
18. K. A. Woodle et al., Phys. Rev. **A41**, 93 (1990).
19. C. J. Oram et. al., Phys. Rev. Lett. **52**, 910 (1984).
20. S. Hatamian, R. S. Conti, and A. Rich, Phys. Rev. Lett. **58**, 1833 (1987).
21. R. G. Beausoleil, D. H. MacIntyre, C. J. Foot, E. A. Hildum, B. Couillaud, and T. W. Hänsch, Phys. Rev. **A35**, 4878 (1987).
22. Steven Chu, A. P. Mills, Jr., A. G. Yodh, K. Nagamine, Y. Miyake, and T. Kuga, Phys. Rev. Lett. **60**, 101 (1988).

23. K. Danzmann, M. S. Fee, and Steven Chu, Phys. Rev. A**39**, 6072 (1989).
24. Allen P. Mills and Steven Chu, in Quantum Electrodynamics (World Scientific, Singapore, 1990), p. 774.
25. Steven Chu, Allen P. Mills, Jr., and John L. Hall, Phys. Rev. Lett. **52**, 1689 (1984).
26. F. G. Mariam, W. Beer, P. R. Bolton, P. O. Egan, C. J. Gardner, V. W. Hughes, D. C. Lu, P. A. Souder, H. Orth, J. Vetter, U. Moser, and G. zu Putlitz, Phys. Rev. Lett. **49**, 993 (1982).
27. E. Fermi, Z. Phys. **60**, 320 (1930).
28. E. Klempt et. al., Phys. Rev. D**25**, 652 (1982).
29. G. Breit, Phys. Rev. **35**, 1447 (1930).
30. M. I. Eides, S. G. Karshenboim, and V. A. Shelyuto, Phys. Lett. B**229**, 285 (1989); 249, 517 (1990).
31. W. E. Caswell and G. P. Lepage, Phys. Rev. Lett. 41, 1092 (1978).
32. M. I. Eides, S. G. Karshenboim, V. A. Shelyuto, Phys. Lett. **202B**, 572 (1988); S. G. Karshenboim, V. A. Shelyuto, and M. I. Eides, Zh. Eksp. Teor. Fiz. **92**, 1188 (1987) [Eng. transl.: Sov. Phys. JETP **65**, 664 (1987).
33. H. Hellwig, R. F. C. Vessot, M. W. Levine, P. W. Zitzewitz, D. W. Allan, and D. J. Glaze, IEEE Trans. Instrum. **IM-19**, 200 (1970).
34. L. Essen, R. W. Donaldson, M. J. Bangham, and E. G. Hope, Nature **229**, 110 (1971).
35. A. C. Zemach, Phys. Rev. **104**, 1771 (1956).
36. S. J. Brodsky and S. D. Drell, Ann. Rev. Nuclear Science **20**, 147 (1970).
37. G. T. Bodwin, and D. R. Yennie, Phys. Rev. D**37**, 498 (1988).
38. V. W. Hughes and J. Kuti, Ann. Rev. Nucl. Part. Sci. **33**, 611 (1983).
39. E. de Rafael, Phys. Lett. **37B**, 201 (1971).
40. P. Gnädig and J. Kuti, Phys. Lett. **42B**, 241 (1972).
41. W. E. Caswell and G. P. Lepage, Phys. Lett. **167B** 437 (1986).
42. M. Ritter, P. O. Egan, V. W. Hughes, and K. A. Woodle, Phys. Rev. A**30**, 1331 (1984).
43. T. Kinoshita and J. Sapirstein, in Atomic Physics **9**, Eds. Robert S. Van Dyck, Jr. and E. Norval Fortson, World Scientific (Singapore, 1984).

This article was processed using Springer-Verlag TEX Z.Physik C macro package 1991
and the AMS fonts, developed by the American Mathematical Society.

Z. Phys. C - Particles and Fields 56, S24-S30 (1992)

Zeitschrift für Physik C Particles and Fields

Electro-weak interaction in muonic atoms

J. Bernabeu

Departement de Fisica Teòrica, Universitat de València, e IFIC, Centre Mixte Univ. València-CSIC, 46100 Burjassot, València SPAIN

01-December-1991

Abstract. The parity non-conserving effective neutral current interaction between charged leptons and nucleons is studied in its implications for atomic physics. Present results on heavy electronic atoms are discussed within the standard electroweak theory and beyond. The new features provided by muonic atoms open the way to the nuclear-spin-dependent parity non-conserving effects. Different observables proposed to study these effects in muonic atoms are reviewed.

1 Introduction

The standard theory of electro-weak interactions has been confirmed by high precision experiments [1] at the CERN electron-positron collider LEP. The measurements of the Z-mass, its width, the leptonic widths, the hadronic width, the forward-backward asymmetries, the longitudinal polarization of the τ-lepton,...., have given a beautiful test of the standard theory and a precise determination of its parameters. Apart from their intrinsic interest, parity non-conserving (PNC) observables in Atomic Physics are complementary to LEP. In this paper I would like to discuss the physics content of the PNC effective neutral current interaction between charged leptons and quarks, its implications for atomic physics and the role played by PNC observables in muonic atoms.

The effective Lagrangian for the PNC lepton-quark interaction is written as

$$L_{PNC} = \frac{G_F}{\sqrt{2}} \sum_i \left[C_{1i}\bar{l}\gamma^\mu\gamma_5 l\bar{q}_i\gamma_\mu q_i + C_{2i}\bar{l}\gamma^\mu l\bar{q}_i\gamma_\mu\gamma_5 q_i\right] \qquad (1)$$

where the index $i = u, d, s...$, runs over the different flavours of quarks. The Standard Model at tree level gives [2] the following couplings:

$$C_{1u} = -\frac{1}{2} + \frac{4}{3}\sin^2\Theta_w$$
$$C_{1d} = \frac{1}{2} - \frac{2}{3}\sin^2\Theta_w \qquad (2)$$

for the axial current of leptons times the vector current of quarks, and

$$C_{2u} = -C_{2d} = -\frac{1}{2} + 2\sin^2\Theta_w \qquad (3)$$

for the vector current of leptons times the axial current of quarks. Using the weak isospin symmetry for the different families of quarks, we have $C_{1s} = C_{1d}$ and $C_{2s} = C_{2d}$.

The matrix elements of the hadronic current for the nucleon are written for $Q^2 = 0$ as

$$< p \mid \bar{q}_i\gamma_\mu q_i \mid p > = G_V^{(i)}\bar{p}\gamma_\mu p$$
$$< p \mid \bar{q}_i\gamma_\mu\gamma_5 q_i \mid p > = G_A^{(i)}\bar{p}\gamma_\mu\gamma_5 p \qquad (4)$$

The conserved vector current and its coherent character, with the vector charge equal to the quark-number, determine the couplings $G_V^{(u)} = 2, G_V^{(d)} = 1, G_V^{(s)} = 0$, for the proton.

In terms of definite $U(3)$ flavour transformation properties, one can introduce the following combination of couplings

$$G_A^{(3)} = G_A^{(u)} - G_A^{(d)}$$
$$G_A^{(8)} = G_A^{(u)} + G_A^{(d)} - 2G_A^{(s)} \qquad (5)$$
$$G_A^{(0)} = G_A^{(u)} + G_A^{(d)} + G_A^{(s)}$$

The charged weak currents tranform, following Cabbibo theory for the light quarks, as an octet under flavour SU(3). The two form factors $G_A^{(3)}$ and $G_A^{(8)}$ can be expressed through the amplitudes F and D known from semi-leptonic decays of baryons

$$G_A^{(3)} = F + D = 1.254 \pm 0.006;$$
$$G_A^{(8)} = 3F - D = 0.68 \pm 0.04 \qquad (6)$$

The EMC measurements [3] of the polarization-dependent structure functions of the proton determines an independent combination of $G_A^{(3)}$, $G_A^{(8)}$ and the singlet $G_A^{(0)}$. One obtains

$$G_A^{(0)} = 0.12 \pm 0.17 \tag{7}$$

It is remarkable that the singlet current coupling $G_A^{(0)}$ is compatible with zero, which seems to be in contradiction with naive expectations from models with constituent u, d quarks for the proton. In such models it would mean that the total helicity carried by all quarks (and antiquarks) in a polarized proton is small. For our considerations we do not explicitly need constituent quark models of the nucleon, since we work instead with the effective nucleonic currents of Eq.(4). The above results lead to

$$G_A^{(s)} = -0.19 \pm 0.06 \tag{8}$$

This important conclusion would be confirmed in the SMC-experiment at CERN, conducted by Prof. V. Hughes.

Using strong isospin as a symmetry of the strong interactions in the limit in which $(m_d - m_u)$ is small, we have that for the neutron the following substitutions have to be made

$$p \to n \qquad \Longrightarrow \qquad \begin{array}{c} G_{V,A}^{(u)} \rightleftharpoons G_{V,A}^{(d)} \\ \\ G_{A,V}^{(s)} \to G_{A,V}^{(s)} \end{array} \tag{9}$$

where $G_{V,A}^{(i)}$ are the form factors defined in Eq.(4).

2 Parity non-conservation in atoms

The study of PNC observables in Atomic Physics provides a test of the fundamental symmmetries of Nature and, specifically, of the standard electro-weak theory, the neutral current lepton-quark couplings and the corresponding electroweak radiative corrections.

Parity Violation effects have been observed in heavy electronic atoms [4]. They are characterized by different experimental methods, atomic elements and transitions observed. All of them use the enhancement, pointed out by Bouchiat and Bouchiat, of the neutral current mixing parameter in heavy atoms. The piece of the effective electron-quark parity violating interaction corresponding to the axial coupling of the electron times the vector coupling of quarks become coherent in the nucleus, so the expectation value is proportional to the number of quarks. The corresponding weak neutral charge induced by the PNC Z-exchange diagram, which interferes with γ-exchange as shown in Figure 1, is given

$$Q_w = -2\{(2Z+N)C_{1u} + (Z+2N)C_{1d}\} \tag{10}$$

where (Z, N) are the (proton, neutron) numbers in the nucleus and $C_{1u,d}$ are the couplings of Eq.(2).

The PNC interactions (with time-reversal symmetry) induces pure-imaginary atomic dipole moments that

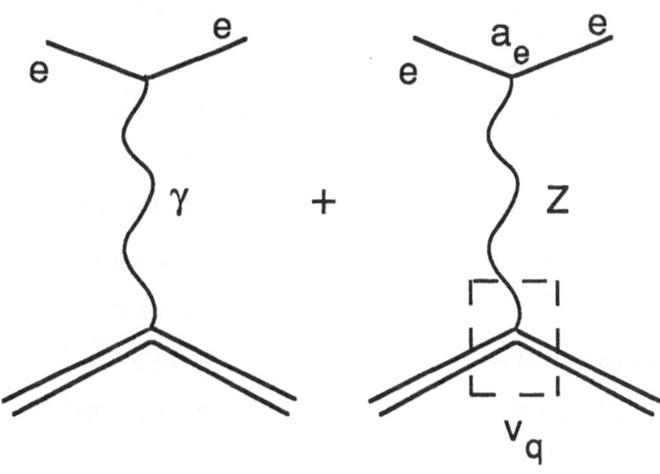

Fig. 1. Parity violation in atoms. The weak neutral charge induced by the PNC Z-exchange interference with the γ-exchange.

are observable in atomic transitions. There are two important classes of such experiments, each involving the interference between an ordinary (but relatively weak, in order to enhance the effect) electromagnetic transition and the PNC-induced E1 transition. In the first (M1) class of experiments, the induced E1 transition is coupled into an M1 transition, producing circularly polarized light or a rotation of plane polarized light. The quantity which is measured by the experiments searching for optical rotation of light in atomic vapour is $R = Im(E_1^{PNC}/M_1)$, which factorizes Q_w from the atomic structure dependent inputs. In Table 1 the experimental values of the quantitiy R for bismuth (two transition lines), lead and thallium are collected.

The difference in the order of magnitude of the effect for thallium is due to the choice of a forbidden magnetic transition in this case.

In the second (Stark) class of experiments, a static electric field $\vec{E_s}$ induces an E1 transition between two states, which interferes with the PNC-induced E1 transition between the same two states. The corresponding transition rate depends then on the relative directions of $\vec{E_s}$ and the polarization vectors of the absorbed and reemitted radiation. The measured quantity is expressed in terms of $Im(E_1^{PNC}/\beta)$, where β is the vector polarizability. The experimental values obtained are shown in Table 2.

The improved measurement [5] of PNC effects in cesium by the Boulder (1988) experiment provides an experimental precision of 2%. This allows the extraction of electroweak radiative correction contributions if the atomic structure calculations are accurate enough. A major theoretical effort for the $6S_{1/2} \to 7S_{1/2}$ transition in Cs has been presented in Ref. [6], with the claim of 1% theoretical uncertainty. The extracted value of Q_w from the Boulder (1988) experimental result of Table 2 and this atomic structure calculation is

Table 2. Results of experiments measuring in a static electric field on induced E1 transitions between two states which interferes with the PNC induced E1 transition.

Experiment	Element	$Im(E_1^{PNC}/\beta)(mV/cm)$
Berkeley (1981)	Tl	-1.80 ± 0.48
Berkeley (1981)	Tl	-1.73 ± 0.27
Paris (1982)	Cs	$+1.33 \pm 0.25$
Paris (1984)	Cs	-1.75 ± 0.27
Boulder (1985)	Cs	-1.65 ± 0.13
Boulder (1988)	Cs	-1.576 ± 0.034

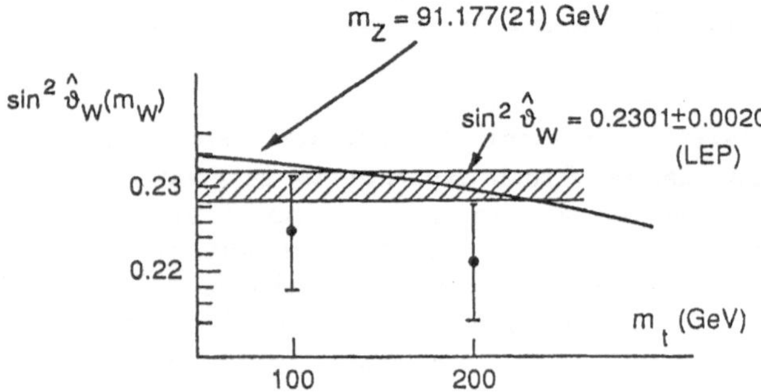

Fig. 2. $sin^2\Theta_w$ as a function of the top quark mass m_t from precise Z mass measurements of the LEP experiments and from the leptonic decay widths of the Z-resonances.

$$Q_w^{exp}(Cs) = -71.04(1.58)[0.88] \quad (11)$$

where the first error is from the experiment and the second from the atomic theory.

When the result (11) is interpreted in the standard electroweak theory, the inclusion of the electroweak radiative corrections brings not only $sin^2\Theta_w$ but also m_t as parameters of the PNC-observable. In the modified minimal-subtraction scheme, one gets [6]

$$sin^2\hat{\Theta}_w(m_w) = \begin{array}{l} 0.2242(65)[36], m_t = 100GeV \\ \\ 0.2215(65)[36], m_t = 200GeV \end{array} \quad (12)$$

This result is compared with the determination provided by the precise Z-mass measurement [7] in the LEP experiments, as shown in Fig.2. The dependence on m_t is almost identical for $Q_w(Cs)$ and for m_Z.

The LEP-value for $sin^2\hat{\Theta}_w$ from the leptonic decay widths of the Z is also included in Fig.2 for comparison.

If, instead of using $sin^2\hat{\Theta}_w$ as independent parameter of the theory, one takes the precisely known Z-mass together with α (fine structure constant), G_F (Fermi coupling constant), m_t (top quark mass) and m_H (Higgs mass), the theoretical prediction of $Q_w(Cs)$ turns out to be almost independent of m_Z.

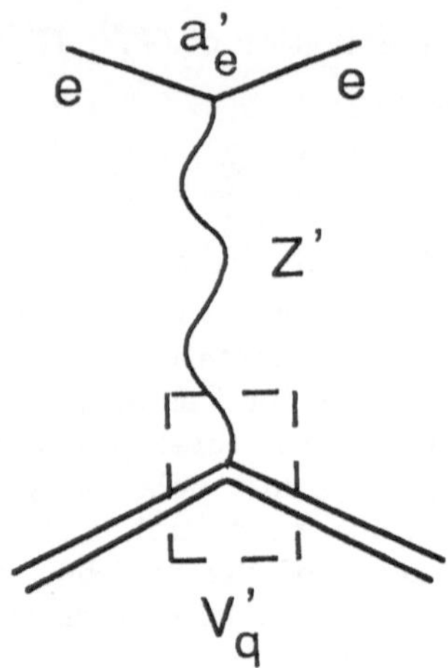

Fig. 3. Additional tree level contribution to the weak charge Q_w from an extra neutral vector boson Z'.

One gets

$$Q_w^{th}(Cs) = -73.10 \pm 0.13 \quad (13)$$

so that, independently of the unknown value of m_t, the difference

$$\delta Q_w(Cs) = Q_w^{exp} - Q_w^{th} = 2.1 \pm 1.8 \quad (14)$$

constitutes a test of the electroweak radiative corrections to the standard theory and a probe sensitive to new physics.

One example of the sensitivity of δQ_w to new physics, appearing through loop corrections, is given by technicolour theories [8]. With the self-energy corrections evaluated in one-loop technifermion approximation, the value of Q_w receives a negative contribution, whose size is proportional both to the number of technicolours and of technidoublets.

In models with an extra neutral vector boson Z', there is an additional tree-level contribution to Q_w coming from Z'-exchange, in the limit of small mixing with the standard Z. This is shown in Fig.3.

One has

$$\delta Q_w^{new} = 16\frac{m_Z^2}{m_{Z'}^2}[(2Z+N)a_e'v_u' + (Z+2N)a_e'v_d'] \quad (15)$$

where v_f', a_f' are the vector and axial vector couplings of the fermion f to the vector boson Z'. These couplings depend [9] on the particular extended gauge theory, like models with extra $U(1)$ or left-right models. To be specific, a superstring-inspired model gives detailed predictions

$$v_u' = 0, \; a_e' = v_d' = \frac{1}{4}\sqrt{\frac{5}{3}}sin\Theta_w \quad (16)$$

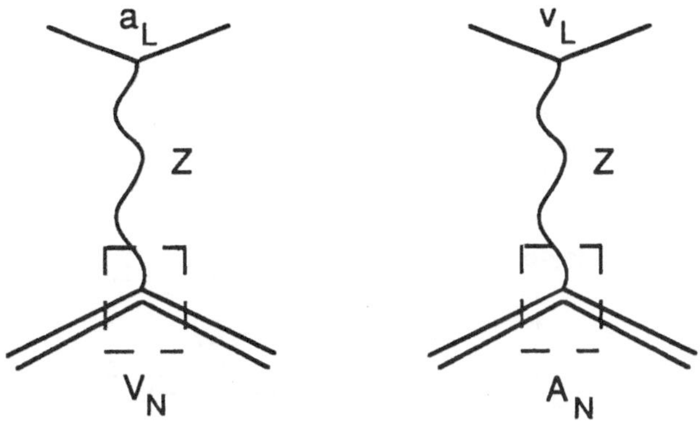

Fig. 4. There are two contributions to the Z-exchange diagrams of charged leptons with nucleons.

so that one would have

$$\delta Q_w^{new} = \frac{5}{3} sin^2\Theta_w (Z+2N) m_Z^2/m_{Z'}^2 \qquad (17)$$

If the limit (14) for $\delta Q_w(Cs)$, whose value is compatible with zero, is used to bound $m_{Z'}$ from Eq.(17) one obtains a significant result $m_{Z'} \gtrsim 400 GeV$.

3 Nuclear-spin-dependent effects

Among the virtues offered by the study of PNC-effects in muonic atoms, one encounters that they are practical tools to obtain a better access to all PNC-couplings of charged leptons to nucleons.

The effective Lagrangian for the PNC interaction of charged leptons with nucleons can be written as

$$L_{PNC} = \frac{G_F}{\sqrt{2}} \sum_{i=p,n} [C_{1i}\bar{l}\gamma^\mu\gamma_5 l\bar{N}_i\gamma_\mu N_i \quad + \quad C_{2i}\bar{l}\gamma^\mu l\bar{N}_i\gamma_\mu\gamma_5 N_i] \qquad (18)$$

associated with the two contributions from the Z-exchange diagram of Fig.4, respectively.

In terms of the neutral current lepton-quark couplings and the form factors of Section 1, one has for the nuclear-spin-independent couplings

$$C_{1p} = 2C_{1u} + C_{1d}$$
$$C_{1n} = C_{1u} + 2C_{1d} \qquad (19)$$

and their associated coherent action on the nucleus, as seen in Eq.(10). For the nuclear-spin-dependent terms one obtains [10]

$$C_{2p} = G_A^{(u)}C_{2u} + [G_A^{(d)} + G_A^{(s)}]C_{2d}$$
$$C_{2n} = G_A^{(d)}C_{2u} + [G_A^{(u)} + G_A^{(s)}]C_{2d} \qquad (20)$$

and they do not add coherently in the nucleus.

In order to single out the nuclear-spin-dependent PNC effect generated by the second term of Eq.(18) one has to measure PNC observables for different hyperfine transitions in atoms. A first indication of this effect has

Table 3. PNC in atomic Cesiums measured in the Boulder experiment for different hyperfine transitions.

	Lines	$Im(E_1^{PNC}/\beta)(mV/cm)$
Boulder (1988), Cs	$F=4 \rightarrow F'=3$	-1.639 ± 0.047
	$F=3 \rightarrow F'=3$	-1.513 ± 0.049

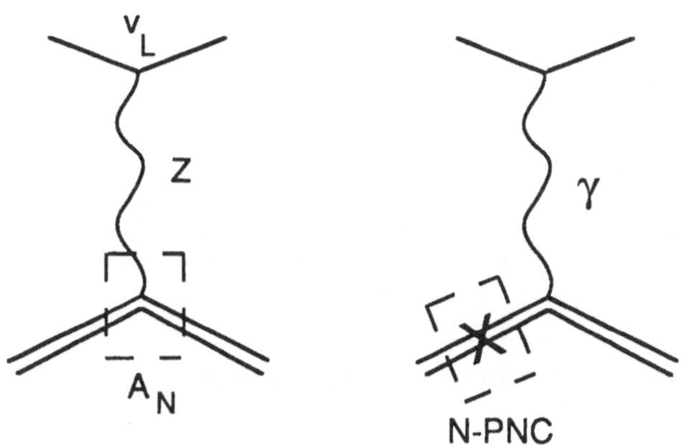

Fig. 5. The weak neutral current and the nuclear anapole moment contributions to PNC effects.

been obtained by the Boulder (1988) experiment [5] in electronic Cs atoms. For different hyperfine levels, the measured quantity $Im(E_1^{PNC}/\beta)$ is different, as shown in Table 3.

The impresssive precision reaching high levels of accuracy in the atomic experiments of the Stark-type can lead to significant nuclear-spin-dependent PNC effects. But the progress is difficult, because the experimental effect is comparatively very small. Furthermore, on the theoretical side the PNC-neutral axial current interaction contribution is contaminated by the effect of the nuclear anapole moment.

The two contributions which compete for the nuclear-spin-dependent PNC effect are given by the two diagrams of Fig.5.

The nuclear anapole moment describes [11] the effect of the parity violating nuclear forces on the nucleus electromagnetic current. This mechanism induces the same effective operator for the lepton-Nucleus amplitude as the neutral current interaction of the vector lepton current with the axial nuclear current. Although formally a higher order α-correction, the nuclear anapole moment contribution has the electromagnetic coherence in heavy nuclei whereas the neutral axial current effect has no coherent effect. Furthermore, the PNC lepton-nucleus interaction associated with the hadronic neutral axial current is suppressed due to the small vector lepton coupling. As a consequence, the nuclear-spin-dependent PNC effect is dominated by the electron interaction with the nuclear anapole moment in heavy atoms.

If one is interested in disentangling the neutral current effect, PNC-observables in light muonic atoms are thus of great value.

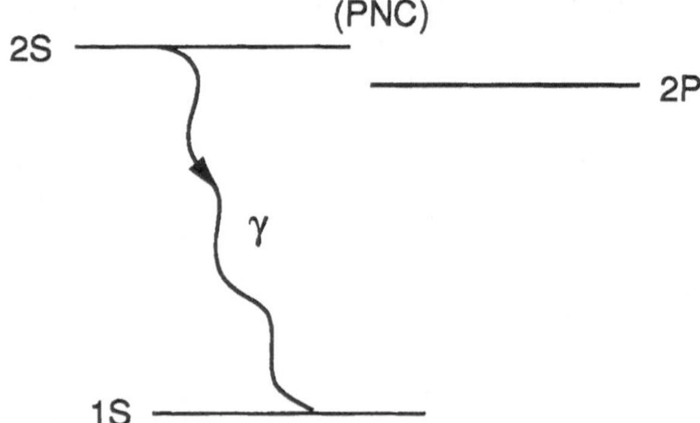

Fig. 6. Transitions between 2S and 1S states of muonic atoms were originally proposed for the detection of neutral currents in muonic atoms.

4 Parity violation in muonic atoms

The original proposal [12] to detect neutral currents in muonic atoms considererd the transition between 2S and 1S states, as shown in Fig.6.

The 2S and 2P states are admixed due to the parity-odd piece of the neutral current interaction between muon and nucleons, as given by Eq.(18). The radiative transition between the 2S and 1S states is then a sum of magnetic and electric dipole amplitudes: M1 + (PNC)E1. The interference of the two amplitudes produces asymmetries in the distribution of the emitted radiation.

Among the interests in the search for PNC-effects in muonic atoms, one should point out the accessibility to the four effective couplings $C_{1p}, C_{1n}, C_{2p}, C_{2n}$ of Eq.(18), the different Q^2-value probed in muonic atoms, the complete control of the atomic physics structure (hydrogen-like), etc. Light muonic atoms offer the additional theoretical perspective of studying the nuclear-spin-dependent PNC-effects and, experimentally, they present important enhancements: 1) of the mixing parameter, due to the near degeneracy of the 2S and 2P levels, and 2) of the amplitude ratio $\mid E1/M1 \mid$, which goes like $(\alpha Z)^{-3}$.

Interesting asymmetries are: 1) a net circular polarization of the radiation; 2) the angular asymmetry of the emitted radiation with respect to the polarization of the muon in the initial 2S- or final 1S-states, and 3) the directional correlation between the momenta of the emitted photon and the decay electron.

J. Missimer and L.M. Simons have proposed [13] the directional correlation as a suitable observable for the two isotopes of boron, ^{10}B and ^{11}B. Muonic boron is unique among the light muonic atoms in fulfilling the conditions necessary for the detection of the M1 transition:

1) Occuring in a gaseous state, the non-radiative 2S decays are suppressed;

2) The radiative lifetime of the 2S-state is much longer than that of other excited states, so it can be prepared by waiting for several ns after the muon stop, and

3) The M1 transition can be distinguished from more probable two-photon (2E1) transition by using an X-ray absorption edge.

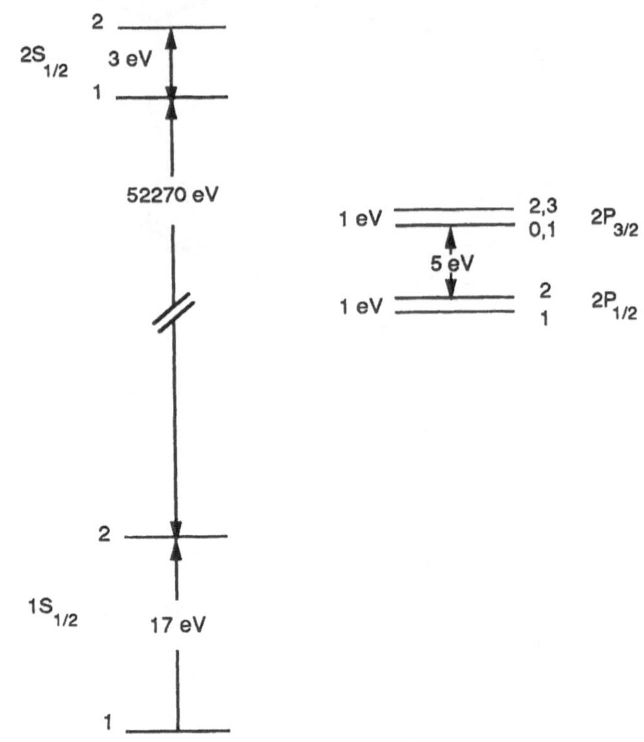

Fig. 7. Energy levels $\mu^{11}B$ which are interesting for parity-mixing experiments. Measurements of the photon-electron directional correlation in the hyperfine components of the initial and final states allow a complete determination of the neutral muon to nucleon couplings.

The two naturally occuring isotopes of boron have nonvanishing spins, so that measurements of the photon-electron directional correlation in the individual hyperfine components of the initial or final state of the radiative transition determines completely the neutral current couplings of the muon to nucleons. The energy levels of $\mu^{11}B$ relevant to parity-mixing experiments are shown in Fig.7.

The different asymmetries discussed above are interconnected in the following way. The partial rate of the hyperfine transition or the angular distribution of photons relative to the muon spin in the final 1S-state is given by

$$d\Gamma^{\gamma}_{F' \to F} \propto \left[d(F', F) + \vec{P}_{\mu}.\hat{k} \quad n(F', F) \right] \frac{d\Omega}{4\,\pi} \tag{21}$$

where $\vec{P}_{\mu}$ is the μ-polarization and $\hat{k}$ the photon direction. We see that the forward-backward angular asymmetry coincides with the circular polarization of photons

$$P_\gamma(F', F) = \frac{n(F', F)}{d(F', F)} \qquad (22)$$

Since only two hyperfine states occur for each S-state, they can be labeled (+) for the $I + 1/2$ and (-) for the $I - 1/2$ components, where I is the nuclear spin. Thus, $P_\gamma(+,-)$ is the asymmetry in the transition between the $F' = I + 1/2$ component of the initial 2S-state and the $F = I - 1/2$ component of the final 1S-state.

The values of P_γ have been calculated for the ^{10}B–^{11}B isotopes [14] and for the $^3He - ^4He$ isotopes [15]. The results for boron are (in percent):

$$\begin{aligned} ^{10}B(I = 3) :& P(+,+) = P(+,-) = \\ & 3.22 \cdot [C_{1p} + C_{1n} + 0.20(C_{2p} + C_{2n})] \\ & P(-,+) = P(-,-) = \\ & 3.52 \cdot [C_{1p} + C_{1n} - 0.26(C_{2p} + C_{2n})] \end{aligned} \qquad (23)$$

$$\begin{aligned} ^{11}B(I = 3/2) :& P(+,+) = P(+,-) = \\ & 3.34 \cdot [C_{1p} + 1.2C_{1n} + 0.20C_{2p}] \\ & P(-,+) = P(-,-) = \\ & 3.95 \cdot [C_{1p} + 1.2C_{1n} - 0.34C_{2p}] \end{aligned}$$

The equality of the asymmetries for identical initial states of the same atom follows from the absence of mixing in the final 1S-state. Inspection of the results (23) shows that the effects are large and that the measurement, if precise enough, of the different hyperfine transitions could determine the four interesting coupling constants. The relation of these effective couplings to the fundamental quark couplings is given in Eqs. (19) and (20), and it involves the flavour form factors $G_A^{(i)}$: the use of the nuclear-spin-dependent PNC effect in muonic atoms as a tool to determine the strangeness form factor $G_A^{(s)}$ has been emphasized in Ref. [10]. As an example, for the strong isoscalar ^{10}B nucleus the relevant couplings in the standard theory are

$$\begin{aligned} C_{1p} + C_{1n} &= 2sin^2\Theta_w \\ C_{2p} + C_{2n} &= [1 - 4sin^2\Theta_w]G_A^{(s)} \end{aligned} \qquad (24)$$

Alternatively, if the standard theory of electroweak interactions is not assumed, the PNC-effects in muonic atoms can be used to test extended gauge models, such as [16] extra Z-bosons, lepto-quarks, compositeness, etc. In this case, the aim is to extract the lepton-quark couplings from this experiment. Using the present results on $G_A^{(i)}$ discussed in Section 1, the information contained in Eq.(20) from C_{2p} and $C2n$ is

$$\begin{aligned} C_{2p} &= (0.78 \pm 0.06)C_{2u} - (0.66 \pm 0.12)C_{2d} \\ C_{2n} &= -(0.47 \pm 0.06)C_{2u} + (0.59 \pm 0.12)C_{2d} \end{aligned} \qquad (25)$$

which in both cases is almost proportional to the combination $C_{2u} - C_{2d}$.

In Ref.[13] the connection between P_γ of Eq.(22) and the directional correlation between the photon and the decay electron is shown:

$$W(\epsilon, x) = \left[2\epsilon^2(3 - 2\epsilon) + 2\epsilon^2(1 - 2\epsilon)P_\gamma x\right] d\epsilon \frac{dx}{2} \qquad (26)$$

where ϵ is the fractional energy of the electron $\epsilon \simeq \frac{E_e}{m_{\mu/2}}$ and $x = cos\Theta$ between the photon and electron momenta. The correlation manifests itself as an asymmetry A in the number of electrons emitted parallel and antiparallel to the photon direction. The measurability is determined by the number of correlated events per second I occurring in given ranges of ϵ and x. The number of correlated events which must be measured to determine the asymmetry to a desired accuracy is inversely proportional to the figure of merit $I \cdot A^2$. The maximum figure of merit for boron is estimated as max $(I \cdot A^2) \simeq 1 \cdot 10^{-11}$, so that $1 \cdot 10^{13}$ μ-stops are required to measure the asymmetry P_γ to a relative accuracy of ten percent.

5 Conclusions

The nuclear-spin-independent PNC-effects in heavy electronic atoms are reaching high precision levels. Within the standard electroweak theory, the experimental results on Atomic Parity Violation constitute a very clean test of the electroweak radiative corrections. If interpreted with models beyond the standard theory, one gets a powerful exploration of additional Z-bosons, technicolour ideas, etc.

Parity Violation in Light Muonic Atoms offers a tool to seperate out the nuclear-spin-dependent PNC-interaction. With appropriate selection of the nuclear isotopes one could seperate the effective axial isovector $(C_{2p} - C_{2n})$ coupling of the nucleon from the effective axial isoscalar $(C_{2p} + C_{2n})$ coupling. They are related to the fundamental quark couplings and the nucleon structure through

$$\begin{aligned} C_{2p} - C_{2n} &= (G_A^{(u)} - G_A^{(d)})(C_{2u} - C_{2d}) \\ C_{2p} + C_{2n} &= (G_A^{(u)} + G_A^{(d)})(C_{2u} + C_{2d}) + 2G_A^{(s)}C_{2d} \end{aligned} \qquad (27)$$

As seen, the accessibility to the $G_A^{(s)}$ form factor is of the highest interest for the understanding of the nucleon structure.

Acknowledgement. I would like to thank the organizers of the Workshop "The Future of Muon Physics" for the warm atmosphere that we had, which triggered lively discussions. This work is dedicated to Profs. V.W. Hughes and G. zu Putlitz, on the occasion of their anniversaries. Fruitful discussions with Prof. S. Bilenky are acknowledged. This research has been supported by CICYT under Grant AEN 90-0040.

References

1. Steinberger, J.: Phys. Rep. 203 (1991) 345.
2. Bernabéu, J. and Pascual, P.: Electroweak Theory, Univ. Autónoma Barcelona, B-21399 (1981).
3. Ashman, J. et al.: EMC experiment, Phys. Letters B206 (1988) 364; Nucl. Phys. B328 (1989) 1.
4. Fortson, E.N. and Lewis, L.L.: Phys. Rep. 113 (1984) 289; Amaldi, U. et al.: Phys. Rev. D36 (1987) 1385; Gosta, G. et al.: Nucl. Phys. B297 (1988) 244.
5. Noecker, M.C., Masterson, B.P. and Wieman, C.E.: Phys. Rev. Letters 61 (1988) 310.
6. Blundell, S.A. Johnson, W.R. and Sapirstein, J.: Phys. Rev. Letters 65 (1990) 1411.
7. Dydak, F.: CERN-PPE / 91-14 (1991).
8. Marciano, W. and Rosner, J.L.: Phys. Rev. Letters 65 (1990) 2963.
9. Altarelli, G. et al.: CERN-TH 6028/91 (1991).
10. Bernabéu, J.: Nucl. Phys. A518 (1990) 317.
11. Zeldovich, Ya.B.: Zh. Eksp. Theor. Fiz 33 (1957) 1531 [JETP 6 (1957) 1184]; Flambaum, V.V. and Khriplovich, I.B.: Zh. Eksp. Theor. Fiz 79 (1980) 1656 [JETP 52 (1980) 835], Bouchiat C. and Piketty, C.A.: Z. Phys. C49 (1991) 91.
12. Bernabéu, J., Ericson, T.E.O. and Jarlskog, C.: Phys. Letters B50 (1974) 467; Feinberg, G. and Chen, M.Y.: Phys. Rev. D10 (1974) 190; Moskalev, A.N.: JETP Letters 19 (1974) 216.
13. Missimer, J. and Simons, L.M.: PR-90-03, PSI (1990).
14. Moskalev, A.N. and Ryndin, R.M.: Sov. J. Nucl. Phys. 22 (1976) 71.
15. Bernabéu, J., Bordes, J. and Vidal, J.: Z. Phys. 41 (1989) 679.
16. Langacker, P.: UPR-0451T (1990): Casalbuoni, R. et al.: UGVA-DPT 1991/05-728.

This article was processed using Springer-Verlag TEX Z.Physik C macro package 1991
and the AMS fonts, developed by the American Mathematical Society.

Z. Phys. C - Particles and Fields 56, S31–S34 (1992)

Zeitschrift für Physik C Particles and Fields

Some Open Problems in Muon Physics

Lali Chatterjee

UGC Research Scientist Department of Physics, Jadavpur University, Calcutta - 700 032, INDIA

01-December-1991

Abstract. Some open problems on different sectors of muon physics have been investigated and the salient features are discussed. These are expected to play a meaningful role in the rich and vibrant future of Muon Physics.

1 Introduction

Spanning a wide range of energies and interactions, the esoteric field of muon physics is of crucial importance both for its rich fundamental physics content and its application potential. Physics evolves through the precision polishing of experimental and theoretical tools. This permits no slackening in our attempts to understand its intricate facts and their subtle nuances. The development of increasingly sophisticated experimental techniques and high precision muon beams provoke a reexamination of our theoretical methodologies and hypotheses to meet the exacting experimental demands. Thus despite the beautiful agreement of the Standard Model with experiment, we have heard in this workshop of the attempts to obtain an aesthetically appealing theory based on sounder physical precepts [1].

Originally a prerogative of the particle and cosmic ray physics community, the muon has estabilshed its position in various disciplines of Science. The field of exotic atoms and molecules provides precision tests of QED and Weak Physics. The subject of subbarrier nuclear fusion catalysed by negative muons generates excitement due to its potential as an emerging energy source and the intricate admixture of the different fundamental interactions it involves. As a nuclear probe high energy muons have proved to be of exceptional value.

We are working on different aspects of the multifaceted domain of Muon Physics. These include the weak characteristics of muons dressed in Coulomb bound states, continuum to bound muon transitions by stochastic radiative modes, exotic atom velocity distributions, the physics of sub-barrier nuclear fusion materialising in a muon cloud, deep inelastic muon studies, muon formation in nuclear matter and muon production methods. We report these in brief and assess their role in the Future of Muon Physics.

2 Muon weak interactions dressed in Coulomb photons

The basic weak decay process

$$\mu^{\pm} \rightarrow e^{\pm} + \bar{\nu}_{\mu}(\nu_{\mu}) + \nu_e(\bar{\nu}_e)\ldots \tag{1}$$

and its capture counterpart

$$\mu^{\pm} + e^{\pm} \rightarrow \bar{\nu}_{\mu}(\nu_{\mu}) + \nu_e(\bar{\nu}_e)\ldots \tag{2}$$

can occur from dressed states of Coulomb bound type. The Coulomb partners are essentially spectators to the basic weak process, but perturb the bare deltafunction constrains by participating in the kinematics. The Coulomb dressing or effects of the continuas exchange of Coulomb photons can be accounted for by use of suitable bound state wave functions. The simplest and most oft-studied process is that of muon decay from the atomic bound state

$$(\mu^- N) \rightarrow e^- + \nu_{\mu} + \bar{\nu}_e + N\ldots \tag{3}$$

μ^- decay from muonic molecules and from the neutral muonic Helium system involve two Coulomb partners in the initial sector, while μ^+ decay from Muonium is hosted by a purely leptonic Coulomb bound state. The respective weak capture reactions involve swallowing of its Coulomb partner by the μ^+, or the capture of the μ^- by its hadronic Coulomb partner. We have studied the processes

$$(\mu^- N_1 N_2) \rightarrow e^- + \nu_{\mu} + \bar{\nu}_e + N_1 + N_2\ldots \tag{4}$$

$$(\mu^+ e^-) \rightarrow e^+ + \bar{\nu}_{\mu} + \nu_e + e^-\ldots \tag{5}$$

and

$$(\mu^+ e^-) \rightarrow \bar{\nu}_u + \nu_e \ldots \quad (6)$$

Henceforth we restrict ourselves to hydrogenic systems for the muomolecular cases so that N_1 and N_2 refer to p, d or t nuclei.

For these light atomic systems, reduction of total available energy due to binding small and Coulomb retardation of the relative motion between the decay lepton and the recoiling spectator is also small. The effect of the wavefunction is possibly most pronounced for the muomolecular cases due to its non-centrality. For reactions (3) and (4), with spectator nuclei, one can neglect their recoil energy, and consider only their recoil momentum for conservation. However, this is not possible for the light spectator for reaction (5). Neglecting the non-local effects of the massive W propagators, as is permissable for these low energy weak processes, one can write the general matrix element for the Coulomb dressed weak decays as

$$M_{fi} = (G/\sqrt{2})[\bar{\psi}(P_e)O_\alpha \psi_\mu(P_\mu)\bar{\psi}_\nu(P_{\nu_1})O_\alpha \psi(p_{\nu_2})\ldots] \quad (7)$$

where

$$I = \int \Phi_\mu(r)\ exp(-i\underline{p}_s \cdot \underline{r}) d^3\underline{r} \ldots \quad (8)$$

and the total decay rate is

$$R = C\int |\ M_{fi}\ |^2 d^3\underline{p}_e d^3\underline{p}_{\nu_1} d^3\underline{p}_{\nu_2} d^3\underline{p}_s/(E_e E_{\nu_1} E_{\nu_2} E_s)$$
$$\delta^4(P_0 - P_e - P_{\nu_1} - P_{\nu_2} - P_s) \ldots\ldots \quad (9)$$

P, E refer to the four momenta and total energies of the subscripted particles and $\underline{p}$ their momenta. 's' refers to the spectator. $\Phi_\mu(r)$ is the initial muon wavefunction for the relevant process. The square of the matrix element is summed over the spins as for free lepton decay and the integration over the neutrino phase space is evaluated in their centre of mass frame [2]. The energy constraint equation for the decay electron-spectator phase space is obtained from the delta function conditions as

$$2[\sqrt{S}E_e + P_e P_S u] + p_e^2 - E_e^2 - 2E_e E_S$$
$$= S + E_s^2 - 2\sqrt{S}E_S - P_S \ldots \quad (10)$$

S is the total available energy and $u = cos\theta$ where θ is angle between the decay lepton and the spectator. Integrating over the angle and the lepton momentum $\underline{p}_e$, one obtains the spectator distribution, which can be finally integrated over the total rate. We consider specifically cases (4) and (5). For the general heteronuclear hydrogenic molecule, one obtains the rate normalised to decay from the $(\bar{\mu}p)$ system as [4].

$$R_m/R_a = (D_m/D_a)[T_m/T_a] \ldots \quad (11)$$

where D_m and D_a are constants

$$T_m = \int p^2 dp\ [W^5 - 2p^2 W^3 + W p^4] J_m$$

and

$$T_a = \int p^2 dp\ [W^5 - 2p^2 W^3 + W p^4] J_a$$

with

$$J_m = [B_1 sin(apR)/(apR) + B_2 sin(bpR)/(bpR)],$$
$$B_i = \eta_6^5/(\eta_i^2 + p_s^2)^4,$$

and W is the available energy.

a and b are the seperations of the nuclei N_1 and N_2 from their centre of mass. The subscript i refers to the nucleus. $J_a = \eta^2/(\eta^2 + p^2)^4$, where η is the inverse Bohr radius of the $(\mu_- p)$ system. The above assumes the BO model for the muomolecule. This simplifies for the homonuclear case and we have shown earlier that for the $(dd\mu)$ case, [3] $R_m/R_a = [W_m^5/W_a^5][1-1.52\alpha^2]/[1-2\alpha^2]$.

For decay of the positive muon from muonium one obtains [4] for the decay in units of the free decay rate (R_0)

$$R/R_0 = C_0 \int_0^1 x^2 dx H^3 K/G_0^2/L^4 \ldots \quad (12)$$

where W_0 is maximum spectator momentum,

$$\eta' = 1/(a_m W_0)$$

$$H = [S/W_0^2 - x^2 - 2E_S W/(W_0^2)], G_0 = [(W/W_0)^2 - x^2]$$
$$W = \sqrt{S} - E_S,$$

and

$$K = [2W^2/W_0^2 - x^2 - S/W_0^2 + 2WEs/W_0^2]$$

C is the relevant constant and a_m the muonium Bohr radius.

$$L = ((\eta')^2 + x^2)$$

These distributions and the exact Coulomb dressed weak rates are expected to be of use in future precision muon experiments.

The muonium (M) annhilitation into neutrinos probed by Pontecorvo [6] is suppressed from the singlet but allowed from the triplet state by helicity constraints.

The vacuum environment of the $M - \bar{M}$ experiments [7] should inhibit triplet $\rightarrow$ singlet flips so that process (6) becomes viable. Despite the obvious experimental difficulties associated with its detection, process (6) could vitiate $M - \bar{M}$ results if the $M - \bar{M}$ coupling were to be very small [5].

The matrix element is obtained from that for process (5) by crossing symmetry. Although reaction (6) is energetically favoured compared to (5), it is severly suppressed due to the low density (N) of available electron states. N is taken as the Muonium wavefunction at contact as for similar problems.

In addition the exitchannel now contains 2 mass less particles in contrast to the 3 particle final state and the phase space integral is hence trivial. We have carried out the explicit calculations [4] for this process. We obtain the rate to be zero for the singlet case as expected from helicity conservation and the branching ration (given by the triplet contribution) to be 1.07 x 10^{-10} compared to the free Muon decay rate, in agreement with Pontecorvo's estimate.

3 Radiative Coulomb capture

We have studied the 'stochastic' population of deep μ^- atom states by direct radiative capture. This channel is generally ignored in the continuum to bound transition sector, which is dominated by the auger capture after thermalisation. We have studied this process, going beyond the soft photon approximation [8]. We are presently reinvestigating the competing thermalisation channels at muon energies where the muon wavelength is smaller than the target atom dimensions and the muon is hence sensitive to target structure. 'Close encounters' with the target nucleus result in 'Stochastic' population of the deep inner orbits [8].

The radiative capture is akin to the bremsstrahlung process and is described by the second order S matrix. In Momentum space, the 2nd order S matrix is

$$S = \left[ie^2/(\sqrt{2}w)\right]\bar{U}_f(\underline{p}_f)\{\left[i/(\not{p}-m)\right]A_0^e(q)\ exp(-i\underline{k}\cdot\underline{x}) + exp(-i\underline{k}\cdot\underline{x}_2)A_0^e(q)\left[i/(\not{p}'-m)\right]\} U_i(\underline{p}_i)\int\chi(x_1)\Phi_{ne}(x_2) exp(i\underline{p}_i\cdot\underline{x}_i - i\underline{p}_f\cdot x_2)d^4x_1d^4x_2d^4p(p')\ldots \quad (13)$$

For the present we have evaluated the radiative capture using the standard non relativistic matrix element [8]

$$M_{fi} = C\int\psi_f(\underline{r})\ \underline{r}\cdot\underline{\epsilon}\ exp(i\underline{k}\cdot\underline{r})\psi_i(\underline{r})d^3r$$

However, we have retained the photon wave function term exp $(i\underline{k}\cdot\underline{r})$ explicitly and thus avoid the soft photon approximation. Since we have computed rates up to 10 keV. incident energy, the use of non relativistic wave function for the muon is justified. We have also used values for the photon momentum as obtained from exact relativistic delta function conditions. While the correction over the soft photon results are small at lower energies, they increase .3% at 10 keV. incident energy [8].

We are presently working out the exact S matrix results for this process along with the competing bremsstrahlung process. We also plan to study the energy loss mechanism in the (1-100 keV.) regime, in a formalism sensitive to target structure and target polarization. This will permit us to ascertain beam depletion effects prior to thermalisation.

4 Exotic atom distributions

In continuation of the Physics discussed in the previous section, it is interesting that radiative capture results in direct formation of near stabilized exotic atoms with velocities larger than those acquired in auger capture.

A compatitive rate for the radiative Coulomb capture provides an appealing panacea for several existing problems [8]. Increased ground state population could be manifested as an altered q_{1s} [9] measure in the μcf scenario aiding resolution of this discrepancy. Perhaps they could account for one component of the dual $(\mu^- p)$ atoms observed in the muon transfer experiments [10].

Finally, carried over to the pion case, radiative capture of 0.7 keV. pions would yield 90 eV $(\pi^- p)$ atoms and could explain their occurence in the $(\pi^- p)$ distribution [11].

These features provoke a deeper and more incisive study of the initial capture sector which cannot be considered as closed and could still yield surprises.

5 Sub barrier nuclear fusion in the muon cloud

We are studying the physics of the Sub-barrier fusion form the muonic molecule [12].

The structure of the fusion vertex, its space time evolution dressed in the muon's Coulomb field, the muon's role during the fusion, its connection to the vertex and its behaviour immediately after the fusion act need to be investigated on a more sound theoretical basis [12,13]. Is the sudden approximation used to compute sticking sudden enough for the Physics? Its basic precept starts from the muomolecule taken at nuclear coalescence and overlaps it with the final state. It therefore starts from stage II, when the barrier has been crossed and ignores the muon's interim role during barrier crossing. It presupposes a compound nucleus routing and also that the muon registers the nuclear collapse and adjusts to it.

In reality the states accessible to observation are the stable muomolecule prior to barrier crossing with the muon in a two centre state and the final fusion products with the muon in bound or free state. Starting from two centre state one would obtain a slightly lower sticking and the consequences for application are appealing.

We have also attempted to incorporate the exit channel kinematics in the calculation of sticking within the SA format [14].

'In flight' muon catalysed fusion for the d-d-case has been computed in an Allis-Morse type screened potential [15] and we obtain $322.6(sec)^{-1}$ for the rate which is somewhat larger than the earlier estimate of Zeldovich although the order of magnitude is the same.

6 Deep inelastic muon studies

We propose to study deep inelastic scattering of muons in a two-stage emulsion experiment. In the first phase, which has been approved by Fermilab [16], a horizontal exposure will allow us to study the particle production characteristics of the three target types (light, medium and heavy) present in the emulsion. In the second phase we plan a vertical 'stack' exposure in an emulsion telescope arrangement that would provide an accurate and sensitive measure of the moment and incorporate 4π detection.

We aim to study the EMC effect, the hadronisation mechanisms in the different targets, and comparative issues dealing with analogous neutral current events.

7 Muon formation in nuclear matter

In neutron stars weak interactions materialises muons when the difference in Chemical potential of n and p exceeds m_μ. These muons are stable as the electron Fermi sea is saturated.

We have investigated the formation of muons in dense asymmetric nuclear matter [17]. Observation of such muons could determine preferences of the equations of State describing such matter as the different models have differing predictions.

Such muons could be found in supernova cores, dense droplets of non-collapsed stars or neutron rich chunks blown away from neutron stars during glitching.

A supernova core of $\sim$ 20 miles diameter should accumulate $\sim 10^{57}$ muons. As the muon lifetime allows the electro magnetic slowing down and capture, the X-rays signalling their cascades or capture could be searched for in astrophysical data.

8 Muon production aspects

We have investigated the possibility of improving muon production efficiency from different angles [18]. These include topological studies of fixed target pion production, including rescattering and double scattering effects. The feasibility of heavy ion collisions providing useful muon sources has also been probed [19], specially for neutron rich collisions as these favour the negative flavour in the produced pions.

9 Discussion

We have glimpsed through the research we are carrying out on the different facts of muon physics from low to high energies - including its application potential as a fusion energy source.

Finally - a physicist proposes - nature disposes.

We can only try to understand the muon's role in the scheme of the universe and we hope that our continued efforts and research will prove fruitful in the future evolution of Muon Physics.

Acknowledgement. L. Chatterjee would like to thank the conference organisers for local hospitality during the conference. Financial support from U.G.C., D.S.T. and DAE for research on different aspects of Muon Physics is gratefully acknowledged.

References

1. Scheck F.: Talk given at Workshop on "Future of Muon Physics" held at Heidelberg, May 7-10 (1991).
2. Chatterjee, L. and Gautam, V.P.: Phys. Rev. D 41, (1990) 1698.
3. Chatterjee, L. and Chakraborty, A.: μ-decay from molecular states. Jadavpur University, preprint JUNHEP/91-LC 4.
4. Chatterjee, L. and Bhattacharya, S.: Physica Scripta 29, (1984) 205.
5. Chatterjee, L. Chakraborty, A. and Das, G.: The weak characteristics of muonium - the standard model and beyond. Jadavpur University Preprint JUNHEP/91-LC 3.
6. Ponte Corvo, B.: Sov. Phys. JETP6 (1958) 429.
7. Jungmann, K. et al.: PSI Proposal R-89-06.
8. Chatterjee, L., Das, G. and Chakraborty, A.: Radiative capture beyond the soft photon approximation. Jadavpur University Preprint JUNHEP/91-LC 2 and: Radiative π/μ capture - a panacea for all ills. Jadavpur University Preprint JUNHEP/91-LC 4; Europhysics Letters 1992 (in press).
9. Anderson, A.N.: AIP Conference proceedings 1st Proceedings μcf '88, Florida, USA, May (1988). Ed. Jones, S.E., Rafelsky, J. and Monkhorst, H.J.
10. Mulhauser, F. et al.: Muon Cat. Fusion 4, (1989) 365.
11. Crawford, J.E. et al.: Phys. Lett. B213, (1988) 391.
12. Chatterjee, L.: Indian. J. Phys. 65A(3), (1991) 175.
13. Chatterjee, L.: μcf '89 Proceedings. Ed: Davies, J.D. (1990). Rutherford Lab. Pub. RAL 90-022.
14. Chatterjee, L.: Phys. Lett. A137, (1989) 4 and Hadronic Journal 13, (1990) 361.
15. Chatterjee, L. and Das, G.: Phys. Lett. A154, (1991) 5.
16. Chatterjee, L., Ghosh, D. and Murphy, T.: Fermilab Experiment No E802.
17. Chatterjee, L. and Dey, J.: Hadronic Journal 13, (1990) 69.
18. Chatterjee, L.: Fusion Technology 12, (1987) 444; Indian Journal Pure and App. Physics 27, (1989) 787
19. Chatterjee, L., Ghosh, D. and Verma, S.D.: μcf 88, AIP Conference Proceedings, 1st. Ed: Jones, S.E., Rafelsky, J. and Monkhorst, H.J., (1989).

This article was processed using Springer-Verlag TEX Z.Physik C macro package 1991
and the AMS fonts, developed by the American Mathematical Society.

Z. Phys. C - Particles and Fields 56, S35-S43 (1992)

Zeitschrift für Physik C Particles and Fields

Muonium

Vernon W. Hughes

Yale University, New Haven, Connecticut 06520

23-July-1992

Abstract. A brief review is given of current and future fundamental topics in muonium research including the ground state hfs and Zeeman effect, the Lamb shift in the n=2 state and spontaneous conversion of muonium to antimuonium. The important 1S→2S transition is discussed by K. Jungmann in this volume. It is emphasized that muonium as an atom composed of two leptons is superior to hydrogen for many precise fundamental studies because of the complexities and uncertainties for hydrogen associated with the proton as a hadron.

1 Introduction

Muonium (M) is the bound atomic state of a positive muon (μ^+) and of an electron (e^-) and hence it is a hydrogenic atom. Muonium was discovered in 1960 through observation of its characteristic Larmor precession in a magnetic field. Since then research on the fundamental properties of M has been actively pursued, as has also the study of muonium collisions in gases, muonium chemistry and muonium in solids.

The principal reason that muonium continues to be important to fundamental physics is that it is the simplest atom composed of two different leptons. The muon retains a central role as one of the elementary particles in the modern standard theory, but we still have no understanding as to "why the muon weights" and in all respects behaves simply as a heavy electron. Muonium is an ideal system for determining the properties of the muon, for testing modern quantum electrodynamics, and for searching for effects of weak, strong, or unknown interactions in the electron-muon bound state. Basically muonium is a much simpler atom than hydrogen because the proton is a hadron and, unlike a lepton, has a structure that is determined by the strong interactions. Thus muonium provides a cleaner system to study than hydrogen for testing QED and the electroweak interactions.

For quantum electrodynamics high energy experiments test the high energy or short distance limits of the theory, whereas low energy high precision experiments test higher order radiative processes characteristic of the renormalized QED field theory. At present the most important low energy tests of QED are the following:

(1) Electron anomalous magnetic moment [1,2] or g_e-2 value ($a_e = \frac{g_e-2}{2}$)

(2) Lamb shift in hydrogen (n=2 state) [3,4] [$S_L = E(2^2S_{1/2} - 2^2P_{1/2})$]

(3) Muonium hyperfine structure interval [5] $\Delta\nu$ (n=1 state)

(4) Muon anomalous magnetic moment [6,7] or g_μ-2 value ($a_\mu = \frac{g_\mu-2}{2}$)

In addition to the muonium hfs interval $\Delta\nu$ - (3) in the above list - muonium studies can make important contributions to the other tests as well. The experimental value for a_e is

$$a_e(\text{exp}) = 1\ 159\ 652\ 188.4(4.3) \times 10^{-12}(4\,\text{ppb}),$$

and the theoretical value can be written

$$a_e(\text{th}) = \frac{\alpha}{2\pi} + A_2(\frac{\alpha}{\pi})^2 + A_3(\frac{\alpha}{\pi})^3 + A_4(\frac{\alpha}{\pi})^4 + \cdots$$

for which the A coefficients have been calculated. At present the most precise value for α to use to evaluate a_e(th) is obtained from condensed matter physics (quantum Hall effect and ac Josephson effect) with an accuracy of 24 ppb, and hence it relies on condensed matter theory. As will be discussed in Section 2, the more precise measurement of $\Delta\nu$ and of the Zeeman effect in the ground n=1 state of muonium being undertaken at LAMPF should determine α to about 25 ppb or better where only QED and atomic theory are involved.

The Lamb shift in the n=2 state (D. Yennie, in this volume) of hydrogen is one of the classic tests of QED. The experimental value is $S_L(\text{exp}) = 1057.845$ (9) MHz (9 ppm) and the theoretical value can be written:

$$S_L(th) = \text{QED terms} + \text{proton structure term}.$$

The proton structure term is proportional to the mean square radius of the proton $< r_p^2 >$ and contributes 140 ppm to S_L(th) with an uncertainty of 9 ppm arising from the error involved in the measurement of $< r_p^2 >$ by high energy elastic electron proton scattering. The error in S_L(th) is due principally to this uncertainty in the proton structure term. The experimental and theoretical values for S_L are in reasonable agreement, and hence until the proton structure term is better known a more sensitive test of QED can not be provided by the Lamb shift.

The muon anomalous magnetic moment or g_μ-2 value provides a precise test of QED and of the behaviour of the muon as a heavy lepton, and indeed because of the higher mass scale involved as compared to the electron can provide an important, sensitive test of electroweak theory and of speculative theories beyond the standard model (F. Farley, T. Kinoshita, B.L. Roberts, in this volume). Determination of g_μ-2 from the muon g_μ-2 experiment requires a value for the muon mass m_μ or equivalently of the muon to proton magnetic moment ratio μ_μ/μ_p. At present μ_μ/μ_p is determined from the measurement of the Zeeman effect and hfs in the muonium ground state with a precision between 0.15 and 0.35 ppm; the new LAMPF experiment should reduce this error to about 50 ppb.

An important general point can be emphasized in comparing tests of QED and of the electroweak interaction with hydrogen and muonium. The effects of strong interactions or hadronic structure, which can not yet be calculated from the basic theory of quantum chromodynamics, are at present limiting importantly the QED tests of the simplest ordinary one-and two-electron atoms including hydrogen and helium. On the other hand high precision spectroscopic measurements can also be made on muonium where no hadronic structure effects are present and very precise theoretical values can be evaluated. In principle the proton structure effects in hydrogen could be evaluated from measurements by laser spectroscopy on muonic hydrogen, μ^-p, but such experiments have not yet been successful.

The energy level diagrams for hydrogen and muonium are shown in Figure 1. Several of the very precise measurements of H energy intervals are listed below in Table 1. [8,9]. The corresponding muonium measurements of considerably less precision are listed as well. For hydrogen the theoretical accuracy for $\Delta\nu$ is limited at the several ppm level by uncertainty in the contributions of proton structure and polarizability [4] and for the 1S-2S interval at 1 part in 2×10^{10} due to uncertainty about proton structure.

The topics on muonium we shall discuss briefly below are the following:

1. Ground state hyperfine structure and Zeeman effect
2. Lamb shift and fine structure in the n=2 state
3. Muonium to antimuonium conversion

HYDROGEN

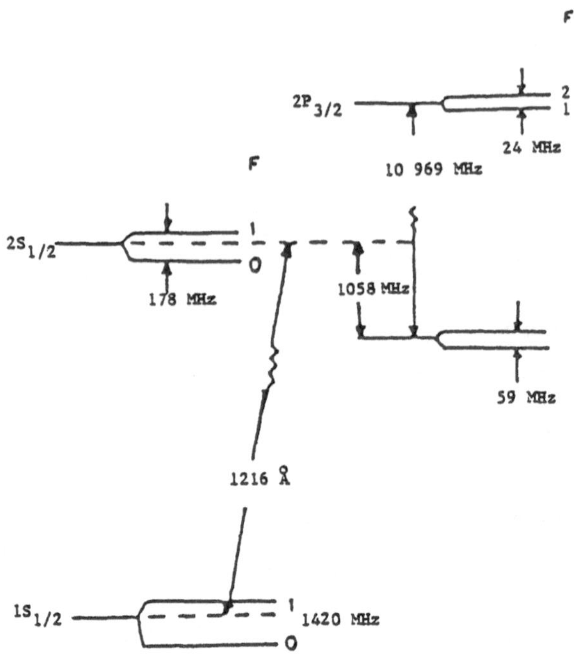

MUONIUM

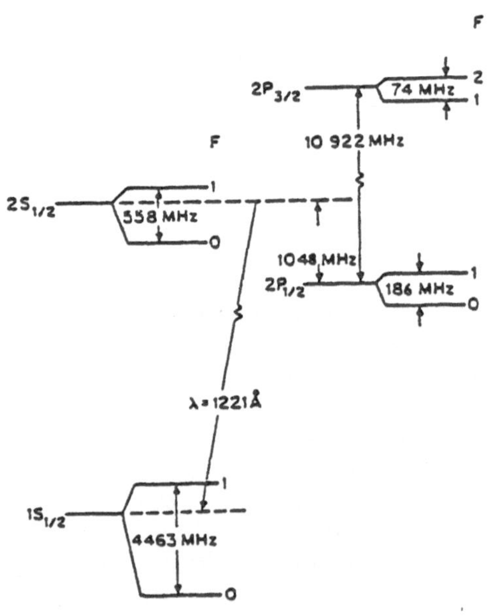

Fig. 1. Hydrogen and Muonium energy level diagrams for n=1 and n=2 states.

2 Ground State Hyperfine Structure and Zeeman Effect

After the discovery of muonium, measurements of its energy levels could be undertaken by microwave magnetic resonance spectroscopy utilizing the facts that the incident μ^+ are polarized so that polarized muonium

Table 1. Precision Measurements.

	Hydrogen
$\Delta\nu$(n=1)	1 420.405 751 766 7 (9) MHz (7 parts in 10^{13})
ν(2S-1S)	2 466 061 414.1 (8) MHz (3 parts in 10^{10})
$\nu(4P_{1/2} - 2S_{1/2})$	616 520 018.02 (7) MHz (1 part in 10^{10}) (determines R_∞)
	Muonium
$\Delta\nu$(n=1)	4 463 302.88 (16) kHz (36 ppb)
ν(2S-1S)	2 455 528 016 (72) MHz (3 parts in 10^8)
$\nu(4P_{1/2} - 2S_{1/2})$	not measured

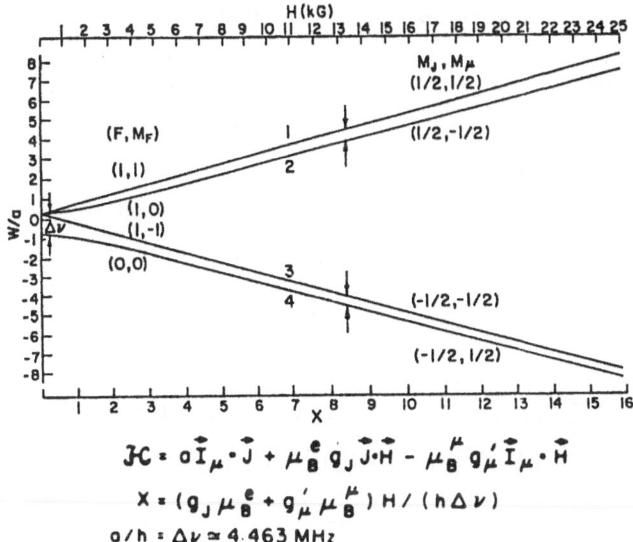

Fig. 2. Breit-Rabi energy level diagram for muonium in its $1^2S_{1/2}$ ground state in a magnetic field. Several of the transitions measured are indicated by the arrows.

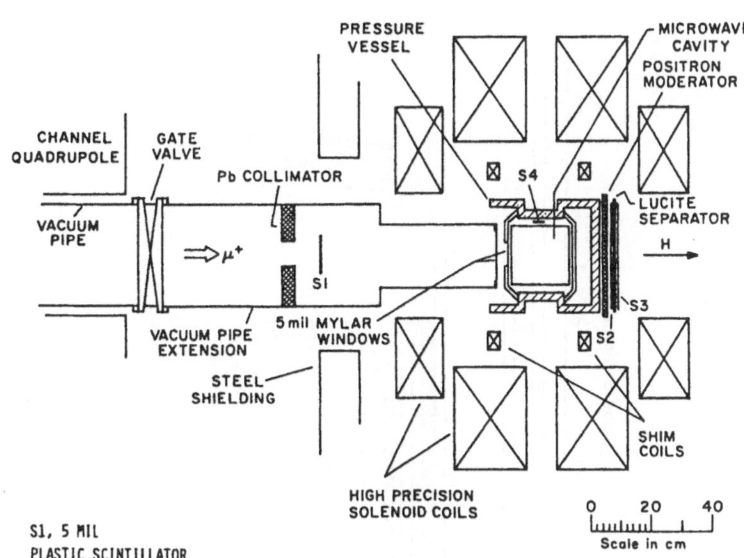

Fig. 3. Experiment at LAMPF in which the latest precision measurement of the hyperfine structure interval $\Delta\nu$ in muonium was made.

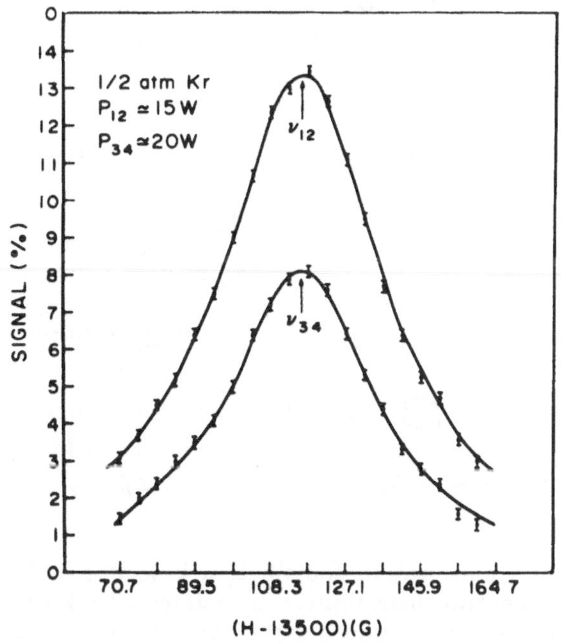

Fig. 4. Resonance lines fitted to the data from the experiment shown in Fig. 3. P_{12} and P_{34} are input powers to the microwave cavity.

is formed and that the decay positrons have an asymmetric angular distribution with respect to the muon spin direction [5]. The Breit-Rabi energy level diagram for the ground state of muonium is shown in Fig. 2. With the aim of determining the hyperfine structure interval $\Delta\nu$ and the muon magnetic moment μ_μ, transitions at both weak and strong magnetic fields have been measured as indicated. Starting in 1962 a series of increasingly accurate measurements were undertaken by both the Yale-Heidelberg and Chicago groups. The latest experiment at the Los Alamos Meson Physics Facility (LAMPF) was a strong field measurement [10]. A schematic diagram of the experimental arrangement is shown in Fig. 3. Typical resonance curves are shown in Fig. 4.

The experimental results for $\Delta\nu$ and μ_μ and the current theoretical value for $\Delta\nu$ are given in Table 2. The radiative and recoil corrections to the leading Fermi value for $\Delta\nu$ have been computed to high order [4]. [D. Yennie, in this volume.] The first error of 1.33 kHz or 0.30 ppm is due mostly to uncertainty in the value of the constant μ_μ/μ_p appearing in the Fermi term E_F. The second uncertainty comes from that of the α value used. The third uncertainty of 1.0 kHz is an estimate of uncalculated terms. Uncertainty in the small hadronic

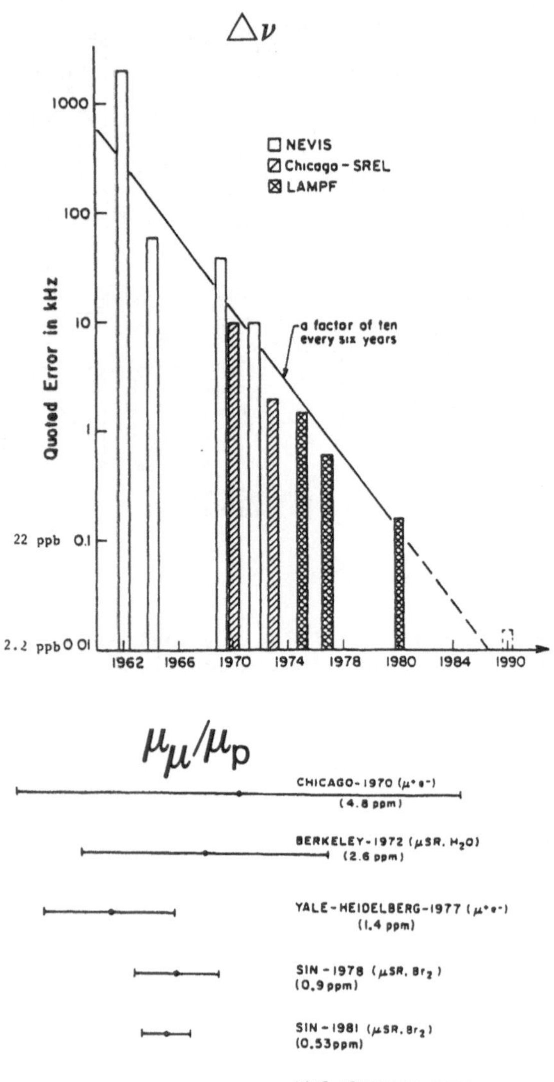

Fig. 5. History of measurements of $\Delta\nu$ and of μ_μ/μ_p.

vacuum polarization contribution is less than 10 ppb. Weak neutral current effects associated with Z exchange in the e-μ interaction are estimated to be 0.07 kHz or 16 ppb and are neglected. The experimental value for $\Delta\nu$ is known to 36 ppb, and the experimental and theoretical values agree well within the theoretical error of 0.4 ppm. This agreement constitutes one of the important sensitive tests of quantum electrodynamics, in particular of the e-μ interaction in their bound state, and of the behaviour of the muon as a heavy electron. Figure 5 displays the history of the experimental precisions in the measurement of $\Delta\nu$ and of μ_μ/μ_p.

A new experiment is being undertaken at Los Alamos (LAMPF 1054) to measure $\Delta\nu$ and μ_μ/μ_p to precisions of 10 ppb and 50 ppb, respectively, thus reducing the present errors by a factor between 5 and 10. The general method of the experiment is the same as that discussed above. The measurement will be done at a strong magnetic field of about 17 kG and will use a high intensity clean muon beam. Resonance line narrowing using the old muonium technique is planned with the muon beam being chopped.

Table 2. Theoretical value for hyperfine structure interval $\Delta\nu$ in muonium ground state.

$$\Delta\nu = \frac{16}{3}\alpha^2 c R_\infty \left(\frac{\mu_\mu}{\mu_p}\right)\left(\frac{\mu_p}{\mu_B^e}\right)\left(1+\frac{m_e}{m_\mu}\right)^{-3} \times$$

$$\left[1 + \text{QED terms } (\alpha, Z\alpha) + \text{Recoil terms } \left(\alpha\frac{m_e}{m_\mu}\right) + \text{Radiative-recoil terms } \left(\alpha^2\frac{m_e}{m_\mu}\right)\right]$$

VALUES OF THE FUNDAMENTAL CONSTANTS

c	= 2.997 924 58 x 10^{10} cm/sec	(exact)
α^{-1}	= 137.035 989 5 (61)	(45 ppb)
R_∞	= 109 737.315 73 (2) cm^{-1}	(0.2 ppb)
a_e	= 1.159 652 193 (10) x 10^{-3}	(9 ppb)
a_μ	= 1.165 923 0 (84) x 10^{-3}	(7.2 ppm)
μ_p/μ_B^e	= 1.521 032 202 (15) x 10^{-3}	(10 ppb)
μ_μ/μ_p	= 3.183 345 5 (9)	(0.3 ppm)
m_μ/m_e	= 206.768 260 (60)	(0.3 ppm)

$\Delta\nu(th)$ = 4 463 303.11 (1.33) (0.40) (1.0) kHz (0.4 ppm)

(the three errors arise from: $\frac{\mu_\mu}{\mu_p}$; α; estimate uncalculated terms)

$\Delta\nu(exp)$ = 4 463 302.88 (0.16) kHz (0.036 ppm)

$\Delta\nu(th) - \Delta\nu(exp)$ = 0.23 (1.7) kHz

$$\frac{\Delta\nu(th)-\Delta\nu(exp)}{\Delta\nu(exp)} = (0.05 \pm 0.4)\text{ ppm}$$

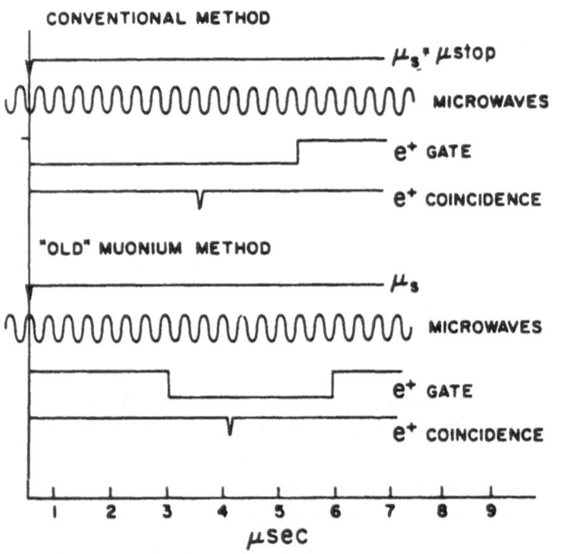

Fig. 6. Line narrowing with a chopped muon beam.

The old muonium method is indicated in Figure 6. A muonium atom will only be studied if it decays during

Fig. 7. Photograph of the superconducting solenoid magnet.

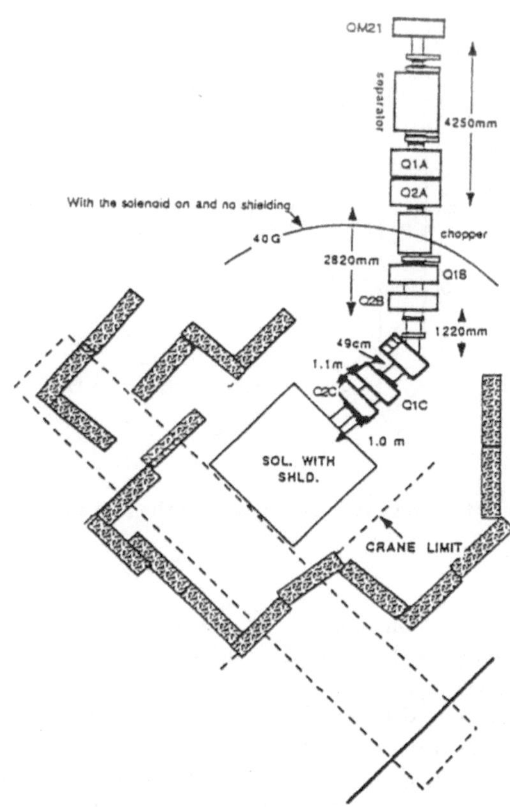

Fig. 8. Muon beam line from stopped muon channel to LAMPF experiment on muonium ground state.

the period of the e^+ gate and hence has lived at least until the time T when the gate is turned on after the arrival of the muon at time t=0. By contrast in the conventional method all muonium atoms are studied from time t=0 until the end of the period of the e^+ gate. In the conventional method the frequency half width of the resonance line is

$$\Delta f_{1/2} = [4 \mid b \mid^2 + \gamma^2]^{1/2}/\pi$$

where $\gamma = 1/\tau_\mu$ (τ_μ = muon lifetime) and $|b|^2 \propto$ microwave power. In the limit of $|b|^2 = 0$, $\Delta f_{1/2} = \gamma/\pi =$ 145 kHz; for $2\,|b|\,\tau_\mu = 1$, $\Delta f_{1/2} = 200$ kHz. In the old muonium method in the limit of $T \gg \tau_\mu$ and $|b| \ll \gamma$, $\Delta f_{1/2} = 2\,|b|\,/\pi$; with $T = 3\tau_\mu$ and $2|b|T = 1$, $\Delta f_{1/2} =$ 60 kHz.

Figure 7 shows a photograph of the magnet which is a superconducting solenoid magnet (originally designed and constructed by Oxford Magnet Technologies as a MRI magnet). The magnet has an inner bore at room temperature 1 m in diameter and 2.2 m in length. It will be surrounded by an iron shield consisting of two side walls and two endcaps. It will be operated in persistent mode with a field of about 1.7 T. The magnet will be shimmed to a homogeneity of better than 1 ppm over a 20 cm diameter spherical volume, and in persistent mode operation should be stable to about 1 part in 10^8 per hour. Knowledge of the magnetic field to about 0.1 ppm in absolute value over the volume of the microwave cavity will be obtained by pulsed NMR with a H_2O probe which is calibrated against a standard spherical H_2O probe. Modulation of the magnetic field by about 200 G to sweep through the resonance line will be provided by an additional pair of solenoidal coils located inside of the main solenoid and operated at room temperature.

Figure 8 shows the beam line from the output of the stopped muon channel at LAMPF into the solenoid where muonium is formed and studied. An $\vec{E} \times \vec{B}$ separator largely removes positrons from the beam, the ratio e^+/μ^+ being about 0.05 for a muon beam momentum p_μ = 25 MeV/c. An important element in the beam line is the electrostatic chopper which when pulsed on deflects the μ^+ beam out of its trajectory so that no μ^+ beam arrives at the gas target. The electrostatic kicker can provide voltages of +20 kV and -20 kV on two parallel plates spaced by 10 cm and 100 cm in length. The rise and fall times of the pulse are about 100 ns and the pulse repetition rate can be as high as 100 kHz with a variable width pulse. The muon extinction factor was measured to be about 0.03. The muon beam is deflected from its initial direction from the stopped muon channel by a dipole magnet to enter the gas target in the solenoid; this deflection should remove further background from the beam.

A more precise theoretical value for $\Delta\nu$ of muonium in its n=1 state should soon be available for comparison with the improved experimental value. (D. Yennie, in this volume). If instead of the α value from the 1986 Adjustment of the Fundamental Constants used in Table 2 a more recent value based on the electron g-2 value is used [11] then the uncertainty in $\Delta\nu$(th) associated with α is reduced to less than 20 ppb. Theoretical calculation of additional higher order terms are in progress (T. Kinoshita, private communication) which should reduce the error associated with uncalculated terms to about 20 ppb.

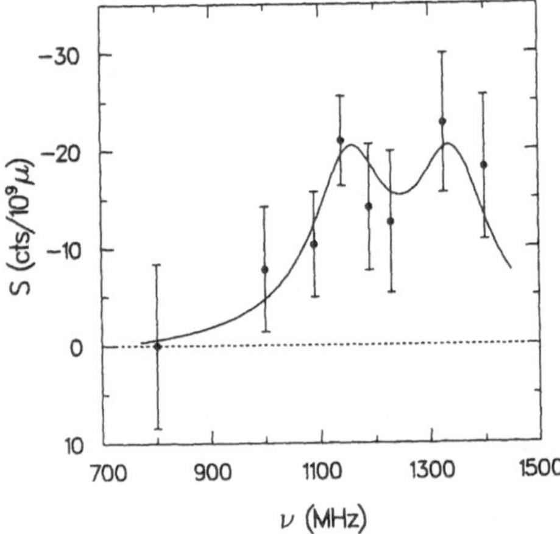

Fig. 9. Microwave-resonance data and the best fit theoretical line shape.

Clearly combination of the expected improved experimental values for $\Delta\nu$ and μ_μ/μ_p with an improved $\Delta\nu$(th) will provide a more sensitive test for muonium $\Delta\nu$ and could determine α to about 25 ppb.

In the future still more precise measurements of the muonium ground state would be possible if a more intense pulsed muon beam from the proton storage ring at LAMPF became available (H. White, in this volume).

3 Lamb Shift in Muonium

With the development of a fast muonium beam in vacuum, [5] measurement of the Lamb shift in muonium became possible. The energy level diagram of the n=1 and n=2 states of muonium is shown in Fig. 1. Two measurements [12,13] have determined the Lamb shift S_L in the n=2 state by observing the $2^2S_{1/2}$ to $2^2P_{1/2}$ transition in a radiofrequency spectroscopy experiment.

The method of the experiments is similar to that of the observation of the fine structure transition $2^2S_{1/2}$ to $2^2P_{3/2}$ discussed below. The resonance line observed in the LAMPF experiment [13] is shown in Fig. 9 in which the lower frequency component corresponds to the transition $2S_{1/2}(F=1)$ to $2P_{1/2}(F=1)$ and the upper to $2S_{1/2}(F=1)$ to $2P_{1/2}(F=0)$.

Table 3 gives the current theoretical value of S_L and the two experimental values. The theoretical value for the muonium Lamb shift differs from that in hydrogen by the absence of a proton structure term and by the relatively greater importance of recoil terms. The experimental values agree with the theoretical value within the limited experimental accuracy of about 1%.

Recently at LAMPF the fine structure transition $2^2S_{1/2}$ to $2^3P_{3/2}$ in muonium has been studied [14]. Figure 10 indicates the experimental method. Muonium is formed in the metastable $2^2S_{1/2}$ state at an $A\ell$ foil just downstream of a low gas pressure MWPC. After collimation the M(2S) beam enters a microwave cavity

Table 3. Muonium Lamb Shift Theoretical Value.

Contribution	Order (mc^2)	Value(MHz)
Self-energy	$\alpha(Z\alpha)^1[ln(Z\alpha)^{-2}, 1, Z\alpha, \cdots]$	1085.812
Vacuum polarization	$\alpha(Z\alpha)^4(1, Z\alpha, \cdots)$	-26.897
Fourth order	$\alpha^2(Z\alpha)^4$	0.102
Reduced mass	$\alpha(Z\alpha)^4(m_e/m_\mu)[ln(Z\alpha)^{-2}, 1]$	-14.493
Relativistic recoil	$(Z\alpha)^5(m_e/m_\mu[ln(Z\alpha)^{-2}, 1, Z\alpha]$	3.159
Higher-order recoil	$(Z\alpha)^4(m_e/m_\mu)^2$	-0.171
Radiative recoil	$\alpha(Z\alpha)^5 m_e/m_\mu$	-0.022
	Total	1047.490(300)

The uncertainty in the theoretical value is due to uncalculated terms of higher order in m_e/m_μ, i.e., terms (m_e/m_μ) (reduced mass term) and α(reduced mass term).

Experimental Value

$$S_L(exp) \quad = (1054 \pm 22) \text{ MHz, LAMPF}$$
$$= (1070^{+12}_{-15}) \text{ MHz, TRIUMF}$$

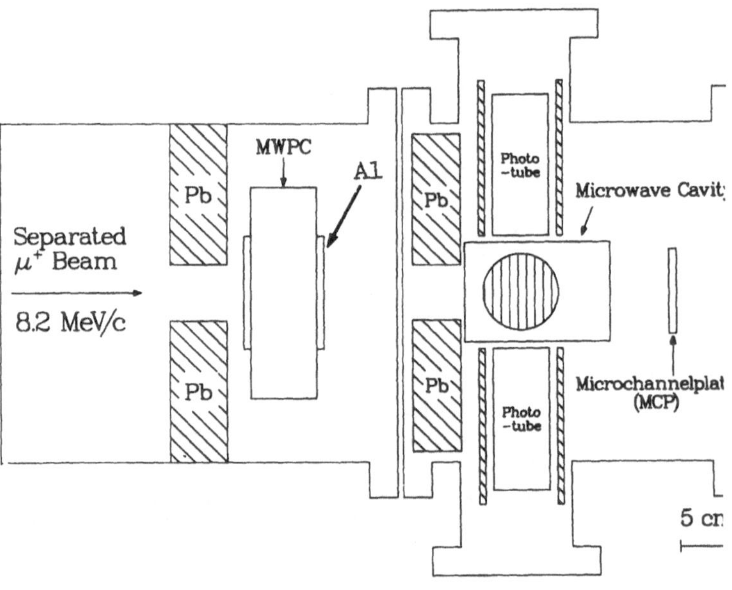

Fig. 10. Diagram of the apparatus used in observation of the muonium fine structure transition $2^2S_{1/2} \to 2^2P_{3/2}$.

operating at a frequency of about 10 GHz which drives the transition $2^2S_{1/2} \to 2^3P_{3/2}$. From the $2^2P_{3/2}$ state M decays to the ground 1S state with a mean life of 1.6 ns, and the Lyman-α 1221 Åphoton is detected by a UV photomultiplier tube, while the resulting M(1S) atom travels to a microchannel plate where it is detected. The signal due to the microwave field, defined as a delayed triple coincidence between a μ^+ count in the MWPC detector, a Lyman-α photon, and a microchannel plate count, is shown in Fig. 11 as a function of the microwave frequency, together with the fitted line shape. Three

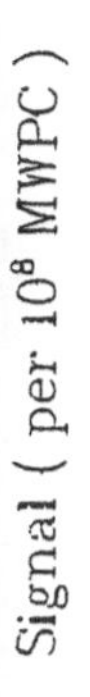

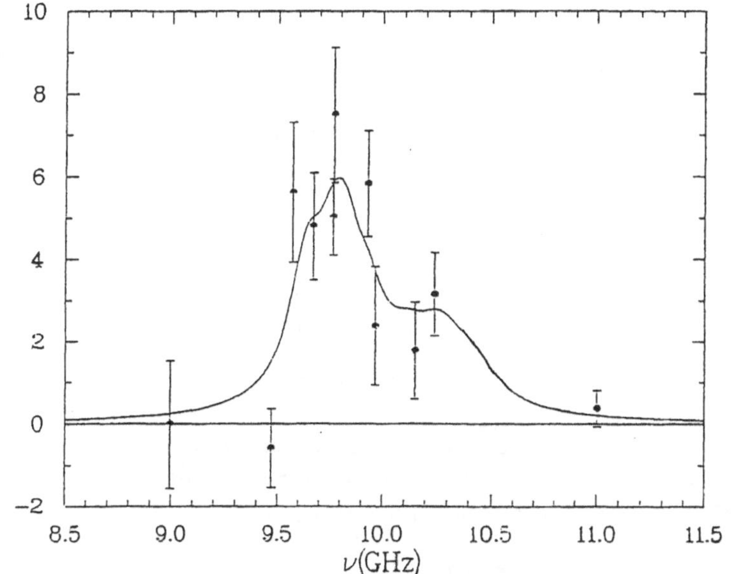

Fig. 11. Muonium $2S^2_{1/2} - 2P^2_{3/2}$ transition showing data and a best fit theoretical line shape.

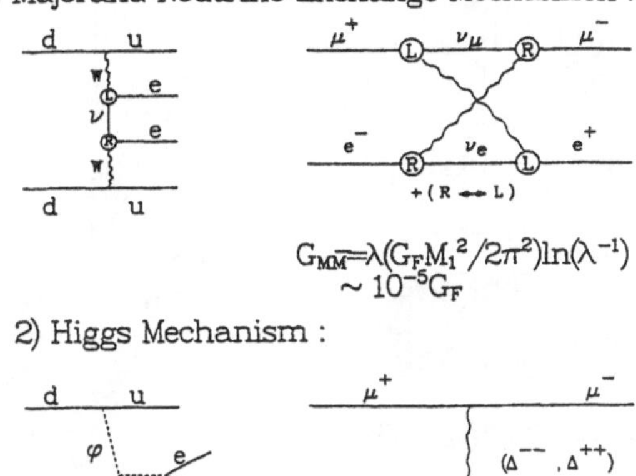

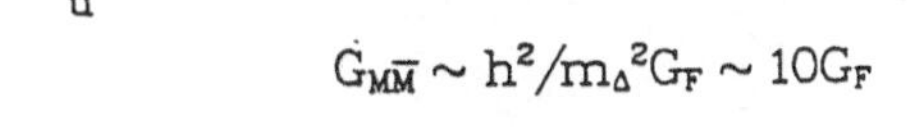

Fig. 12. Majorana neutrino exchange mechanism: Higgs mechanism.

resonance transitions are involved, but they are not resolved. From the lowest to the highest frequencies these transitions are $2S_{1/2}(F{=}1)$ to $2P_{3/2}(F{=}1)$, $2S_{1/2}(F{=}1)$ to $2P_{3/2}(F{=}2)$ and $2S_{1/2}(F{=}0)$ to $2P_{3/2}(F{=}1)$ [Fig. 1]. Analysis gives the value 9783^{+35}_{-30} MHz for the transition frequency $2S_{1/2}(F{=}1)$ to $2P_{3/2}(F{=}2)$. Correcting for the hfs this gives the value 9895^{+35}_{-30} for the fine structure interval $2S_{1/2}$ to $2P_{3/2}$ in agreement with the theoretical value of 9874.3(3) MHz. This $2S_{1/2}$ to $2P_{3/2}$ resonance can also be used to determine the Lamb shift assuming the theoretical value for the fine structure interval and gives $S_L = 1027^{+30}_{-35}$ MHz.

Future improvement in the accuracy of determining the Lamb shift and fine struture interval in the n=2 state could be considered using the $2^2S_{1/2}$ to $2^2P_{3/2}$ transition. It should involve the suppression of the first-order Doppler effect by the choice of cavity mode relative to the M velocity direction, the use of M(2S) production from the foil at about 30° to the direction of the incident μ^+ to reduce the background associated with energetic μ^+ [15], and use of UV photon detectors with higher acceptance and efficiency (perhaps using highly reflecting UV mirrors and a small photon gas counter). With these features the experiment might determine S_L to about 0.1%.

This experiment is severely limited by signal rate which at present amounts to about 10 counts/hr, and additional future progress would require higher M(2S) beam intensities. Fundamentally we need a major increase in the phase-space density of μ^+ beams (L. Simons, P. Taqqu, in this volume).

4 Muonium → Antimuonium Conversion

The muon and the electron may be considered to belong to two different generations of leptons, which thus far appear to remain separate or obey independent conservation laws of muon number, ($L_\mu = +1(-1)$ for $\mu^-(\mu^+)$ and for $\nu_\mu(\bar{\nu}_\mu)$, 0 for other particles) and of electron number ($L_e = +1(-1)$ for $e^-(e^+)$ and for $\nu_e(\bar{\nu}_e)$, 0 for other particles) as required in the standard theory. Any connection between the muon and the electron, such as a process which would violate muon number conservation, would be an important clue to the relationship between the two generations and to physics beyond the standard model. Speculative modern theories which seek a more unified theory of particles and their interactions, such as the left-right symmetric theory, predict muon number violating processes. Figure 12 shows possible processes allowed in the left-right symmetric theory (R. Mohapatra, P. Herczeg, in this volume). As yet no such rare decay process has been observed (H.K. Walter and K.Jungmann, in this volume), and with our present knowledge theory has little useful predictive power.

The conversion of muonium (μ^+e^-) to its antiatom antimuonium (μ^-e^+) would be an example of a muon number violating process, and like neutrinoless double beta decay would involve $\Delta L_e = 2$. The M-$\bar{M}$ system also bears some relation to the $K^\circ - \bar{K}^\circ$ system, since the neutral atoms M and $\bar{M}$ are degenerate in the absence of an interaction which couples them.

Usually the M to $\bar{M}$ conversion is discussed with the following postulated Hamiltonian which is of the four-Fermion, V-A type with the coupling constant $G_{M\bar{M}}$.

$$H_{M\bar{M}} = \frac{G_{M\bar{M}}}{\sqrt{2}}\bar{\mu}\gamma^\lambda(1-\gamma_5)e\bar{\mu}\gamma_\lambda(1-\gamma_5)e + \text{h.c.}$$

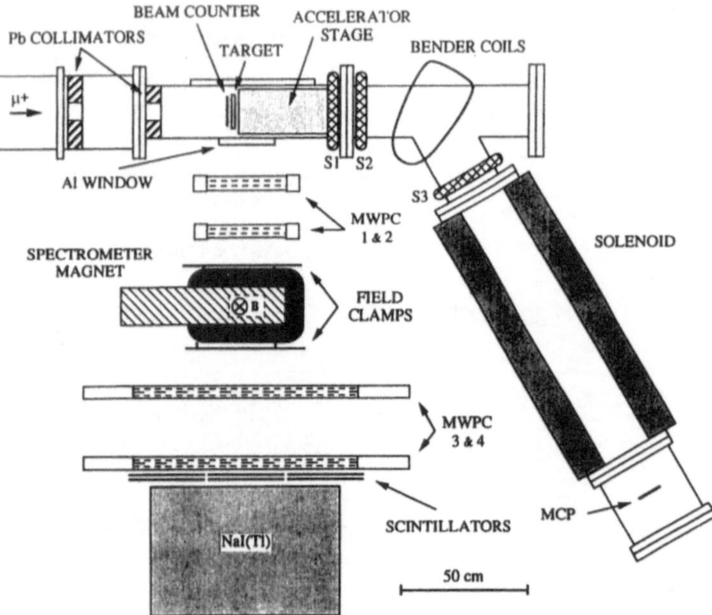

Fig. 13. Top view of the experiment to search for $M \to \bar{M}$ conversion at LAMPF.

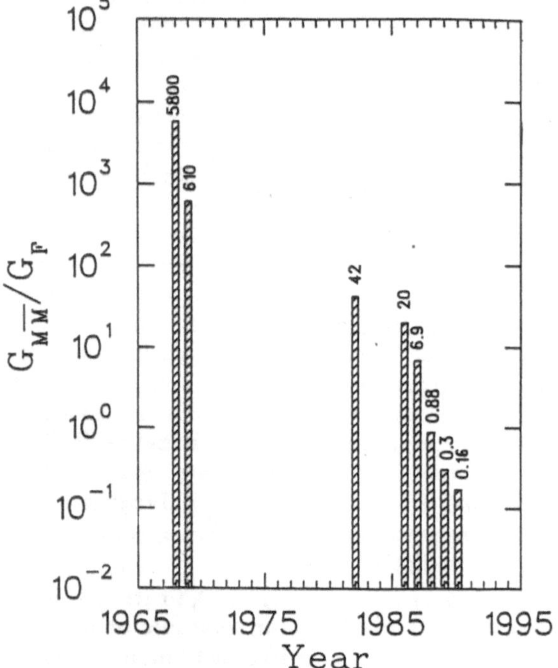

Fig. 14. History of experimental upper limits on $G_{M\bar{M}}$.

Table 4. Muon Properties.

MASS

$\frac{m_{\mu^+}}{m_e} = 206.768\,259(62)(0.3\text{ppm})$

$\frac{m_{\mu^-}}{m_e} = 206.765(10)(50\text{ppm})$

SPIN

$I_\mu = 1/2$

MAGNETIC MOMENT

$\frac{\mu_{\mu^+}}{\mu_p} = 3.183\,345\,47(95)\,(0.3\text{ppm})$

$\frac{\mu_{\mu^-}}{\mu_p} = 3.183\,4(9)(300\text{ppm})$

G-VALUE

$(g_{\mu^+}-2)/2 = 1\,165\,911(11) \times 10^{-9}(10\text{ppm})$

$(g_{\mu^-}-2)/2 = 1\,165\,937(12) \times 10^{-9}(10\text{ppm})$

ELECTRIC DIPOLE MOMENT

$\mu_e \leq 7 \times 10^{-19}$ e-cm (95% confidence level)

STATISTICS

Fermi - Dirac

LIFETIME

$\tau_{\mu^+} = 2\,197.03(4)\,\text{ns}\,(18\text{ppm})$

MUON NEUTRINO MASS

$m_{\nu_\mu} < 0.25$ MeV

Rabi, *"Who Ordered That?"*
(the muon)

where standard notation is used. The M→ $\bar{M}$ conversion violates the additive laws of muon and electron number conservation, but satisfies the multiplicative laws of muon and electron conservation for which $(-1)^{\sum L_\mu}$ = constant and $(-1)^{\sum L_e}$ = constant. The probability $P(\bar{M})$ that a muonium atom formed at t=0 will decay from the $\bar{M}$ form due to the action of the Hamiltonian $H_{M\bar{M}}$ is given by:

$$P(\bar{M}) = 2.5 \times 10^{-5} \left(\frac{G_{M\bar{M}}}{G_F}\right)^2$$

where G_F is the Fermi coupling constant.

Thus far no spontaneous conversion of M→ $\bar{M}$ has been observed and the most sensitive search was done at LAMPF [16] with the apparatus shown in Fig. 13. Thermal muonium is formed by μ^+ stopped in a SiO_2 powder target and diffuses out into vacuum. The signal for an M→ $\bar{M}$ conversion would be a coincident high energy $e^-(\geq 30MeV)$ from a μ^- decay detected with the high energy spectrometer arm and a low energy atomic $e^+(\sim 10$ eV) detected with the low energy spectrometer arm. No candidate events were observed and the limit $G_{M\bar{M}} < 0.16\,G_F$ (90% C.L.) was established. The history of the limits on $G_{M\bar{M}}$ in the various experimental searches since 1968 are shown in Fig. 14.

A new search for the M → $\bar{M}$ conversion is being undertaken at PSI [17]. The method is the same as the latest LAMPF experiment just discussed and the experimental setup is shown in Fig. 15. The cw beam at PSI will be used and the spectrometer has a large solid angle and will also require the observation of the 2γ annihilation of e^+. The goal is a sensitivity $G_{M\bar{M}} \simeq 10^{-3} G_F$.

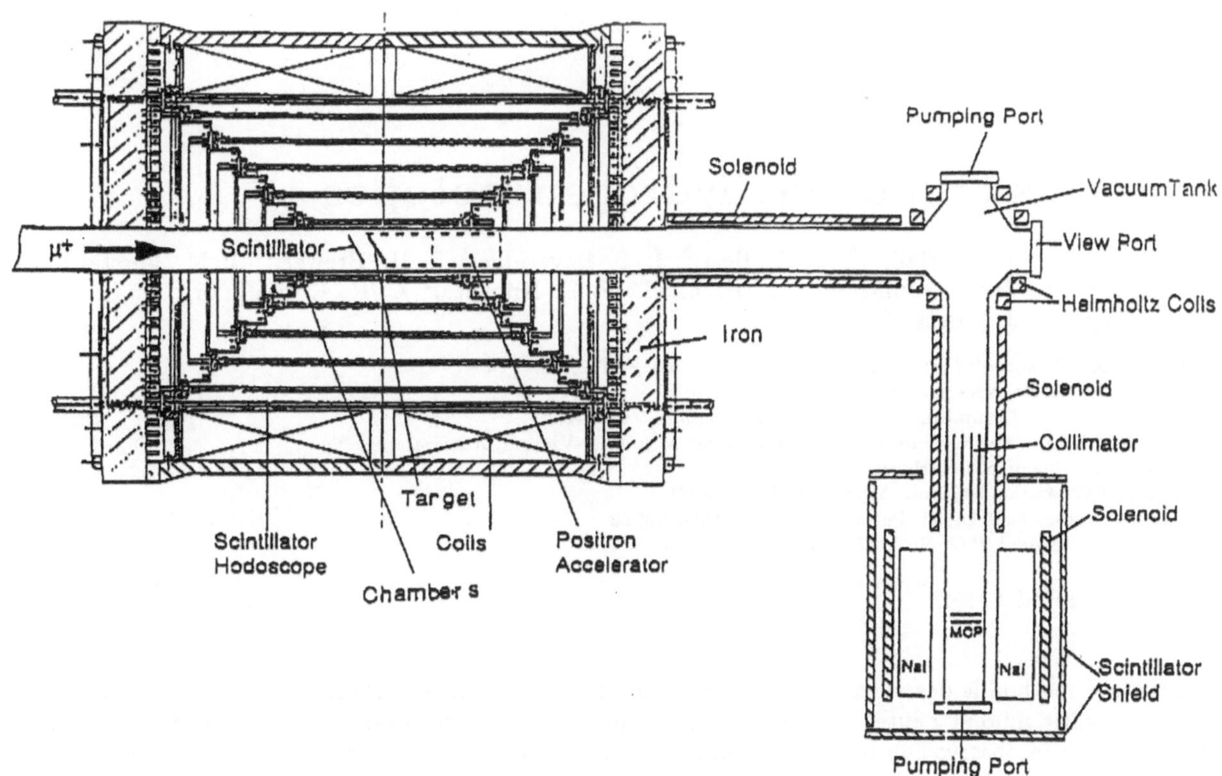

Fig. 15. MAC's a new Muonium to Antimuonium Conversion Spectrometer

5 Summary

Table 4 summarizes our present knowledge of muon properties. As a concluding remark we emphasize that research on the fundamental properties of muonium is flourishing with many important recent advances and with bright prospects for the future.

Research supported in part by DOE under contract DE-AC02-76ERO3075.

References

1 R.S. van Dyck, Jr., Quantum Electrodynamics, ed. by T. Kinoshita (World Scientific, Singapore, 1990) p. 322.
2 T. Kinoshita, Quantum Electrodynamics, ed. by T. Kinoshita (World Scientific, Singapore, 1990) p. 218.
3 F.M. Pipkin, Quantum Electrodynamics, ed. by T. Kinoshita (World Scientific, Singapore, 1990) p. 696.
4 J.R. Sapirstein and D.R. Yennie, Quantum Electrodynamics, ed. by T. Kinoshita (World Scientific, Singapore, 1990) p. 560.
5 V.W. Hughes and G. zu Putlitz, Quantum Electrodynamics, ed. by T. Kinoshita (World Scientific, Singapore, 1990) p. 822.
6 F.J.M. Farley and E. Picasso, Quantum Electrodynamics, ed. by T. Kinoshita (World Scientific, Singapore, 1990) p. 479.
7 T. Kinoshita and W.J. Marciano, Quantum Electrodynamics, ed. by T. Kinoshita (World Scientific, Singapore, 1990) p. 419.
8 N.F. Ramsey, Quantum Electrodynamics, ed. by T. Kinoshita (World Scientific, Singapore, 1990) p. 673.
9 M.G. Boshier et al., Phys. Rev. A40, 6169 (1989).
10 F.G. Miriam, et al., Phys. Rev. Lett. 49, 993 (1982).
11 T. Kinoshita and D.R. Yennie, Quantum Electrodynamics, ed. by T. Kinoshita (World Scientific, Singapore, 1990) p. 1.
12 C.J. Oram et al., Phys. Rev. Lett. 52, 910 (1984).
13 A. Badertscher et al., Phys. Rev. Lett. 52, 914 (1984); K.A. Woodle et al., Phys. Rev. A41, 93 (1990).
14 S.H. Kettell, Ph.D. Thesis, Yale University (1991); S.H. Kettell et al., Abstracts of Twelfth Inter. Conf. on Atomic Physics (ICAP), eds. W.E. Baylis, G.W.F. Drake and J.W. McConkey, (1990) p. I-9; S.H. Kettell et al., Bull. Am. Phys. Soc. 36, 1258 (1991).
15 H. Ahn, Ph.D. Thesis, Yale University (1992); H. Ahn et al., Abstracts of Twelfth Inter. Conf. on Atomic Physics (ICAP), eds. W.E. Baylis, G.W.F. Drake and J.W. McConkey, (1990) p. XII-4.
16 B.E. Matthias, et al., Phys. Rev. Lett. 66, 2716 (1991).
17 K. Jungmann et al., Proposal for an Experiment to Search for Spontaneous Conversion of Muonium to Antimuonium (1989).

This article was processed using Springer-Verlag TEX Z.Physik C macro package 1991
and the AMS fonts, developed by the American Mathematical Society.

Z. Phys. C - Particles and Fields 56, S44-S47 (1992)

Zeitschrift für Physik C Particles and Fields

Reactions of muonic hydrogen isotopes

G. M. Marshall,[1] J. M. Bailey,[8] G. A. Beer,[2] J. L. Beveridge,[1] J. H. Brewer,[3] B. M. Forster,[3] W. N. Hardy,[3] T. M. Huber,[7] R. Jacot-Guillarmod,[3] P. Kammel,[4] P. E. Knowles,[2] A. R. Kunselman,[6] G. R. Mason,[2] A. Olin,[2] C. Petitjean,[5] and J. Zmeskal[4]

[1] TRIUMF, 4004 Wesbrook Mall, Vancouver, B.C., Canada V6T 2A3
[2] University of Victoria, Victoria, B.C., Canada V8W 3P6
[3] University of British Columbia, Vancouver, B.C., Canada V6T 2A6
[4] Institute for Medium Energy Physics, Boltzmanngasse 3, A-1090 Vienna, Austria
[5] PSI, CH-5234 Villigen, Switzerland
[6] University of Wyoming, Laramie, Wyoming, USA 82071
[7] Gustavus Adolphus College, St. Peter, Minnesota, USA 56082
[8] University of Liverpool, P.O. Box 147, Liverpool, UK L69 3BX

Submitted 15 September 1991

Abstract. A method is described for the study of reactions of energetic muonic deuterium and tritium atoms. It is based on the observation of emission into vacuum of muonic deuterium from thin layers of solid hydrogen. A simple model for emission is accompanied by some rate calculations. Finally, an experiment is proposed to take advantage of the process to study muonic hydrogen reactions.

1 Introduction

A great deal of effort, both experimental and theoretical, has been put toward the understanding of muonic molecular reactions. This is especially true for the case of muonic hydrogen (μ^-p, μ^-d, or μ^-t), where the practical application of energy production via muon catalyzed fusion has encouraged the construction of very detailed models of the processes responsible for high fusion yields.[1] Naturally, the research which remains to be done to confirm theoretical predictions has become increasingly difficult with traditional methods, and alternatives are being sought to circumvent difficult technical problems.

Muon catalyzed fusion (μCF) has been observed to occur in mixtures of deuterium and tritium[2] with a yield of over 100 fusions per muon. This yield is only possible if the formation of a muonic molecule of two hydrogen isotopes occurs at a rate much higher than expected from calculations involving "simple" interactions. It has been shown that a resonance mechanism is necessary.[3] In fact, detailed calculations have been made of the energy dependence of resonance formation.[4] In order to validate precisely the calculations by experiment, it is necessary to create muonic tritium atoms with kinetic energy in the range of 0.1–1.0 eV (10^3–10^4 K). It is also important to control or measure the energy of the atoms. Thermal activation is possible but difficult as the required temperatures are high, and one achieves a broad Maxwellian distribution from which it may be difficult to extract precise energy information.

An alternative to thermal activation is discussed in this paper. It is the application of a phenomenon observed many years ago with muons in hydrogen isotope mixtures,[5] which results in energetic muonic deuterium (μ^-d) or tritium (μ^-t) atoms without the need for extremely high temperature target systems.

2 Principle of muonic hydrogen production

The first experimental observation of muon catalyzed fusion was in a liquid hydrogen bubble chamber[5] containing protium with small atomic concentrations of deuterium. Some events were characterized by the ionization track of an incoming muon and a track corresponding to a regenerated muon with 5.3 MeV energy, following the reaction

$$p\mu d \longrightarrow {}^3\mathrm{He}(0.2\mathrm{MeV}) + \mu(5.3\mathrm{MeV}). \qquad (1)$$

A striking feature of many such events was a clearly discernible separation of order 1 mm between the tracks.

The explanation of the separation was that a deuteron "robbed" the muon from muonic protium, obtaining some 45 eV in kinetic energy due to the difference in reduced mass and therefore binding energy. It then penetrated the hydrogen, leaving no ionization track, by a measurable distance before forming the muonic molecule and emitting the fast muon after fusion. That the distance of penetration could be so large was curious, until it was explained in terms of a fortuitous reduction in the cross section for scattering of μ^-d on protons in hydrogen.[6] It is a manifestation of what is known as the Ramsauer-Townsend effect, first observed for electrons in rare gases, and results from a strong attractive potential with short range. In scattering terminology, the lowest partial wave is shifted by $\pi/2$ and the corresponding scattering amplitude vanishes; the higher order scattering amplitudes are small, so the cross section is also unusually small. The scattering interaction has been calculated in greater detail,[7,8] and the transparency of protium to μ^-d as well as μ^-t in the range of 0.1–10 eV kinetic energy has been theoretically confirmed.

The effect has been observed recently in an experiment using solid films of hydrogen with various deuterium atomic concentrations and thicknesses (up to $\sim$ 1 mm).[9] The results show clearly that it is possible to produce μ^-d in vacuum in the energy range of interest. The partial experimental yield is observed to be approximately 1% under optimum conditions. The total yield, which includes emission of μ^-d of energy below the detection threshold, is expected to be higher by a factor of about two.

3 Estimating the yield

Given the appropriate theoretical cross sections, the emission of muonic deuterium (or tritium) from a hydrogen layer can be calculated as a function of deuterium atomic concentration and layer thickness. A precise calculation requires a numerical simulation, but with certain simplifying assumptions an analytical estimate can be made. We will describe here the ingredients of this estimation procedure, and give some results. Please note that approximations will be made to illuminate the gross dependence of the important processes, and to prevent obfuscation with small corrections; the absolute values of the results are therefore not exact.

For the emission of muonic deuterium to occur via the proposed mechanism, three sequential processes must take place. Each process can be calculated independently, and the fraction $\mathcal{F}$ of the initial incident muon beam which is emitted as μ^-d will be the product of the fractions $\mathcal{F}_i$ of the three processes.

The first, $\mathcal{F}_1$, is the fraction of incident μ^- which stops and captures initially on a proton in the hydrogen layer. We assume that the thickness x of the layer is small compared with the stopping range distribution ΔR of the incoming muon beam. Typically a beam of 30 MeV/c μ^- with a momentum width of 5% full width at half maximum (FWHM) has a range and range spread equivalent to 150 mg cm^{-2} and 40 mg cm^{-2} FWHM respectively in carbon. The low energy stopping power of protium is about 2.5 times higher than carbon, so a density of $\rho = 88\,\mathrm{mg\,cm^{-3}}$ for solid hydrogen at 2.5 K gives $\Delta R = 1.8\,\mathrm{mm}$. The maximum hydrogen layer thickness is 1 mm. Note that the stopping power of protium is nearly the same as for deuterium when expressed *per atom* rather than per unit thickness (expressed as either x or ρx); in this case one can safely ignore concentration-dependent differences in density, and consider the thickness x to be the equivalent protium value scaled by atoms per cm^2. It is also assumed that protons and deuterons are the only atoms present (i.e., $c_p + c_d = 1$), so that the probability of capture on a proton is $(1 - c_d)$, where c is the fractional atomic concentration of the indicated isotope in the hydrogen layer. One then has

$$\mathcal{F}_1 = (1 - c_d)x/\Delta R. \tag{2}$$

The second factor, $\mathcal{F}_2$, is the probability that a muon in a μ^-p atom will transfer to a deuteron in its lifetime. This is estimated by the ratio of the rate for transfer (λ_d) to the sum of the rates for all modes of disappearance of muonic protium. Apart from muon transfer to a deuteron, the muon may decay with rate λ_0, or the muonic atom may form a muonic molecule with another proton with rate λ_{pp}. The result for the second factor is

$$\mathcal{F}_2 = \frac{\lambda_d}{\lambda_0 + \lambda_{pp} + \lambda_d}. \tag{3}$$

Perhaps a less precise approach is taken to the estimation of $\mathcal{F}_3$, the probability of emission of μ^-d from a layer of thickness x, after the muon has transferred from μ^-p. The assumptions are made that *all* μ^-d atoms are created with kinetic energy *above* the penetration energy (i.e., an energy in the range for which the cross section for elastic scattering with protium is within the Ramsauer-Townsend minimum), and furthermore are slowed so as to travel with the penetration energy. It is also assumed that any further scattering process will lower the kinetic energy below the penetration value, leading to stopping of the muonic atom within the hydrogen layer. In this way the problem becomes one of finding the probability that the μ^-d with the penetration energy will *not* scatter before reaching one boundary of the hydrogen layer. Given a value Σ for the probability of interaction per unit length (sometimes called the "macroscopic cross section"), the calculation is an integration of the survival probability $\exp(-\Sigma r)$. The variable $r = z/\cos\theta$ is the distance from the initial point to a point on the surface, z is the perpendicular distance to the surface, and θ the angle of emission with respect to a perpendicular from the surface. The integration is taken over all possible straight paths from the initial point to one surface of a layer, and over all possible distances of the initial point from the surface. With the correct normalization, one has

$$\begin{aligned}\mathcal{F}_3 &= \frac{1}{x}\int_0^x \left[\frac{1}{2}\int_0^{\pi/2} \exp(-\frac{\Sigma z}{\cos\theta})\sin\theta\, d\theta\right] dz \\ &= \frac{1}{2\Sigma x}\left(\frac{1}{2} - \int_1^\infty \frac{\exp(-\Sigma x q)}{q^3} dq\right).\end{aligned} \tag{4}$$

The integral in the latter expression can be evaluated numerically and is often found in computer function libraries.

The interaction probability can be expressed in terms of the conventional cross sections by

$$\Sigma = N_0\phi[(1 - c_d)\sigma^{el}_{\mu dp} + c_d\sigma^{el}_{\mu dd}] \tag{5}$$

where $N_0 = 4.25 \times 10^{22}\mathrm{cm}^{-3}$ is the atomic density of liquid hydrogen and $\phi = 1.24$ is the density of the solid

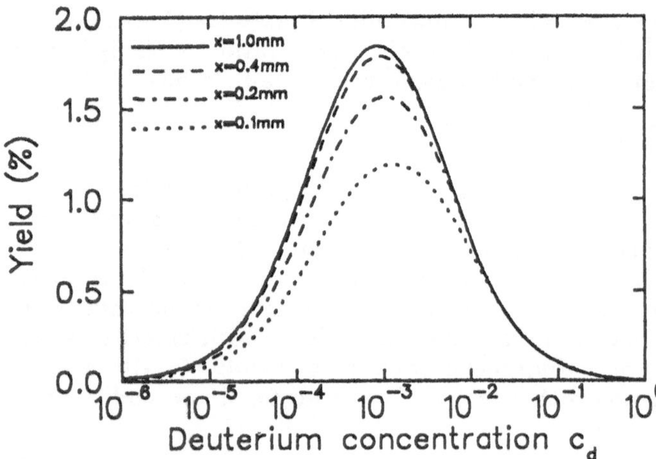

Fig. 1. Yield of muonic deuterium from a layer of thickness x as a function of deuterium atomic concentration

layer relative to liquid hydrogen. Values of the cross sections for elastic scattering of μ^-d by protons and by deuterons are chosen to be $\sigma^{el}_{\mu dp} = 1 \times 10^{-21}\,\text{cm}^{-2}$ and $\sigma^{el}_{\mu dd} = 2.3 \times 10^{-19}\,\text{cm}^{-2}$, which represent typical theoretical values at the penetration energy.[7,8] Corrections for electron screening for atomic targets are calculated to be small.[8] The rates for transfer and muon molecular formation depend on density and concentration by

$$\lambda_d = c_d\phi\Lambda_d, \lambda_{pp} = (1 - c_d)\phi\Lambda_{pp} \tag{6}$$

where the reduced rates $\Lambda_d = 1.7 \times 10^{10}\,\text{s}^{-1}$[10] and $\Lambda_{pp} = 2.5 \times 10^6\,\text{s}^{-1}$[11] are consistent with experiment, assuming no strong temperature dependence.

The product $Y = \mathcal{F}_1\mathcal{F}_2\mathcal{F}_3$ is shown in Fig. 1 as a function of deuterium atomic concentration c_d for several values of the thickness x. The normalization is to the number of incident muons, and will depend on the properties of the muon beam; the scale of Y should not be assumed to be precise. The maximum value occurs slightly above the natural concentration of deuterium atoms in hydrogen from sea water, $c_d = 1.4 \times 10^{-4}$.

4 Reaction studies with emitted hydrogen isotopes

The emission of μ^-d and μ^-t due to the penetration mechanism is the basis of a proposal to measure the resonant energy dependence of the cross section for muon molecular formation

$$\mu t + \text{dxee} \rightarrow (\text{d}\mu\text{t})\text{xee} \tag{7}$$

where x is p, d, or t. Recent calculations have been made[4] for the effective rate of the reaction in a mixture of isotopes at different temperatures and for different μ^-t spin states, for the most interesting vibrational states of target and product molecules. At 30 K, the resonances for each spin state are distinct, with the lowest

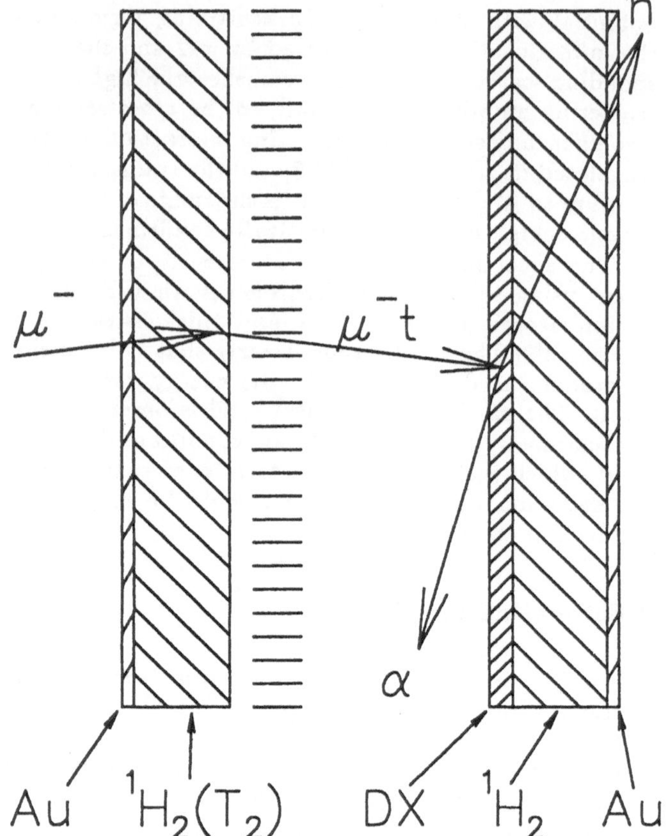

Fig. 2. Schematic geometry for an experiment which uses adjacent solid hydrogen layers and emission of energetic muonic tritium

and strongest structures occurring at μ^-t kinetic energies from 0.2 to 0.6 eV. In a conventional target at 30 K, however, this is two orders of magnitude greater than kT, and one must rely on epithermal energies for resonance effects. At 1000 K, when the resonance energy region is thermally accessible, the structures are broadened by the distribution in energy of the reactants.

The practical difficulties of hot tritium targets can be overcome in principle by utilizing the penetration mechanism in a geometry consisting of an emitting layer and a target layer (Fig. 2). The emitting layer consists of protium with a small atomic concentration of tritium, $c_t \sim 1 \times 10^{-3}$. Because the cross sections for elastic scattering of μ^-d and μ^-t are very similar in the penetration region, muonic tritium should be emitted with essentially the same properties as muonic deuterium, i.e. with kinetic energy in the range of 0.1-1.0 eV, matching very well the calculated resonance energy region. An adjacent target layer is composed of the desired hydrogen molecules, such as D_2 or HD, with a thickness corresponding to about one interaction length for muon molecular formation. A sublayer consisting of pure protium beneath the target layer allows muonic atoms which pass through the target layer to stop without any interaction which could emulate the signature for molecular formation. The signature is the observation of a fusion product such as a neutron or alpha particle, since fusion follows molecular formation at a rate of order $10^{12}\,s^{-1}$.

The energy dependence is determined by time of flight. Experimentally one measures the interval between the time at which a muon enters the target and the time at which a fusion product is detected. The energy of the muonic atom is calculated from the time of flight between the adjacent layers and the distance travelled. The time at which the muonic atom escapes from the emitting layer after entering the target is determined by the transfer rate of the muon to the heavier isotope, and is of the order of 100 ns. Because the rate of fusion from the $d\mu t$ system is so high ($\sim 10^{12}\,s^{-1}$), the arrival time at the target layer is essentially the time of detection of a fusion product. The energy resolution is therefore limited by the time of the transfer process and by the uncertainty in the distance travelled, since muonic atoms are emitted with an angular distribution. If the angles can be restricted, for example by means of a collimation system, then an energy resolution of better than 20% might be possible. It is also essential that the possibility is minimized for energy loss in the target layer before molecular formation, in order that the time of flight correctly represents the energy. Fortunately the resonant cross section for molecular formation is not small compared with elastic cross sections which might compete, but it emphasizes the importance of the choice of the minimum target layer thickness.

The event rate for the proposed experiment has been estimated. Assuming detection of 14 MeV neutrons from dt fusion with an efficiency of 0.1 in an area of $250\,cm^2$ at an effective distance of 15 cm, a muon rate of order $10^4\,s^{-1}$ will give an event rate of order 0.1-1 s^{-1}. While this is certainly large enough to justify the measurement, a distinct improvement is possible if a detector for the charged fusion products can be placed close to the target layer, perhaps even serving as the surface on which the hydrogen is frozen. Measurements to determine the feasibility of this method are underway at TRIUMF.

References

1. For recent reviews, see L.I. Ponomarev, Contemporary Physics 31, (1990) 219; W.H. Breunlich, P. Kammel, J.S. Cohen, and M. Leon, Ann. Rev. Nucl. Part. Sci. 39, (1989) 311
2. S.E. Jones et al., Phys. Rev. Lett. 56, (1986) 588; W.H. Breunlich et al., Phys. Rev. Lett. 58, (1987) 329
3. S.S. Gershtein and L.I. Ponomarev, Phys. Lett. 72B, (1977) 80
4. M.P. Faifman and L.I. Ponomarev, Phys. Lett. B 265 (1991) 201
5. L.W. Alvarez et al., Phys. Rev. 105, (1957) 1127
6. Stanley Cohen, David L. Judd, and Robert J. Riddell, Jr., Phys. Rev. 119, (1960) 397
7. M. Bubak and M.P. Faifman, JINR preprint E4-87-464, Dubna (1987)
8. James S. Cohen and Michael C. Struensee, Phys. Rev. A 43, (1991) 3460; James S. Cohen, Phys. Rev. A 43 (1991), 4668
9. B.M. Forster et al., Hyp. Int. 65, (1990) 1007; G.M. Marshall et al., Proc. μCF-89 (Oxford, 1989), report RAL-90-022, Rutherford Appleton Laboratory, (1990) 45
10. A. Bertin et al., Lett. Nuovo Cimento 4, (1972) 449
11. V.M. Bystritskii et al., Sov. Phys. JETP 43, (1976) 606 [original Zh. Eksp. Teor. Fiz. 70, (1976) 1167]

This article was processed using Springer-Verlag TEX Z.Physik C macro package 1991
and the AMS fonts, developed by the American Mathematical Society.

Z. Phys. C - Particles and Fields 56, S48-S58 (1992)

Zeitschrift für Physik C Particles and Fields

Muonic atoms spectroscopy

L.A. Schaller

Institut de Physique, Université de Fribourg, CH-1700 Fribourg, Switzerland

1-July-1991

Abstract. A review is given of the physics of muonic atoms. The results of spectroscopic experiments on μ^- atomic systems are discussed.Special emphasis is given to the determination of nuclear charge moments, the nuclear polarization correction, and QED corrections.

1 Introduction

Using the tool of muonic atoms spectroscopy, there is a rich physics domain to be covered, if the high fluxes available at the medium energy muon factories are combined with the high energy resolutions of semiconductor detectors and crystal spectrometers. Some examples shall be mentioned in the following:

Due to the large overlap of the innermost muonic wave functions with the nuclear charge density in all but the lightest nuclei, nuclear ground state moments and their 'fine structures' in terms of isotope or isotone shifts can be determined with high precision [1,2]. By means of hyperfine structure splittings, magnetic dipole moments and electric quadrupole moments of the nuclear ground states can be measured [3]. Transitions to the muonic 1s level probe the Bohr-Weisskopf effect, i.e. the finite nuclear magnetization distribution, while hyperfine splittings between higher-lying levels yield the nuclear ground state moments model-independently [4,5]. Magnetic and electric moments of excited nuclear states are also amenable to experiment [6,7,8]. Dynamic muon-nucleus interactions lead furthermore to nuclear polarization effects, whose determinations are a challenge to both theoreticians and experimentalists [9,10]. While K-series transitions in heavy muonic atoms may lead to prompt muon-induced fission [11,12,13], scattering and transfer processes in mixtures containing hydrogen isotopes help to elucidate the phenomenon of muon-catalyzed fusion [14]. If muonic atom transitions between intermediate levels are studied, valuable information about electron screening effects can be extracted [15]. In particular however, high precision tests of QED corrections can then be performed, and the first order vacuum polarization can be tested with accuracies of the order of 1000 ppm [16].

At high principal muon quantum numbers $n_\mu \gtrsim 10$, atomic, molecular and chemical capture effects become important. Such effects may be studied by looking at the sum of the K-series intensities, or by studying relative intensities [17]. Intensities also depend on different electron refilling times. From line enlargements in transitions between low-lying levels in heavy muonic atoms, the natural line widths can be extracted.

Besides energies and intensities, direct rate measurements become possible in the ns to μs time ranges. In the following, we again enumerate some examples:

Interesting quantities are e.g. the disappearance rates of muonic protium by transfer to heavier hydrogen isotopes and in particular to helium [18]. The latter rate is poorly understood and may possibly limit the muon-catalyzed fusion cycle. In addition, there are still many open questions concerning transfer to larger Z nuclei. Nuclear muon capture rates and in particular capture rates to different isotopes can be accurately determined by looking at the time span between the stopped muon and the appearance of electrons, neutrons, delayed nuclear γ-rays or fission fragments. Using scintillation counters for the stopped muons and a series of parallel-plate avalanche counters to detect the fission products, time resolutions of the order of 500 ps can be obtained [19]. Such resolutions may lead to the detection of short-lived fission isomers in muon-induced fission [20]. Finally, the fission channel with its excellent time resolution may in the future also be useful when studying transfer from the μp system to fissionable nuclei.

The next chapter contains a brief theoretical desription of the muonic atom system. We then concentrate on new measurements of nuclear charge radii and isotope and isotone shifts (chapter 3), on the nuclear polarization corrections (chapter 4), on precision measurements of the first order (e^+e^-)-vacuum polarization and also

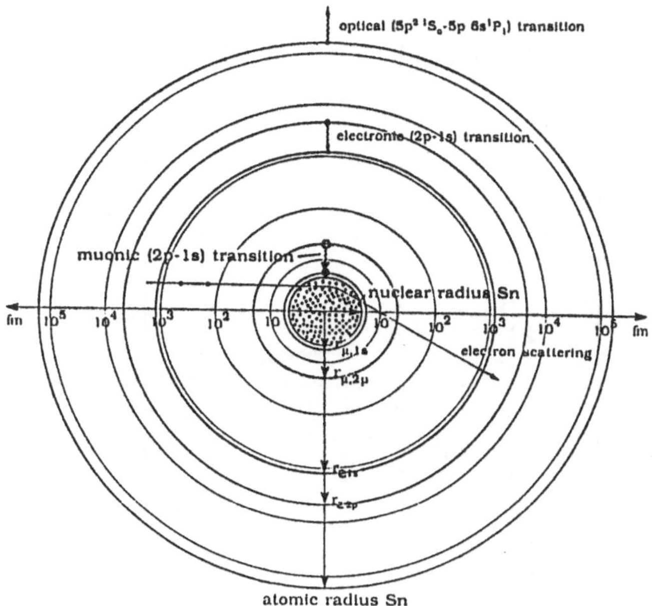

Fig. 1. Sensibilities of elastic electron scattering, muonic atom x-rays, electronic x-rays and optical transitions to the finite nuclear charge extension for the example of the tin atom. Note the logarithmic length scale.

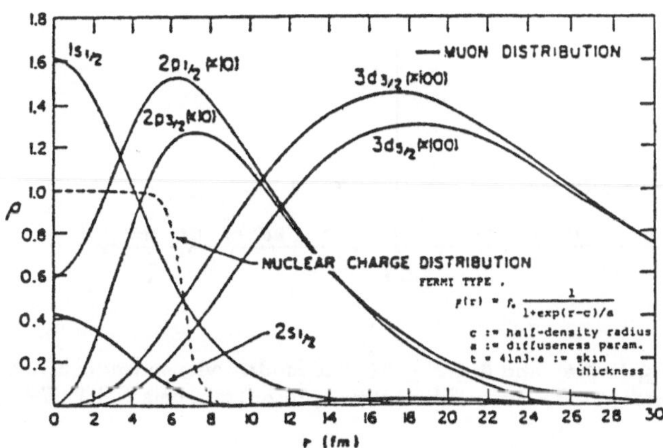

Fig. 2. Overlap of the muon probability densities $|\psi|^2$ of the innermost states with the nuclear charge density ρ in the case of $\mu^- - {}^{208}Pb$.

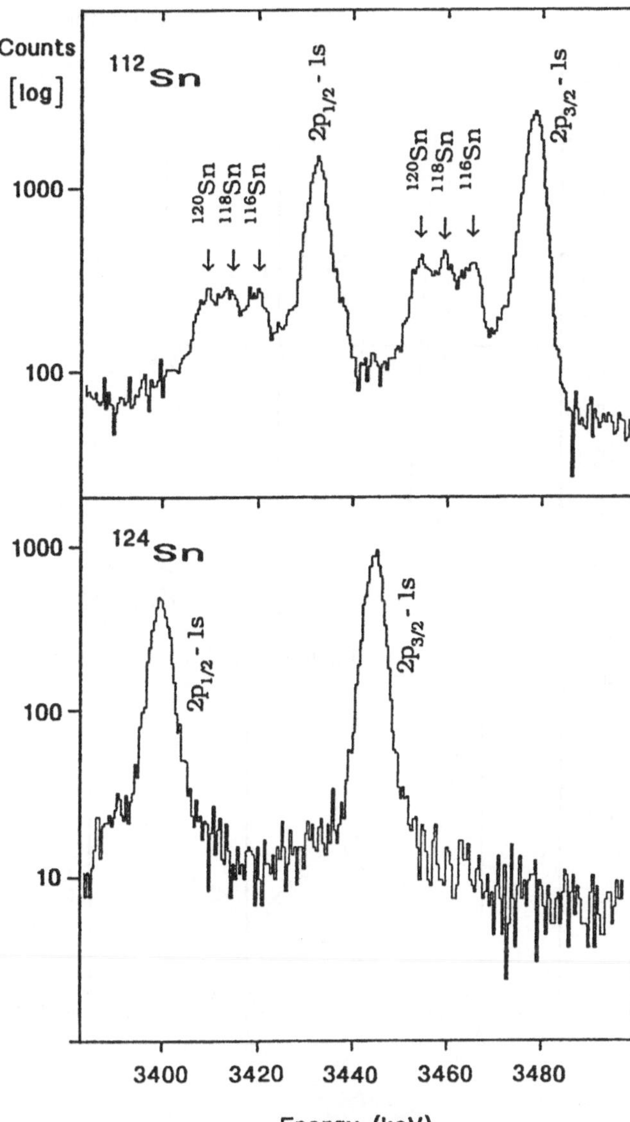

Fig. 3. Prompt muonic x-ray spectra showing the $2p_{1/2} - 1s$ and $2p_{3/2} - 1s$ transitions in the two tin isotopes at the extreme ends of stability, ${}^{112}Sn$ and ${}^{124}Sn$ (ref.2). The isotopic purity of ${}^{112}Sn$ was 68%, which explains the appearance of further tin isotopes in the upper half of this figure.

on measurements of QED corrections of higher order (chapter 5). Finally, a short outlook towards unsolved problems is given (chapter 6).

2 Theory

Starting point for the calculations of all muonic atom energy levels is the Dirac point nucleus approximation, which may be written in short as the Hamiltonian equation

$$H\Psi = E\Psi. \tag{1}$$

In intermediate muonic states, the muon-nucleus system is almost hydrogenlike, and the approximation is good. Near the nucleus however, the finite nuclear charge size may lower the muonic binding energy by as much as 50%. Hence, the Dirac equation has to be numerically solved, using a static central potential with adjustable nuclear charge parameters. Due to the double integration procedure, the exact form of the chosen nuclear charge distribution (usually a two-parameter Fermi distribution) does not influence the final result in an appreciable way. This is particularly true if equivalent radii

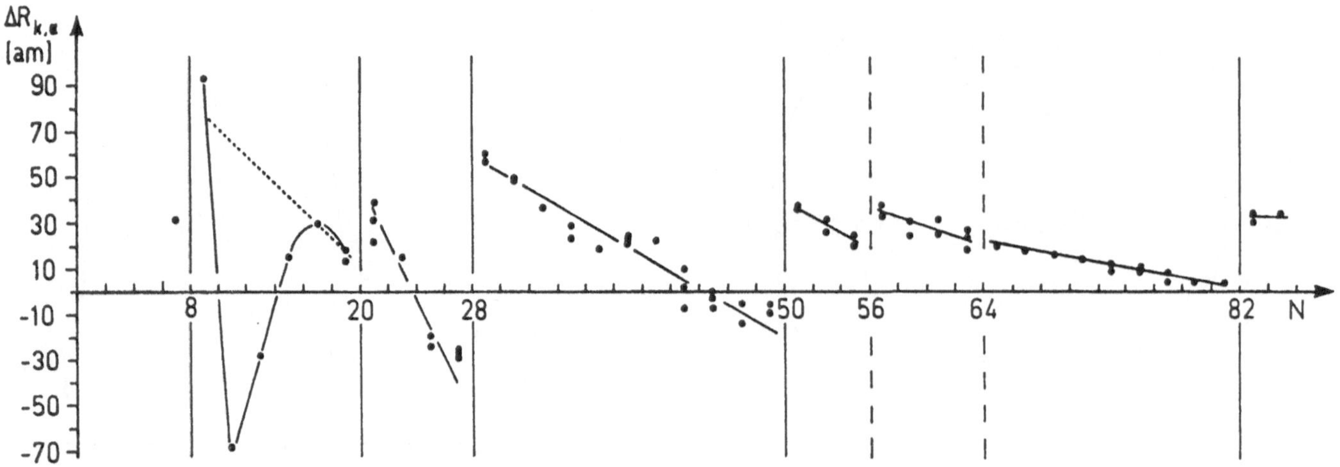

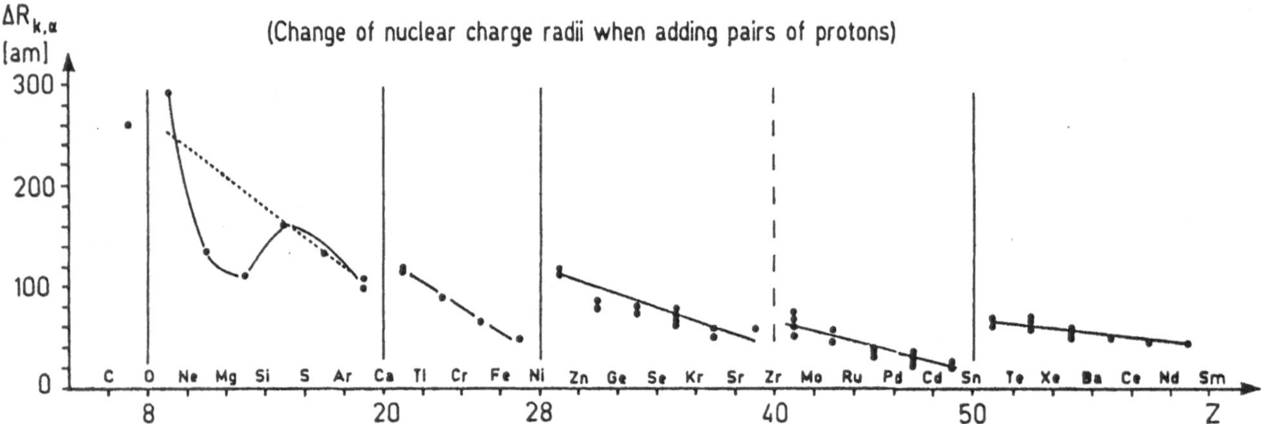

Fig. 4. Isotope and isotone shifts as measured by our collaboration for $6 \leq N \leq 86$ and $6 \leq Z \leq 62$. The model-independent equivalent charge radii differences $\Delta R_{k\alpha}$ between two even isotopes (isotones) are plotted at the respective neutron (proton) numbers $N+1$ $(Z+1)$.

$R_{k\alpha}$ are deduced (see chapter 3). In high precision work, also the first order e^+e^--vacuum polarization correction, the socalled Uehling potential, is included in the numerical integration of the Dirac equation. Such a procedure automatically also includes all ladder graphs of order $(\alpha(Z\alpha))^n$ besides the basic $\alpha(Z\alpha)$ graph, a fact which has to be taken into account when adding higher-order vacuum polarization corrections as perturbations. Explicitly then, the Hamiltonian H of eq.(1) and the energy E may be written as

$$H = H_0 + H', E = E_0 + E', \tag{2}$$

where

$$E_0 = E_{Dirac,point\ nucleus} + E_{finite\ size} + E_{Uehling\ pot.} \tag{3}$$

All other corrections, namely higher order e^+e^- as well as $\mu^+\mu^-$ and hadronic vacuum polarization corrections (VP), Lamb shift of first and second order and anomalous magnetic moment (Lamb), the relativistic reduced mass term (rel. mass), electron screening (el. scr.) and nuclear polarization (NP) corrections are added as perturbations H' or E', resp:

$$E' = E_{VP} + E_{Lamb} + E_{rel.mass} + E_{el.scr.} + E_{NP}. \tag{4}$$

Of all corrections, the least known are the nuclear polarization corrections, whose discussion we dedicate an own chapter (chapter 4). In cases where the QED corrections are by far the most important corrections in muonic atom transitions, muonic atom spectroscopy provides a means to test quantum electrodynamics and in particular the Uehling potential to high precision (chapter 5).

3 Nuclear charge moments

Figure 1 presents in a spherically symmetric and logarithmic way typical dimensions in a typical medium-heavy muonic atom ($\mu^- - Sn$). While electrons in elastic electron scattering and muons in the 1s level overlap strongly with the nuclear charge density, very high precision is demanded, if the finite nuclear charge extension is to be tested using electronic X-rays (crystal spectrometers) or optical transitions (laser spectroscopy). In terms of the squares of the muonic wave functions $|\psi|^2$, figure 2 shows their overlap with the nuclear charge density ρ for the heavy muonic standard atom $\mu^- - ^{208}Pb$. The muonic 2p intensities are blown up by a factor of 10, the 3d intensities by a factor of 100. Note for later, that the $2p_{1/2}$-wave functions overlap more strongly with ρ than the $2p_{3/2}$-intensities.

Muonic atoms probe strictly speaking not the rms radius $< r^2 >$, but a generalized nuclear charge moment, first introduced by Barret [21]:

$$< r^k e^{-\alpha r} >= \int_0^\infty \rho(r) r^k e^{-\alpha r} 4\pi r^2 dr. \qquad (5)$$

Each muonic atom transition determines a different and model-independent generalized moment. The data are usually plotted in terms of the equivalent radius $R_{k\alpha}$, which is the radius of a homogeneously charged sphere having the same generalized moment as the real charge distribution:

$$3[R_{k\alpha}]^{-3} \int_0^{R_{k\alpha}} r^k e^{-\alpha r} r^2 dr =< r^k e^{-\alpha r} > . \qquad (6)$$

For light nuclei, the moment k is close to 2 and the parameter α close to 0, so that Barrett moment and rms radius are almost the same.

As an example, figure 3 shows on a logarithmic scale the energy spectra in the region of the $2p_{1/2} - 1s$ and $2p_{3/2} - 1s$ transitions for the tin isotopes ^{112}Sn and ^{124}Sn measured with a high-resolution Ge detector [2]. The employed 'prompt' time gate of 20 ns after each stopped muon considerably reduces the background. The high muon fluxes at the PSI muon channels allow precise measurements with small amounts of rare isotopes like ^{112}Sn, whose natural abundance is only 1%. In the corresponding spectrum, the isotopic impurities $^{116,118,120}Sn$ are seen with respective intensities of about 10%.

The next picture (fig. 4) presents the results of our Fribourg-Mainz-Los Alamos collaboration for isotopes and isotones from carbon up to samarium [1,2,22-29]. Plotted are the changes of the equivalent nuclear charge radii $\Delta R_{k\alpha}$ when adding pairs of neutrons and protons at the neutron numbers $N+1$ or the proton numbers $Z+1$, respectively. If we first consider only the results for $N, Z \geq 20$, we obtain the 'normal' systematics of nuclear charge radii changes, which can be described by the following characteristica:

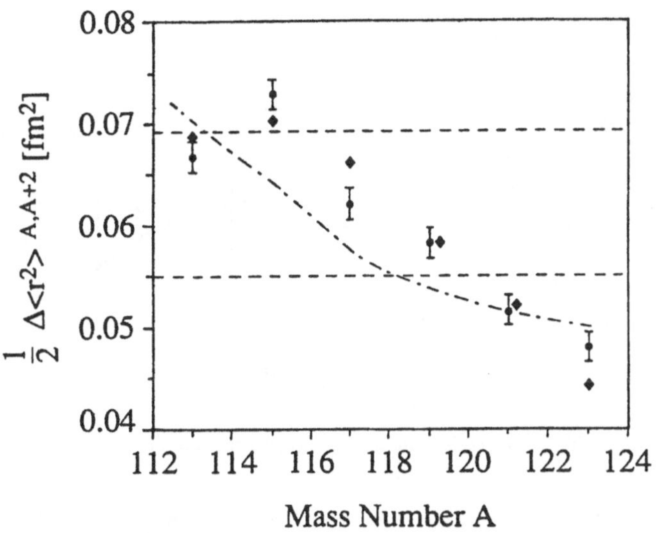

Fig. 5. Brix-Kopfermann diagram for the tin isotopes [2]. Plotted are the rms radii differences $1/2\Delta < r^2 >^{A,A+2}$ as a function of mass number A, drawn-in at A+1. Our results (black circles) are presented with errors. Also shown are optical data (black diamonds) as well as theoretical predictions (see text).

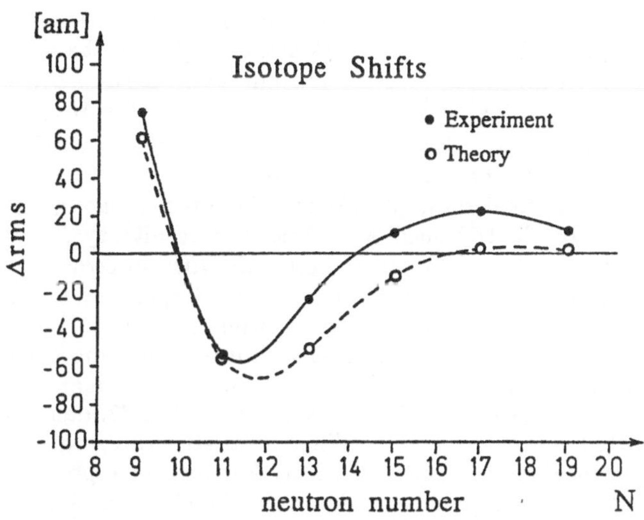

Fig. 6. Comparison of theory and experiment for the even-even isotopes in the s-d shell [29]. The theoretical points (open circles) are Skyrme force calculations by Friedrich [33] including deformation corrections (see text). The black dots are our results. The two curves are drawn-in to guide the eye.

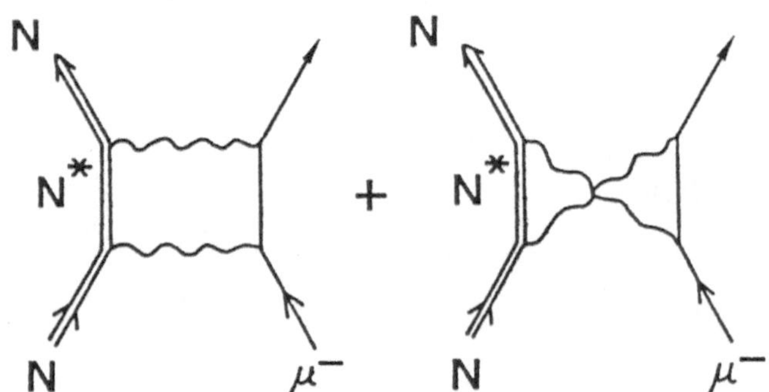

Fig. 7. The two leading Feynman graphs for nuclear polarization.

(1) For the same neutron number, the change of nuclear charge radii when adding pairs of neutrons is only weakly dependent on the respective proton configuration.

(2) After the closing of the magic neutron shells at $N =$ 20, 28, 50 and 82, there is a sudden increase of the radii differences reflecting the nuclear shell structure.

(3) In-between magic numbers, the addition of a pair of neutrons results in an almost linear decrease of the radii differences.

(4) The behaviour of isotone shifts is analogous to the behaviour of isotope shifts.

A detailed look within a given shell may also reveal subshell effects. As an example, fig.5 shows our measured charge radii differences (black circles with error bars) in the even Sn isotopes. The 65th and 66th neutron begin to occupy a higher-lying subshell (probably the $1h_{11/2}$-subshell). The radius difference $^{114,116}Sn$ is therefore increased as compared to the difference $^{112,114}Sn$. The newest optical data [30,31] are not precise enough to uniquely identify such an effect. The results by Eberz et al. are plotted as black diamonds with no error bars (their uncertainties are about 2-3 times larger than ours). The simpler theories like Myers droplet model [32] only yield limits (horizontal dashed lines), while more sophisticated Hartree-Fock calculations including ground state correlations and the Skyrme force (ref.33, dashed-dotted line) show indeed differential structure, but fail to reproduce the subshell effect. For the corresponding $\Delta R_{k\alpha}$ radii differences (not shown here), accuracies of the order of 0.5 am are obtained.

If we now go back to fig.4 and look at the isotope and isotone shifts below $N, Z = 20$, we notice immediately that the systematics applied to heavier nuclei fails. There are two reasons for such a failure. First, the nuclei with $8 \leq N, Z \leq 20$ occupy the s-d shell. These nuclei are known to be strongly deformed, changing their deformation from a strong oblate to a strong prolate shape. Diminishing deformations towards the end of a particular shell have formerly been evoked as a reason for smaller nuclear charge radii when adding pairs of neutrons (see e.g. the Kr isotopes below $N = 50$, ref.34). Second, the

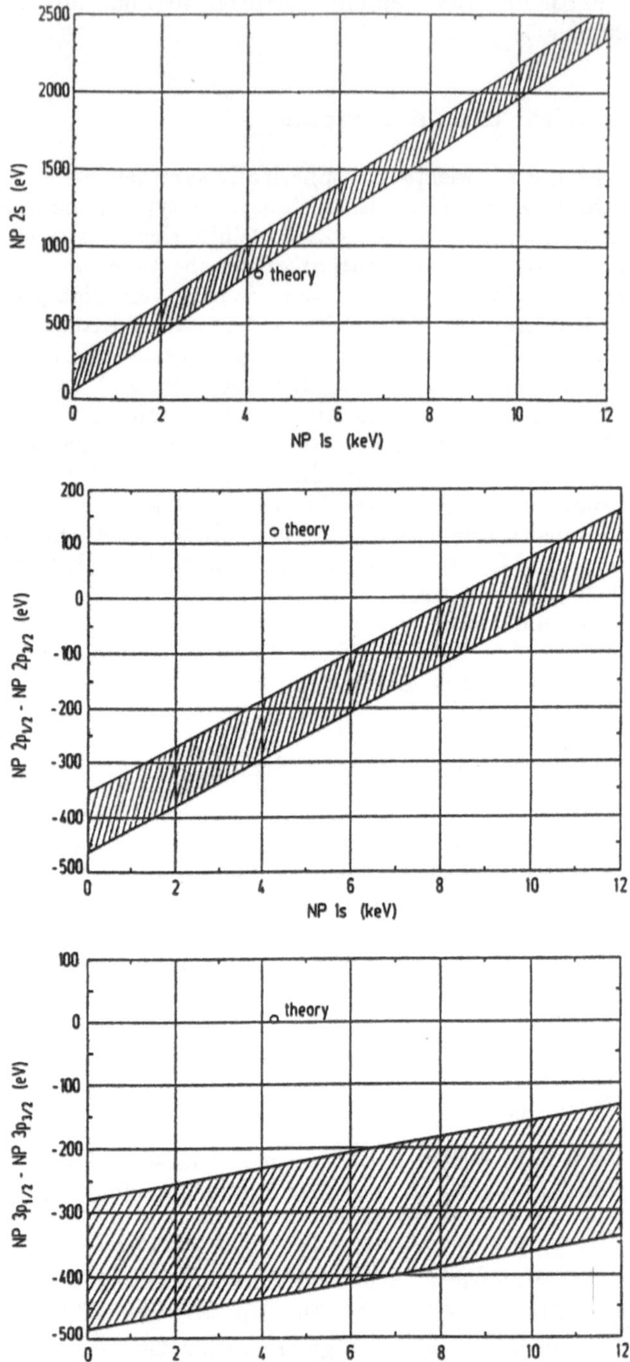

Fig. 8. Nuclear polarization (NP) correlations in $\mu^- - {}^{208}Pb$ [37]. Shown are plots of NP2s versus NP1s, of $\Delta NP2p$ vs. NP1s and of $\Delta NP3p$ vs NP1s. The shaded areas correspond to the experimental uncertainties, the theory is basically taken from the work of Rinker and Speth [38].

s-d shell occupation numbers as calculated by Brown et al. [35] are such that for the lighter nuclei the innermost $1d_{5/2}$ subshell is preferentially filled up before the outer subshells $2s_{1/2}$ and $1d_{3/2}$ are populated. As an extreme example, the two additional neutrons in ^{22}Ne as com-

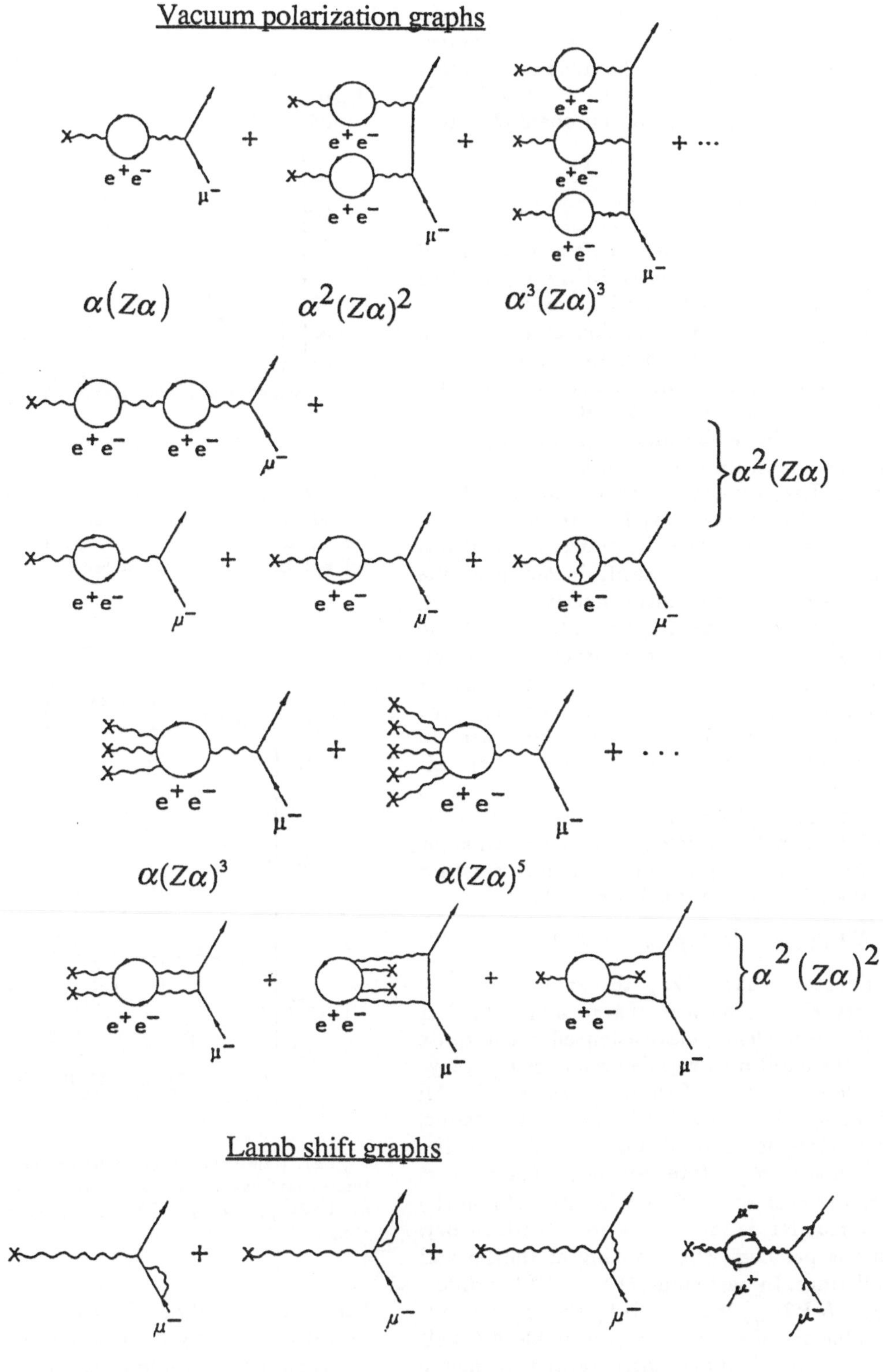

Vacuum polarization graphs

$\alpha(Z\alpha)$ $\alpha^2(Z\alpha)^2$ $\alpha^3(Z\alpha)^3$

$\alpha^2(Z\alpha)$

$\alpha(Z\alpha)^3$ $\alpha(Z\alpha)^5$

$\alpha^2(Z\alpha)^2$

Lamb shift graphs

Fig. 9. Feynman graphs for vacuum polarization corrections of different order and for Lamb shift.

pared to ^{20}Ne are mostly added to the $1d_{5/2}$ subshell pulling by means of strong interaction with the protons nuclear charge into the interior of the nucleus. Hence, the ^{22}Ne rms radius becomes 51 am smaller than the ^{20}Ne radius. Similar reasoning is applicable for $^{24,26}Mg$. For nuclei beyond Si however, the outer subshells are filled up leading again to 'normal' systematics towards the end of the s-d shell (see dashed line in fig.4). If we take e.g. the spherical Hartree-Fock calculations of Friedrich [33] and add-on deformation contributions by employing the pairing-plus-quadrupole model of Reehal and Sorensen [36], theory and experiment come to a reasonable agreement, as can be seen in fig.6, where the rms radii changes $\Delta < r^2 >^{1/2} \;=\; < r^2 >^{1/2} (N+2) - < r^2 >^{1/2} (N)$ are plotted at the respective neutron numbers $N+1$.

4 Nuclear polarization

The accuracy of nuclear charge moments as determined by the method of muonic atoms is ultimately limited by the nuclear polarization (NP) corrections. These corrections are analogous to the dispersion corrections in electron scattering. The leading Feynman graphs of the nuclear polarization correction are shown in fig.7. This dynamical effect leads to virtual excitations of nuclear states by the muon. Since the muon mass is large, the nuclear excitation spectrum extends well into the giant resonance region. The deexcitation of such nuclear states to the ground state transfers energy back to the muon increasing its binding energy. The theoretical calculations are rather cumbersome and necessarily incomplete, due to a lack of knowledge of the entire nuclear excitation spectrum. The corresponding energy shifts are of the order of 200 ppm for the muonic 1s level in medium-heavy muonic atoms and hence in principle measurable to about 10%. However, it is not easy to disentangle the NP corrections from the finite nuclear charge size. In order to obtain reliable results, all muonic atom transitions sensitive to the nuclear charge extension have to be measured with the highest possible precision. It turns out that correlations between NP corrections of different muon states are particularly predicative, that is a plot of NP(2s) versus NP(1s) or NP splittings like

$$\Delta NP(np) = NPnp_{1/2} - NPnp_{3/2} \quad (n \geqq 2). \tag{7}$$

Regarding for instance $\Delta NP(2p)$, the $2p_{1/2}$ and $2p_{3/2}$ states yield very similar nuclear charge moments, but their sensitivity to nuclear polarization effects is quite different, since the giant quadrupole resonance e.g. plays a significant role in the case of the $2p_{3/2}$ state only. All current NP theories yield for $\Delta NP(2p)$ a positive value, since the $2p_{1/2}$ state is more strongly affected by the nuclear size than the $2p_{3/2}$ state (see fig.2). Fig.8 shows our results for the case of μ^--^{208}Pb (ref.37). While the plot of NP2s versus NP is shows agreement with theory, there is a serious problem when looking at the 2p and also the 3p splittings. In particular, the sign of the difference $NP2p_{1/2} - NP2p_{3/2}$ is reversed by the experiment, if the NP1s value is taken to be a reasonable 4-6 keV as predicted by theory [38] and corroborated by elastic electron scattering data [39]. In order to obtain a positive difference, a value of around 12 keV for NP1s from the $\Delta NP(2p)$ data and even about twice as much from the $\Delta NP(3p)$ data has to be assumed. Similar results have been obtained in μ^--^{90}Zr [40]. More indications for a sign reversal in $\Delta NP(2p)$ stem from the $2p_{3/2} - 1s$ and the $2p_{1/2} - 1s$ transitions of isotope and isotone chains measured by our group [27/28]. These data are however not conclusive, since only two parameters, namely the

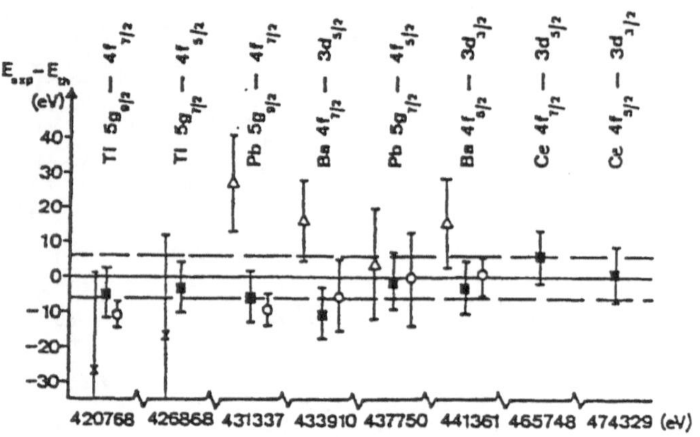

Fig. 10. Differences between measured and calculated transition energies for heavy muonic atom experiments. The crosses refer to ref.41, the triangles to ref.42 and the open circles to ref.43. The full rectangles represent our work [16].

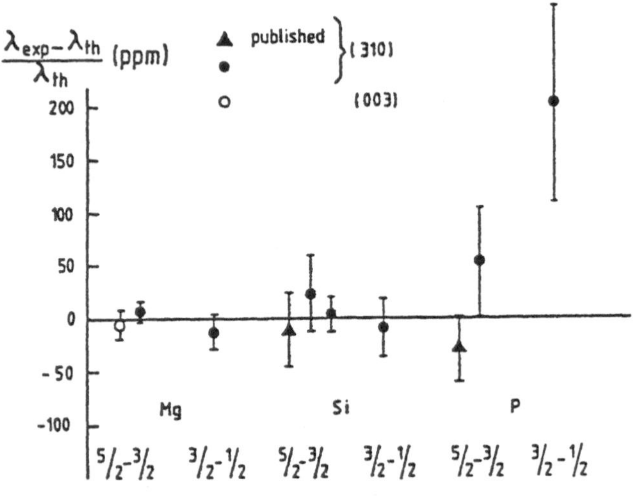

Fig. 11. Relative differences between experimental and theoretical transition wavelengths for crystal spectrometer measurements [44] on the $3d_{5/2} - 2p_{3/2}$ and $3d_{3/2} - 2p_{1/2}$ transitions in muonic ^{24}Mg, ^{28}Si and ^{38}P.

Fermi half-density radius c and the NP2p splitting, could be varied, making the errors in general too large.

What effects could possibly cause such discrepancies? Magnetic polarization effects, not considered in theory, are expected to have the same sign for both np states. In addition, they should not contribute more than about 10% to the two NPnp values. Uncertainties in the calculated NP shifts of higher levels or in higher-order vacuum polarization corrections would influence the μ^--^{208}Pb data considerably more than the μ^--^{90}Zr data, but the observed effect is in both cases the same. Further high precision muonic atom work combined with accu-

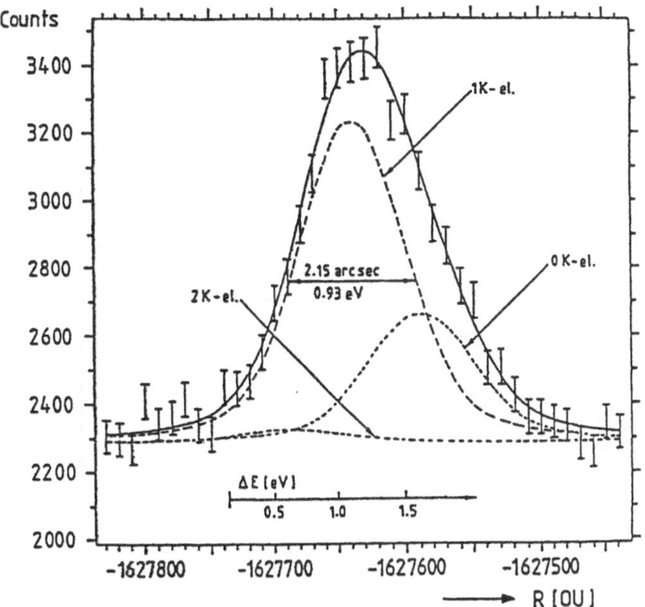

Fig. 12. Bragg reflection of the 4f-3d transition in pionic ^{24}Mg measured with a 2.1 mm thick bent quartz crystal in 3rd order [45]. The fitted function is the sum of 3 individual peaks corresponding to two, one or zero K electrons present during the pionic transition.

rate (e,e)-measurements as well as new theoretical efforts are necessary in order to clarify the nuclear polarization problem.

5 QED corrections

We have already mentioned that QED corrections are important corrections when calculating muonic atom energies. Fig.9 shows the Feynman graphs for the main vacuum polarization contribution and the ladder graphs, for the Källen-Sabry terms of order $\alpha^2(Z\alpha)$, for graphs of order $\alpha(Z\alpha)^{2n+1}$, for the Delbrück terms of order $\alpha^2(Z\alpha)^2$ as well as for the main Lamb shift graphs. The largest QED correction is the first-order (e^+e^-)-vacuum polarization (VP), which is by roughly the fine structure constant α smaller than the muonic binding energy. For the 2p-1s transitions in μ^--^{12}C (75262 eV), first order vacuum polarization (372 eV) is of the same order of magnitude than the finite size effect (-398 eV), but opposite in sign. For the $5g_{7/2} - 4f_{5/2}$ transition in μ^--^{208}Pb on the other hand, vacuum polarization of order $\alpha Z\alpha$ (2191 eV) is by far the largest correction to the Dirac point nucleus energy of 435664 eV. The next-sized correction for the latter transition is electron screening (-81 eV) followed by (e^+e^-)-vacuum polarization of order $\alpha(Z\alpha)^3$ (-44 eV). In the energy range from 400 to 500 keV, accurately known calibration lines are plentyful and centre-of-gravity line precisions of better than 20 ppm can be obtained using high purity Ge detectors. Hence, we have performed precision measurements of 4f-3d and 5g-4f transition energies in medium-heavy to heavy muonic atoms in order to test both first order vacuum polarization corrections as well as higher order corrections [16]. Note that in heavy exotic atoms QED is tested at very high field strengths, of the order of 10^{18} to 10^{19} V/cm. Fig.10 shows the results for 8 transitions in $\mu^- - Ba$, $\mu^- - Ce$, $\mu^- - Tl$ and $\mu^- - Pb$ with energies between 420 and 475 keV. Plotted are the differences of experimental minus calculated energies. Within ± 6 eV, the two energies are equal. Since the vacuum polarization corrections are by far the largest corrections, this result corresponds to a test of first order VP corrections to 0.3% and of higher-order VP corrections to about 20%.

A few years ago, a group at SIN [44] has performed bent-crystal spectrometer measurements to check vacuum polarization corrections in the $3d_{5/2} - 2p_{3/2}$ and the $3d_{3/2} - 2p_{1/2}$ transitions in μ^--^{24}Mg, μ^--^{28}Si and $\mu^- -^{31}P$. The corresponding energies are situated between 56 and 88 keV. Fig.11 presents the obtained results in terms of experimental minus theoretical wavelength differences. The observed precisions in the centre-of gravity positions of about 1 eV or 15 ppm translate to a test of the $\alpha Z\alpha$ VP corrections of the order of 0.2%. However, no higher order QED corrections are amenable to a check in these light nuclei. If one measures at higher orders n of the Bragg reflections, the resolutions of the crystal spectrometer gets proportionally better with n, but the intensities become weak. A very nice example of a recent high precision crystal spectrometer measurement at a meson factory is the determination of the mass of the negative pion by B. Jeckelmann et al. [45]. Fig.12 shows that at 26 keV (4f-3d transition in $\pi^- -^{24}Mg$) the obtained resolution in 3rd order Bragg reflection is 1.6 eV leading to an accuracy for the π^--mass of 3.8 ppm if QED is assumed to hold exactly. However, due to a lack of knowledge of the electronic K-shell occupation number during the pionic cascade, the absolute mass might have to be shifted by as much as 19 ppm [46].

6 Outlook

For future decisions which QED or nuclear polarization tests should be performed using the method of muonic atoms spectroscopy, one should consult diagrams like the ones presented in fig.13 (ref.9). If several transitions with energies larger than 100 - 200 keV are demanded, high resolution semiconductor detectors are preferable to crystal spectrometers. They have the advantage of presenting the whole energy spectrum 'at once' and with high efficiency. Examples are nuclear polarization tests in all muonic atom levels sensitive to the finite nuclear charge extension [37,40] or vacuum polarization tests between intermediate muon states in medium-heavy to heavy muonic atoms [16]. However, if a particular line at energies below 100 - 200 keV has to be determined with the highest possible accuracy, or if very small energy shifts are involved, DuMond type bent-crystal spectrometers should be employed. Such spectrometers are presently in use at PSI while measuring strong interaction shifts in pionic atoms or performing nuclear γ

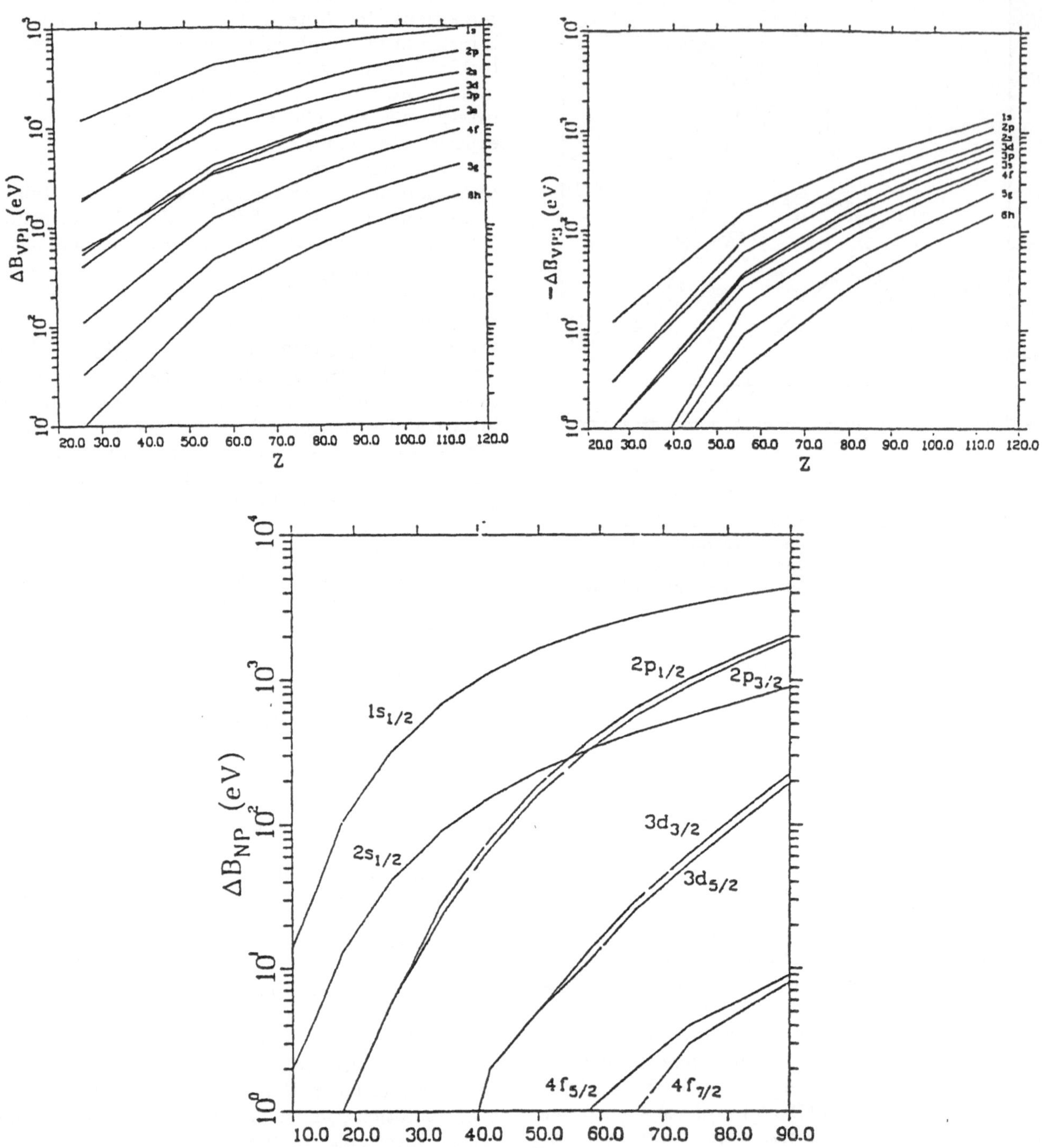

Fig. 13. Order $\alpha Z\alpha$ and $\alpha(Z\alpha)^n$ ($n \geq 3$) vacuum polarization energy shifts and nuclear polarization energy shifts as a function of Z for muonic atoms, as calculated by Borie and Rinker [9].

-ray spectroscopy [47]. In order to eliminate electron screening effects, which ultimately limit the accuracy of mass measurements or QED precision tests (see chapter 5), one should either consider low Z gas targets or employ double-flat crystal spectrometers. In the former case, complete stripping during the muonic cascade has been observed [48]. In the latter, the troublesome electron screening shifts could in principle be resolved due to the higher resolution [49]. However, the line intensities are then considerably weaker than for a bent-crystal spectrometer.

Before the interest turns to possible anomalous muon-nucleus interactions mediated by some type of Higgs scalars [50], independent high-precision results, for instance from elastic electron scattering data on the nuclear charge radius of interest, should become available. In addition, all corrections to the Dirac point nucleus energies have then to be strictly believed.

Acknowledgement. Last but not least, the author wishes to thank all members of the muonic atom collaboration and in particular his colleagues P. Bergem, R. Jacot-Guillarmod, C. Piller, L. Schellenberg and H. Schneuwly at Fribourg. The experiments described here have been partially supported by the Swiss National Foundation.

References

1. Schaller L.A., Barandao D.A., Bergem, P., Boschung M., Phan T.Q., Piller G., Rüetschi A., Schellenberg L., Schneuwly H., Fricke G., Mallot G. and Sieberling H.G.: Phys.Rev. C31, (1985), 1007
2. Piller C., Gugler C., Jacot-Guillarmod R., Schaller L.A., Schellenberg L., Schneuwly H., Fricke G., Hennemann T. and Herberz J.: Phys.Rev. C42, (1990), 182
3. Weber R., Kern J., Kiebele U., Pinston J.A., Aas B., Beltrami I., Bernold T., Bongardt K., Ledebur T.V., Leisi H.J., Ruckstuhl W., Strassner G. and Vacchi A.: Phys.Lett. 98B, (1981), 343
4. Schaller L.A. and Dey W.: Helv.Phys.Acta 47, (1974), 483
5. Dey W., Ebersold P., Leisi H.J., Scheck F., Walter H.K. and Zehnder A.: Nucl.Phys. A326, (1979), 418
6. Rüetschi A., Schellenberg L., Phan T.Q., Piller G., Schaller L.A. and Schneuwly H.: Nucl.Phys A422, (1984), 461
7. Zumbro J.D., Naumann R.A., Hoehn M.V., Reuter W., Shera E.B., Bemis C.E. and Tanaka Y.: Phys Lett. 167B, (1986), 383
8. de Laat C.T.A.M., Taal A., Duinker W., Konijn J., d'Achard van Enschut J.F.M., David P., Hartfiel J., Janszen H., Mayer-Kuckuk T., von Mutius R., Piller C., Schaller L.A., Schellenberg L., Krogulski T., Petitjean C., Reist H.W. and Müller W.: Phys.Lett. 189B, (1987), 7
9. Borie E. and Rinker G.A.: Revs.Mod.Phys. 54, (1982), 67
10. Schaller L.A.: Proc.Int.Conf. on Dynamics of Collective Phenomena in Nuclear and Subnuclear Long-Range Interactions in Nuclei, Bad Honnef, May 1987, ed. by P. David, World Scientific, Singapore 1988, p.145
11. David P., Hartfiel J., Janszen H., Mayer-Kuckuk T., von Mutius R., Petitjean C., Reist H.W., Polikanov S.M., Konijn J., de Laat C.T.A.M., Taal A., Krogulski T.,Johansson T., Tibell G. and d'Achard van Enschut J.F.M.: Z.Phys. A328, (1987), 37
12. David P., Hänscheid H., Hartfiel J., Janszen H., Mayer-Kuckuk T., von Mutius R., Petitjean C., Reist H.W., Polikanov S.M., Duinker W., Konijn J., de Laat C.T.A.M., Taal A., Krogulski T.,Johansson T., Tibell G. d'Achard van Enschut J.F.M., Theobald J.P., Trautmann N., Gugler C., Schaller L.A. and Schellenberg L.: Z.Phys. A330, (1988), 397
13. Schrieder W., David P., Hänscheid H., Konjin J., de Laat C.T.A.M., Paganetti H., Petitjean C., Reist H.W., Risse F., Rösel Ch., Schaller L.A., Schellenberg L., Sinha A.K., Taal A. and Trautmann N.: Z.Phys. A339,(1991),445
14. Breunlich W., contribution to this conference
15. Schneuwly H. and Vogel P.: Phys.Rev. A22, (1980), 2081
16. Dubler T., Kaeser K., Robert-Tissot B., Schaller L.A., Schellenberg L. and Schneuwly H.: Nucl.Phys. A294, (1978), 397
17. Schneuwly H., Boschung M., Kaeser K., Piller G., Rüetschi A., Schaller L.A. and Schellenberg L.: Phys.Rev. A27, (1983), 950
18. Jacot-Guillarmod R., Bienz F., Boschung M., Piller C., Schaller L.A., Schellenberg L., Schneuwly H., Reichart W. and Torelli G.: Phys.Rev. A38, (1988), 6151
19. Hänscheid H., David P., Konjin J., Krogulski T., de Laat C.T.A.M., Mayer-Kuckuk T., Petitjean C., Polikanov S.M., Reist H.W., Risse F., Rösel C.F.G., Schaller L.A., Schellenberg L., Schrieder W., Sinha A.K. and Taal A.: Z.Phys. A335, (1990), 1
20. Hartfiel J.: Thesis, Physics Dept., University of Bonn, 1985, unpublished
21. Barrett R.C.: Phys.Lett. B33, (1970), 388
22. Schaller L.A., Dubler T., Kaeser K., Rinker G.A., Robert-Tissot B., Schellenberg L. and Schneuwly H.: Nucl.Phys. A300, (1978), 225
23. Schellenberg L., Robert-Tissot B., Kaeser K., Schaller L.A., Schneuwly H., Fricke G., Glückert S., Mallot G. and Shera E.B.: Nucl.Phys. A333, (1980), 333
24. Schaller L.A., Schellenberg L., Rüetschi A. and Schneuwly H.: Nucl.Phys. A343, (1980), 333
25. Emrich H.J., Fricke G., Hoehn M.V., Kaeser K., Mallot M., Miska H., Robert-Tissot B., Rychel D., Schaller L.A., Schellenberg L., Schneuwly H., Shera E.B., Sieberling H.G., Steffen R.M., Wohlfahrt H.D. and Yamazaki Y.: Proc. 4th Int. Conf. on Nuclei far from Stability, Helsingor 1981, p.33
26. Schaller L.A., Schellenberg L., Phan T.Q., Piller G., Rüetschi A., Schneuwly H.: Nucl.Phys. A379, (1982), 523
27. Piller G., Jacot-Guillarmod R., Schaller L.A., Schellenberg L., Schneuwly H., Fricke G., Hennemann T., Herberz J., Klein R., Reutter M. and Shera E.B.: Nucl.Phys. Spring Meeting Strasbourg 1990, Verh.DPG(VI) 25, (1990), H3.1
28. Fricke G., Bernhardt C., Hack T., Hennemann T., Herberz J., Jansen J., Klein R., Mazanek P., Schellenberg L., Jacot-Guillarmod R., Piller C., Schaller L.A. and Schneuwly H.: Nucl.Phys Spring Meeting Strasbourg 1990, Verh.DPG(VI) 25, (1990), H3.2
29. Fricke G., Herberz J., Hennemann T., Mallot G., Schaller L.A., Schellenberg L., Piller C. and Jacot-Guillarmod R.: Phys.Rev. C, (1992),in press
30. Anselment M., Bekk K., Hanser A., Hoeffgen H., Meisel G., Göring S., Rebel H. and Schatz G.: Phys.Rev. C34, (1986), 1052
31. Eberz J., Dinger U., Huber G., Lochmann H., Menges R., Ulm G., Kirchner R., Klepper O., Kühl T.U. and Marx D.: Z.Phys. A326, (1987), 121
32. Myers W.D. and Schmidt K.H.: Nucl.Phys. A410, (1983), 61
33. Friedrich J. and Reinhard P.G.: Phys.Rev. C33, (1986), 335
34. Fricke G., Mallot G., Phan T.Q., Piller G., Rüetschi A., Schaller L.A., Schellenberg L. and Schneuwly H.: 9th Int. Conf. on High Energy Physics and Nuclear Structure, Versailles 1981, 231
35. Brown B.A., Radhi R. and Wildenthal D.H.: Phys.Rep. 101, (1983), 313
36. Reehal B.S. and Sorensen R.A.: Nucl.Phys. A161, (1971), 385
37. Bergem P., Piller C., Rüetschi A., Schaller L.A., Schellenberg L., and Schneuwly H.: Phys.Rev. C37, (1988), 2821
38. Rinker G.A. and Speth J.: Nucl.Phys. A306, (1978), 397
39. Yamazaki Y., Wohlfahrt H.D., Shera E.B., Hoehn M.V. and Steffen R.M.: Phys.Rev.Lett. 42, (1979), 1470
40. Phan T.Q., Bergem P., Rüetschi A., Schaller L.A. and Schellenberg L.: Phys.Rev. C32, (1985), 609
41. Vuilleumier J.L., Dey W., Engfer R., Schneuwly H., Walter H.K. and Zehnder A.: Z.Phys. A278, (1976), 109
42. Tauscher L., Backenstoss G., Fransson K., Koch H., Nilsson A. and de Raedt J.: Phys.Rev.Lett. 35, (1975), 410

43. Hargrove C.K., Hincks E.P., McKee R.J., Mes H., Carter A.L., Dixit M.S., Kessler D., Wadden J.S., Anderson H.L. and Zehnder A.: Phys.Rev.Lett. 39, (1977), 307
44. Aas B., Beer W.,Beltrami I., Ebersold P., Eichler R., Ledebur T.V., Leisi H.J., Ruckstuhl W., Sapp W.W., Vacchi A., Kern J., Pinston J.A., Schwitz W. and Weber R.: Nucl.Phys. A375, (1982), 405
45. Jeckelmann B., Nakada T., Beer W., de Chambrier G., Elsenhans O., Giovanetti K.L., Goudsmit P.F.A., Leisi H.J., Rüetschi A., Piller O. and Schwitz W.: Phys.Rev.Lett. 56, (1986), 1444
46. Frosch R., contribution to this conference
47. Perny B., Dousse J.-Cl., Gasser M., Kern J., Lanners R., Rhême Ch. and Schwitz W.: Nucl.Instr.Meths. A267, (1988), 120
48. Simons L.M.: Z.Phys. C46, (1990), S183
49. Hughes V.W., this conference, private communication
50. Ruckstuhl W., Aas B., Beer W., Beltrami I., de Boer F.W.N., Bos K., Goudsmit P.F.A, Kiebele U., Leisi H.J., Strassner G., Vacchi A. and Weber R.: Phys.Rev.Lett. 49, (1982), 859

This article was processed using Springer-Verlag TEX Z.Physik C macro package 1991
and the AMS fonts, developed by the American Mathematical Society.

Z. Phys. C - Particles and Fields 56, S59-S69 (1992)

Zeitschrift für Physik C Particles and Fields

Laser spectroscopy of muonic atoms

Klaus Jungmann

Physikalisches Institut der Universität Heidelberg, Philosophenweg 12, D-6900 Heidelberg, Germany

24-March-1992

Abstract. Laser spectroscopy of muonic atoms is well suited for testing the Quantum Electrodynamics of bound states, for precise measurements of fundamental constants, for investigating properties of hadronic nuclei, and for searching for anomalous muon interactions. Successful experiments have been reported for the 2S-2P transitions in muonic helium ion and for the 1S-2S two-photon transition in muonium. Experiments on muonic hydrogen are coming up. A source of very slow positive muons is being developed which employs laser photoionization of muonium. Polarized muonic helium was obtained by spin transfer from laser polarized Rubidium. The field is open for addressing a variety of interesting physics problems.

PACS: 36.10Dr;32.30.-r;06.20.Jr

1 Introduction

Today laser spectroscopy offers a variety of well established techniques for studying electronic atoms and electronic molecules as well as for investigating solids and liquids [1,2]. Bright laser light sources are available for exciting even very weak transitions. Methods have been developed for studying atoms and molecules that are very rare or available in very small quantities [3]. Optical frequencies can be measured with a precision of parts in 10^{10} using secondary frequency standards, for example He-Ne lasers which are stabilized to transitions in molecular iodine [4]. Much higher accuracy can be expected in the near future from a frequency synthesis chain directly linking optical laser frequencies to the Cs frequency standard [5]. With interferometric techniques optical frequency differences can be determined to parts in 10^{11} [6]. At todays most intense muon sources one has now the possibility to investigate muonic atoms with laser spectroscopic methods.

In this article we will discuss some problems that have been addressed by laser spectroscopy in the past or which are conceivable in the near future. Some interesting problems are listed together in Table 1. Three different kinds of experiments are distinguishable:

(i) There is spectroscopy of muonic atoms ($\mu^- {}^A_Z X$) and ions ($(\mu^- {}^A_Z X)^{n+}$). In these atomic systems one or more electrons are replaced by a negative muon (μ^-).

The mass of the muon (m_μ) is about 207 times larger than the mass of the electron(m_e) [7]

$$m_\mu = 206.768262(30) \cdot m_e \qquad (1).$$

The Bohr-radius in muonic atoms is

$$a_\mu = m_e/m_\mu \cdot a^e_B \qquad (2)$$

where $a^e_B = 0.529177249(24) \cdot 10^{-10}\ m$ is the electronic Bohr radius [7]. Bound muonic states are therefore much more sensitive to size and structure of the atomic nuclei compared to corresponding electronic states. In particular s-states show large shifts due to the nuclear effects. With their Bohr-radius being of the order of magnitude of the electrons Compton wavelength $\lambda_e/2\pi = 3.86159323(35) \cdot 10^{-13}\ m$ [7] muonic atomic states experience also stronger electronic vacuum polarization effects than electronic atoms [9]. Also one may look for anomalous muon interactions, i.e. any difference between muon-nucleus and electron-nucleus interactions except for those due to the mass difference. Such deviations would be expected from a non-point-like muon.

(ii) The hydrogen-like muonium atom ($\mu^+ e^-$) [10] is one of the simplest atomic systems. The positive muon (μ^+) replaces the hadronic nucleus (proton) in a hydrogen atom. The system must be compared to other natural and exotic hydrogen isotopes. It consists of two "point-like" leptons of different generations. The energy levels of the system can be extremely well calculated because of the absence of any hadronic constituent. Muonium is an excellent object for testing Quantum Electrodynamics(QED) and the behaviour of the muon as a pointlike heavy leptonic particle.

Table 1. Laser spectroscopic experiments in muon physics that have been carried out in the past or are conceivable in the near future.

System	Transition	Frequency [THz]	Physics Problem	Remarks
muonium (μ^+e^-)	1S-2S	2 x 1228.5	Lamb shift, μ^+ mass	finished at KEK, QED contribution to 2,5%; progressing at RAL QED contribution to 1%
	1S-2S-continuum	2457 + 366	Low energy μ^+ beams	under development at KEK
	1S-2P-2S	2457 + 0.00105	2S Lamb shift	laser and microwave transition; future possibility
	1S-3P-2S	2912	2S Lamb shift	optical pumping of the 2S state; future possibility
muonic helium(4) ion ($\mu^{-4}He)^4$	2S-2P	369.6, 334.2	QED vacuum polarization, α particle structure	finished successfully at CERN, vacuum polarization at 0,2%, $\langle r_\alpha \rangle^{1/2}$ to 1.8%;failed at PSI
	3D-3P	30.4	QED vacuum polarization	finished at BNL, no clear signal
muonic helium(3) ion ($\mu^{-3}He)^+$	2S-2P	282.7 to 365.1	^3He nuclear size and nuclear polarizability	future possibility
muonic helium(3) ($\mu^{-3}He\ e^-$)	n=1,G=0-G=1	332	^3He nuclear size	future possibility
muonic hydrogen (μ^-p)	2S-2P	48.84	p nuclear size and nuclear polarization, QED vacuum polarization	future possibility
	3D-3P	1.6	QED vacuum polarization	in preparation at PSI
	n=1, F=0-F=1	43.9	p nuclear size and nuclear polarizability	future possibility
	Rydberg transitions	visible	μ^- mass, test of CPT	future possibility
heavy muonic atoms	Rydberg transitions	near UV, visible or IR	anomalous μ^- interactions, vacuum polarization	future possibility
($\mu^-{}^A_ZX$)	Hyperfine transitions	near UV, visible or IR	anomalous μ^- interaction, vacuum polarization	future possibility
laser polarized muonic helium ($\mu^{-4}He - Rb$)	5S-5P in Rb	377.5	μ-capture from polarized target, test of T-invariance	7(1)% muon polarization at LAMPF

(iii) Finally, there are experiments where laser spectroscopy is employed as an auxiliary tool, i.e. the laser field does not directly interact with the muonic system itself.

Some differences are noticeable between laser spectroscopy in muon physics and laser spectroscopy of electronic systems:

(i) Muonic atoms can be produced only at moderate rates of the order of a few atoms per second at most. For obtaining reasonable signal strength one needs efficient excitation of the transitions which requires in most cases very powerful lasers. Also one needs an effective detection scheme for the transitions. In some cases (see below) the muon decay $\mu^- \rightarrow e^- + \nu_\mu + \bar{\nu}_e$ (respectively $\mu^+ \rightarrow e^+ + \nu_e + \bar{\nu}_\mu$) releases an energetic electron (respectively positron) which can be used as part of the signature for reducing background. Due to the parity violation in the weak interaction, the muon decay carries information on the muon polarization in the angular distribution of the decay electrons.

(ii) The lifetime of the muon [8]

$$\tau_\mu = 2.19703(4) \cdot 10^{-6}\ s \tag{3}$$

sets a lower limit to the natural linewidth of any transition of

$$\Delta\nu_{nat} \geq (\pi \cdot \tau_\mu)^{-1} = 145kHz. \tag{4}$$

For μ^- muonic atoms the lifetime is even shorter due to nuclear muon capture. The interaction times $\tau_{interact.}$ of the systems with the laser field are restricted to periods of the order of

$$\tau_{interact.} \approx \tau_\mu, \tag{5}$$

which makes pulsed laser sources very attractive for such experiments.

(iii) The excitation probability in μ^- muonic systems is reduced compared to the excitation of electronic transitions with the same frequencies ω by a factor of m_μ^2, because of the reduced scale of the electric dipole matrix element [11]. We have for the electric dipole transitions the transition probability

$$A_{E1} \propto m_\mu^{-2} \cdot \omega^3. \tag{6}$$

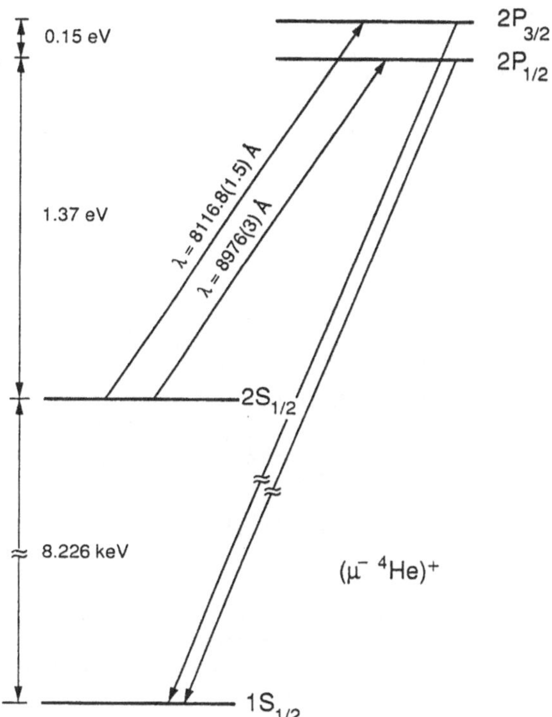

Fig. 1. Energy level diagram of the $(\mu^{-4}He)^+$ ion. Laser excitations of the $2S_{1/2} - 3P_{1/2}$ transition at $\lambda = 5976(3)$ Å and of the $2S_{1/2} - 2P_{3/2}$ transition at $\lambda = 8116.8(1.5)$ Å have been observed in the pioneering experiment at CERN [13].

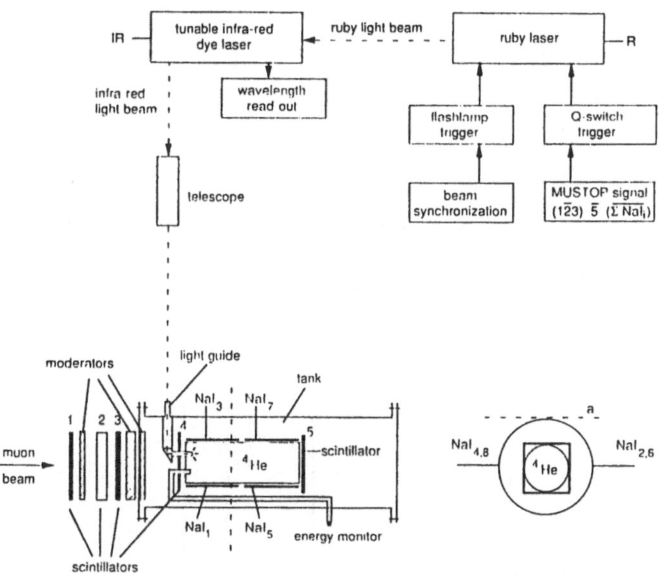

Fig. 2. Schematic view of the apparatus employed in the excitation of the 2S-2P transitions in the muonic helium(4) ion at CERN. The transition is indicated by the observation of a 8.23 keV x-ray from the subsequent spontaneous 2P-1S transition in coincidence with a pulse from the ruby laser pumped infrared dye laser [13].

(iv) The energies of atomic transitions scale with the muon mass. Therefore only transitions between excited states of μ^--muonic atoms, fine structure and hyperfine transitions are accessable for available lasers in the visible, the near ultraviolet and infrared range of the electromagnetic spectrum. Excitations from the ground state (1S) to excited states of μ^--atoms with keV transition frequencies are beyond the capabilities of present laser technology. However, the γ-rays from transitions between the excited states and the ground state can be detected efficiently and with high resolution using modern semiconductor devices. Such transitions are of interest for detection schemes, if they follow laser transitions between excited states.

2 Muonic helium

2.1 Muonic helium ion

2.1.1 2S-2P transitions in the muonic helium ion

The pioneering experiment in the field of laser spectroscopy of muonic atoms has been performed already in the 1970's at the muon beam of the CERN synchrocyclotron by E. Zavattini and his co-workers [12,13,14]. Fine-structure transitions between the $2S_{1/2} - 2P_{1/2}$ and the $2S_{1/2} - 2P_{3/2}$ states of the muonic helium(4) ion $(\mu^{-4}He)^+$ have been induced. (see Fig. 1).

In the experiment negative muons are slowed down and are stopped in a helium $(^4He\,e^-e^-)$ gas target at a pressure of 40 atmospheres. The 2S-state of the muonic helium ion $(\mu^{-4}He)^+$ is formed with 4% probability [15]. The metastable state was observed to decay mainly by two-photon decay at a rate of $\lambda_{2\gamma} = 5.8(5)\cdot 10^5 s^{-1}$ [15]. The ion is excited from the 2S to the 2P-state with pulsed infrared laser radiation from a tunable dye laser system which is pumped by a ruby laser operated at 0.25 Hz. The laser is fired 500 ns after the arrival of a negative muon in the apparatus. The P-states of the $(\mu^{-4}He)^+$ ion decay to the 1S ground state through allowed electric dipole radiation by emitting x-rays of 8.23 keV energy which are observed in NaI(Tl) counters surrounding the pressure vessel (Fig. 2). By observing these x-rays in coincidence with laser pulses, the experiment has observed resonances at 8116.8(1.5) Å and 8976(3) Å corresponding to the $2S_{1/2} - 2P_{3/2}$ and the $2S_{1/2} - 2P_{1/2}$ transitions (see Fig. 3).

The 2S-2P transitions are sensitive to both the mean square charge radius of the ^4He nucleus (α-paricle) and the QED vacuum polarization contribution. On one hand, the results can be interpreted as check of the vacuum polarization contribution at the $1.7\cdot 10^{-3}$ level, if one uses a value for $\langle r_\alpha^2\rangle^{1/2}$ obtained from electron scattering [16]. This value is superior to the results from Lamb shift measurements in natural hydrogen and demonstrates the potential of the method. On the other hand, with confidence in the QED calculations one obtains information about the α particle's charge radius. The experiment yields

$$\langle r_\alpha^2\rangle^{1/2} = 1.6733(30)\ fm \qquad (7)$$

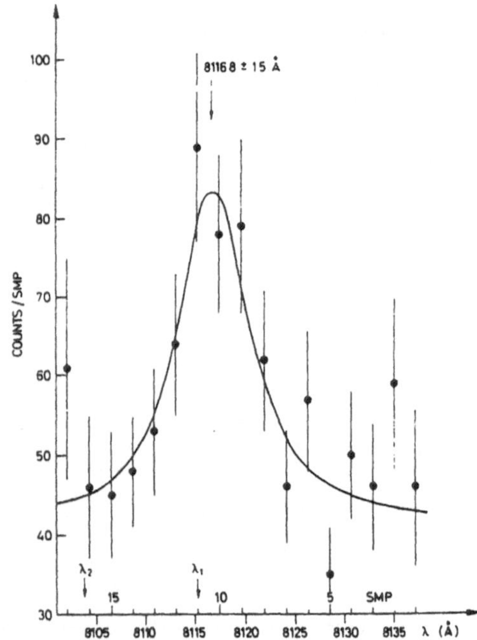

Fig. 3. Experimental signal of the laser excited $2S_{1/2}-2P_{3/2}$ transition in the $(\mu^{-4}He)^{+}$ ion. The data points have been normalized to the number of stopped muons and fit well to a Lorentzian line shape with a line width (FWHM) of $\Gamma = 7.2(2.8)$ Å [13].

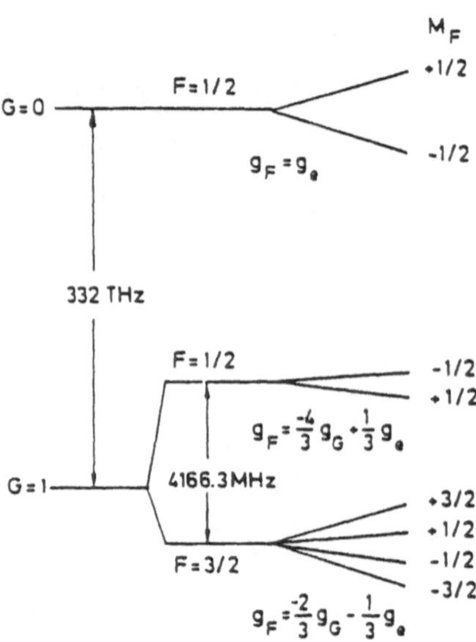

Fig. 4. Energy levels of the ground state of muonic helium(3). The transition between the G=1 and G=0 states is of interest for a laser experiment at $\lambda_1 = 0.9\ \mu m$ which can provide information on the ^3He nucleus.

in fair agreement with electron scattering results.

The method developed for the $(\mu^{-4}He)^{+}$ ion would be ideal for a measurement of the charge radius of the ^3He nucleus [17]. Unfortunately, an attempt to reproduce the experimental results described above failed at PSI [18]. In addition, the metastability of the 2S state at high pressures, which is a prerequisite to the laser experiment carried out, could not be confirmed in independent efforts [19,20,21]. With the observation of long lifetime component for antiprotons in liquid helium [22] some new aspects have been introduced which may help to solve the puzzle of the long lifetime in the $(\mu^{-4}He)^{+}$ ion [23].

2.1.2 3D-3P transitions in the muonic helium ion

The overlapp with the nucleus of P-states and D-states in atomic systems is smaller than for S-states. Therefore P-D transitions show less sensitivity to nuclear size and structure than S-P transitions [24]. They are better suited for tests of QED, in particluar of QED vacuum polarization effects. For the muonic helium ion the 3D-3P transition at $\lambda = 9.85\ \mu m$ appears to be ideally suited, since its width of $\delta\lambda = 0.047\ \mu m$ promises a high quality of $\lambda/\delta\lambda$ of 208 and, fortunately, from the experimentalists point of view the wavelength happens to coincide with powerful CO_2 laser lines. For a detailed discussion of a test experiment carried out at BNL we would like to refer to the article of L.Braci et al. in this volume [25]. A high precision experiment would largely benefit from the availibility of an intensive pulsed source of negative muons.

2.2 Laser polarized muonic helium(3)

Polarized neutral muonic helium(3) $(\mu^{-3}He\ e^{-})$ is desirable in order to investigate the spin dependence of nuclear muon capture in ^3He and for a measurement of the induced pseudoscalar coupling. In addition, polarized muonic atoms are of interest for tests of fundamental symmetries, for example T-invariance [26]. In a recent experiment at LAMPF [27] polarized muonic helium(3) was produced with a $7 \pm 1\%$ muon polarization by spin exchange with polarized Rb atoms which had been optically pumped on the Rb D_1 line by a Ti-Sapphire laser. Details were presented by P.Souder at this workshop [28]. This experiment is a demonstration of an application of laser spectroscopy as an auxiliary tool in muon physics.

2.3 Muonic helium(3) atom

The muonic helium(3) atom $(\mu^{-3}He\ e^{-})$ is a simple system consisting of three different spin 1/2 particles. The coupling of the muon to the ^3He nucleus causes a hyperfine splitting of the order of 332 THz between two states with angular monentum G=0 and G=1 (see Fig. 4). In a laser experiment one can measure precisely the elctromagnetic form factor of the ^3He nucleus. If polarized muonic helium(3) will used, the transition can be observed as a change of the spatial asymmetry of the electrons from the muon decay.

Table 2. Fundamental constants used in the calculations of the contributions to the muonium 1S-2S energy level difference.

Constant		Value	Ref.
speed of light $[m/s]$	c	299792458	[35]
fine structure constant	α^{-1}	137.0359895(61)	[7]
Rydberg constant $[cm^{-1}]$	R_∞	109737.315709(18)	[4]
muon-electron mass ratio	m_μ/m_e	206.768259(62)	[7]
muon g-factor	g_μ	2.002331846(17)	[7]
electron g-factor	g_e	2.002319304386(20)	[7]

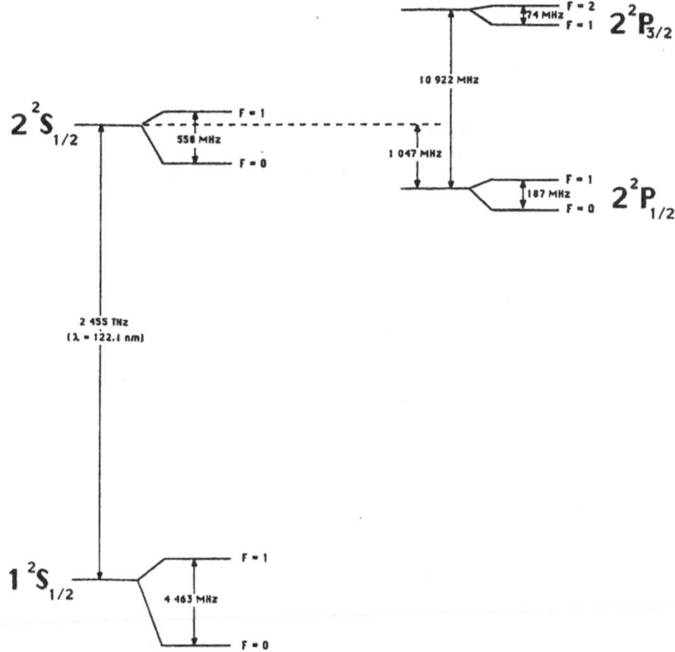

Fig. 5. Muonium energy levels of the $n = 1$ ground state and the $n = 2$ first excited state. The transition between the $1^2S_{1/2}, F = 1$ and $2^2S_{1/2}, F = 1$ states is excited by two photons of $\nu = 1228.5$ THz corresponding to $\lambda_1 = 244$ nm.

3 Muonium

3.1 1S-2S two-photon transition

The 1S-2S two-photon transition in atomic hydrogen has been used for testing calculations of the Lamb shift of the 1S state and for most precise measurements of the Rydberg constant [29]. Recent experiments have reached a resolution of 1 part in 10^{11}. Now the experiments approach a level of precision where the theoretical calculations are limited by the knowledge of the rms charge radius of the proton [30]. Unless new and independent measurements of the proton's form factor exceed the present precision obtained from electron scattering, further improvements of measurements of the hydrogen 1S-2S interval cannot be used for testing QED calculations.

The hydrogen-like muonium atom, being the bound state of two leptons of different generations, is an excellent test object for QED. Since the system consists of two "point-like" particles for which no finite size or internal structure is known so far, the atom is theoretically simpler than other hydrogen-like atoms with hydronic nuclei. Its energy levels (Fig. 5) can be calculated almost exclusively by QED [31]. Weak contributions are of the order [32]

$$G_F \cdot m_e^2 \alpha^3 m_e c^2 / n^3 \simeq 6 \cdot 10^{-13} eV/n^3, \tag{8}$$

and contributions due to the strong interactions arising from vacuum polarization loops containing hadrons are of the order of magnitude of [34]

$$0.1 \cdot m_e c^2 \alpha^5 (m_e/m_\pi)^2 \simeq 10^{-11} eV. \tag{9}$$

In contrast to positronium (e^+e^-) [33], a second purely leptonic system, muonium has the advantage of having a heavy "nucleus" which allows the description in the Furry picture. Complications due to virtual and real annihilation are absent. The QED corrections due to relativistic reduced mass and relativistic recoil are about ten times larger compared to atomic hydrogen and heavier hydrogen like atoms. The system allows a precise determination of the fine structure constant α in a measurement of the hyperfine structure of the ground state or accurate measurements of the magnetic moment μ_μ of the positive muon in a measurement of the Zeeman-effect of the hyperfine structure in the ground state[36,10]. The 1S-2S interval is sensitive to QED contributions and to the mass of the positive muon. All major contributions to the muonium n=1 and n=2 energy level are given in Table 3. They have been calculated using the fundamental constants of Table 2. The reduced mass term is the most uncertain one due to the limited knowledge of the muon mass. By a measurement of the 1S-2S interval to a precision of the order of magnitude of 1 MHz or better one can obtain an improved value for the muon mass. Alternatively the experiment is a test of QED, if the muon mass can be taken from more accurate measurements of the ground state hyperfine structure splitting and the muon magnetic moment which are presently under way at LAMPF [37] and from improved calculations of the ground state hyperfine structure.

In a first attempt at the pulsed muon beam source at KEK a glimpse of a muonium 1S-2S transition has been observed by S. Chu et al. [38]. The precise measurement of the frequency interval is presently on the way and carried out by a collaboration of the universities of Heidelberg, Oxford, Strathclyde, Southampton and Yale and the Rutherford Appleton Laboratory (RAL) [39].

The pulsed muon channel at RAL [40] is operated at 50 Hz and delivers 3000 positive muons per pulse at a momentum of 20 MeV/c and with 7.5 % momentum bite. The muons are stopped in a SiO_2 powder target and form muonium by electron capture [41]. The atoms

Table 3. Contributions to the muonium $n = 1$ and $n = 2$ energy levels of muonium. The evaluation has been according to Ref. 25 using the fundamental constants listed in Table 2. Apparently the most uncertain term is reduced mass contribution indicating that a precise measurement may be interpreted in terms of an improved value for the muon mass.

		Leading order	$1S_{1/2}$ [MHz]	$2S_{1/2}$ [MHz]
Dirac energy		R_∞	-3289885760.0(5)	-822474177.4(1)
Reduced mass		$R_\infty \frac{m_e}{m_\mu + m_e}$	15834191.9(4.7)	3958600.3(1.2)
Lamb shift			8091.1(1.1)	1034.8(1)
Self energy		$R_\infty Z^4 \alpha^3$	8396.46(8)	1.072.90(1)
Vacuum polarization		$R_\infty Z^4 \alpha^3$	-215.19	-26.90
Relativistic reduced mass		$R_\infty Z^4 \alpha^3 \frac{m_e}{m_\mu}$	-112.6(1.1)	-14.42(14)
Radiativ recoil		$R_\infty Z^5 \alpha^4 \frac{m_e}{m_\mu}$	-0.179	-0.022
Relativistic recoil		$R_\infty Z^4 \alpha^3 (\frac{m_e Z}{m_\mu})$	21.40(21)	3.03(3)
Higher order QED			1.06	0.13
Hyperfine structure	F=0	$R_\infty \alpha^2$	-3347.5	-418.4
	F=1	$R_\infty \alpha^2$	1115.8	139.5
Total energy	F=0		-3274046824.6(4.9)	-818514960.7(1.2)
	F=1		-3274042361.3(4.9)	-818514402.8(1.2)

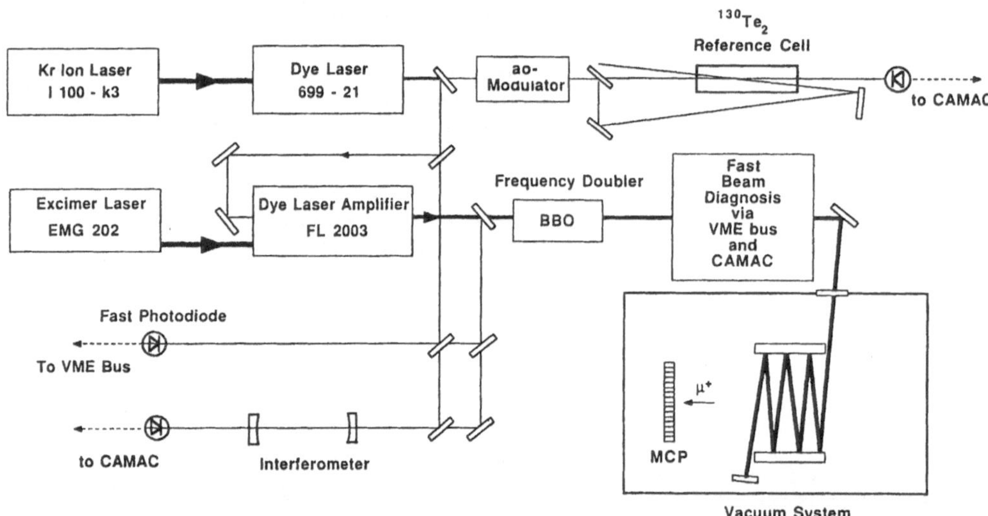

Fig. 6. The laser system employed in the muonium 1S-2S experiment at RAL. The beam of a cw ring dye laser at 488 nm wavelength is amplified in an excimer laser pumped four stage dye laser amplifier and frequency doubled in a BBO crystal. Part of the cw light is frequency downshifted by an acousto optic modulator (AOM) and locked to the d_4 line of the $^{130}Te_2$ molecule. Systematic shifts of the frequency in the dye amplifier are observed by simultaneously recording the fringes of an interferometer.

diffuse to the surface and leave the target for the surrounding vacuum at thermal energies with a Maxwell-Boltzmann velocity distribution. The average velocity has been found to be 7.43(2) $mm/\mu s$. The atoms interact about 1 cm above the target with two counter-propagating pulsed laser beams at 244.2 nm wavelength. A single laser beam is reflected back and forth between two dielectric mirrors with 93 % reflectivity which are separated by 10 cm. After three bounces the beam is reflected back onto itself by a third mirror. For every laser shot there are 1 to 2 atoms in the laser beam volume.

The laser system (Fig. 6) consists of a Krypton ion laser pumped cw ring dye laser at 488 nm wavelength. It feeds a pulsed dye amplifier which is pumped by an excimer laser running at 25 Hz repetition rate. The pulsed amplified light is frequency doubled in a β-BBO crystal. Pulses of 17 nsec length and up to 5 mJ energy were obtained at 244 nm wavelength. The laser was fired 1.5 μs after the arrival of the muons from every second muon pulse. For the purpose of frequency calibration a fraction of the cw laser ligth is frequency down shifted by between 766 and 868 MHz by doublepassing through an acousto-optic modulator. The cw laser is frequency

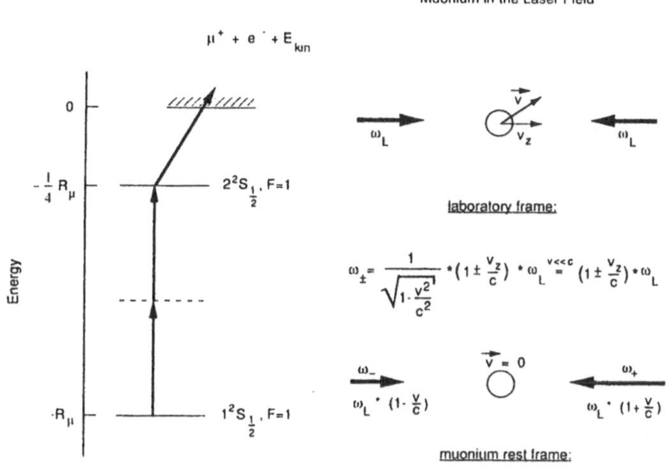

Fig. 7. The principle of the experiment for exciting the two-photon 1S-2S transition in muonium. The atom interacts with two counter-propagating laser beams of the frequency ω_L. The absorbtion of one photon from each laser field is Doppler-free to first order. The two-photon transition is detected by observing the positive muon released after photoionization of the excited state in the same laser field.

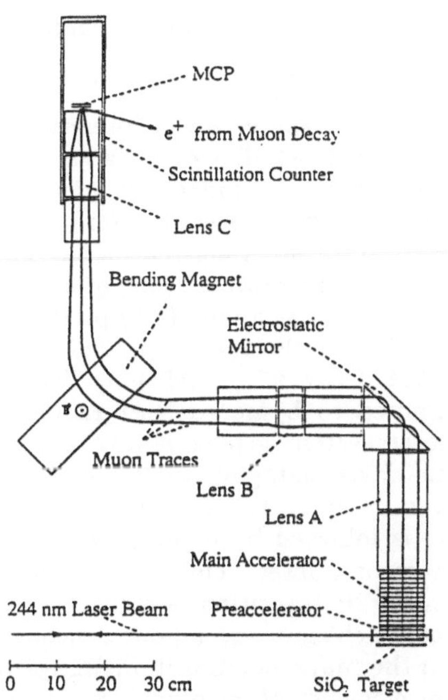

Fig. 8. Muons released by the photoionisation of the 2S-state are guided in an energy and momentum selective path on to a microchannel plate (MCP) detector. Decay positrons from the muons are detected in a plastic scintillator tube surrounding the MCP.

offset locked to the d_4 resonance of molecular $^{130}Te_2$ using saturation spectroscopy. The line center is currently known to an accuracy of 0.8ppb [42].

The two-photon transition is detected via the photoionization of the 2S state with a third photon from

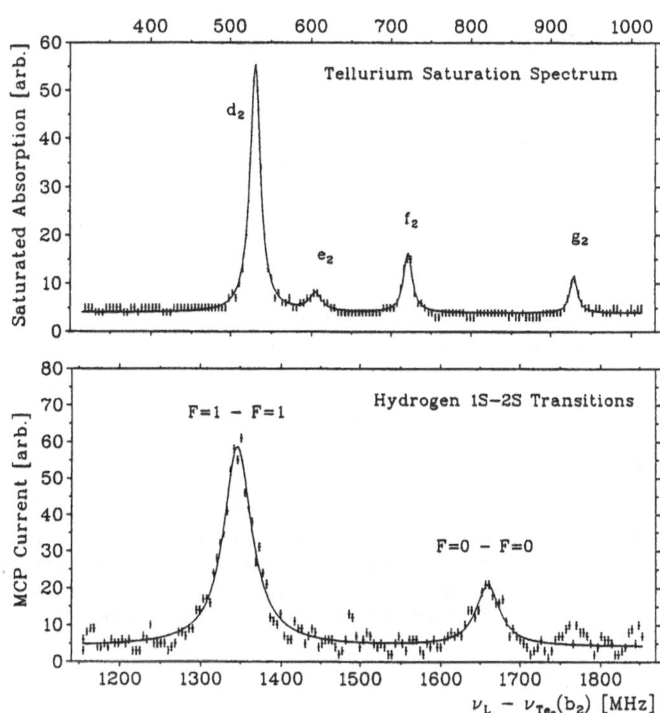

Fig. 9. Doppler-free spectrum of the 1S-2S transitions in atomic hydrogen together with a simultaneously recorded Doppler-free saturation spectrum of $^{130}Te_2$ molecules for frequency calibration. The reference is shifted by 840 MHz. The spectra were recorded within 20 seconds.

the same laser beams (see Fig. 7). The muon set free in this process is accelerated in a two stage electro-static accelerator up to 1 keV and guided along an energy and momentum selective path to a microchannel plate detector (MCP), as shown in Fig. 8. The muon is identified by the detection of the Michel positron from the decay $\mu^+ \rightarrow e^+ + \nu_e + \bar{\nu}_\mu$ in a plastic scintillation counter surrounding the MCP.

The apparatus has been tested and optimized using Doppler-free two-photon signals from atomic hydrogen (H) which has been produced by photodetachment from H_2O and organic molecules on the mirror surfaces close to the SiO_2 target (see Fig. 9). Protons were detected instead of muons on the MCP.

The muonium signal shown in Fig. 10 was obtained within 19 hours of data taking at 3.7(4) mJ of UV laser pulse energy. The laser was stepped hourly to one of five selected freuqencies. On resonace one counts 4.5(1.8) muons per 10^5 laser shots. The background count level when the laser is off is 0.94(38) per 10^5 accelerator pulses in good agreement with 0.49(49) counts per 10^5 pulses for the laser being tuned 2.5 GHz below resonace. A Lorentzian can be fitted to the data. The line center is 843(15) MHz above the d_4 reference line. The line width (FWHM) is 49(31) MHz. It is mainly due to the laser bandwidth which has been independently measured using an interferometer and the signals from atomic hydrogen to 40(7) MHz.

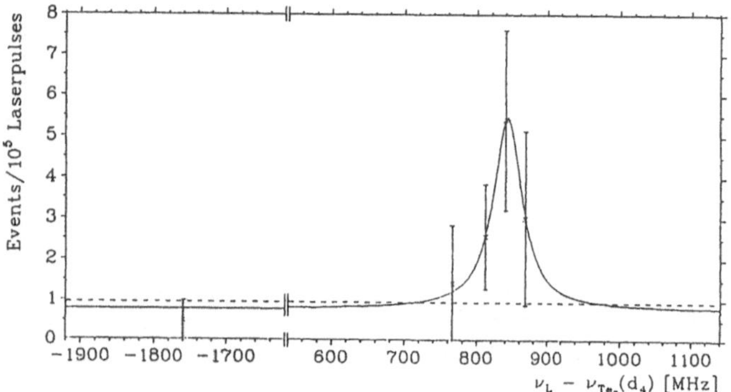

Fig. 10. Doppler-free signal of the muonium 1S-2S transition. The background countlevel of 0.94(38) per 10^5 accelerator pulses in the laser off phase is indicated by a dashed line. The off resonant counting rate of 0.49(49) per 10^5 laser pulses has been measured by tuning the laser by 2.5 GHz below resonance. The frequency scale corresponds to the offset of one quarter of the 1S-2S transition frequency from the d_4 reference line in the molecular $^{130}Te_2$ spectrum.

The two-photon transition is subject to numerous systematic effects. There is a linear Doppler shift contribution of ±11 MHz due to the finite crossing angle between the counter-propagating laser beams and the asymmetric velocity distribution of the muonium atoms. The second order Doppler-effect amounts to -0.9 MHz. Residual electric fields in the interaction region cause a dc-Stark shift of less than 0.12 MHz. The dynamic Stark-effect in the intense laser field was numerically calculated to 32(10) MHz taking into account measured temporal and spatial beam profiles. Rapid changes of the refractive index in the pulsed laser amplifier (frequency chirp) cause at λ=488 nm an average blue shift of 20(10) MHz. They have been carefully investigated by an optical spectrum analyzer and by comparison with the atomic hydrogen resonances.

The corrected 1S-2S transition frequency for muonium is

$$\Delta\nu_{1S-2S} = 2455528016(58)(43)\ MHz \tag{10}$$

where the first error is statistical and the second one arises from systematic effects. The value is in agreement with theoretical calculations and the previous experiment at KEK (see Table 4). The Lamb shift contribution to the 1S-2S interval can be calculated to be 7056.2(1.2) MHz. It has been tested in this experiment at the 1 % level. The experiment represents the most precise Lamb shift measurement in muonium so far. Significant improvements are expected in the near future. A precision of 1 MHz or better for the 1S-2S frequency interval appears feasible. A hydrogen – muonium isotope shift measurement in the same apparatus has the best perspectives for a muon mass measurement, because of the expected cancellation of systematic effects and also because of the lower sensitivity of the isotope shift to uncertainties in the Rydberg constant.

Table 4. Breakdown of the corrections applied in the experimental determination of the $1S_{1/2}, F = 1 - 2S_{1/2}, F = 1$ two-photon transition frequency in muonium at RAL. The final result is listed together with the KEK experimental result [31] and the latest theoretical calculation [26]. The figures are listed for the fundamental blue laser frequency ν_{1S-2S} which is one quarter of the atomic transition frequency and corresponds to λ = 488.4 nm laser wavelength.

	Frequency [MHz]	Statistical Error [MHz]	Systematic Uncertainty [MHz]
d_4 Te_2-reference line	613881150.8	-	±0.6
AOM-frequency at centre of resonance	+842.7	±14.5	-
dc-Stark shift	−0.03 < ... < 0.0	-	-
ac-Stark shift	−8.0	-	±2.5
linear Doppler shift	0.0	-	±2.8
second order Doppler shift	+0.23	-	-
dye amplifier shift	+20	-	±10
$\frac{1}{4}\nu_{1S-2S}(exp., RAL)$	613882005.8	±14.5	±10.7
$\frac{1}{4}\nu_{1S-2S}(exp., KEK)$	613881984	±30	±35
$\frac{1}{4}\nu_{1S-2S}(theor.)$	613881989.8	±0.25†	±0.88‡

† upper limit of the theoretical error
‡ uncertainty arising from the reduced mass

3.2 Ultra slow muon beams from laser photoionization of muonium

Ultra slow muon beams are of interest to the μSR community for precise studies of surfaces, for example. Presently such a beam is under development at KEK [43]. Thermal muonium atoms in vacuum are formed by electron capture from hot tungsten metal [44] which is located in the primary proton beam of an accelerator. The tungsten also acts as a pion (π^+) production target and a moderator for slowing down and stopping the muons from the decay $\pi^+ \rightarrow \mu^+ + \nu_\mu$. The muonium atoms are exposed to Lyman-α laser radiation. An all solid state laser system generates 122 nm VUV light pulses by four-wave-mixing of two 212 nm photons and a 820 nm photon in a Kr/Ar gas mixture. The 1S-2P excitation is followed by photoionization of the 2P state with a second laser field at λ = 355 nm. The single photon 1S-2P transition is subject to the linear Doppler-effect which causes an inhomogeneous linewidth of 156 GHz for thermal muonium atoms at a temperature of 2000 K. The efficiency of the process depends therefore strongly on the effective laser bandwidth. A muon beam with ultra low momentum distribution is extracted at energies of several keV. The project is described by K.Nagamine in this volume.

3.3 2S-2P Lamb shift of muonium

Measurements of the n=2 Lamb shift in muonium [45,46,47] have reached the 10^{-2} level of precision. Significant improvements are necessary in order to challenge experiments which were performed in natural hydrogen [48,49]. The precision of those experiments has reached the accuracy of theortical calculations which are limited by knowledge of the rms radius of the proton. In the case of muonium which is free of nuclear size effects the major difficulty is the lack of an intense source of slow (thermal) metastable ($n = 2$) muonium. Up to date metastable muonium can be produced in noticable quantities only by a beam foil method. For muons of 7.3 MeV/c momentum passing through a 200 $\mu g/cm^2$ aluminum foil one finds a fraction of 12 % neutral muonium with 10 % of the atoms in the metastable 2S state [38]. The atoms from a beam foil technique have at least several keV of kinetic energy due to the resonant character of the process. With such a source Lamb shift ($2S_{1/2} - 2P_{1/2}$) and fine structure ($2S_{1/2} - 2P_{3/2}$) transitions could be observed in experiments at TRIUMPF [36] and LAMPF [37,38]. The rate of 1S-2S two-photon transitions is too low for quantitatively populating the 2S state. In addition, photoionization reduces the n=2 population.

The situation has changed with the availability of powerful laser systems that can produce kW laser pulses at the Lyman-α (L_α) wavelength of $\lambda_{L_\alpha} = 122.1$ nm using sum frequency mixing in Kr gas [50]. The 1S-2P transition in thermal muonium in vacuum can be excited with high probability. By inducing a pulsed microwave transition at 1047 MHz one can prepare muonium in the 2S state. This state can be probed by delayed photoionization with a second laser which is followed by the detection of the muon or by delayed Stark quenching of the 2S state in an external electric field and the observation of Lyman-α photons. Precision Lamb shift measurements can start. Such an experiment fits best in the environment of an intensive pulsed surface muon beam. The feasibility of such an experiment will be demonstrated with the success of the ultra slow positive muon beam at KEK [43].

An even more attractive way of obtaining reasonable 2S population would be optical pumping using a laser at the Lyman-β frequency of $\lambda_{L_\beta} = 103.0\ nm$ for exciting the 1S-3P transitions. The 3P state has a lifetime of 5.4 ns and decays with 11.8 % probability into the metastable 2S state. High efficiencies could be obtained within a few pumping cycles.

4 Muonic hydrogen

The two-body system muonic hydrogen (μ^-p) can be treated theoretically similar to one-electron atoms [11,51]. In this system the muon is much closer to the nucleus compared to electronic hydrogen isotopes. The QED vacuum polarization is the dominant radiative correction. Even higher order vacuum polarization corrections are non-negligible. Muonic hydrogen renders the possibility of testing QED vacuum polarization in the absence of perturbations from further electrons and with minimal perturbation from other radiative corrections [51]. Particularly S-states are very sensitive to the rms charge radius, the polarizability of the proton [52] and to anomalous muon interactions. By choosing the proper transition one can either focus on QED tests or one can measure nuclear properties.

As in the case of the muonic helium ion (see 2.1.2) the P-D transitions are almost free of nuclear size corrections. A clean test of QED vacuum polarization can be expected. At PSI recently an experiment has been proposed by E. Zavattini et al. [53] which aims for a laser excitation of the 3D$-$3P transitions. There are 10 allowed electric dipole transitions in range from $\lambda = 120\ \mu m$ and $\lambda = 234\ \mu m$. They are planed to be induced by a free electron laser which is under development in Frascati. The system should have 20 kW of output power and macro pulses of 1 μs duration [53].

Accurate measurement of the 2S$-$2P transitions at $\lambda \approx 6\ \mu m$ will yield the charge radius of the proton and can help to distinguish between two earlier electron scattering experiments at Stanford and Mainz [30]. It would also be of advantage to the interpretation of precision laser spectroscopy results from natural atomic hydrogen. To some lesser extend these transitions are sensitive to the polarizability of the proton. The corresponding transitions in muonic deuterium (μ^-d) should provide information on the deuteron's electromagnetic size and polarizability.

The n=1 state is most sensitive to the nuclear effects. The ratio of the size of proton to the size of the muons orbit is about 10^{-2}. The hyperfine transition between the F=1 and F=0 states can be driven with radiation at $\lambda \approx 6.8\mu m$ which is reachable with todays lasers. For these investigations one would largely benefit from the availability of polarized muonic hydrogen at low energies. In that case the signal could be observed as a change in the spatial anisotropy of the decay electrons from the μ^- decay in the muonic atom analogous to the hyperfine structure signals from muonium. With all the current attempts for the development of slow muon sources [54,55] one can hope, that a source of polarized muonic hydrogen atoms can be realized in the not to distant future. The effort will be paid off by the almost unique possibility of precise measurements of the proton charge radius and the proton's polarizability.

If muonic hydrogen is formed from electronic hydrogen by capture of a negative muon, the system starts with principal quantum numbers of about n $\approx$ 14 from where the muon cascades down. These Rydberg states have reasonable long lifetimes, especially if they have large angular momenta. They are not very much affected by the proton's size and structure. Single- or two-photon transitions between Rydberg states can have frequencies in the visible region (e.g. n=14 to n=18) which is convenient for precise measurements. Measurements of the Rydberg constant in a muonic system accurate to parts in 10^8 or better can be expected. The value must be compared to the Rydberg constant measured in electronic hydrogen. Alternatively, the mass of the negative muon can be extracted and compared to the mass of the positive muon as a test of CPT invariance.

5 Conclusions

Precision laser spectroscopy of muonium and selected transitions in μ^--atoms has been demonstrated to be feasible. For most of the experiments pulsed accelerators are of great advantage since mostly high laser intensities are necessary which can be provided easily from pulsed laser sources. At pulsed muon beam lines one has a larger chance of having a sufficient number of atoms available at the time the laser is fired. Background which is uncorrelated with the accelerator beam can be reduced significantly by suitable gating. A pulsed lepton source at LAMPF [56] which would have a factor of 20 higher pulse intensity compared to the presently brightest pulsed muon source at the Rutherford Appleton Laboratory would be a large boost for laser experiments in muon physics.

Laser spectroscopy of muonium and of muonic atoms can provide us with precise values for fundamental constants, accurate tests of QED, information on nuclear sizes and polarizabilities, tests of the CPT theorem, and limits on anomalous muon interactions. Last but not least, the nature of the muon as point-like heavy leptonic particle is tested in all the experiments. One can expect interesting results within the next decade from the experiments which are presently carried out and from new ones that will come.

Acknowledgement. We would like to acknowledge fruitful discussions on the subject with P.E.G. Baird, J.R.M. Barr, M.G. Boshier, A.I. Ferguson, V.W. Hughes, F. Kottmann, F. Maas, B.E. Matthias, G. zu Putlitz, W. Schwarz, P.G.H. Sandars, D. Taqqu, W.T. Toner and K.A. Woodle. This work was supported in part by the Federal Minister of Science and Technology of Germany.

References

1. W. Demtröder: Laser Spectroscopy, Springer-Verlag, Berlin Heidelberg, New York (1982).
2. S. Svanberg: Atomic and Molecular Spectroscopy, Springer Verlag, Berlin Heidelberg New York (1991).
3. J.J. Snyder, R.A. Keller (eds.): Ultrasensitive Laser Spectroscopy, Special issue J. Opt. Soc. Am. **B2** (1985), V.S. Letokhov: Sci. Am. **259**/3, 44 (1988), and G.S. Hurst and M.G. Payne eds.: Principles and Application of Resonance Ionization Spectroscopy, Adam Hilger, Bristol (1988).
4. F. Biraben, J. C. Garreau, L. Julien, M. Allegrini, Phys. Rev. Lett. **62**, 621 (1989).
5. V.P. Chebotayev, private communication (1991)
6. R.G. DeVoe and R.G. Brewer, Phys.Rev. **A30**, 2827 (1984), and R.G. DeVoe, C. Fabre, K. Jungmann, J. Hoffnagle, R.G. Brewer, Phys. Rev. **A37**, 1802 (1988).
7. E.R. Cohen and B.N. Taylor, Rev.Mod.Phys. **59**, 1121 (1987)
8. Particle data group, Phys.Lett.**239B**, 5 (1990)
9. L. Schaller, this volume
10. V.W. Hughes, this volume, see also: V.W. Hughes and G. zu Putlitz, in: Quantum Electrodynamics, T. Kinoshita (ed.), World Scientific, Singapore, p. 822. (1990)
11. H.A. Bethe and E.E. Salpeter: Quantum Mechanics of One- and Two- electron Atoms, Springer, Berlin Göttingen Heidelberg (1957)
12. G. Carboni, U. Gastaldi, G. Neri, O. Pitzurra, E. Polacco, G. Torelli, A. Bertin, G. Gorini, A. Placci, E. Zavattini, A. Vitale, J. Duclos, J. Picard, Nuov.Cim. **34A**, 493 (1976).
13. G. Carboni, G. Gorini, G. Torelli, L. Palffy, F. Palmonari, E. Zavattini, Nucl. Phys. **A278**, 381 (1977).
14. G.Carboni, G.Gorini, E. Iacobini, L. Palffy, F. Palmonari, G. Torelli, and E. Zavattini, Phys.Lett. **73B**, 229 (1978)
15. A. Bertin, G. Carboni, J. Duclos, U. Gastaldi, G. Gorini, G. Neri, J. Picard, O. Pizurra, A. Placci, E. Polacco, G. Torelli, A. Vitale, and E. Zavattini, Nuov.Cim. **26B**, 433 (1975)
16. I. Sick, Phys.Lett. **116B**, 212 (1982)
17. PSI proposal R-82-16 "Measurement of the 2S-2P Transition Energies of muonic Helium(3) using an Excimer-Dye-Laser", H. Orth, spokesman, (1982)
18. PSI proposal R-83-20 "Measurement of the $2^1S_{1/2} - 2^3P_{1/2}$ Energy Difference in muonic Helium(3) at Low Gas Density", F. Kottmann, spokesman (1983), see also: A. Bianchetti, P. Hauser, H. Hofer, F. Kottmann, Ch. Lüchinger, R. Schären, and L. Simons, PSI Newslett. 1989, 67 (1990)
19. H.P. von Arb, F. Dittus, H. Heeb, H. Hofer, F. Kottmann, S. Niggli, R,. Schaeren, D. Taqqu, J. Unternährer, and P. Egelhof, Phys.Lett. **136B**, 232 (1984)
20. M. Eckhause, P. Guss, D. Joyce, J.R. Kane, R.T. Siegel, W. Vulcan, R.E. Welsh, R. Whyley, R. Dietlicher, and A. Zehnder, Phys.Rev. **A33** 1743 (1986)
21. J. Rosenkranz, K.P. Arnold, M. Gladisch, J. Hofmann, H.J. Mundinger, H. Orth, G. zu Putlitz, M. Stickel, W. Schäfer, W. Schwarz, and V.W. Hughes, Ann.Phys. **47**, 667 (1990)
22. M.Iwasaki, S.N. Nakamura, K. Shikagi, Y. Shimizu, H. Tamura, T. Ishikawa, R.S. Hayano, E. Takada, E. Widmann, H. Outa, M. Aoki, P. Kitching, and T. Yamazaki, Phys.Rev.Lett.**67**, 1246 (1991)
23. L. Bracci and E. Zavattini, Phys. Rev A41, 2352 (1990).
24. BNL proposal 745 An experiment to measure vacuum polarization in 3D-3P transitions in muonic helium atoms",A.M. Sachs, J. Fox, R. Cohen, and E. Zavattini, spokesmen (1979)
25. L. Bracci, A. Vacchi, and E. Zavattini, this volume
26. J. Deutsch, in: Proceedings of the Workshop on Fundamental Muon Physics: Atoms, Nuclei, and Particles, C.M. Hoffmann, V.W. Hughes, and M. Leon (eds.), Los Alamos , p. 47 (1986)
27. N.R. Newbury, A.S. Barton, P. Bogorad, G.D. Cates, M. Gatzke, B. Saam, L. Han, R. Holmes, P.A. Souder, J. Xu, and D.Benton, Phys.Rev.Lett.**67**, 3219 (1991)
28. P.A. Souder, this volume
29. M.Weitz, F. Schmidt-Kaler, and T.W. Hänsch, Phys.Rev.Lett. **68**,1120 (1992), C. Zimmermann, R. Kallenbach, and T.W. Hänsch, Phys.Rev.Lett. **65**, 571 (1990), M.G. Boshier, P.E.G. Baird, C.J. Foot,E.A. Hinds, M.D. Plimmer, D.N. Stacey, J.B. Swan, D.A. Tate, D.M. Warrington, and K.A. Woodgate, Phys.Rev. **A40**, 6169 (1989), and J.R.M Barr, J.M. Girkin, J.M. Tolchard, and A.I. Ferguson, Phys.Rev.Lett. **56**, 580 (1986)
30. L.M. Hand, D.G. Miller, and R. Wilson, Rev.Mod.Phys. **35**, 335 (1963), and G.G. Simon, Ch. Schmitt, F. Barkowski, and V.W. Walther, Nucl.Phys. **A333**, 381 (1980)
31. J. R. Sapirstein and D. R. Yennie, in: Quantum Electrodynamics, T. Kinoshita (ed.), World Scientific, Singapore, p. 560 (1990)
32. M.A.B. Bég and G. Feinberg, Phys.Rev.Lett. **33**, 606 (1974), and Phys.Rev.Lett. **35**, 130 (1975)
33. S.J. Brodsky and S.D. Drell, Ann.Rev.Nucl..Sci. **20**,147 (1977)
34. Steven Chu, Allen P. Mills, Jr., and John L. Hall, Phys. Rev. Lett. **52**, 1689 (1984), see also: Allen P. Mills and Steven Chu, in: Quantum Electrodynamics, T. Kinoshita (ed.), World Scientific, Singapore, p. 774 (1990)
35. Comité Consultatif pour la definition du Mètre (20.Oct.1983)
36. F. G. Mariam, W. Beer, P. R. Bolton, P. O. Egan, C. J. Gardner, V. W. Hughes, D. C. Lu, P. A. Souder, H. Orth, J. Vetter, U. Moser, and G. zu Putlitz, Phys. Rev. Lett. **49**, 993 (1982).

37. LAMPF proposal 1054 "Ultrahigh Precision Measurements on Muonium Ground State: Hyperfine Structure and Muon Magnetic Moment", V.W. Hughes, G. zu Putlitz, P.A. Souder, spokesmen (1986)
38. Steven Chu, A. P. Mills, Jr., A. G. Yodh, K. Nagamine, Y. Miyake, and T. Kuga, Phys. Rev. Lett. **60**, 101 (1988), see also: K. Danzmann, M. S. Fee, and Steven Chu, Phys. Rev. A**39**, 6072 (1989)
39. K. Jungmann, P.E.G. Baird, J.R.M. Barr, C. Bressler, P.F. Curley, R. Dixson, G.H. Eaton, A.I. Ferguson, H. Geerds, V.W. Hughes, J. Kenntner, S.N. Lea, F. Maas, M.A. Persaud, G. zu Putlitz, P.G.H. Sandars, W. Schwarz, W.T. Toner, M. Towrie, G. Woodman, L. Zhang, and Z. Zhang, Z.Phys.D **21**, 241 (1991)
40. G.H. Eaton, this volume, see also: G.H. Eaton, A. Carne, S.F.J. Cox, J.D.Davies, R. de Renzi, O. Hartmann, A. Kratzer, C. Ristori, C.A. Scott, G.C. Sterling, and T. Sundquist, Nucl.Instr.and Meth. A**269**, 483 (1988)
41. K.A. Woodle, K.P. Arnold, M. Gladisch, J. Hofmann, M. Janousch, K. Jungmann, H.J. Mundinger, G. zu Putlitz, J. Rosenkranz, W. Schäfer, G. Schiff, W. Schwarz, V.W. Hughes, and S.H. Kettell, Z.Phys. D**9**, 59 (1988)
42. G.P. Barwood, W.R.C. Rowley, P. Gill, J.L. Flowers, and B.W. Petley, Phys.Rev. A**43**, 4783 (1991)
43. K. Nagamine, this volume
44. A.P. Mills, J. Imazato,Y. Kawashima, S. Saitoh, A. Uedono, and K. Nagamine, Phys.REv.Lett. **56**, 1463 (1986)
45. C.J. Oram, J.M. Bailey, P.W. Schmor, C.A. Fry, R.F. Kiefl, J.B. Warren, G.M. Marshall, and a. Olin, Phys.Rev.Lett. **52**, 910 (1984)
36. A. Badertscher, S. Dhawan, P.O. Eagen, V.W. Hughes, D.C. Lu, M.W. Ritter, K.A. Woodle, M. Gladisch, H.Orth, G. zu Putlitz, M. Eckhause, J. Kane, F.G. Mariam, and J. Reidy, Phys. Rev. Lett. bf 52, 914 (1984), and K.A. Woodle, A. Badertscher, V.W. Hughes, D.C. Lu, M.W. Ritter, M. Gladisch, H.Orth, G. zu Putlitz, M. Eckhause, J. Kane, and F.G. Mariam, Phys. Rev. A**41**, 93 (1990)
47. S.H. Kettell, Ph.D. thesis, Yale, unpublished (1990)
48. S. R. Lundeen and F. M. Pipkin, Phys. Rev. Lett. **46**, 232 (1981), and S. R. Lundeen and F. M. Pipkin, Metrologia **22**, 9 (1986).
49. F.M. Pipkin, in: Quantum Electrodynamics, T. Kinoshita (ed.), World Scientific, Singapore, p. 696 (1990)
50. R.Brünger, A. Borsutzky, and R. Wallenstein, Verhandl.DPG. (VI) **27**, 1445 (1992)
51. E. Borie and G.A. Rinker, Rev.Mod.Phys. **54**, 67 (1982)
52. V.W. Hughes and T. Kinoshita, in: Muon Physics ,V.W. Hughes and C.S. Wu (eds.), Academic Press, p. 11 (1977)
53. PSI proposal R-92-06 "Spectroscopy of Muonic Hydrogen", E. Zavattini, spokesman (1992)
54. D. Taqqu, this volume
55. E. Morenzoni, this volume
56. D.H. White, this volume

This article was processed using Springer-Verlag TEX Z.Physik C macro package 1991
and the AMS fonts, developed by the American Mathematical Society.

Z. Phys. C - Particles and Fields 56, S70–S73 (1992)

Zeitschrift für Physik C Particles and Fields

Time-Fourier spectroscopy of muonium

V.P. Chebotayev*

* Permanent address: Institute of Laser Physics of the USSR Acadamy of Sciences, Novosibirsk 630090, Russia
* Present address: Max-Planck-Institut für Quantenoptik, 8046 Garching Postfach 1513, FRG

15-September-1991

Abstract. A new method is proposed for precise measurements of optical transition frequencies. Time Fourier spectroscopy uses two short laser pulses separated by a well known time difference for exciting atomic transitions. The precision is limited by the accuracy of the time separation of the two laser pulses.

PACS: 36.10.Dr;32.30-r

At present a precise measurement of the 1S-2S transition frequency in muonium can be performed only in the pulsed regime. The accuracy, which can be obtained in such an experiment, is the same as in frequency measurements in hydrogen atoms which can be performed in the continuous regime. At least, there are two most important circumstances which have direct influence on the accuracy of the experiment. The first one is connected with the process of obtaining the narrow resonance in the center of 1S-2S muonium transition. The second one is connected with the process of an absolute transition frequency measurement. In this article we shall consider a new method of nonlinear laser spectroscopy: *Time Fourier spectroscopy.* As we think, it will allow to obtain ultimate resolution as well as to carry out an absolute frequency measurement.

All the known methods of high resolution spectroscopy are based on resonant interaction between particles and optical fields. Hence, highly monochromatic laser radiation is essential in these methods. The observation time of a resonance is much longer than the oscillation period of the quantum transition under investigation. A new spectroscopic method based on using pairs of pulses of laser radiation with a linewidth which is broader than the measured quantum-transition frequency interval, but with a stable interpulse time, has been considered in [1]. The duration of the interaction between a particle and the single pulse is shorter than the oscillation period of the quantum transition. The result of the atomic interaction with two short pulses is determined by the phase of the free oscillations of the dipole moment at the time of the arrival of the second pulse. In other words, the particle makes the transition from one energy level into the other one synchronously with the atomic oscillation. The synchronized quantum transition is very accurately determined in time, and may be applied to direct precision measurements of time and frequency. The phenomenon of /synchronization of quantum transitions is not critically dependent on the nature of the pulse's perturbation. The method has been considered in details in [2]. The probability of a particle interaction upon the excited state is the oscillating time-delay function with the period $2\pi\omega_{os}^{-1}$, where ω_{os} is the frequency of transition. If we have to consider the ensemble of levels, the total absorption probability under the action of two identical pulses may be presented as a sum

$$W(T) = \sum_{i=2}^{n} W_{1i} cos(w_{1i}T) \tag{1}$$

where ω_{1i} is the frequency of the transition $1 - i$, T is a delay time between pulses, W_{1i} is the transition probability per one pulse.

The Fourier transformation of equation(1) will give an absorption spectrum. The time Fourier spectroscopy of hydrogen Rydberg atoms has been considered in [3].

The application of the described method is possible if the pulse separation instability is much less than the oscillation period $2\pi/\omega_{os}$. Therefore, it is of great importance to produce very short pulses which are delayed with respect to each other with high accuracy. In [4] a method to obtain the stable pulses on the basis of a self-mode-locked He-Ne laser has been developed. The time repetition frequency was synchronized with an external frequency generator by modifying the optical length of a resonator. In this case, the stability of the pulse frequency corresponds to that of the rf generator and, hence, may be as good as 1 part in 10^{-12} to 10^{-13}.

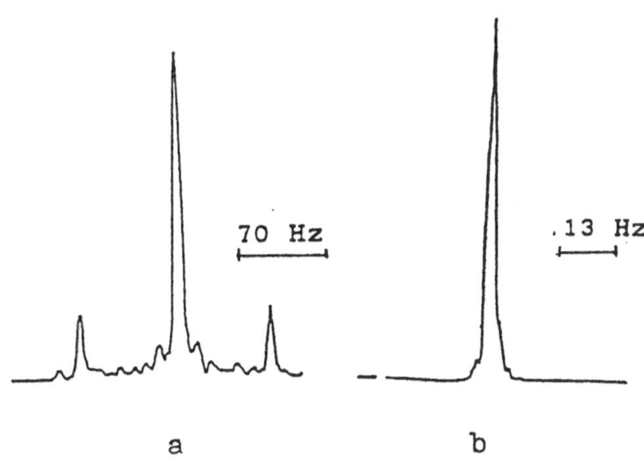

Fig. 1. A comparison of the intermode beat spectra (a) and (b) shows the improvement achieved both in the interpulse time stability and the pulse parameters in the phase-locking regime.

A sample of the a spectrum obtained in the free-running self-mode-locking regime is shown in Fig.1(a) from which can be seen that the spectral width is not more than 10 Hz. Results such as those in Fig.1(a) imply that self-mode-locking is accompanied by a modulation spectrum generally due to external perturbations — primarily, instability in the discharge current.

Our investigations showed that the frequency change in the intermode beats varies as $\Delta f = \Delta L/L$, where ΔL is the change in effective cavity length L. This conclusion led us to consider the possibility of stabilizing the beat frequencies (and, hence, the laser frequencies) by phase-locking the cavity fluctuations against an external oscillator. Fig.1(b) shows the intermode beat frequency under this phase-locked conditions. Here the spectral width is less than 0.02 Hz, which is the resolution of the spectrum analyzer used. The relative stability of the pulse frequency under these conditions corresponds to that of the external oscillator (10^{-10} over 10 sec). In the He-Ne laser the perturbation of intermode beat frequency due to the fluctuations of the discharge current is strong. Their compensation by changing the distance between mirrors increases the instability of the optical frequencies. Another possibility of phase locking of the frequency repetition rate has been investigated. In this case the POL error signal was connected to the discharge current. A narrowing of spectrum fluctuations of the optical frequencies has been observed. The comparison of the intermode beat spectra in Figs.1a and 1b shows a great improvement both in the interpulse time stability and the pulse parameters in the phase-locking regime. The stabilization of the intermode beat frequencies implies an equivalent stabilization of the optical length of the cavity and of the optical frequencies.

This stable short-pulse regime can be used to obtain a set of frequencies with a very stable frequency interval Ω. One may consider such a system as a frequency synthesizer for both optical and microwave frequencies. The range of the frequency synthesizer is determined by the pulse time and can be more than 10 THz. Synchronization of pulses from different lasers permits making an efficient nonlinear transformation to increase the number of synchronized frequencies. In consequence, there is a real possibility of obtaining a continuous distribution of frequencies through the infrared and optical regions with constant intervals Ω. Therefore, the experiments described here may be important for obtaining extremely short pulses and for the development of the method [1,2].

In the near future it is problematic to apply the method of synchronized quantum transitions[1,2] in the optical region. Presently obtainable pulse durations of $\cong 10^{-14}s$ do not allow us to use the method described in [1] in the optical part of the spectrum . Here we shall examine the application of Fourier spectroscopy in the case of pulse durations which are much longer than the period of oscillations of the atomic oscillator. Obtaining the time-stable pulses gives us this possibility. Interaction between light pulses and atoms for $\tau_{imp} \gg \omega_{os}$ was investigated quite well during the development of the method of separated fields, quantum echo, and other techniques of optical spectroscopy (see [5,6]). It is known, that the realization of the method of separated fields, which is well elaborated for the microwave region [7], demands the removal of the influence of the Doppler effect. In [8] different variants of realization of this technique for two- and many-level systems were considereded. For two-level atoms the physical nature of the Doppler shift compensation is analogous to photon echo [8]. The nonlinear part of the transition probability for two-level atoms under the action of three standing-wave fields (the same amplitude) is drawing up with the expression at the frameworks of perturbation theory

$$W = 2W_1(1 + cos(2\Omega T + \alpha_1)) \quad , \qquad (2)$$

here W_1 is the transition probability under the action of one field, α_1 is the time phase difference between fields, and Ω is the detuning of the carrier frequency from the line center.

Like in the case considered above, the transition probability is an oscillating function, too. Unlike in reference [1], the period of the oscillations depend on the frequency detuning Ω. The location of the maximum depends on the optical phase. It limits very much the application of this method in the optical range in experiments connected with a measurement of the line center position. The Fourier spectroscopy variant, which is considered here, allows one to get over this difficulty. Unlike traditional spectroscopy which is based on the

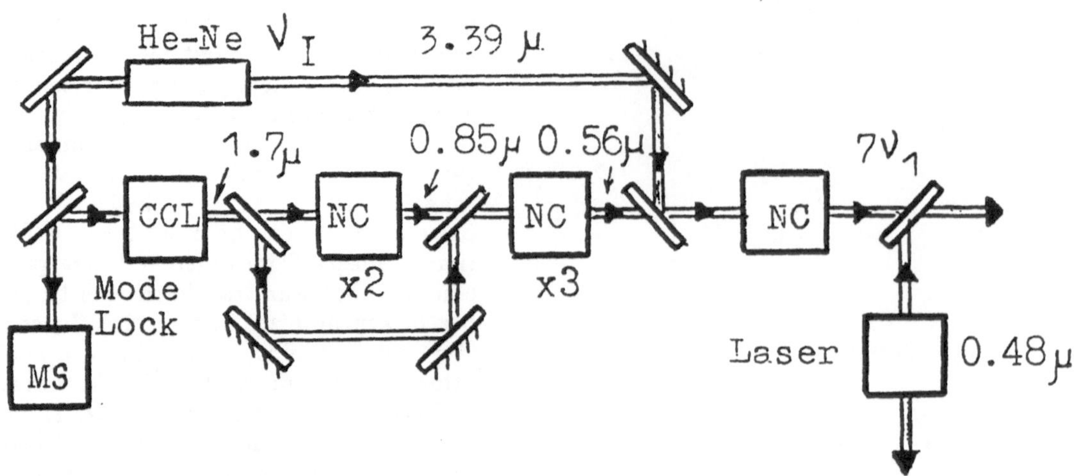

Fig. 2. Schematic setup for an absolute frequency measurement of a color center laser at $\lambda = 1.7\mu$.

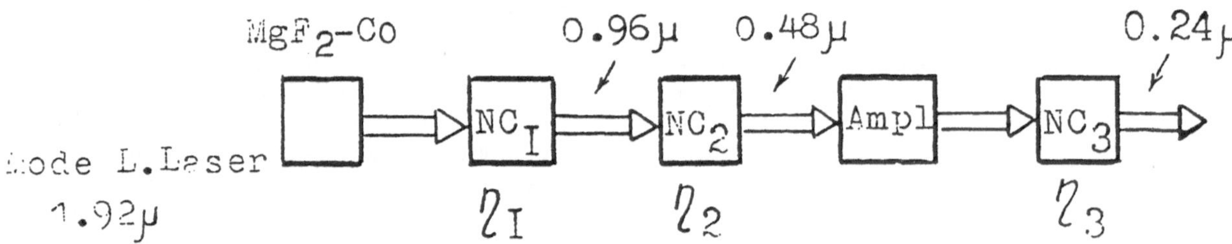

Fig. 3. Frequency synthesis chain for producing a stable freqeuncy reference in a hydrogen 1S-2S two-photon excitation experiment. The setup is similar to a setup needed in a muonium experiment.

measurement of the absorption as a function of the frequency detuning Ω, in the described method here , the dependence of the probability on the delay time T with a fixed frequency detuning Ω is observed.

The process of a precision measurement of the frequency detuning consists in a precise determination of the oscillating period of the transition probability W by varying the delay time T. In this method of measurement the position of the probability maximum is not important. Therefore, the dependence of the results of measurements on the phase of the optical field is expected. It consists of the positive difference of the described method from traditional method of two fields separated in space and time, which are based on the frequency measurements. But the described method does not permit to increase the resolution capability as the delay time T is defined by the homogeneous transition width γ.

Now let us consider the most important part of the method, which deals with the process of precision frequency measurements in pulsed regime. Pulsed radiation with relatively stable optical phase may be obtained in different ways. The simplest one is connected with the pulsed amplification of the continuous laser radiation. It is known that in the process of pulse amplification with a duration of 10^{-8} sec not only phase modulation but also the frequency modulation take place. Additional frequency measurements are needed. The most perspective way is using mode-locked laser pulses a stabilized frequency repetition rate δ. If a radiation spectrum of a mode-locked laser covers the known frequency ω_{ref} of a reference laser then the measured beat frequencies are $\Delta\omega = \omega_{ref} - \omega_r$, where ω_r is the frequency of a longitudinal mode r. One may define the frequencues of all optical modes

$$\omega_i = (N_i - N_r)\delta + \Delta\omega + \omega_{ref} \quad , \tag{3}$$

where N_i and N_r are the numbers of modes i and r. Fig.2 shows the scheme for an absolute frequency measurement of a color center laser at $\lambda = 1.7\mu$.

It is well known that the efficiency of frequency doubling is much higher in the pulsed regime compared to the cw case. It gives the possibility to develop new frequency chain systems for absolute frequency measurements in the visible and VUV regions. Fig.3 shows the scheme of a frequency synthesis chain for the absolute measurement of the frequency of the hydrogen Balmer-β line. It permits not only to carry out the frequency measurements, it can also be used for obtain the resonances in pulsed regime. In this scheme it is preferable to use the fourth harmonic of the MgF_2-Co laser at $\lambda = 1.92\mu$ rather than the fifth harmonic of CO_2 laser. The fourth harmonic of MgF_2-Co laser may be used as the main signal for the pulsed amplifier which is synchronized with the pulse repitition frequency. If the pulse duration $\tau_{imp} \cong 10^{-11} sec$ and the length of amplifier $L_{amp} \cong .1 cm$ the change of optical phase of pulse during of amplification is negligible. It permits to use the amplified signal for both the excitation of the resonance and the precise measurement of the frequency.

Acknowledgement. I am grateful to Prof. T.Hänsch for valuable discussions.

References

1 Chebotayev V.P.: Pis'ma Zh.E.T.F. 49, (1989), 429
2 Chebotayev V.P. and Ulybin V.A.: Appl.Phys. B50, (1990), 1
3 ibid: B52, (1991), 347
4 Chebotayev V.P., Klementyev V.M., Pyltsin O.I. and Zakhar V.F.: Appl.Phys., in press
5 Coherence in Spectroscopy and Modern Physics, ed. by Arecchi F.T., Bonifacio R. and Scully N.O. (Plenum, New York, London)
6 Multiphoton Resonant Processes in Atoms: Topics in Current Physics, Vol. 21, ed. by Feld M.S. and Letokhov V.S. (Springer, Berlin a.o.)
7 Ramsey N.F.: Molecular Beams (Oxford University Press, New York, London 1955)
8 Chebotayev V.P.: Super High Resolution Spectroscopy,Laser Handbook, Vol.5, ed. by Bass M. and Stitch M.L. (North-Holland, Amsterdam, Oxford, New York, Tokyo 1985)

This article was processed using Springer-Verlag TEX Z.Physik C macro package 1991
and the AMS fonts, developed by the American Mathematical Society.

Z. Phys. C - Particles and Fields 56, S74-S79 (1992)

Zeitschrift für Physik C Particles and Fields

Laser induced transition 3D-3P in muonic helium

L.Bracci[1] , A. Vacchi[2] , E. Zavattini[2]

[1] University of Pisa Italy

[2] University of Trieste and INFN Trieste Italy

6-December-1991

Abstract. In recent years quantum electrodynamics (QED) vacuum polarization corrections (at low momentum transfer) has been tested in several physical situations using muonic atoms: this has permitted a direct check for different values of $Z\alpha$ [1- 6]. The vacuum polarization test obtained by measuring the 2S- 2P energy difference, by means of laser-induced transition in muonic helium, has reached the 0.17% level. Such a test is limited by the uncertainty of the helium electric form factor; in order to overcome this limit, a scheme to measure the 3D-3P energy difference in muonic helium has been suggested and a test measurement performed. In this article we wish to make some comments on this experimental method and discuss its future possibilities.

1 Laser induced Transitions in muoniuc helium

Some years ago, at the CERN Syncro-cyclotron, a CERN-Pisa Collaboration [1] devised a scheme to perform a laser-induced transition experiment on a muonic-helium ion $(\mu^{-4}He)_{2S}$. At that time this rather complex set-up, connected with the operation of the Synchro-cyclotron itself, allowed the energy level differences D_i^2 =2S-2P, to be measured with an accuracy of $1.5 \cdot 10^{-4}$. In table 1 the experimental results and the expected theoretical values given by the quantum electrodynamics (QED) calculations [2] are reported.

It is important to note from the table that about 20% of D_i^2 is, in this case, due to corrections induced by the nuclear electric form factor $< r_e^2 >^{1/2}$. Taking for this form factor the experimental value [3]

$$< r_e^2 >^{1/2} = (1.676 \pm 0.008) fm,$$

(as seen by an electron probe), and assuming muon-electron universality, one sees that the QED correction is tested to the level of 0.17%; the result of this experiment is among the best direct tests so far performed, see table 2.

Later a CERN-Columbia Collaboration [7] looked into the possibility of performing an experiment of the same type (i.e. laser-induced transition in muonic ions), but on transitions between levels not belonging to the S states in order to avoid the influence of the electromagnetic form-factor uncertainties. A set-up to measure the 3D-3P energy-level differences, via laser stimulation, in muonic helium was realized. In this paper as well as giving a brief account of the apparatus built at BNL and its performances [8,9] some comments are also reported on the possibilities of this experimental method.

In fig. 1 the lower energy levels of the free muonic-helium ion, together with some of their characteristics, are shown. The principle of the experiment consists in observing a variation in the K_β line intensity caused by the laser- stimulated 3D-3P transition when negative muons are stopped in a multipass optical cavity filled with helium and where high- density electromagnetic radiation is stored. It is crucial to realize that the lifetimes of the muonic levels involved are all around $10^{-12}s$.

The main physical background, from which the small increase due to the laser-stimulated emission has to be extracted, is the natural emission of K_β X-rays during the cascade of the negative muons.

In table 3 are presented the results [7,9] of a calculation for the various contributions to the 3D-3P energies obtained under the assumption that the $(\mu^{-4}He)^+$ ion is free. The wavelength of the radiation to match the resonance conditions is obtained with CO_2 lasers.

From fig. 1 one sees that the width Γ of the line is

$$\Gamma = 5 \cdot 10^{-4} eV = 0.0473 \mu m.$$

If N_D is the number of negative muons passing the D levels of the muonic ion (about 60%), and E/V is the energy density of the radiation at the side of the

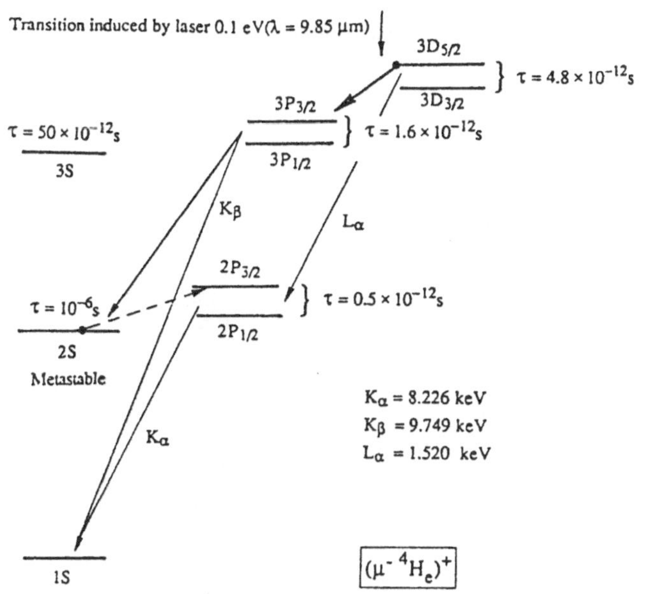

Fig. 1. Scheme of the first energy levels of the $(\mu^{-4}He)^+$ muonic ion.

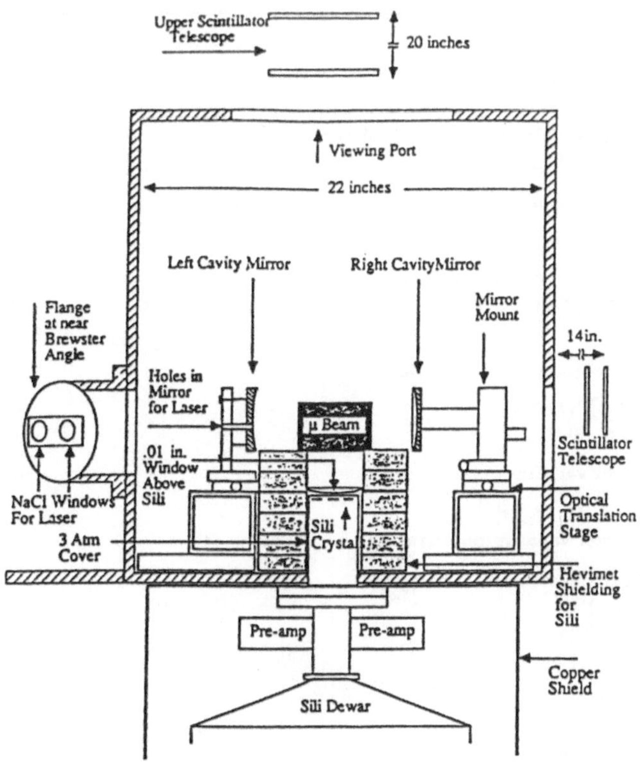

Fig. 2. Layout of the target and optical cavity-system.

stopping muon, then the fraction of transitions is given by [8,9]:

$$\epsilon(\omega) = \frac{N_{stim}}{N_D} = \frac{\Gamma/2}{\hbar^2(\omega-\omega_0)^2 + \Gamma^2/4} \cdot \frac{1}{\gamma_{3D}} \cdot |F_{3D-3P}|^2 E/V \,, \tag{1}$$

where $\Gamma = (\gamma_{3D} + \gamma_{3P})$ is the width of the transiton, γ_{3D} and γ_{3P} are the radiative decay rate of the D and P levels of interest, respectively (see fig. 1 ; $\gamma = \hbar/\tau$), and $|F_{3D-3P}|^2 = 2.43(ea_\mu)^2$ is the square of the electric-dipole transition- matrix element. Taking $V = 0.4(cm^3)$ and $E = 4(Joules)$ in Eq. (1), at the resonance frequence one gets $\epsilon = 0.6\%$.

Figure 2 is a sketch of the target set-up; the effective target is represented by a multipass optical cavity [10], where the CO2 laser burst was stored (for about 100 ns). The cavity is assembled in 3 atm of helium and, in the small cavity region, the burst of negative muons is stopped during the presence of the radiation.

The X-ray detecting system was composed of three Si(Li) detectors each having a sensitive area of about $1cm^2$. An isotopic $^{13}C^{18}O_2$ laser delivered the 4 J of radiation stored in the optical cavity. A particular frequency was selected by means of a grid to locate the centre of the resonance (98595Å for $3D_{5/2} - 3P_{3/2}$) to better than 10 Å.

At the Single Burst Extraction (SBE) Beam of the BNL AGS (28 GeV/c), an isolated burst of negative muons (about 50 ns wide) was brought to stop in the helium gas target. The SBE operation was synchronized with the laser firing so that muons and laser radiation would be present at the same time in the cavity target (see fig. 3). The relevant figures for the muon beam (10^{12} protons on target, 1 burst each 1.4s) are given below:

- momentum	25 MeV/c
- μ^- stops in cavity	300/burst
- e/μ	8
- Si(Li) total counts per burst	0.65/counter

It is important to note that since the muon beam was instantaneous compared to the integration time of the detectors, only one count could be accepted during a single burst. The optimum integration time was a compromise between the need for good energy resolution and the minimization of the pile-up. In fig. 4 is shown the K_β X-ray yield [8] obtained in the experiment (laser off). It is easy to see that a useful quantity is the ratio K_β/K_α: from the experiment, the following value was deduced for this ratio:

$$Y_{exp} = \frac{\left(\frac{K_\beta}{K_\alpha}\right)_0 - \left(\frac{K_\beta}{K_\alpha}\right)_f}{\left(\frac{K_\beta}{K_\alpha}\right)_f} = 0.012 \pm 0.014$$

where 0 means laser on, f laser off. The expected value for Y, assuming the use of the correct wavelength, is

$Y_{th} = 0.017$.

Table 1. Contributions to n = 2 energy splittings in the $(\mu^{-4}\mathrm{He})^+_{2S}$ system.

Contribution	Transition energies (meV) $2P_{3/2} - 2S_{1/2}$	$2P_{1/2} - 2S_{1/2}$
Dirac contribution with Coulomb potential and point like charges	145.70	0
Nuclear polarizability	3.1 ± 0.6	3.1 ± 0.6
Finite size*	-289.5 ± 2.8	-289.5 ± 2.8
Electronic vacuum polarization		
Uehling term: first iteration	1664.44	1664.17
higher iteration	1.70	1.70
Kallen-Sabry term $(\alpha^2 Z\alpha)$	11.55	11.55
$\alpha(Z\alpha)^n, n > 3$	-0.02	-0.02
$\alpha^2(Z\alpha)^2$	0.02	0.02
Muon vacuum polarization	0.33	0.33
$\mu - e$ vacuum polarization	0.02	0.02
Hadron vacuum polarization	0.15	0.15
Vertex corrections and (g-2)		
$\alpha(Z\alpha)$	-10.52	-10.85
$\alpha(Z\alpha)^n, n > 1$	-0.16	-0.16
$\alpha^2 Z\alpha$	-0.03	-0.03
Recoil terms		
Breit	0.28	0.28
Two photons	-0.44	-0.44
Weak contribution	0.00002	0.00002
Sum theory	1526.6 ± 2.8	1380.3 ± 2.8
Experiment	1527.5 ± 0.3	1381.3 ± 0.5

* Value obtained from the results of Ref.[3] $\langle r_e^2\rangle^{1/2} = (1.676 \pm 0.008)$ fm.

Table 2. Summary of the most accurate tests of QED vacuum polarization correction.

Experiment	Total Effect	Vacuum Polarization	Total Uncertainty	Major source of Uncertainty	Relative Uncertainty in Vacuum Polarization	Ref.
Lamb shift H	$1\mathrm{x}10^3$ MHz	26 MHz	0.02 MHz 0.05 MHz	Calculation	$\sim 1.7\mathrm{x}10^{-3}$	[17]
$(g\text{-}2/2)_e$	$1.1\mathrm{x}10^{-3}$ eV	$1\mathrm{x}10^{-7}$ eV	$2\mathrm{x}10^{-10}$ eV	Value of α	$\sim 1.7\mathrm{x}10^{-3}$	[18]
$(g\text{-}2/2)_\mu$	$1.1\mathrm{x}10^{-3}$ eV	$6\mathrm{x}10^{-6}$ eV	$2\mathrm{x}10^{-8}$ eV	Strong Int. Cont. + Expt. error	$\sim 1.7\mathrm{x}10^{-3}$	[19]
Muonic Atoms High Z	$4\mathrm{x}10^5$ eV	$2\mathrm{x}10^3$ eV	8 eV	Expt. error + Electron screening	$4\mathrm{x}10^{-3}$	[20]
Muonic Atoms, He Laser induced	1.5 eV	1.7 eV	$4\mathrm{x}10^{-3}$ eV	$\langle r^2\rangle^{1/2}$	$1.7\mathrm{x}10^{-3}$	[1]
Muonic Atoms Low $Z \leq 13$	^{24}Mg $5.6\mathrm{x}10^4$ eV ^{28}Si $7.7\mathrm{x}10^4$ eV	$1.8\mathrm{x}10^2$ eV $2.8\mathrm{x}10^2$ eV	3 ppm; 0.16 eV 3 ppm; 0.25 eV	Electron screening + Calculation	$0.95\mathrm{x}10^{-3}$	[6]

Table 3. Contributions to the 3D-3P energy level difference for the free $(\mu^{-4}He)^+$ system.

Transition	Vacuum polarization α (Uehling-Serber) (meV)	α^2 (Kallen-Sabry) (meV)	Fine structure (Dirac) (meV)	Hyperfine structure (meV)	Total (meV)	λ (μm)
$3D_{3/2}$-$3P_{1/2}$	110.458	0.905	43.164	0	154.528	8.0235
$3D_{5/2}$-$3P_{3/2}$	110.458	0.905	14.389	0	125.751	9.8595
$3D_{3/2}$-$3P_{3/2}$	110.458	0.905	0	0	111.363	11.1334

$\Gamma = 5x10^{-4}$ eV $\simeq 0.047\mu m$

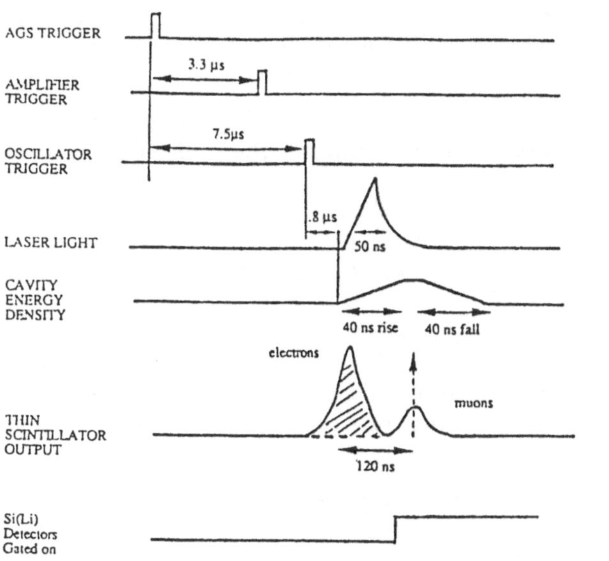

Fig. 3. Timing sequence of the experiment.

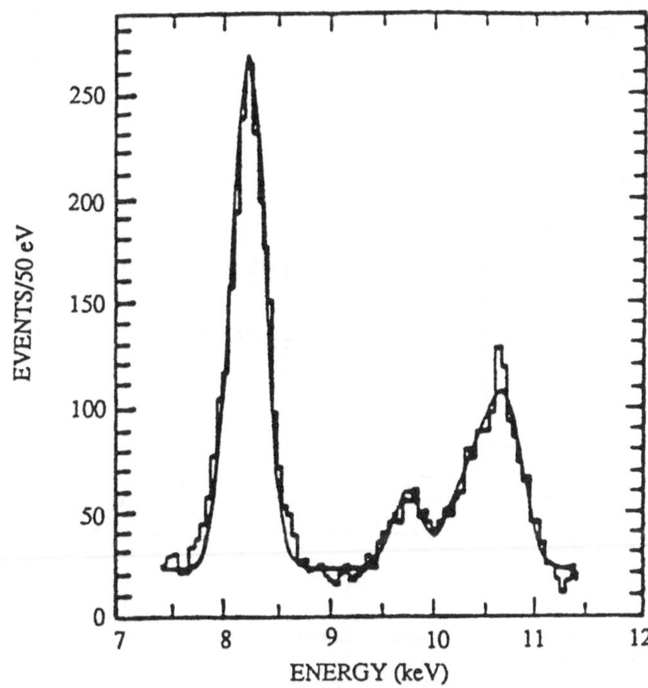

Fig. 4. Energy spectrum recorded in a Si(Li) detector of the prompt K X-rays from the negative muons stopped in the He target at 3 atm. The smooth curve is a fit to the data, using Gaussian shapes with fixed centres and widths for theK_α, K_β,K_γ, including also a combination of higher-K X-rays and a flat background, [9].

It appears evident that in order to perform a significant measurement, i.e. comparable or competitive with the 2S-2P difference measurement [1], much higher statistics would have been required. The experiment faced pratical limits owing to the size of the sensitive area of the Si(Li) detectors. A large array of X-ray detectors having high segmentation, good energy resolution, and fast response would allow the experiment to be performed.

Meanwhile, the processes occurring when a negative muon is stopped in a helium gas target at a pressure of few atmospheres were studied (see Refs. [11-13]). These studies have shown that most probably at these densities, very soon (i.e. within the cascade time), the single charged system $(\mu^{-4}He)^+ = M_n^+$ remains bound in a helium-ionic molecular system: i.e. the M_n^+ formed in the gas (which appears as a heavy hydrogen ion), does not remain free and forms bound muon-helium molecular ions $[(M_n^+He)^+He]_i^+$. As a consequence, electron screening effects on the muonic energy levels should be expected.

All possible bound muon-helium molecular ions (ground and excited ones) from Ref. [13] are shown in fig. 5. The possible prompt formation of various bound muonic molecular ions, unfortunately, changes the prospects of using the 3D-3P energy difference measurements, suggested in Ref. [7], as a high- precision QED vacuum-polarization test.

We wish to discuss and point out the substantial correction $\Delta_{corr,3}$ due to the electrons screening in the molecules $[(M_n^+He)^+He]_i^+$, to the 3D-3P energy level difference. The correction will, in general, depend on the molecular-ion formed : however the correction can be written

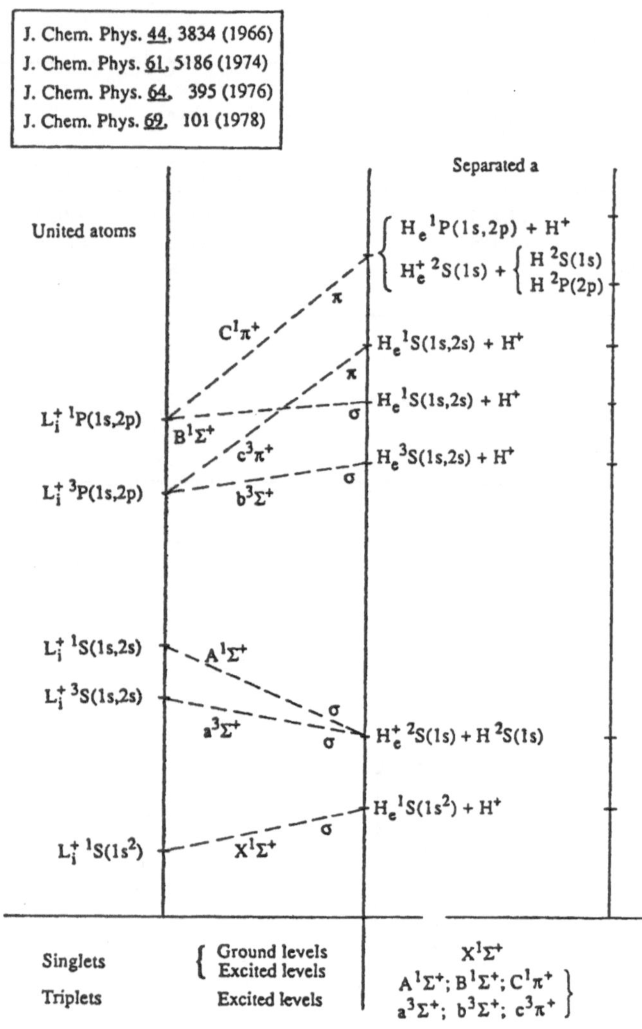

Fig. 5. Possible bound muonic-helium molecular ions $[(\mu^{-4}He)^{+}He]_i^{+}$.

$$\Delta_{corr,3} = -|\rho|\frac{e^2}{a_0}\left(\frac{a_{\mu He}}{a_0}\right)^2 \frac{2^3}{3\cdot 5}\int e^{-2/3r_\mu} r_\mu^4 d^3\vec{r}_\mu$$

$$= -4\cdot 10^{-3}|\rho|(eV)$$

where $|\rho|$ is the electron screening density at the muon's site in e/a_0^3 units, a_0 is the Bohr radius of the hydrogen atom, $a_\mu He$ is the Bohr radius of the muonic helium; e is the electron's charge. An estimation of $\Delta_{corr,3}$ taking for the muonic molecular ion the ground system $X'\Sigma^+$(see fig. 5) gives as a correction to the values of table 3

$$\Delta_{corr,3} = -10^{-3}eV = 0.0860\mu m \approx 2\Gamma.$$

Interpolating the data given in Ref.[14] for the $[H^+He]_i^+$ molecular excited ions, values for $\Delta_{corr,3}$ below Γ were obtained for the excited muonic systems, depending on the level considered.

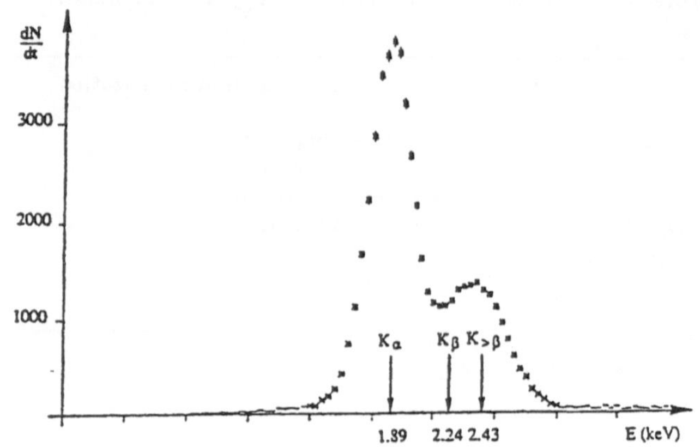

Fig. 6. K transition of the m-p system as measured by a Xe gas scintillation detector; the FWHM is about 20% at 2 keV.

The correction to the 2S-2P energy difference, $\Delta_{corr,2}$, due to the screening of the molecular electrons can be written

$$\Delta_{corr,2} = -|\rho|\frac{e^2}{a_0}\left(\frac{a_{\mu He}}{a_0}\right)^2 \frac{1}{3*2^2}\int e^{-r_\mu} r_\mu^2 d^3\vec{r}_\mu$$

$$= -10^{-3}|\rho|(eV),$$

and on the assumption that the molecular ion is the $X'\Sigma^+$ ground system one gets the following corrections to the values in table 1:

$$\Delta_{corr,2} = -2.4\cdot 10^{-4} = 1.3\text{Å},$$

i.e. very small compared with the uncertainty caused by the form factor and a fraction 1/7 of the width ($\approx 8\text{Å}$).

2 Conclusions

Before carrying on with the experiment discussed in Ref. [7,8,9], as a QED vacuum polarization test, further work is required. At first glance it appears that the formation at early times of muon-molecular ions makes this experimental method unlikely to give results competitive with the 2S-2P experiment.

However looking at the experimental results of the Columbia test is has to be stressed that: i) the clean observation of the K_β line in an extremely intense pulsed low energy muon beam (25 MeV/c) has been a success, ii) with the multipass cavity-target technique it has been possible to make a (fast) muon-beam laser-pulse coincidence.

These facts suggest that it is worth while employing this technique to study the neutral μ^-p system formed by stopping negative muons in hydrogen gas: in particular to look at the 3D-3P energy difference. In fig. 6 the energy spectrum of the μ^-p K transitions is presented as measured by a Xe gas scintillation proportional detector [15,16]. For orientation we have calculated the energy

Table 4. Various contributions to the 3D-3P energy level differences for the $(\mu^- p)$ neutral system. Width $\Gamma \simeq 2x10^{-5}$ eV.

Transition	Dirac (meV)	Hyperfine (meV)	Vehling (meV)	Kallen (meV)	Total (meV)	λ (mm)
$3D^3_{3/2} - 3P^5_{3/2}$	0	-0.6757171	4.649347	0.0442757	4.017960	0.308582
$3D^5_{3/2} - 3P^5_{3/2}$	0	-0.1351433	"	"	4.558480	0.271988
$3D^3_{3/2} - 3P^3_{3/2}$	0	+0.2252390	"	"	4.918862	0.252061
$3D^5_{5/2} - 3P^5_{3/2}$	0.831152	-0.5405736	"	"	4.984201	0.248756
$3D^7_{5/2} - 3P^5_{3/2}$	0.831152	-0.1930620	"	"	5.331713	0.232543
$3D^5_{3/2} - 3P^3_{3/2}$	0	+0.7650126	"	"	5.459436	0.227103
$3D^5_{5/2} - 3P^3_{3/2}$	0.831152	+0.3603822	"	"	5.885157	0.210674
$3D^3_{3/2} - 3P^3_{1/2}$	2.493455	-0.9009561	"	"	6.286122	0.197363
$3D^5_{3/2} - 3P^3_{1/2}$	2.493455	-0.360382	"	"	6.826695	0.181618
$3D^3_{3/2} - 3P^1_{1/2}$	2.493455	+1.351434	"	"	8.538511	0.145270

Notation, nL_J^{2F+1}; $\vec{J} = \vec{l} + \vec{s}$; $\vec{F} = \vec{J} + \vec{I}$; I = nuclear spin, s = lepton spin.

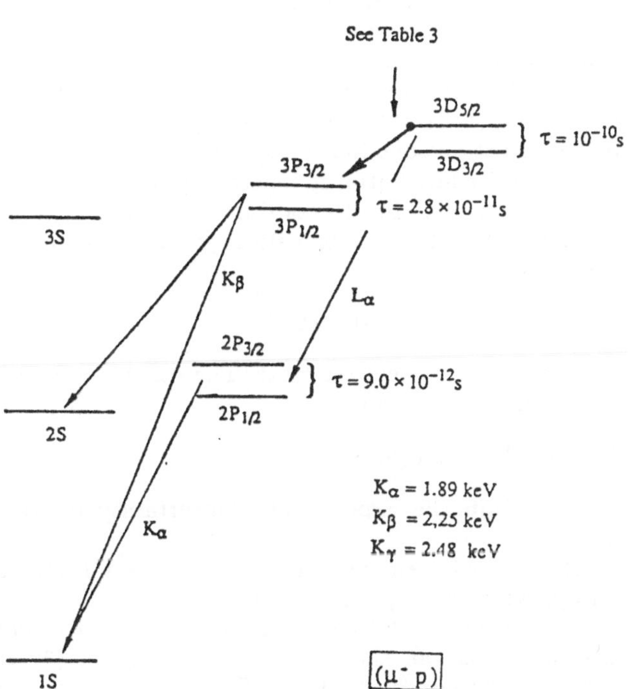

Fig. 7. Scheme of the first energy levels of the $(\mu - p)$ muonic atoms.

differences of the 3D-3P levels (see fig. 7) and found the values shown in table 4: the corresponding line width Γ is

$$\Gamma = \frac{\hbar}{\tau_P} + \frac{\hbar}{\tau_D} = 2 \cdot 10^{-5} eV,$$

i.e. we have transitions at frequencies around 1000 GHz with a width of about 6 GHz. We note that for the 3D-3P levels the transition probabilities in the $\mu^- p$ case are, for the same energy density, about 100 times higher than for the $(\mu^{-4}He)^+$.

References

1. G. Carboni et al., Nucl. Phys. A278, 381 (1977).
2. E. Borie and G.A. Rinker, Phys. Rev. A18, 324 (1978).
3. I. Sick, Phys. Lett. B116, 212 (1982)
4. E. Zavattini, Proc. First Course of Int. School of Physics of Exotic Atoms, Erice 1977, eds. G Fiorentini and G. Torelli (Lab. Naz. di Frascati, 1977), p.43.
5. W.G. Bauer and H. Salecker, Found. Phys. 13, 115 (1983).
6. B.Aas et al., Nucl. Phys. A451 (1986) 679; A375 (1982) 405 ; A429 (1984) 381.
7. A.M. Sachs, J. Fox, R. Cohen and E. Zavattini: An experiment to measure vacuum polarization in 3D-3P transitions in muonic helium atoms. Proposal exp. 745, BNL 1979.
8. J.S. French et al , Phys. Rev. A40, 158 (1989).
9. J.S. French, Thesis, Columbia University, Nevis 263 (1987).
10. D. Herriot et al., Appl. Opt. 3, 523 (1964).
11. J.S. Cohen, Phys. Rev. A25, 1791 (1982)
12. L.I. Menshikov et al., Z. Phys. D7, 203 (1987).
13. L. Bracci and E. Zavattini, Phys. Rev A41, 2352 (1990).
14. H.H. Michels, J. Chem. Phys. 44 3834 (1966).
15. J. Böcklin et al., Nucl. Instruments and Methods, 176, 105 (1980).
16. J. Böcklin Thesis, ETHZ 7161 (1982).
17. E. Borie, Phys. Rev. Lett. 47 (1981) 568, and references therein.
18. E. Borie et al.,Proc. Mainz Conf. on present status and aims of QED. Lecture Notes in Physics, Vol. 143 (Springer, Berlin, 1981) p. 68.
19. J. Baily et al., Nucl. Phys. B150 (1979) 1.
20. L. Tauscher, Proc. Mainz Conf. on present status and aims of QED. Lecture Notes in Physics, Vol. 143 (Springer, Berlin, 1981), and Z. Phys. A283, 139 (1978).

This article was processed using Springer-Verlag TEX Z.Physik C macro package 1991
and the AMS fonts, developed by the American Mathematical Society.

Z. Phys. C - Particles and Fields 56, S80–S87 (1992)

Zeitschrift für Physik C Particles and Fields

Muon g − 2: Theory

Toichiro Kinoshita

Newman Laboratory of Nuclear Studies, Cornell University, Ithaca, NY14853 U.S.A.

15-September-1991

Abstract. Current status of the theory of the muon anomalous magnetic moment is reviewed. The QED part of the theory is in a very good shape at present and the accuracy of its prediction far exceeds the expected precision of the new muon g - 2 experiment at the Brookhaven National Laboratory. On the other hand, further improvement is urgently needed for the hadronic contribution, which requires new precision measurements of hadron production cross sections in e^+e^- collisions. A complete standard-model calculation of the two-loop electroweak contribution will also be very useful in reducing the remaining theoretical uncertainties.

1 Introduction

The most accurate measurements of the muon anomalous magnetic moment a_μ available at present are those carried out in the last CERN experiment [1]. The details of these experiments are discussed in the article by Farley in this volume and in a recent review article by Farley and Picasso [2]. They give the values

$$a_{\mu^-} = 1\ 165\ 936(12) \times 10^{-9},$$
$$a_{\mu^+} = 1\ 165\ 910(11) \times 10^{-9}. \tag{1}$$

Combining these results they obtain

$$a_\mu^{expt} = 1\ 165\ 923(8.5) \times 10^{-9}. \tag{2}$$

Although the theory of a_μ has been reviewed recently [3], it will be useful to examine and update it from time to time in preparation for the new muon g - 2 experiment E821 at the Brookhaven National Laboratory.

The theory of the anomalous magnetic moment of the muon looks very similar to that of the electron. However, the physics of the muon anomalous moment is quite different from that of the electron due to the fact that the internal momenta of the muon's structure scale as the muon mass rather than the electron mass. This makes the electron-loop vacuum polarization effect very important, and leads to logarithms of m_μ/m_e in the coefficients of the second and higher powers of α/π. For the same reason, the hadronic vacuum polarization and the weak interaction contribute significantly to a_μ. Thus the theory of muon anomalous magnetic moment divides itself into three parts: pure QED contribution a_μ(QED), hadronic contribution a_μ(had), and weak interaction contribution a_μ(weak). The current status of these contributions, together with their possible improvements, will be discussed in the following sections.

Collecting all the theoretical results available at present one finds

$$a_\mu^{th} = 116\ 591\ 917(176) \times 10^{-11}, \tag{3}$$

as the best theoretical estimate. The difference between theory and experiment

$$a_\mu^{th} - a_\mu^{expt} = -3.8(8.7) \times 10^{-9} \tag{4}$$

is well within the experimental uncertainty quoted in (2).

The Brookhaven muon g - 2 experiment E821 aims at improving the experimental precision by about a factor of 20 over the CERN result (2). Once this goal is achieved, it is the theoretical uncertainty in (3) that must be reduced substantially. This uncertainty comes predominantly from that of the hadronic contribution a_μ(had), which in turn results mainly from the rather poorly controled systematic errors in the measurements of R, the ratio of the total cross sections for hadron production and $\mu^+\mu^-$ production in e^+e^- collisions. The only way to reduce the error in a_μ(had) is to perform new measurement of R, particularly in the energy range from the two pion threshold to about 3 GeV, with strong emphasis on reduction of systematic errors. In order to match the expected precision of the forthcoming measurement of a_μ, it is necessary to cut down the error of a_μ(had) to less than 30×10^{-11}.

When the theoretical error in the hadronic contribution is reduced to this size, the new g - 2 experiment will enable us to test the validity of the standard model prediction for a_μ(weak), which is about 195×10^{-11}, to a precision of better than 30 percent. This is in fact the primary motivation for this experiment. Note that the weak interaction contribution arises from intermediate virtual states of one-loop order and hence is potentially sensitive to the physics beyond the standard model as well as the range of current high energy accelerators.

It has recently been pointed out [4] that two-loop electroweak diagrams may contribute as much as 10×10^{-11} to a_μ. It is desirable to complete the two-loop calculation and remove this theoretical uncertainty before the experimental result becomes available.

In view of the recent measurements [5] of the mass and width of the Z^0 boson, which give convincing confirmation of the standard model with three generations of lepton families whose neutrinos are (almost) massless, the calculated electroweak contribution to a_μ is very likely to be confirmed in the experiment E821. If this happens, comparison of theory and experiment on a_μ may impose a very strict lower bound on the mass of not-yet-discovered particles in the energy range of SSC. In this sense, the relevance of the Brookhaven muon g - 2 experiment to high energy physics is further enhanced by recent results from LEP.

2 The QED contribution

The QED contribution can be written in the general form

$$a_\mu(\text{QED}) = A_1 + A_2(m_\mu/m_e) + A_2(m_\mu/m_\tau) + A_3(m_\mu/m_e, m_\mu/m_\tau), \quad (5)$$

where m_e, m_μ, and m_τ are the masses of the electron, muon, and tauon, respectively.

Renormalizability of QED guarantees that the functions A_1, A_2, and A_3 can be expanded in power series in α/π with finite calculable coefficients:

$$A_i = A_i^{(2)}\left(\frac{\alpha}{\pi}\right) + A_i^{(4)}\left(\frac{\alpha}{\pi}\right)^2 + A_i^{(6)}\left(\frac{\alpha}{\pi}\right)^3 + \dots, \quad i = 1, 2, 3. \quad (6)$$

Evaluation of the mass-independent term A_1 is identical with that of the electron anomaly a_e. Thus far it has been evaluated [6] to order α^4:

$$\begin{aligned} A_1^{(2)} &= \ \ 0.5, \\ A_1^{(4)} &= -0.328\ 478\ 965\dots, \\ A_1^{(6)} &= \ \ 1.176\ 13(42), \\ A_1^{(8)} &= -1.434(138). \end{aligned} \quad (7)$$

Here $A_1^{(6)}$ takes into account the recently completed analytic calculation of the diagrams containing a light-by-light scattering subdiagram [7]. Its numerical value is

$$A_1^{(6)}(\text{light} - \text{light}) = 0.371\ 005\ 292\cdots, \quad (8)$$

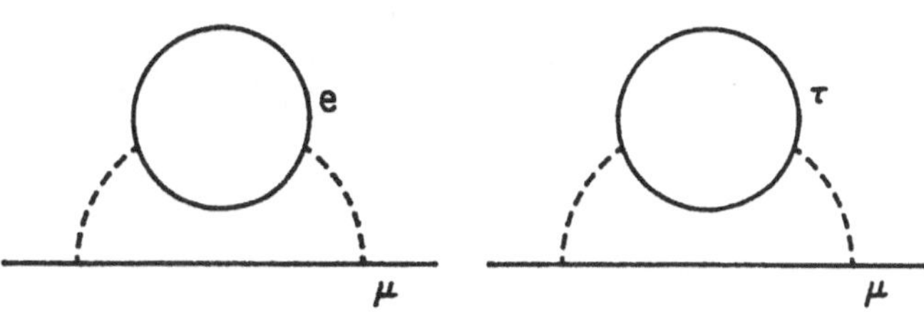

Fig. 1. Fourth-order Feynman diagrams contributing to the muon anomalous magnetic moment. In this and following figures fermions are often assumed to propagate in a constant magnetic field. Thus no external magnetic field vertex is shown explicitly.

which is in very good agreement with previous results obtained by numerical integration:

$$\begin{aligned} A_1^{(6)}(\text{light} - \text{light}) &= 0.370\ 986(20), \quad &(\text{Ref. [8]}) \\ &= 0.371\ 12(28). \quad &(\text{Ref. [9]}) \end{aligned} \quad (9)$$

Some parts of $A_1^{(6)}$ and all of $A_1^{(8)}$ in (7) are evaluated by the numerical integration routine VEGAS [10]. The uncertainties are also estimated by VEGAS.

If one chooses as α the most recent value based on the quantized Hall effect [11]:

$$\alpha^{-1} = 137.035\ 997\ 9(32), \quad (10)$$

one obtains from (6) and (7) the result

$$A_1 = 1\ 159\ 652\ 136.2(5.3)(4.1)(27.1) \times 10^{-12}, \quad (11)$$

where the first and second uncertainties come from the numerical uncertainties in $A_1^{(6)}$ and $A_1^{(8)}$, respectively, while the third reflects the uncertainty in α quoted in (10). Further improvement of $A_1^{(6)}$ and $A_1^{(8)}$ is in progress.

As for A_2 and A_3, it is easy to see that $A_2^{(2)} = A_3^{(2)} = A_3^{(4)} = 0$. In the fourth order there are contributions to $A_2(m_\mu/m_e)$ and $A_2(m_\mu/m_\tau)$ from the Feynman diagrams shown in Fig. 1. Both are known analytically [12]. Thus their numerical errors come only from the uncertainties in the measurements of the mass ratios m_e/m_μ and m_μ/m_τ. To the accuracy of our interest these contribution are [13]

$$A_2^{(4)}(m_\mu/m_e) = \frac{1}{3}\ln\frac{m_\mu}{m_e} - \frac{25}{36} + \frac{3}{2}\frac{m_e}{m_\mu}\zeta(2) - 4\left(\frac{m_e}{m_\mu}\right)^2 \ln\frac{m_\mu}{m_e} + 3\left(\frac{m_e}{m_\mu}\right)^2 - \frac{15}{2}\left(\frac{m_e}{m_\mu}\right)^3 \zeta(2) + \ldots \tag{12}$$

$$= 1.094\ 258\ 28(5)$$

and

$$A_2^{(4)}(m_\mu/m_\tau) = \frac{1}{45}\left(\frac{m_\mu}{m_\tau}\right)^2 + \frac{1}{70}\left(\frac{m_\mu}{m_\tau}\right)^4 \ln\frac{m_\mu}{m_\tau} + \ldots \tag{13}$$

$$= 7.745(4) \times 10^{-5},$$

where $\zeta(n)$ is the Riemann zeta function of argument n. The best mass values available at present [14] are $m_e =$ 0.510 999 06(15) MeV (0.30 ppm), $m_\mu =$ 105.658 389(34) MeV (0.32 ppm), and $m_\tau =$ 1784.1(+2.7/ − 3.6) MeV. Note, however, that I used in (12) the value 206.768 262 (30) (0.15 ppm) for the ratio m_μ/m_e, which is more accurately known than m_e or m_μ separately.

In the sixth order there are 24 Feynman diagrams contributing to $A_2^{(6)}(m_\mu/m_e)$. These include 18 diagrams containing electron vacuum polarization loops and 6 diagrams containing light-by-light scattering subdiagrams. Typical diagrams of both kinds are shown in Fig. 2 and Fig 3, respectively.

Unfortunately these contributions are not known analytically at present. Until recently, the most accurate values have been those obtained by numerical integration. For the diagrams of Fig. 2 the most recent values are as follows[1]:

$$\begin{aligned} A_2^{(6)}[Fig.2(a)] &= 1.493\ 73(13), \\ A_2^{(6)}[Fig.2(b)] &= 2.718\ 63(5), \\ A_2^{(6)}[Fig.2(c)] &= 0.100\ 519(5), \\ A_2^{(6)}[Figs.2(d,e)] &= -2.392\ 38(43). \end{aligned} \tag{14}$$

These results are obtained by VEGAS [10]. The number of iterations and the number of sampling points per iteration are $(30, 12\times 10^7)$ for Fig. 2(a) and $(40, 12\times 10^7)$ for a related diagram no shown, and $(24, 12 \times 10^7)$, $(20, 12 \times 10^6)$, $(45, 24 \times 10^7)$, $(48, 12 \times 10^7)$, for Figs. 2(b), 2(c), 2(d), and 2(e), respectively.

[1] These results are newly evaluated for this paper. They are not only 3 to 6 times more accurate than the old values reported in Ref.3 but also remove the shifts of 0.1 percent or less in the latter caused by one numerical constant which was inadvertently assigned a single precision instead of the intended double precision.

As is well-known, $\ln(m_\mu/m_e)$ terms and mass-independent terms of $A_2^{(6)}$[Fig.2] can be determined algebraically by the renormalization group consideration [15]. It is found that

$$\begin{aligned}
A_2^{(6)}[\text{Fig.2(a)}] &= \frac{1}{4}\ln\frac{m_\mu}{m_e} + \frac{1}{2}\zeta(3) - \frac{5}{12} + \ldots \\
&= 1.517\ 26 + \mathcal{O}(m_e/m_\mu), \\
A_2^{(6)}[\text{Fig.2(b)}] &= \frac{2}{9}\ln^2\left(\frac{m_\mu}{m_e}\right) - \frac{25}{27}\ln\frac{m_\mu}{m_e} + \frac{317}{324} + \frac{2}{9}\zeta(2) + \ldots \\
&= 2.724\ 15 + \mathcal{O}(m_e/m_\mu), \\
A_2^{(6)}[\text{Fig.2(c)}] &= \left(\frac{119}{27} - \frac{8}{3}\zeta(2)\right)\ln\frac{m_\mu}{m_e} + \frac{2}{9}\zeta(2) - \frac{61}{162} + \ldots \\
&= 0.100\ 52 + \mathcal{O}(m_e/m_\mu), \\
A_2^{(6)}[\text{Figs.2(d, e)}] &= \left(-\frac{31}{12} + \frac{10}{3}\zeta(2) - 4\zeta(2)\ln 2 + \zeta(3)\right)\ln\frac{m_\mu}{m_e} - \frac{79}{9}\zeta(2) + \frac{115}{24} + 10\zeta(2)\ln 2 - \frac{7}{2}\zeta(3) + 3C_4 + \ldots \\
&= -2.397\ 89 + \mathcal{O}(m_e/m_\mu),
\end{aligned} \tag{15}$$

where C_4 is a known constant with the value 0.833 768 09 ... .

Clearly, these results do not have enough accuracy to check (14). For this purpose terms linear or higher order in m_e/m_μ must be evaluated. Recently, the situation has been improved significantly by Samuel and Li [13] who have evaluated all coefficients of terms up to $(m_e/m_\mu)^2$ analytically. As a consequence, the last lines of (15) are now replaced by

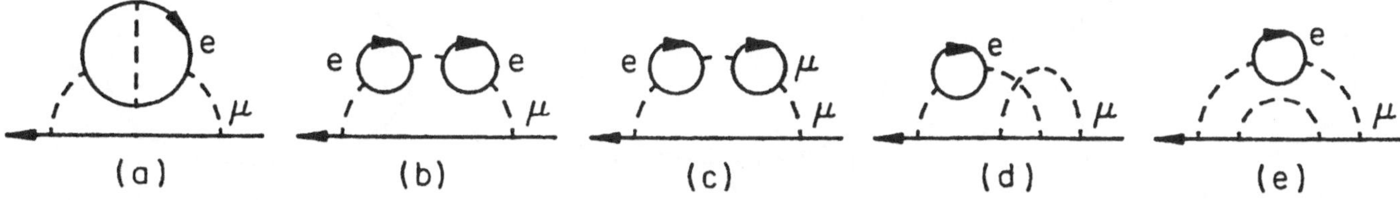

Fig. 2. Sixth-order muon vertices obtained by inserting electron vacuum-polarization loops in the photon lines of second- and fourth-order muon vertices. The number of vertex diagrams represented by the diagrams (a), (b), (c), (d), and (e) are 3, 1, 2, 6, and 6, respectively.

$$\begin{aligned}
A_2^{(6)}[\text{Fig.2(a)}] =\ & 1.493\ 46 \\
& + \mathcal{O}\left(\left(\frac{m_e}{m_\mu}\right)^3 \ln^2 \frac{m_\mu}{m_e}\right), \\
A_2^{(6)}[\text{Fig.2(b)}] =\ & 2.718\ 57 \\
& + \mathcal{O}\left(\left(\frac{m_e}{m_\mu}\right)^3 \ln^2 \frac{m_\mu}{m_e}\right), \\
A_2^{(6)}[\text{Fig.2(c)}] =\ & 0.100\ 519 \\
& + \mathcal{O}\left(\left(\frac{m_e}{m_\mu}\right)^3\right), \\
A_2^{(6)}[\text{Figs.2(d, e)}] =\ & -2.392\ 396 \\
& + \mathcal{O}\left(\left(\frac{m_e}{m_\mu}\right)^3 \ln \frac{m_\mu}{m_e}\right).
\end{aligned} \tag{16}$$

The result (16) is very close to the numerical result (14), leaving no room for doubt about the correctness of both results[2].

By far the most important sixth-order contribution numerically, however, comes from six diagrams containing light-by-light scattering subdiagrams represented by Fig. 3 [3]:

$$A_2^{(6)}[\text{Fig.3}] = 20.947\ 1(29). \tag{17}$$

From (14) and (17) we obtain

$$A_2^{(6)}(m_\mu/m_e) = 22.867\ 6(30). \tag{18}$$

The only other non-negligible sixth-order contributions are [3,13]

$$\begin{aligned}
A_2^{(6)}(m_\mu/m_\tau) &= 6.9 \times 10^{-5}, \\
A_3^{(6)}(m_\mu/m_e, m_\mu/m_\tau) &= 5.24 \times 10^{-4}.
\end{aligned} \tag{19}$$

The eighth-order term $A_2^{(8)}(m_\mu/m_e)$ has contributions from 469 Feynman diagrams. As was reported previously [3], numerical evaluation of these integrals leads to

$$A_2^{(8)}(m_\mu/m_e) = 126.92(41), \tag{20}$$

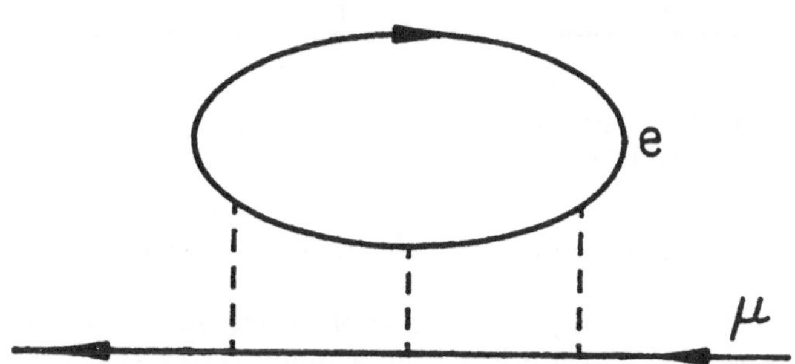

Fig. 3. Sixth-order muon vertex obtained by insertion of an electron loop light-by-light scattering subdiagram. This diagram represents six vertex diagrams in which the external magnetic field vertex is inserted in the electron lines in all possible ways.

where the dominant contribution comes from the diagrams of Fig. 4 with $(l_1, l_2) = (e, e)$.

For diagrams containing vacuum-polarization subdiagrams consisting of only one electron loop, the leading $\ln(m_\mu/m_e)$ terms as well as mass-independent terms may be determined from renormalization group considerations [15,16]. Meanwhile, only logarithmic terms were known for diagrams whose vacuum-polarization subdiagram contains further electron-loop subdiagrams. Recently, however, it has been shown that mass-independent terms of such diagrams may also be determined by a renormalization group approach [17]. Applied to the diagrams of Fig. 5 (with $(l_1, l_2) = (e, e)$), this method yields

$$\begin{aligned}
A_2^{(8)}[\text{Fig.5(e, e)}] &= \frac{1}{12}\ln^2\left(\frac{m_\mu}{m_e}\right) \\
&\quad + \frac{1}{3}(\zeta(3) - 2)\ln\frac{m_\mu}{m_e} + \frac{1531}{1728} \\
&\quad + \frac{5}{12}\zeta(2) - \frac{1025}{1152}\zeta(3) + \dots \\
&= 1.452\ 570 \dots + \mathcal{O}\left(\frac{m_e}{m_\mu}\right).
\end{aligned} \tag{21}$$

This is in good agreement with the numerically evaluated result in [3], or the more accurate result

$$A_2^{(8)}[\text{Fig.5(e, e)}] = 1.440\ 62(10) \tag{22}$$

newly evaluated by VEGAS for this paper, where the number of iterations and the number of sampling points

[2] It is not known to me why the remainders are not as large as $(m_e/m_\mu)^3 \ln^3(m_\mu/m_e)$.

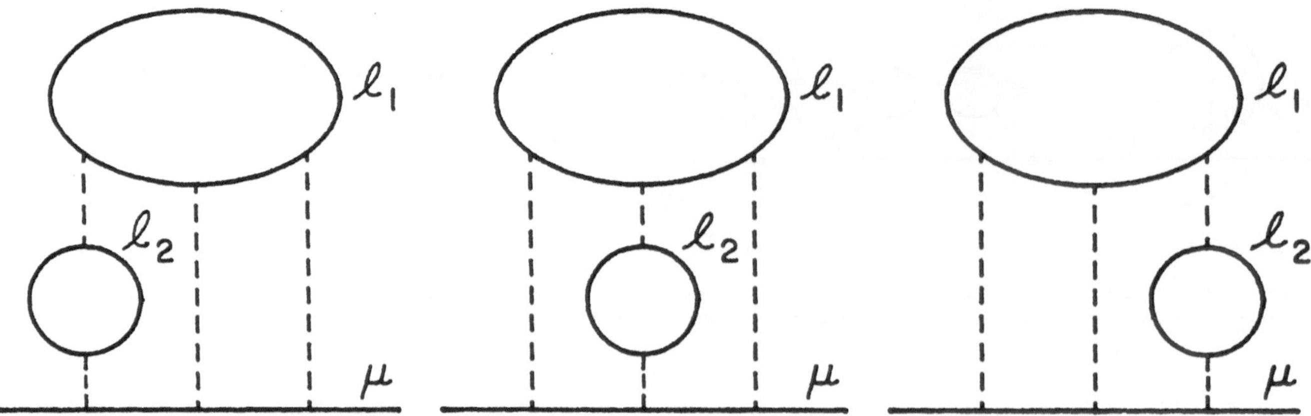

Fig. 4. Muon vertex diagrams obtained by inserting a second-order electron vacuum polarization loop in the diagrams of Fig.3, where $(l_1, l_2) = (e, e)$.

per iteration are (40, 6×10^7) for both diagrams of Fig. 5.

The only other nonnegligible contribution to the eighth-order term arises from a muon vertex that contains an electron light-by-light scattering subdiagram and a tauon vacuum-plarization loop and another in which the role of electron and tauon are interchanged. (See Fig. 4 with $(l_1, l_2) = (e, \tau), (\tau, e)$). Their values have been evaluated numerically. Their sum is given by [3]

$$A_3^{(8)}(m_\mu/m_e, m_\mu/m_\tau) = 0.079(3). \qquad (23)$$

The tenth-order contribution has also been estimated. The result is [3]

$$A_2^{(10)}(m_\mu/m_e) = 570(140). \qquad (24)$$

Collecting the results (11), (12), (13), (18), (20), (23), and (24), we obtain

$$a_\mu(\mathrm{QED}) = 1\ 165\ 846\ 951(44)(28) \times 10^{-12}. \qquad (25)$$

This result is in good agreement with the value quoted in Ref. 3. Note, however, that some of the contributing terms have undergone somewhat larger changes.

3 The hadronic contribution

The sensitivity of the muon anomaly to the hadronic structure of the photon was first pointed out by Bouchiat and Michel [18] and Durand [19], who showed the dramatic enhancement effect of low-lying resonances such as the ρ resonance. There are several recent re-evaluations of hadronic vacuum-polarization contribution to the second-order QED vertex diagram shown in Fig. 6(a):

$$\begin{aligned} a_\mu(\mathrm{Fig.6(a)}) &= 7068(59)(164) \times 10^{-11}, && \text{(Ref. 20)} \\ &= 7100(105)(49) \times 10^{-11}, && \text{(Ref. 21)} \\ &= 684(11) \times 10^{-10}, && \text{(Ref. 22)} \quad (26) \\ &= 7052(60)(46) \times 10^{-11}, && \text{(Ref. 23)} \\ &= 7048(105)(46) \times 10^{-11}. && \text{(Ref. 23)} \end{aligned}$$

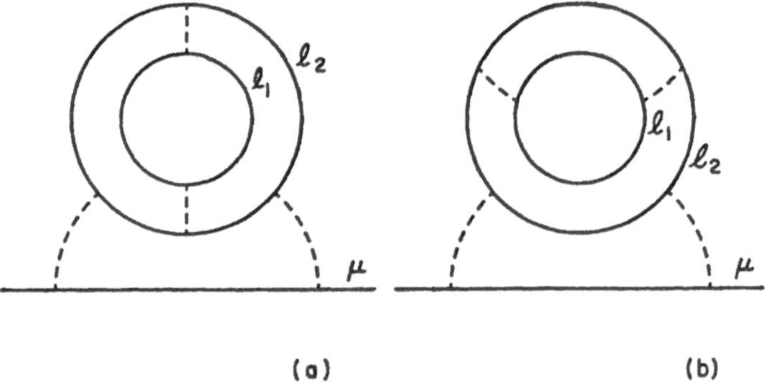

Fig. 5. Muon vertex diagrams containing vacuum-polarization diagrams which have two electron loops. $(l_1, l_2) = (e, e)$.

The agreement between these estimates seems to be good. However, it is somewhat misleading. It is a consequence of the fact that they are mostly dealing with the same set of data on $R = \sigma_{total}(e^+e^- \to \mathrm{hadrons})/\sigma_{total}(e^+e^- \to \mu^+\mu^-)$.

What demands our close attention is the considerable difference in error estimates among different evaluations in (26). Although it is partly due to different definitions of errors, it reflects primarily the different treatments of systematic errors, which are only poorly understood at best and very difficult to evaluate from the available information. The difficulty of dealing with the systematic error was previously noted in footnote *a* to Table III of Ref. 20.

The first and second errors in the first two lines of (26) are statistical and systematic, respectively. The first errors in the last two lines of (26) are combinations of statistical and systematic errors with varying degree of optimism, while the second errors are referred to as model errors. Although the evaluation of the last two lines of (26) relies on a more elaborate treatment of analytic structure of the pion charge from factor and utilizes some data not available earlier, the apparent improvement is mostly a result of rather optimistic estimates of systematic errors.

Very recently, another estimate of a_μ(Fig.6(a)) has been made [24]:

$$a_\mu(\text{Fig.6(a)}) = 7244.7(66.3)(258.7) \times 10^{-11}. \tag{27}$$

This evaluation emphasizes the presence of large systematic errors in the multi-hadron production cross section in the energy range of 1 to 3 GeV.

Clearly, a real progress in the evaluation of the hadronic contribution must wait for better measurements of R, the ratio of $\sigma_{total}(e^+e^- \to \text{hadrons})$ and $\sigma_{total}(e^+e^- \to \mu^+\mu^-)$. The most important contribution of a_μ(Fig.6(a)) comes from the energy range below 1 GeV, which is dominated by the ρ resonance. The challenge in this region is to improve the already-well-measured R even further. New measurements of R at VEPP-2M at Novosibirsk are expected to contribute significantly to this improvement. However, the most promising method in reducing the error in this energy range will be along the line of CERN NA7 experiment in which stationary e^- targets are bombarded with an e^+ beam of several hundred GeV. A preliminary but very encouraging result [25] of this approach has been incorporated in (26). Further measurement along this approach is urgently needed. A comparable precision may also be achievable by another scheme using a uranium-liquid argon calorimeter as a detector [26]. This method might work for multi-pion production measurements, too. Use of asymmetric collider has also been explored [27].

The second major source of systematic error in a_μ(Fig.6(a)) is of different nature. It is due to the relatively poor quality of the current experimental data in the $1-3$ GeV range. (In fact, how to handle systematic errors in this region is the cause of difference between (26) and (27).) The large systematic error there reflects the difficulty of measuring exclusive multi-hadron production cross sections. Computing R as a sum over exclusive channels results in systematic errors of about 20 percent of their contribution to a_μ. Here, what is needed is a brand new experiment in which $\sigma_{total}(e^+e^- \to\text{hadrons})$ is measured inclusively rather than summing up the contributions of exclusive channels. Currently, the only collider available for such a measurement is VEPP-2M, which is designed to cover energies up to 1.4 GeV. In view of the fact that there is no other suitable e^+e^- collider operating or being constructed in the higher energy range, it is very important to stretch the energy of VEPP-2M upwards as much as possible.

There are also small contributions of the higher-order hadronic terms [20] arising from the diagrams of Figs. 6 (b), (c), and (d):

$$a_\mu(\text{Figs.6(b)} - \text{(d)}) = -90(5) \times 10^{-11}. \tag{28}$$

Theoretically most difficult to evaluate is the contribution of hadronic light-by-light scattering contribution shown in Fig. 7. No way has been found to relate this contribution to some observable hadronic amplitude or cross section. The best estimate thus far, obtained by saturating low-lying hadronic states with mesons, is [20]

$$a_\mu(\text{Fig.7}) = 49(5) \times 10^{-11}. \tag{29}$$

In view of the rather fragile theoretical basis of this calculation, it is highly desirable to find a better and more reliable estimate[3].

From (28), (29), and the first line of (26), one finds

$$a_\mu(\text{had}) = 7\,027(175) \times 10^{-11}, \tag{30}$$

as the best current estimate of the hadronic contribution.

4 The weak interaction contribution

As is well known, the one-loop weak correction to a_μ in the standard model consists of three parts

$$a_\mu^{weak}(\text{one}-\text{loop}) = a_\mu^W + a_\mu^Z + a_\mu^H \tag{31}$$

where W, Z, and H refer to W, Z, and Higgs bosons, respectively. Each contribution is separately finite and given by [29]

$$\begin{aligned} a_\mu^W &= \frac{10}{3}\frac{G_F m_\mu^2}{8\sqrt{2}\pi^2} + \mathcal{O}\left(\frac{m_\mu^4}{m_W^4}\right), \\ a_\mu^Z &= -\left(\frac{5}{3} - \frac{(3-4\cos^2\theta_W)^2}{3}\right)\frac{G_F m_\mu^2}{8\sqrt{2}\pi^2} \\ &\qquad + \mathcal{O}\left(\frac{m_\mu^4}{m_W^4}\right), \\ a_\mu^H &= 3F\left(\frac{m_H^2}{m_\mu^2}\right)\frac{G_F m_\mu^2}{8\sqrt{2}\pi^2}, \end{aligned} \tag{32}$$

where $G_F = 1.166\ 37 \times 10^{-5}\text{GeV}^{-2}, \sin^2\theta_W \simeq 0.233$ and F is a known function. The current experimental information on the Higgs mass indicates that the a_μ^H is completely negligible [30]. Thus we obtain [3]

$$a_\mu^{weak}(\text{one}-\text{loop}) = 195(1) \times 10^{-11}. \tag{33}$$

Recently, it has been pointed out that the $\gamma\gamma Z$ electron triangle anomaly diagrams induce a relatively large two-loop contribution [4]

$$-\frac{3\alpha^2}{8\pi^2\sin^2\theta_W}\left(\frac{m_\mu^2}{m_W^2}\right)\ln\left(\frac{m_Z}{m_\mu}\right) \simeq -10 \times 10^{-11}$$

to the muon anomalous moment. This effect will be compensated in part by analogous u and d quark triangle diagrams which cancel the short distance anomaly. Summing over the three generations of leptons and quarks

[3] Ref. 28 gives a lower bound to a_μ(Fig.7) which is larger than (29). However, their estimate is based on a perturbative quark-loop formula for a_μ(Fig.7) which may not be consistent with their nonperturbative determination of quark mass. A nonperturbative analysis of the hadronic light-by-light scattering amplitude will be needed to resolvs this issue.

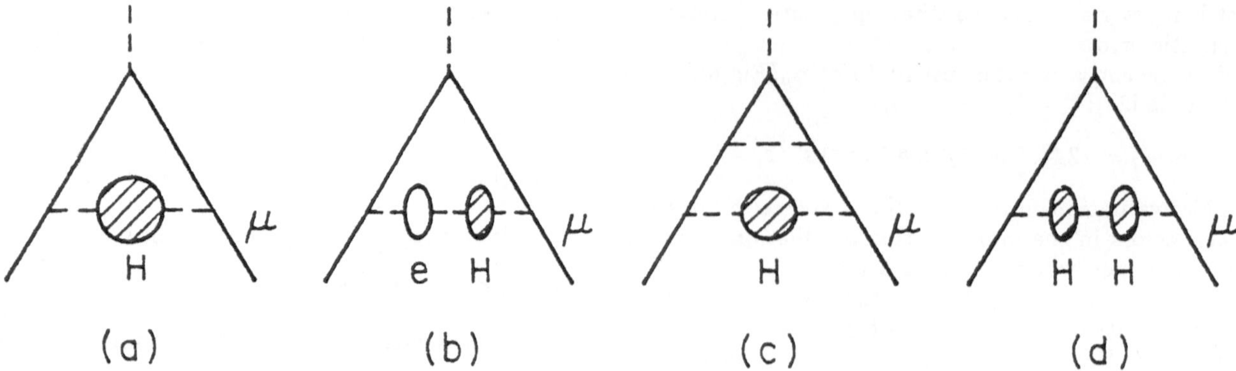

Fig. 6. Various hadronic vacuum-polarization contributions to a_μ.

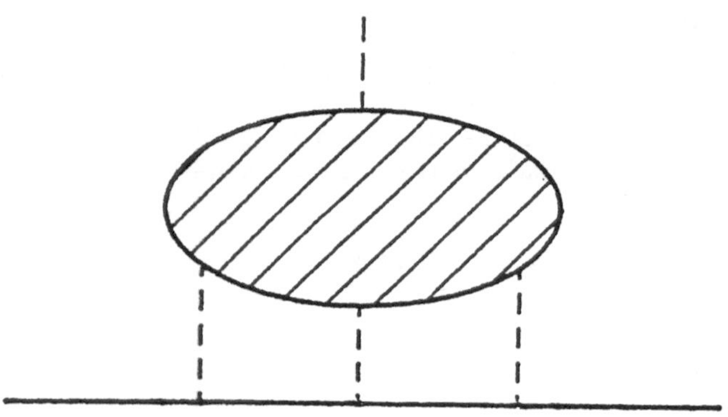

Fig. 7. Hadronic light-by-light scattering contribution to a_μ.

might bring the total contribution of the fermion triangle anomaly diagrams to $\sim 1 \times 10^{-10}$. There are in addition two-loop weak contributions of relative order $\alpha/\pi \sin^2\theta_W \simeq 0.01$ with respect to the one-loop effects in (32). They could potentially add up to $\mathcal{O}(10 \times 10^{-11})$. At present we have therefore no choice but to accept

$$a_\mu^{weak} = 195(10) \times 10^{-11} \tag{34}$$

as the best estimate of the electroweak contribution to a_μ [3]. It is important to complete the two-loop calculation and remove this uncertainty before the new experimental result becomes available.

5 "New physics" contributions

One of the main purposes of the new muon $g-2$ experiment is to obtain strict constraints on "new physics" possibilities, such as extra gauge bosons, additional Higgs scalars, supersymmetry, and compositeness of leptons. Provided that a_μ is measured to the precision of 40×10^{-11} and that theory of a_μ(had) is improved similarly, one can obtain much more stringent bounds than those available at present on "new physics" mass scales and couplings. For a review of these topics see Ref. 3. Additional work based on an extension of supersymmetric standard model can be found in Ref. 31.

Acknowledgement. I should like to thank E. Remiddi and M. A. Samuel for communicating their results prior to publication. This work is supported in part by the U. S. National Science Foundation. Part of the numerical work was conducted at the Cornell National Supercomputing Facility, which receives major funding from the U. S. National Science Foundation and the IBM Corporation, with additional support from New York State and members of the Corporate Research Institute.

References

1. J. Bailey *et al.*, Phys. Lett. B **68** (1977) 191.
2. F. J. M. Farley, this volume and F. J. M. Farley and E. Picasso, in "Quantum Electrodynamics", ed. by T. Kinoshita (World Scientific, Singapore,1990), pp. 479 – 559.
3. T. Kinoshita and W. J. Marciano, in "Quantum Electrodynamics", ed. by T. Kinoshita (World Scientific, Singapore,1990), pp. 419 – 478.
4. E. A. Kuraev, T. V. Kukhto, and A. Schiller, Yad. Fiz. **51** (1990) 1631 [Sov. J. Nucl. Phys. **51** (1990) 1031].
5. G. S. Abrahams *et al.*, Phys. Rev. Lett **63** (1989) 2173; L3 Collab., B. Adeva *et al.*, Phys. Lett. B **231** (1989) 509; ALEPH Collab., D. Decamp *et al.*, Phys. Lett. B **231** (1989) 519; OPAL Collab., M. Z. Akrawy *et al.*, Phys. Lett. B **231** (1989) 530; DELPHI Collab., P. Aarnio *et al.*, Phys. Lett. B **231** (1989) 539.
6. T. Kinoshita, in "Quantum Electrodynamics", ed. by T. Kinoshita (World Scientific, Singapore,1990), pp. 218 – 321.
7. S. Laporta and E. Remiddi, Phys.Lett. B **265** (1991) 182.
8. T. Engelmann and M. J. Levine (unpublished), quoted in M. J. Levine, H. Y. Park, and R. Z. Roskies, Phys. Rev. D **25** (1982) 2205.
9. T. Kinoshita, Phys. Rev. Lett. **61** (1988) 2898.
10. G. P. Lepage, J. Comput. Phys. **27** (1978) 192.
11. M. E. Cage et al., IEEE Trans. Instrum. Meas. **38** (1989) 284.
12. H. H. Elend, Phys. Lett. **20** (1966) 682; **21** (1966) 720.
13. M. A. Samuel and G. Li, Phys.Rev. **D44** (1991) 3935.
14. E. R. Cohen and B. N. Taylor, Rev. Mod. Phys. **59** (1987) 1121.
15. T. Kinoshita, Nuovo Cimento **51B** (1967) 140.
16. B. E. Lautrup and E. de Rafael, Nucl. Phys. **B70** (1974) 317.
17. T. Kinoshita, H. Kawai, and Y. Okamoto, Phys. Lett. B **254** (1991) 235; R. N. Faustov, A. L. Kataev, S. A. Larin, and V. V. Starshenko, Phys. Lett. B **254** (1991) 241; H. Kawai, T. Kinoshita, and Y. Okamoto, Phys. Lett. B **260** (1991) 193.
18. C. Bouchiat and L. Michel, J. Phys. Radium **22** (1961) 121.
19. L. Durand, III., Phys. Rev **128** (1962) 441.

20. T. Kinoshita, B. Nizic, and Y. Okamoto, Phys. Rev. D **31** (1985) 2108.
21. J. A. Casas, C. Lopez, and F. J. Yndurain, Phys. Rev. D **32** (1985) 736.
22. L. M. Kurdadze *et al.*, Yad. Fiz. **40** (1984) 451 [Sov. J. Nucl. Phys. **40** (1984) 286].
23. L. Martinovic and S. Dubnicka, Phys. Rev. D **42** (1990) 884.
24. F. Jegerlehner, private communication to V. W. Hughes.
25. S. R. Amendolia *et al.*, Phys. Lett. B **138** (1984) 454.
26. W. Willis, private communication.
27. M. Greco, Frascati Report No. LNF-88/24(P), 1988 (unpublished).
28. J. Cole, G. Penso, and C. Verzegnassi, Trieste preprint ISAS 19/85/EP, 1985, unpublished.
29. R. Jackiw and S. Weinberg, Phys. Rev. D **5** (1972) 2396; I. Bars and M. Yoshimura, Phys. Rev. D **6** (1972) 374; K. Fujikawa, B. W. Lee, and A. I. Sanda, Phys. Rev. D **6** (1972) 2923; W. A. Bardeen, R. Gastmans, and B. E. Lautrup, Nucl. Phys. B46 (1972) 319.
30. J. Gunion, H. Haber, G. Kane, and S. Dawson, "Higgs Hunter's Guide", (Addison-Wesley, New York, 1990).
31. R. M. Francis, M. Frank, and C. S. Kalman, Phys. Rev. D **43** (1991) 2369.

This article was processed using Springer-Verlag TEX Z.Physik C macro package 1991
and the AMS fonts, developed by the American Mathematical Society.

Z. Phys. C - Particles and Fields 56, S88–S96 (1992)

Zeitschrift für Physik C Particles and Fields

The Cern (g-2) measurements

F.J.M. Farley

Yale University, New Haven, Ct 06511, USA

27-August-1991

Abstract. Measurements of the anomalous magnetic moment of the muon are reviewed. The emphasis is put on the first one of a series of three CERN experiments which led to an experimental value of $a = 1\,165\,922.9\,(8.5) \cdot 10^{-9}$ ($\pm 7.2 ppm$), and which compares well with the present theoretical result $a = 1\,165\,919.2\,(1.8) \cdot 10^{-9}$ ($\pm 1.5 ppm$).

1 Introduction

This paper reviews the three muon (g-2) measurements carried out at CERN from 1958 to 1975. More emphasis is placed here on the first experiment because it is less well known. Many other reviews are available[1].

First recall the principle of the experiment in non-relativistic approximation. In a uniform magnetic field B the momentum turns at the cyclotron frequency $\omega_c = eB/mc$, while the spin precession frequency is $\omega_s = g(e/2mc)B = (1+a)\omega_c$, where $a = (g-2)/2$ is the anomalous magnetic moment to be measured. So if polarized muons are injected into the field the spin turns slightly faster than the momentum vector and the polarization angle of the spin relative to the direction of motion turns at the difference frequency

$$\omega_a = \omega_s - \omega_c = a(eB/mc) \tag{1}$$

As a result the anomaly a can be measured directly. Note that eq.(1) is valid also at relativistic energies, though to prove this would take a whole lecture on its own [2]. The (g-2) precession is not slowed down by time dilation!

In practice one measures ω_a relative to the proton precession frequency ω_p in the same magnetic field. With a little algebra on can derive

$$a = R/(\lambda - R) \tag{2}$$

where $R = \omega_a/\omega_p$ is the quantity measured by this experiment, and $\lambda = \omega_s/\omega_p = \mu_\mu/\mu_p$ is the ratio of muon to proton magnetic moments. λ is known from hyperfine structure measurements [3] in muonium and from muon precession at rest in bromine [4]. The values are in good agreement and the current mean value adopted by Cohen and Taylor [5] is $\lambda = 3.18334547(95)(0.3ppm)$.

The problem then is to store polarized muons in a magnetic field so that they make many turns, and then to measure the change in spin angle as a function of time.

2 Cyclotron experiment

In the first experiment the polarized muons came from the cyclotron and were stored in a 6m long magnet. To get the particles into the magnet is not difficult; they are passed through a beryllium degrader as shown in fig. 1, lose momentum and turn in a smaller circle completely inside the field. To make them miss the degrader after one turn, the field has a gradient in the y-direction. This makes the orbit 'walk' to the right (x-direction) by 2 cm per turn. Fig. 2 shows the first evidence of muons trapped for many turns in a small experimental magnet.

Getting the muons out of the field, so that they can decay in a field-free region and indicate their spin direction, is more difficult; many doubted that this could be achieved because in slowly varying fields the flux through the drifting orbit is an invariant of the motion. The trick is to increase the gradient adiabatically so that the step size is about 11 cm per turn (for an orbit of radius 20 cm). The particles then arrive at the end of magnet unexpectedly, forget the invariant, and leave the field.

In the overall design [5], shown in fig. 3, there is another adiabatic transition to reduce the step size to 4 mm/turn in the central part of the magnet, thus increasing the storage time. The rather complex shim patterns needed to achieve the correct fields are shown in fig. 4. (Note that the vertical scale is magnified).

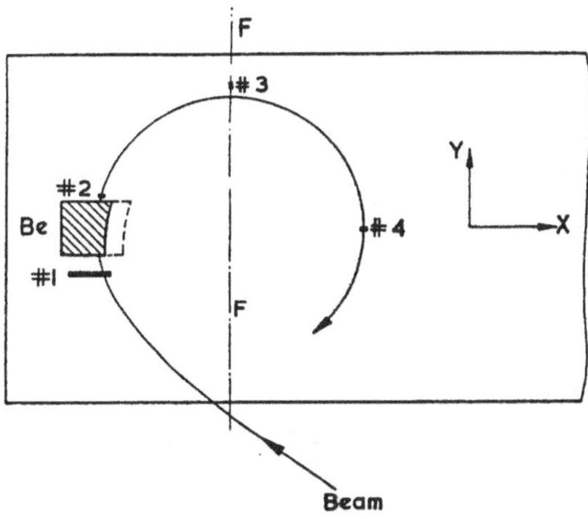

Fig. 1. Injecting muons. The first turn is located by coincidence 1234 with the beryllium moved forward (dotted lines) to symmetrize the scattering.

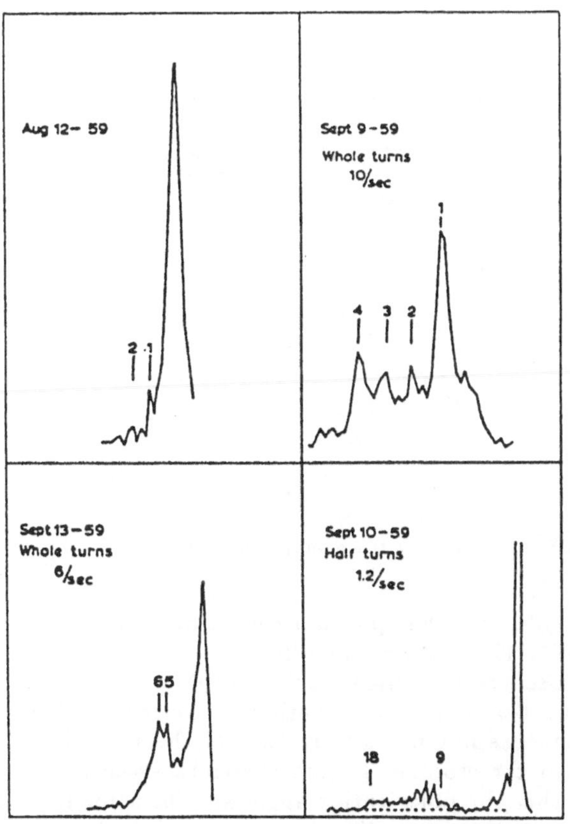

Fig. 2. Time of flight data showing muons turning inside a small experimental magnet.

To avoid losing the particles by sideways drifting there should be no gradient in the x-direction. This means that the flux through an orbit drifting along the centre line of the magnet must be constant. This was checked by running a 40 cm diameter search coil along the x-axis.

Fig. 5 shows the tracks of muons emerging from the end of the magnet, as measured with a floating wire system. The 'polarization analyser', seen in fig. 4, was positioned so that the majority of the ejected muons stopped in the methylene iodide target. Backward and forward decay electrons were then detected in counter telescopes 66' and 77'. When a stopping muon was signaled, a short pulse of vertical magnetic field was applied to the target, to flip the spin through $\pm 90°$ in alternate runs. Independent of solid angle factors, one can then calculate for each telescope an asymmetry

$$A = (n_+ - n_-)/(n_+ + n_-) = A sin \Theta_s \tag{3}$$

Here n_+, n_- are the decay counts for flipping through $+90°$, $-90°$, and Θ_s is the spin angle relative to the polarization analyzer. Plotting A as a function of the time the muon has spent in the field (the 'storage time') gave fig. 6 for the first (1.7 %) measurement [6], later improved to fig. 7 for the 0.4 % measurement [7].

From this data one can read off the (g-2) frequency ω_a, and using Eqs. (1) and (2) calculate some result for $a = (g-2)/2$. To get the correct result required an array of supplementary measurements [7].

In Eq. (3) the spin angle Θ_s relative to the polarization analyser has several components,

$$\Theta_S = a(e/mc)\bar{B}t + \Theta_P + \Theta_S^o - \Theta_P^o. \tag{4}$$

Here a is the quantity to be determined, $\bar{B}$ is the mean magnetic field seen by the muons, t is the time spent in the magnet ('storage time'), Θ_P is the directions of arrival at the polarization analyser, while Θ_S^o and Θ_P^o are the corresponding spin and momentum directions for the beam before it entered the storage magnet. Θ_S^o and Θ_P^o were measured by putting the same polarization analyser into the incoming magnet. The parameters $\bar{B}$, Θ_P, and the initial polarization angle $\Theta_S^o - \Theta_P^o$ can all be functions of the storage time t, and of the orbit radius (which correlates with the muon range).

Inevitably the stored muons had a range of orbit radii, and the orbit centres were not all on the magnet centre-line. Fig. 8 shows the field vs transverse coordinate y measured in the centre of the storage region. Muons with different orbit radii and different orbit centres see a different mean gradient, and so advance at different rates and have different storage times. They also have different mean fields. The orbits corresponding to each one-microsecond range of t were located on the y-axis by moving brass flaps into the storage region until they cut the beam. The mean field could then be calculated for each storage time. Because the orbit follows contour lines of the magnetic field, the value determined at one point is valid for the whole magnet.

The direction of arrival at the polarization analyser was obtained with a 'venetian blind' counter. This consisted of an array of parallel sheets of scintillator, adjustable in angle, which vetoed the event if a muon touched one of the sheets. By rotating the sheets one

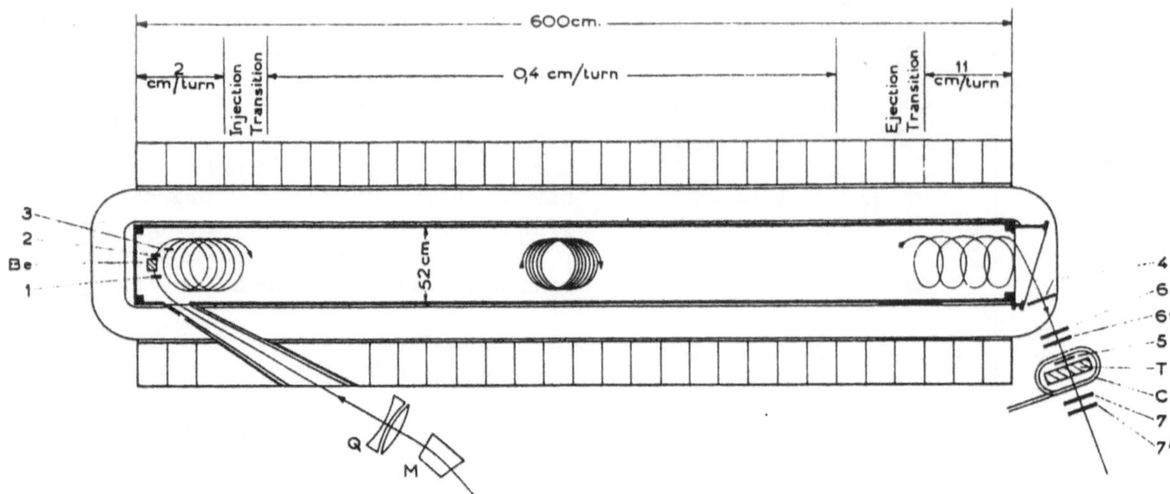

Fig. 3. 6m bending magnet used for storing muons for up to 2000 turns. The field gradient makes the orbit walk to the right. At the end a very large gradient is used to eject the muons so that they stop in the polarization analyser.

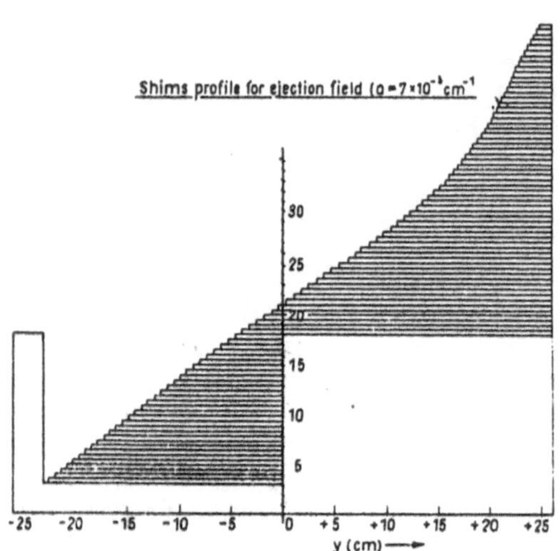

Fig. 4. Shim profiles used for the ejection region. (Expanded vertical scale).

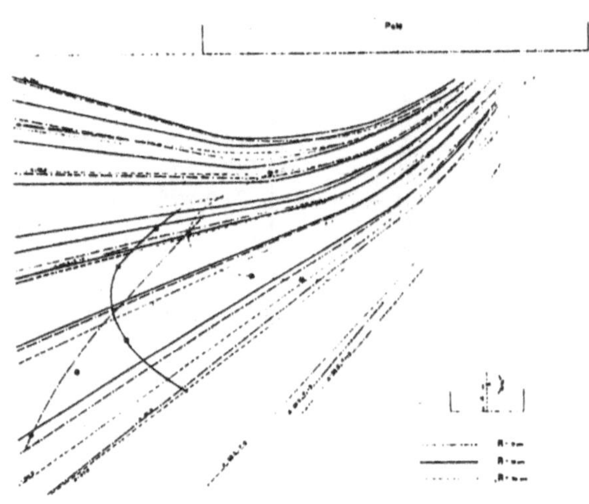

Fig. 5. Floating wire trace of ejected muons.

could select muons with a well defined direction of arrival. Fig. 9 shows the observed distribution, which was moreover found to be independent of storage time.

With the polarization analyser in the input beam, the polarization angle Θ_S^0 - Θ_P^0 was found to change rapidly with increasing muon range. The large slope (10° per cm range in carbon) implied that the mean range of the muons selected by the storage system would need to be very well measured. This spin-range correlation had its origin in the left-right asymmetry of π and μ tracks in the magnetic field of the cyclotron. No such asymmetry should exist in the vertical plane; therefore the effect could be reduced by passing the muon beam through a solenoid, fig. 10, which rotated the transverse components of spin 90° around the beam axis, effectively interchanging horizontal and vertical components. With the solenoid, the spin-range correlation was reduced to $-(19 \pm 12)\ mR/cm$ of carbon.

Because the magnet only accepts one per cent of the input beam, the polarization measured for the whole beam was not necessarily that of the partial beam selected for storage. It was therefore essential to check whether the polarization angle was the same in different parts of the beam. The beam traversing the cylindrical pipe through the cyclotron shield wall was subdivided by placing brass stops at its entrance and exit; each stop had only one quadrant open, as shown in fig. 10. The four possible positions of each open quadrant were combined to subdivide the beam into 16 sub-beams. Strong variations (of order 20°) were found in the polarization angle from cell to cell.

To eliminate this effect the beam structure was scrambled by placing a 1 mm lead scattering foil close to

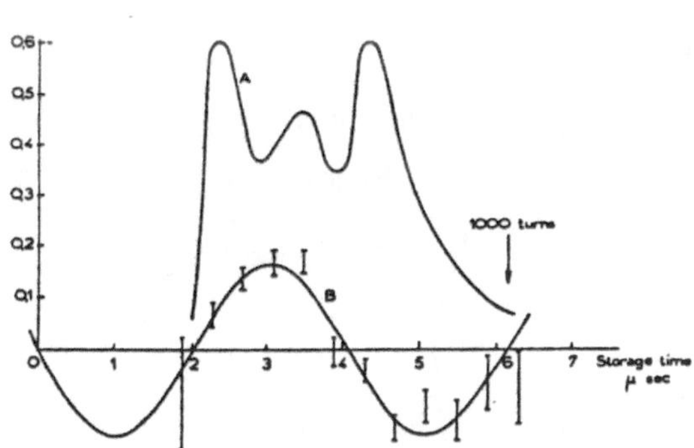

Fig. 6. Asymmetry A vs storage time t, combined data for the 1.7% experiment.

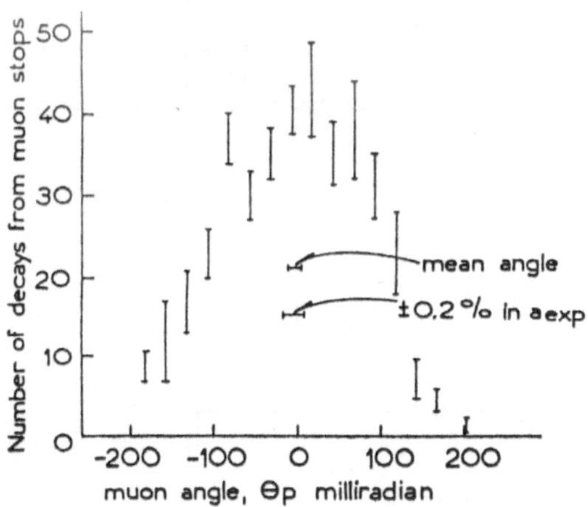

Fig. 9. Arrival direction of ejected muons at the polarization analyser.

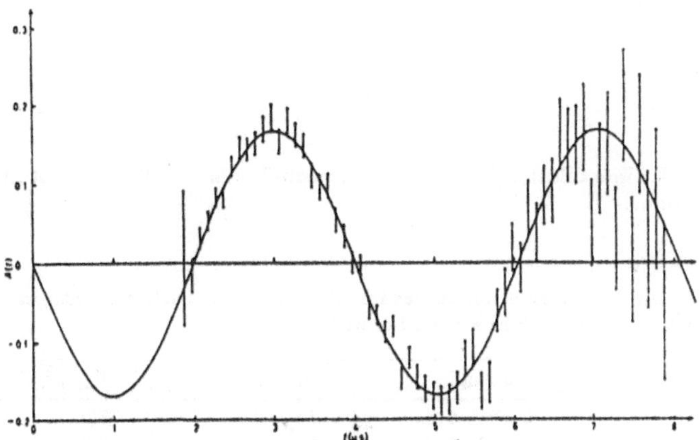

Fig. 7. Asymmetry A vs storage time t, combined data for the 0.4% experiment.

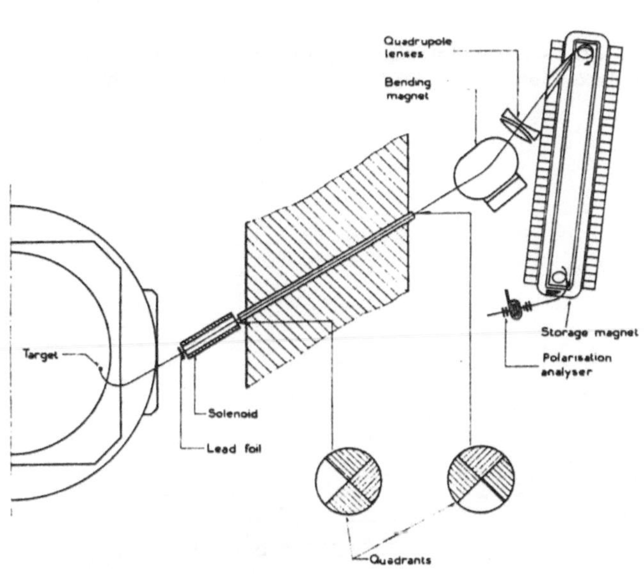

Fig. 10. Overall layout showing cyclotron target 1, solenoid 2, lead foil 3, polarization analyser 4, storage magnet 5, bending magnet 6, quadrupoles 7 and brass stops 8.

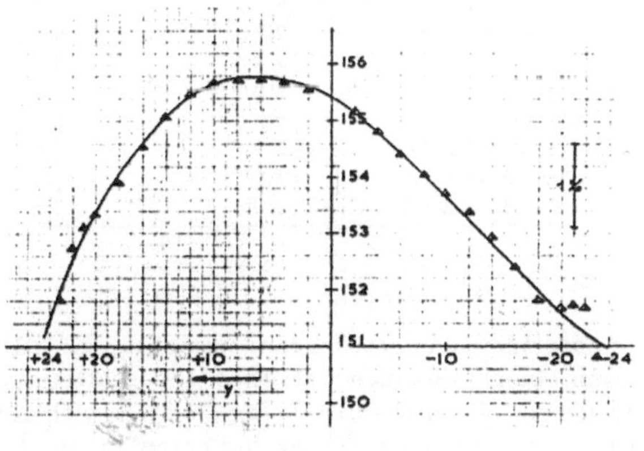

Fig. 8. Magnetic field ve transverse (y) coordinate at centre of magnet.

the cyclotron. Muons which had originally belonged to one cell were thus distributed over many cells. Table 1 shows the spin angles measured for the 16 cells, together with the fraction each contributed to the input beam, and to the stored beam.

This data was used to compute a small correction to the input polarization to compensate for the selective nature of the storage process.

A further correction arose from asymmetrical left-right scattering in the beryllium degrader at injection: scattering through angle Θ changes the polarization angle by Θ/γ, and left and right do not have equal probabilities of injecting a muon. The scattering angle for the

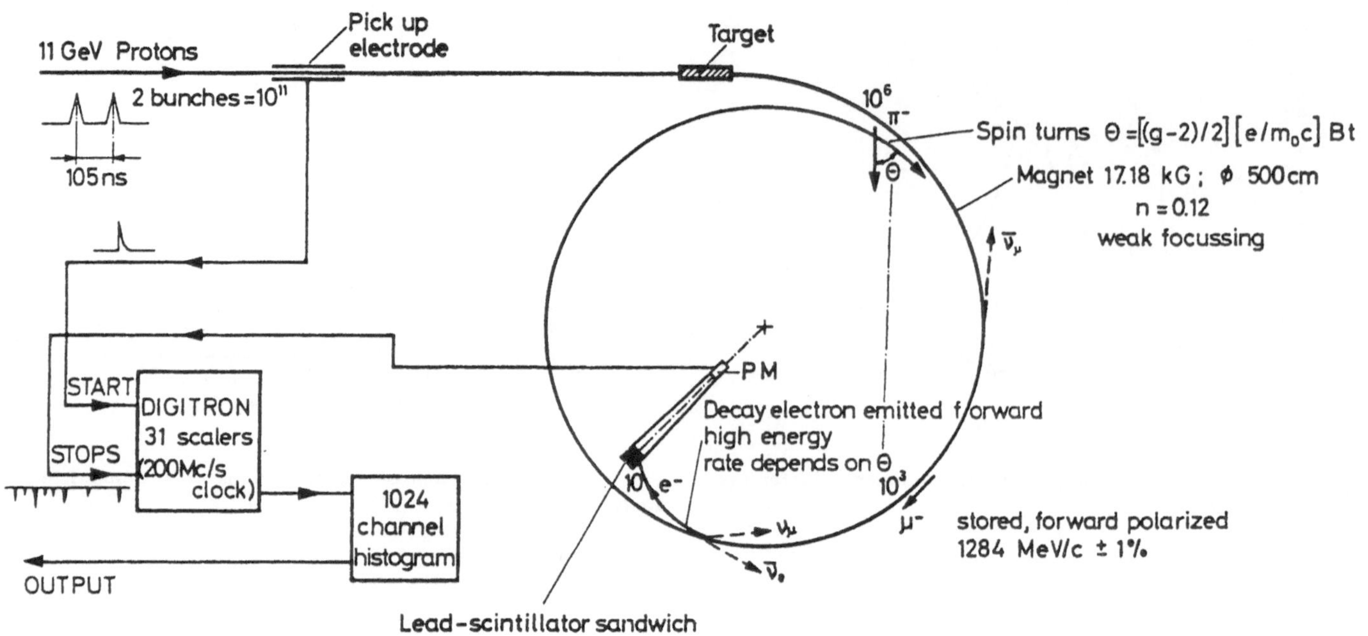

Fig. 12. First muon storage ring, diameter 5m, muon momentum 1.3 GeV, time dilation factor 12. An injected pulse of 10 GeV protons produces pions at the target, which then decay in flight to give stored muons.

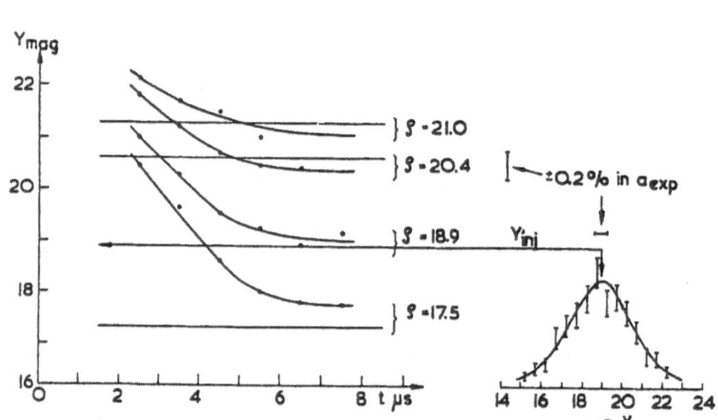

Fig. 11. Horizontal scattering of muons in beryllium at injection, from measurements of the first turn: ·· y-coord for stored muons, – –y-coord for injected (unscattered) muons. On right: scatter distribution of injected muons from scan of # 3 position in fig. 1.

stored particles was related to the y-coordinate of the orbit after one quarter turn, obtained by probing with an anti-coincidence counter. The corresponding position of the unscattered injected beam was obtained with the beryllium block advanced so that the scattering was symmetrized, the first turn being located with several counters in coincidence [7], as indicated in fig. 1. The results, given in fig. 11 as a function of storage time and orbit radius, show that this scattering correction was an important effect.

Table 1. Polarization angles for 16-sub-beams with percentages F in raw beam and in stored beam.

Cell	Spin angle	$F_{beam}[\%]$	$F_{stored}[\%]$
1	-1.64 ± 1.4	11.6	8.7
2	$+2.80 \pm 2.3$	3.1	5.4
3	$+2.00 \pm 2.0$	2.2	2.9
4	-1.14 ± 1.5	15.0	7.9
5	$+1.20 \pm 1.6$	12.3	10.1
6	$+4.46 \pm 1.8$	8.0	9.4
7	$+4.00 \pm 1.2$	4.1	4.7
8	$+1.26 \pm 2.1$	8.1	5.3
9	-0.22 ± 1.6	5.6	6.5
10	$+2.41 \pm 2.6$	2.7	5.3
11	$+4.81 \pm 1.3$	2.2	4.9
12	-2.50 ± 2.5	5.6	7.3
13	$+1.16 \pm 1.3$	6.2	6.6
14	$+1.67 \pm 2.6$	1.1	2.6
15	$+3.46 \pm 1.2$	1.9	3.5
16	$+3.30 \pm 1.0$	9.9	8.5

Including all these effects in fitting Eqs. (3) and (4) to the data, the mean field $\bar{B}$ and the mean phase $\Theta_P + \Theta_S^0 - \Theta_P^0$ was entered separately for each 1 microsecond range of storage time. The fit shown in fig. 7 is therefore

Fig. 13. Photograph of the 5m muon storage ring showing counters being installed.

not an exact sine curve. The result for the anomaly was $a = (1162 \pm 5)10^{-6}$.

After all these precautions it is encouraging to find, in retrospect, that the value was correct. This is apparent, not from the theory, which may not always be an infallible guide (see below), but because it agreed with the later measurements.

3 First muon storage ring

By this time the CERN proton synchrotron was operating and there was a possibility of making a more accurate measurement by using relativistic muons with a dilated life time. Because there is no factor γ in Eq. (1) the (g-2) frequency remains the same, so more cycles of $(g-2)$ precession can be observed before the muons decay. The new experiments were made possible by four miracles of Nature! (One approach to science, perhaps, is first to discover your miracle; then use it for what you want to do).

The first miracle is that it is easy to inject muons into a storage ring. If a burst of relativistic pions pass round the ring, they decay in flight; the forward decay muons have almost exactly the same momentum, and so follow the same orbit: they will usually hit something and be lost. But the muons from almost forward decay have slightly lower momentum; the orbit shrinks and some of them fall onto permanently stored orbits. They remain after the pions and associated background have

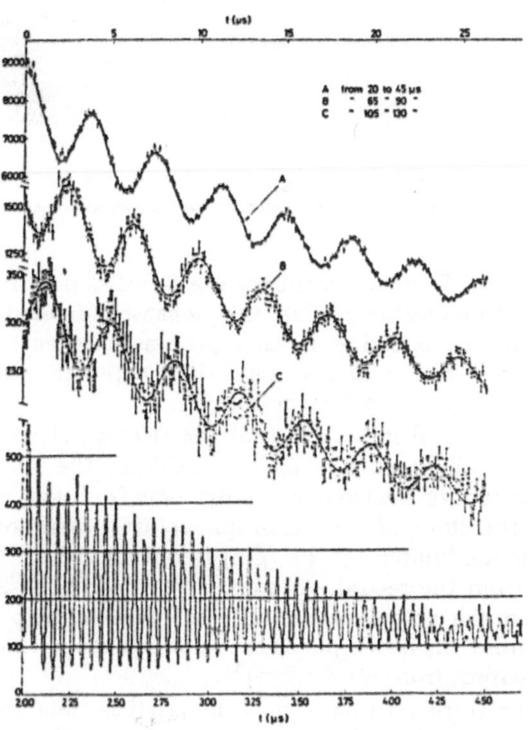

Fig. 14. Decay electron counts vs time after the injected pulse for muon storage ring I. The upper three curves show the g-2 precession out to $130\mu s$ (upper time scale). The lower curve shows the 19 MHz modulation due to the rotation of muon bunch (lower scale).

Fig. 15. Second muon storage ring, diameter 14 m, momentum 3.094 GeV/c, muon life-time $64\mu s$. A pulse of preselected pions of momentum 3.15 GeV/c was inflected and decayed to stored muons during the first run.

died away. The stored muons are forward polarized. In the first muon storage ring, the pions were created inside the ring, by injecting 12 GeV protons onto an internal target. In the second storage ring, a preselected pion beam was inflected for one turn.

The second miracle is that the electrons from muon decay in flight emerge on the inside of the ring. The higher energy electrons can only come from forward decays; therefore, as the muon spin rotates, their counting rate is modulated by the (g-2) frequency, which can be read from the record. As many cycles of the (g-2) precession can now be observed, the complex factors that determine the initial phase need not be known; the phase is apparent from the first cycle.

The third miracle is that the electron detectors can only count when the muons are passing. The injected pulse corresponding to one rf-bunch of the proton synchrotron was only 10 ns long, but the rotation period for the 5m diameter ring was 52 ns. Therefore at early times the electron counts were modulated by the rotation frequency of the muon bunch. By reading off this frequency the mean radius of the muons could be calculated, thus locating their mean position in the weak-focusing gradient field. This was needed to determine the average field seen by the muons. In fact, a Fourier transform of the electron count rate gave the radial distribution of the muons in some detail.

Fig. 12 gives an overview of the experiment [8]. Fig. 13 is a photograph of the ring magnet, with the counters being installed. The primary proton target, used to create the pions, is behind the concrete shielding. The (g-2) precession data out to $130\mu s$ is shown in fig. 14, which also includes the rotation pattern of the muons at early times (lower curve).

The result of this experiment [8] was $a = 116616(31) \cdot 10^{-8}$. Again, in retrospect, this value was correct because it agreed with the next experiment. But it disagreed with the theory by $450 \pm 270 ppm$. This stimulated theorists to re-examine the contribution of the light-by-light scattering diagrams in the α^3-term of the QED expansion. It had previously been postulated that the six diagrams might cancel, giving a very small or zero contribution. Aldins et al [9], however, found that the coefficient of $(\alpha/\pi)^3$ 'should be 18.6!! With this correction the agreement between experiment and theory became good, $240 \pm 270 ppm$.

4 Second muon storage ring

In the first storage ring the uncertainty in muon radius, coming from the observed rotation frequency, was estimated as $\pm 3mm$, and in the weak focusing gradient field this contributed $\pm 160ppm$ to the final error. To improve further it would be necessary to reduce, or eliminate, the dependence of the (g-2) frequency on radius, without losing the vertical focusing. This is possible with a combination of magnetic and electric fields. The forces that hold the muon in its orbit and give focusing for small deviations from equilibrium arise from what appears in the muon rest frame as a pure electric field. But the forces that determine the spin precession appear there as a pure magnetic field! Therefore, in principle, it should be possible to control these effects independently, using an appropriate combination of electric and magnetic fields in the laboratory frame.

In the second CERN storage ring, 14 m diameter, the magnetic field was uniform; vertical focusing was achieved by electric quadrupoles, all of the same sign. The combination of uniform magnetic field with the vertically focusing electric quadrupoles is equivalent, for the particle motion, to a weak focusing magnetic ring. The idea is that the spin motion should depend on the uniform magnetic field, and thus be independent of the muon radius. In general the electric quadrupole field has horizontal components which vary with radius, and also affect the spin motion: so nothing is gained. However at one particular energy, called the 'magic energy', defined by $\gamma = (1 + 1/a)^{1/2} = 29.3$, the horizontal electric field does not affect the (g-2) frequency [10]. The fourth miracle is that this occurs at an experimentally accessible energy 3.096 GeV. This was the principle used in the 14m diameter storage ring [11], shown in fig. 15; the position of the average muon in the storage volume did not then have to be know so precisely, in spite of the large increase in statistical accuracy. A small sub-set of the data, reproduced in fig. 16 shows both the (g-2) precession and the bunch rotation frequency. The muon life time was now dilated to $64\mu s$ and the (g-2) precession was visible as far as $640\mu s$.

The experimental result for $R = \omega_a/\omega_p$, combined with the latest value for λ given above, leads to

$$a = 1\,165\,922.9\,(8.5) \cdot 10^{-9}\ (7.2ppm),$$

compared with the current theoretical value [12] of

$$a = 1\,165\,919.2\,(1.8) \cdot 10^{-9}\ (1.5ppm)$$

implying agreement with theory to $3.2 \pm 7.5ppm$.

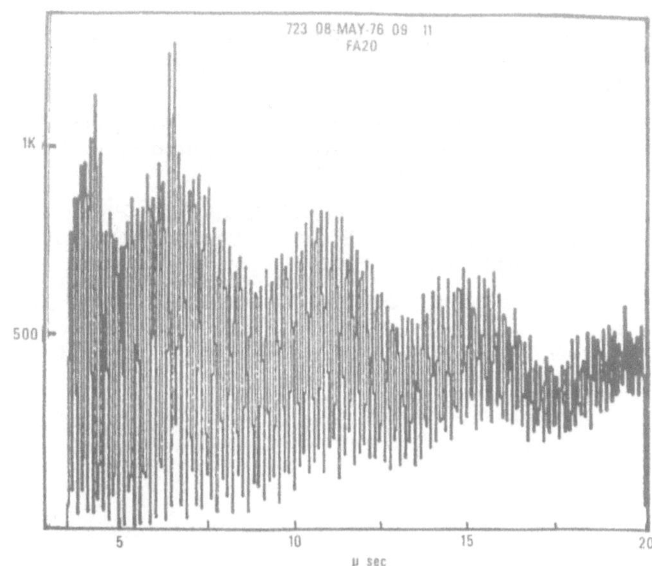

Fig. 16. Decay electrons vs storage time for storage ring II, showing the first cycles of g-2 modulation, and the rotation of the muon bunch. This is a small sub-set of the data.

5 Conclusions

The main result of this series of experiments has been to establish the muon firmly as a heavier version of the electron, obeying the laws of QED for a particle of that mass. Experimentally the muon g-factor is an essential input for calculating the muon mass, and for interpreting the measurements on muonium.

The muon is point-like with a cut-off parameter $\Lambda > 31GeV$. Hadronic vacuum polarization is confirmed to 14 %. No new couplings of the muon are apparent. There can be no undiscovered fermions with mass less than $2m_\mu$.

As a by-product, Einstein's time dilation in a circular orbit at $\gamma = 30$ is correct to 0.1 %.

References

1. Farley F.J.M., and Picasso E. In: Kinoshita T. (ed.) Quantum Electrodynamics. World Scientific, Singapore (1990), p. 479
2. Mendlowitz H., and Case K.M. (1955) Phys. Rev. 97: 33; Bargmann V., Michel L., Telegdi V.L. (1959) Phys. Rev. Lett. 2: 435; Farley F.J.M. (1968) In: Levi M. (ed.) Cargese Lectures in Physics, Gordon and Breach , p. 55; Farley F.J.M. (1975) Contemp. Phys. 16: 413
3. Mariam F.G., Beer W., Bolton P.R., Egan P.O., Gardner C.J., Hughes V.W., Lu D.C., Souder P.A., Orth H., Vetter J., Moser U., zu Putlitz G. (1982) Phys. Rev. Lett. 49: 993
4. Klempt E., Schulze R., Wolf H., Camani M., Gygax F.N., Ruegg W., Schenk A., Schilling H. (1982) Phys. Rev. D25: 652
5. Cohen E.R., and Taylor B.N. (1987) Rev. Mod. Phys. 59: 1121
6. Charpak G., Farley F.J.M., Garwin R.L., Muller T., Sens J.C.,Telegdi V.L., Zichichi A. (1961) Phys. Rev. Lett. 6: 128
7. Charpak G., Farley F.J.M., Garwin R.L., Muller T., Sens J.C., Zichichi A. (1962) Phys. Rev. Lett. 1: 16; Charpak G., Farley F.J.M., Garwin R.L., Muller T., Sens J.C., Zichichi A. (1965) Nuovo Cimento 37: 1241

8. Bailey J., Bartl W., von Bochmann G., Brown R.C.A., Farley F.J.M., Giesch M., Joestlein H., van der Meer S., Picasso E., Williams R.W. (1977) Phys. Lett.A9: 369
9. Aldins J., Kinoshita T., Brodsky S.J. Dufner A.J. (1969) Phys. Rev. Lett. 23: 441;Aldins J., Kinoshita T., Brodsky S.J. Dufner A.J. (1970) Phys. Rev.D1: 2378
10. Bailey J., and Picasso E. (1970) Progr. Nucl. Phys. 12: 43 Combley F., and Picasso E. (1974) Phys. Rep. C14:1 Farley F.J.M. (1975) Contemp. Phys. 16: 413
11. Bailey J., Borer K., Combley F., Drumm H., Eck C., Farley F.J.M., Field J.H., Field J.H., Flegel W., Hattersley P.M., Krienen F., Lange F., Lebee E., McMillan E., Petrucci G., Picasso E., Runolfsson O., von Rueden W., Williams R.W., Wojcicki S. (1979) Nucl. Phys. B150: 1
12. Kinoshita T., and Marciano W.J. (1990) In: Kinoshita T. (ed.) Quantum Electrodynamics. World Scientific, Singapore (1990), p. 419

This article was processed using Springer-Verlag TEX Z.Physik C macro package 1991
and the AMS fonts, developed by the American Mathematical Society.

Z. Phys. C - Particles and Fields 56, S97-S100 (1992)

Zeitschrift für Physik C Particles and Fields

The Electric Dipole Moment of the Muon

W. Bernreuther

Institut für Theoretische Physik, Universität Heidelberg, D-6900 Heidelberg, F.R.G.

15-November-1991

Abstract. The forthcoming experiment [1] on the anomalous magnetic moment of the muon will be able to search also for an electric dipole moment d_μ of the muon with increased sensitivity. A brief survey is given on what various models of CP violation predict for d_μ.

1 Introduction

The experiment [1] being developed at Brookhaven National Laboratory to measure the muon's anomalous magnet moment $a_\mu = (g_\mu - 2)/2$ aims at improving the previous measurement [2] of a_μ by about a factor of 20. The principal motivations for this effort are the weak interaction contribution to a_μ and new physics effects which would manifest itself in a deviation of a_μ from its Standard Model (SM) value [3]. Moreover, the measurement sensitivity of this experiment to a further form factor, the electric dipole moment (EDM) d_μ of the muon — which, if nonzero, would signal T violation — probably increases also by about an order of magnitude. Here I shall discuss how large d_μ is expected to be in various models of CP violation in view of the recent experimental upper bounds on the EDM of the neutron and especially of the electron. As rather detailed accounts of particle EDMs in various models of CP violation were recently published [4-8], I shall give only a brief overview.

2 Particle EDMs

It is well known that a stationary particle state (of a particle having spin and being non-selfconjugate) with well-defined parity cannot have a permanent EDM unless parity (P) and time-reversal invariance (T) are violated, which means CP violation assuming CPT invariance. For a spin-1/2 fermion f the EDM d_f is defined by

$$\langle f(p')|J_\mu(0)|f(p)\rangle = \bar{u}\left[F_3(q^2)\sigma_{\mu\nu}\gamma_5 q^\nu/2m_f + ...\right]u, \quad d_f = -F_3(0)/2m_f \tag{1}$$

where $q = p' - p, m_f$ denotes the mass of the fermion, and the dots indicate the three T-conserving form factors which are not exhibited in (1).

So far a nonzero EDM of a particle has not been found. For the muon the best bounds on d_μ come from the CERN experiment on a_μ more than a decade ago. If $d_\mu \neq 0$, the precession of the muon spin in the external magnetic and electric fields seen by the muon would be affected: It would result in a tilt of the spin precession plane, the effect being proportional to d_μ. From a direct search for this effect the result [9]

$$d_\mu = (3.7 \pm 3.4) \times 10^{-19} e\ cm \tag{2}$$

was obtained. The tilt of the precession plane would increase the precession frequency which would be noticed through an "increase" by $\Delta a_\mu(\mathrm{EDM}) \sim |d_\mu|^2$ of the anomalous magnetic moment. Comparison of the SM prediction for a_μ with its measured value and blaming the difference $\delta(a_\mu^{exp} - a_\mu^{SM})$ on $\Delta a_\mu(\mathrm{EDM})$ resulted in the bound [2]

$$|d_\mu| < 7.3 \times 10^{-19} e\ cm (95\% CL). \tag{3}$$

Note that this procedure is, however, not without caveat. CP-conserving new physics effects could decrease a_μ with respect to its SM value. This would leave a somewhat larger margin for $\Delta a_\mu(\mathrm{EDM})$. Yet this margin is limited by the direct measurement (2).

The present upper bounds on the EDMs of various fermions are collected in Table 1. In particular for the τ lepton another T- respectively CP-violating form factor is of interest: its weak dipole form factor $\tilde{d}_\tau$ which may be present in the $Z\tau\bar{\tau}$ vertex (with Lorentz structure $\tilde{d}_\tau \bar{\tau} i\sigma_{\mu\nu}\gamma_5\tau(\partial^\mu Z^\nu - \partial^\nu Z^\mu)$). So far the bounds on d_τ and $\tilde{d}_\tau$ are indirect ones: They are obtained from the possible contributions of these form factors to the cross sections for $e^+e^- \to \tau^+\tau^-$ off and on the Z resonance. These bounds are subject to caveats analogous to the

Table 1. Upper bounds (95 % CL) on the EDMs of various fermions ($d_\tau \equiv d_\tau(35\ \mathrm{GeV})$) and on the weak dipole form factor $\tilde{d}_\tau \equiv \tilde{d}_\tau(m_z)$ of the τ lepton. The references should be consulted for a discussion of experimental errors, theoretical uncertainties involved, etc.

Particle	Upper bound [$e\ cm$]
Electron	$\lvert d_e\rvert < 1.9 \times 10^{-26}$ [10]
Muon	$\lvert d_\mu\rvert < 7.3 \times 10^{-19}$ [2]
Tau	$\lvert d_\tau\rvert \lesssim 2 \times 10^{-16}$ [5,11,12,13]
	$\lvert \tilde{d}_\tau\rvert < 4.2 \times 10^{-17}$ [12,15]
Neutron	$\lvert d_n\rvert < 1.2 \times 10^{-25}$ [16]
Proton	$\lvert d_p\rvert < 1.6 \times 10^{-22}$ [17]
Lambda	$\lvert d_\Lambda\rvert < 1.5 \times 10^{-16}$ [18]

one discussed below eq. (3). However, CP-odd correlations exist [14,15] with which these form factors can be determined unambiguously. These correlations are being measured for $Z \to \tau^+\tau^-$ at LEP. It is expected that this eventually leads to an increase of the sensitivity to $\tilde{d}_\tau$ by about an order of magnitude.

It seems instructive to compare the precision with which d_e and d_μ are known with the present measurement precision of their magnetic counterparts. The experimental errors on a_e and a_μ translate into

$$\begin{aligned}\delta(F_2^e(0)/2m_e) &= 2 \times 10^{-22} e\ cm\\ \delta(F_2^\mu(0)/2m_\mu) &= 8 \times 10^{-22} e\ cm.\end{aligned} \tag{4}$$

That is, the EDM of the electron is known four orders of magnitude more precisely than its magnetic moment, whereas for the muon the situation is reverse.

3 Models

If the Kobayashi-Maskawa (KM) phase δ_{KM} in the mixing matrix of the charged weak quark currents is the only source of CP violation in Nature, then particle EDMs are too small to be observable in the forseeable future. In particular, for leptons one has (cf. e.g. [4-6])

$$|d_{lepton}|_{SM} << |d_n|_{KM} < 10^{-30} e\ cm. \tag{5}$$

However, if the SM is embedded into a larger gauge theory, various other CP-violating couplings (in particular couplings not connected to the mixing of fermion generations) involving quarks and leptons are possible in a natural way. As to EDMs the following generic features should be recalled: (a) In a renormalizable (gauge) theory, the EDM coupling $d_f \bar{f} i\sigma_{\mu\nu}\gamma_5 f F^{\mu\nu}$ cannot be present at Born level, but must be induced by loops. (b) The EDM coupling flips chirality, $f_L \leftrightarrow f_R$; therefore, $d_f \neq 0$ requires fermion masses in addition to CP violation. The question is how do the EDMs, in particular those of leptons, scale: $d_\ell \sim (m_\ell)^p$, $d_\ell \sim (m_f)^q, \ldots$? The generic diagrams which can generate a nonzero muon EDM are depicted in Fig. 1. The boson B must couple both to μ_L and μ_R with complex couplings g_L and g_R, respectively, such that $Im(g_L g_R^*) = 0$. The mass of the

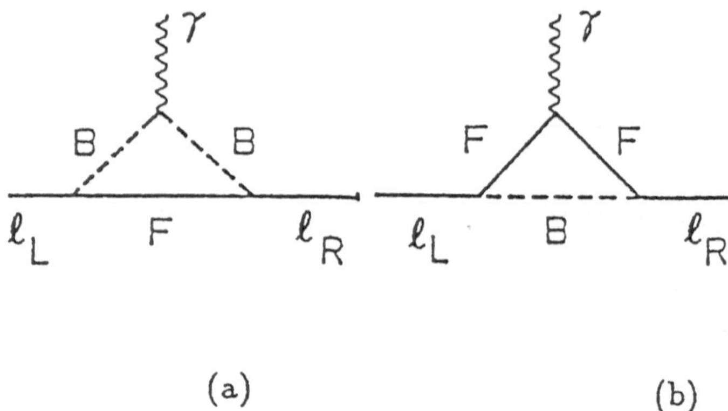

Fig. 1. Generic one-loop diagrams that can generate a nonzero lepton EDM. F denotes a fermion and B a boson of spin zero or one.

intermediate fermion F provides the necessary chirality flip.

If $g_{L,R}$ are non-gauge couplings (e.g. Yukawa couplings), they may also be proportional to some fermion mass. Are there "reasonable" models of CP violation which predict $|d_\mu/d_e| >> m_\mu/m_e \sim 100$? An inspection of some widely discussed models of CP violation yields the following:

Supersymmetric Models: CP violation, which is specific to supersymmetric models and which generates dipole moments of charged leptons, can be induced by the lepton-slepton-photino (or other neutralino) couplings. A muon (electron) EDM is generated through diagram Fig. 1(a) where F=photino, B=scalar muon (selectron). If the mixing of the two types of scalar leptons $\tilde{\ell}_{L,R}$ is proportional to the lepton mass of the respective generation, then one obtains (cf., e.g., [6])

$$d_\mu/d_e \sim m_\mu/m_e. \tag{6}$$

It is unlikely that this source of CP violation generates $|d_\mu/d_e| >> 100$.

Left-right symmetric models: CP violation in this model which is of relevance here can be assigned (in a certain phase convention and neglecting a possible mixing of lepton generations) to complex Dirac neutrino mass terms D_ℓ connecting the light $\nu_{\ell L}$ with the heavy $N_{\ell R}$ neutrinos [19]. An EDM d_ℓ arises from the amplitude Fig. 1(a) where $B = W$ boson (i.e. a mixture of $W_L \leftrightarrow W_R$) and F=neutrino (mass eigenstate) of the respective generation, arising from the mixing of $\nu_{\ell L}$ and $N_{\ell R}$ [20]. From the amplitude Fig. 1(a) one obtains

$$d_\ell = \frac{eG_F}{2\sqrt{2}\pi^2}\xi f(M_i^2/M_{W_R}^2) Im D_\ell, \tag{7}$$

where $|\xi| \lesssim .004$ is the W_L, W_R mixing parameter, $1 \geq f(x) \gtrsim 1/4$ for $0 \leq x \leq \infty$, and $M_i \sim M_{W_R} =$

$O(1TeV)$. In the simplest version of these models $|D_\ell|$ cannot be very large because of the "seesaw" relation $m_{\nu_\ell} \simeq |D_\ell|^2/M_i$ for the light neutrino masses. Using the experimental upper limit $m_{\nu_\mu} < 25MeV$, one has $|D_\mu| \sim .5GeV$. Even if $\arg D_\mu = O(1)$ the muon EDM will than not be larger than

$$|d_\mu| \simeq 1.7 \times 10^{-23} e\ cm. \tag{8}$$

In an extension [21] of the simple left-right models, the constraints on $|D_\ell|$ can be diluted and one may argue [8] that d_μ can be one or two orders of magnitude larger than (8).

Higgs models: If the Higgs sector is more complicated than in the SM, it may provide an additional source of CP violation. This can occur already in an extension of the SM using two Higgs doublets $\Phi_{1,2}$ which are coupled such that no flavour-changing neutral currents appear at the Born level (cf. e.g. [6]). The particle spectrum contains three neutral Higgs particles. If the scalar self-interactions are CP-violating, these states mix and the resulting mass eigenstates have no definite CP parity. That is, they couple both to scalar ($f\bar{f}$) and pseudoscalar ($\bar{f}i\gamma_5 f$) quark and lepton currents with couplings being proportional to the fermion masses. These couplings then generate quark and lepton EDMs at one-loop (through Fig. 1(b)) which behave like $d_f \simeq e\sqrt{2}G_F m_f^3/4\pi^2 M_\varphi^2$ (M_φ is the mass of the Higgs particle being exchanged), where the scalar and pseudoscalar coupling contributes each a factor m_f and the additional factor of m_f arises from the necessary chirality flip. If this were the whole story, one would expect a muon EDM being drastically larger than d_e. However, it was pointed out in [22]that there are some two-loop contributions to d_f being linear in m_f. For the electron EDM these contributions are in fact $10^7 - 10^8$ times larger than the one-loop term given above. It is a simple exercise, using the formulae of [22,23], to see that for the muon EDM the two-loop amplitudes still dominate; only for the τ and heavier fermions the EDM contributions to one loop eventually become larger than the two-loop effect. This means that at least in the simplest Higgs models of CP violation one has the scaling relation

$$d_\mu/d_e \simeq m_\mu/m_e \tag{9}$$

which yields $|d_\mu| \simeq 4 \times 10^{-24} e\ cm$.

Leptoquark models: Another exotic possibility are (scalar) leptoquarks with CP-violating couplings to quarks and leptons (cf. e.g. [24,8] and references contained in [6]). For the sake of illustration, let us consider the following toy model. Suppose there is a scalar charged leptoquark ϕ which couples charged leptons to $Q = 2/3$ quarks such that generation-changing transitions are severely suppressed or absent (this can be imposed by a discrete symmetry). This interaction is given by the Lagrangian

$$L_I = \sum_{i=1}^{3} (\lambda_i \bar{u}_{iR} \ell_{iL} + \lambda_i' \bar{u}_{iL} \ell_{iR})\phi \ + \ \text{h.c.} \tag{10}$$

where $(u_1, u_2, u_3) = (u, c, t)$, $(\ell_1, \ell_2, \ell_3) = (e, \mu, \tau)$ and $Im(\lambda_i^* \lambda_i') \neq 0$ in order to get non-zero EDMs. The leptonic EDMs are obtained from Figs. 1(a), (b) where $B = \phi, F = u_i$. If $Im(\lambda_i^* \lambda_i')$ is of the same order of magnitude for the three generations then

$$d_\tau : d_\mu : d_e \approx m_t : m_c : m_u \tag{11}$$

where the quark masses arise from the chirality flip along the internal quark lines in the amplitudes. Using $m_u \simeq 5$ MeV, $m_c \simeq 1.4$ GeV yields $d_\mu \approx 300 d_e$. However, if the phases of $\lambda_i^* \lambda_i'$ are all of the same order, but the magnitudes of these couplings are of the Higgs type, i.e. $|\lambda_i| \sim m_{u_i}/m_\phi, |\lambda_i'| \sim m_{\ell_i}/m_\phi$, one obtains

$$d_\tau : d_\mu : d_e \approx m_t^2 m_\tau : m_c^2 m_\mu : m_u^2 m_e. \tag{12}$$

In this case one hase $d_\mu \approx 10^7\ d_e$, but one also gets with $m_t > 89$ GeV that the electric and weak dipole form factors $d_\tau, \tilde{d}_\tau \approx 10^5 d_\mu$.

4 Conclusions

Some popular "non-standard" models of CP violation predict that the muon EDM is not very much bigger than about a few $\times 100 d_e$. A substantially larger d_μ is possible in specific versions of left-right symmetric models and leptoquark models. However, if d_μ is of order $10^{-19} - 10^{-20} e\ cm$, one would expect even larger electric and weak dipole form factors of the τ which should eventually be measurable at LEP. Nevertheless, future experimental determinations of d_μ will be of interest. It seems also worthwhile to search for experimental methods which would substantially increase the sensitivity to this form factor.

References

1. B. L. Roberts: talk given at this symposium
2. J. Bailey et al.: Nucl. Phys. B 150 (1979) 1
3. T. Kinoshita and W. Marciano, in: Quantum Electrodynamics. T. Kinoshita (ed.), p. 419. Singapore; World Scientific 1990
4. X. G. He, B. H. J. McKellar, S. Pakvasa: Int. J. Mod. Phys. A4 (1989) 501
5. S. M. Barr, W. Marciano, in: CP violation, C. Jarlskog (ed.), p. 445. Singapore: World Scientific 1989
6. W. Bernreuther, M. Suzuki: Rev. Mod. Phys. 63 (1991) 313
7. W. Grimus: Univ. Wien preprint UWThPh-1990 (1990)
8. C. Q. Geng, J. N. Ng: Phys. Rev. D42 (1990) 1509
9. J. Bailey et al.: J. Phys. G4 (1978) 345
10. K. Abdullah, C. Carlsberg, E. D. Commins, H. Gould, S. B. Ross: Phys. Rev. Lett. 65 (1990) 2347
11. J. Körner et al.: Z. Phys. C49 (1991) 447
12. A. de Rujula et al.: Nucl. Phys. B 357 (1991) 311
13. F. delAguila, M. Sher: Phys. Lett. B 252 (1990) 116
14. W. Bernreuther, O. Nachtmann: Phys. Rev. Lett. 63 (1989) 2787
15. W.Bernreuther, G. W. Botz, P. Overmann, O. Nachtmann: preprint HD-THEP-91-7 (1991), to be published in Z. Phys. C
16. K. F. Smith et al.: Phys. Lett. B 234 (1990) 191
17. D. Cho, K. Sangster, E. A. Hinds: Phys. Rev. Lett. 63 (1989) 2559
18. L. Pondrom et al.: Phys. Rev. D 23 (1981) 814

19. R. N. Mohapatra, in: CP violation. C. Jarlskog (ed.), p. 384. Singapore: World Scientific 1989
20. J. F. Nieves, D. Chang, P. B. Pal: Phys. Rev. D 33 (1986) 3324; J. Liu: Nucl. Phys. B 271 (1986) 531
21. J. M. Frère, J. Liu: Nucl. Phys. B 324 (1989) 333
22. S. M. Barr, A. Zee: Phys. Rev. Lett. 65 (1990) 21
23. R. G. Leigh, S. Paban, R. M. Xu: Nucl. Phys. B 352 (1991) 45; J. F. Gunion, R. Vega: Phys. Lett. B 251 (1990) 157; D. Chang, W. Y. Keung, T. C. Yuan: Phys. Rev. D 43 (1991) 14
24. S. M. Barr, A. Masiero: Phys. Rev. Lett. 58 (1987) 187

This article was processed using Springer-Verlag TEX Z.Physik C macro package 1991
and the AMS fonts, developed by the American Mathematical Society.

Z. Phys. C - Particles and Fields 56, S 101-S 108 (1992)

Zeitschrift für Physik C Particles and Fields

The new muon *(g-2)* experiment at Brookhaven

B. Lee Roberts

Department of Physics, Boston University, Boston, Massachusetts 02215

2-January-1992

Abstract. A new measurement of the muon anomalous magnetic moment (BNL E821) aims to obtain a factor of 20 in precision over the previous CERN measurement which had an accuracy of 7.3ppm. This increase in precision will provide a sensitivity to virtual particles into the TeV mass range and opens a substantial window for the discovery of new physics beyond the standard model. The CERN measurements established the muon as a point-like lepton obeying QED with additional corrections from virtual pions. It also provided constraints on the virtual production of new particles with masses on the order of tens of GeV. By increasing the precision of this measurement by a factor of 20, the experiment will be sensitive to the single-loop radiative contribution from the W and Z bosons with an experimental error $\sim$ 20% of the magnitude of the effect predicted by the standard model. In addition to experimentally testing electroweak renormalization, this measurement will be sensitive to the presence of "new physics" such as muon substructure to several TeV, supersymmetric particles, an anomalous W-boson magnetic moment, a CP violating muon electric dipole moment, and other physics beyond the standard model. In several cases, the sensitivity of E821 goes well beyond that which will be available at Fermilab, LEPII or the SSC. The construction and preparation of the precision storage ring has begun at Brookhaven. Storage ring completion is anticipated in 1993 and data collection will commence in late 1994.

1 Introduction

At this meeting, Professor Farley has given a thorough introduction to the previous muon *(g-2)* measurements at CERN [1]. These precision measurements took advantage of the relatively long muon lifetime [2] of 2.2 μs ($c\tau = 658.65m$) which made it possible to perform such experiments. Although its mass of 105.658387 (34) MeV is 209 times the electron's, to the accuracy of current experiments it appears to be a point-like lepton. However, since its mass is so much heavier than the electron's there are measurable contributions to its anomalous moment from loops with virtual μ, τ, π (hadrons), and in principle from those with virtual $W^{\pm}$ and Z^0.

We have learned a great deal from the CERN measurements, and at present we see no deviation from the value expected in the standard model. The earlier measurements have verified the presence of contributions to the muon anomalous moment from such QED processes as light on light scattering, as well as contributions from virtual muons and hadrons.

The current muon *(g-2)* value has served as an important hurdle for new theories of subatomic particles. Any theory which predicted a muon anomaly at variance with the CERN result was automatically discarded by its proponent without troubling the rest of the community. Nevertheless, the 7.3 parts per million (ppm) precision of the present value is not adequate to confirm the predicted contributions of the $W^{\pm}$ and Z^0 bosons to the muons anomalous magnetic moment.

Since this single-loop electroweak contribution to the muon *(g-2)* is so fundamental to the standard model, and since this experiment provides one of the cleanest tests of single-loop radiative corrections, a new collaboration [3] is preparing the next generation experiment at the Brookhaven National Laboratory Alternating Gradient Synchrotron (AGS). The goal of the new measurement is to obtain a factor of 20 improvement in precision, with an expected total error of 0.35 ppm.

Before discussing the theoretical motivation in detail, we need to review a few basic ideas about spin precession in a magnetic field. The gyromagnetic ratio g is defined by the relationship between the magnetic dipole moment and the spin angular momentum

$$\mu_s = g_s \frac{e}{2mc} \mathbf{S}.$$

The formulation of the Dirac equation, and its prediction that the gyromagnetic ratio of a spin 1/2 particle is exactly 2 (in the absence of a Pauli term) was a major step in the development of the theory of subatomic physics.

The nonrelativistic Larmor precession frequency (for a positive particle) is given by:

$$\omega_s = \frac{g_s}{\hbar}(\frac{e\hbar}{2mc})B = \left[1 + \frac{g_s - 2}{2}\right]\frac{e}{mc}B.$$

where the second term in brackets is the anomalous moment,

$$a = \frac{g_s - 2}{2}$$

and the (total) magnetic moment is given by

$$\mu = [1 + a]\,\frac{e\hbar}{2mc}$$

which is the number quoted in the Particle Data Tables [2]. The first term in the square bracket is the Dirac moment (which is +, 0 or − depending on the charge), and the quantity a is the anomalous (Pauli) moment. We will drop the subscript s on g in the discussion below.

The relativistic spin precession for a muon travelling through a region which has both magnetic and electric fields present is given by

$$\omega_a = \frac{d\Theta_R}{dt} = \frac{e}{mc}\left[a_\mu \mathbf{B} - \left(a_\mu - \frac{1}{\gamma^2 - 1}\right)\beta \times \mathbf{E}\right] \quad (1)$$

where $\Theta_R = (\mathrm{s}, \beta)$ is the angle between the muon spin direction in its rest frame and the muon velocity direction in the laboratory frame. The other quantities refer to the laboratory frame.

At the "magic γ" where

$$\left(a_\mu - \frac{1}{\gamma^2 - 1}\right) = 0$$

the electric field does not contribute to the precession. One can thus achieve (weak) vertical focusing with electrostatic quadrupoles, and build the storage ring magnet to have a uniform dipole magnetic field. This condition is satisfied if $\gamma = 29.3$ and $p_\mu = 3.094$ GeV/c.

2 Theoretical Considerations and the Electron (g-2)

Professor Kinoshita has given a detailed discussion on calculations of *(g-2)* at this meeting. Nevertheless, I will take a few minutes to review why we wish to re-measure the the muon *(g-2)* value. The simplest Feynman diagram for the interaction of a lepton with an external field is shown in Fig. 1 and represents $g = 2$.

An anomalous moment can result from radiative corrections to Fig. 1 and from internal structure if it is present. For example the proton's anomalous moment of $+1.79\ \mu_N$ is a reflection of the its quark structure. For leptons, even in the absence of internal structure, radiative corrections are significant. The lowest order radiative correction to Fig. 1 is shown in Fig. 2.

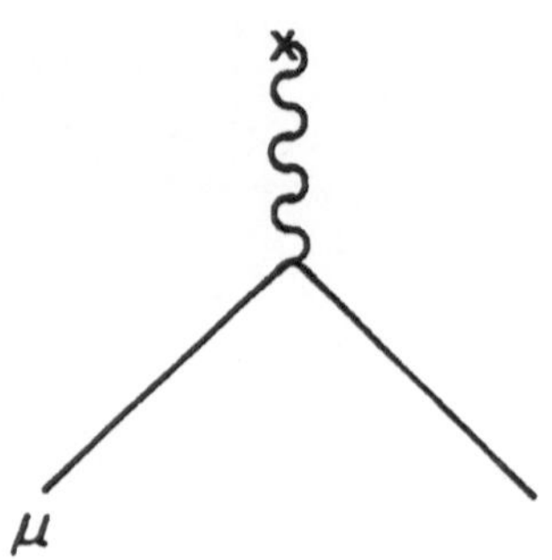

Fig. 1. The interaction of a spin 1/2 particle with an external magnetic field.

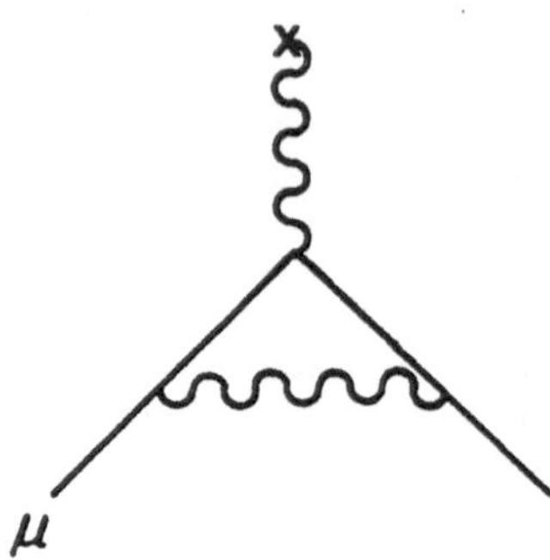

Fig. 2. The lowest order QED correction to $g - 2$.

The first measurement of the anomalous moment for the electron was made in 1947 by Kush and Foley [4], who determined that

$$g_e = 2(1.00119 \pm 0.00005),$$

consistent with the first-order correction of $\frac{\alpha}{2\pi}$ calculated by Schwinger [5].

Since this seminal work, the anomalous magnetic moment of the electron, a_e, has been measured with ever-increasing precision. To the level of current measurements, QED diagrams alone suffice to determine its theoretical value, which can be written [6] as an expansion in powers of α:

$$a_e^{theory} = C_1\left(\frac{\alpha}{\pi}\right) + C_2\left(\frac{\alpha}{\pi}\right)^2 + C_3\left(\frac{\alpha}{\pi}\right)^3 + C_4\left(\frac{\alpha}{\pi}\right)^4 + \ldots + \delta a$$

where $C_1 = 0.5$, $C_2 = -0.328\ 478\ 965\ \ldots$, $C_3 = 1.17611(42)$, $C_4 = -1.434(138)$ and $\delta a = 4.46 \times 10^{-12}$. The correction δa comes from contributions of virtual muons, hadrons, $W^\pm$ and Z^0. The theoretical prediction is [6]

$$a_e^{theory} = 1\ 159\ 652\ 140\ (28) \times 10^{-12}$$

where the value for the fine structure constant

$$\alpha_{QH}^{-1} = 137.035\ 997\ 9\ (32)$$

is taken from measurements of the quantum Hall effect.

The values of $a_{e^\pm}$ from the Penning trap experiments are [7]

$$a_{e^-} = 1\ 159\ 652\ 188.4(4.3) \times 10^{-12}$$

and

$$a_{e^+} = 1\ 159\ 652\ 187.9(4.3) \times 10^{-12}.$$

There is a 2σ agreement with QED at the level of 10^{-12} and, to the extent that the particle and anti-particle properties are the same, CPT invariance is verified at the 2×10^{-12} level. One should note that the contribution from heavier particles to the electron *(g-2)* value is at the level of the current experimental errors.

The muon anomalous magnetic moment can be calculated in the same fashion as that of the electron. However, because the mass of the muon is much greater than that of the electron, diagrams with quark loops, or to a lesser extent, weak bosons, contribute measurably to the final result. The relative contribution of heavier particles to the muon's anomalous moment goes as $\sim (m_\mu/m_e)^2$ which makes the contribution of such particles on the measured muon anomaly much larger ($\sim 4 \times 10^4$) than for the electron. The theoretical value for a_μ can be expressed [8]

$$a_\mu^{theory} = a_\mu^{QED} + a_\mu^{hadronic} + a_\mu^{weak} \qquad (2)$$

where

$$a_\mu^{QED} = 1\ 165\ 846\ 955(46)(28) \times 10^{-12} \qquad (3)$$

The first uncertainty is from calculation, the second from α.

Loops where virtual hadrons are emitted and then re-absorbed contribute to the value of a_μ at a measurable level (60 parts per million). The lowest order hadronic diagrams are shown in Fig. 3.

The contribution from these hadronic loops cannot be calculated reliably in QCD. However, the dominant contribution, shown in Fig. 3a, can be related, through dispersion theory, to the experimentally measured cross-section annihilation for e^+e^- into hadrons. The single hadron loop contribution to $g-2$ is given by

$$a_\mu^{hadronic} = (\frac{\alpha m_\mu}{3\pi})^2 \int_{4m_\pi^2}^{\infty} \frac{ds}{s^2} K(s)R(s) \qquad (4)$$

where the quantity $R(s)$ is defined by

$$R(s) = \frac{\sigma_{tot}(e^+e^- \to hadrons)}{\sigma_{tot}(e^+e^- \to \mu^+\mu^-)}$$

is measured experimentally, and

$$K(s) = \frac{3s}{m_\mu^2}\Big[x^2(1-\frac{x^2}{2}) + (1+x)^2(1+\frac{1}{x^2})\left[ln(1+x) - x + \frac{x^2}{2}\right] + \frac{1+x}{1-x}x^2 lnx\Big]$$

Table 1. Evaluations of the Hadronic Contribution to the Muon $(g-2)$ Value. The several evaluations of $a_\mu^{hadronic}$ were not carried out with the exact same data sets. The errors on the hadronic contribution are given in parentheses. In the cases where two are listed the first is statistical and the second is systematic. The two were added in quadrature to obtain the total uncertainty in ppm.

$a_\mu^{hadronic} \times 10^{11}$	Error in ppm	Reference
7068 (59) (164)	1.5	Kinoshita et al. [8,9]
6840 (110)	0.94	Barkov et al. [10]
7100 (105) (49)	0.99	Casas et al. [11]
7245 (66) (259)	2.3	Burkhardt et al. [12], Jegerlehner [13]
7048 (115)	0.99	Martinovič et al. [14]

with

$$x = \frac{1-\beta}{1+\beta}, \qquad \beta = \sqrt{1 - \frac{4m_\mu^2}{s}}$$

The integral is dominated by the kinematic regions from threshold where the low-energy $\pi\pi$ resonances such as the ρ are important up to the J/ψ. The contributions from the virtual hadron loops in Fig. 3 is [8]

$$a_\mu^{had} = 7027(175) \times 10^{-11} \qquad (5)$$

The uncertainty is dominated by the knowledge of $R(s)$ (see Eq. (4)).

There are four recent evaluations [8,9,10,11,12,13,14] of the hadronic contribution to the muon *(g-2)* value and these are listed in Table I.

The contribution of higher order diagrams have been estimated by Kinoshita [8] to be

$$a_\mu^{hadronic}(higher\ order) = -41(7) \times 10^{-11} \quad (0.06ppm)$$

Although there is good agreement between these five evaluations, there is a difference of opinion on how to handle the experimental errors which results in the differing errors on $a_\mu^{hadronic}$. In all cases, the uncertainty in the hadronic contribution is dominated by the experimental errors on $R(s)$ from threshold to the J/ψ, with the regions below and above $\sqrt{s} = 1.4$ GeV contributing roughly equally.

A new experiment is being mounted at INP in Novosibirsk to measure $R(s)$ over the ρ resonance (up to $\sqrt{s} = 1.4$ GeV) to the necessary accuracy. The improved storage ring, VEPP 2M, is now operational, as is a new detector, CMD 2. Data collection should begin in late 1991. These new data from INP should halve the uncertainty on $a_\mu^{hadronic}$ [15].

Diagrams with the intermediate vector bosons $W^\pm$, Z^0 will contribute to the value of a_μ at the few ppm level. The lowest order $W^\pm$ and Z^0 contributions are shown in Fig. 4 and their sum is given by [8]

$$a_\mu^{weak} = 195(10) \times 10^{-11} \qquad (6)$$

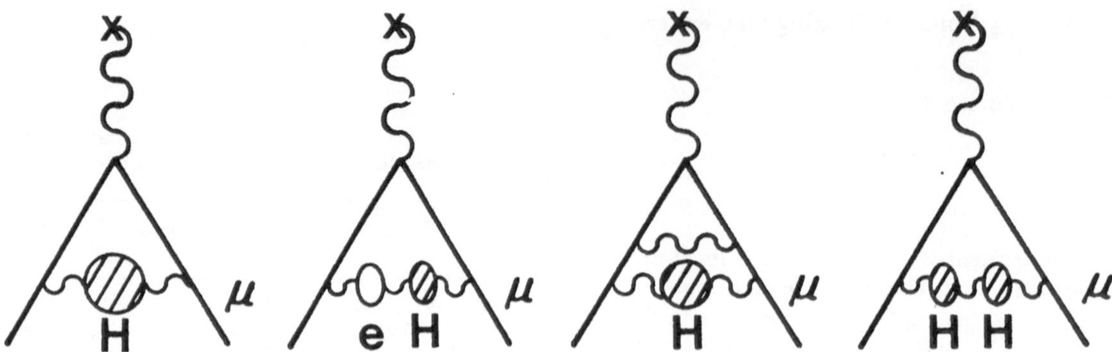

Fig. 3. Lowest order hadronic vacuum polarization diagrams.

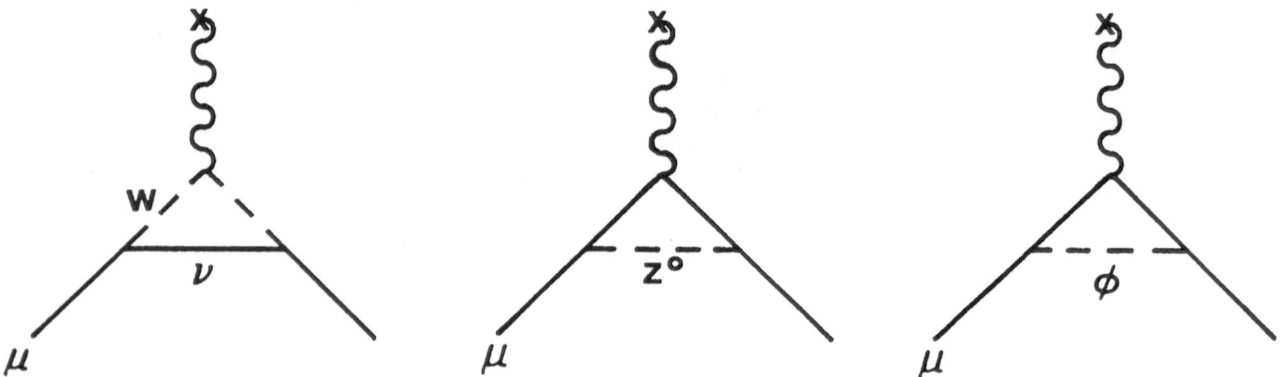

Fig. 4. Single loop weak contributions to the muon $(g-2)$ value.

The uncertainty is dominated by the estimated second order weak contribution. Provided its mass is considerably greater than 300 MeV, the contribution of the Higgs to $(g-2)$ is negligible.

The total theoretical value of a_μ is

$$a_\mu^{theory} = 116\ 591\ 918\ (176) \times 10^{-11} \tag{7}$$

We note that the uncertainty on the hadronic term (taken from Ref. [8] (Eq. (5)) is of the same order as that on the weak term(Eq. (6)). If we are to learn much about the weak term alone, it is essential that the error on the hadronic term be reduced.

The current experimental values of the muon anomaly are

$$a_{\mu^-}^{exp} = 1\ 165\ 937(12) \times 10^{-9}$$

and

$$a_{\mu^+}^{exp} = 1\ 165\ 911(11) \times 10^{-9}$$

(for a combined accuracy of 7.3 ppm), in good agreement with the calculated value. The goal of the new experiment (E821) is a measurement of $(g-2)$ to the level of 0.35 ppm, a factor of twenty improvement over the CERN precision. This sensitivity is $\sim 20\%$ of the magnitude of the expected electro-weak contribution to $(g-2)$ from single loop diagrams involving the intermediate vector bosons $W^\pm$, Z^0.

A measurement at the proposed accuracy would thus provide an essential test of the renormalization of the unified electroweak theory. In addition, the experiment will produce an order of magnitude better limit on the muon electric dipole moment.

The muon anomaly is sensitive to the presence of new fields, as well as those associated with known particles. As a result, E821 is sensitive to a range of new physics, and can serve to put new and more stringent limits on masses of new particles, muon substructure and other phenomena outside of the standard model.

The sensitivity to new physics has been discussed by Kinoshita and Marciano [8] assuming that the new experimental error will be 0.35 ppm and that the theoretical uncertainty on the hadronic contribution is reduced substantially by new measurements of $R(s)$.

In Table 2 we show limits on various parameters which will come from E821, LEP II and the SSC [8].

We emphasize that a precision measurement of $(g-2)$ is complementary to other high energy measurements. In some cases, such as composite W bosons, the limit which will be placed by the new $(g-2)$ experiment is considerably better that that which will come from LEP

Table 2. Sensitivity to New Physics of the New Muon $(g-2)$ Experiment

New Physics	E821 Sensitivity	Comments
μ substructure[a]	$\Lambda \geq 5\ TeV$	SSC domain
excited muon	$m_{\mu^*} \geq 400\ GeV$	LEPII comparable
$W^{\pm}$ substructure	$\Lambda \geq 2\ TeV$	LEPII $\sim 100-200\ GeV$
$\frac{(g_W-2)}{2} \neq 0$	$\geq$ 0.02	LEPII, SSC $\sim$ 0.2
Light Higgs	$\leq$ 300	$O(10^{-3}g)$ coupling
Heavy Higgs	$\leq 500GeV$	$O(g)$ coupling
Supersymmetry	$\leq 130\ GeV$	FNAL $\bar{p}p$ collider
$W_R^{\pm}$	$\leq 250\ GeV$	FNAL $\bar{p}p$ collider
$Z'(E6)$	$\leq 100\ GeV$	FNAL $\bar{p}p$ collider

[a]For substructure $\Delta a_{\mu} \sim m_{\mu}^2/\Lambda^2$

II or the SSC. Even before the measurement of $R(s)$ is completed (see Eq.(5) there is a window of opportunity for new physics in going from the 7.3 ppm measurement at CERN to a 1 ppm measurement at Brookhaven. To exploit fully the additional factor of three in precision available at the AGS, we will need an improved value for the hadronic contribution to (g-2).

While it is true that indirect evidence of new physics is not as satisfying as a direct observation, the discovery of a $(g-2)$ value not predicted in the standard model would be viewed as a fundamental and exciting discovery, similar to the excitement which would result from the discovery of a rare kaon decay mode not allowed by the standard model. Unlike a rare decay experiment, a deviation from the standard model prediction would contain two pieces of information, a magnitude and a sign. For example, supersymmetry in the mass range accessible at the Tevatron would cause a value of $(g-2)$ smaller than the standard model value.

3 The New Experiment

At Brookhaven National Laboratory we are constructing a new muon storage ring to perform the next generation muon *(g-2)* measurement. The experimental technique is a refined version of the last CERN measurement [1], *viz.* a weak focussing storage ring (n=0.135) operating at the magic γ. In order to improve on the systematic errors of 1.5 ppm present in the CERN measurement, substantial improvements have been made in all areas of the experiment.

To perform the *(g-2)* experiment, one injects a short bunch of muons into a storage ring and then counts decay electrons with detectors placed just outside the vacuum chamber of the storage region. Since the decay electrons are lower in energy, they will spiral into the detectors. For the arrival of each electron one records an energy and a time. If one constructs a histogram of number of electrons *vs.* time, the exponential muon lifetime is observed, modulated by the *(g-2)* frequency ω_a. At the magic γ the lifetime is dilated to 64.4 μs.

At CERN, pions just above the magic momentum were injected into the storage ring and the $\pi \to \mu\nu$ decay was used to kick the (polarized) muons onto orbit. The efficiency for this mode of injection is $\sim 10^{-4}$ which limited the number of muons they could store to a few hundred per fill. With the AGS booster accelerator in operation, each of the twelve micro-bunches will contain 5×10^{12} protons. The secondary beamline will contain a 72 m long pion decay channel which will permit pion injection or direct muon injection into the ring. We expect around 1×10^4 (6×10^4) muons stored per fill for pion (muon) injection. In addition to the increase in flux, muon injection will eliminate the problems caused by the large hadronic background created by pions which do not decay and are lost into the storage ring material. The development of a full aperture kicker to kick the muon bunch onto a stable orbit is underway.

The muon decay $\mu^+ \to e^+\nu_e\bar{\nu}_\mu$ is self analyzing. The highest energy electrons are emitted preferentially along the muon spin direction. The decay asymmetry varies from $-1/3$ to $+1$ for the lowest to highest energy electrons respectively. For a detailed discussion of the decay kinematics see the article by Farley and Picasso[16].

The pion/muon beam produced by the AGS will be introduced into the storage region by a superconducting inflector. This truncated, double cosine theta, iron-free septum magnet [17] cancels the storage ring field and permits the π/μ to enter the storage region. The use of a static device to cancel the main field removes the problem of transient effects from a pulsed device, such as the pulsed coaxial line used in the CERN experiment, and permits the field leaking from the end of the inflector into the storage region to be shimmed out.

The storage ring magnetic field will be provided by a monolithic superferric "C" magnet which is under construction. The continuous yoke will be made up of 30° sectors of 1006 steel. A cross section of the magnet is shown in Fig. 5, and it can be seen that the yoke consists of 6 separate steel plates. The pole pieces are made of high quality 10005 steel and will be mounted with an air gap between them and the yoke. This air gap will serve to decouple the field in the storage region from any in the yoke. The storage aperture will be circular (9 cm diameter) and its center lies at a radius of 711.2 cm. The four $\sim$ 14 m diameter superconducting coils which excite the $\sim$ 1.5 T field are shown along with their cryostats. The two coils at the outer radius (inside the "C") share a common cryostat. Field uniformity of 1 ppm integrated over azimuth will be achieved by a combination of active and passive shimming.

The **B** field will be monitored by fixed NMR probes positioned around the ring as well as by pickup coils. Both systems will be fed back to the power supply. Periodically the field in the entire storage region will be mapped by a trolley which travels inside the storage region without having to turn off the magnet or remove the vacuum chamber. We expect to monitor the field to the 0.1ppm level.

The relevant **B** field which must be inserted in Eq. (1) is the field in the storage region averaged over the muon

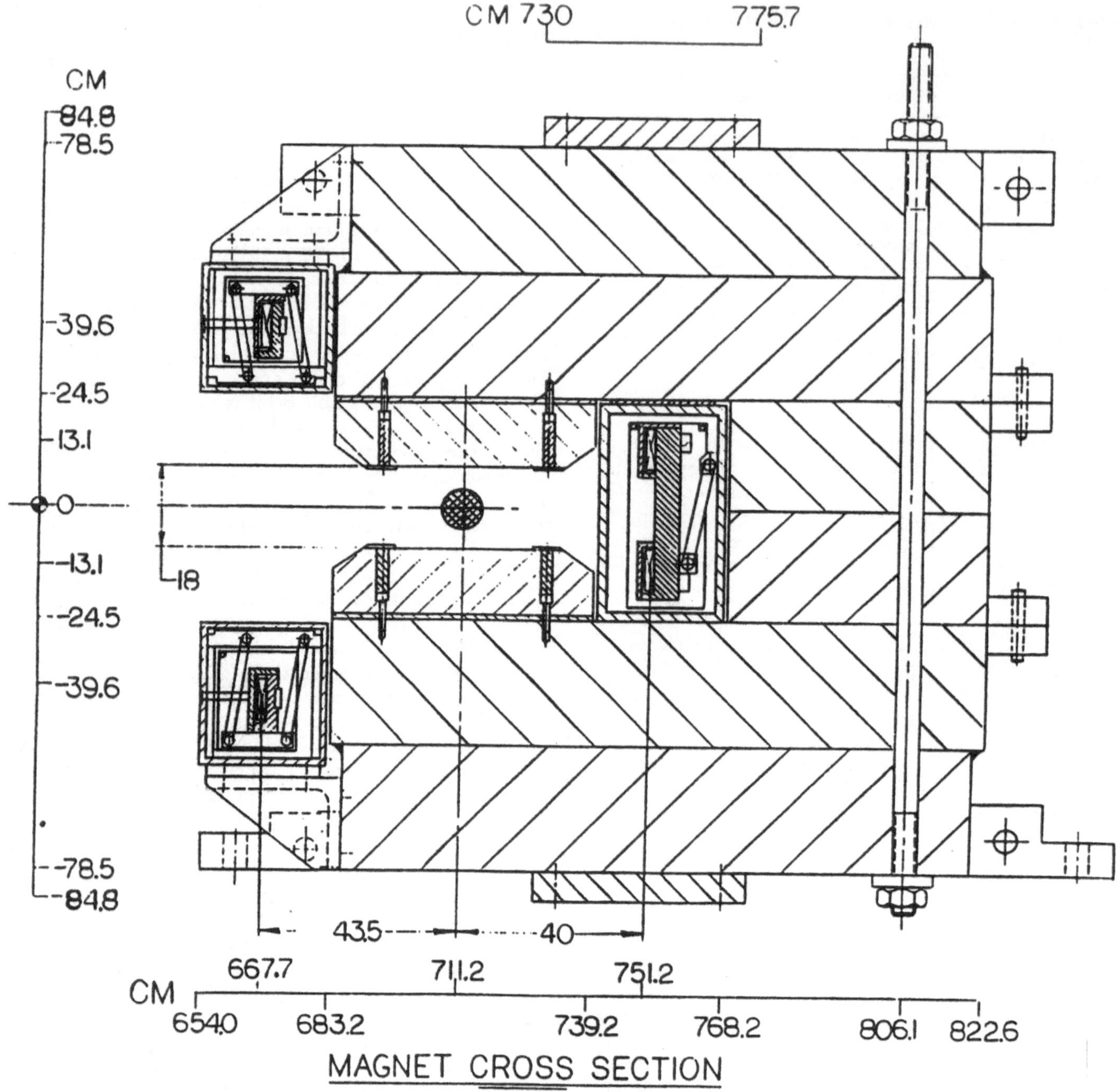

Fig. 5. A cross section of the storage ring magnet.

distribution. This breaks down into a sum of multipoles of the magnetic field times the appropriate multipole of the muon distribution. The circular aperture in the storage region greatly decreases the presence of higher order multipoles in the muon distribution, and thus decreases our sensitivity to the higher multipoles in the magentic field. In addition to measuring the fast rotation time at early times, (which gives the average radius) we will measure the muon distribution directly during data collection. Monte Carlo studies have shown that by measuring the trajectories of decay electrons with wire chambers, and tracking them back to a point tangent to the storage ring gives a good reconstruction of the the parent muon distribution. One or two stations in the ring will be outfitted with the appropriate array of chambers, etc. to make this measurement.

The ring will have four-fold symmetry with the horizontal focussing provided by four sets of electrostatic quadrupoles. This symmetry has the advantage that $\beta(max) \simeq \beta(min)$ in the ring, so the beam does not

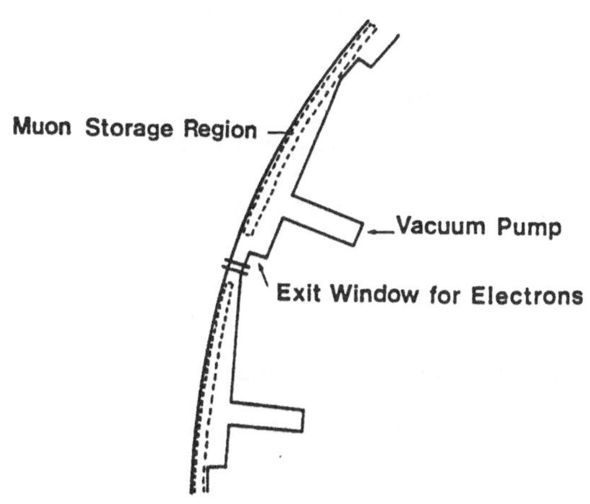

Fig. 6. A portion of the vacuum tank showing detector locations.

breathe as it did in the two-fold symmetry of the CERN ring. In addition, this geometry leaves over two thirds of the ring free of material where muon decay electrons could pre-shower. The vacuum chamber will be scalloped as shown in Fig. 6, which eliminates electron pre-showering in the vacuum chamber walls.

We plan to stage the experiment, running first with pion injection, and then with muon injection.

The electron detectors will consist of a position sensitive detector immediately downstream of the vacuum window, followed by an electron calorimeter which will be made of scintillating fiber embedded in a Pb-Bi eutectic alloy which has a radiation length of $\sim$ 1 cm and a resolution of $\sigma/\sqrt{E} = 15\%$ or better. The counters in front of the calorimeters will facilitate pileup rejection at the high rates expected. With the availability of fast clocks, and cheap memory, the electronic queueing losses and other electronic deadtime which contributed to the CERN systematic errors will be negligible in E821.

4 Conclusions

The new muon *(g-2)* experiment at Brookhaven will represent a significant improvement over the CERN measurement. With the booster accelerator the improvement of a factor of 400 in statistics will be easy. The continuous "C" magnet rather than 40 separate dipoles, plus better monitoring and mapping should eliminate systematic errors from the B field. Modern electron calorimeters and electronics will facilitate data collection at the high instantaneous rates expected, especially with muon injection.

The experiment will provide a window of opportunity for the discovery of new physics as well as a sensitive test of single loop electroweak radiative corrections. With the beam intensity available at the AGS, we will be able to carry out substantial tests of systematic uncertainties and minimize their influence on the final result. The theoretical error in the hadronic correction will be reduced roughly a factor of two by the new data from Novosibirsk. Even without these new data there is a window for the discovery of new physics of a factor of seven without any additional information on $R(s)$. The detailed electroweak tests will depend on the existence of new e^+e^- data.

If the new measurement of *(g-2)* were to provide a number which was not in agreement with the standard model, it would be exciting indeed. With the standard model now firmly in place, there is common agreement on what signatures would indicate physics beyond it. The muon *(g-2)* experiment, along with the program of rare kaon decay experiments at Brookhaven will provide the most sensitive probes of new physics in the TeV mass range during the pre-SSC era.

5 Appendix

E821 Collaboration Members

D. Brown, R. Carey, E. Hazen, F. Krienen, J.P. Miller, B.L. Roberts‡, L.R. Sulak, C. Wang, W. Worstell - *Boston University;* H.N. Brown, G. Bunce§, G.T. Danby, C. Gardner, J.W. Jackson, R. Larsen, Y.Y.Lee, S. Mane, W.-Z Meng, W.M. Morse‡, A. Prodell, R. Shutt, K. Woodle - *Brookhaven National Laboratory;* K. Becker, M.S. Lubell - *City College of New York;* T. Kinoshita, Y. Orlov - *Cornell University;* D. Winn - *Fairfield University;* U. Haeberlen, K. Jungmann, G. zu Putlitz - *University of Heidelberg*; W.P. Lysenko - *Los Alamos National Laboratory;* L.M. Barkov, B.I. Khazin, D.N. Grigorev, E.A. Kuraev, S.I. Redin, Yu.M. Shatunov, E. Solodov - *Institute of Nuclear Physics, Novosibirsk;* K. Nagamine, M. Iwasaki - *University of Tokyo; K. Endo, H. Hirabayashi, S. Kurokawa, Y. Mizumachi, T. Sato, A. Yamamoto - KEK;* L. Ishida - *Riken;* H. Ahn, P. Cushman, S.K. Dhawan, F.J.M. Farley, X. Fei, V.W. Hughes‡, - *Yale University.*

‡Spokesmen
§Project Manager

Acknowledgement. I wish to thank Jim Miller and Rob Carey for reading this manuscript and for a number of helpful discussions and suggestions. I wish to thank Bill Marciano for discussions on the new physics potential of the (g-2)experiment. This work is supported in part by the U.S. National Science Foundation, the U.S. Department of Energy, the Federal Ministry for Science and Technology of Germany ant the Japan-U.U.Japan High Energy Physics Cooperation.

References

1. J. Bailey et al., Nucl. Phys. B150, 1 (1979)
2. Particle Data Group, Phys. Lett. B239 (1990)
3. Brookhaven National Laboratory Experiment E82 1, *A New Precision Measurement of the Muon (g-2) Value at Level of*

0.35 ppm, V.W. Hughes, W.M. Morse and B.L. Roberts - Spokesmen
4. P. Kush and H.M. Foley, Phys. Rev.72, 250 (1948)
5. J. Schwinger, Phys. Rev. 73, 416 (1948), 75, 898 (1949)
6. T. Kinoshita, Quantum Electrodynamics, (Directions in High Energy Physics Vol. 7) T. Kinoshita ed., World Scientific, 1990,
7. R.S. Van Dyck et al., Phys. Rev. Lett. 59, 26 (1987)
8. T. Kinoshita and W.J. Marciano, *Quantum Electrodynamics*, (Directions in High Energy Physics Vol. 7) T. Kinoshita ed., World Scientific, 1990, p.
9. T. Kinoshita, B. Nižić and Y. Okamoto, Phys. Rev. D31, 2108 (1985)
10. L.M. Barkov et al., Nucl. Phys. B256 , 365 (1985)
11. J.A. Casas, C. López and F.J. Yndurán, Phys. Rev. D32, 736 (1985)
12. H. Burkhardt, F. Jegerlehner, G. Penso, C. Verzegnassi, Z. Phys. C43, 497 (1989)
13. F. Jegerlehner, private communication.
14. Ľubomír Martinovič and Stanislav Duvnička, Phys. Rev. D42, 884 (1990)
15. V.W. Hughes, private communication
16. F.J.M. Farley and E. Picasso, *Quantum Electrodynamics*, (Directions in High Energy Physics Vol. 7) T. Kinoshita ed., World Scientific, 1990, p 479
17. F. Krienen et al. Nucl. Inst. and Meth. A238, 5 (1989)

This article was processed using Springer-Verlag TEX Z.Physik C macro package 1991
and the AMS fonts, developed by the American Mathematical Society.

Z. Phys. C - Particles and Fields 56, S 109-S 113 (1992)

Zeitschrift für Physik C Particles and Fields

Muon Decay

Wulf Fetscher

Institut für Mittelenergiephysik, ETH Zürich, CH-5232 Villigen PSI

13-August-1991

Abstract. Muon decay, after more than fifty years of experimental and theoretical investigations, has been established as a firm basis for the Standard Model of Electroweak Interactions. The decay products of the muon have been identified, and the Lorentz structure of the decay interaction has been determined completely from existing experiments in a model-independent way. Together with the completely anologous leptonic τ decays it continues to be an important and sensitive testing ground for deviations from the Standard Model.

1 Introduction

The muon, as an electron-like fermion, is the second of the three charged leptons e^-, μ^- and τ^-. It decays purely leptonic via the charged weak interaction. Thus it offers the unique possibility to determine the Lorentz structure of the weak interaction in a single reaction (including inverse muon decay) without being hampered by the effects of the strong interaction. Although the muon is now known for more than fifty years, it was only recently that the nature of the neutral decay products and the Lorentz structure of the decay interaction have been determined by experiment.

2 Some historical remarks

The muon has been discovered more than fifty years ago by Neddermeyer and Anderson [1] and by Street and Stevenson [2,3]. It was called 'mesotron' at that time and was considered to be the particle postulated by Yukawa [4] as the mediator of strong interactions. (What is not so well known is that the muon has actually been seen five years earlier, in 1932, by Paul Kunze [5] in Rostock using a Wilson cloud chamber with a 1.84 T magnetic field. He selected pictures with double tracks and found a muon which presumably stopped in the chamber wall and emitted a 37 MeV positron which was identified by Kunze. About the upper track be noted: '.... für ein Proton ionisiert sie wohl zu wenig, und für ein positives Elektron zu viel'.) It took ten more years until the experiments of Conversi et al. [6] led to the conclusion [7] that the new particle could not be the Yukawa particle, since it was able to penetrate 7 mm of nuclear matter without interaction. It was then realized by Pontecorvo [8] that the muon could be a fermion or, more specific, a heavy electron which should be able to decay electromagnetically via $\mu \rightarrow e\gamma$. A little later, in 1948, Steinberger [9] found that the electron from μ decay was not monoenergetic and suggested a decay similar to nuclear beta decay with an electron and two neutral particles in the final state. From this result we can draw two important conclusions:

(1) The muon does not have the same lepton number as the electron

(2) Muon decay is a weak decay.

We emphasize that both nuclear beta decay and muon decay are the basis of the V-A hypothesis of the charged weak interaction which in turn is incorporated into the Standard Glashow-Salam-Weinberg Model, and that the universal Fermi coupling constant is derived from μ decay. Thus lepton number conservation, V-A and universality are put into the Standard Model. It is therefore essential to verify these assumptions by experiment.

3 Nature of neutrinos in normal decay

3.1 Test of $\overline{\nu_e}$

From the lack of the decay $\mu \rightarrow e\gamma$ and other forbidden processes one finds that muon number is conserved separately. One can design additive and multiplicative schemes:

$$(a) \sum_i (\mathcal{L}_e^{(i)} + \mathcal{L}_\mu^{(i)}) = const., \sum_i L_\mu^{(i)} = const.$$

Table 1. Muon decay for two different lepton number conservation schemes.

Decay	Allowed by	
$\mu^+ \to e^+\nu_e\bar{\nu}_\mu$	a	b
$\mu^+ \to e^+\bar{\nu}_e\nu_\mu$	–	b

$$(b)\sum_i(\mathcal{L}_e^{(i)}+\mathcal{L}_\mu^{(i)}) = const., (-1)^{\Sigma L_\mu^{(i)}} = const.$$

The process $\mu^+ \to e^+\bar{\nu}_e\nu_\mu$ (see table 1) has been excluded by an experiment at LAMPF [10].

Principle: Stopped μ^+, targets H_2O and D_2O. Possible reactions are $\bar{\nu}_e p \to ne^+$ and $\nu_e d \to ppe^-$. It has been found that neutrinos from μ^+ decay at rest produce e^- in D_2O, **not** e^+ in H_2O.

$$\to \quad \frac{\sigma(\mu^+ \to e^+\bar{\nu}_e\nu_\mu)}{\sigma(\mu^+ \to all)} < 0.10 \quad (90\% c.l.)$$

3.2 Test of ν_μ via the inverse decay (CHARM Coll.[11])

Principle: $\bar{\nu}_\mu$ from $\pi^- \to \mu^- + \bar{\nu}_\mu$ and ν_μ from $\pi^+ \to \mu^+ + \nu_\mu$

$$\to \quad \frac{\sigma(\bar{\nu}_\mu + e^- \to \mu^- + \bar{\nu}_e)}{\sigma(\nu_\mu + e^- \to \mu^- + \nu_e)} < 0.05\ (90\% c.l.)$$

The test of the inverse reaction shows that the muon neutrino in the decay of the μ^- is the ν_μ, not the $\bar{\nu}_\mu$.

4 Lorentz structure of the decay $\mu \to e\bar{\nu}_e\nu_\mu$

Muon decay can be described by the most general, local, derivative-free, lepton-number conserving four-fermion point interaction [12] (L. Michel, 1950). The Hamiltonian contains ten complex coupling constants corresponding to 19 independent parameters to be determined by experiment. The status of the knowledge of the interaction as of 1965 was discussed at the 'International Conference on Weak Interactions' in Argonne [13]. There seemed to be no hope to experimentally determine the complete set of the nineteen real parameters. However, in the same year, Cecilia Jarlskog [14] has shown that relations without interference terms of the type $\Sigma \mid X_i \mid^2=0$ hold, if some particular experiments yield the V-A values. This shows that the determination of the 19 decay parameters does not necessarily require 19 independent measurements! Furthermore a neutrino-electron correlation experiment seemed necessary. The use of a chiral projection hamiltonian as proposed by Mursula and Scheck [15] in 1983 led to most transparent relations between coupling constants and experiments, with a minimum of interference terms. Finally, in 1986 Fetscher, Gerber and Johnson [16] succeed to completely **determine** the interaction describing normal and inverse muon decay, based on a set of key experiments and **without** any additional model assumptions.

Their matrix element is given by

$$M = \sqrt{8}G_F \sum_{\substack{\gamma = S,V,T \\ \varepsilon,\mu = R,L}} g^\gamma_{\varepsilon\mu} \cdot \qquad (1)$$

$$< \bar{e}_\varepsilon|\Gamma^\gamma|(\nu_e)_n >< (\bar{\nu}_\mu)_m|\Gamma_\gamma|\mu_\mu >$$

γ labels the type of interaction: Γ^S, Γ^V, Γ^T (4-scalar, 4-vector, 4-tensor). The indices ε and μ indicate the chiral projection (lefthanded, righthanded) of the spinors of the experimentally observed particles, $\varepsilon\hat{=}$electron, $\varepsilon\hat{=}$muon. The helicities n and m for the $\bar{\nu}_e$ and the ν_μ, respectively, are uniquely determined for given γ, ε, μ. In this picture, the standard model corresponds to $g^V_{LL}=1$, all other couplings being zero.

The strength of the interaction is determined by the μ lifetime. Since one is only interested in the relative weights of the different couplings, the $g^\gamma_{\varepsilon\mu}$ are normalized:

$$\sum_{\varepsilon,\mu=R,L} Q_{\varepsilon\mu} = 1 \qquad (2)$$

with

$$Q_{RR} \equiv \frac{1}{4}|g^S_{RR}|^2 + |g^V_{RR}|^2 \qquad (3)$$

$$Q_{LR} \equiv \frac{1}{4}|g^S_{LR}|^2 + |g^V_{LR}|^2 + 3|g^T_{LR}|^2 \qquad (4)$$

$$Q_{RL} \equiv \frac{1}{4}|g^S_{RL}|^2 + |g^V_{RL}|^2 + 3|g^T_{RL}|^2 \qquad (5)$$

$$Q_{LL} \equiv \frac{1}{4}|g^S_{LL}|^2 + |g^V_{LL}|^2 \qquad (6)$$

Here $Q_{\varepsilon\mu}$ is the probability to obtain an electron of handedness ε from a muon of handedness μ. It is possible to derive experimental values for Q_{RR}, Q_{LR} and Q_{RL} from the following complete set of measurements [16]:

Shape of positron energy spectrum

$$\rho = 0.752 \pm 0.0027\ [17] \qquad (7)$$

Decay asymmetry between μ spin and e momentum
At spectrum end point:

$$\xi\delta/\rho = 0.9989 \pm 0.0023[18] \qquad (8)$$

Differential:

$$\delta = 0.7486 \pm 0.0038[19] \qquad (9)$$

The polarization vector (P_L, P_{T_1}, P_{T_2}) of the positron yields the remaining six parameters:

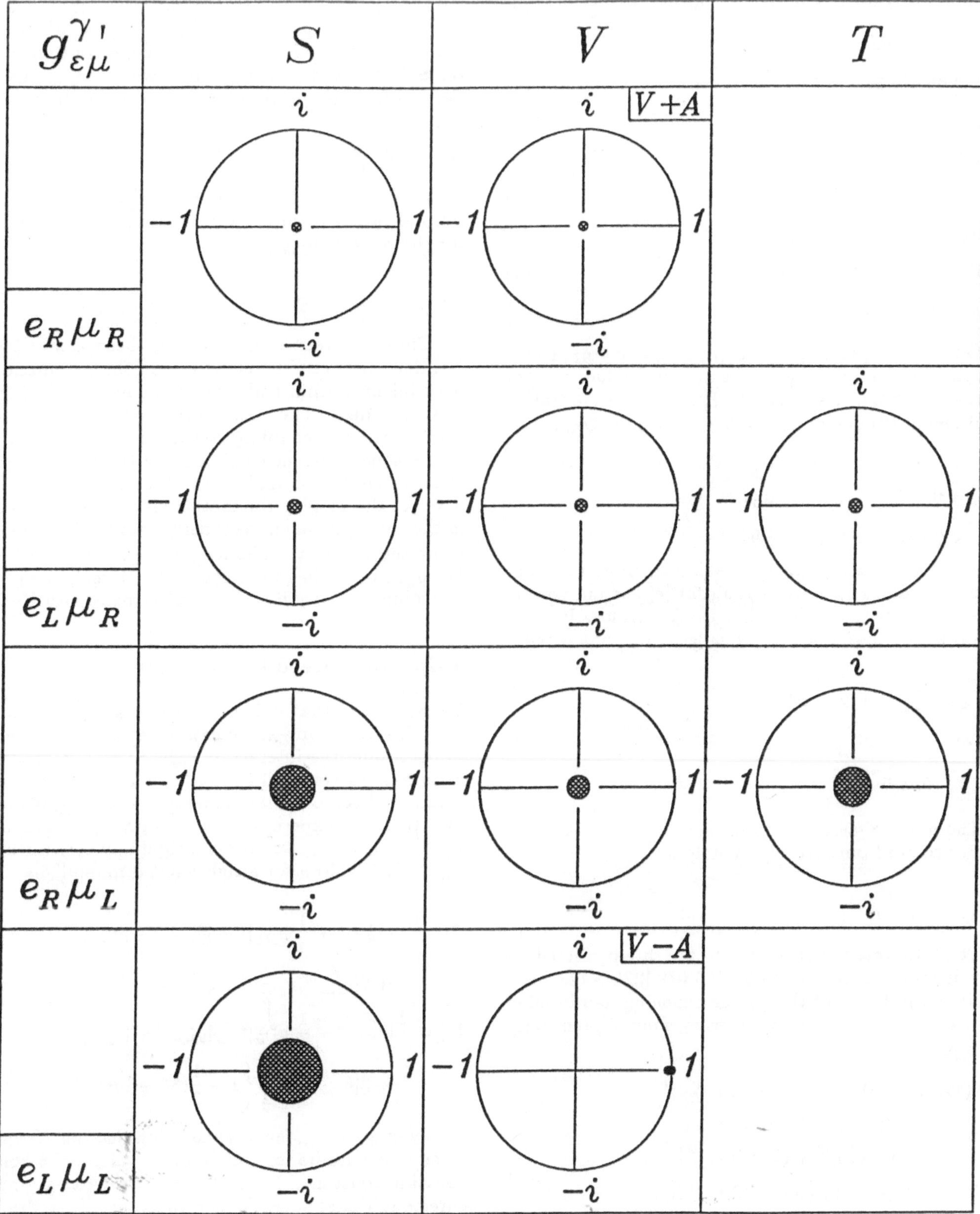

Fig. 1. 90% c.l. limits [16] for the reduced coupling constants $g^{\gamma'}_{\varepsilon\mu} = g^{\gamma}_{\varepsilon\mu}/max(g^{\gamma}_{\varepsilon\mu})$ for the decay $\mu^- \to e^-\bar{\nu}_e\nu_\mu$. Each coupling is uniquely determined by the handednesses ε and μ of the electron and the muon, respectively, and the type of interaction $\gamma = S, V$ or T. The maximal possible values $max(g^{\gamma}_{\varepsilon\mu})$ of the coupling constants are 2, 1 and $1/\sqrt{3}$ for $\gamma = S, V$ resp. T.

Longitudinal polarization P_L

$$< P_L > \equiv \xi' = 0.998 \pm 0.045 \quad [20] \tag{10}$$

Angular dependence of P_L

$$\xi'' = 0.65 \pm 0.36 \quad [20] \tag{11}$$

Energy dependence of P_{T_1}

$$\alpha/A = 0.015 \pm 0.052 \quad [21] \tag{12}$$

$$\alpha'/A = -0.047 \pm 0.052 [21] \tag{13}$$

Energy dependence of P_{T_2}

$$\beta/A = 0.002 \pm 0.018 \quad [21] \tag{14}$$

$$\beta'/A = 0.017 \pm 0.018 \quad [21] \tag{15}$$

One finds upper limits for Q_{RR}, Q_{LR} and Q_{RL}, which in turn yield upper limits for the absolute values of eight complex coupling constants $g^\gamma_{\varepsilon\mu}$. From the normalization requirement (2) one obtains a lower limit for Q_{LL}:

$$Q_{RR} < 0.002, \qquad Q_{LR} < 0.004 \quad ,$$

$$Q_{RL} < 0.045, \qquad Q_{LL} > 0.949 \quad .$$

Since Q_{LL} is bounded by a lower limit, it is not possible to deduce an upper limit for $|g^S_{LL}|$ from normal muon decay without detecting the neutrinos. In fact, with the data from normal muon decay we cannot tell if

$$g^S_{LL} = 0, g^V_{LL} = 1 (V - A) \quad ,$$

or

$$g^S_{LL} = 2, g^V_{LL} = 0$$

The missing piece of information to resolve this ambiguity comes from inverse muon decay:

$$\nu_\mu + e^- \rightarrow \mu^- + \nu_e \quad .$$

The total rate S, normalized to the rate predicted by V–A, is found to be S = 1.054 ± 0.079 [22]. S has been calculated in terms of the charge changing hamiltonian in the parity representation. In terms of our $g^\gamma_{\varepsilon\mu}$ (chirality representation) one gets

$$\begin{aligned} S = (1/2)(1-h)\{&|g^V_{LL}|^2 + (3/8)|g^V_{RL}|^2 \\ &+ (3/32)|g^S_{LR} - (10/3)g^T_{LR}|^2 \\ &+ (3/32)|g^S_{RR}|^2 + (4/3)|g^T_{LR}|^2\} \\ + (1/2)(1+h)\{&|g^V_{RR}|^2 + (3/8)|g^V_{LR}|^2 \\ &+ (3/32)|g^S_{RL} - (10/3)g^T_{RL}|^2 \\ &+ (3/32)|g^S_{LL}|^2 + (4/3)|g^T_{RL}|^2\} \quad , \end{aligned} \tag{16}$$

where h is the helicity of the ν_μ from pion decay. The deviation of |h| from 1 is known very precisely: 1 − |h| < 4.1 x 10^{-3} [23,18]; the sign of h has been determined by **electromagnetic** interaction for ν_μ and $\bar{\nu}_\mu$ [24,25]. Thus S gives information about the first 5 coupling constants g^V_{LL}, g^V_{RL}, g^S_{LR} g^T_{LR} and g^S_{RR}, all of which couple to lefthanded ν_μ. The influence of four of them on S is found to be negligible with the upper limits derived from **normal** muon decay. One obtains

$$S = |g^V_{LL}|^2 \quad , \tag{17}$$

which yields a **lower** limit for $|g^V_{LL}|$, and through the normalization requirement (1) one gets an upper limit for the remaining $|g^S_{LL}|$:

$$|g^S_{LL}|^2 < 4(1-S) \quad . \tag{18}$$

Thus the weak interaction has been completely determined between the electron and the muon and their neutrini in normal and inverse muon decay using only **leptonic** data. The results are shown in Fig. 1, where each of the ten coupling constants is given within one of the squares defined uniquely by the handednesses of electron and muon and by the type of interaction. The outer circles display the mathematical limits for the $g^\gamma_{\varepsilon\mu}$ in the complex plane, the inner circles for nine of the $g^\gamma_{\varepsilon\mu}$ show the areas still allowed by experiment (90% c.l.). For g^V_{LL}, which has been chosen to be real, one gets the small line close to $g^V_{LL} = 1$ in agreement with V–A.

5 How to proceed in the future

We have seen that it is not necessary to perform all of the 19 possible measurements in muon decay in order to determine the interaction. The Standard Model predicts V-A corresponding to $g^V_{LL} = 1$. We can then find a minimal set of measurements needed to **prove** this hypothesis by experiment by calculating the probability $P^e_R \equiv Q_{RR} + Q_{RL}$ to obtain a righthanded electron and $P^\mu_R \equiv Q_{RR} + Q_{LR}$ for a righthanded muon [26]:

$$\begin{aligned} P^e_R = \frac{1}{4}|g^S_{RR}|^2 + \frac{1}{4}|g^S_{RL}|^2 + |g^V_{RR}|^2 + |g^V_{RL}|^2 \\ + 3|g^T_{RL}|^2 = \frac{1}{2}(1-\xi') \end{aligned} \tag{20}$$

$$\begin{aligned} P^\mu_R = \frac{1}{4}|g^S_{RR}|^2 + \frac{1}{4}|g^S_{LR}|^2 + |g^V_{RR}|^2 + |g^V_{LR}|^2 \\ + 3|g^T_{LR}|^2 = \frac{1}{2}\left\{1 + \frac{1}{9}(3\xi - 16\xi\delta)\right\} \end{aligned} \tag{21}$$

By measuring the positron polarization ξ' one obtains upper limits for the absolute values of 5 complex coupling constants, by measuring the decay asymmetry three additional ones. We can conclude: If the Standard Model hypothesis is correct, then only the following five experiments are necessary to determine the interaction: τ_μ(→ Fermi coupling constant), δ, ξ and ξ' (→ 16 decay parameters) and $S(\nu_\mu + e^- \rightarrow \mu^- + \nu_e)$(→ 'V-A')!

Thus the most efficient way to improve our knowledge of the decay interaction is to measure those quantities with higher precision which determine the chiralities of the leptons: Positron polarization, muon decay asymmetry and cross section for the inverse decay with neutrinos of negative helicity. If these measurements agree with the V-A hypothesis, then there is no need to measure the other decay parameters; if they do not agree one can think of specific additional measurements needed to differentiate between the different additional couplings.

Experimentally, however, it seems difficult to improve the existing measurements: $P_\mu\xi$ is now known with a precision of $8\cdot10^{-3}$ [27]; data are being evaluated and the limit will be improved by a factor of 2 by the same group. Most challenging is the measurement of the positron polarization with a precision of <1%; the difficulty there is to find a target with sufficiently large ($\gg$7%) electron polarization.

We conclude by noting that the decay interaction can and should be studied also for the two leptonic decays of the τ lepton $\tau^- \to \mu^-\bar{\nu}_\mu\nu_\tau$ and $\tau^- \to e^-\bar{\nu}_e\nu_\tau$. Most of the decisive experiments of muon decay can also be performed for leptonic τ decays [28]. Even for the same type of coupling the coupling constants can be different from those of μ decay. Scalar couplings, for example, induced by charged Higgs bosons, would be proportional to the mass of the charged lepton ('weak universality') and thus effects would show up in the muonic of the decay τ with a factor $\approx 10^7$ larger than in muon decay! Maybe it is the heaviest lepton that hides some surprises.

References

1. S.H. Neddermeyer and C.D. Anderson, Phys. Rev. **51** (1937) 884.
2. J.C. Street and E.C. Stevenson, 'Washington Meeting', April 29, 1937.
3. A more detailed account of the discovery of the muon is given in the review article by H.-J. Gerber, 'Lepton Properties', Proceedings of the International Europhysics Conference on High Energy Physics, July 1987.
4. H. Yukawa, Proc. Phys. Math. Soc. Jpn. **17** (1935) 48.
5. P. Kunze, Z. Phys. **83** (1933) 1.
6. M. Conversi et al., Phys. Rev. **71** (1947) 209.
7. E. Fermi et al., Phys. Rev. **71** (1947) 314.
8. B. Pontecorvo, Phys. Rev. **72** (1947) 246.
9. J. Steinberger, Phys. Rev. **74** (1948) 500.
10. S.E. Willis et al., Phys. Rev. Lett. **44** (1980) 522.
11. F. Bergsma et al., CHARM Collaboration, Phys. Lett. **B122** (1983) 465.
12. L. Michel, Proc. Phys. Soc. **A63** (1950) 514.
13. Proceedings: ANL-7130, Argonne National Laboratory, Illinois 60440.
14. C. Jarlskog, Nucl. Phys. **75** (1966) 659.
15. K. Mursula and F. Scheck, Nucl. Phys. **B253** (1985) 189.
16. W. Fetscher, H.-J. Gerber and K.F. Johnson, Phys. Lett. **B173** (1986) 102 W. Fetscher, H.-J. Gerber, 'Note on μ decay parameters', in Rev. of Part. Prop., Phys. Lett. **239B** (1990) VI. 11'.
17. S.E. Derenzo, Phys. Rev. **181** (1969) 1854.
18. J. Carr, G. Gidal, B. Gobbi, A. Jodidio, C.J. Oram, K.A. Shinsky, H.M. Steiner, D.P. Stoker, M. Strovink and R.D. Tripp, Phys. Rev. Lett. **51** (1983) 627; H.M. Steiner, private communication.
19. B. Balke et al., Phys. Rev. **D37** (1988) 587.
20. H. Burkard, F. Corriveau, J. Egger, W. Fetscher, H.-J. Gerber, K.F. Johnson, H. Kaspar, H.J. Mahler, M. Salzmann and F. Scheck, Phys. Lett. **150B** (1985) 242.
21. H. Burkard, F. Corriveau, J. Egger, W. Fetscher, H.-J. Gerber, K.F. Johnson, H. Kaspar, H.J. Mahler, M. Salzmann and F. Scheck, Phys. Lett. **160B** (1985) 343.
22. D. Geiregat et al., CHARM II Collaboration, Phys. Lett. **247B** (1990) 131.
23. W. Fetscher, Phys. Lett. **140B** (1984) 117.
24. L.Ph. Roesch et al., Helv. Phys. Acta **55** (1982) 74.
25. A.I. Alikhanov et al., JETP **11** (1960) 1380; G. Backenstoss et al., Phys. Rev. Lett. **6** (1961) 415; M. Bardon et al., Phys. Rev. Lett. **7** (1961) 23; A. Possoz et al., Phys. Rev. Lett. **B70** (1977) 265; R. Abela et al., Nucl. Phys. **A39** (1983) 413.
26. W. Fetscher, Complete Determination of the Weak Interaction in Muon Decay and Comparison with the Standard Model, in "12th International Conference on Neutrino Physics and Astrophysics", Sendai, Japan, June 3-8, 1986.
27. I. Beltrami, H. Burkard, R.D. von Dincklage, W. Fetscher, H.-J. Gerber, K.F. Johnson, E. Pedroni, M. Salzmann and F. Scheck, Phys. Lett. **194B** (1987) 326.
28. W. Fetscher, Phys. Rev. **D42** (1990) 1544.

This article was processed using Springer-Verlag TEX Z.Physik C macro package 1991
and the AMS fonts, developed by the American Mathematical Society.

Z. Phys. C – Particles and Fields 56, S 114–S 116 (1992)

Zeitschrift für Physik C Particles and Fields

Measurements of the muon-neutrino mass

M. Daum, R. Frosch, D. Herter, M. Janousch, P.-R. Kettle

Paul Scherrer Institute, CH-5232 Villigen-PSI, Switzerland

06 September 1991

Abstract. The most precise upper limits for the muon-neutrino mass m_{ν_μ} have been obtained by measuring the muon momentum p_{μ^+} in pion decay $\pi^+ \rightarrow \mu^+ \nu_\mu$ at rest. The current result for the squared neutrino mass, ${m_{\nu_\mu}}^2 = -0.154 \pm 0.045\, MeV^2$, negative by 3.4 standard deviations, indicates that the experimental value for at least one of the three quantities used in the $({m_{\nu_\mu}}^2)$-derivation $(p_{\mu^+}, m_{\pi^+}, m_{\mu^+})$ is wrong.

1 Introduction

The question whether the neutrinos have non-vanishing rest masses is one of the main problems of particle physics. The newest compilation of the Particle Data Group [1] gives the following upper limits:

$$m_{\nu_e} < 17\, eV \quad (CL = 0.95); \tag{1}$$

$$m_{\nu_\mu} < 0.27\, MeV \quad (CL = 0.90); \tag{2}$$

$$m_{\nu_\tau} < 35\, MeV \quad (CL = 0.95). \tag{3}$$

Here, the 'weak eigenstates' ν_e, ν_μ, and ν_τ are assumed to be also mass eigenstates. If this should turn out to be wrong, Eqs. (1)–(3) would be approximately valid for the mass eigenstate which occurs most frequently in the mentioned weak eigenstate.

The limit of m_{ν_μ} quoted above was obtained by measuring the momentum p_{μ^+} of the muon from the decay $\pi^+ \rightarrow \mu^+ \nu_\mu$ of pions at rest [2]. Several other methods of measuring m_{ν_μ} have been used or discussed:

a) $\pi^+ \rightarrow \mu^+ \nu_\mu$ decay of pions in flight. The measured π^+ and μ^+ momenta lead to a precise m_{ν_μ} result if one selects decays in which the μ^+ flies in approximately the same direction as the initial π^+. The best $({m_{\nu_\mu}}^2)$ value obtained by this method is [3]

$${m_{\nu_\mu}}^2 = -0.14 \pm 0.20\, MeV^2. \tag{4}$$

The corresponding upper limit for m_{ν_μ} is $0.50\, MeV$ $(CL = 0.90)$. This result has the advantage of depending only weakly on the π^+ and μ^+ masses used in the analysis.

b) $K_L^0 \rightarrow \pi^- \mu^+ \nu_\mu$ and $\rightarrow \pi^+ \mu^- \overline{\nu}_\mu$ decay. The maximal invariant mass of the $\pi^\mp \mu^\pm$ system strongly depends on m_{ν_μ}. The best upper limit obtained by this method is [4]

$$m_{\nu_\mu} < 0.65\, MeV \quad (CL = 0.90). \tag{5}$$

c) Radiative stopped pion decay $\pi^+ \rightarrow \mu^+ \nu_\mu \gamma$ [5]. The maximal energy of the photons from this decay strongly depends on m_{ν_μ}. However, no good method of subtracting the large background due to photons from radiative stopped muon decay $\mu^+ \rightarrow e^+ \nu_e \overline{\nu}_\mu \gamma$ has been found.

d) Muon capture, *e.g.* $\mu^-\, {}^3\mathrm{He} \rightarrow d\, n\, \nu_\mu$ or $\mu^-\, {}^6\mathrm{Li} \rightarrow t\, t\, \nu_\mu$. The maximal deuteron or triton energy strongly depends on m_{ν_μ}. Theoretical predictions concerning the first of these processes $(\mu^-\, {}^3\mathrm{He})$ are expected to be available soon [6]. For the second process $(\mu^-\, {}^6\mathrm{Li})$, the triton rate in the relevant energy region has been calculated to be too low for the present purpose [7].

In the last few years, progress has been made in the determination of m_{ν_μ} by the 'classical' method, using $\pi^+ \rightarrow \mu^+ \nu_\mu$ decay at rest. For this decay, 4-momentum conservation leads to the equation

$${m_{\nu_\mu}}^2 = {m_{\pi^+}}^2 + {m_{\mu^+}}^2 - 2 m_{\pi^+} \sqrt{{m_{\mu^+}}^2 + {p_{\mu^+}}^2}\,. \tag{6}$$

It appears possible, in the next few years, to determine the squared neutrino mass $({m_{\nu_\mu}}^2)$ by this method to a precision of $\sim\ \pm 0.01\, MeV^2$. The corresponding maximal uncertainties required for the quantities on the right-hand side of Eq. (6), i.e., for m_{π^+} ($\sim$139.57 MeV), m_{μ^+} ($\sim$105.66 MeV), and p_{μ^+} ($\sim$29.79 MeV), are:

$$\Delta m_{\pi^+} \approx \pm 0.17\, keV\ (1.2\, ppm)\,, \tag{7a}$$

$$\Delta m_{\mu^+} \approx \pm 0.18\, keV\ (1.7\, ppm)\,, \tag{7b}$$

$$\Delta p_{\mu^+} \approx \pm 0.13\, keV \; (4.4\, ppm) \, . \tag{7c}$$

These uncertainties are further discussed in the following sections.

2 The charged pion mass

In order to determine $(m_{\nu_\mu}{}^2)$ from Eq. (6), one needs the positive pion mass m_{π^+}. According to the CPT theorem, m_{π^+} is equal to the negative pion mass m_{π^-}, which can be determined precisely from energies of X-rays emitted in the de-excitation of pionic atoms. The current world average from these measurements is [1]

$$m_{\pi^-} = 139.56737 \pm 0.00033\, MeV \, . \tag{8}$$

This average is dominated by the result obtained by Jeckelmann *et al.* [8] from the $4f$–$3d$ X-ray energy of pionic ^{24}Mg,

$$m_{\pi^-} = 139.56752 \pm 0.00037\, MeV \, . \tag{9}$$

The analysis leading to this result has recently been subjected to an experimental test by Thomann *et al.* [9], concerning the value of Γ_{3d}, the strong nuclear absorption width for the 3d state of pionic ^{24}Mg, which is needed to derive the number of K-electrons present during the $4f$–$3d$ transition of pionic ^{24}Mg atoms [8].

In their original analysis [8] Jeckelmann *et al.* used optical potential parameters to calculate

$$\Gamma_{3d} = 0.011 \pm 0.011\, eV \, . \tag{10}$$

In the recent test [9], involving observation of coincidences between $4f - 3d$ and $3d - 2p$ X-rays, the strong absorption width was directly measured. Under certain assumptions concerning dead-time corrections, the result is [9]

$$\Gamma_{3d} = 0.074 \pm 0.018\, eV \, . \tag{11}$$

Possible changes of the dead-time corrections would lead to even larger values. If this new result for Γ_{3d} is used in the analysis [8], one finds that an alternative π^- mass value, higher than that given in Eq. (9) by a factor of ~1.00002, *i.e.*, a mass value near 139.570 MeV, can no longer be excluded. More precise studies of the cascade in pionic ^{24}Mg will be needed to determine reliably which of the two values is correct. The uncertainty of the $m_{\pi^\pm}$ world average after those studies is anticipated to be approximately as that given in Eq. (8), which is higher than required in Eq. (7) by a factor of ~2. This requirement could be met by new m_{π^-} measurements similar to that of Ref. [8].

3 The muon mass

This mass, which is the second of the three quantities needed to calculate $(m_{\nu_\mu}{}^2)$ from Eq. (6), has the current world average [1]

$$m_{\mu^\pm} = 105.658387 \pm 0.000034\, MeV \, . \tag{12}$$

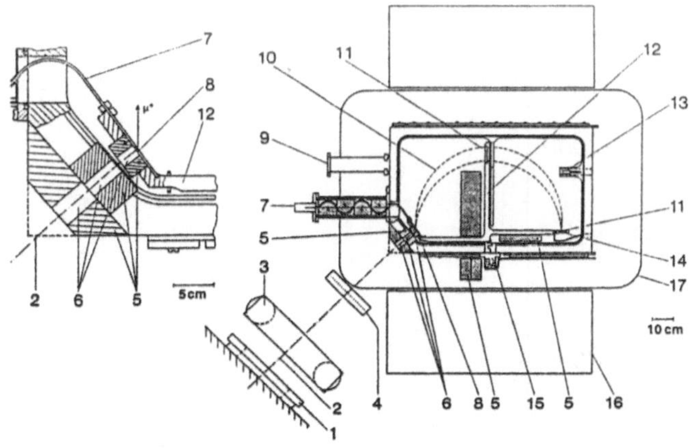

Fig. 1. Experimental apparatus of the new p_{μ^+} measurement of Daum *et al.* [10]: (1) exit vacuum window of pion channel, (2) central π^+ trajectory, (3) beam steering magnet, (4) multi-wire proportional chamber, (5) lead collimators and shielding, (6) graphite degrader, (7) pion-stop scintillation counter light guide, (8) scintillator, (9) port of vacuum pump, (10) accepted muon trajectories, (11) copper collimators, (12) aluminium T-arm supporting scintillator, collimators and silicon counter, (13) NMR probe for magnetic field measurement and stabilization, (14) silicon surface barrier detector, (15) water cooling spiral for temperature control of T-arm, (16) yoke of spectrometer magnet, (17) magnet coils. In the left part of the figure, the pion entry region is shown on a larger scale.

This uncertainty ($\pm 0.3\, ppm$) is better than the requirement given in Eq. (7) by a factor of five.

4 The muon momentum in pion decay at rest

This momentum, p_{μ^+}, is the third quantity required in Eq. (6). The apparatus used in the new p_{μ^+} measurement of Daum *et al.* [10,11] is shown in Fig. 1.

Positive pions from a high-intensity channel of PSI are stopped in a small organic scintillator. Muons from their decay are momentum-analyzed in a homogeneous-field magnetic spectrometer and identified in a silicon surface-barrier detector. The muon spectrum obtained after background subtraction (Fig. 2) has a cutoff at a magnetic field of ~2760 Gauss, which is due to muons originating near the downstream face of the π^+-stop scintillator.

Muons from decays further inside the scintillator lose some of their initial momentum on the way out and lead to the flat part of the spectrum in Fig. 2. The Monte-Carlo generated curve shown in the figure was fitted to the data points with two free parameters only, namely the initial muon momentum p_{μ^+} and a normalization factor for the ordinate.

The result of this new measurement [10], $p_{\mu^+} = 29.79206 \pm 0.00068\, MeV$, agrees with that of the previous experiment [2]. Combination of the two results leads to the new average [10]

$$p_{\mu^+} = 29.79179 \pm 0.00053\, MeV. \tag{13}$$

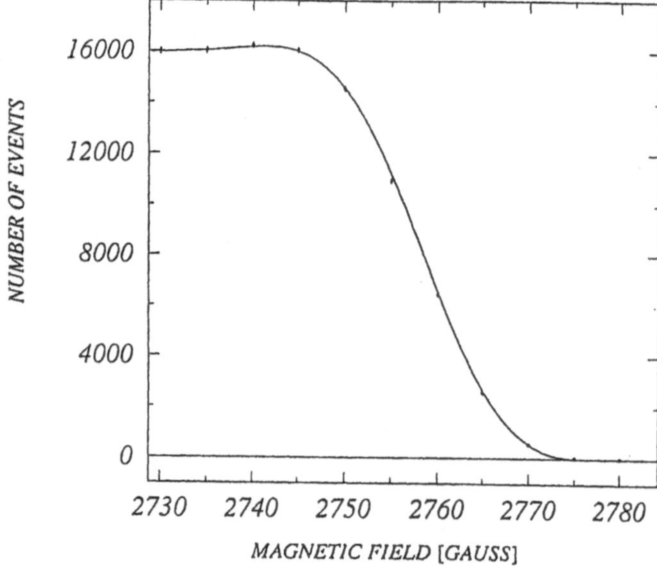

Fig. 2. Muon spectrum determined with the apparatus shown in Fig. 1. Points with error bars: dependence of the good event rate on magnetic field. The solid curve was calculated by a Monte-Carlo program and fitted to the experimental points with two free parameters, namely the muon momentum p_{μ^+} and the normalization factor for the ordinate. χ^2=8.5 for 9 degrees of freedom.

This uncertainty ($\pm 18\,ppm$) is higher than that required in Eq. (7) by a factor of ~4. A clear improvement is expected from p_{μ^+} measurements using a surface muon channel, *i.e.*, a channel which transports muons from stopped π^+ decays at the surface of a pion production target. First tests for such a measurement are about to start at PSI [12].

5 The muon-neutrino mass

Insertion of the values given in Eqs. (8), (12) and (13) into Eq. (6) gives the squared neutrino-mass result

$$m_{\nu_\mu}{}^2 = -0.154 \pm 0.045\,MeV^2\,, \tag{14}$$

which, being negative by ~3.4 standard deviations, indicates that at least one of the three mentioned values is wrong. If the m_{π^-} result of Eq. (8) is replaced by the alternative value discussed in section 2 above, one obtains a $(m_{\nu_\mu}{}^2)$-value which is consistent with zero and corresponds to an upper limit similar to that given in Eq. (2), *i.e.*, ~$0.3\,MeV$.

Acknowledgements: The authors wish to thank Dr. P.F.A. Goudsmit for helpful discussions.

References

1. Particle Data Group, Phys. Lett. B 239 (1990).
2. R. Abela *et al.*, Phys. Lett. B 146 (1984) 431.
3. H.B. Anderhub *et al.*, Phys. Lett. B 114 (1982) 72.
4. A.L. Clark *et al.*, Phys. Rev. D 9 (1974) 533.
5. J. Missimer *et al.*, Nuc. Phys. B 188 (1981) 29.
6. J. Egger, W. Glöckle, private communications, 1991.
7. K. Junker, Nuc. Phys. A 407 (1983) 460.
8. B. Jeckelmann *et al.*, Nuc. Phys. A 457 (1986) 709.
9. S. Thomann, Ph. D. thesis, ETH Zurich, 1990.
10. M. Daum *et al.*, Phys. Lett. B 265 (1991), 425.
11. M. Janousch, Ph. D. thesis, University of Zurich, in preparation.
12. M. Daum *et al.*, proposal R-87-01.2, PSI, 1990.

This article was processed using Springer-Verlag TEX Z.Physik C macro package 1991
and the AMS fonts, developed by the American Mathematical Society.

Z. Phys. C - Particles and Fields 56, S117-S128 (1992)

Zeitschrift für Physik C Particles and Fields

Rare muon decays and physics beyond the standard model

Rabindra N. Mohapatra

Department of Physics, University of Maryland, College Park, Maryland 20742

15-November-1991

Abstract. Rare muon processes such as muonium to antimuonium conversion, $\mu \to e\gamma$ and $\mu \to 3e$ provide sensitive tests of new physics beyond the standard model. Specifically the left-right symmetric models of weak interaction with a low scale for the right-handed W_R boson (in the TeV range) provide a whole range of rare muon processes which are experimentally accessible. In this talk, I discuss the implications of the left-right symmetric models for rare muon processes and also briefly touch on the SUSY models with R-parity violation and their implications for these processes.

1 Introduction

The standard $SU(2)_L \times U(1)_Y$ model of electroweak interaction has proved extremely successful in the description of low energy ($E \leq 100\ GeV$) electroweak phenomena. Yet, there are a number of aspects of the model, which are arbitrarily put in to fit observations, that cry out for a more satisfactory theory. Among them, are the origin of weak mass scale, mass and mixing hierarchy among fermions, strong and weak CP problem, origin of parity violation etc. Several approaches [1] beyond the standard model advocated to solve some of these problems are (i) supersymmetry, that addresses the gauge hierarchy problem; (ii) Left-right symmetry that addressed the problem of parity violation, neutrino masses, quark-lepton mass hierarchies and the strong CP-problem; (iii) Grand-unification, which provides further unification of gauge symmetries as well as the unification of quarks and leptons etc., (iv) Technicolor and compositeness that probe new dynamics at shorter distance scales which may throw light on the origin of electroweak symmetry breaking.

In this paper, I will focus primarily on the left-right symmetric models [2,3] and show how the existing precision experiments allow parameters and mass scales of the model to be such that they lead to measurable rates for several rare muon processes. In the end, I will also briefly discuss the supersymmetric models with R-parity violation and its implications.

The specific version of the left-right model which will be of interest is that of ref.3, which incorporates the see-saw mechanism for neutrino masses, with right-handed neutrinos being heavy Majorana particles. The processes we will focus on are:

i) $\mu \to e\gamma$ and $\mu \to 3e$ decay;
ii) $\mu^-(Z,A) \to e^+(Z-2,A)$,
iii) Muonium to Anti-muonium ($M \to \bar{M}$) transition and exotic muon decay, $\mu^+ \to e^+\bar{\nu}_e\nu_\mu$

To begin with, let us recall that, at present there exist stringent experimental bounds on these processes [4]:

$B(\mu \to e\gamma)$	$\leq 4.9\times10^{-11}$	(ref.5)
$B(\mu \to 3e)$	$\leq 10^{-12}$	(ref.6)
$B(\mu^- Ti \to e^- Ti)$	$\leq 4.9\times10^{-12}$	(ref.7)
$B(\mu^- Ti \to e^+ Ti)$	$\leq 1.7\times10^{-10}$	(ref.8)
$G_{\mu^+ e^- \to \mu^- e^+}$	$\leq .1\ G_F$	(ref.9)
$G_{\mu^+ \to e^+ \bar{\nu}_e \nu_\mu}$	$\leq .18\ G_F$	(ref.10)

It is worth remarking that all these rare processes do break some leptonic quantum numbers. If we assign separate lepton numbers L_e and L_μ to the electron and muon type leptons, then, the first three processes conserve $L_e + L_\mu$ but break $L_e - L_\mu$ by 2 units ($i.e. \Delta(L_e - L_\mu) = 2$). The $\mu^- \to e^+$ conversion on the other hand breaks $L_e + L_\mu$ by two units but conserves $L_e - L_\mu$. The last two processes conserves $L_e + L_\mu$ but break $L_e - L_\mu$ by four units.

To appreciate the significance of rare muon decays for new physics, let us recall the salient features of the leptonic sector of the standard model. Under the $SU(2)_L \times U(1)_Y$ group, the left handed quarks (u_L, d_L) transform as $SU(2)_L$ doublets with $Y = \frac{1}{3}$ whereas u_R and d_R transform as $SU(2)_L$ singlets with $Y = \frac{4}{3}$ and $-\frac{2}{3}$ respectively. On the other hand, the left-handed leptons (ν_L, e_L) transform as weak isodoublets with Y = −1

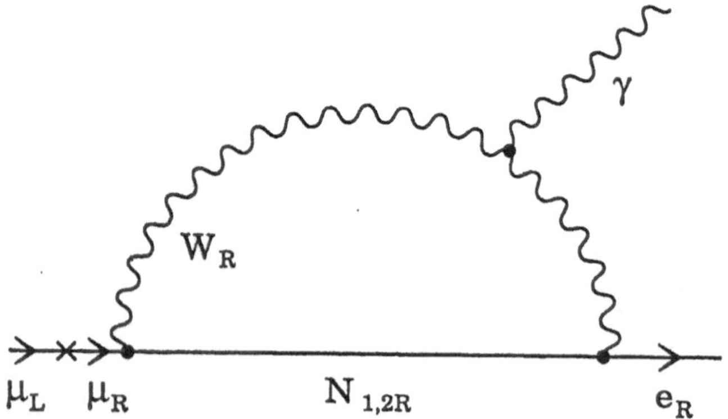

Fig. 1. W_R-induced one loop graph for the $\mu \to e\gamma$ decay.

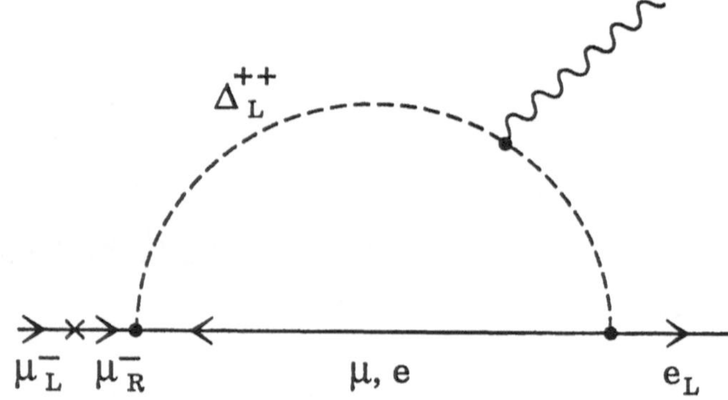

Fig. 2. Δ_L^{++} induced one loop graph for $\mu \to e\gamma$ decay. An analogous graph involving Δ_L^{+}-intermediate states exists.

and e_R transforms as a Y = −2 iso-singlet and no ν_R is included in the spectrum,leading to massless neutrinos. This basic asymmetry between the quarks and leptons built into the construction of the standard model has very significant implications for the rare muon decays since it implies that in the lepton sector, L_e, L_μ and L_τ are separately conserved (in the absence of electroweak anomalies) whereas in the quark sector only the total baryon number is conserved. As a result, all rare muon decays are forbidden in the standard model. Clearly, therefore, any evidence for any of the rare muon processes is an indication of physics beyond the standard model and by the same token, stronger and stronger upper bounds on the rare muon processes increase the domains of validity of the standard model by postponing the scale of new physics further and further. Both are significant for our understanding of the fundamental laws of nature.

2 Quark-Lepton Symmetry, B-L Violation and Rare Muon Processes

In the previous section, we argued that quark-lepton asymmetry of the standard model is intimately connected with the absence of rare lepton number violating process. If we extend the standard model to include the right-handed neutrinos (one per generation), the theory has quark-lepton symmetry. This has two important implications: first, the neutrinos can have mass and since neutrinos are believed to be electrically neutral, this mass can be both Dirac type that conserves lepton number ($\Delta L = 0$) or Majorana type ($i.e. \Delta L = 2$). Secondly, the inclusion of right-handed neutrinos leads to the appearance of B-L as a local anomaly free gaugeable symmetry. If the neutrino mass is Dirac type, B-L remains an exact symmetry; on the otherhand; if it is Majorana type, it is broken by two units.

Let us now turn to the implication of general quark-lepton symmetric theories to rare muon processes. Obviously, if the neutrino mass is Dirac type $\Delta(B-L) = 0$; the only rare processes allowed are the ones that conserve $L_e + L_\mu + L_\tau = 0$ (i.e. $\mu \to e\gamma$, $\mu \to 3e$ and $\mu^- \to e^-$ conversion). It however, turns out that, in this case, a GIM type mechanism is operative in the lepton sector so that all flavor violating processes such as $\mu \to 3e$, $\mu \to e\gamma$ and $\mu^- \to e^-$ conversion are proportional to neutrino masses. Since, the neutrino masses have very stringent upper limits (i.e.$m_{\nu_e} \leq 9.4$ eV; $m_{\nu_\mu} \leq 290$ keV and $m_{\nu_\tau} \leq$ 35 MeV), typical GIM suppression factors in the amplitude are of type ($\Delta\ m_\nu^2/M_{W_L}^2 \leq 10^{-7}$), which makes all the above processes unobservable. For instance, we get, $B(\mu^- \to e^-\gamma) \leq 10^{-26}$.

On the other hand, if the neutrinos are Majorana particles, a popular way to understand their small masses is to use the see-saw mechanism [11], which relies on a mass matrix of the following type:

$$M_\nu = \begin{pmatrix} 0 & m_D \\ m_D & M_R \end{pmatrix} \tag{1}$$

where m_D and M_R are 3× 3 matrices.

In eqn.(1), $M_R \gg m_D$. This matrix can be block diagonalized to lead in lowest order to a light-neutrino mass matrix with

$$M_{\nu_L} \simeq M_D^T M_R^{-1} M_D \tag{2}$$

and a heavy neutrino matrix

$$M_{\nu_R} \simeq M_R \tag{3}$$

We clearly see that the elements (and hence eigenvalues) of the ν_L matrix are much smaller than the typical entries in m_D, the Dirac mass of the neutrino; for m_D, of the order of typical charged fermion (quark or lepton) masses, m_ν is naturally very light without any fine tuning. The see-saw mechanism has an important implication for rare processes. Since the see-saw mass matrix mixes left and right-handed neutrinos, we lose the GIM mechanism for flavor changing radiative loop induced processes, giving rise to the possibility that rare muon

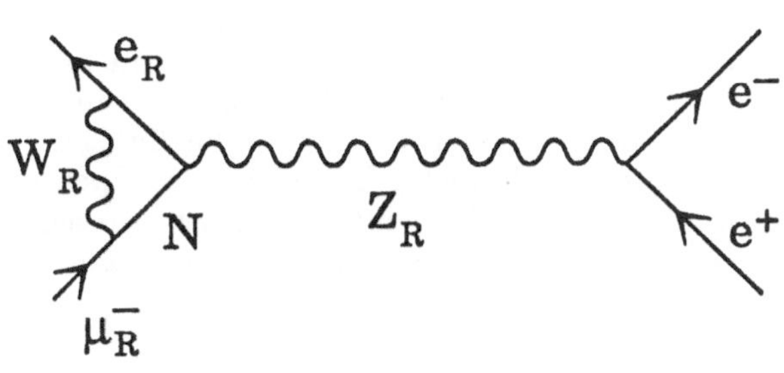

Fig. 3. One loop graphs for $\mu \rightarrow 3e$ decay.

processes can get enhanced. They indeed do as we shall see subsequently.

Let us now discuss if there are any compelling arguments for the two theoretical reasons that can lead to enhanced rare processes. In a complete unification scheme where one considers quarks and leptons to be different manifestations of one fermionic state (multiplet), a right-handed quark would naturally be accompanied by a right-handed neutrino. Besides, a completely quark-lepton symmetric theory is aestetically much more appealing, which also provides a physical meaning to the U(1) generator of the electroweak symmetry group by making it identical to the B-L symmetry [12]. This then leads to the electric charge formula

$$Q = I_{3_L} + I_{3_R} + \frac{B-L}{2} \tag{4}$$

instead of $Q = I_{3_L} + \frac{Y}{2}$ of the standard model.Note that unlike in the standard model,where Y is not identifiable as any physical quantum number,all generators appearing in eq.(4) are physical quantum numbers.

Next, let us see if there is any compelling reason for neutrinos to be Majorana rather than Dirac type particles.

We give two arguments: we see from equation (4) that in the energy regime $E \gg M_{W_L}$, both I_{3_R} and B-L symmetries are violated together since ΔQ and $\Delta I_{3_L} = 0$ i.e.

$$\Delta I_{3_R} = -\frac{1}{2}\Delta(B-L) \tag{5}$$

Thus, as I_{3_R} is broken, we have breaking of lepton number that leads to Majorana neutrinos. The manifest left-right symmetry of the electric charge formula in eqn.(4) suggests that the entire electroweak group is perhaps left-right symmetric i.e. $G_{WK} = SU(2)_L \times SU(2)_R \times U(1)_{B-L}$. In such a case parity violation (i.e.$\Delta I_{3_R} \neq 0$) automatically implies Majorana neutrinos.

A second argument has to do with understanding the quantization of electric changes observed in nature. The point is that in the standard model without right-handed neutrinos, constraints of anomaly freedom and vectorlike QED lead to quantization of electric charges [13]. However, once the right-handed neutrinos are included, this nice property is lost. However, again charge quantization is restored, if we demand that neutrinos are Majorana particles [14]. There exist, therefore, strong theoretical arguments for the existence of right-handed neutrinos as well as Majorana character of neutrinos with heavy right-handed neutrinos.

Experimentally, there is no direct evidence at the moment for the neutrino mass. However, if solar neutrino puzzle is genuine, then any particle physics resolution requires a massive neutrino. Similarly, if the recent evidences for a 17 keV neutrino from beta decay anomalies hold up, that will constitute a direct evidence for massive neutrinos. There also exist interpretations of the dark matter problem and structure formation in the universe, where neutrino mass plays a role. In any case, throughout this talk, I will assume massive neutrinos and look for rare muon processes in theories such as the left-right symmetric models where neutrino masses arise naturally.

3 The Left-Right Symmetric Model of Weak Interactions

Let us now describe the basic features of the left-right symmetric models [3] based on the gauge group $SU(2)_L \times SU(2)_R \times U(1)_{B-L}$. Denoting the quark doublet, (u, d) $\equiv Q^T$ and the lepton doublet $(\nu, e) \equiv \psi^T$ (omitting the generation index);we have the following transformation rules for them under the gauge group:

$$\begin{aligned} Q_L &\equiv (2,1,\tfrac{1}{3}) \quad ; \quad Q_R \equiv (1,2,\tfrac{1}{3}) \\ \psi_L &\equiv (2,1,-1) \quad ; \quad \psi_R \equiv (1,2,-1) \end{aligned} \tag{6}$$

Denoting the $SU(2)_L$, $SU(2)_R$ and $U(1)_{B-L}$ gauge bosons by $(W_L^{\pm}, W_L^3)$, $(W_R^{\pm}, W_R^3)$ and B respectively, the gauge interactions of quarks and leptons can be written as:

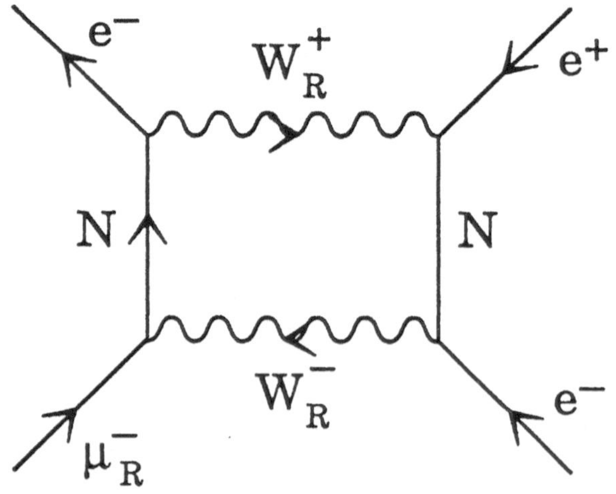

Fig. 4. Box graph for the process $\mu \rightarrow 3e$ decay.

$$
\begin{aligned}
L_{wk} = ig/\sqrt{2}\left[J^{\dagger}_{\mu_L} W^{\mu-}_L + J^{\dagger}_{\mu_R} W^{\mu-}_R + h.c.\right] \\
+ g\left[J^3_{\mu_L} W^{\mu 3}_L + J^3_{\mu_R} W^{\mu 3}_R\right] \\
+ g'\left[\frac{1}{6}(\bar{u}\gamma_\mu u + \bar{d}\gamma_\mu d) - \frac{1}{2}(\bar{\nu}\gamma_\mu \nu + \bar{e}\gamma_\mu e)B^\mu\right]
\end{aligned} \tag{7}
$$

Note that prior to symmetry breaking all gauge bosons are massless; therefore, the theory is parity conserving. In order to explain the near maximal parity violation observed in low energy weak processes, we must have the gauge symmetry breaking such that $M_{W^\pm_R} \gg M_{W^\pm_L}$ and the corresponding massive neutral gauge bosons to satisfy $M_{Z_R} \gg M_{Z_L}$. The process of gauge symmetry breaking also induces mixings between quark and lepton families. Changing to a basis where the quarks and leptons are mass eigenstates introduces the quark and lepton mixing angles θ_L and θ_R in the charged current $W_{L,R}$ interactions for quarks and leptons. As far as the neutral current interactions go, the quark and charged lepton coupling are completely flavor diagonal at the tree level; however for the neutrinos, due to the see-saw mechanism, the neutral couplings are flavor violating at the tree level to order $O(M_{W_L}/M_{W_R})$. For details, see ref. 15.

Finally, the process of symmetry breaking in the minimal model also induces a mixing between W_L and W_R. We will denote the mixing angle by ζ. We will ignore CP-Violating effects.

There exist a variety of low energy observations as well as high energy neutral current scattering data involving neutrinos and collider data on the production of heavy charged bosons, which imply that [15] $M_{W_R} \geq$ 800 GeV to 2 TeV (the former if we ignore the constraints of $K_L - K_S$ mass difference) and $M_{Z_R} \geq 800$ GeV for $M_{\text{HIGGS}} \simeq 1$ TeV from precision measurements of Z-width at LEP. [16] In deriving the above limit on M_{W_R}, it was assumed that the quark mixing angles in the left and right-handed W interactions are the same. In principle, they could be different in which case, the limit on W_R becomes much weaker and we get [17], $M_{W_R} \geq 300$ GeV. Whether the mixing angles in the left and right-handed sector are same or different depends on the detailed choice of the Higgs mechanism. We will work in a minimal version of the model where $\theta_L = \theta_R$ but since the detailed nature of symmetry breaking is an open question even for the standard model, we will present the results for the $\theta_L \neq \theta_R$ as well when appropriate.

3.1 The Higgs sector

The Higgs Sector of the minimal left-right model [3] consist of triplet Higgs bosons: $\Delta_L(3,1,+2) \oplus \Delta_R(1,3,+2)$ and a bidoublet Higgs boson $\phi(2,2,0)$. The Higgs potential of the model has been analyzed in several papers [3, 18]. It was shown in ref. 3 that, there exists a range of parameters for which, the minimum of the Higgs potential corresponds to

$$
< \Delta^\circ_R > = v_R \neq 0 \qquad < \phi > = \begin{pmatrix} \kappa & 0 \\ 0 & \kappa' \end{pmatrix}
$$

and

$$
< \Delta^\circ_L > = v_L \tag{8}
$$

with

$$
v_L \ll \kappa, \kappa' \ll v_R \quad ,
$$

with

$$
v_L \simeq (\gamma_1 \kappa^2 + \gamma_2 \kappa\kappa' ... + \gamma_3 \kappa'^2)/v_R \quad . \tag{9}
$$

At the first stage of the symmetry breaking, $SU(2)_L \times SU(2)_R \times U(1)_{B-L}$ breaks down the $SU(2)_L \times U(1)_Y$ (with $Y = 2I_{3R} + (B-L)$), the gauge group of the standard model. The $W^{\pm_R}$ and Z_R pickup mass at this stage. The second stage of the symmetry breaking is implemented by $< \phi > \neq 0$, when $W^{\pm_L}$ and Z_L pickup mass leaving $U(1)_{em}$ unbroken and a massless photon.

To discuss the fermion masses, let us write down the Yukawa Coupling:

$$
\begin{aligned}
L_Y = \Sigma_{a,b}\Big(h_{ab}\bar{Q}_{La}\phi Q_{R_b} + \tilde{h}_{ab}\bar{Q}_{La}\tilde{\phi} Q_{Rb} \\
+ h^\ell_{ab}\bar{\psi}_{La}\phi\psi_{Rb} + \tilde{h}^\ell_{ab}\bar{\psi}_{La}\tilde{\phi}\psi_{Rb} \\
+ f_{ab}\psi_{la} C^{T-1}\tau_2 \Delta_L \psi_{lb}\Big) + (L \rightarrow R) + h.c.
\end{aligned} \tag{10}
$$

Left-right symmetry implies that $h = h^\dagger$ and $\tilde{h} = \tilde{h}^\dagger$, $h^\ell = h^{\ell\dagger}$ and $\tilde{h}^\ell = \tilde{h}^{\ell\dagger}$. Note that the constraint of that left-right symmetry reduces the number of Yukawa Couplings considerably. For instance in the quark sector, in the absence of CP-violation, for three generations, there are 18 arbitrary parameters in the standard model. On the other hand, the corresponding number for the left-right model is 12, as can be seen from eqn. (10).

It is now easily seen from eqn. (10) that at the first stage, (i.e. when $< \Delta^\circ_R > \neq 0$), the right-handed neutrino picks up a heavy Majorana mass or order f v_R.

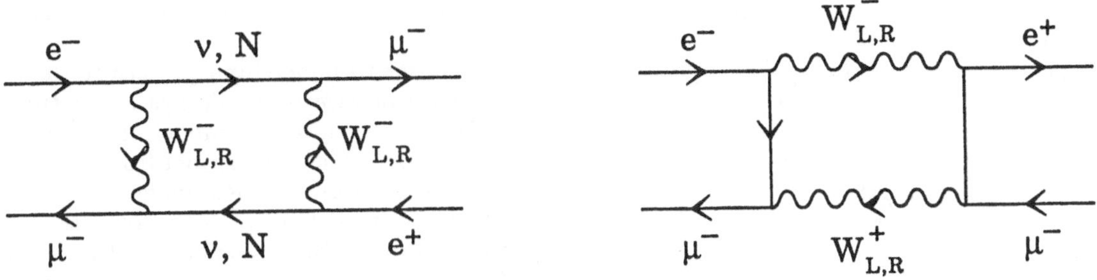

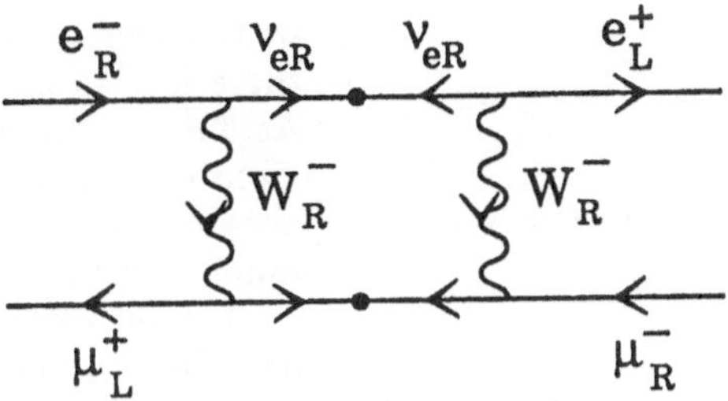

Fig. 5. $W_{L,R}$-induced box graphs for $\mu^+e^- \rightarrow \mu^-e^+$ oscillation.

At the second stage, the neutrinos acquire Dirac masses of order $h^\ell\kappa + \tilde{h}^\ell\kappa'$ and the left-handed neutrinos aquire a small Majorana mass of order v_L. If we set $v_L = 0$, we get the see-saw mass matrix, eqn. (1). There are ways to extend the theory so that this is justified. [1] In any case, we will assume $v_L = 0$, in what follows. There are lower bounds on the masses of the right-handed neutrino, which follow from different observations. First of all if we assume that generation mixing effects are small, we get the following formula for light neutrino masses. Assuming $\kappa' \ll \kappa$ and all Yukawa couplings are of the same order of magnitude, one finds

$$m_{\nu_a} \simeq m^2_{\nu^D_a}/m_{N_a} \tag{11}$$

where the Dirac mass of the neutrino, $m^D_{\nu a} \simeq (h^\ell_{aa}\kappa + \tilde{h}^\ell_{aa}\kappa')$ and N_a denotes the heavy Majorana (right-handed) neutrino. This formula follows from the diagonalization of the see-saw neutrino matrix in eqn. (1) and setting generation mixing parameters $h^\ell_{ab}, \tilde{h}^\ell_{ab}, f_{ab}$ $(a \neq b)$ to small. A prior m^D_ν are of course arbitrary parameters and will depend on further details of the theory. One plausible assumption motivated by unified gauge theories is to assume that $m^D_{\nu a} \simeq m_{la}$ (i.e. the masses of the charged lepton of the corresponding generation). In the bulk of the talk, we will make this assumption, but comment on other possibilities at the end.

Within this assumption, the present upper limits on m_{ν_a} $(a = e, \mu, \tau)$ imply lower limits on N_a as follows:

$$M_{N_e} \geq 6\ GeV\ ;\quad M_{N_\mu} \geq 9\ GeV\ ;\quad M_{N_\tau} \geq 23\ GeV\ . \tag{12}$$

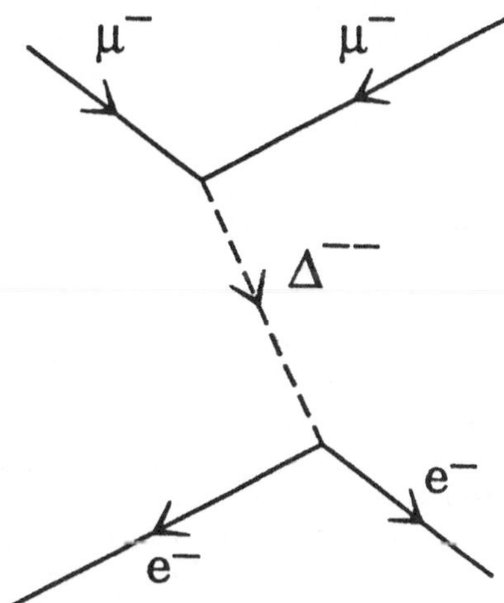

Fig. 6. Δ^{++}_L induced tree-level graph for$\mu^+ e^- \rightarrow \mu^-e^+$ oscillation.

As the upper limits on light neutrino masses go down, the lower limits in eqn.(11) go up. On the mass of N_e, a stronger bound can be derived from a combination of double beta decay and vacuum stability of the left right model.[19] This value is $m_{N_e} \geq 800\ GeV$. Of course, if the Higgs boson mass is very precisely fine tuned it can lower the above bound. There is however, no such bound on N_μ and N_τ.

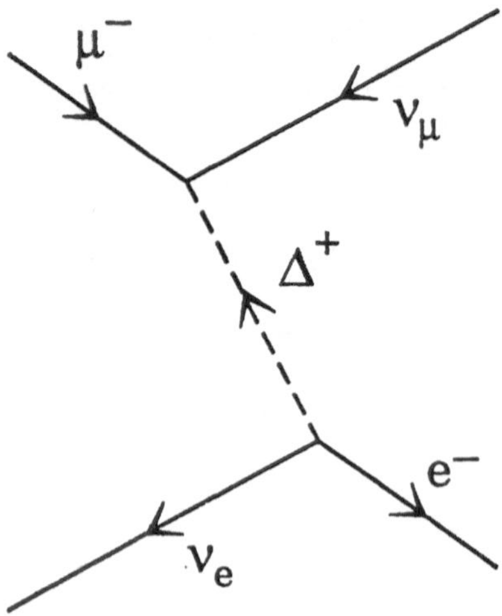

Fig. 7. Δ_L^+ induced graph for exotic muon decay $\mu^- \to e^- \nu_e \bar{\nu}_\mu$.

3.1.1 Mass of the Higgs Bosons

As far as the Higgs bosons are concerned, if they are charged or are $SU(2)_L$ non-siglets, they would contribute to the Z-width if their masses are lighter than 45 GeV. Since, the measured Z-width agrees very well with the predictions of the standard model, it implies that all the Higgs particles have masses heavier than 45 GeV. In most estimates, we will use, Higgs masses of order 50 GeV - 100 GeV. It must however be pointed out that, the physical Higgs boson ϕ_2^o (in the limit $\kappa' = 0$) must be heavier than 3 - 5 TeV since they mediate flavor changing neutral interactions.[20] Since ϕ_2^+ is the $SU(2)_L$ partner of ϕ_2^o, it must also have a mass close to ϕ_2^o (i.e. in the multiTeV range). In the next section, we consider some further constraints on the Δ_L-type Higgs bosons.

Finally, I wish to mention that, while the fermion sector of our model is quark-lepton symmetric, the Higgs sector is not and the phenomenological consistency of the model does not require it to be. However, it is not implausible, to surmise that, if the present scenerio is embedded into dynamical schemes, new Higgs bosons with diquark quantum numbers will emerge, which will carry color and couple to two quarks. They could be either triplets or singlets under the electroweak SU(2) group. But since the di-leptonic Higgs ($\Delta_{L,R}$) are chosen to be triplets to enable them to break $SU(2)_R$ symmetry, quark lepton symmetry would dictate that the diquark Higgs bosons also be $SU(2)_{L,R}$ triplets. Therefore, we will choose $\Delta_L^{qq}(3,1,6) + \Delta_R^{qq}(1,3,6)$ with following Yukawa Couplings:

$$L = f_{ab}^q Q_{La}^T C^{-1} \tau_2 \Delta_L^{qq} Q_{Lb} + L \to R + h.c. \tag{13}$$

The Higgs potential will then contain powers of Δ^{qq}'s.

3.1.2 Couplings of the Fermions and Higgs bosons

In order to understand physically the reason for the occurence of rare muon decays at an enhanced rate in the left-right symmetric model, let us note the features of the model which are distinct from the standard model: a) There are right-handed gauge bosons and the heavy right-handed neutrinos which give rise to new induced contribution to rare muon decays; b) Secondly, there are 14 extra Higgs bosons (four doubly charged, four singly charged and six neutral) in the unitary gauge, some of which can mediate rare muon processes.

Due to the see-saw mechanism, the gauge couplings of right-handed neutrinos are in general flavor violating at the tree-level; furthermore, due to the same reason, there is also no GIM cancellation at the one-loop level. As far as the Higgs bosons are concerned, they also violate flavor in the tree-level couplings. As a result, one obtains, as we will see, highly enhanced rates for rare muon processes, which are necessarily flavor violating.

Let us write down gauge couplings for the physical neutrinos. Defining $\rho = m_D M^{-1}$ andU_+ as the orthogonal matrix that diagonalizes $m_D M_R^{-1} m_D$,V_+ as the matrix that diagonalizes the matrix M_R we find as the matrix that diagonalizes the change lepton mass matrix.

$$\begin{aligned} L_{wk} = & \frac{g}{\sqrt{2}} \left(\bar{\nu}_L K_L \gamma_\mu E_L + \bar{N}_L K' \gamma_\mu E_L \right) W_L^\mu \\ & + \frac{g}{\sqrt{2}} \left(\bar{N}_R K_R \gamma_\mu E_R - \bar{\nu}_R K'' \gamma_\mu E_R \right) W_R^\mu \\ & + 0\left(\rho^2\right) \end{aligned} \tag{14}$$

where

$$\begin{aligned} K_L = U_+^T U_- \; ; \quad K' = V_+^T \rho U_- \; ; \\ K_R = V_+^T U_- \; ; \quad K'' = U_+^T \rho U_- \end{aligned} \tag{15}$$

and $\nu^T = (\nu_e, \nu_\mu, \nu_\tau)^T$ and $E^T = (e, \mu, \tau)^T$ (we have ignored CP-Violation).

Similarly, the neutral current couplings also acquire flavor changing pieces, which we do not display here since, we will not need them in what follows. Incidentially, note that, typically for the muon electron sector, $\rho_{ij} \sim 3\times10^{-3}$.

Turning now to the the Δ-Couplings, it is clear that they induce corrections to several precisely measured parameters of the electrons and muons such as (g-2) and $\mu \to 3e$ decay, Bhabha scattering etc.[21] To discuss this it is more convenient to work in a basis where the charged lepton mass matrix is diagonal. In that basis, we can rewrite the Δ_L Couplings in eqn.(10) as:

$$\begin{aligned} L_\Delta = & E_L^T C^{-1} f E_L \Delta_L^{++} \\ & + \left(1/\sqrt{2}\right) E_L^T C^{-1} f K_L^T \nu_L \Delta_L^+ \\ & + \left(1/\sqrt{2}\right) \nu_L^T K_L f C^{-1} E_L \Delta_L^+ \\ & + \nu_L^T C^{-1} K_L f K_L^T \nu_L \Delta_L^o + h.c. \end{aligned} \tag{16}$$

It is important to note that K_L is an observable matrix involving only the the neutrino mixing angles, which

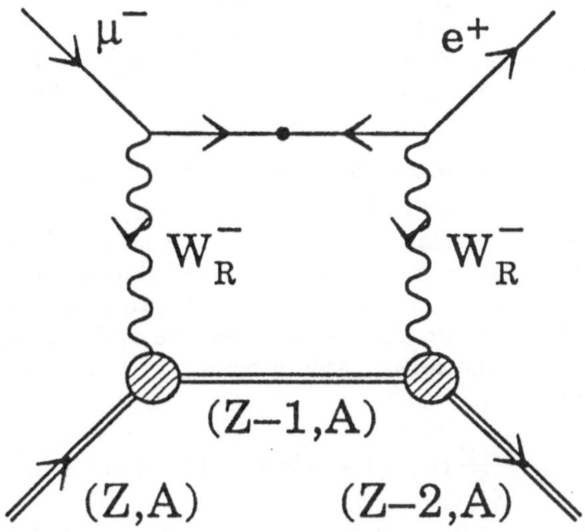

Fig. 8. W_R induced graph for $\mu^- \to e^+$ conversion.

are measurable in the neutrino oscillation parameters. For large mass differences, existing neutrino oscillation searches imply,

$$K_{L,12} \leq .03$$
$$K_{L,13} \leq .17$$
and
$$K_{L,23} \leq .03$$

$$\mu \to 3e \text{ decay: } f_{\mu e} f_{ee} \leq 5 \times 10^{-7} \left(\frac{M_{\Delta}^{++}}{100\ GeV}\right)^2$$

$$(g-2)_\mu = \frac{f_{\mu\mu}^2}{6\pi^2}\left(\frac{m_\mu}{M_{\Delta}^{++}}\right)^2 \leq 2\times10^{-8} \qquad (17)$$

Eqn.(17) implies for instance that for $M_{\Delta}^{++} \simeq 100\ GeV$, $f_{\mu\mu} \leq .25$. Similarly, Bhabha scattering implies $f_{ee} \leq .3$ or so. There are two ways to satisfy the $\mu \to 3e$ constraint: one way is to consider $f_{\mu e} = 0$. In the bulk of this review, we do not consider the possibility, where we leave $f_{\mu e}$ arbitrary but set $f_{\mu\mu} = f_{ee} = 0$. This case is interesting since, $L_e - L_\mu$ becomes an exact symmetry of the Lagrangian and forbids $L_e - L_\mu$ violating processes such as $\mu \to 3e$, $M - \bar{M}$ transition etc.

There are also limits on the mass difference between $M_{\Delta_L^0}$ and $M_{\Delta_L^{++}}$ from the fact that existing data restrict the electroweak ρ - parameter to be very close to unity: (1) Requiring $\Delta\rho_{EW} \leq .01$, and using the formula:

$$\Delta\rho_{EW} = \frac{G_F}{4\sqrt{2}\pi^2}\left[f_{0,+} + f_{+,++}\right] \qquad (18)$$

where

$$f_{a,b} = \left(m_a^2 + m_b^2 - \frac{2m_a^2 m_b^2}{m_a^2 - m_b^2}\ell n\frac{m_a^2}{m_b^2}\right) \qquad (19)$$

we get, for $m_{\Delta_L^0}^2 = (50\ GeV)^2, M_{\Delta_L^+}^2 \leq (230\ GeV)$ and $M_{\Delta_L^{++}}^2 \leq (320\ GeV)^2$ and for $m_{\Delta_L^0}^2 = (100\ GeV)^2$ we get $m_{\Delta_L^+}^2 \leq (245\ GeV)^2$ and $m_{\Delta_L^{++}}^2 \leq (330\ GeV)^2$.

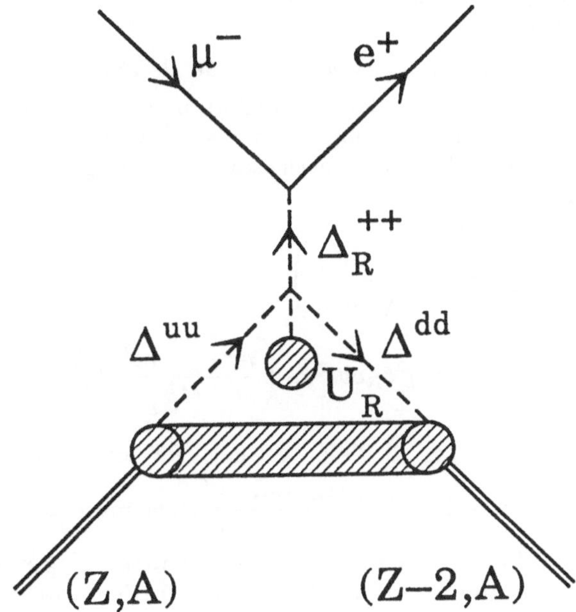

Fig. 9. Diquark Higgs induced graph for $\mu^- \to e^+$ conversion.

Let us now briefly discuss the constraints on the masses and couplings of the diquark Higgs bosons Δ^{qq}. The most stringent constraint on their masses and couplings come from flavor-changing processes such as $K^0 - \bar{K}^0$ and $D^0 - \bar{D}^0$ transition induced by them at the tree level. Let us work in a basis where the down-quarks are eigenstates of quark mass matrices to start with. In this basis, we choose $f_{a,b}^q = 0$ for a≠b so that $K^\circ - \bar{K}^\circ$ transitions are not generated at the tree level. This will however, generate $D^\circ - \bar{D}^\circ$ transitions with strengths proportional to:

$$G_{D^\circ - \bar{D}^\circ} \simeq \frac{(f_{11}^q - f_{22}^q)^2 sin^2\theta_c}{M_{\Delta^{uu}}^2} \qquad (20)$$

The present upperlimit is $G_{D^\circ - \bar{D}^\circ} \leq G_F \times 10^{-6}$.If we allow for fine-tunings between the f-couplings,we can have $M_{\Delta^{uu}} \simeq 50\ GeV, f_{11}^q \leq 1.5\times10^{-3}$. Similar constraints also exist on f_{22}^q.

3.2 Purely Leptonic Rare Processes:

The four leptonic rare processes we consider in this section are:

i) $\mu \to e\,\gamma$
ii) $\mu \to e\,\bar{e}e$
iii) $\mu^+ e^- \to \mu^- e^+$ and
iv) $\mu^+ \to e^+ \bar{\nu}_e \nu_\mu$.

The first two have been discussed in several papers [22,23];$M\bar{M}$ conversion was the subject of ref.24 and 25 and exotic muon decay was considered in ref.26 and 27.

There are two classes of contributions to the above four processes: a) Contribution of W_R and N_R mediated loops; b) Exchange of Δ-Higgs bosons at the tree and loop level.

(i) and (ii): $\mu \to e\gamma$ *and* $\mu \to$ *3e Decays*

The leading one loop contribution to $\mu \to e\gamma$ decay comes from the diagram of fig.1 involving virtual W_R and N_R. This has been calculated [22] to yield branching ratio:

$$B(\mu \to e\gamma) = \frac{3\alpha}{32\pi}\left(\frac{M^2_{W_L}}{M^2_{W_R}}\right)^4 \cdot sin^2\theta_R cos^2\theta_R \left(\frac{m^2{}_{N_2} - m^2{}_{N_1}}{M^2_{W_L}}\right)^2 \quad (21)$$

For $M_{W_R} \simeq 2\ TeV$ we get, for maximal θ_R, $B(\mu \to e\gamma) \simeq \frac{1}{4}10^{-11}\left(\frac{m^2_{N_1} - m^2_{N_2}}{M^2_{W_L}}\right)^2$. The present upper limit of $B(\mu \to e\gamma) \le 10^{-12}$ implies $\Delta M^2_N \le \frac{2}{3}M^2_{W_L}$. Therefore, improvement of this limit further will limit the mass differences between the different heavy right-handed neutrinos.

There are also contributions to $\mu \to e\gamma$ from $\Delta-$exchange graphs (see fig.2). The intermediate states also contain a $\Delta^+\nu$ state. Adding these contributions together, we get [28]

$$B_\Delta(\mu \to e\gamma) \simeq \frac{2\alpha}{3\pi}\frac{m^4_{W_L}}{g^4}\left[\frac{(f'^+f')_{12}}{m^2_{\Delta^{++}}} + \frac{(f^+f)_{12}}{m^2_{\Delta^+}}\right]^2 \quad (22)$$

where $f'_{ij} = 2f_{ij}$ if $i \neq j$ and $f'_{ij} = f_{ij}$ for i = j, where again, we work in the basis where the charged leptons are diagonal. It is important to note that the neutrino mixing matrix K does not appear in eqn. (22). Therefore, if we decouple the (see discussion following eqn.(17)) third generation and set $f_{e\mu} = 0$, then this contribution vanishes.

Turning now to $\mu \to 3e$, there are several contributions: one where an e^+e^- pair is attached to the photon lines in fig. 1 and 2. The contribution of fig. 1 is $B(\mu \to 3e) \simeq \frac{\alpha}{sin^2\theta_W} B(\mu \to e\gamma)$, which is below the present upper bound. Those of fig.2 vanish, if $f_{e\mu} = 0$ as already argued. Then there are the usual flavor changing one loop contributions involving Z_R-exchange (fig.3) and box graphs (fig.4). These graphs do not vanish when $f_{e\mu} = 0$.

Typically, these contributions to $B(\mu \to 3e)$ are of order

$$B(\mu \to 3e) \approx 5\times10^{-5}\left(\frac{M_{W_L}}{M_{W_R}}\right)^4 \cdot \left(\frac{\Delta M^2_N}{M^2_{W_R}}\right)^2 \ell n^2\left(\frac{M^2_{W_R}}{M^2_N}\right) . \quad (23)$$

For $M_{W_R} \simeq 2$ TeV, this implies $\Delta M^2_N \le \frac{1}{3} M^2_{W_R}$, which is a much weaker bound than the one obtained from $\mu \to e\gamma$ decay.

Finally, there is the Δ^{++} contribution discussed in sec. 4, which vanishes for $f_{e\mu} = 0$.

Of course, if we did not assume $f_{e\mu} = 0$, these processes will seriously constrain $f_{e\mu}$ to less than 10^{-6} or so for $M'_\Delta s$ in the 100 GeV range.

(iii)*Muonium-Antimuonium Transition* ($M \leftrightarrow \bar{M}$).

$M - \bar{M}$ transition violates L_μ and L_e by two units i.e. $\Delta L_\mu = -\Delta L_e = 2$. Since the first stage of symmetry breaking in the left-right models violate all lepton flavors by two units, one will expect processes such as $M - \bar{M}$ to occur. As before, there are both gauge and Higgs contributions to this process. It is convenient to parameterize,

$$H_{M\bar{M}} = \frac{G_{M\bar{M}}}{\sqrt{2}}(\bar{\mu}\gamma^\alpha(1+\gamma_5)e)(\bar{\mu}\gamma_\alpha(1+\gamma_5)e) + h.c. \quad (24)$$

The muonium-antimuonium transition probability is given by [29]

$$P(M) \simeq \frac{\delta^2}{2\Gamma} \quad (25)$$

where Γ is the inverse of the muon life time ($\Gamma = .5\times 10^6 sec^{-1}$) and

$$\delta \equiv 2 < M \mid H_{M\bar{M}} \mid \bar{M} > \equiv \frac{16G_{M\bar{M}}}{\sqrt{2}\pi a^3} \quad (26)$$

(a is the Bohr radius of muonium). Several experiments have been conducted over the years [30] to measure P(M). The recent experiments from the TRIUMF [31] and by the Yale-Heidelberg-Los-Alamos group [32] has led to the following bounds on $G_{M\bar{M}}$:

$G_{M\bar{M}} \le .3\ G_F$ (ref.31)
$G_{M\bar{M}} \le .16 G_F$ (ref.32)

The gauge contribution to $G_{M\bar{M}}$ arise from the box graphs depicted in fig. 5. The graphs with a right-handed neutrino intermediate state (fig.5c) is the dominant graph, which gives

$$G_{M\bar{M}} \simeq \frac{G^2_F}{4\pi^2}\left(\frac{M^2_{W_L}}{M^2_{W_R}}\right)^2 M^2_N \ell n\left(\frac{M^2_{W_R}}{M^2_N}\right) \quad (27)$$

For $M_{W_R} \simeq 2\ TeV$, this implies $G_{M\bar{M}} \simeq 10^{-6}G_F$ for $M_N \simeq 400\ GeV$, which is for below the present experimental bounds.

The Higgs contribution to this process arises from the exchange of the doubly charged Higgs boson $\Delta^{++}_{L,R}$ (fig.6), leading to

$$G_{M\bar{M}} \simeq \frac{f_{ee}f_{\mu\mu}}{4\sqrt{2}M^2_{\Delta^{++}}} \quad (28)$$

for $M^{++}_\Delta \simeq 50\ GeV, f_{ee}, f_{\mu\mu} \simeq .1, G_{M\bar{M}} \simeq 7 \times 10^{-2} G_F$, which is not far from the present upper bound and is measurable.

An interesting property of the minimal left-right symmetric model with an eV-keV-MeV type neutrino

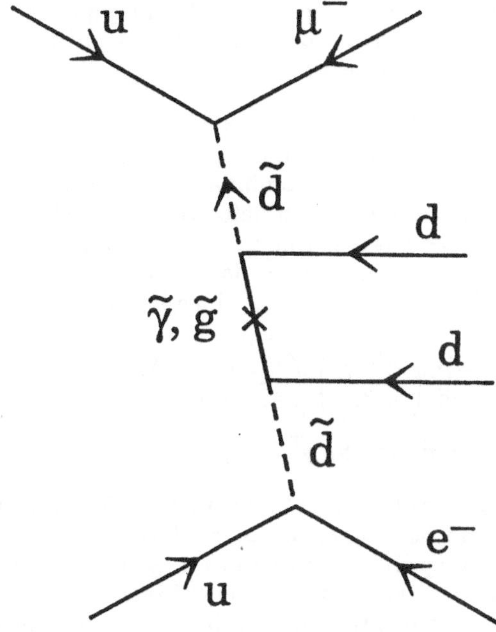

Fig. 10. Photino and gluino induced graph for $\mu^- \rightarrow e^+$ conversion in SUSY models with R-parity violation.

spectrum is that, cosmological constraints on the muon neutrino lifetime can be converted into a lower bound [27] on $G_{M\bar{M}}$. Let us briefly sketch the argument.

Cosmological mass density constraints imply that neutinos heavier than $100\Omega h^2 eV$ must be unstable with life-time obeying the bound:

$$\tau_{\nu_H} \leq \left(\frac{100\Omega h^2 eV}{m_{\nu_H}}\right)^2 t_U \tag{29}$$

where $\frac{1}{2} \leq h \leq 1$; $\Omega \leq 2$; m_{ν_H} is its mass and t_U is the age of the universe ($\simeq 3.15\times10^{17} sec.$). In the left-right model, the ν_μ can decay only via the decay mode [33]$\nu_\mu \rightarrow 3\nu_e$. This process arises via the exchange of Δ_L° bosons and involves the same set of parameters as $M - \bar{M}$ oscillation. Eqn. 29 then implies that,

$$(K_L f K_L^T)_{ee}(K_L f K_L^T)_{e\mu} \geq 3 \times 10^{-12}\,\beta \tag{30}$$

where

$$\beta = \frac{\left(M_{\Delta_L^\circ} inGeV\right)^2}{\left(m_{\nu_\mu} inGeV\right)^{3/2}}$$

If we decouple the third generation and assume as before, $f_{e\mu} = 0$, then we get using eqn. (15)

$$f_{ee}(f_{ee} - f_{\mu\mu}) \geq 10^{-10}\,\beta \tag{31}$$

It is therefore clear from eq. 31 that f_{ee} and $f_{\mu\mu}$ must satisfy a lower bound. From eqn.(31) it might appear that, we could set $f_{\mu\mu} = 0$ and satisfy eqn.(31) and in that case, $G_{M\bar{M}}$ will will vanish. Note, however, that, if $f_{\mu\mu} = 0$, then there is no see-saw mechanism

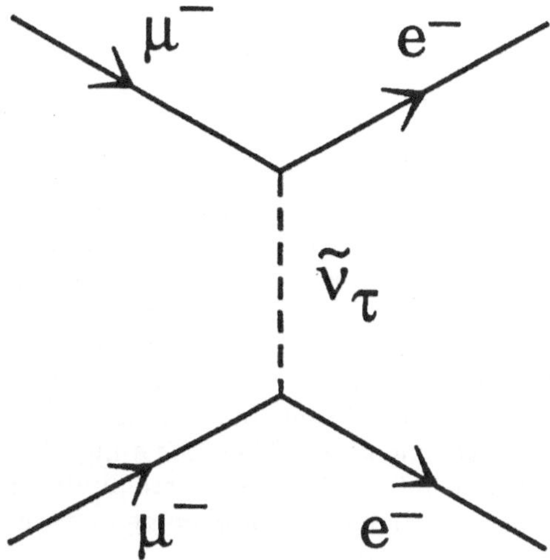

Fig. 11. $\tilde{\nu}_\tau$ mediated graph for $M - \bar{M}$ transition in R-parity violating theories.

for ν_μ andm_{ν_μ} would be of order m_μ (if we assume hierarchical pattern for the leptonic mixing angles and flavor changing Yukawa couplings)in contradiction with observations.

In the see-saw picture, since $m_{\nu_\mu} \simeq m_\mu^2/(2f_{\mu\mu}v_R)$ for v_R in the TeV range,

$$f_{\mu\mu} \geq 1.7\times10^{-2} \tag{32}$$

Note furthermore, $M_{\Delta_L^{++}}$ cannot be too heavy either due to ρ-parameter constraints described earlier. Putting all these together, Herczeg and I [27] have argued that, there must be the following lower bound on $G_{M\bar{M}}$:

$$G_{M\bar{M}} \geq 2 \times 10^{-4}\, G_F \tag{33}$$

The lower bound increases with decreasing muon neutrino mass.

(iv) *Exotic Muon Decay* $\mu^+ \rightarrow e^+\, \bar{\nu}_e \nu_\mu$

In the V-A theory, μ^+ decays to $e^+\nu_e\bar{\nu_\mu}$ which conserves both $L_e and L_\mu$. However, in the left-right symmetric models, presence of the $\Delta^\dagger$-Higgs leads to the new decay mode $\mu^+ \rightarrow e^+\bar{\nu_e}\nu_\mu$ via the Feynman diagram of fig.7(ref.26). The strength of the four-fermi interaction describing this process is given by:

$$G_{\mu^+ \rightarrow e^+\bar{\nu}_e\nu_\mu} \simeq \frac{f_{ee} f_{\mu\mu}}{4\sqrt{2}M^2_{\Delta_L^+}} \tag{34}$$

The same arguments that led to a lower bound for $G_{M\bar{M}}$ then lead to a lower bound for this process: [27]

$$G_{\mu^+ \rightarrow\, e^+\bar{\nu}_e\nu_\mu} \geq 8\times10^{-4}\, G_F \tag{35}$$

3.3 $\mu^- \rightarrow e^\pm$ Conversion in Nuclear Muon Capture

Within the standard model, a nuclear muon capture [34] results in the following processes:

$$\mu^- + p \rightarrow \nu_\mu + n$$

$$\mu^- + p \rightarrow \nu_\mu + n + \gamma \tag{36}$$

$\mu^- \rightarrow e^{\mp}$ conversion can however occur in left-right models and in this section, we will discuss under what conditions, these processes can be observable. It is important to note that the muon conversion process, a slow muon gets captured into an orbit of the nucleus forming a muonic atom. Due to the higher mass, Bohr orbit of the muonic atom is 200 times larger than the corresponding electronic atom. Thus, for a high Z-atom, the muon captured into an S-wave is actually inside the nucleus so that, its chance of undergoing point interaction with a proton goes up.

Let us now discuss the prediction of left-right models for this process:

(i) $\mu^- + (Z, A) \rightarrow e^- + (Z, A)$:

The Feynman diagrams leading to these processes are very similar to those in fig. 4 and 5 except that, the electron line on the right is replaced by the nuclear line. Evaluation of the box graph requires putting a complete set of nuclear states. We will simplify this and consider only the process [22] $\mu^- + p \rightarrow e^- + p$ with a single neutron intermediate state. Summing the box graph and the vertex correction graph (the analogs of fig 4 and 3), we get

$$\begin{aligned} H_{eff} = & -\frac{G_F}{\sqrt{2}} \frac{\alpha\varepsilon}{8\pi} (sin\theta_R cos\theta_R) \bar{e}\gamma_\alpha \\ & \cdot (1-\gamma_5)\mu\bar{p}\left[C_V\gamma_\alpha - C_A\gamma_\alpha\gamma_5\right]p \end{aligned} \tag{37a}$$

where

$$\begin{aligned} C_V &= 4 - a(1 - 4sin^2\theta_W) \\ C_A &= 4 - a(1 - 2sin^2\theta_W) \end{aligned} \tag{37b}$$

where

$$a = \frac{\frac{1}{2}cos^2\theta_W}{cos^2 2\theta_W} \simeq \frac{3}{2}$$

and

$$\begin{aligned} \varepsilon = & \left(\frac{\Delta M_N^2}{sin^2\theta_W M_{W_L}^2}\right) \\ & \cdot \left(\frac{M_{W_L}^2}{M_{W_R}^2}\right)^2 \left(\ell n\left(\frac{M_{W_R}^2}{M_N^2}\right) - 2\right) \end{aligned} \tag{37c}$$

(Here M_N is the average mass of $N_e and N_\mu$). For maximal mixing angles $\theta_R \simeq \pi/4$ and $M_{W_R} \simeq 2\ TeV$ we find,

$$\begin{aligned} \frac{\Gamma(\mu^- \rightarrow e^-)}{\Gamma(\mu^- \rightarrow \nu_\mu)} & \simeq 5 \times 10^{-16} \left(\frac{\Delta M_N^2}{M_{W,L}^2}\right) \\ & \simeq 3\times 10^{-16} \end{aligned} \tag{38}$$

where we used the bound on $\Delta M_N^2/M_{W_L}^2 \simeq \frac{2}{3}$ from $\mu \rightarrow e\gamma$ decay. This is four orders of magnitude smaller than the current best limits.

(ii) $\mu^-(Z, A) \rightarrow e^+(Z-2, A)$:

In this process, the total lepton number changes by two units whereas the combination of lepton numbers $L_e - L_\mu$ is conserved. In left- right models, without diquark Higgs bosons, this process can arise only from the box graph in Fig. 8. This diagram leads to an effective Hamiltonian for the process, which is [22]

$$\begin{aligned} H_{eff} = & \frac{G_F}{\sqrt{2}}\left(\frac{M_{W_L}^2}{M_{W_R}^2}\right)\left(\frac{\alpha}{8\pi sin^2\theta_W}\right) \\ & \cdot sin\theta_R cos\theta_R \mu^T C\gamma_\alpha\gamma_\beta(1-\gamma_5)eM_{\alpha\beta} \end{aligned} \tag{39}$$

where $M_{\alpha\beta}$ involve the muclear matrix elements of weak currents and an integration over internal momenta flowing in the loop. It was noted in ref. 22 that if the nuclei have spin zero, two extra powers of internal momentum appear in the q integral, thereby increasing the higher momentum contribution to the integral. Using this property, in ref. 22, it was shown that,

$$\begin{aligned} B(\mu^- \rightarrow e^+) \simeq & \frac{1}{32} sin^2\theta_R cos^2\theta_R \\ & \cdot \left(\frac{\alpha^2}{8\pi sin^2\theta_W}\right)^2 \left(\frac{M_{W_L}^2}{M_{W_R}^2}\right)^2 \left(\frac{\Delta M_N}{M_A}\right)^2 \end{aligned} \tag{40}$$

For $\theta_R \simeq \frac{\pi}{4}, M_{W_R} \simeq 2$ TeV, this gives

$$B(\mu^- \rightarrow e^+) \simeq 10^{-14} \left(\frac{\Delta M_N}{M_A}\right)^2 \tag{41}$$

This is four orders of magnitude (if $\Delta M_N \simeq M_A$) below the present limit [8] but could perhaps be accessible in a future experiment. In making this estimate, we have simply used an order of magnitude estimate for nuclear matrix elements.

In the presence of the diquark Higgs bosons, new contributions to $\mu^- e^+$ conversion arise from diagrams of type shown in fig. 9.

Typical strength of these graphs using $f^{uu} \leq 10^{-3}$, $f^{dd} \simeq 10^{-1}$ and $M_\Delta^{qq} \simeq 50\ GeV \simeq M_{\Delta^{++}}$:

$$M(\mu^- \rightarrow e^+) \simeq \frac{f_{e\mu} f_{uu} f_{dd} v_R}{M_{\Delta^{++}}^2 M_{\Delta_{qq}}^4} M_{Nuc.} \tag{42}$$

It is clear that, this case is viable if $L_e - L_\mu$ is a good symmetry so that $f_{ee} = f_{\mu\mu} = 0$. To estimate the branching ratio $B(\mu^- \rightarrow e^+)$,one needs the nuclear matrix elements.Several recent estimates [35] suggest a big suppression due to nuclear matrix elements leading to

$$B(\mu^- \rightarrow e^+) \simeq 10^{-17} \tag{43}$$

which is far below the reach of any contemplated experimet.

4 Impact of Solar Neutrino Puzzle on Low Scale W_R Models

An important area of particle physics where hints of new physics are accumulating is the solar neutrino puzzle [36]. The two particle physics solutions to this puzzle are: A) Mikheyev-Smirnov-Wolfenstein matter oscillation mechanism, which requires a neutrino spectrum like $m_{\nu_e} \ll m_{\nu_\mu} \simeq 10^{-3}eV$; b) Voloshin-Vystotskii-Okun magnetic moment mechanism, where also both $\nu_e and \nu_\mu$ masses are small (less than 1 eV or so). The question is, does this fit into a low scale W_R picture. The conventional wisdom is derived from the minimal left-right or SO(10) type pictures where neutrino Dirac masses are of order the charged fermion masses because of which an MSW spectrum world require, $M_{W_R} \simeq 10^{11}$GeV or so. But it is possible to construct radiative pictures [37] for fermion masses, where $m_{\nu_i^D} \ll m_{\ell_i}$. In such pictures an MSW like spectrum emerges, even with M_{W_R} in the TeV range. Therefore, we would expect all our expectations for rare muon processes to hold except the lower bounds for $M - \bar{M}$ and $\mu^- \to e^- \nu_e \bar{\nu}_\mu$ process.

5 R-Parity Violating SUSY Models

In this section, we will consider an alternative class of extensions of the standard model based on supersymmetry, which provides a solution to the Higgs mass problem (i.e. why the Higgs boson mass is not close to the Planck scale). (For a review of the basic ideas of supersymmetry and its application to particle physics, see ref. 1). In the supersymmetric standard models, every fermion is part of a supermultiplet and is accompanied by a boson and every boson is accompanied by a fermion. Thus, the particle spectrum is doubled. Collectively the fermions and bosons together form a chiral supermultiplet if it is to represent a matter or a Higgs particle and a vector supermultiplet, if it is to represent a gauge particle. A superpartner in what follow will be denoted by a twidle over the corresponding particle. The particle content of the SUSY standard model is therefore given by: Quark doublet $Q = (Q, \tilde{Q})$, Lepton doublet $L = (L, \tilde{L})$. Similarly for Singlets and Higgs and gauge bosons using the standard notation. One can assign an R-parity quantum number which is +1 for particles and -1 for their super partners.

The dynamics of supersymmetric theories [38] is specified by three parts: (i) A gauge interaction part, which in addition to the usual gauge interactions of the standard model, interactions of particles with gauginos and their superpartners. (ii) Superpotential part W, which replaces the Yukawa interactions and the Higgs potential (iii) An explicit Soft Supersymmetry breaking part V_{SB} which could arise from Planck scale physics.

The gauge part is automatically fixed once the superfields are assigned to multiplets of the gauge groups. On the otherhand, the superpotential part can be chosen to fit observations as is the V_{SB}. In the SUSY extension of the standard model, one can shoose the superpotential in such a way that R-parity is either conserved or it is violated. In an R-conserving SUSY model R-parity can also be broken spontaneously. It is the R-parity violating models [39] that lead to some interesting rare muon processes at observable rates.

To discuss these effects, let us write the complete superpotential W with the R-conserving part W_o and R-violating part W_1, together:

$$W = W_o + W_1 \tag{44a}$$

$$W_o = h_u Q H_u U^c + h_d Q H_d D^c + h_e L H_d E^c + \mu H_u H_d \tag{44b}$$

$$W_1 = \lambda_{ijk} L_i L_j E_k^c + \lambda_{ijk} Q_i L_j D_k^c + \mu' H_u L_\tau \quad . \tag{44c}$$

We omitted the baryon violating part by assuming Baryon number conservation to hold exactly. W_1 induces corrections to $e - \mu - \tau$ universality, $\pi \to e\nu$ decay, neutrino masses, which have been analyzed by Barger, Giudice and Han [40] who list the bounds on λ andλ' from present observations.

Two rare muon processes, which occur at significant rates in R-violating SUSY theories are $\mu^- \to e^+$ conversion and $\mu^+ e^- \to \mu^- e^+$ transitions. Let us estimate their strength.

a) $\mu^- \to e^+$ *Conversion*

This process is similar to the neutrinoless double beta decay which has been discussed in R-parity violating theories by this author [41]. The dominant diagram is the one shown in fig.10, which leads to an amplitude with strength:

$$G_{\mu^- \to e^+} \simeq \frac{4\pi\alpha_s}{M\tilde{g}} \times \frac{\lambda'_{211}\lambda'_{111}}{M_{\tilde{d}}^4} \tag{45}$$

For $M_{\tilde{g}} \approx 50\, GeV$, $M_{\tilde{d}} \simeq 100\, GeV$, $\lambda'_{111} \leq 3\times10^{-2}$ and $\lambda'_{211} \leq 10^{-1}$, (from ref.40), we get, $G(\mu^- \to e^+) \simeq 10^{-13}$, which due to nuclear suppression effects will most likely lead to an unobservable branching ratio for $B(\mu^- \to e^+)$ as in the previous case.

b) $M - \bar{M}$ *Transition*

This can arise from the diagram in fig. 11, if Re $\tilde{\nu_\tau}$ and Im $\tilde{\nu_\tau}$ have different masses, which will be the case if $\mu' \neq 0$ in W_1.We expect this to lead to observable amplitudes for muonium-anti-muonium transition.A crude estimate for generic SUSY models would give:

$$G_{M\bar{M}} \simeq \frac{\lambda'^2_{321}}{4}\left(\frac{1}{M^2_{Re\nu_\tau}} - \frac{1}{M^2_{Im\nu_\tau}}\right) \tag{46}$$

The present bound on λ_{321} is 10^{-1}; choosing slepton masses of order 100 GeV, we expect, $G_{M\bar{M}} \approx 10^{-2}G_F$ which is certainly within the reach of the next round of experiments. [42]

6 Conclusion

In this review, we have argued that some rare muon processes can occur at measurable rates in left-right symmetric models with see-saw mechanism for neutrino masses. Observation of any of the processes would test a very deep theoretical idea that at the next level of unification quark-lepton symmetry, and left-right symmetry manifest themselves. The obvious question is are there other extensions of the standard model that can lead to similar predictions for rare muon process? I have briefly discussed the supersymmetric models with R-parity violation where the processes $\mu^- \rightarrow e^+$ conversion and $M - \bar{M}$ transition can arise at a rate comparable to that in the left-right supersymmetric models due to photino exchange.

Acknowledgement. I like to acknowledge useful discussions with P. Herczeg, L. Kisslinger, K. Jungmann, B. Matthias and would like to thank Prof. G. zu Putlitz and Dr. K. Jungmann for a wonderful symposium on Muon Physics and for their hospitality. Finally, I want to wish a very happy seventieth birthday to Prof. V. Hughes and sixtieth birthday to Prof. G. zu Putlitz both of whom have added a considerable amount to our understanding of the Muon. This Work is supported by a Grant from the National Science Foundation.

References

1. See R. N. Mohapatra, "Unification and Supersymmetry," Springer-Verlag, (1986).
2. J. C. Pati and A. Salam, Phys. Rev.**D10**, 275 (19740. R. N. Mohapatra and J. C. Pati, Phys. Rev.D11, 566, 2558 (1975). G. Senjanovic and R. N. Mohapatra, Phys. Rev. **D12**, 1502 (1975).
3. R. N. Mohapatra and G. Senjanovic, Phys. Rev. Lett.**44**, 912 (1980) Phys. Rev. **D23**, 165 (1981).
4. R. Engfer and H. K. Walter, Ann. Rev. Nucl. Part. Sci.**36**, 327 (1986).
5. A. Bolton et. al. Phys. Rev. **D38**, 2077 (1988).
6. F. Bellgardt et. al., Nucl. Phys. **B299**, 1 (1988).
7. S. Ahmad et.al. Phys. Rev. **D38**, 2102 (1988).
8. S. Ahmad et.al. Phys. Rev. Lett **59**, 970 (1987).
9. B. E. Matthias et. al., Phys. Rev. Lett. **66**, 2716 (1991). T. M. Huber et. al., Phys. Rev. **D41**, 2709 (1990).
10. D. Krakauer et. al., Phys.Lett.**B263**,535 (1991).
11. M. Gell-Mann, P. Ramond and R. Slansky, in "Supergravity" ed. D. Freedman and P. van Nieuenhuizen, North Holland (1979); T. Yangida, in Proceedings of KEK Workshop on "The Baryon Number of the Universe," ed. R. N. Mohapatra and G. Senjanovic, Phys. Rev. Lett. **44**, 912 1980
12. R. E. Marshak and R. N. Mohapatra, Phys. Lett. **91B**, 222 (1980).
13. N. G. Deshpande, Oregon Preprint OITS-107 (1979). R. Foot, G. C. Joshi, H. Lew and R. Volkas, Mod. Phys. Lett.**A5**, 95 (1990).
14. K. S. Babu and R. N. Mohapatra, Phys. Rev. Lett. **64**, 938 (1989).
15. For a recent summary, see R. N. Mohapatra, Prog. Part. and Nucl. Phys. **26**, 1 ed. A. Faessler (Pergamon Press) (1990).
16. G. Altarelli, R. Casalbuoni and S. DeCurtis, CERN-TH 6051 /91.
17. P. Langacker and S. Uma Sankar, Phys. Rev. **D40**, 1569. F. I. Olness and M. E. Ebel, Phys. Rev. **D30**, 1034 (1989) (1984). A. Datta and A. Raichoudhury, Phys. Rev. **D28**, 1170 (1983).
18. A. Masiero, R. N. Mohapatra and R. D. Peccei, Nucl. Phys. **B192**, 66 (1981). J. Baseq, J. Liu, J. Milutinovic, L. Wolfenstein, Nucl. Phys. **B272**, 145 (1986).
19. R. N. Mohapatra, Phys. Rev. **D34** 909 (1986).
20. R. N. Mohapatra, G. Senjanovic and M. D. Trahn, Phys. Rev. **D28**, 546 (1983); G. Ecker, W. Grimus and H. Neufeld, Nucl. Phys.**B229**, 421 (1983).
21. J. F. Gunion, J. Grifols, A. Mendez, B. Kayser and F. I. Olness, Phys. Rev.**D40**,1546 (1989)
22. Riazuddin, R. E. Marshak and R. N. Mohapatra, Phys. Rev. **D24**, 1310 (1981).
23. P. B. Pal, Nucl. Phys. **227**, 237 (1983) M. Schwarz,Phys.Rev. **D40**,1521 (1989)
24. A. Halprin, Phys. Rev. Letter **48**, 1313 (1982).
25. R. N. Mohapatra, in "Quarks, Lepton and Beyond" ed. H. Fritzech et.al. (Plenum, 1985).
26. P. Herczeg and R.N. Mohapatra,unpublished;Reported in P. Herczeg, "Rare Decay Symposium",ed.D.Bryman,et al (World Scientific,1989) p.24.
27. P. Herczeg and R. N. Mohapatra, Los Alamos Preprint (1991).
28. K. S. Babu and R. N. Mohapatra, Phys. Rev. Lett **64**, 9 (1990).
29. G. Feinberg and S. Weinberg, Phys. Rev. **123**, 1439 (1961).
30. G. M. Marshall et. al. Phys. Rev. **D25**, 1174 (1982). G. W. Beer et. al. Phys. Rev. Lett. **57**, 671 (1986). B. Ni et. al. Phys. Rev. Lett. **59**, 2716 (1987).
31. T. Huber et. al. Ref. 9.
32. B. Matthias et. al. Ref.9.
33. M. Roncadelli and G. Senjanovic, Phys. Lett. **107B**, 59 (1983).
34. For an excellent review, see N. Mukhopadhyay, "Weak and Electromagnetic Interactions in Neclei," ed. P. Deppomier (Edition Frontier, 1989), P. 52.
35. A. Kamal and J. Ng, Phys. Rev. **D20**, 2269 (1979). J. Vergados, Phys. Rep. **133**, 1 (1986);L.S.Kisslinger and R.N.Mohapatra, University of Maryland Preprint UMD-PP-91-021 (1991).
36. For recent reviews, see R. N. Mohapatra and P. B. Pal, "Massive Neutrinos in Physics and Astrophysics," (World Scientific, 1991). J. N. Bahcall, "Neutrino Astrophysics" (Cambridge University Press, 1989).
37. K. S. Babu and R. N. Mohapatra, University of Maryland Report UMD-PP-91-286 (1991).
38. H. Haber and G. Kane, Phys. Rep. **117**, 75 (1985).
39. C. S. Aulakh and R. N. Mohapatra, Phys. Lett. **119B**, 136 (1983). F. Zwirner, ibid **132B**, 103 (1983). L. Hall and M. Suzuki, Nucl, Phys. **B231**, 419 (1984). I. H. Lee, Nucl. Phys. **B246**, 120 (1984). G. G. Ross and J. W. F. Valle, Phys. Lett. **151B**, 375 (1985). J. Ellis et. al., Phys. Lett. **150B**, 142 (1985). S. Dawson, Nucl. Phys. **B261**, 297 (1985). R. Barbieri and A. Masiero, Nucl. Phys. **B267**, 679 (1986).
40. V. Barger, G. F. Giudice and T. Han, Phys. Rev. **D40**, 2987 (1989).
41. R. N. Mohapatra, Phys. Rev. **D34**, 3457 (1986).
42. K.Jungmann and W.Bertl et al, PSI expt. R-89-06.1.

This article was processed using Springer-Verlag TEX Z.Physik C macro package 1991
and the AMS fonts, developed by the American Mathematical Society.

Z. Phys. C - Particles and Fields 56, S129–S134 (1992)

Zeitschrift für Physik C Particles and Fields

The neutrinos in muon decay

Peter Herczeg

Theoretical Division, Los Alamos National Laboratory
Los Alamos, New Mexico 87545

Submitted 21 October 1991

Abstract. We review the available information on the identity of the neutrino states emitted in muon decay, and discuss the exotic decay $\mu^+ \to e^+ \bar{\nu}_e \nu_\mu$.

1 Introduction

The main decay mode of the muon is the decay into two neutrinos [1]: $\mu^+ \to e^+ + n + n'$. In the standard model $n = \nu_{eL}$, $n' = \bar{\nu}_{\mu L}$, where ν_{eL} and $\nu_{\mu L}$ are massless left-handed neutrinos which accompany the corresponding left-handed charged leptons in doublets of $SU(2)_L$. The interaction responsible for this decay is due to W–exchange and has the V-A form

$$H^{(\mu)}_{V-A} = \frac{G_F}{\sqrt{2}} \bar{\mu}\gamma_\lambda(1-\gamma_5)\nu_\mu \bar{\nu}_e \gamma^\lambda (1-\gamma_5) e \; + \; H.c. \quad , (1)$$

where $G_F = (g^2/8m_W^2)(1+\Delta r)$; Δr represents radiative corrections [2].

In extensions of the standard model there may be new decay modes of the type $\mu^+ \to e^+ +$ *neutrinos*, and new decay interactions may be present. Among the decays $\mu^+ \to e^+ +$ *neutrinos* there could be some which violate the conservation of lepton family numbers and possibly also the conservation of the total lepton number. In the presence of the new interactions the neutrinos are expected to be massive, and the gauge group eigenstates are not expected to coincide with the mass-eigenstates. The mixing of the neutrinos may involve also heavy neutrino states, which cannot be emitted in the decays.

In this talk we shall review the existing information on the identity of the neutrinos in the main decay mode of the muon, and then discuss the particular exotic decay mode $\mu^+ \to e^+ \bar{\nu}_e \nu_\mu$.

2 The identity of the muon decay neutrinos

The most general local nonderivative four-fermion interaction that allows for lepton family number and total lepton number violation can be written in the helicity projection form [3] as [4]

$$\begin{aligned} H = 4 \sum_{i,j} [& (g^V_{LL})_{ij} \bar{e}_L \gamma^\lambda n_{iL} \bar{n}_{jL} \gamma_\lambda \mu_L \\ & + (g^V_{LR})_{ij} \bar{e}_L \gamma^\lambda n_{iL} \bar{n}^c_{jR} \gamma_\lambda \mu_R \\ & + (g^V_{RL})_{ij} \bar{e}_R \gamma^\lambda n^c_{iR} \bar{n}_{jL} \gamma_\lambda \mu_L \\ & + (g^V_{RR})_{ij} \bar{e}_R \gamma^\lambda n^c_{iR} \bar{n}^c_{jR} \gamma_\lambda \mu_R \\ & + (g^S_{LL})_{ij} \bar{e}_L n^c_{iR} \bar{n}^c_{jR} \mu_L \\ & + (g^S_{LR})_{ij} \bar{e}_L n^c_{iR} \bar{n}_{jL} \mu_R \\ & + (g^S_{RL})_{ij} \bar{e}_R n_{iL} \bar{n}^c_{jR} \mu_L + (g^S_{RR})_{ij} \bar{e}_R n_{iL} \bar{n}_{jL} \mu_R \\ & + (g^T_{LR})_{ij} \bar{e}_L t^{\alpha\beta} n^c_{iR} \bar{n}_{jL} t_{\alpha\beta} \mu_R \\ & + (g^T_{RL})_{ij} \bar{e}_R t^{\alpha\beta} n_{iL} \bar{n}^c_{jR} t_{\alpha\beta} \mu_L] + H.c. \end{aligned} \quad (2)$$

The fermion fields in Eq. (2) are mass-eigenstates. The indices i, j run over all the neutrino states that can be emitted in the decay. For a fermion field $(f) f_L = \frac{1}{2}(1-\gamma_5)f$, $f_R = \frac{1}{2}(1+\gamma_5)f$; $t_{\alpha\beta} = \frac{i}{2\sqrt{2}}[\gamma_\alpha, \gamma_\beta]$. n_{iL} includes all the left-handed neutrino states ($n_{1L} \equiv \nu_{eL}$, $n_{2L} \equiv \nu^c_{eL}$, $n_{3L} = \nu_{\mu L}$ etc.), and the set n^c_{Ri} all the right-handed ones ($n^c_{1R} \equiv \nu^c_{eR}$, $n^c_{2R} \equiv \nu_{eR}$, $n^c_{3R} \equiv \nu^c_{\mu R}$, etc.).[1]

A special case of the Hamiltonian (2) is the one ($H^{(\mu)}_{LC}$), which contains all the possible interaction types (V,A,S,...), but allows only decay modes which conserve the individual lepton family numbers, and includes only ν_{eR} and $\nu_{\mu R}$ in addition to ν_{eL} and $\nu_{\mu L}$. $H^{(\mu)}_{LC}$ contains 10 coupling constants [$(g^V_{LL})_{13} \equiv g^V_{LL}$, $(g^V_{LR})_{14} \equiv g^V_{LR}$, $(g^V_{RL})_{23} \equiv g^V_{RL}$, $(g^V_{RR})_{24} \equiv g^V_{RR}$, $(g^S_{LL})_{24} \equiv g^S_{LL}$, $(g^S_{LR})_{23} \equiv g^S_{LR}$, $(g^S_{RL})_{14} \equiv g^S_{RL}$, $(g^S_{RR})_{13} \equiv$

[1] In Ref. [4] the constant $(g^V_{LL})_{ij}$ is denoted as $(G_o/\sqrt{2}) g^{LL}_{ij}$; the correspondence between the notations for the other constants is analogous.

g^S_{RR}, $(g^T_{LR})_{23} \equiv g^T_{LR}$, and $(g^T_{RL})_{14} \equiv g^T_{RL}$]. In Ref. [5] limits have been set on all of these using the available experimental results on the muon lifetime, the positron energy spectrum and polarizations, and the inverse muon decay cross-section. One of the results of the analysis is the lower bound

$$Q_{LL} \equiv \left(\frac{1}{4} \mid g^S_{LL} \mid^2 + \mid g^V_{LL} \mid^2\right)(G_\mu/\sqrt{2})^{-2} > 0.949 \text{ (90\% } c.l.), \tag{3}$$

obtained from muon-decay data alone on the quantity Q_{LL} which contains the standard model contribution. In Eq. (3) G_μ is the muon decay constant ($G_\mu = 1.16637(2) \times 10^{-5} GeV^{-2}$). For the contribution of the remaining coupling constants to the decay rate upper limits have been obtained (also from muon decay measurements), which are smaller than $(G_\mu/\sqrt{2})^2$ by factors of about 20 to 500. However, some of the coupling constants could still be quite large. For example the limit on $\mid g^S_{RL} \mid$ is $\mid g^S_{RL} \mid < 0.424(G_\mu/\sqrt{2})$ [5].

In the general case the muon decay parameters can be expressed through a set of quadratic functions of the coupling constants, which are generalizations of those for the lepton family number conserving case [4]. There is a one-to-one correspondence between the two sets [4], and consequently it is possible to use the results for the lepton family number conserving case to obtain constraints for the Hamiltonian (2). Thus, since [4]

$$\begin{aligned} \frac{1}{4} \mid g^S_{LL} \mid^2 &\longleftrightarrow \sum_{i>j} \mid (g^V_{LL})_{ij} + \frac{1}{2}(g^S_{LL})_{ji} \mid^2 , \\ \mid g^V_{LL} \mid^2 &\longleftrightarrow \sum_{i\leq j} \mid (g^V_{LL})_{ij} + \frac{1}{2}(g^S_{LL})_{ji} \mid^2 , \end{aligned} \tag{4}$$

the constraint (3) becomes

$$Q_{LL} \equiv \sum_{i,j} \mid (g^V_{LL})_{ij} + \frac{1}{2}(g^S_{LL})_{ij} \mid^2 > 0.949(G_\mu/\sqrt{2})^2 . \tag{5}$$

In the analysis it has been assumed that the masses of the neutrinos that can be emitted in the decay are small enough that their effect on the positron spectrum can be neglected.

Information on one of the neutrino states in muon decay comes from the inverse muon decay process $\nu_\pi e^- \rightarrow \mu^- n_i$ where ν_π is the neutrino state emitted in $\pi^+ \rightarrow \mu^+ \nu_\pi$ decay, and n_i are some neutrino states. The cross-section for this reaction has been measured recently by the CHARM II collaboration [6] and by the CCFR collaboration [7], obtaining

$$S = 1.054 \pm 0.079 \qquad (CHARMII) , \tag{6}$$

$$S = 0.981 \pm 0.057 \qquad (CCFR) , \tag{7}$$

where S is the total cross-section for $(\nu_\pi e^- \rightarrow \mu^- n_1) + (\nu_\pi e^- \rightarrow \mu^- n_2) + ...$ relative to the cross-section predicted by the standard model. As ν_π is to an excellent approximation left-handed [8], S is given by (taking into account the limits on the coupling constants)

$$S \simeq \sum_i \mid (g^V_{LL})_{i3} + \frac{1}{2}(g^S_{LL})_{3i} \mid^2 (G_\mu/\sqrt{2})^{-2} , \tag{8}$$

where the neutrino state $\nu_\pi = \sum_j c_j n_{jL}$ has been denoted as n_{3L}. ($S \simeq\mid g^V_{LL} \mid^2 (G_\mu/\sqrt{2})^{-2}$ in the case of the Hamiltonian $H^{(\mu)}_{LC}$ [5]).

The experimental value of S enables one to set a lower bound on the term which includes the standard model contribution, and using $Q_{LL} \leq 1$ an upper bound on the remaining part of Q_{LL} [5]. From (7) one obtains [7]

$$\sum_i \mid (g^V_{LL})_{i3} + \frac{1}{2}(g^S_{LL})_{3i} \mid^2 > 0.925(G_\mu/\sqrt{2})^2 \text{ (90\% } c.l.), \tag{9}$$

$$\sum_{\substack{i,j \\ j\neq 3}} \mid (g^V_{LL})_{ij} + \frac{1}{2}(g^S_{LL})_{ji} \mid^2 < 0.075(G_\mu/\sqrt{2})^2 \text{ (90\% } c.l.) \tag{10}$$

The limits from (6) are only slightly weaker.

The bound (9) implies that at least one of the μ^+–decay modes which involves the neutrino $\bar{\nu}_\pi$ produced in $\pi^- \rightarrow \mu^- \bar{\nu}_\pi$ decay dominates the μ^+–decay rate [4].

Regarding the nature of the state ν_π there is some experimental information from a search [9] for $e^\pm$–production by ν_π on nucleons. The experiment yielded $\Gamma(\pi^+ \rightarrow \mu^+ \bar{n}_e)/\Gamma(\pi^+ \rightarrow \text{all}) < 1.5 \times 10^{-3}$ (90% c.l.) and $\Gamma(\pi^+ \rightarrow \mu^+ n_e)/\Gamma(\pi^+ \rightarrow \text{all}) < 8 \times 10^{-3}$ (90% c.l.), where $\bar{n}_e$ and n_e are neutrino states capable of producing e^+ and e^-, respectively.[2] This indicates that ν_π is not the state which accompanies the positron or the electron in nuclear beta decay.

Information on the second neutrino in muon decay follows from the experiment of Ref. [10], where neutrinos (n_e) from μ^+–decay have been observed through the reaction $n_e D \rightarrow ppe^-$. The good agreement of the measured $n_e D \rightarrow ppe^-$ cross-section and the calculated one in the standard model indicates that the total muon decay rate contains a substantial contribution from muon decay into a final state in which one of the neutrinos is the one accompanying the positron in nuclear beta decay.

Experimental results [10,11,12,13] are available also on decays of the type $\mu^+ \rightarrow e^+ \bar{n}_e n_x$, where n_x is some neutrino state and $\bar{n}_e$ is a neutrino state capable of

[2] We remind the reader that we use the term "neutrino state" to refer to either a "neutrino" or an "antineutrino." Also, e.g. the state $n_e(\bar{n}_e)$ could be an antineutrino (neutrino) state. For Majorana neutrinos there is of course no distinction. On the other hand $\nu_e(\bar{\nu}_e)$ denotes the electron neutrino (electron antineutrino).

producing positrons on protons.[3] The best limit on the branching ratio

$$R \equiv \Gamma(\mu^+ \to e^+ \bar{n}_e n_x)/\Gamma(\mu^+ \to \text{ all}) \tag{11}$$

is

$$R < 0.018 \quad (90\% \; c.l.) \quad , \tag{12}$$

from the experiment of Ref. [13].[4]

It is evident from the above discussion that the experimental information regarding the muon decay interaction and the nature of the neutrinos involved is consistent with the standard model picture. Searches for non-standard contributions and non-standard decay modes continue to be of great importance. In the next section we shall consider the exotic decay $\mu^+ \to e^+ \bar{\nu}_e \nu_\mu$.

3 The decay $\mu^+ \to e^+ \bar{\nu}_e \; \nu_\mu$

The exotic decay mode $\mu^+ \to e^+ \bar{\nu}_e \nu_\mu$[5] was first considered [15] before the advent of gauge theories, in connection with the question regarding the nature of the suspected invariance principle which was supposed to account for the apparent absence of processes like $\mu \to e\gamma$, or $\mu^- N \to e^- N$. In the scheme of Ref. [15] the decay $\mu^+ \to e^+ \bar{\nu}_e \nu_\mu$ is one of the processes (muonium to antimuonium conversion is another) which would be allowed if the conservation of a multiplicative quantum number (muon parity) would be involved, but forbidden if the conservation law concerned additive quantum numbers (muon number and electron number). In the standard model the lepton family numbers are conserved and therefore $\mu^+ \to e^+ \bar{\nu}_e \nu_\mu$ (like $\mu \to e\gamma$, etc.) is forbidden. Beyond the standard model the presence of conserved lepton numbers (additive or multiplicative) is generally not expected.[6] We should note also that since the strength of the $\mu^+ \to e^+ \bar{\nu}_e \nu_\mu$ interaction is not related to the weak interactions, the existence of a conserved multiplicative quantum number cannot be ruled out by the absence of $\mu^+ \to e^+ \bar{\nu}_e \nu_\mu$ (or muonium to antimuonium conversion, etc.) at a certain level.

The decay $\mu^+ \to e^+ \bar{\nu}_e \nu_\mu$ could be mediated at the tree-level by non-standard Higgs bosons, or new gauge bosons. A simple extension of the standard model which allows $\mu^+ \to e^+ \bar{\nu}_e \nu_\mu$ can be obtained by adding to the Higgs doublet a singlet charged Higgs boson (h) and including singlet right-handed neutrinos. A coupling of the form[7]

$$L = g_{ee} \overline{\nu^c_{eR}} e_R h + g_{\mu\mu} \overline{\nu^c_{\mu R}} \mu_R h \; + \; H.c. \tag{13}$$

is then possible, which (if the right-handed neutrinos are sufficiently light) gives rise to $\mu^+ \to e^+ \bar{\nu}_e \nu_\mu$. The corresponding interaction after a Fierz transformation can be written in the form

$$H = \frac{g_{ee} g^*_{\mu\mu}}{8 m_h^2} \bar{\mu} \gamma_\lambda (1+\gamma_5) \nu_e \bar{\nu}_\mu \gamma^\lambda (1+\gamma_5) e \; + \; H.c. \tag{14}$$

Denoting $\overline{G} = \sqrt{2} g_{ee} g^*_{\mu\mu}/8m_h^2$, the branching ratio R (see (11), where now $\bar{n}_e = \bar{\nu}_e$, $n_x = \nu_\mu$) is given by $R = \mid \overline{G} \mid^2 / (G_F^2 + \mid \overline{G} \mid^2) = \mid \overline{G}/G_\mu \mid^2$. The experimental limit (12) does not apply for this case, since the right-handed neutrinos do not couple to the W. There are however several indirect constraints on $\overline{G}$.

One constraint follows from the limits on the coupling constants of the general Hamiltonian (2). The Hamiltonian consisting of the standard model contribution and the interaction (14) is a special case of (2) with

$$(g^V_{LL})_{13} = G_F/\sqrt{2} \;\; , (g^S_{RR})_{24} = 2\overline{G}^*/\sqrt{2} \;\; , \tag{15}$$

and all the other coupling constants set to zero. From the analysis of muon decay data one has the limit $\mid g^S_{RR} \mid < 0.066 (90\% \; c.l.)$ [5] for $H^{(\mu)}_{LC}$, which translates in the general case to

$$\sum_{i>j} \mid (g^V_{RR})_{ij} \; + \; \frac{1}{2} (g^S_{RR})_{ji} \mid^2 < \; 0.0011 (G_\mu/\sqrt{2})^2 \tag{16}$$
$$(90\% \; c.l.) \; .$$

Since the left-hand side of Eq. (16) is simply $\mid \frac{1}{2} (g^S_{RR})_{24} \mid^2$, we obtain from (15) and (16) the bound

$$\mid \overline{G} \mid < 0.032 \; \mid G_\mu \mid \quad (90\% \; c.l.) \; . \tag{17}$$

Limits on $\overline{G}$ are implied also by the experimental value of the W−mass, and by charged current universality.

The muon decay constant in the presence of the interaction (14) is given by $G_\mu = G_F(1+$

[3] Further sources of positrons could be neutrinos from ordinary μ^+−decay due to oscillations, or even without oscillations if the weak eigenstate neutrinos contain heavy mass-eigenstates which cannot be produced in the decay[14]. A Majorana ν_{eL} can also produce a positron, but the amplitude is proportional to the neutrino mass.

[4] It should be noted that the limit (12) holds only for such decays $\mu^+ \to e^+ \bar{n}_e n_x$ where the spectrum of $\bar{n}_e$ (or the spectrum of n_x, if n_x can produce positrons) is the same as the spectrum of $\bar{\nu}_\mu$ in μ^+−decay. I am grateful to R. L. Burman for calling my attention to this aspect of the experiment. The experiment of Ref. [12], in which the reaction $\bar{\nu}_\pi + e^- \to \mu^- \bar{n}_e$ rather than muon decay was searched for, sets a limit only for the branching ratio of $\mu^+ \to e^+ \bar{n}_e \nu_\pi$.

[5] The $\bar{\nu}_e$ and ν_μ in this decay mode are by definition identical or nearly equal to the weak eigenstates $\bar{\nu}'_e$ and ν'_μ. If ν_e is a Majorana neutrino, then for left-handed (right-handed) couplings $\bar{\nu}_e$ is the right-handed (left-handed) component of ν_e.

[6] It is interesting to mention however the model of Ref. [16], where it was shown that the three-family standard model for the leptons can be extended in such a way that a multiplicative quantum number is conserved, while the conservation of the lepton family numbers is broken. This is achieved by requiring invariance of the unbroken theory under the permutation group S_3, and introducing three Higgs doublets, one for each lepton family. The decay $\mu^+ \to e^+ \bar{\nu}_e \nu_\mu$ is mediated by one of the new Higgs bosons. If the quarks are treated in this model in the same way as the leptons, the model runs into contradiction with experiment.

[7] This is analogous to the coupling of a doubly charged singlet Higgs boson to right-handed charged leptons, considered in Ref. [17].

$| \overline{G}/G_F |^2)^{1/2}$. Since G_μ is known experimentally and G_F can be evaluated using the experimental values of m_W, $\sin^2\theta_W$ and Δr, a constraint follows for $| \overline{G}/G_F |$. With $m_W = (79.91 \pm 0.39)$ GeV [18], $\sin^2\theta_W = 0.2291 \pm 0.0034$ [19], and $\Delta r = 0.056^{+0.006}_{-0.010}$ [19], we find

$$| \overline{G}/G_\mu | < 0.23 \quad (90\%\ c.l.)\ . \tag{18}$$

The experimental value of the ud-element U_{ud} of the Kobayashi-Maskawa matrix is defined as the ratio of the experimental value of the beta decay vector constant G_β and the muon decay constant G_μ. In the standard model $G_\beta = G_F\widehat{U}_{ud}$ (where $\widehat{U}_{ud}$ is the true KM matrix element), and $G_\mu = G_F$, so that $U_{ud} = G_\beta/G_\mu = \widehat{U}_{ud}$. In the presence of the interaction (14) we have $| U_{ud} |^2 = | \widehat{U}_{ud} |^2 (1 + | \overline{G}/G_F |^2)^{-1}$. Analogous relations hold for $| U_{us} |^2$ and $| U_{ub} |^2$, so that using the unitarity relation for the 3-family case one obtains.

$$| U_{ud} |^2 + | U_{us} |^2 + | U_{ub} |^2 = (1 + | \overline{G}/G_F |^2)^{-1}\ . \tag{19}$$

A recent analysis [20] yields $| U_{ud} |^2 + | U_{us} |^2 + | U_{ub} |^2 = 0.9989 \pm 0.0012$, implying

$$| \overline{G}/G_\mu | < 0.053 \quad (90\%\ c.l.)\ . \tag{20}$$

The example (14) of an interaction that can give rise to $\mu^+ \to e^+\bar{\nu}_e\nu_\mu$ is just a possibility, without a particular motivation. Mohapatra and I have investigated [21] $\mu^+ \to e^+\bar{\nu}_e\nu_\mu$ and also muonium to antimuonium ($M \to \overline{M}$) conversion in the left-right symmetric $SU(2)_L \times SU(2)_R \times U(1)_{B-L}$ model of Ref. [22]. In this model $\mu^+ \to e^+\bar{\nu}_e\nu_\mu$ and $M \to \overline{M}$ conversion arise naturally, and moreover turn out to play a distinctive role. We have pointed out that with reasonable assumptions concerning some of the parameters of the model there is a lower bound in these models for the $\mu^+ \to e^+\bar{\nu}_e\nu_\mu$ rate and the $M \to \overline{M}$ conversion rate for the range of the muon neutrino mass m_{ν_μ} for which the constraint from cosmology requires ν_μ to be unstable. Below I shall give a brief sketch of this work, referring the reader to Ref. [21] for details and complete references.

The Higgs sector of the model contains the bidoublet field ϕ (2,2,0) and the triplet fields $\Delta_R(1,3,2)$ and Δ_L (3,1,2). Left-right symmetric models provide an attractive framework for understanding the origin of parity violation in the weak interactions. The class of $SU(2)_L \times SU(2)_R \times U(1)_{B-L}$ models with triplet Higgs bosons can also provide an explanation of the smallness of the masses of the observed neutrinos.

The observed energy density of the universe implies that neutrinos which are heavier than about 40 eV must be unstable, and that there is an upper bound on their lifetimes, which is a decreasing functin of their mass. For the muon neutrino the only decay mode that can satisfy the cosmological constraint is the decay $\nu_\mu \to \nu_e\nu_e\bar{\nu}_e$ mediated by Δ_L^0-exchange. The cosmological constraint gives a lower bound (proportional to $m_{\nu_\mu}^{-3/2}$) on the strength $| G_0/\sqrt{2} |$ of the $\nu_\mu \to \nu_e\nu_e\bar{\nu}_e$ interaction, which in turn implies an upper bound (proportional to $m_{\nu_\mu}^{3/4}$) on the mass m_o of the Δ_L^0. This bound combined with the lower bound $m_o \gtrsim 43$ GeV on m_o provided by the experimental value of the invisible width of the Z dictates that if ν_μ is unstable, its mass has to be larger than $\sim$ 36 keV. It follows that the model is viable for $m_{\nu\mu} \lesssim$ 40 keV and for $m_{\nu\mu}$ in the range 36 keV $\lesssim m_{\nu\mu} \lesssim$ 270 keV.

The decay is mediated by the exchange of the singly-charged Higgs boson Δ_L^+ [23]. The corresponding interaction can be written in the form

$$H = 2\frac{G_+}{\sqrt{2}}\bar{\mu}\gamma_\lambda(1-\gamma_5)\nu_e\bar{\nu}_\mu\gamma^\lambda(1-\gamma_5)e \ + \text{H.c.}\ , \tag{21}$$

where $G_+ \simeq \sqrt{2}f_{ee}f^*_{\mu\mu}/8m_+^2$; m_+ is the mass of the Δ_L^+; f_{ee} and $f_{\mu\mu}$ are lepton$-\Delta_L$ Yukawa couplings.

The constant G_+ is related to G_o as

$$G_+^* = \frac{1}{2}G_oK_{e\mu}^{-1}\frac{f_{\mu\mu}}{f_{ee}-f_{\mu\mu}}\frac{m_o^2}{m_+^2}\ , \tag{22}$$

where $K_{e\mu}$ is the $e\mu$-element of the mixing matrix K in the charged-current interactions of the light neutrinos. Eq. (22) yields a lower bound on $| G_+ |$, since it can be shown that not only $| G_o |$ but also $| f_{\mu\mu} |$ and m_o^2/m_+^2 are bounded from below.

Muonium to antimuonium conversion arises in the model at the tree level through Δ_L^{++}-exchange [24]. The resulting effective $M \to \overline{M}$ interaction is given by

$$H = \frac{G_{++}}{\sqrt{2}}\bar{\mu}\gamma^\lambda(1-\gamma_5)e\bar{\mu}\gamma_\lambda(1-\gamma_5)e \ + \text{H.c.}\ , \tag{23}$$

where $G_{++} \simeq \sqrt{2}f_{ee}f^*_{\mu\mu}/8m^2_{++}$. G^*_{++} is related to G_o in the same way as G_+^* except for the replacement $m_+ \to m_{++}$ in Eq. (22). Since m_o^2/m_{++}^2 is, like m_o^2/m_+^2, bounded from below, a lower bound follows also for $| G_{++} |$.

We find $| G_+ | \gtrsim 4 \times 10^{-4}G_F$ and $| G_{++} | \gtrsim 2 \times 10^{-4}G_F$ for 36 keV $\lesssim m_{\nu_\mu} \lesssim$ 270 keV [21]. The lower bounds increase with decreasing m_{ν_μ}. Thus, as the experimental limits on $| G_+ |$ and/or $| G_{++} |$ become more and more stringent, the allowed range of m_{ν_μ} for which the model is viable becomes increasingly smaller. For $m_{\nu_\mu} \simeq 36$ keV we obtain $| G_+ | \gtrsim 2 \times 10^{-2}$ and $| G_{++} | \gtrsim 10^{-2}$.

The branching ratio R in Eq. (11) (with $\bar{n}_e = \bar{\nu}_e$, $n_x = \nu_\mu$) is given by

$$R = 4 | G_+/G_\mu |^2\ . \tag{24}$$

The experimental limit (12) implies

$$| G_+ | < 0.067G_\mu\ . \tag{25}$$

The Hamiltonian consisting of the interaction (21) and the standard model contribution corresponds to the general Hamiltonian (2) with

$$(g^V_{LL})_{13} = G_F/\sqrt{2}\ ,\ (g^S_{LL})_{13} = 4G_+^*/\sqrt{2} \tag{26}$$

and all the remaining constants absent. Eq. (10) in this case implies

$$| G_+ |< 0.14 G_\mu \quad (90\% \; c.l.) \tag{27}$$

From the experimental value of the W−mass and from charged current universality we obtain

$$| G_+ |< 0.12 \; G_\mu \quad (90\% \; c.l.) \tag{28}$$

and

$$| G_+ |< 0.026 \; G_\mu \quad (90\% \; c.l.) \; , \tag{29}$$

respectively. These are the same constraints as for $| \overline{G}/2 |$ before, since the muon decay constant is now given by $G_\mu = G_F(1 + 4 \mid G_+/G_F \mid^2)^{1/2}$. From the upper limits (25), (27), (28) and (29) the most stringent at present is the one from charged current universality. It should be noted however that this bound may be affected by theoretical uncertainties.

An experiment in preparation at LAMPF [25] plans to search for $\mu^+ \rightarrow e^+ \bar{\nu}_e \nu_\mu$ with a sensitivity corresponding to $| G_+ | \simeq 6 \times 10^{-3} G_F$. The LAMPF experiment will improve simultaneously the limit on $K_{e\mu}$ by a factor of ~ 3.5. This will increase the lower bound on $| G_+ |$ and $| G_{++} |$ by a factor of ~ 2.

The present experimental upper limit on $| G_{++} |$ is $| G_{++} |< 0.16 G_F$ (90% c.l.) [26]. The experiment under way at PSI [27] is expected to lower the upper limit to $10^{-3} G_F$.

4 Conclusions

In this talk we discussed two subjects in the field of muon decay: the status of our knowledge regarding the identity of the neutrinos emitted in muon decay, and the exotic decay mode $\mu^+ \rightarrow e^+ \bar{\nu}_e \nu_\mu$.

Concerning the identity of the muon decay neutrinos experiment indicates that among the decays $\mu^+ \rightarrow e^+ +$ *neutrinos* the decay mode which dominates the $\mu^+ \rightarrow e^+ +$ *neutrinos* rate involves the neutrino species of the standard model scenario: one of the neutrinos in this decay mode is the state $\bar{\nu}_\pi$ emitted in $\pi^- \rightarrow \mu^- \bar{\nu}_\pi$ decay, and the other the state n_e which accompanies the positron in nuclear beta decay; experiment indicates also that ν_π and n_e are not the same states. Muon decay and inverse muon decay data constrain the interaction that governs this decay mode to be the standard model interaction and/or a scalar type interaction. The experimental values of m_W and $\sin^2\theta_W$ indicate that the standard model contribution dominates.

The contribution of other decay interactions and decay modes to the $\mu^+ \rightarrow e^+ +$ *neutrinos* rate is constrained to be less than about 10%. Searches for non-standard contributions continue to be of great importance. From non-standard decay modes we discussed the exotic decay $\mu^+ \rightarrow e^+ \bar{\nu}_e \nu_\mu$. More and more sensitive searches for this decay (as also searches for muonium to antimuonium conversion) will provide important information on an attractive class of left-right symmetric models.

Acknowledgement. I am grateful to R. L. Burman, V. W. Hughes, K. Jungmann, W. C. Louis, B. E. Matthias, R. N. Mohapatra and D. H. White for valuable conversations. I would also like to thank K. Jungmann and G. zu Putlitz for a most enjoyable symposium and wonderful hospitality. This work was supported by the United States Department of Energy.

References

1. Recent reviews of aspects of muon decay include R. Engfer, H. K. Walter: Ann. Rev. Nucl. Part. Sci. 36(1986)327; S. P. Rosen: Los Alamos National Laboratory preprint LA-UR-88-4013, invited talk at the Workshop on new directions in neutrino physics at Fermilab, Fermilab, Sept. 14-16, 1988; P. Herczeg: In: Rare Decay Symposium, eds. D. Bryman, J. Ng, T. Numao, J. Poutissou, p. 24. Singapore: World Scientific 1989
2. A. Sirlin: Phys. Rev. D22(1980)970
3. F. Scheck: Leptons, Hadrons and Nuclei. Amsterdam: North-Holland 1983; K. Mursula, F. Scheck: Nucl. Phys. B55(1985)189
4. P. Langacker, D. London: Phys. Rev. D39(1989)266
5. W. Fetscher, H.-J. Gerber, K. F. Johnson: Phys. Lett. 173B(1986)102; W. Fetscher, H.-J. Gerber, Note on muon decay parameters, In: Review of particle properties: Phys. Lett. 239B(1990)VI.11
6. D. Geiregat et al.: Phys. Lett. 247B(1990)131
7. S. R. Mishra et al.: Phys. Lett. 252B(1990)170
8. W. Fetscher: Phys. Lett. B140(1984)117
9. A. M. Cooper et al.: Phys. Lett. 112B(1982)97
10. S. E. Willis, et al.: Phys. Rev. Lett. 44(1980)522
11. T. Eichten et al.: Phys. Lett. 46B(1973)281; J. Blietschau et al.: Nucl. Phys. B133(1978)205
12. F. Bergsma et al.: Phys. Lett. 122B(1983)465
13. D. A. Krakauer et al.: Phys. Lett. 263B(1991)534
14. P. Langacker, D. London: Phys. Rev. D38(1988)907
15. G. Feinberg, S. Weinberg: Phys. Rev. Lett. 6(1961)381
16. E. Derman, D. R. T. Jones; Phys. Lett. 70B(1977)449; E. Derman: Phys. Lett. 78B(1978)497; E. Derman: Phys. Rev. D19(1979)317
17. R. N. Mohapatra: In: Eighth Workshop on Grand Unified Theory, Syracuse, ed. K. C. Wali, p. 200. Singapore: World Scientific 1987; D. Chang, W.-Y. Keung: Phys. Rev. Lett. 62(1989)2583
18. S. Lloyd: In: Proceedings of the 1991 Aspen Winter Conference in Elementary Particle Physics (unpublished)
19. P. Langacker, M. Luo: Phys. Rev. D44(1991)817
20. D. H. Wilkinson: Isospin and Quarks in Nuclear Beta-Decay, paper presented at the International Conference on Spin and Isospin in Nuclear Interactions, Telluride, Colorado, March 11-15, 1991, TRIUMF preprint TRI-PP-91-9
21. P. Herczeg, R. N. Mohapatra: Los Alamos National Laboratory preprint, Oct. 1991. A brief version of this paper is given in P. Herczeg, R. N. Mohapatra: Los Alamos National Laboratory preprint LA-UR-91-3038, to appear in the proceedings of the Particles and Fields '91 Conference of the APS, Vancouver, Canada, Aug. 18-22, 1991
22. R. N. Mohapatra, G. Senjanović: Phys. Rev. Lett. 44(1980)912; R. N. Mohapatra, G. Senjanović Phys. Rev. D23(1981)165
23. P. Herczeg, R. N. Mohapatra (unpublished). Reported in P. Herczeg: In: Rare Decay Symposium, eds. D. Bryman, J. Ng, T. Numao, J. Poutissou, p. 24. Singapore: World Scientific 1989

24. A. Halprin: Phys. Rev. Let. 48(1982)1313

25. X.-Q. Lu et al.: A Proposal to Search for Neutrino Oscillations with High Sensitivity in the Appearance Channels $\nu_\mu \rightarrow \nu_e$ and $\bar{\nu}_\mu \rightarrow \bar{\nu}_e$, LAMPF Proposal LA-11842-P, Aug. 1990.

26, B. E. Matthias et al.: Phys. Rev. Lett. 66(1991)2716

27. K. Jungmann, W. Bertl et al.: Search for Spontaneous Conversion of Muonium to Antimuonium, PSI Experiment R-89-06.1

This article was processed using Springer-Verlag TEX Z.Physik C macro package 1991
and the AMS fonts, developed by the American Mathematical Society.

Z. Phys. C - Particles and Fields 56, S135-S142 (1992)

Zeitschrift für Physik C Particles and Fields

Rare Decays: Experiments

Hans-Kristian Walter [1]* and Klaus Jungmann [2]**

[1] Paul Scherrer Institute, CH-5232 Villigen PSI, Switzerland
[2] Physikalisches Institut der Universität Heidelberg, Philosophenweg 12, D-6900 Heidelberg, Germany

6 December 1991

**Representing the SINDRUM II Collaboration: (ETH Zürich, Univ. Zürich, Paul Scherrer Institute Villigen, RWTH Aachen, Univ. Tbilisi)*

***Representing the M-$\overline{M}$ Collaboration: (Univ. Heidelberg, ETH Zürich, Univ. Zürich, Paul Scherrer Institute Villigen, RWTH Aachen, Univ.Swierk, JINR Dubna, Univ. Tbilisi, Yale University)*

Abstract. The experimental status of generation number violating muon and kaon decays is reviewed. We concentrate on three experiments with a large PSI contribution, neutrinoless muon-electron conversion, muonium-antimuonium conversion and the decay $K^+ \rightarrow \pi^+\mu e$. The SINDRUM II experiment at PSI searches for the coherent conversion of a muon into an electron in titanium. The design and construction of the SINDRUM II detector and the results of the first data-taking period in 89 are presented. In total 3.2 million events have been analyzed and no candidate for the process $\mu^- Ti \rightarrow e^- Ti$ has been found. Using two independent determinations of the total number of muon stops an upper limit of $B_{\mu e} < 4.4 \times 10^{-12}$ (90% C.L.) is obtained. The plans to lower the sensitivity by two orders of magnitude are discussed. A muonium-antimuonium conversion experiment is being set up at PSI using the refurbished SINDRUM I detector. The goal of the experiment is to improve the sensitivity for the effective coupling constant $G_{M\overline{M}}$ by a factor of $\sim$ 100. Finally a new proposal E865 has been accepted at BNL to improve the sensitivity for the branching ratio of the decay $K^+ \rightarrow \pi^+\mu e$ by a factor 70 to 3×10^{-12}.

1 Introduction

The standard model of particle physics, although most successful and uncontradicted by experiment, is believed to be an effective low energy approximation of a more fundamental theory unbroken at higher energies. The driving force for this belief is the quest for simplicity and a minimum of ad hoc assumptions, i.e. mainly aesthetical. The most unsatisfactory aspects of the standard model is the lack of understanding of the multitude of interactions and flavors and of mass generation. In particular the origin of a triplet of families is mysterious and experiments establishing connections between these families are of fundamental importance. Neutrino masses and mixings and lepton flavor changing effective neutral currents belong to this category. The processes $\mu^+ \rightarrow e^+\gamma$, $\mu^+ \rightarrow e^+e^+e^-$, μ^- + (A,Z)$\rightarrow e^-$ + (A,Z), $K_L \rightarrow \mu e$ and $K^+ \rightarrow \pi^+\mu e$ are very sensitive probes since high intensity beams and unique signatures allow to measure very small branching ratios, which in turn probe very high energy scales for new physics.

Table 1 lists upper limits for branching ratios of flavor-violating muon-, pion-, and kaon-decays. New experiments in progress at the meson factories and at Brookhaven and their preliminary results have been included, whereas other recent results on rare kaon decays like $K_L \rightarrow \pi^o e^+e^-$, γe^+e^-, 4e and decays of heavy bosons have been omitted.

2 Muon electron conversion

Neutrinoless muon-electron conversion is especially sensitive to new physics since the constituents of the nucleus (A,Z), say Ti, add coherently to the conversion rate. The aim of our new experiment at PSI is to search for $\mu - e$ conversion in Ti with a sensitivity which is 2 - 3 orders of magnitude better than achieved previously [24].

2.1 Backgrounds and beam

The experimental signature is a single mono-energetic electron emitted from the ground state of the muonic Ti-atom which has a lifetime of 329 ns [25] . The energy of the electron is $E_{e^-} = m_\mu c^2 - B_\mu - R_{nucl.}$, where m_μ is the muon mass, B_μ the binding energy of the muonic atom ground state and $R_{nucl.}$ the recoil energy of the nucleus; it equals E_{e^-} = 104.27 MeV. The sensitivity may be limited by potential backgrounds which are:

Beam muons: i) Muon decay from the atomic ground state (MIO), which has an endpoint energy of the signal itself. ii) Radiative capture of the muon followed by internal or external production of an asymmetric e^+e^-

Table 1. Upper limits for branching ratios of flavor violating μ, π and K-decays

Process	B [90% C.L.]	Ref.	Sensitivity of new proposal	Ref.	Preliminary Results	Ref.
$\mu \to e\gamma$	$< 4.9 \times 10^{-11}$	[1]	10^{-13}	[2,3]		
$\mu \to e\gamma\gamma$	$< 7.2 \times 10^{-11}$	[4]	10^{-12}	[2]		
$\mu \to 3e$	$< 1 \times 10^{-12}$	[5]	2×10^{-14}	[6]		
$\mu^- N \to e^- N$	$< 4.6 \times 10^{-12}$	[7]	5×10^{-14}	[8]	4.4×10^{-12}	[9]
			10^{-16}	[10]		
$\mu^+ e^- \to \mu^- e^+$	$< 6.5 \times 10^{-7}$	[11]	10^{-11}	[12]		
$\pi^o \to \mu e$	$< 1.6 \times 10^{-8}$	[13]		[14]		
$\pi^o \to \nu\bar{\nu}$	$< 6.5 \times 10^{-6}$	[15]	10^{-9}	[16]	8.3×10^{-7}	[17]
$K_L^o \to \mu e$	$< 2.2 \times 10^{-10}$	[18]	10^{-12}	[19]	7.1×10^{-11}	[20]
			10^{-11}	[21]	4.16×10^{-11}	[22]
$K^+ \to \pi^+ \mu e$	$< 2.1 \times 10^{-10}$	[13]	3×10^{-12}	[14]		
$K^+ \to \pi^+ \nu\bar{\nu}$	$< 1.4 \times 10^{-7}$	[23]	2×10^{-10}	[16]	5×10^{-9}	[17]

pair, of which the positron remains undetected. The endpoint energy depends on the initial and final state mass difference ($\Delta E = 3.99$ MeV for Ti). Both decays can be suppressed by good energy resolution, the former can be used as an independent normalization.

Beam pions: Radiative pion capture with asymmetric pair production. This background must be suppressed by a combination of low beam contamination and a prompt veto, where the former must be more and more effective, as larger intensities and smaller branching ratios are attempted. With 200 μA proton beam intensity and at a muon momentum of 100 MeV/c a μ^- flux of $\sim 10^7\ s^{-1}$ with a π/μ contamination of $\sim 10^{-7}$ could be achieved at the μE1 channel of PSI, of which ~25% was stopped in a 140 mm $\phi \times 300$ mm long Ti target with effective density of 0.103 g/cm^3. A prompt beam veto further reduced the pion contamination.

Beam electrons: Beam electrons and electrons from muon decay in flight scattered off the target with a momentum around 100 MeV/c could fake a $\mu - e$ conversion signal. The choice of a small beam momentum, a prompt beam veto, pulse height selection in the beam counter, and a veto on a RF-correlated phase segment can suppress these electrons.

Cosmic rays: A large variety of background types is induced by cosmic rays. Electrons of ~100 MeV can be produced by showers in the yoke iron, knock-out of electrons or pair production from the target or other detector material by charged cosmic particles or photons or neutrons. Active and passive shielding as well as event topologies showing associated track activity must be used to suppress this background.

2.2 The detector SINDRUM II

The setup of the spectrometer is shown in Fig. 1. The beam is analyzed in time and pulse height by a beam counter, read out via a plastic fiber light guide into a wave form digitizer for optimal double pulse resolution. A superconducting solenoid produces a magnetic field of up to 1.5 T (here 1.2 T) parallel to the beam axis in a cylindrical volume of 1350 mm ϕ and 1800 mm length. Since most of the electrons from MIO are confined to a radius of 330 mm, the various detectors are placed outside this radius. Times are measured with a 64 element (3 mm thick) plastic scintillator hodoscope ($\sigma = 0.7$ ns), which is also used in the trigger. Two concentric drift chambers measure the trajectories of the emitted electrons outside the hodoscope. The curvature of the first and because of the damping outermost revolution determines the momentum. Ionization from the outermost revolution arrives first at the anode wires since the electric drift field points radially outwards to the anode wire cylinder located at the chambers outer cylinders. The inner drift chamber is filled with CO_2/iC_4H_{10} (70/30) and has 384 anode and focus - and 768 grid wires. With a gradient of 1 kV/cm and at a field of 1.2 T the drift

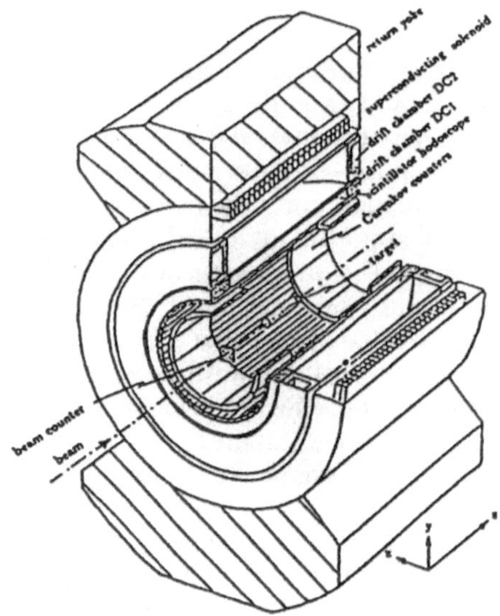

Fig. 1. Cut-away view of the SINDRUM II detector (not to scale). Also shown is the definition of the coordinate system $x - y - z$. The axis corresponds to the beam axis. The front and back mirror plates are not shown

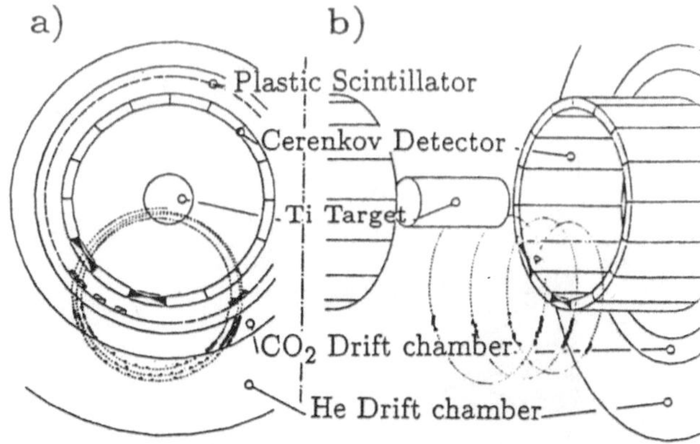

Fig. 2. Electron track a) the full spectrometer information in the x-y view, b) z-y projection of the 3-dimensional hits measured by the inner drift chamber.

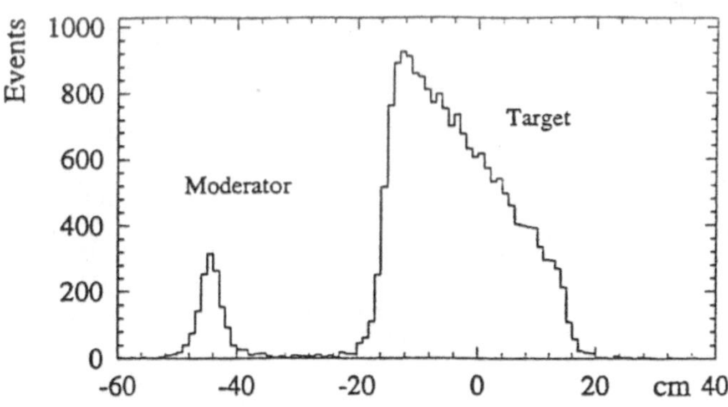

Fig. 3. Distribution of the position of the track origin along the beam axis which reflects the muon stop distribution

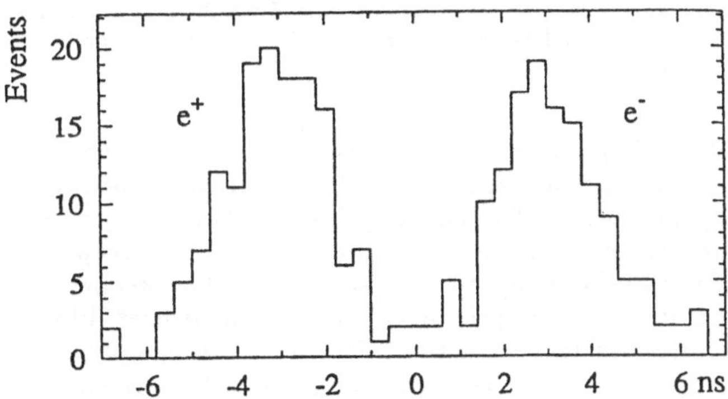

Fig. 4. Time-of-flight between two successive crossings of the scintillating hodoscope for cosmic ray events. The charge separation obtained from this information is about 98%

velocity is 0.95 cm/μs, the Lorentz angle is 6° and the measured resolution is $\sigma = 170$ μm. Coordinates along the axis are determined by measuring signals induced on 384 cathode strips with an accuracy of $\sigma = 3.8$ mm, which will be improved in the future. A sandwich structured cylinder of 0.8×10^{-3} radiation length thickness separates the gas volumina and the electric fields of the inner and the outer drift chambers. The outer chamber is filled with He/iC$_4$H$_{10}$ (88/12) and has 96 wire cages. With a gradient of $\sim$ 350 V/cm the drift velocity is 1.3 cm/μs, the Lorentz angle 32° (B = 1.2 T) and the measured resolution $\sigma = 1.7$ mm. The gas system has to keep the gas mixtures constant, monitor temperature, pulse height and drift velocity and avoid accidents (e.g. $p_{DC1} - p_{DC2} < 5$ mbar). Two Cerenkov hodoscopes at both ends each consisting of 16 plexiglass counters of thickness 30 mm complete the setup. Details about the electronics, the trigger system, the data acquisition and monitoring systems as well as a more detailed description of the apparatus and its performance can be found in refs. [26-29]. The performance was tested using cosmic ray muons and positrons from the decay $\pi^+ \to e^+\nu_e$. A momentum resolution of $\sigma = 0.6\%$ was measured not including target absorption, which increases it to 1% for the $\mu - e$ conversion signal. The spectrometer was brought into operation a few months before the 1990-91 shutdown of the PSI accelerator. A data taking period of one month mainly in the 50 MHz mode followed and during 2.15×10^6 s about 3.2×10^6 events were written on tape. Test and calibration measurements completed this data taking period. In the following the data evaluation and results of a preliminary analysis are discussed.

2.3 Analysis of the 1989 run

First the helical trajectory of the particle which triggered the detector readout is reconstructed and translated into the momentum vector at the track origin which is defined as the point of closest approach to the detector axis. An example is shown in Fig. 2. In a second step the presence of additional tracks is checked which is mainly used to recognize cosmic ray background. In a third step the information from the beam counter is analyzed to identify scattered beam electrons and electrons from pion capture.

Using the results from the computer reconstruction events are classified according to the track origin (Fig. 2), the charge (Fig. 3) and the type of beam particle (Fig. 4).

Two thirds of the reconstructed events are scattered beam electrons which have been strongly suppressed by rejecting prompt events with 9.25 ns $< \Delta t_{rf} <$ 12.75 ns. The final analysis is based on the momentum spectra of different classes of the events (Figs. 6 and 7). In the distribution of the delayed electrons shown in Fig. 6a no event is found in the region 98-105 MeV/c where 86% of the hypothetical conversion events are expected.

The 90% C.L. upper limit for the branching ratio thus is:

$$B_{\mu e} < \frac{2.3}{N_{stop}\, f_{cap}\, A_g^{\mu e} \times A_t^{\mu e} \times \varepsilon_{tot}^{\mu e}} \tag{1}$$

whereas the number of MIO events detected is:

$$N^{MIO} = N_{stop}(1 - f_{cap}) A_g^{MIO} A_t^{MIO} \varepsilon_{tot}^{MIO} \tag{2}$$

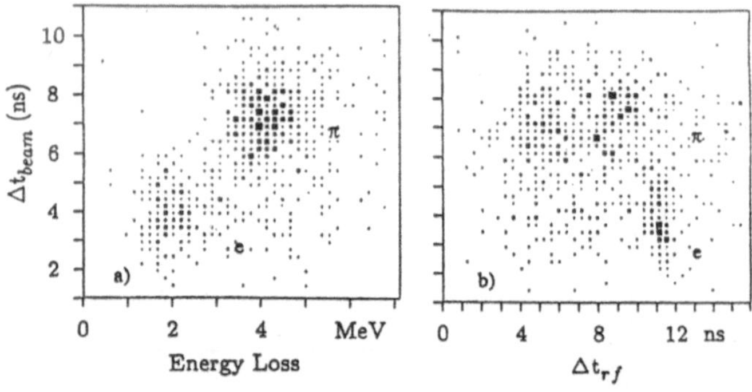

Fig. 5. Time-of-flight Δt_{beam} between counter and track origin versus: a) the amplitude in the beam counter, b) the phase Δt_{rf} relative to the cyclotron rf signal. In order to enhance the pion contribution events were selected with a momentum above 97 MeV/c and a track origin inside the moderator. In distribution b) two sources of pion events can be recognized which have been associated with particles scattering either in the injection or in the extraction of the muon channel

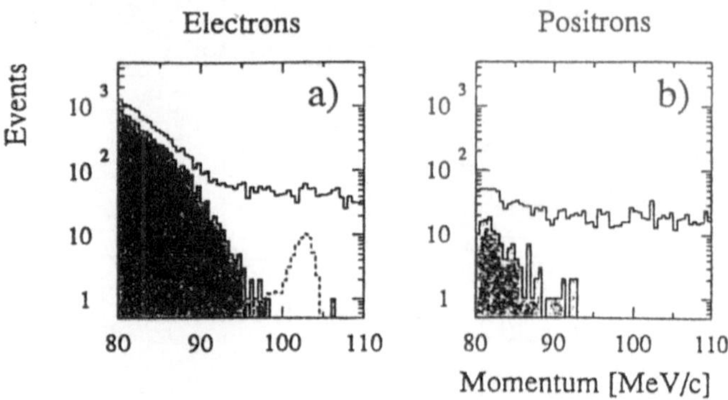

Fig. 6. Momentum distributions from delayed electrons (a) and positrons (b) which originate in the target with (gray) and without (white) rejection of cosmic ray background. The dashed histogram in part a is the distribution for conversion events generated by a Monte Carlo simulation under the assumption of a branching ratio of 10^{-10}

where N_{stop} = number of muons stopped in the target during the experiment, f_{cap} = fraction of muons captured by the Ti nucleus (85.3 ± 0.3%) [2], A_g = geometrical acceptance of spectrometer, A_t = acceptance of the online trigger conditions on the hodoscopes and chambers. ε_{tot} = product of various efficiencies, like intrinsic and reconstruction efficiencies, cuts, losses, etc. Table 2 gives the corresponding quantities which have been obtained by measurement and/or Monte Carlo simulations. The number of muon stops was evaluated using the beam counter signals and the stop fraction from measurement and Monte Carlo simulations of the beam to be $(4.24 \pm 0.35)\times10^{12}$. This number was cross-checked using ~ 16000 measured MIO events and the theoretical branching ratio as $(4.20 \pm 0.80)\times10^{12}$. The value $(4.23 \pm 0.32)\times10^{12}$ was used and the corresponding upper limit on the branching ratio is:

$$B_{\mu e} < 4.4 \times 10^{-12} \text{ (90\% C.L.)}. \tag{3}$$

As shown by a Monte Carlo simulation the measured spectrum can be described both in shape and in rate by normal muon decay in orbit. It has been demonstrated that in the final analysis of this first run the sensitivity can be improved by about 20%.

The distribution in Fig. 7a of prompt electrons surviving the cut in Δt_{rf} consists mainly of electrons from muon decay in the channel which scatter in the target. The pion spectrum shown in the same figure is dominated by accidental coincidences for momenta below 90 MeV/c. The positron spectra in this region (Figs. 6b and 7b) are dominated by misidentified electrons.

The processes contributing background in the momentum region of interest are in order of their rate: cosmic ray events, scattered beam electrons, scattered electrons from muon decay in flight and pion induced events. One type of cosmic ray events could only be rejected at the cost of a reduction in acceptance. These are asymmetric e^+e^- pairs produced in the target by isolated high energy photons.

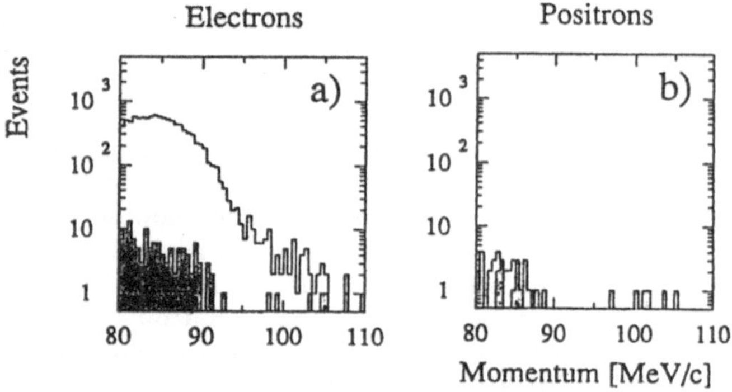

Fig. 7. Momentum distributions from prompt electrons (a) and positrons (b) which originate in the target, separated in to contributions from electrons (white) and pions (gray)

The events show pronounced peaks at angles which are not shielded by either concrete area walls or the return yoke and could be rejected by 16% cuts in the acceptance. In future experiments additional shielding will eliminate this problem. The rejection of prompt background introduced a similar inefficiency. In the next run, which is scheduled for end 1991, a reduction of the dead time of the beam counter from 15 ns to 10 ns will be achieved by the use of a faster waveform digitizer.

A rejection improvement by a factor of three of asymmetric pair production background will be attempted by introducing a second hodoscope at a radius of ~130 mm with which associated positrons can be vetoed. Finally

Table 2. Acceptances and efficiencies for $\mu - e$ conversion and MIO

	$\mu - e$	MIO
A_g	0.500	22 $\times$ 10^{-8}
A_t	0.754	0.269
$A = A_g \times A_t$	0.376 ± 0.022	$(5.9 \pm 0.9) \times 10^{-8}$
ϵ_{tot}	0.386 ± 0.032	0.441 ± 0.048
$A \times \epsilon$	0.145 ± 0.015	$(2.60 \pm 0.49) \times 10^{-8}$

a rate increase by a factor of 2 to 4, a stop fraction improvement by a different moderator geometry, an improvement of wire chamber efficiency by a new intermediate cylinder, and a longer measuring time should contribute to allow a sensitivity gain of one order of magnitude. In addition, studies for a pion to muon convertor (PMC) at the new high intensity beam line πE5 have been carried out, which would allow to lower the final sensitivity by yet another order of magnitude.

3 Muonium-antimuonium conversion

The spontaneous conversion of muonium ($M = \mu^+e^-$) into antimuonium ($\overline{M} = \mu^-e^+$) violates lepton generation number by two units. Although not provided in the standard model, it can be expected in many extensions to it. For historic reasons the strength of the new interaction is described by a coupling constant $G_{M\overline{M}}$ from an effective $V - A$ type Hamiltonian. It is connected to the probability $P_{\overline{M}}$ of observing an $\overline{M}$ atom decay from a system which started as pure M through

$$P_{\overline{M}} = (2.57 \times 10^{-5})\,(G_{M\overline{M}}/G_F)^2 \quad , \tag{4}$$

where G_F is the Fermi coupling constant of the weak interaction.

The M-$\overline{M}$-conversion is of particular interest for left-right symmetric models and supersymmetric theories. Within the framework of left-right symmetry Mohapatra and Herczeg [30,31] have calculated a lower bound on the effective coupling constant of $G_{M\overline{M}} \gtrsim 2 \times 10^{-4} G_F$. The interaction is mediated by the exchange of a doubly charged Higgs boson $\Delta^{++}(\Delta^{--})$. This model allows the prediction of M-$\overline{M}$-conversion with an even larger coupling constant $G_{M\overline{M}}$, if the present upper limit on the muon neutrino mass will be lowered in future experiments. A coupling strength of $\lesssim 10^{-5}$ has been estimated for models in which M-$\overline{M}$-oscillation is mediated by massive Majorana neutrinos [32]. Within generic SUSY models a value of $G_{M\overline{M}} \approx 10^{-2} G_F$ can be expected for slepton masses of the order of 100 GeV [30].

3.1 Signature of the new generation of M-$\overline{M}$ conversion experiments

The recent development of intense sources of thermal M in vacuum [33,34] has enabled and stimulated very sensitive searches for spontaneous M-$\overline{M}$ conversion [11,35]. Positive muons at subsurface momentum are

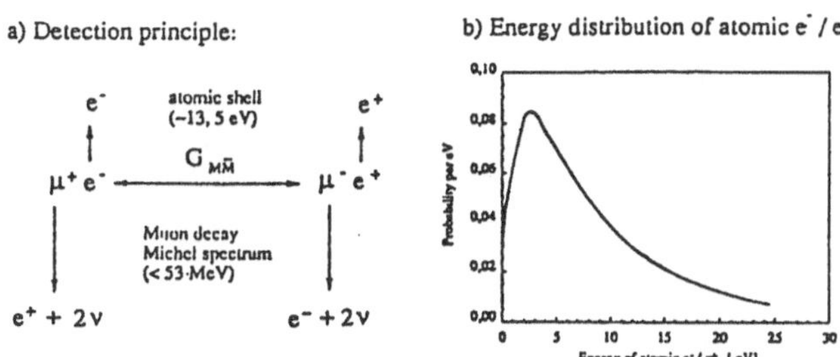

Fig. 8. Signature (a) of the new generation of searches for spontaneous muonium to antimuonium conversion. The decay of the positive muon in a muonium atom releases an energetic positron with an energy spectrum (Michel spectrum) ranging up to 53 MeV. The energy distribution (b) of the electron from the muonium atomic shell, which is left behind, averages to 13.5 eV. In case of an antimuonium atom decay we have a fast Michel electron and a slow atomic positron.

stopped in a SiO_2 powder target and form M by charge exchange. The atoms diffuse to the surface and leave the target with Maxwell-Boltzmann velocity distribution at thermal ($T \approx 300K$) energies into the surrounding vacuum.

Muonium atoms decaying at rest in vacuum liberate fast positrons with a well known energy spectrum (Michel spectrum) which has a maximum at 52.83 MeV. The electron in the atomic shell of the M atom is left behind (Fig. 8). These electrons have a distribution of kinetic energies which averages to 13.5 eV with a rapidly falling tail towards higher energies (see Fig. 8b). In the case of a decaying antimuonium atom there is a fast (Michel) electron and a slow positron from the atomic shell. Both particles can be uniquely identified and they can be observed in coincidence.

This signature for M-$\overline{M}$-conversion which looks for both the constituents of the anti-atom has been employed for the first time in a recent experiment at LAMPF [11]. It is by far superior to earlier efforts, firstly, because of low sensitivity to experimental background and, secondly, because of the observation of $P_{\overline{M}}$ rather than the much smaller probability $P'_{\overline{M}}$ of finding an $\overline{M}$ atom at a fixed time after the creation of a M atom. The probability $P'_{\overline{M}}$ was the only quantity that could be measured in all previous experiments which were looking for μ^- reactions with atomic matter (μ^- atomic cascade or μ^- nuclear capture) as part of the $\overline{M}$-event signature.

The M-$\overline{M}$-oscillations are sensitive to external magnetic fields which are applied to the system. The dependence of the conversion probability on an external magnetic field has been calculated for an effective $V - A$ type Hamiltonian and is shown in Fig. 9 [36]. At fields

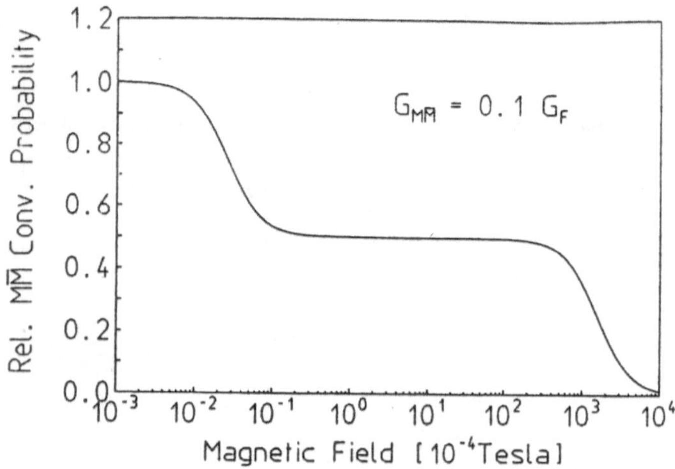

Fig. 9. Dependence of the muonium-antimuonium conversion probability on the strength of an external magnetic field [36]. In case of a V–A type interaction the conversion is expected to be completely suppressed for fields above several 100 mT.

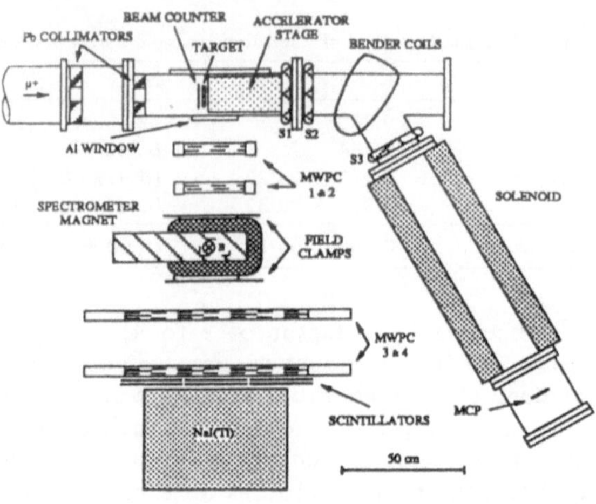

Fig. 10. Setup of the latest LAMPF experiment. The fast Michel particles are detected in a magnetic spectrometer consisting of two pairs of multiwire proportional chambers with a C-magnet in between. The atomic electrons /positrons are accelerated to 5.7 keV and transported onto a microchannel plate (MCP) detector. The view is from the top. The Michel particle detectors are mounted under 50° with respect to the incident muon beam.

as low as 2.6 μT the conversion probability is reduced by a factor of two due to the removal of degeneracy of corresponding M and $\overline{M}$ magnetic hyperfine structure sublevels. The conversion is expected to be almost completely suppressed for fields above several 100 mT.

3.2 LAMPF experiment

A 20 MeV μ^+beam from the LAMPF stopped muon channel (SMC) with 7.5% momentum bite was counted in a 150μm thick plastic scintillator. A fraction of $(56 \pm 2)\%$ was stopped in a SiO_2 powder target of 9 mg/cm^2 thickness. Muonium was formed at a yield of $(5.02 \pm 0.06)\%$. The residual external magnetic field in the conversion region was 1 mT causing a suppression of any possible M-$\overline{M}$-conversion probability by a factor of two. The apparatus (see Fig. 10) could be operated to observe either M or $\overline{M}$ decays in the vacuum region near the SiO_2 target. Positrons or electrons from decaying M or $\overline{M}$ could leave the vacuum chamber through a 100 μm aluminum window. They were detected in a set of four multiwire proportional chambers aligned on an axis perpendicular to the incident muon beam. A C-yoke dipole magnet with 52.2 mT central field, located between the second and third wire chambers, served as a charge and momentum selecting device. The timing signals were derived from a double layer of plastic scintillators behind the last wire chamber. The Michel particle energy was determined in a 76 cm diameter cylindrical NaI(Tl) crystal. The acceptance of the high energy particle spectrometer was mainly limited by a rather small solid angle. Michel particles were detected with an overall efficiency of $(2.50 \pm 0.02) \times 10^{-3}$. For $\overline{M}$ data taking a curvature sensitive trigger was installed by grouping wires oriented parallel to the C-magnet field in all four wire chambers. 98.9% of the e^+ were rejected and the apparatus had a e^- acceptance of 66%.

The atomic electron/positron was accelerated to 5.7 keV in a electrostatic three stage device. The sign of charge as well as the momentum of the particle were selected by a 1.5 mT iron free dipole magnet. The particles were guided through a 2m long solenoid onto a 75 mm diameter microchannel plate detector (MCP). The time of flight distribution for atomic e^- from M atoms decaying in the vacuum region downstream of the SiO_2 target (Fig. 11) has a width (FWHM) of only 7 nsec and allows a narrow coincidence window. The atomic electron/positron could be transported and detected with an overall efficiency of $(15.5 \pm 0.8)\%$.

A maximum likelihood analysis yielded zero $\overline{M}$ events out of $(6.17 \pm 0.28) \times 10^6$ M atoms with a 90% C.L. upper limit of two events. The experiment sets an upper bound for the M-$\overline{M}$ conversion probability of $P_{\overline{M}} < 6.5 \times 10^{-7}(90\%\ C.L.)$. This corresponds to a coupling constant of $G_{M\overline{M}} < 0.16 G_F (90\%\ C.L.)$. This value supersedes the final result of $G_{M\overline{M}} < 0.29\, G_F (90\%\ C.L.)$ from an experiment carried out at TRIUMF which had reached its background limit [35].

3.3 New PSI experiment

The goal of the new experiment [12] presently under way at PSI (Fig. 12) is a sensitivity for the coupling constant of $G_{M\overline{M}} < 10^{-3}$. This can be achieved by using the same clean signature as in the LAMPF experiment and by introducing various improvements:

- The refurbished SINDRUM I detector will be used as high energy spectrometer operated at 0.1 T magnetic field. The field strength is a compromise between good e^- identification and moderate suppression of the M-$\overline{M}$ -

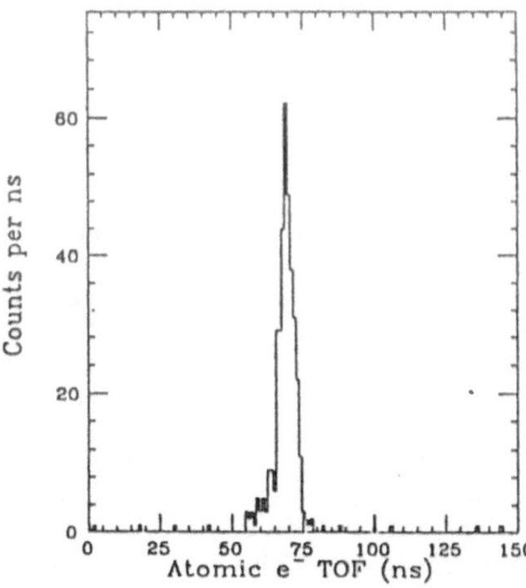

Fig. 11. Time of flight spectrum of atomic electrons from muonium atoms decaying in vacuum.

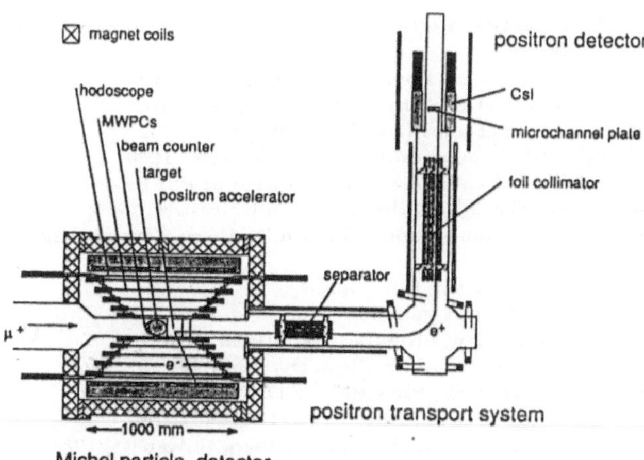

Fig. 12. Setup of the new muonium-antimuonium experiment currently under way at PSI. The refurbished SINDRUM I detector is used to identify the fast Michel particles. Atomic positrons are accelerated to approximately 10 keV and guided in a magnetic field of 0.1 T to a position sensitive multichannel plate detector. 511 keV γ's from positron annihilation are detected in a set of twelve CsI crystals surrounding the MCP.

conversion probability. This detector has an acceptance for the Michel particle of about 60.

- After a possible $\overline{M}$ decay the atomic e^+ will be accelerated electrostatically to 10 keV in a two stage device. They are then transported adiabatically in a 0.1 T magnetic field onto a position sensitive multichannel plate detector (MCP). The magnetic transport system has a 90° bend for the separation of high energy particles from the atomic e^+. The electrode configuration of the accelerator and the applied voltages will be chosen for a minimum width of the e^+ time of flight distribution. The e^+ trace can be projected back into the M cloud region allowing the reconstruction of the $\overline{M}$decay vertex.
- A barrel of twelve CsI crystals will surround the MCP for the detection of the two 511 keV γ's which will be emitted back-to-back when the e^+ annihilates with an e^- in the MCP. The efficiency times solid angle acceptance of the CsI detector is expected to be 60%.

The major source of expected background in the SINDRUM I detector is the misidentification of e^+ from M atom decays due to multiple scattering in the wirechambers or due to Bhabha scattering of Michel e^+ in the vacuum beam tube. With 0.5 mm C-fiber for the tube one expects a probability of 1.4×10^{-5} for both processes combined .

Various different sources of potential background in the atomic positron detector have to be considered.

- The allowed decay $\mu^+ \rightarrow e^+\nu\bar{\nu}e^+e^-$ with a low energy e^+ in the final state could fake an $\overline{M}$ event, if the second e^+ is not detected because of the finite acceptance of the SINDRUM I detector. With stringent cuts on the energies of the detected particles and the e^+ time of flight distribution this effect can be kept below 10^{-11}.
- Highly energetic positrons from the incident beam or from decaying muons will not follow the 90° bend of the transport system. Positrons from the low energy end of the muon decay which will be guided in the 0.1 T magnetic field can be efficiently suppressed in a set of collimators consisting of 1cm spaced Al plates. The positron contamination in the incident muon beam is reduced to about 5% in a crossed-field electrostatic and magnetostatic separator.
- Low energy secondary electrons (δ-rays) from the target will be suppressed by the collimator system as well. They are expected to contribute a few kHz to the count rate of the MCP.
- Internal Bhabha scattering where the e^+ from a Matom decay transfers significant energy to the e^- from the M atomic shell will have the ideal $\overline{M}$ signature. The probability of such a process is expected to be below 10^{-11}.
- Cosmic rays will not produce any severe background. The positron detector will be passively shielded by 5cm lead and actively shielded by a plastic scintillation counter enclosure.

No correlated background between Michel particle and slow positron detectors is expected above the 10^{-11} level. The rate of accidental coincidences R_{acc} is estimated to be of the order of

$$R_{acc} \approx 10^{-13} \times R_\mu^2 \cdot \Delta t \tag{5}$$

where R_μ is the incident muon rate and Δt is the width of the coincidence gate.

The experiment has been started in the πE3 beam area which delivers $1 \times 10^6 \mu^+/sec$ with $100\mu A$ proton current at $21MeV/c$. After the realization of the high proton current plans at PSI one can expect at least $5 \times 10^6 \mu^+/sec$. Within 10^7 sec of data taking one expects 0.13 background events with a 5 nsec coincidence window. Thus, the experiment will be sensitive to spontaneous M-$\overline{M}$-conversion at a level which is interesting for comparison with the model of Mohapatra and Herczeg [30,31].

4 The decay $K^+ \rightarrow \pi^+ \mu e$

A new experiment (E865) to search for the decay $K^+ \rightarrow \pi^+ \mu e$ is being set up at Brookhaven Nat. Lab. by a collaboration of four American institutes, three institutes from the former USSR, the Universities of Basel and Zürich and PSI. The aim of the experiment is to improve the sensitivity of the preceding experiment E777/851 by a factor of 70 to 3×10^{-12}. In detail this factor is composed as follows:

- Improvement in the K beam to a long unseparated 6 GeV/c beam requiring 1.2×10^{13} p/pulse from the new booster (factor 7).
- Larger apparatus by symmetrizing the old setup and using a magnet with larger aperture (factor 3).
- $\approx$ 40 weeks of running time compared to $\approx$ 14 weeks (factor > 2.3).
- Improved hardware and software efficiency (factor 1.5).

A number of improvements have to be done. First the number of muon identifier modules has to be tripled. Assembling and testing will be done at PSI in collaboration with the University of Tbilisi. A high rate beam chamber will be built by DUBNA in collaboration with PSI, which will build similar chambers also for internal use. Furthermore,four large low-mass MWPC's are needed for efficiency improvement, for which R&D is being done together with the University of Zürich. A new scintillator-lead calorimeter with wave length shifter optical fiber readout is mainly being developed by the INR Moscow. Finally, the on-line and off-line computing tools have to be provided with a strong participation of the University of Basel.

In summary,we emphasize that a healthy mixture of in-house and external experiments on one hand and of precision low-energy and highest energy experiments on the other hand is necessary to optimize the chance of finding new and unexpected results, to offer students a spuring and most scientific atmosphere and to exploit fully all possible synergy effects.

References

1. R.D. Bolton et al., Phys. Rev. Lett. 56 (1986) 2461
2. LAMPF Exp. 969, M.D. Cooper, spokesman
3. PSI Letter of Intent R-85-15.0, H.K. Walter, spokesman
4. D. Grosnick et al., Phys. Rev. Lett. 57 (1986) 3241
5. U. Bellgardt et al., Nucl. Phys. B299 (1988) 1 and W. Bertl et al., Nucl. Phys. B260 (1985) 1.
6. A.F. Kornyushkin et al., Proposal at Moscow Meson Factory, 1987
7. S. Ahmad et al., Phys. Rev. Lett. 59 (1987) 970
8. PSI Exp. R-87-03, A. van der Schaaf, spokesman
9. A. Badertscher et al., J. Phys. G 17 (1991) 47
10. A.I. Bochkarev et al., Proposal at Moscow Meson Factory, 1987
11. B.E. Matthias et al., Phys. Rev. Lett. 66 (1991) 2716
12. PSI Exp. R-89-06.1, W. Bertl and K. Jungmann, spokesmen
13. A.M. Lee et al., Phys. Rev. Lett. 64 (1989) 165
14. AGS Exp. E865, M.E. Zeller, spokesman
15. J. Dorenbosch et al., Z. Phys. C40 (1988) 497
16. AGS Exp. E787, I.H. Chiang, spokesman
17. M.S. Atiya, Particles and Fields 1991, Vancouver, August 18-22, 1991
18. C. Mathiazhagan et al., Phys. Rev. Lett. 63 (1989) 2181
19. AGS Exp. E791, S.G. Wojcicki, spokesman
20. S. Kettell, Particles and Fields 1991, Vancouver, August 18-22, 1991
21. KEK Exp. E137, T. Inagaki, spokesman
22. T. Shinkawa, Particles and Fields 1991, Vancouver, August 18-22, 1991
23. Y. Asano et al., Phys. Lett. 107B (1981) 159
24. S. Ahmad et al., Phys. Rev. D38 (1988) 2102
25. T. Suzuki et al., Phys. Rev. C35 (1987) 2212
26. SINDRUM collaboration, PSI newsletter 1989
27. SINDRUM collaboration, PSI preprint PSI-PR-90-41 (1990)
28. F. Muheim, Ph.D. thesis, Zurich University, 1990
29. M. Grossmann-Handschin, Ph.D. thesis, Zurich University, 1990
30. R.N. Mohapatra, this volume
31. P. Herczeg, this volume
32. A. Halprin, Phys.Rev.Lett. 48 (1982) 1313
33. K.A. Woodle et al., Z.Phys. D9 (1988) 59
34. G.A. Beer et al., Phys.Rev.Lett. 57 (1986) 671
35. T.M. Huber et al., Phys.Rev D41 (1990) 2709
36. W. Schäfer, Ph.D. thesis, Heidelberg University, 1988

This article was processed using Springer-Verlag TEX Z.Physik C macro package 1991
and the AMS fonts, developed by the American Mathematical Society.

Z. Phys. C - Particles and Fields 56, S 143–S 145 (1992)

Zeitschrift für Physik C Particles and Fields

The future of muon physics: Nuclear muon capture

Jules Deutsch

Institut de Physique, Université Catholique de Louvain, Chemin du Cyclotron, 2, B-1348 Louvain-la-Neuve, Belgium

18-September-1991

Abstract.

1) Overview and Perspectives :
2) Two cases-studies :
2.1) Precision recoil-spectroscopy in ^{3}He muon-capture,
2.2) T-violation tests in muon-capture.

1 Overview and perspectives

Let us recall that in the Standard Model the basic constituants of matter are a number of fermions and their anti-particles. They are grouped into generations, the number of which seems to be limited to three. Within each generation the masses and interactions allow to distinguish two quarks (e.g. u and d) and two leptons, one being electically charged, the other neutral (e.g. μ and ν_μ).

The basic fermions acquire mass by their coupling to a scalar boson still to be observed (the Higgs-particle) ; they interact exchanging some or all of the various vector bosons which characterize the various interactions.

In nuclear muon capture we are concerned with the coupling of the first-generation quarks u and d to the second-generation leptons μ and ν_μ by the exchange of a charged vector-boson W.

It is customary to write the effective hamiltonian of this process as :

$$GU_{ud}[\tilde{u}_\nu \gamma_\lambda (1+\gamma_5) u_\mu] < f|V_\lambda + A_\lambda|i> .$$

Here G stands for the Fermi coupling constant (G $= g^2/8$ in terms of the lepton-W coupling-constant g), duly renormalized by U_{ud}, first element of the Cabibbo-Kobayashi-Maskawa mixing matrix, u_ν and u_μ for the corresponding spinors and γ for the Dirac matrixes. $|i>$ and $|f>$ are the initial and final nuclear wave functions, V_λ and A_λ the most general vector and axial-vector Lorentz covariants. Assuming invariance under G-parity inversion, they can be expressed in terms of the momentum-transfer q as :

$$V_\lambda = F_V(q^2)\gamma_\lambda + F_M(q^2)\sigma_{\nu\lambda}q_\nu$$

and

$$A_\lambda = F_A(q^2)\gamma_\lambda\gamma_5 - iF_p(q^2)q_\lambda\gamma_5$$

Here the various form-factors $F_i(q^2)$ reflect the effect of the strong-interaction : we do not observe muon-capture on elementary u-quarks but on nucleons or nuclei. Constraints on these various form factors are provided by the Conserved Vector Current hypothesis and various soft-pion theorems.

The aims of nuclear muon-capture experiments are either the exploration of the particle-physics aspects, the nucleus serving so-to-say as a laboratory, or that of the nuclear-physics aspects, the muon serving as a probe. For illustration's sake we list briefly some of the topics which are or were under active consideration in the past ; we will be more specific discussing below two particular experiments.

The experiments dealing with the particle-physic aspects test predictions of the Standard Model and search for deviations which could provide hints of more satisfactory descriptions of nature.

In this perspective, let us mention the experiments searching for lepton-number violation as appearing in the $\mu^- + A \rightarrow e^- + A$ reaction, forbidden by the Standard Model, or related searches for generation-mixing in the neutrino-sector by charged-particle spectroscopy in two-body final states produced by muon-capture as will be discussed below. Precision-experiments were also considered to measure the ν_μ mass or to improve the constraints upon it. Experiments are in preparation to test muon-electron universality in the charged- and neutral current couplings. Finally, we mention the tests of the weak hadron current symmetry-properties : tests of the conservation of the isovector current (CVC) and those of the absence of time-reversal violation beyond the Standard Model we shall also discuss in more detail.

The nuclear physics issues first investigated were the response of the nuclear medium to the muon-probe in terms of single-particle or collective excitations, its comparaison to excitation by other probes, the measurement of transition-radii and other more-or-less classical nuclear spectroscopy studies. The high-momentum tail of the nuclear response-function continues to attract attention revealing nucleon pair-correlation and the anticipated "pion-like" muon capture. Most of the ungoing effort is directed, however, toward the investigation of the pionic degrees of freedom and the influence the nuclear medium exerts on them. The induced pseudoscalar form-factor is particularily sensitive to this influence and its predicted quenching still remains to be established unambigously.

The experimental approaches to these challenges, both in ordinary and radiative muon capture, extend from the more classical capture-rate measurements (total and partial) to the observation of the rare break-up channels and to the various more modern correlation-experiments. These correlation experiments involve generally the muon-spin, the target-spin in the case of J $\neq$ 0 target-nuclei (hyperfine conversion and the use of polarized targets), the spin of the product-nucleus observed through its decay-pattern and the neutrino recoil-direction.

Amongst the principal experiments on-way or planed let us mention the investigation of radiative- and ordinary muon capture in hydrogen and/or He^3 (TRIUMF, PSI), various correlation-experiments making use of the Doppler-effect in the decay of the recoiling product-nucleus (Dubna and TRIUMF), the attempts to use polarized He^3-targets in muon-capture (LAMPF and TRIUMF), the search for neutrino-mixing by precision recoil-spectroscopy in He^3 muon-capture (PSI) and various ideas to search for T-odd corelations in muon-capture.

2 Two case-studies

2.1 Precision recoil-spectroscopy in ^3He muon-capture

Extensions of the Standard Model allow for non-zero neutrino-masses and consequently for generation-mixing similar to the one observed in the quark-sector. In these scenarios, the muon, for example, couples to neutrinos of various masses with a coupling-strength characteristic of each corresponding neutrino mass eigen-state. The coupling-strength of the various charged leptons to the various massive neutrinos are expressed than by the elements of a neutrino mixing-matrix similar to the Cabbibo-Kobayashi-Maskawa mixing-matrix of the quark-sector.[1)]

Searches for neutrino oscillation and neutrino decay attempt to observe neutrino-mixing but up to now failed to observe it and produced only upper limits to elements of the mixing-matrix, if we except the possible deficit in solar neutrinos which could be ascribed to neutrino mixing in the Sun.

Schrock called attention [2)] to the possibility of an approach complementary to the above mentioned ones, featuring generally a higher sensitivity, albeit for more massive neutrinos.

This approach consists in a precision-spectroscopy of the charged particle emitted together with the neutrino in a two-body final state reaction. The admixture of a massive neutrino into the dominant channel would produce a second peak in the energy-spectrum of the charged particle in addition to the dominant peak corresponding to the emission of a light neutrino.

The challenge was to build a detector able to observe the 1.9 MeV triton-recoil with good energy-resolution and to reject with good efficiency the charged particles from muon-decay and from the various triton break-up channels. This challenge was met constructing a helium-3 filled scintillating proportional chamber allowing the spatial reconstruction of the charged particle tracks and featuring a better than 1 % energy-resolution for tritons in the central region of the fiducial volume.

The device, its performance and the first results obtained were described with some detail in the oral presentation of this talk. As, however, a written account of the work appeared in the meantime[3)], we do not repeat it here but refer the interested reader to ref. 3.

It may be noted that the device could be used for other muon-capture experiments on ^3He : for a precision-measurement of the partial capture rate and that of the triton recoil-asymmetry. These developments are actually under consideration ; because of the low residual muon-polarization in ^3He, the asymmetry-experiment may however have to await the development of a polarized ^3He-target, actively persued at LAMPF[4)]

A more speculative line of research will be discussed in the following chapter.

2.2 T-violation tests in muon-capture

As no experiments or even worked-out proposals exist in this field, the aim of this brief comment is solely to trigger further interest in this issue, under active consideration [5,6,7,8)].

CP-violation was observed up to now only in the K_L-system. It was searched for also in semi-leptonic weak decays such as K-decay[9)] or nuclear (nucleon) beta-decay[10)].

If the only source of CP-violation would be the one offered by the Standard Model, effects in beta-decay (and muon-capture) would be of second order only and vanishingly small. So any observation of T-violation in these processes would indicate new physics beyond the Standard Model.

New experiments are actually designed to push further the precision-limits of the searches in K^+-decay and nuclear (nucleon) beta-decay. In the following we would like to advocate a similar effort in muon-capture.

In the muon-sector results, even less precise than the ones already obtained in the electron-sector, could be of interest. Phenomenologically there is no need indeed to have effects similar in the two sectors, as exemplified by the Higgs-coupling which is trivially stronger in the muon-sector than in the electron-sector : if such a (charged Higgs) coupling contributes to CP-violation, it is expected to be stronger in muon-capture than in beta-decay[11].

Experimentally, one could search for both P-even and P-odd T-odd correlations.

A P-even T-odd correlation is examplified by $((k_\gamma x k_{recoil}).\sigma_\mu)$. $(k_\gamma \,.\, k_{recoil})$ as discussed in ref. 8. Here k_γ stands for the direction-vector of the gamma-ray which de-excites the nucleus formed in muon-capture : it measures the alignement of the final state. k_{recoil} stands for the nuclear recoil direction and σ_μ for the muon spin. The possibility to measure such a correlation in the $\mu + {}^{16}O \rightarrow \nu_\mu + {}^{16}N$ (1^-, 397 keV) transition was explained with some detail in the oral presentation ; we refer the interested reader to ref. 8 for these details.

A P-odd T-odd correlation is examplified by the search for $(J_{final} x k_{recoil}).\sigma_\mu$ correlation, where J_{final} is observed by the preferential direction of the subsequent beta-decay. The principle of this method, applied to muon-capture in ${}^{12}C$, was pioneered by L. Grenacs ; for more detail we refer to his ref. 12.

These interesting challenges clearly qualify for the title of our workshop "The Future of Muon Physics".

It is a pleasure for me to express at this occasion my appreciation for the warm and most stimulating friendship of Guisbert zu Putlitz and Vernon Hughues.

References

1. cfr. e.g. F. Boehm and P. Vogel, Physics of the massive neutrinos, Cambridge Univ. Press, 1987.
2. R. Schrock, Phys. Lett. 96B (1980) 159 and Phys. Rev. D24 (1981) 1232 and 1273.
3. B. Tasiaux et al., Particle World 2 (1991) 81.
4. P. Souder, these proceedings.
5. J. Deutsch, A comment on T-violating triple correlations in mon-capture, Workshop on Fundamental Muon Physics, Los Alamos Nat. Lab., Jan. 20-22, 1986.
6. St. Chiezanovicz and J. Deutsch, Low-Energy Muon Science Workshop 90, PSI, April 1990, unpublished.
7. N. Mukhopadhyay and P. Herczeg, PANIC 1990.
8. A.S. Carnoy, UCL Ph. Thesis annex, 1991, unpublished.
9. M.K. Campbell et al., Phys. Rev. Letters 47 (1981) 1032.
10. cfr. e.g. : F. Calaprice, Hyperf. Int. 22 (1985) 83 and refs. cited.
11. J.-M. Gérard, private communication.
12. L. Grenacs, Ann. Rev. Nucl. Part. Sci. 45 (1985) 455 and refs. cited.

This article was processed using Springer-Verlag TEX Z.Physik C macro package 1991
and the AMS fonts, developed by the American Mathematical Society.

Z. Phys. C - Particles and Fields 56, S146-S149 (1992)

Zeitschrift für Physik C Particles and Fields

Laser Polarized Muonic Atoms

Paul A. Souder

Syracuse University, Syracuse, NY 13244

26-October-1991

Abstract. Lasers are an important tool in the field of muon physics. A new application of lasers, namely producing polarized muonic atoms, is the subject of a new program at LAMPF. One technique already demonstrated is stopping unpolarized muons in a laser polarized ^{3}He target. A more promising idea is to polarize neutral muonic helium by collisions with laser polrized Rb vapor. These methods for producing polarized muonic helium will be useful for measuring the spin dependence of nuclear muon capture and for determining the induced pseudoscalar coupling.

PACS: 36.10.Dr,29.25.Kf,32.80.Bx

1 Introduction

Although the use of lasers in muon physics has a relatively long history, recent developments in both laser technology and muon beams should make a new series of exciting experiments possible. A variety of physics topics may be addressed by these methods. One topic is laser driven transitions in muonic atoms, such as the 2S-2P transition in helium studied by Zavatini and collaborators in their pioneering work.[1,2] There is still active interest in the issues raised by these experiments.[3] Another topic, which has been described at this workshop, is laser spectroscopy of the electron in the muonium atom.[4]

At the Los Alamos Clinton P. Anderson Meson Physics Facility (LAMPF), we have initiated a program using lasers for a new purpose, namely the polarization of muonic atoms. At present, we see three physics goals for this work:

1. Learn about the atomic physics involved in the complex processes that influence the polarization of negative muons stopped in matter.
2. Use polarized atoms to study the spin-dependence of the part of the weak Hamiltonian involved in nuclear muon capture. One focus of this work will be the determination of g_p, the induced pseudoscalar coupling.
3. Polarized atoms may be important for experiments testing symmetries, in particular time-reversal invariance. Although this is the least developed part of our program, the possibilities have been discussed by others.[5]

This paper will be arranged in three parts. The first will describe a successful experiment performed in the summer of 1990 in which we measured the polarization of initially unpolarized muons stopped in a polarized ^{3}He target. The second will describe the spin observables in muon capture in ^{3}He. The third will describe a new technique, direct spin exchange (DSE), that is especially well suited for studies of nuclear capture in ^{3}He.

2 Muons Stopped in Polarized ^{3}He

Muons bound to polarized nuclei become polarized via the hyperfine interaction, a technique denoted "repolarization" and first applied to ^{209}Bi.[6] We have successfully used this method by stopping a beam of unpolarized muons in a ^{3}He target [7] which was polarized via spin exchange with laser-optically pumped Rb atoms.[8]

The core of the apparatus is a 2.5 cm diameter spherical glass cell containing $\sim$8 atm of ^{3}He, $\sim$70 torr of N_2, and several mg of metallic Rb. Operation at 200°C provided ample Rb vapor density. The ^{3}He was polarized by spin exchange with Rb vapor optically pumped by a 5 W Ti:Sapphire laser. Average polarizations of $\sim$35% were achieved during the run. The targets were polarized in a "pumping station" outside of the beam cave. When transported to the apparatus in the muon beam, the targets decayed with a lifetime of about 30 hours. In a uniform field, these cells have lifetimes on the order of 80 hours; in our apparatus we were limited by magnetic field gradients produced by the last beam quadrupole.

In order to stop muons cleanly in this small, thin target, a special beam tune was developed for the Stopped Muon Channel at LAMPF.[9] The main features were a narrow momentum spread and, through a combination of focusing and collimation, a small beam spot with an area of $\sim$1 cm^2. At the usual operating momentum of 22.5 Mev/c, the beam provided an average muon rate of $\sim$1 kHz. Extreme care was taken to minimize the material in the beam; the beam telescope was a pair of 40μ plastic scintillation counters, and the target cell walls were $\sim$ 100μ thick.

The direction of the spins in the target followed the direction of a 1–2 G magnetic field which was reversed every few minutes. Decay electons were detected by scintillation counter telescopes, and an asymmetry was formed by comparing the number of electrons that decayed parallel $N\uparrow\uparrow$ and antiparallel $N\uparrow\downarrow$ to the magnetic field. The experimental asymmetry

$$A_{exp} = \frac{N\uparrow\uparrow - N\uparrow\downarrow}{N\uparrow\uparrow + N\uparrow\downarrow} \tag{1}$$

is shown in Figure 1. A large asymmetry is apparent which corresponds in aggregate to a 20 standard deviation effect. The sign of the asymmetry reverses, as expected, when the direction of the spins relative to the magnetic field is changed at the pumping station. The ability to flip the asymmetry in two independent ways (flipping the magnetic field and flipping the spins relative to the field) is a powerful technique for measuring small asymmetries with minimal systematic errors. It is one of the key advantages lasers bring to this field.

The muon polarization normalized to the target polarization, P_μ^N, is obtained from A_{exp} by correcting for the fraction of stops in the gas, target polarization, etc. We obtain a preliminary result of $P_\mu^N = 7 \pm 1\%$, which is about a factor of 2 smaller than predicted by calculations that neglect the effects of collisions during the cascade.[10,11] We note that when polarized muons are stopped in ^{4}He, the asymmetry is a factor of 3 or so smaller than predicted.[12,13]

3 Theory of Muon Capture by ^{3}He

Measurements of the spin dependence of negative muons captured by ^{3}He into the ground state of ^{3}H, $\mu^- + N_i \rightarrow \nu_\mu + N_f$ can provide unique information about the charged weak current,[14,15] given by

$$J_\mu = \bar{u}_f \Big(g_V \gamma_\mu + \frac{g_M}{2M_P} \sigma_{\mu\nu} q^\nu + g_A \gamma_5 \gamma_\mu - g_P q_\mu \gamma_5 \frac{2M}{m_\pi^2} \Big) u_i. \tag{2}$$

Here g_V, g_M, g_A and g_P are isovector form factors, which depend upon the four momentum transfer Q^2, and $M = \frac{1}{2}(M_f + M_i)$. The form factor g_P is important for muon capture because the mass of the initial state lepton, the muon, is significantly different from that of the final state lepton. Since there is no pseudoscalar coupling in the Standard Model for free quarks, g_P arises from the fact that the quarks are confined in the proton by QCD forces. Indeed, the primary motivation of many muon capture experiments is to obtain information about this elusive but fundamental form factor.

The numerical values of the form factors are given by $g_V = 0.82$, $g_M = -4.73$, $g_A = -1.06$, and $g_P = 0.68$. The first two form factors are obtained from CVC and electron scattering data, the third extrapolated to our Q^2 value from beta decay of ^{3}H, and the last equation comes frpm PCAC:

$$g_P(Q^2)(1 - Q^2/m_\pi^2) \approx -g_A(Q^2). \tag{3}$$

The transition matrix element for muon capture may be written in the form

$$T = \chi_f^\dagger \chi_\nu^\dagger \Big(1 - \sigma \cdot \hat{k}\Big) \cdot \Big[G_V + G_A(\sigma_l \cdot \sigma_h) + G_P(\sigma_l \cdot \hat{k})(\sigma_h \cdot \hat{k})\Big] \chi_i \chi_\mu, \tag{4}$$

where the χ's are Pauli spinors, $\sigma_{h(l)}$ operates on the hadronic (leptonic) spinors, and $\hat{k}$ is the direction of the neutrino in the final state. The new constants G_V etc. may be given in terms of the original form factors evaluated at the appropriate Q^2 as follows

$$\begin{aligned} G_V &\approx g_V \big(1 + \frac{E_\nu}{2M_f}\big) - g_M \frac{m_\mu}{2M_P} \frac{E_\nu}{2M_f} \approx 0.84 \\ G_A &\approx -g_A - g_V \frac{E_\nu}{2M_f} - g_M \frac{E_\nu}{2M_P} \approx 1.31 \\ G_P &\approx g_P \frac{m_\mu E_\nu}{m_\pi^2} + \big(g_A - g_V\big) \frac{E_\nu}{2M_f} - g_M \frac{m_\mu}{2M_P} \frac{E_\nu}{2M_f} \\ &\approx 0.63 \end{aligned} \tag{5}$$

where E_ν is the neutrino energy in the final state.

The reaction is highly spin dependent. This fact may be exploited experimentally by determining the initial hyperfine state described by the quantum numbers F and M_F, and the direction of the recoil ^{3}H, which is exactly opposite $\hat{k}$. It is convenient to define a vector polarization P_V, a tensor polarization P_T, and a departure from the statistical singlet-triplet mixture Δ based on the initial state populations $P(F, M_F)$ as follows:

$$\begin{aligned} P_V &= P(1,1) - P(1,-1) \\ P_T &= P(1,1) + P(1,-1) - 2P(1,0) \\ \Delta &= P(1,1) + P(1,0) + P(1,-1) - 3P(0,0) \end{aligned} \tag{6}$$

Then we may define a decay rate

$$\frac{d^2\omega}{d\hat{k}} = \frac{R}{4\pi} \big(A + B \cdot P_V(\hat{k} \cdot \hat{z}) + C \cdot P_T[(\hat{k} \cdot \hat{z})^2 - \frac{1}{3}] + D \cdot \Delta\big). \tag{7}$$

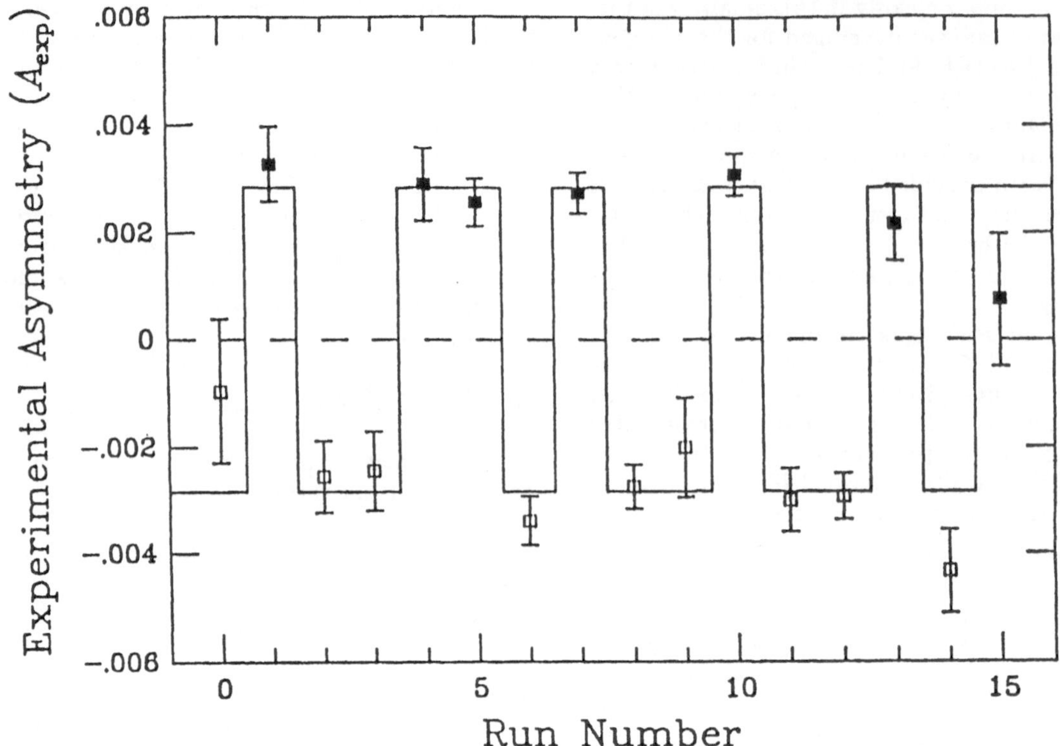

Fig. 1. Raw asymmetries adjusted to a constant helium polarization of 36%. As indicated by the solid line, the sign of the asymmetry conforms to that expected from the relative orientation of the target spin to the applied magnetic field.

Here the average rate $\Gamma = RA$ for an unpolarized initial state defines the constant R, and the constants

$$\begin{aligned} A &= G_V^2 + 3G_A^2 + G_P^2 - 2G_AG_P \approx 4.60 \\ B &= -(G_V + G_A - G_P)^2 \approx -2.31 \\ C &= -2G_P(G_V + G_A) \approx -2.71 \\ D &= 2\{G_A(G_V + G_P - G_A) - \frac{1}{3}G_P(G_V + G_A)\} \\ &\approx -0.48 \end{aligned} \tag{8}$$

are simple functions of the three constants G_V, G_A, and G_P. The one constraint on the four constants, $A + B + \frac{2}{3}C + D = 0$, corresponds to the fact that angular momentum conservation forbids the neutrino to decay in the direction of the polarization of a triplet state.

The constant A has been measured [16,17] by using unpolarized ^{3}He, but little other experimental information exists on this system. It is highly desirable to measure at least one of the other coefficients, since each is very sensitive to g_P. Perhaps the easiest choice is B, which can be isolated in an experiment providing for vector polarization by using either polarized beams or targets. Unfortunately, the polarizations are tiny for polarized beams and, as shown by our work described above, still rather modest with polarized beams. Such experiments are challenging.

It also would be useful to measure C and D to provide redundancy or to check some of the underlying assumptions such as the absence of second class currents. Under naive assumptions, however, both P_T and Δ are *quadratic* in the relevant polarizations and therefore are likely to be small when presently established methods are used.

4 Direct Spin Exchange

To overcome the problem of small polarizations, we have conceived of a new approach, direct spin exchange (DSE).[18] The DSE method will yield high polarizations, in principle as high as 75% for atoms that have been in the target sufficiently long. An important effect of high vector polarizations is that moderate values of P_T become possible, providing a practical way of measuring the constant C.

The basic idea of DSE is to transfer the spin of the Rb atoms directly to the muonic ^{3}He in the ground state. This may be accomplished if the muonic ^{3}He is in a paramagnetic state, namely as a neutral "hydrogen" atom. Thus a two step process is required. First, the muonic ion, which is produced when muons are captured by He, must be neutralized. This may be accomplished by charge exchange with some donor, such as Xe or CH4, before the muonic ion thermalizes. Thermalized muonic ions form molecular ions, which do not react with the above donors. The second problem is to provide a sufficient density of Rb to polarize the atoms during the 2.2μsec lifetime of the muon. Measurements of Rb densities possible in practical cells as well as estimates of the spin-exchange cross section scaled from hydrogen data [19] suggest that reasonable polarizations are possible; $\sim$ 25% averaged over all times and ~45% for muons that have been in the target for 2 μsec. Another possibility for neutralization is charge exchange of the molecular ions containing the muonic ion with Rb atoms. This process, which will also polarize the muon because the captured electron is polarized, has a cross section similar to that for spin exchange.

The atomic physics issues in the neutralization of muonic helium are quite interesting in themselves. The large concentrations of Xe ($\sim$ 1%) which were observed to be required in previous experiments forming neutral muonic He will significantly depolarize the Rb. Thus cells incorporating Xe will require a compromise between the low concentrations required to maintain highly polarized Rb and the high concentrations required for efficient neutralization. In contrast, CH_4 has no unwanted interactions with the Rb. CH_4 also has been demonstrated to efficiently neutralize positive muons.[20] However, studies of hot tritium chemistry in He-CH4 mixtures [21,22] indicate that unwanted diamagnetic species such as neutral H_2 (where one of the H is muonic He) will also be formed at some level. We are planning to study cells this summer at LAMPF to empirically determine the merits of Xe, CH4, and Rb at various concentrations.

Finally, we comment on the issue of incorporating a drift chamber into one of our cells. Polarized targets, which must support ~80 hour lifetimes for the ^{3}He, are made of special glass because most other materials cause significant depolarization. If metal electrodes are added, they must have the smallest possible area and probably will still degrade the cell performance. In contrast, the DSE approach, which requires no long time constants, permits a more flexible design. Larger electrodes are possible, although they must still be compatible with the Rb. Another interesting advantage with DSE is that only the part of the cell where the muons stop need to have polarized Rb; the rest of the cell used to stop the recoil tritons need not absorb laser power. We are optimistic that the DSE approach will lead to a practical experiment which will measure g_P.

References

1. Carboni, G., Gastaldi, U., Neri, G., Pitzurra, O., Polacco,E., Torelli, G., Bertin, A., Gorini, G., Placci, A., Zavattini, E., Vitali, A., Duclos,J., Picard, J.: Nuovo Cimento **34A**, 493 (1977).
2. Carboni, G., Gorini, G., Torelli, G., Palmonari, F., Zavattini, E.,: Nucl. Phys. **A278**, 381 (1977).
3. von Arb, H. P., Dittus, F., Heeb, H., Hofer, H., Kottmann, F., Niggli, S., Schaeren, R., Taqqu, P., Unternahner, J., Egelhof, P.: Phys. Lett. **36B**, 232 (1984).
4. Chu, S. Mills, A. P., Yodh, A. G., Nagamine, K., Miyake, Y., Kuga, T.: Phys. Rev. Lett. **57**, 1847 (1986).
5. Deutsch, J. G.: Proceedings of the Workshop on Fundamental Muon Physics: Atoms, Nuclei, and Particles, ed. by Hoffman, C. M., Los Alamos, New Mexico, (1986).
6. Kadono, R., Imazato, J., Ishikawa, T., Nishiyama, N., Nag-7)amine, K., Yamazaki, T., Bosshard, A., Döbeli, M., van Elmbt, L., Schaad, M., Truöl, P., Bay, A., Perroud, J. P., Deutsch, J., Tasiaux, B., Hagn, E.: Phys. Rev. Lett. **57**, 1847 (1986).
7. Newbury, N. R., Barton, A. S., Bogorad, P., Cates, G. D., Gatzke, M., Saam, B., Han, L., Holmes R., Souder, P. A., Xu J., Benton, D,: to be published in Phys. Rev. Lett.
8. Happer, W., Miron, E., Schaefer, S., Schreiber, D., van Wijngaarden, W. A., Zeng, X.:Phys. Rev. A **29**, 3092 (1984).
9. Holmes, R., Kim, D.-H., Kumar, K. S., Souder, P. A., Benton, D., Cates, G. D., Newbury, N., Pillai, C.: Nucl. Instrum. and Meth. in Phys. Res. A **303**, 226 (1991).
10. Kuno, Y., Nagamine, K., Yamazaki, Nucl. Phys. A **475** 615 (1987).
11. Bukhvostov, A. P., Popov, N. P.: JETP **19**, 1240 (1964).
12. Souder, P. A., Casperson, D. E., Crane, T. W., Hughes, V. W., Lu, D. C., Orth, H., Reist, H. W., Yam, M. H., zu Putlitz, G.: Phys. Rev. Lett. **34**,1417 (1975).
13. Souder, P. A., Crane, T. W., Hughes, V. W.,Lu, D. C. Orth, H., Reist, H. W., Yam, M. H., zu Putlitz, G.: Phys. Rev. A **22**, 33 (1980).
14. B. R. Holstein Phys. Rev. C **4**, 764 (1971)
15. W-Y. P. Hwang, Phys. Rev. C **17**, 1799 (1978).
16. Auerbach, L. B., Esterling, R. J., Hill, R. E., Jenkins, D. A., Lach, J. T., Lipman, N. H.: Phys. Rev. **138**, B127 (1965).
17. Clay, D. R., Keuffel, J. W. Wagner, R. L., Edelstein, R. M.,: Phys. Rev. **140**, B586 (1965).
18. Cates, G. D., Souder, P. A., Barton, A. S., Benton, D., Holmes, R., Newbury, N. R., Xu, J.: LAMPF proposal 1231 (1991) (unpublished).
19. Redsun, S. G., Knize, R. J., Cates, G. D., Happer, W.: Phys Rev. **42**, 1293 (1990).
20. Fleming, D. G., Mikula, R. J., Garner, D. M., Phys. Rev. A **26**, 2527 (1982).
21. Seewald D., Wolfgang, R.: J. Chem. Phys. **47**, 143 (1967). Wolfgang, R., Prog. Reaction Kinetics, **3**, 97 (1965)
22. Estrup P. J., Wolfgang, R., J. Am. Chem. Soc., **82**, 2661 (1960).

This article was processed using Springer-Verlag TEX Z.Physik C macro package 1991
and the AMS fonts, developed by the American Mathematical Society.

Z. Phys. C - Particles and Fields 56, S150–S155 (1992)

Zeitschrift für Physik C Particles and Fields

Radiative muon capture on hydrogen

W. Bertl[3], S. Ahmad[1], D.S. Armstrong[4], G. Azuelos[2,5], M. Blecher[4], C.Q. Chen[1], P. Depommier[5], P. Gumplinger[1], T.P. Gorringe[7], M.D. Hasinoff[1], R. Henderson[2,6], G. Jonkmans[5], A.J. Larabee[1], J.A. Macdonald[2], S.C. McDonald[6], J.-M. Poutissou[2], R. Poutissou[2], B.C. Robertson[8], D.G. Sample[1], W. Schott[1], G.N. Taylor[6], T. von Egidy[2], D.H. Wright[1], N.S. Zhang[2]

[1] University of British Columbia, Vancouver, B.C., Canada V6T 2A6, [2] TRIUMF, Vancouver, B.C., Canada V6T 2A3, [3] PSI/SIN, CH-5232, Villigen, Switzerland, [4] Virginia Polytechnic Inst. and State U., Blacksburg, VA, USA 24061, [5] Université de Montréal, Montréal, P.Q. Canada H3C 3J7, [6] University of Melbourne, Parkville, Victoria, Australia, 3052, [7] University of Kentucky, Lexington, KY, USA 40506, [8] Queen's University, Kingston, Ontario, Canada K7L 3N6

28 October 1991

Talk presented by W. Bertl at the Symposium "The Future of Muon Physics" in Heidelberg, Germany, May 7–9, 1991

Abstract. The radiative capture of negative muons by protons can be used to measure the weak induced pseudoscalar form factor. Brief arguments why this method is preferable to ordinary muon capture are given followed by a discussion of the experimental difficulties. The solution to these problems as attempted by experiment #452 at TRIUMF is presented together with preliminary results from the first run in August 1990. An outlook on the expected final precision and the experimental schedule is also given.

1 Physics motivation

Weak semi-leptonic interactions can be described as a point interaction of leptonic and hadronic currents as long as the momentum transfer involved is small compared to the intermediate vector boson mass, M_W. The vector and axial vector components of the hadronic current are given by [1]

$$V_\mu = g_V\gamma_\mu + ig_M\sigma_{\mu\nu}\frac{q^\nu}{2m_N} + \frac{q_\mu}{m_l}g_S \qquad (1)$$

$$A_\mu = g_A\gamma_\mu\gamma_5 + i\frac{q_\mu}{m_l}g_P\gamma_5 + g_T\sigma_{\mu\nu}\gamma_5\frac{q^\nu}{2m_N} \qquad (2)$$

respectively, where q is the momentum transfer, g are the form factors[1] which can themselves depend on q, m_N is the mean nucleon mass and m_l the lepton mass. Terms explicitly depending on q are called *induced* weak currents. Scalar and tensor contributions (so called second class currents) vanish for the case of unbroken charge symmetry and time reversal invariance [1]. The present experiment aims to improve our knowledge about g_P, the weak induced pseudoscalar form factor.

The expected value[2] for g_P, assuming the axial vector current is partially conserved (PCAC–hypothesis) with the pion field being the source of the inhomogeneity, is given by [2]

$$g_P(q^2) = \frac{\sqrt{2}f_\pi g_{\pi NN}(-m_\pi^2)m_\mu}{m_\pi^2 - q^2} + \text{correction terms} \qquad (3)$$

For ordinary muon capture this results in (f_π = 131.7 MeV the pion decay constant, $g_{\pi NN}$ = 13.5 the pion-nucleon coupling constant)

$$g_P(q^2 = -0.88\,m_\mu^2) = -8.4\pm0.25 \text{ in units of} 10^{-5}/m_p^2 \qquad (4)$$

The cited error is to be understood as the maximum possible deviation due to unknowns in the momentum dependence of g_A and the dispersion integral. In radiative muon capture processes the momentum transfer is not fixed but varies as $q^2 = 2m_\mu k - m_\mu^2$ (k the photon momentum). Not included are effects from Δ-contributions to the RMC-rate which might be as large as 8 % [3]

Several experiments attempting a measurement of g_P have been undertaken in the past [4]. The most accurate results achieved so far are from muon capture on hydrogen using a lifetime measurement method ($g_P = -7.1 \pm 3.0$ [5]) and from recoil polarization measurements following muon capture on carbon. ($g_P = -11.3 \pm 2.4$ [6][3]). The world averaged value (hydrogen experiments only) is $g_P = -8.7\pm1.9$ in agreement with the PCAC prediction. However, the error of 22 % is to be compared

[1] V,M,S,A,P,T stand for vector, magnetic, scalar, axial vector, pseudoscalar and tensor components.

[2] the momentum transfer is conveniently chosen at the value valid for ordinary muon capture

[3] the paper actually gives $g_P(q^2 = -0.88m_\mu^2)/g_A(q^2 = 0) = 9.0 \pm 1.9$. The most recent measurement of g_A gives [7] $g_A(q^2 = 0) = -1.282 \pm 0.002$

with the 3 % accuracy in the theoretical prediction and the 0.3 % error currently achieved in measurements of g_V and g_A using free neutron decay [7]. A substantial improvement is obviously desirable.

Radiative muon capture on nuclei has also been a tool to determine g_P. The results seem to indicate that g_P is a function of nuclear charge. Models exist which explain the deviation from the PCAC predicted value determined from hydrogen measurements by a renormalization of the pion field inside the nucleus. For a detailed discussion see e.g. the review article by Gmitro and Truöl [8]. An accurate measurement of g_P in hydrogen would set the reference point for discussing any possible renormalization effects in nuclei.

2 Why use Radiative Muon Capture (RMC) to determine g_P ?

The RMC process $\mu^- p \rightarrow n\gamma\nu$ is a radiative subprocess to ordinary muon capture (OMC; $\mu^- p \rightarrow n\nu$) and it is therefore suppressed by $(m_e/m_\mu)^2$, making RMC measurements very difficult. Nevertheless, there are several reasons why the extremely rare RMC processes are more favourable than OMC processes for the study of g_P:

- Higher sensitivity of RMC rate to g_P (see figure 1), therefore less vulnerable to systematic errors.
- The shape of the photon spectrum (figure 2) is affected by the q^2 - dependence of g_P near the pion pole, particularly in the high energy part of the spectrum which is the experimentally accessible region. In OMC no q^2–dependence of g_P can be determined.
- The detection of high energy photons from RMC is experimentally easier and more precise than the detection of the low energy neutrons from OMC.

3 Experimental difficulties

A notorious problem with all muon capture experiments in hydrogen is what is sometimes called "muon chemistry" (see fig. 3).

The muon, initially bound in a neutral $(\mu^- p)$-atom after having lost its energy, is eventually transferred to impurity atoms or deuterium [4] as these transitions are energetically favoured. Since the capture probability increases with Z^4, muons bound in non-hydrogenic atoms $(\mu^- Z)$ pose a considerable background problem. Roughly speaking, all elements other than hydrogen must be reduced to a concentration of less than 10^{-9} in order to keep the associated high energy photon background safely below 10 % of the hydrogen RMC rate ($\approx 10^{-8}/\mu$-stop). Such a small concentration is achieved by cleaning the hydrogen gas with a palladium diffusion purifier and maintaining a leak-tight target cell with a low outgassing rate. For the palladium purifier the through-going flux of N_2 was measured to be less than 10^{-11} of the hydrogen flux at room temperature. An outgassing rate of $\approx 5 \times 10^{-11}$ l/min for the gold-walled target cell was achieved, eventually accumulating to 2×10^{-6} liter in 30

[4] natural concentration of tritium is negligible

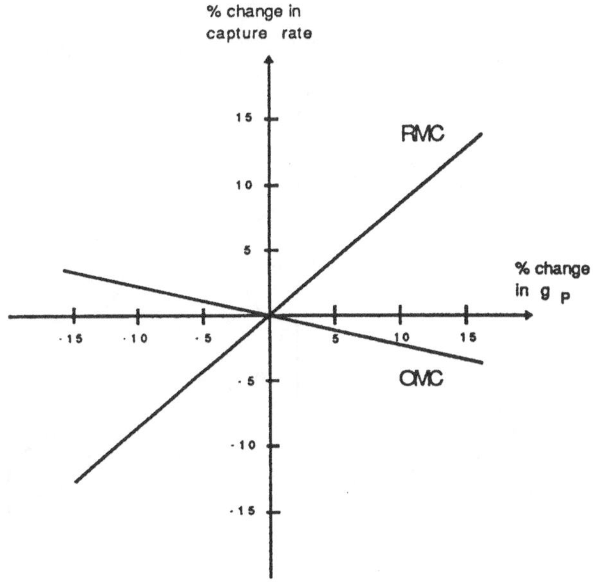

Fig. 1. The sensitivity in determining g_P by OMC and RMC rate measurements respectively assuming the induced tensor form factor $g_T = 0$.

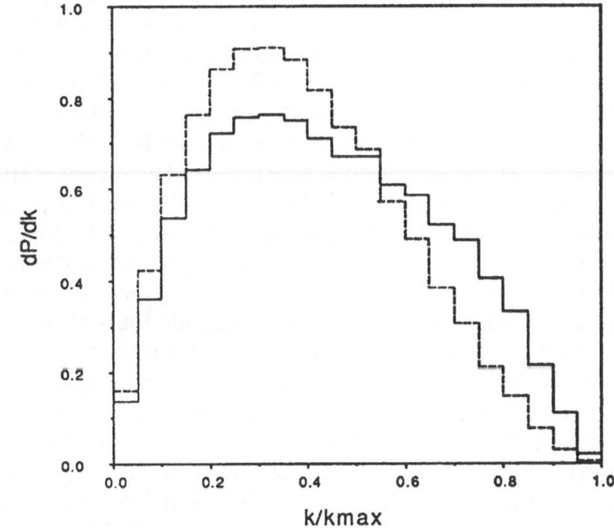

Fig. 2. The photon spectrum of $\mu^- p \rightarrow n\gamma\nu$ as expected from PCAC using formulas for the capture rate calculated by Opat [12] (the max. photon momentum is kmax = 99.7 MeV/c). g_P is taken in the form given by eq. 3 with $q^2 = 2m_\mu k - m_\mu^2$. The dashed line shows the photon distribution assuming a *constant* form factor with no q^2 dependence. The value taken for g_P is -8.4 which reproduces the PCAC predicted value for OMC. However, the differences in the corresponding RMC photon spectra are obvious.

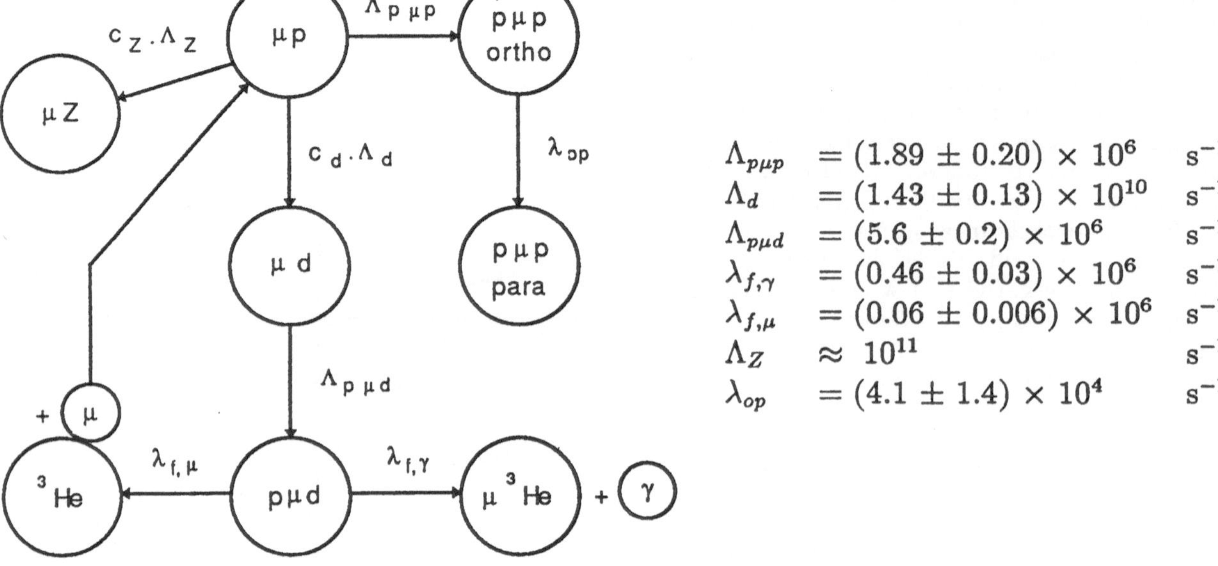

$\Lambda_{p\mu p}$	$= (1.89 \pm 0.20) \times 10^6$	s^{-1}	[16]
Λ_d	$= (1.43 \pm 0.13) \times 10^{10}$	s^{-1}	[16]
$\Lambda_{p\mu d}$	$= (5.6 \pm 0.2) \times 10^6$	s^{-1}	[19]
$\lambda_{f,\gamma}$	$= (0.46 \pm 0.03) \times 10^6$	s^{-1}	[19]
$\lambda_{f,\mu}$	$= (0.06 \pm 0.006) \times 10^6$	s^{-1}	[19]
Λ_Z	$\approx 10^{11}$	s^{-1}	[18]
λ_{op}	$= (4.1 \pm 1.4) \times 10^4$	s^{-1}	[19]

Fig. 3. Simplified picture of reactions catalyzed by stopped μ^- in hydrogen with small deuterium admixtures (concentration c_d) and impurities Z (concentration c_Z). All reactions shown compete with ordinary muon decay and capture processes. Rates Λ are density dependent and are normalized to liquid hydrogen density ($\rho_0 = 4.25 \times 10^{22}$ atoms/cm^3). Rates λ are independent of density.

days[5]. This corresponds to less than 10^{-9} of the liquid hydrogen filling of 2500 l (NTP).

Muonic deuterium atoms together with adjoining hydrogen atoms eventually form the $(p\mu d)^+$ molecule with subsequent fusion to ^{3}He. The muon is either recycled or remains in the ^{3}He orbit with a chance to produce radiation from RMC on ^{3}He. Numerical estimates show that $\approx 3 \times 10^{-6}$ deuterium contamination is sufficient to generate a 10 % background effect. As natural deuterium concentrations in water are between 100 and 150 ppm, deuterium depleted water was purchased[6] and hydrogen gas was produced by electrolysis. This manufactured target gas, as well as samples of the original water,are currently being mass analyzed by various laboratories.

The high probability of muon transfer from hydrogen to heavy elements with subsequent radiative capture on the latter makes the design of the target cell an extremely critical issue. Muons may capture in the wall material by stopping in the front wall of the target, by scattering into the side walls or by diffusion of neutral $\mu^- p$ atoms. Monte-Carlo simulations, verified by muon scattering experiments with an equivalent weight foam target replacing the hydrogen and scintillators replacing the target walls, led to a cylindrical target of 16 cm ø ×15 cm length with a hemispherical front dome. The effect from direct muon stops is further reduced by choosing gold as the wall material, since the muon lifetime in gold is only 73 ns. A blanking time of 350 ns after each μ-stop is used in the analysis to eliminate events due to the decay of the (μ^-Au)-system. Diffusion of $\mu^- p$ atoms to the wall is of less concern for liquid targets, because after $\approx$ 500 ns all $\mu^- p$-atoms have been transformed into $(p\mu p)^+_{ortho}$ ions which have only restricted freedom to move due to their positive charge. Possible effects due to a nonthermalized $\mu^- p$ system at formation time (e.g. evidence for nonthermalized $\mu^- d$-systems have been reported in [9], giving indications for a "hot" $\pi^- p$-system in [10]) as well as effects on the mean free path due to the Ramsauer-Townsend effect [13] have also been investigated. However, given the size of the target cell the maximum mean free path of order 10^{-2} to 10^{-1} mm are insignificant.

Apart from the difficulties inherent to muonic processes in hydrogen and its contaminants itself, there are more problems to be solved:

(i) Radiative muon decay produces a background photon spectrum with a high energy cut off at $m_\mu/2$. The finite energy resolution of the detector results in a tail of radiative decay events reaching to energies greater than the kinematical limit. The effective threshold used for the RMC analysis is therefore determined by investigating the photon spectrum measured with a μ^+-beam, where no capture process can occur (see Fig. 7 and discussion below).

(ii) A severe source of background is due to pions in the beam. A negative pion stopped in hydrogen induces the following reactions:

$$\pi^- p \rightarrow n + \gamma \text{ with } E_\gamma = 129\ MeV \quad (5)$$

$$\pi^- p \rightarrow n + \pi^0 \text{ and } \pi^0 \rightarrow \gamma\gamma \text{ with } E_\gamma = 55 - 83\ MeV \quad (6)$$

[5] The outgassing rate measured at room temperature is several orders of magnitudes larger than at liquid hydrogen temperature. Therefore this estimate is a very conservative upper limit for the impurities.

[6] manufactured by AECL-Canada and sold by ISOTEC Inc., Miamisburg, Ohio, USA

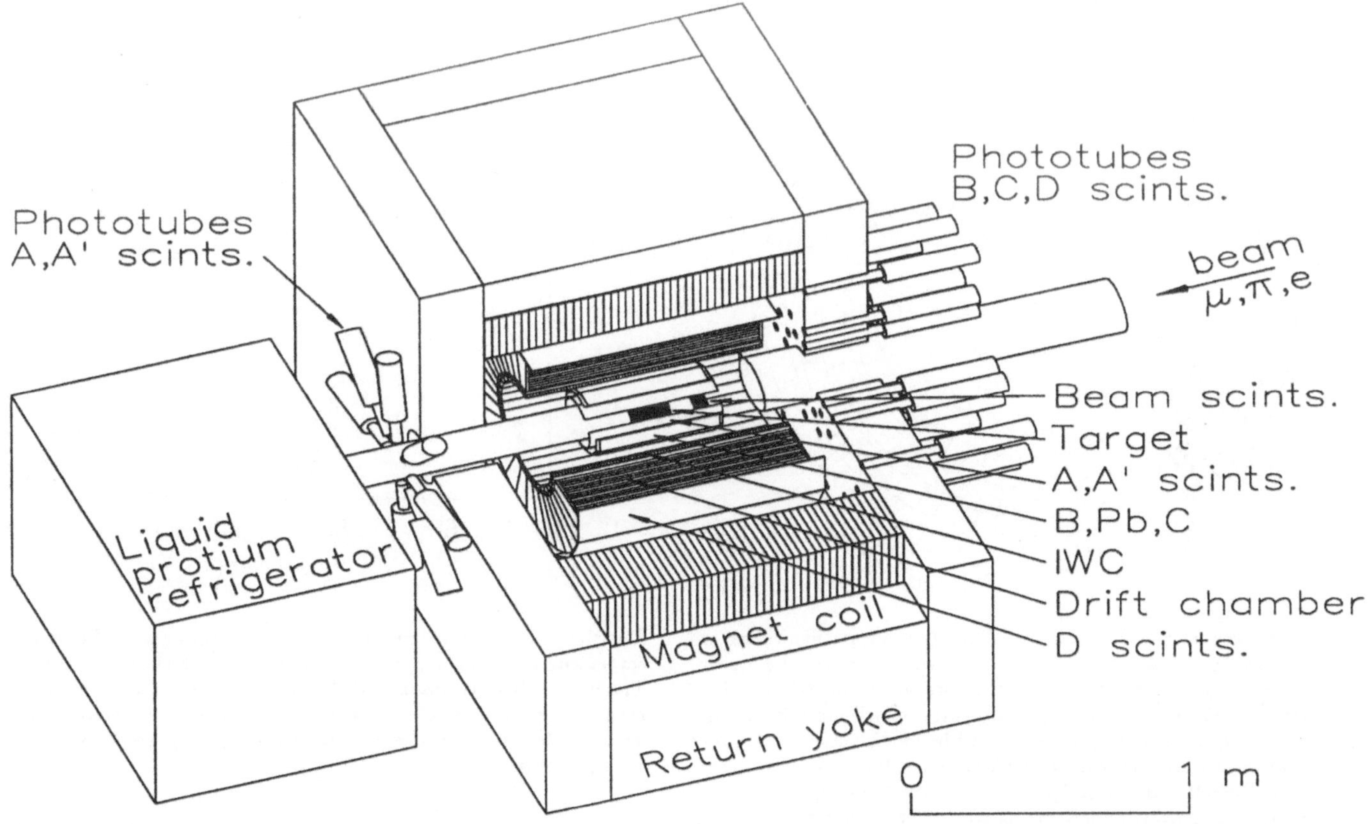

Fig. 4. Schematic view of the TRIUMF RMC facility (see text).

The branching ratio of reaction 5 over 6, known as the Panofsky ratio, is 1.546 ± 0.009 [11] and there are no other reactions of significance. Therefore pions must be suppressed to less than 10^{-9} per incoming muon to avoid contributions from this background. This reduction is achieved in three steps:

- using an RF–separator in the beam reduces the pion contamination by a factor of 1000 to a level of π/μ of $\approx 10^{-3}$.
- taking advantage of the pions shorter range by properly chosen degrader and target thickness.
- using the beam counter timing and pulse-height signals as a prompt veto in the analysis, since $\pi^- p$ formation and nuclear capture are prompt (1 ns).

(iii) Last but not least, cosmic ray background is also a concern. The detector is shielded with both drift chambers and scintillators to detect incoming radiation, but photon induced background events are unavoidable. Measurements without beam and analysis of the photon spectrum above 100 MeV are used to characterize this background.

4 The Detector

The small RMC branching ratio, of the order of 10^{-8}, makes the use of a 4π detector mandatory. The target is surrounded by a cylindrical setup consisting of several concentric layers of scintillators (A, A', B, C, D), a lead photon conversion layer (Pb), a proportional wire chamber (IWC) and a multilayered drift chamber (Figure 4). This pair-spectrometer technique, albeit lower in intrinsic efficiency than a crystal detector (NaI, BGO, CsI etc.) is favoured because of the extremely clear photon signature, which is essential to eliminate the copious potential neutron background and reduce many of the various systematic errors. The scintillator hodoscopes A, A' and B serve as veto counters to eliminate charged particles from ordinary muon decay. A combination of spatially related counters of the C and D hodoscopes creates a trigger signal to test for photon candidates. A typical event is displayed in figure 5.

The detector was constructed at TRIUMF and commissioned in 1989 to serve as a general RMC facility. Measurements on non-hydrogenic nuclei have been reported already [14]. A detailed description of the facility is being published elsewhere [15].

5 Preliminary Results

The acceptance of the spectrometer can be determined by measuring the photons from pionic reactions 5 and 6, respectively. Figure 6 shows the measured photon spectrum compared to a Monte-Carlo simulated distribution.

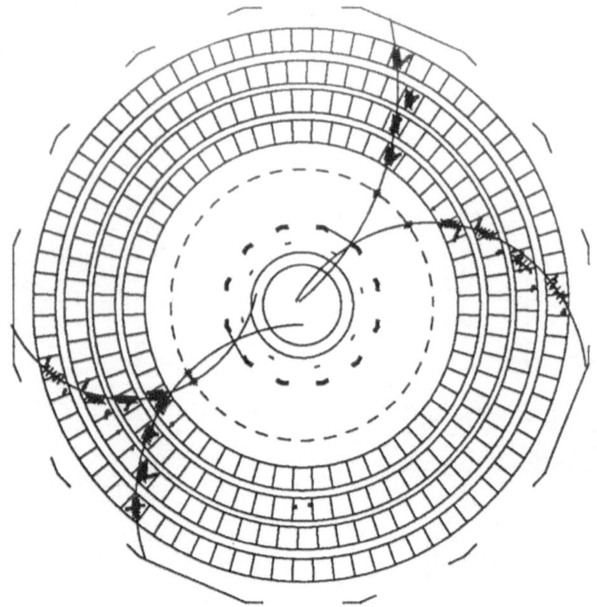

Fig. 5. Shown is the $r - \phi$ projection of an event identified as a $\pi^0 \rightarrow \gamma\gamma$ decay with both photons converted to an e^+e^- pair by the lead converter. Struck scintillators are indicated by line segments, IWC hits are displayed as stars. Scintillators from hodoscopes A, A' and B, which are located inside the converter show no signal as expected. Hits in the drift chamber layers are displayed by crosses (small squares indicate mirror hits due to the left-right ambiguity). The third superlayer is constructed as a stereo layer to determine the z - coordinate. The obvious displacement of the hits are due to the projection onto the center plane of the detector and not real if looked at it in a three dimensional way.

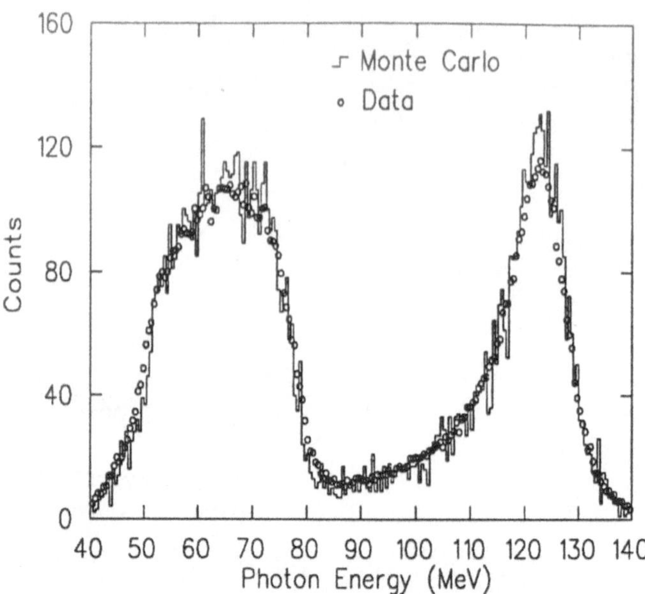

Fig. 6. A photon spectrum acquired by stopping a π^- beam in the target and requiring a prompt coincidence with the beam particles. The comparison is made with Monte-Carlo simulated yields from reactions $\pi^- p \rightarrow n + \gamma$ and $\pi^- p \rightarrow n + \pi^0$ respectively and is used to determine the acceptance and energy resolution of the spectrometer.

The agreement is generally good except in the threshold region, where tracks curling up in the drift chamber are difficult to analyze properly. Pion runs are taken periodically to monitor the acceptance throughout the RMC data collection. The preliminary value for the acceptance is 0.74 ± 0.02 % for 58 MeV $\leq E_\gamma \leq$ 130 MeV.

Figure 7 shows photon spectra taken with μ^- and μ^+ beam, respectively. The raw spectrum in fig. 7a) obtained with a μ^--beam recording 2.14×10^{11} stops still shows a very significant contribution from the pionic reactions 5 and 6. Removing all events coincident with incoming beam particles within 10 ns leads to spectrum fig. 7b), which is dominated by the radiative muon decay spectrum above the detection threshold of 35 MeV. In figure 7c) the 350 ns blanking time after the μ-stop is applied to the data. The events due to RMC on gold which are rejected by the blanking cut can be visualized by comparing Fig. 7c) with Fig. 7b), where an increase in the number of events at the high energy end of the spectrum in figure 7b) is very noticeable. Events from RMC on hydrogen are shown in a magnified display of the high energy region of spectrum 7c). Of these high energy events 3 are expected to be induced by cosmic rays. In figure 7d) the photon spectrum obtained with 5×10^{10} μ^+ stopping in the target is shown – the high energy part of the spectrum is very clean. The high energy tail from events due to badly fitted radiative muon decay events contributes no events above 58 MeV. The single event at 100 MeV is interpreted as a cosmic ray induced event in agreement with estimates based on background measurements. Because the analysis is in a very preliminary stage we do not yet have a result to report for g_P.

6 Conclusion and Future Work

Currently (September 1991) about 2×10^{12} μ^- have been stopped in pure hydrogen (protium) from which one would expect about 200 RMC events to be seen. In addition, runs with ordinary hydrogen with approximately 100 ppm deuterium contamination have been performed ($\approx 3.9 \times 10^{11}$ stops) to check the expected increase in high energy photons due to RMC on ^{3}He following $p\mu d$ - fusion. A third measurement with a different admixture of deuterium is under consideration. The goal of 400 RMC events from the protium target should be achieved by mid-1992. However, the 5 % statistical error for the RMC rate will be increased by the systematic errors, in particular by the 4 % uncertainty of the ortho-para transition rate Λ_{op}. A combined error of 8 % would result in an final accuracy of 10 % for the value of g_P.

Acknowledgement. This work is supported by NSERC & NRC (Canada), the U.S. NSF, the Australia Research Council, and PSI (Switzerland).

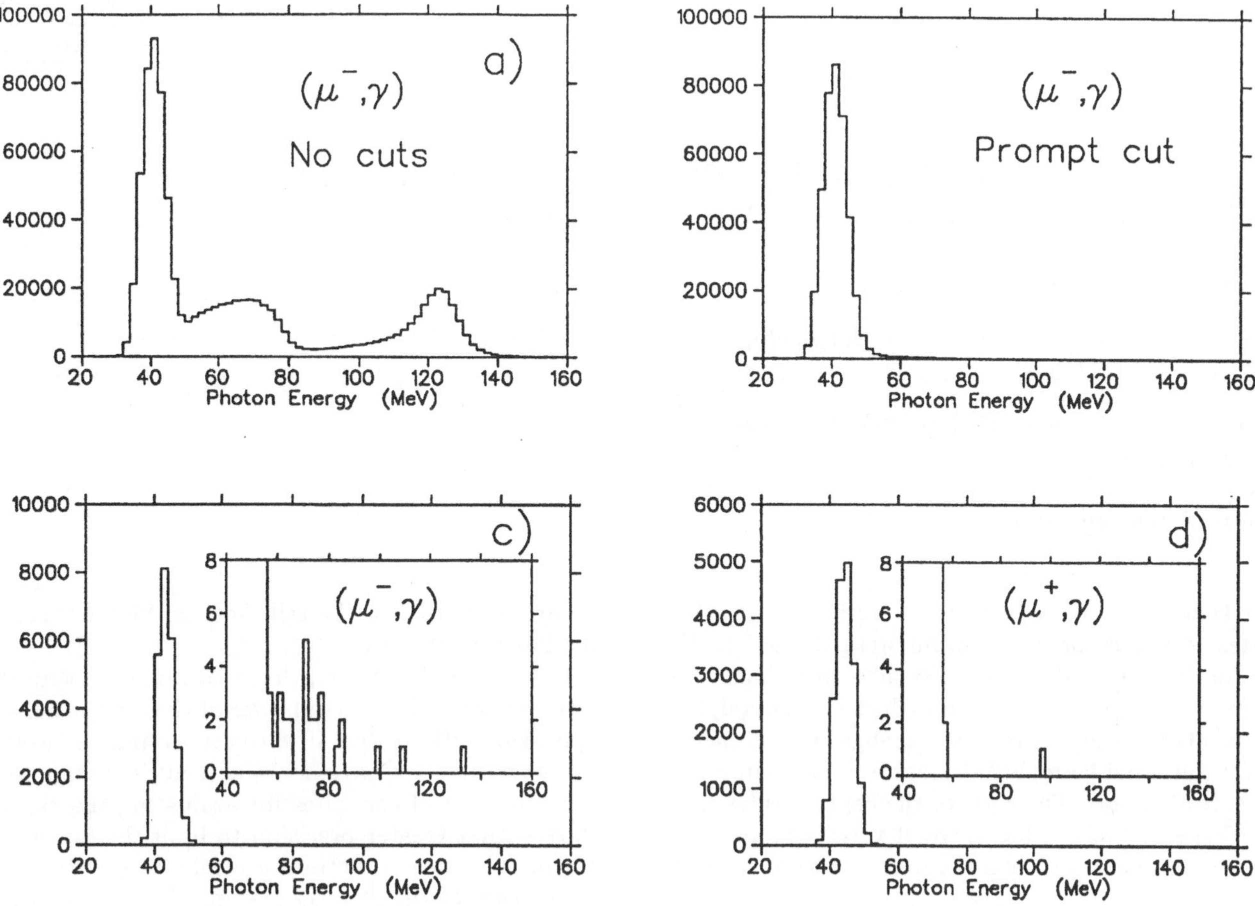

Fig. 7. Photon spectra taken with μ^- and μ^+ beam respectively. The interpretation of the histograms and the applied cuts is given in the text.

References

1. R.E. Marshak, Riazuddin, C.P. Ryan in "Theory of Weak Interactions in Particle Physics", Wiley, New York, 1969. For the notation see e.g. J.D. Bjorken, S.D. Drell, "Relativistic Quantum Fields", Mc Graw-Hill, New York, 1965.
2. L. Wolfenstein, in High Energy Physics and Nucl. Structure ed. S. Devons, Plenum, New York 1970.
3. D.S. Beder and H.W. Fearing, Phys. Rev. D 35 (1987) 2130.
4. For a recent review see e.g. L. Grenacs, Ann. Rev. Nucl. Part. Sci. 35 (1985) 455.
5. G. Bardin et al., Nucl. Phys. A352 (1981) 365.
6. L. Roesch et al., Phys. Rev. Lett. 46 (1981) 1507.
7. D. Dubbers et al. Europhys. Lett. 11 (3) (1990) 195.
8. M. Gmitro and P. Truöl, Adv. Nucl. Phys. 18 (1987) 241.
9. W.H. Breunlich et al; PSI Nucl. Part. Phys. Newsl. 44 (1978).
10. J. Crawford et al; Phys. Lett. 213B, (1988) 391.
11. J. Spuller et al., Phys. Lett. 67B (1977) 479.
12. G.I. Opat, Phys. Rev. 134 (1964) B428.
13. L. Bracci et al., Preprint IFUP-TH 21/90 Pisa, June 1990.
14. A. Serna-Angel et al., XII Int. Conf. on Particles and Nuclei (PANIC) MIT, Boston, June 1990, Abstracts Book XII – 17.
15. D. H. Wright et al., Submitted to Nucl. Instr. and Meth.
16. E.J. Bleser et al., Phys. Rev. 132 (1963) 2679.
17. C. Petitjean et al., Muon Catalyzed Fusion 5/6 (1990/91) 199.
18. S.S Gershtein and L.I. Ponomarev in "Muon Physics" Vol. 3, eds. V.W. Hughes and C.S. Wu, Acad. Press, N.Y. and London 1975.
19. G. Bardin et al., Phys. Lett. 104B (1981) 320.

This article was processed using Springer-Verlag TEX Z.Physik C macro package 1991
and the AMS fonts, developed by the American Mathematical Society.

Z. Phys. C - Particles and Fields 56, S156–S158 (1992)

Zeitschrift für Physik C Particles and Fields

Muon Capture in Hyperfine States of Muonic Deuterium and Induced Pseudoscalar Form Factor

M. Morita[1] and R. Morita[2]

[1] Faculty of Science, Josai University, Sakado, Saitama 350-02, Japan

[2] Josai Women's Junior College, Sakado, Saitama 350-02, Japan

Received June 30, 1991

Abstract. Muon capture rates from the hyperfine states of the muonic deuterium were studied as the functions of the induced pseudoscalar form factor g_P. These capture rates vary also with the $\pi N \Delta$ coupling constants through the meson exchange current effects. Variation is not large, but still appreciable to limit the magnitude of g_P. The ratio of the capture rates from two hyperfine states has a small variation, and it is more sensitive to g_P than the muon capture rates.

1 Introduction

One of the important problems in the study of nuclear weak processes is to find the weak nucleon currents accurately. In fact, these currents are reasonably well established by careful studies on various physical quantities obtained from nuclear beta decays. The form factors of the main parts of the vector and axial vector currents are summarized together with those of their induced currents, see, e. g., [1-2]. This is, however, not the case for the induced pseudoscalar form factor g_P. The reason is the effective strength of this coupling, which is proportional to the lepton mass involved in each process. Therefore, the detection of the effect of the induced pseudoscalar term is difficult in beta decays, while it is possible in muon capture reactions. The effect can be understood mainly due to the pion-pole term.

In principle, the magnitude of g_P can be studied in the muon capture on hydrogen with no ambiguity for atomic, molecular, and nuclear physics. Experimental works of this reaction have been performed repeatedly. Averaging over the latest data by several groups, we have a value of g_P in good agreement with the predicted value by PCAC, although individual data are distributed in a relatively wide range. Statistics in experiments seem to be still not too high because of the low capture rate.

In the complex nuclei, the total muon capture rate increases with the fourth power of the atomic number approximately so that it is easier to find the events. Unfortunately, we have no reliable nuclear wave functions in most of the cases for evaluating the capture rates with a greater precision to limit the magnitude of g_P. The wave functions are relatively well known in the case of the $A = 12$ system. It seems to us almost impossible to predict the partial muon capture rate in ^{12}C leading to the ground state of ^{12}B within a 5 % accuracy [3]. We found, however, appropriate physical quantities which are less model dependent, but reasonably sensitive to g_P. These are the longitudinal and average polarizations of the recoil nucleus [3]. We investigated these quantities, by taking into account the general $0p$-shell wave functions with the first- and second-order core polarization effects, the meson exchange current effects, and the full formula of the muon capture reaction. We obtained the results of g_P which are about 15 % higher than the canonical value, with about 20 % of the error bar due to the experimental uncertainty [4]. A further analysis is being made by taking into account the all-order core polarization effects, the meson exchange current effects, and the full formula. A preliminary result with a simplified formula shows us a tendency of reducing the magnitude of g_P [5]. From this series of investigations, we noticed that the induced pseudoscalar form factor g_P in ^{12}C seems to be close to the canonical value, if we make an effort to evaluate nuclear polarizations by adopting reasonable nuclear wave functions together with the meson exchange current effects.

Since the muon capture in ^{12}C is a little too complicate in nuclear physics, we come back to the work on muon capture in the deuterium, where the nuclear model dependence is very weak [6]. Furthermore, a new experimental data became available to us [7]. There is also a long history of the experimental and theoretical investigations of this reaction, see, e. g., references cited in [6-8]. And the problem was to evaluate the meson exchange current effects. The calculated capture rate is too low in the impulse approximation, and a sizable effect of the exchange currents is required to explain the experimental data.

2 Muon capture in deuterium

We have been interested in the new experimental data of the partial muon capture rate from the doublet hyperfine state of the muonic deuterium [7],

$$\Gamma_{1/2}^{\mathrm{exp}} = 409 \pm 40\ \mathrm{s}^{-1}, \qquad (1)$$

and also that of the neutron energy spectrum in this reaction [9]. The latter has a bump at the high energy side which might be the effect of the exchange current. We adopted the Paris potential for describing the two-nucleon system. We took into account the S- and D-waves of the deuteron and the two-neutron system with angular momenta up to four. For the isobar current, effects of the pion and rho-meson exchange are carefully studied because of possible cancellations. We made a numerical work on the neutron energy spectrum and neutrino asymmetry coefficient. The meson exchange current effects enhance the energy spectrum by a factor of 10.6 at the neutron energy of around 50 MeV in the case of the doublet state. This is, however, too small to explain the experimental data at high energies. We had the capture rate,

$$\Gamma_{1/2} = 402\ \mathrm{s}^{-1}, \qquad (2)$$

if g_P is the canonical value [6]. Adam et al. [8] also obtained a similar value as above, and discussed the capture rate with the experimental data in (1) and another new value [10],

$$\Gamma_{1/2}^{\mathrm{exp}} = 470 \pm 29\ \mathrm{s}^{-1}, \qquad (3)$$

which is far from our expectation in (2). If this is the case, we have to take into account some other processes which are not discussed here.

In our latest work [11], the induced pseudoscalar form factor, $g_P(normal)$, with PCAC is given by

$$g_P(normal) = \frac{2Mm_\mu g_A}{k_\lambda^2 + m_\pi^2}, \qquad (4)$$

where M, m_μ, and m_π are the nucleon, muon, and pion mass, respectively. The k_λ^2 dependency of the axial vector form factor g_A is taken into account as usual. Furthermore, we varied the ratio $g_P/g_P(normal)$ only, by assuming the same k_λ^2 dependency of g_P as that of $g_P(normal)$.

The partial muon capture rates from the doublet and quartet hyperfine states were carefully studied as the functions of the $\pi N \Delta$ coupling constants, c_0 and d_1, and the pseudoscalar form factor g_P. The former parameters can be determined in the pion-nucleon scattering at low energies, and there are different sets of these parameters proposed by different groups. All of these can reproduce the capture rate within the error bar, in conformity with the experimental data in (1). For example, one of the set of parameters, which are by a factor of two smaller than those adopted in (2), reduces the muon capture rate $\Gamma_{1/2}$ by about 3.5 %. This is not large compared with the error bar in the experimental value (1), but appreciable in the case where we discuss the magnitude of g_P. On the other hand, the ratio, $\Gamma_{1/2}/\Gamma_{3/2}$, of the muon capture rates from doublet and quartet states is insensitive to the $\pi N \Delta$ coupling constants as is seen in Fig. 1.

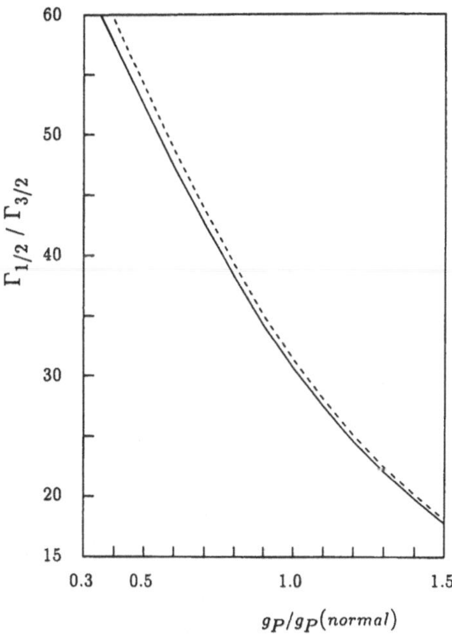

Fig. 1 - Ratio $\Gamma_{1/2}/\Gamma_{3/2}$ of the muon capture rates from the doublet and quartet states of the deuterium as a function of g_P [11]. The solid (dashed) line refers to $c_0/m_\pi^3 = 0.188$ (0.0969).

In fact, if the ratio $\Gamma_{1/2}/\Gamma_{3/2}$ is measured with 10 % uncertainty, we can limit g_P up to 10 % accuracy, in the vicinity of $g_P = g_P(normal)$. This is by about five times more sensitive to g_P than the capture rate $\Gamma_{1/2}$. Therefore, this is a very useful quantity to test PCAC in future. The detail of this investigation is published elsewhere [11].

Acknowledgement

The authors would like to express their sincere thanks to Professor H. Ohtsubo, Dr. T. Sato, Dr. K. Koshigiri, and Dr. M. Doi for their continual collaboration.

References

1. D. H. Wilkinson: Nuclear Weak Process and Nuclear Structure, ed. M. Morita et al., pp.1-45, Singapore, World Scientific 1989
2. M. Morita: Progress in Nuclear Physics, ed. W-Y. P. Hwang et al., pp.272-282, New York, Elsevier Pub. Co. 1991
3. M. Fukui, K. Koshigiri, T. Sato, H. Ohtsubo, and M. Morita: Prog. Theor. Phys. **70** (1983) 827
4. M. Fukui, K. Koshigiri, T. Sato, H. Ohtsubo, and M. Morita: Prog. Theor. Phys. **78** (1987) 343
5. K. Koshigiri, K. Kubodera, H. Ohtsubo, and M. Morita: Nuclear Weak Interaction and Nuclear Structure, ed. M. Morita et al., pp.52-58, Singapore, World Scientific 1989
6. M. Doi, T. Sato, H. Ohtsubo, and M. Morita: Nucl. Phys. **A511** (1990) 507
7. M. Cargnelli, Ph.D. Thesis, Tech. Univ. Wien 1987, unpublished
8. J. Adam, Jr., E. Truhlik, S. Ciechnowicz, and K. -M. Schmitt: Nucl. Phys. **A507** (1990) 675
9. Y. K. Lee, T. J. Hallman, L. Madansky, S. Trentalange, G. R. Mason, A. J. Caffrey, E. K. Mcintyre, Jr., and T. R. King: Phys. Lett. **B188** (1987) 33
10. G. Bardin, J. Duclos, J. Martino, A. Bertin, M. Capponi, M. Piccinini, and A. Vitale: Nucl. Phys. **A453** (1986) 591
11. M. Doi, T. Sato, H. Ohtsubo, and M. Morita: Prog. Theor. Phys. **86** (1991) No. 1, in press

Z. Phys. C - Particles and Fields 56, S 159-S 168 (1992)

Zeitschrift für Physik C Particles and Fields

Photon-hadron interactions revisited in deep inelastic muon scattering

D. F. Geesaman

Argonne National Laboratory, Argonne, Il 60439, USA

4 September 1991

Abstract. New results for deep inelastic muon scattering are reviewed. These include shadowing results to $x_{bj} \sim 10^{-5}$, nucleon structure functions measurements, the Gottfried sum rule and the use of deep inelastic scattering to obtain information on gluon distributions. The future of muon scattering in the HERA era is discussed.

The title of this talk was deliberately chosen to bring to mind Feynman's classic book *Photon-Hadron Interactions* [1]. Muon scattering has played a prominent role in understanding the interaction of the photon with hadrons and understanding the structure of the nucleon. In this talk some recent developments in deep inelastic scattering are reviewed. One of the most interesting developments is that the time has come for a reexamination of the relationship between the two primary themes of Feynman's book, the hadronic component of the photon and the parton model.

Deep inelastic scattering was pivotal in changing our picture of the photon-hadron interaction. It was through the pioneering deep inelastic scattering measurements at SLAC that the first evidence emerged for light point-like constituents inside of the nucleon, the partons that are today clearly identified with the quarks of the standard model. This evidence was the observation of scaling, and the deviations from scaling have provided a quantitative test of the model of strong interactions, Quantum Chromodynamics (QCD), and a significant measure of the strong coupling constant, α_s. As will be discussed below, in the infinite momentum frame deep inelastic scattering measures the one body longitudinal momentum distributions of the quarks in the nucleon.

While the first measurements were done with electrons, it was quickly realized that muon beams provided significant advantages over the existing electron beams. As secondary beams at hadron accelerators, muons were available at substantially higher energies than electrons. This allowed studies at greatly extended ranges of momentum and energy transfers. Today, the final state hadronic energies available in deep inelastic muon scattering are comparable to those of the extensive $e^+ e^-$ results at PEP and PETRA, but deep inelastic scattering brings to these studies the knowledge of the kinematics of the virtual photon, a powerful addition to the physics. Higher energy is by no means the only virtue of muon beams. The large ratio of the muon to electron masses significantly reduces the radiative corrections which are required in muon scattering and allows thicker targets to help compensate for the lower intensity of the muon beams. Finally, the method of production of muon beams from meson decay provides a natural polarization of the muon spin, making polarized beam and target experiments possible. The next two talks will concentrate on this last aspect of deep inelastic muon scattering.

The full power of deep inelastic scattering was revealed by choosing to analyze the data in the correct coordinate frame, the infinite momentum frame. This is illustrated in Figure 1a. The motivation is to measure a one-body distribution of electromagnetic charges which requires the time between quark interactions, τ_{SQ} to be large compared to the lifetime of the virtual photon, τ_{int}. This is accomplished by Lorentz dilation, in a system where the momentum of the proton (P) becomes infinite:

$$\tau_{int} \sim \frac{2xP}{Q^2} \ll \tau_{SQ} \sim \frac{2xP}{m^2 + k_\perp^2} \tag{1}$$

This equation gives the scale by which we require Q^2, the four momentum of the virtual photon to be large in the parton model. Another scale, to be discussed later, is the dependence of the QCD coupling constant α_s on the scale parameter Q^2. The major advantage of an infinite momentum frame is that certain classes of diagrams, especially those involving quark-antiquark splittings can be shown to contribute little compared to diagram 1a (e.g. reference [2]). Figure 2 defines the

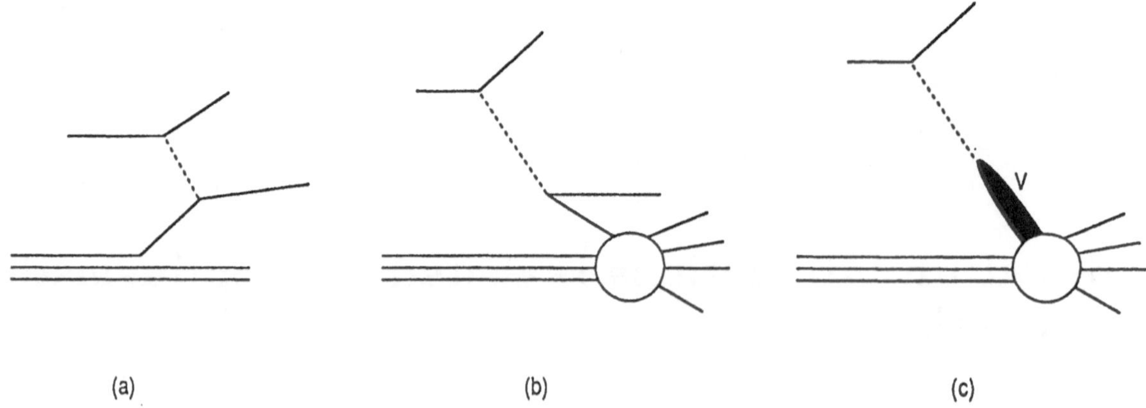

Fig. 1. Diagrams representing three physical descriptions of deep inelastic scattering. a) The parton model in which the virtual photon is absorbed by one quark from the target. This is particularly appropriate for an infinite momentum frame description of deep inelastic scattering. b) The virtual photon fluctuates into a $q\bar{q}$ pair and either the q or $\bar{q}$ interacts with the target. This is particularly appropriate for a lab frame description of deep inelastic scattering. c) The virtual photon fluctuates into a vector meson such as the ρ. The vector meson interacts with the target with typical hadron-hadron interactions. This is the vector dominance model.

Kinematics

$q = k - k'$

$Q^2 = 4E_l E_l' sin^2(\theta_l/2)$

$\nu = \frac{q\cdot P}{M} = E_l - E_l'$

$y = \frac{q\cdot P}{k\cdot p} = \nu/E_l$

$x = Q^2/(2M\nu)$

$\equiv Fraction\ of\ P\ carried\ by\ quark\ as\ P \longrightarrow \infty$

$W^2 = (P+q)^2 \equiv Mass^2\ of\ recoiling\ hadronic\ system$

$s = (P+k)^2 \equiv Total\ Energy^2$

Cross Sections and Structure Functions

$$\frac{d\sigma^e}{dx dQ^2} = \frac{4\pi\alpha^2}{Q^4 x}\left[(1 - y - \frac{M^2}{s-M^2}xy)F_2^e + xy^2 F_1^e\right]$$

$$\frac{d\sigma^{\nu,\bar{\nu}}}{dx dQ^2} = \frac{G_F^2}{2\pi x}\frac{M_W^2}{(Q^2+M_W^2)^2}\left[(1 - y - \frac{M^2}{s-M^2})F_2^\nu + xy^2 F_1^\nu \pm (y - \frac{y^2}{2})xF_3^\nu\right]$$

$$\sigma_L/\sigma_T = (F_2 - 2xF_1)/(2xF_1)$$

Parton Model of Structure Functions

$$F_2^e = 2xF_1^e = x[\tfrac{4}{9}(u+\bar{u}+c+\bar{c}+t+\bar{t}) + \tfrac{1}{9}(d+\bar{d}+s+\bar{s}+b+\bar{b})]$$

$$F_2^\nu = 2xF_1^\nu = x[d+s+b+\bar{u}+\bar{c}+\bar{t}]$$

$$F_3^\nu = 2[d+s+b-\bar{u}-\bar{c}-\bar{t}]$$

$$F_2^{\bar{\nu}} = 2xF_1^{\bar{\nu}} = x[u+c+t+\bar{d}+\bar{s}+\bar{b}]$$

$$F_3^{\bar{\nu}} = 2[u+c+t-\bar{d}-\bar{s}-\bar{b}]$$

Fig. 2. Definitions of the kinematic variables, structure functions, and quark distributions in deep inelastic lepton scattering.

kinematic variables and the relationships between the measured cross sections and the quark distributions in this reference frame. The important kinematic variables are Q^2, ν, the energy loss of the incident lepton in the laboratory frame and $x_{bj} = Q^2/(2m_p\nu)$ which measures the fraction of the momentum of the proton carried by the struck quark in the infinite momentum frame. A major advantage of this description is the universality of charged lepton and neutrino scattering.

An alternative reference frame for viewing high energy photon-hadron reactions is the lab frame. Here, the same old fashion perturbation theory arguments that suggest diagram 1a dominates in the infinite momentum frame suggest that diagram 1b dominates in the lab frame at low x_{bj}[3]. Now the phase space of the quark-antiquark pair is an important kinematic and dynamic constraint. At low energies, this phase space leads naturally to vector meson dominance as illustrated in Figure 1c and back to Feynman's book. The basis idea is to make use of what knowledge we have of hadron-hadron reactions, in this case Regge theory. The electromagnetic current is given directly in terms of a hadronic current:

$$e\langle N|j_\mu^{em}|M\rangle = \frac{e}{f_v}\frac{m_v^2}{Q^2+m_v^2}\langle N|J_\mu^v|M\rangle \quad (2)$$

The vector mesons are the well-known ρ, ω, and ϕ, the more poorly determined higher mass mesons, and the isoscalar pomeron which ensures that the hadron-hadron cross sections at high energy are energy independent. Regge theory then suggests that the parton distributions vary as $x_{bj}^{-\alpha}$ at low x, where α is 0.5 for the low mass mesons and 1.0 for the pomeron.

In this lab frame description, the important kinematic parameter is the lifetime of the quark-antiquark fluctuation which, from the uncertainty principle, is:

$$L \sim c\triangle t \sim c/\triangle E \sim \frac{2*\nu}{Q^2+m_v^2} \sim \frac{0.4fm*\nu(GeV)}{Q^2+m_v^2} \quad (3)$$

For low ν (< 1.0 GeV), L is small compared to the finite size of the proton. At large Q^2, L becomes $\sim 0.2/x_{bj}$ so that at very low x_{bj} values such as 10^{-2} to 10^{-5}, L is many fermis and is much larger than the diameters of atomic nuclei.

This lab frame description provides an alternative description of deep inelastic scattering at low x_{bj} in terms of the spectral function of the photon (measured in $\frac{\sigma(e^+e^-\to hadrons)}{\sigma(e^+e^-\to\mu^+\mu^-)}$) and the interaction of a quark with a nucleon [4,5,6] and can equally well describe the nucleon structure functions at low x_{bj}. The physics is, of course, independent of the reference frame chosen to analyze the problem, but these two reference frames provide rather different insights into the space-time picture of deep inelastic scattering. One could ask the question as to whether the $q-\bar{q}$ pair should be identified with the photon or the nucleon. In the context of the photon-nucleon interaction this is not a well posed question. However, if we impose an external length scale on the problem, it becomes more interesting. We know that hadron-nucleon interactions are "shadowed", the cross section on a nucleus is less than $A^*\sigma_p$. In the lab frame this is understood as interference of a second scattering with the first scattering, or that the hadron flux is absorbed on the front face of the nucleus, a surface effect. Shadowing is a "natural phenomena" in the lab frame and early studies avoided low x_{bj} for exactly this reason. There is, however, an infinite momentum frame description of shadowing, espoused especially by Nicolaev [7] and Mueller and Qiu [8] that describes shadowing as a modification of the nuclear parton distributions due to the spatial overlap of the parton distributions of neighboring nucleons. Here the key feature is that the partons at low momentum are not well localized in the nucleon:

$$\triangle r \sim 1/x_{bj} \quad (4)$$

At low x_{bj}, a quark or gluon from one nucleon can recombine with a gluon or quark from an adjacent nucleon and the resulting parton distributions may be different in a nucleus and a free proton.

In both descriptions, x_{bj} determines the relevant length scale and two physical length scales are important. The first is the separation between nucleons. Once L or $\triangle r$ become comparable to the nucleon separation, shadowing should begin to be observed. Once L or $\triangle r$ become comparable to the diameter of the nucleus, shadowing should saturate and eventually, the virtual photon would be expected to behave like a real photon as $Q^2 \to 0$. In that case, ν becomes the only relevant scale of the probe.

Previous results on the shadowing of nuclear cross sections have clearly identified the onset of shadowing. The most recent high statistics data come from the NMC collaboration [9]. At very low x_{bj}, radiative corrections and muon-electron scattering have always limited the analysis. In Figure 3, new preliminary results from FNAL E665 are shown which for the first time present the ratio of cross sections on a heavy nuclear target (Xenon) to those on deuterium over the range $10^{-5} < x_{bj} < 10^{-1}$. The open circles show a conventional analysis using only the scattered muons and applying standard radiative corrections. The full circles are the result of an analysis which uses an electromagnetic calorimeter to reject muon-electron and bremsstrahlung events. An independent analysis requiring two same sign coincident hadrons gives consistent results, with larger statistical errors. Each analysis has an overall systematic error of 8% which is not shown in Figure 3.

The saturation of the ratio of Xe to Deuterium cross sections at a value of 0.70 +/- 0.03 +/- 0.08 (where the first error is statistical and the second is systematic) is clearly seen in this analysis. This is completely consistent, within the overall systematic errors with the real photon value (0.60 +/- 0.07) obtained from extrapolation [10]. The onset of shadowing is seen to occur at $x_{bj} \sim 0.05$ and the saturation occurs at $x_{bj} \sim 0.002$. All the results (NMC [9], NA28 [13] and E665 show little, if any, dependence on Q^2. What is interesting is that the ratio of the x_{bj} value where shadowing starts to the saturation value, $0.05/0.002 \sim 25$ is considerable larger than the ratio of the diameter of Xenon to the spacing

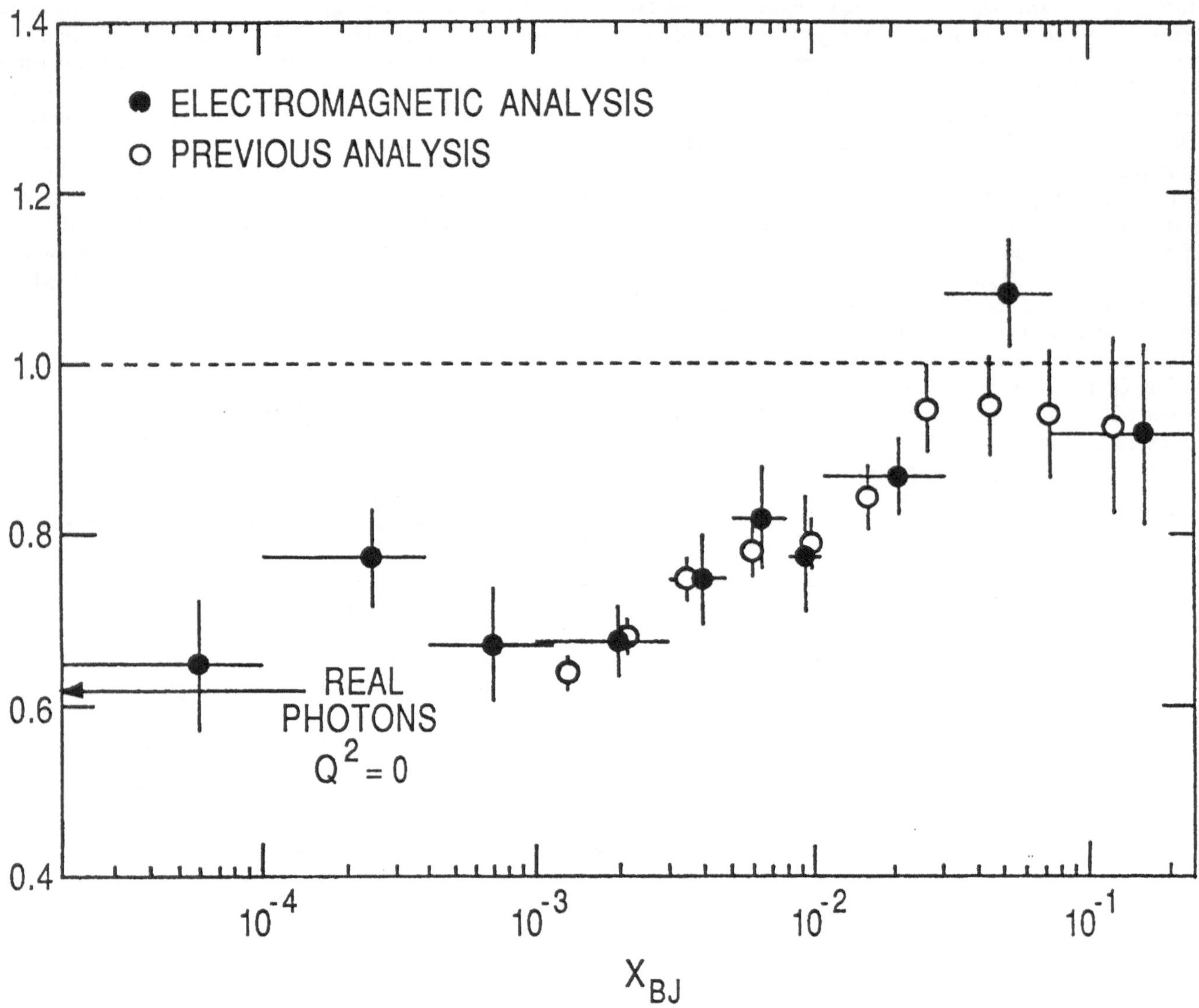

Fig. 3. Preliminary results from FNAL E665 for the x_{bj} dependence of the ratio of Xenon to Deuterium deep inelastic cross sections. The error bars shown are statistical. There is an additional 8% overall uncertainty in the normalization. The open circles are results using only the scattered muons and making radiative corrections (ref [10]). The solid circles are a new analysis using an electromagnetic calorimeter to eliminate radiative and $\mu - e$ events (ref [11]). The real photon point is extrapolated from the results of reference [12].

between nucleon centers: $11/1.8 \sim 6$. It is much closer to the ratio of the diameter of Xenon to the average spacing between the nucleon surfaces in a nucleus: $11/0.4 \sim 27$. However if the relevant distance scale for the onset of shadowing is so small, one would have expected shadowing to start at a considerably larger x value.

From this point on, only the infinite momentum frame description will be considered and I will turn my attention to what we have learned about the structure of the proton. There are now excellent electron, muon and neutrino data and several global fits to these data and related hadronic reaction data provide tight constraints on the nucleon structure functions over a significant kinematic range. Two of the most recent global fits are those of Harriman, Martins, Roberts and Stirling [14] and those of Tung and Morfin [15] Attention continues to center on the small inconsistencies of the various data sets, particularly the discrepancies between BCDMS and EMC. To an outside party, it is extremely difficult to evaluate the relative merits of the two measurements. Fortunately, a reanalysis [16] of the entire SLAC data set seems to provide a partial answer. In recent publications [17] by members of the BCDMS collaboration, it is seen that if the normalizations of the muon experiments are allowed to float relative to the electron scattering measurements, the best agreement at low x_{bj} is obtained if the BCDMS results are renormalized by 0.99 and the EMC results are renormalized by 1.08. At larger x_{bj}, the agreement is improved by a 1.3 standard deviation modification of the BCDMS data by their main systematic

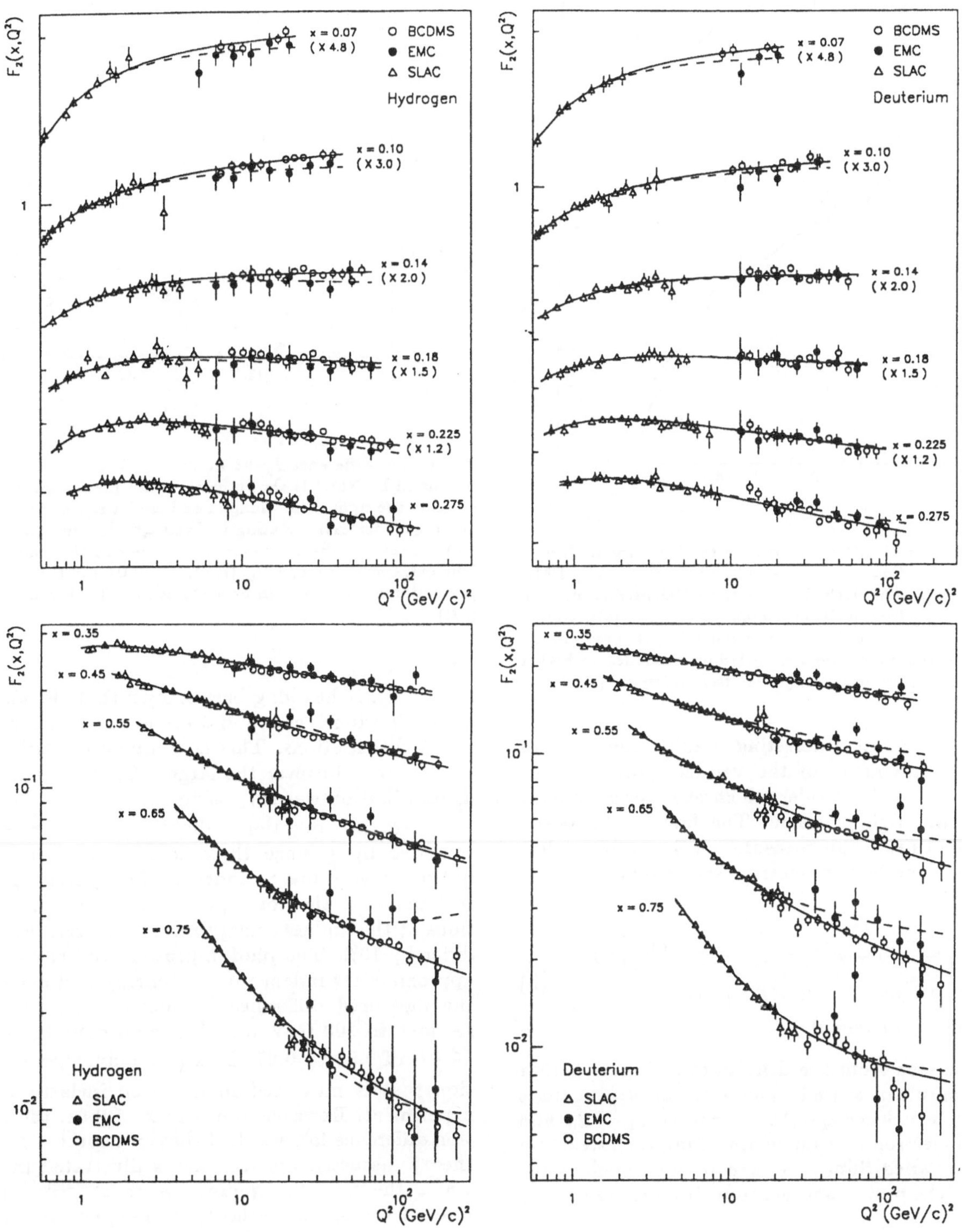

Fig. 4. Results from SLAC [16], EMC [18], and BCDMS [19] taken from a combined fit to the three data sets [17] The EMC data have been renormalized by 1.08. The BCDMS data have been renormalized by 0.99 and modified by the 1.3 standard deviations in their main systematic error. The solid (dashed) curve is a phenomenological fit to the SLAC and BCDMS (EMC) data. See reference [17]

error (essentially the field integral calibration of the iron toroidal magnet). The results with these modifications are shown for hydrogen and deuterium in Figure 4 [17].

The excellent comparison of QCD with the Q^2 dependence of the structure functions is shown in Figure 5 [20].

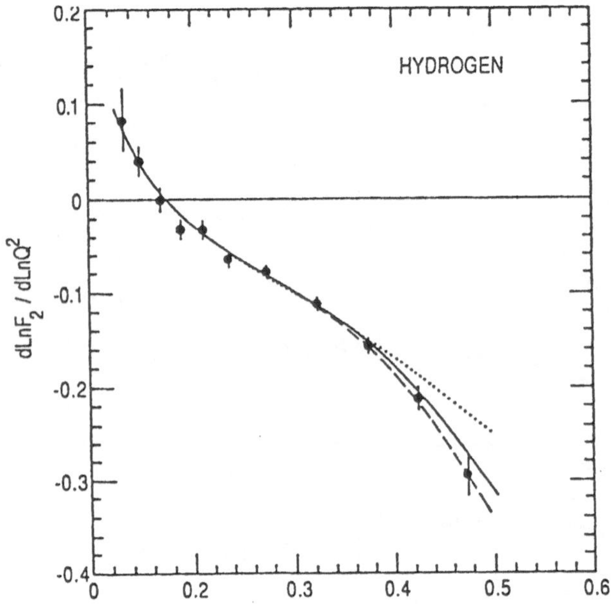

Fig. 5. The results of Milsztajn [20] for the logarithmic derivative of F_2, $dlnF_2/dlnQ^2$ from a combined fit to SLAC, EMC and BCDMS data. The dashed line is the QCD prediction with λ_{QCD} = 250 MeV from a fit with higher twist and target mass corrections included. The solid line includes target mass corrections but not higher twist and the dotted line includes no higher twist or target mass corrections. (Taken from reference [20].)

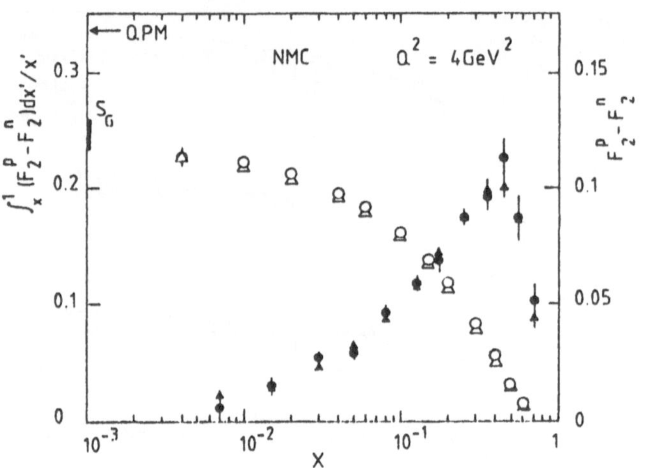

Fig. 6. The difference F_2^n-F_2^p at Q^2 = 4 GeV^2 as a function of x_{bj} measured by NMC is shown as the solid points and refers to the scale on the right. The triangles and circles represent two different techniques for extrapolating to fixed Q^2. The integral from x_{min} to 1 is shown as the open points and refers to the scale on the left. The extrapolated result S_G (for x_{min} of 0) and the parton model prediction are also shown on the left axis. (Taken from reference [22]).

One of the most interesting results comes from sum rule measurements of the valence quark properties. There are two sum rules which have recently been measured with high precision. The first is the Gross Llewellyn-Smith sum rule measured in the parity violating structure functions in neutrino scattering:

$$\begin{aligned} S_{GLS} &= \int \frac{1}{x} x F_3(x, Q^2) dx = 3[1 - \frac{\alpha_s}{\pi} + \mathcal{O}(\frac{1}{Q^2})] \\ &= 2.68 \pm 0.08 \qquad (CCFR \quad [21]) \\ &= 2.63 \qquad (Theory) \end{aligned} \tag{5}$$

It can be seen from the definitions in Figure 2 that the GLS sum rule measured on an isoscalar target counts the number of valence quarks as long as $\bar{u}_p = \bar{d}_n$ and $\bar{d}_p = \bar{u}_n$. In electron and muon scattering a related sum rule, the Gottfried Sum Rule can be measured in the difference of the proton and neutron structure functions:

$$\begin{aligned} S_{GSR} &= \int \frac{F_2^p - F_2^n}{x} dx \\ &= \frac{1}{3} + \frac{2}{3} \int (\bar{u}_p(x) - \bar{d}_p(x)) dx \\ &= 0.240 \pm 0.016 \qquad (NMC \quad [22]) \end{aligned} \tag{6}$$

The integrand and the integral from x to 1 are shown in Figure 6 as the closed and the open circles respectively. There is a clear discrepancy from the value $\frac{1}{3}$ expected if $\bar{u}_p = \bar{d}_p$. It has long been known that the strange sea quark distributions s and $\bar{s}$ are about half the light sea quark distributions. This is commonly attributed to a mass effect. However the large difference in the $\bar{u}_p$ and $\bar{d}_p$ distribution was surprising.

There are questions about any sum rule that is weighted by $\frac{1}{x}$ since these could have a large contribution at very low x. Indeed, NMC point out that for real photons, the total proton and neutron cross sections at the highest energy measured precisely, 17 GeV, differ by 10%. The photon-proton total cross section is approximately independent of energy and extrapolating the measured difference in neutron and proton cross sections to 200 GeV would give an expected difference $\frac{\sigma_d - \sigma_p}{\sigma_p}$ of 0.93 to 0.97. This is the analogous cross section that is measured in the deep inelastic scattering experiment. However, down to x of 0.01, the cross section difference follows the behavior expected from Regge theory discussed above. This is illustrated in Figure 7 where the solid line represents an $x^{0.62 \pm 0.05}$ behavior. Even if something pathological happens at lower x, we are forced to accept that the x dependence of the sea distributions cannot be the same.

There are three "standard" mechanisms which might give rise to the difference in the $\bar{u}$ and $\bar{d}$ distributions. The first is the expectation that the extra valence u quarks Pauli block the occupation of some $u\bar{u}$ states compared to $d\bar{d}$ states. While this sounds intuitive, it is by no means trivial. Calculations in 1 space and 1 time dimension indicate that exactly the opposite result will emerge [23]. The second explanation focuses on a

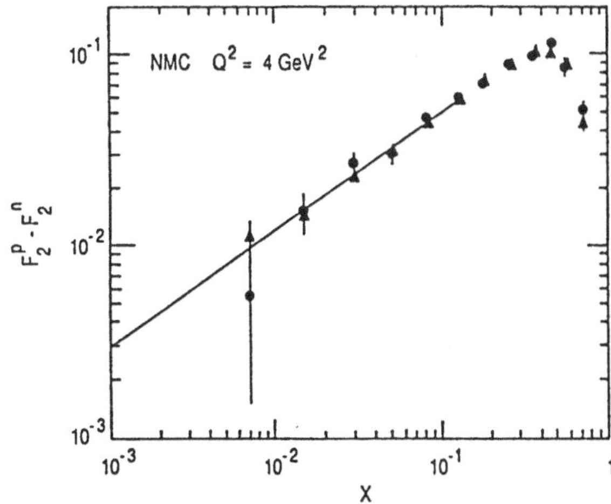

Fig. 7. The difference F_2^n-F_2^p at $Q^2 = 4$ GeV^2 (the triangles and the circles represent two different techniques for extrapolating to fixed Q^2) and the fitted function ax^b used in the extrapolation to x=0. (Taken from reference [22]).

hadronic picture. A proton will spend part of its time as a neutron and a π^+ ($uud + uddu\bar{d}$) giving an extra $d\bar{d}$ contribution. The third "story" which also has not provided any quantitative description attributes the difference to the small mass difference in the u and d quarks.

What the Gottfried Sum Rule result does point out is that, even for the light quarks, the sea does not have the same symmetry as the gluon field. It is not possible to devolve an asymmetric sea at low Q^2 to pure glue. Once again, just as in the spin structure function results discussed in Dr. Voss's talk, we see the limitations of the insight of the constituent quark model in the high momentum regime.

Now it is time to turn to effects which are beyond the leading order in α_s, what I like to call "Sniffing the Glue". Figure 8 shows the lowest order and the next order contributions to deep inelastic scattering, gluon bremsstrahlung (which is well studied in e^+e^- collisions) and photon-gluon fusion. At this level deep inelastic scattering becomes directly sensitive to the gluon distributions. One must take care with the renormalization scale and scheme dependence since the relative contributions of the three graphs and the parton distributions themselves are to some extent a matter of arbitrary convention. It is certainly possible to study deep inelastic scattering in a scheme where photon-gluon fusion is minimized, but there are a number of physical effects in the final state which are much more naturally understood in terms of graphs 8b and 8c. One which has been exploited by the NMC collaboration [24] is the production of charmed quarks. The static charmed quark content of the nucleon is expected (and measured) to be very small. However a high energy gluon can easily create $c\bar{c}$ pairs. A second final state observable is the detection

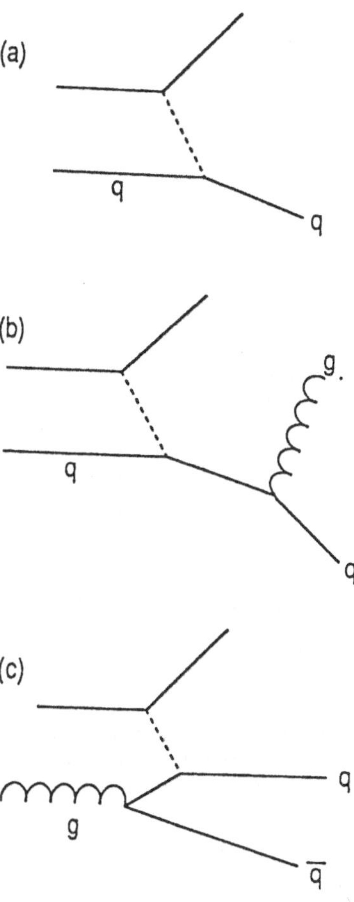

Fig. 8. Leading and next to leading order diagrams for deep inelastic scattering. Only a single time ordering has been shown in each case. a) The lowest order parton model as in Figure 1a) and Figure 2. b) Gluon Bremsstrahlung. c) Photon-Gluon Fusion.

of two forward jets of hadrons in the reaction. This has been studied at FNAL E665 and will be a major effort at HERA.

NMC measures the charmed quark production by detecting the J/Ψ mesons through their decays to two muons. The gluon distribution is extracted with a model for meson formation, the color-singlet model [25]. Care is taken to eliminate the coherent J/Ψ's from the event sample. The measured rapidity distributions and azimuthal asymmetry distributions of the J/Ψ provide important consistency checks for the model and are well reproduced. Figure 9 shows the x dependence of the measured gluon distribution and a fit of the form:

$$xG(x) = c\frac{\eta + 1}{2}(1 - x)^\eta \tag{7}$$

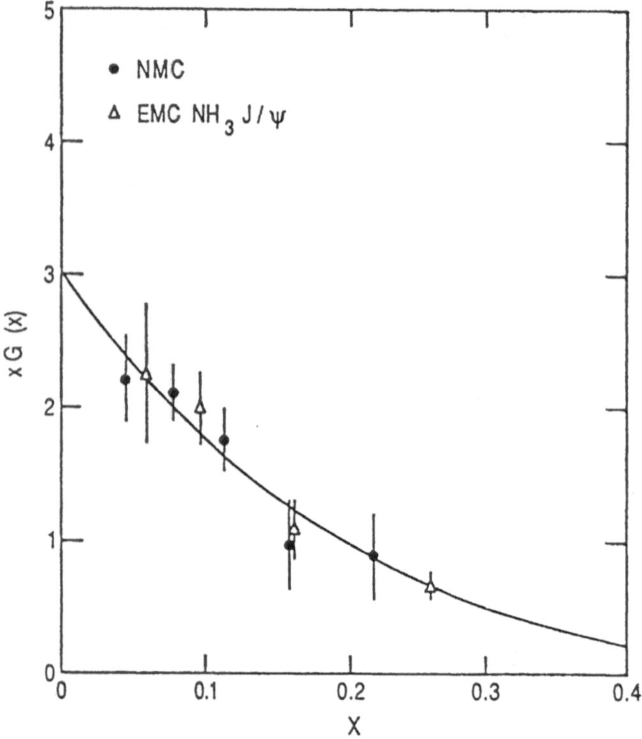

Fig. 9. The x dependence of the gluon distribution xG(x) extracted from J/ψ production in deep inelastic muon scattering using the color singlet model. (Taken from ref. [24]).

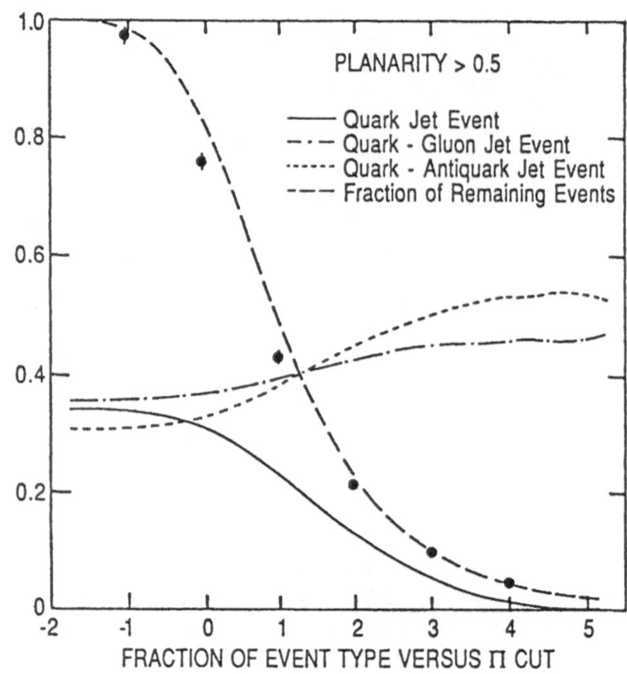

Fig. 10. The solid points show the fraction of the total deep inelastic scattering events with 4 or more charged tracks, $W^2 > 300$ GeV^2, $Q^2 > 3$ and planarity > 0.5 as a function of minimum Π and the Monte Carlo calculation [27] of the fraction (long dashed line). The other lines show the calculated relative fraction of quark jet events (diagram 8a, solid), quark-gluon jet events (diagram 8b, dash-dot), and quark-antiquark jet events (diagram 8c, short dashed). (Taken from reference [26])

with $\eta = 5.1 \pm 0.9$ and $c = 2.4 \pm 0.4$. The one problem with the result is that the overall normalization of the gluon distribution, c, should be 1.0. This last parameter is sensitive to one of the free parameters in the model, the charm quark mass, which is not well constrained by other data.

To identify the photon-gluon fusion effects in events with light quarks, it is necessary to look at the outgoing hadrons and study the event structure. If the $q\bar{q}$ or qg both carry a significant fraction of the energy of the virtual photon, then each will hadronize in a jet of hadrons. The momenta of the virtual photon and the two partons define a natural plane in space and two forward jet events will tend to be "pancake shaped", rather than cylindrical about the virtual photon axis. E665 has investigated a number of algorithms to identify two-forward jet events. One of the simplest involves looking at high charged multiplicity events ($n_{ch} \geq 4$) at high total hadronic energy W ($W^2 > 300$ GeV^2), finding the plane containing the virtual photon which minimizes the transverse momentum out of the plane, and selecting events on the basis of two transverse momentum variables, the planarity $\mathcal{P}$ and a normalized transverse momentum Π:

$$\begin{aligned} \mathcal{P} &= \frac{\sum[p_t^2(in) - p_t^2(out)]}{\sum[p_t^2(in) + p_t^2(out)]} > 0.5 \\ \Pi &= \frac{4}{\sqrt{n_{ch}}} \sum(|p_t| - .32) \end{aligned} \tag{8}$$

where (in) and (out) refer to transverse momentum in and out of the event plane. Figure 10 shows the fraction of events from diagram 8a-8c which pass a given Π cut. Using these event selection criteria, preliminary results [26] for the resulting flow of particles in the event plane are shown in Figure 11 and compared to a Monte Carlo calculation [27]. The evolution of the particle and energy flow (not shown) from a single-lobed one-jet structure to a distinct two-lobed structure is evident, and well reproduced by the calculation including QCD effects. We are now studying the sensitivity of the results to the event selection criteria and the underlying parton distributions. The calculations indicate that the majority of the two forward jet events at low x_{bj} values are due to photon-gluon fusion and the cross sections are sensitive to the choice of gluon distribution. It will be fascinating to extend these studies to look for A dependent effects and shadowing.

In the near term future we can expect considerably more exciting results from these two experiments. NMC will provide high statistics structure functions and

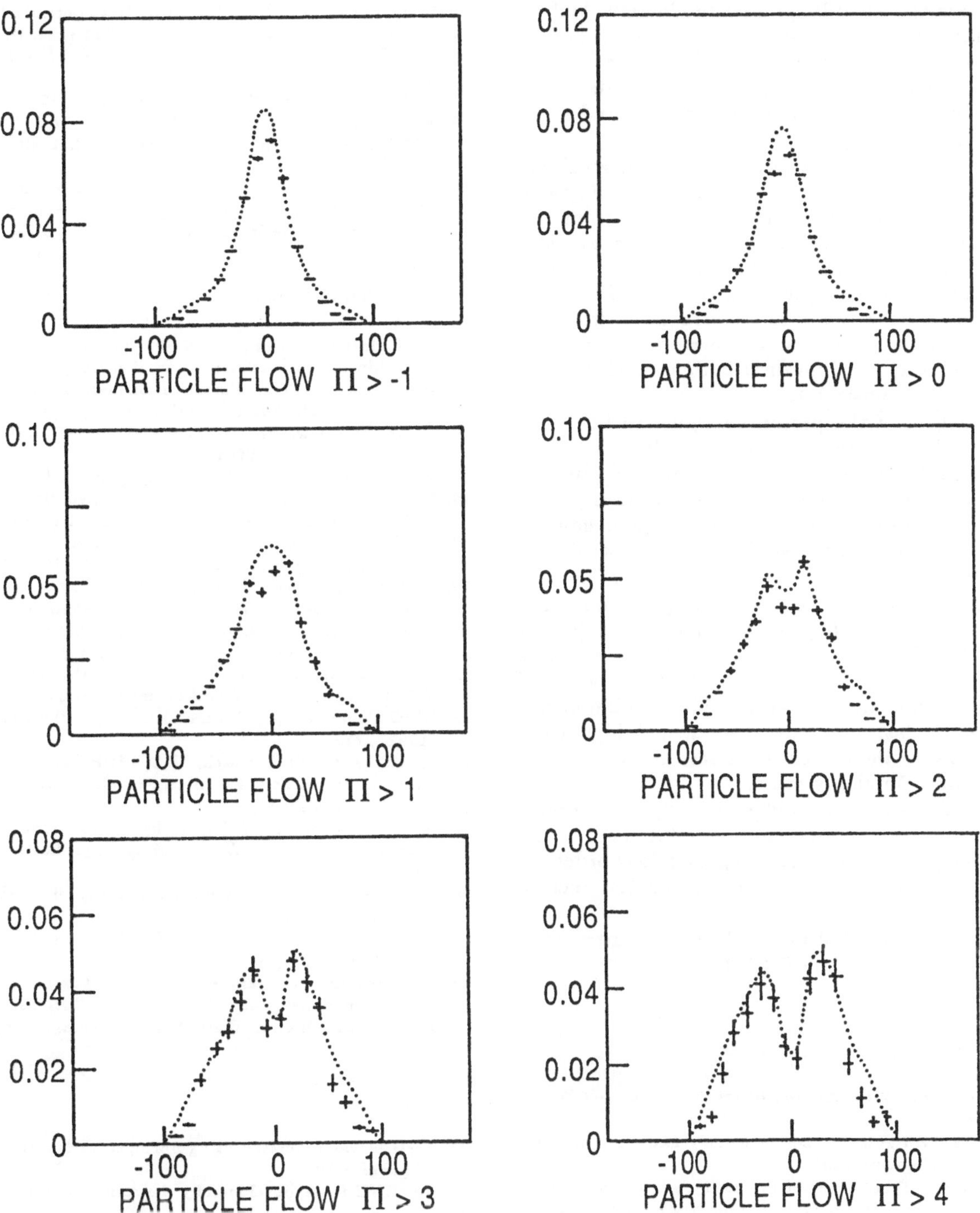

Fig. 11. The flow of particles in the event plane versus angle for six different minimum values of Π. The curves are the Monte Carlo calculations [27] of the particle flow. The event selection criteria are the same as those for Figure 10. (Taken from reference [26])

multi-muon results on hydrogen, deuterium and nuclear targets. The data taking phase of this experiment is now complete and a number of publications have appeared in the past year and are expected in the near future. FNAL E665 complements this with a higher beam energy (490 GeV) and systematic studies of the final state hadrons concentrating on the W>23 GeV regime which is not accessible at CERN. The first data on hydrogen, deuterium and Xenon were taken in 1988. In 1990, E665 used hydrogen, deuterium, carbon, calcium and lead targets using a target system which allowed to targets to be interchanged every minute to reduce the time dependent systematic errors. In 1991, E665 will complete data taking with higher statistics measurements on hydrogen and deuterium concentrating on jet production. Following in the scientific legacy of CHIO, EMC and BCDMS, these experiments will continue to highlight the key role of muon deep inelastic scattering. The first results shown here are only whetting our appetite for more and much more is to come.

In summary, I have shown some of the new results of deep inelastic muon scattering and tried to relate them to our understanding of photon-hadron interactions and the structure of the nucleon. It is clear we are still grasping at the non-perturbative structure in each case. Measurements at low x_{bj} open new horizons. From one simple point of view we must expect new effects to become important. At low x, the parton distributions behave as $q(x) \sim 1/x$ and the distributions must saturate to avoid unitarity constraints.

The structure of the proton continues to hold surprises. From the Gottfried Sum Rule results, we cannot devolve the proton into three constituents and glue because the quark sea does not have the same symmetry as the glue. At the same time we are getting new handles on the gluon distributions from deep inelastic scattering experiments.

All of these results make us want to go even further to lower x values and higher Q^2 values. That is just what HERA will do when it begins operating later this year, increasing our range in x_{bj} and Q^2 by two orders of magnitude. Even in this era, there will be an important role for deep inelastic muon scattering in bridging the large gap between the current data sets and the HERA results. As Marc Virchaux has shown [28], a new muon scattering experiment can bridge this gap and provide much more accurate results at $x_{bj} > 0.1$ This would provide a unique data set spanning over four orders of magnitude in Q^2 and x_{bj} to confront the photon-nucleon interaction and QCD.

This work is supported in part by the Department of Energy, Contract W-31-109-ENG-38.

References

1. R. P. Feynman, *Photon-Hadron Interactions*, Addison-Wesley (Reading) 1972
2. F. E. Close, *An Introduction to Quarks and Partons*, Academic Press (London) 1979
3. J. D. Bjorken, *Proceedings of the 1971 Symposium on Electron and Photon Physics at High Energies*, ed. by N. Mistry, Cornell (1971)
4. S. J. Brodsky and H. J. Lu, Phys. Rev. Lett. **64** (1990) 1342
5. G. Piller and W. Weise, Phys. Rev. C**42** (1990) 1834
6. L. L. Frankfurt and M. I. Strikman, Nucl. Phys. B **316** (1989) 340
7. N. N. Nicolaev and V. I. Zakhorov, **B 55** (1975) 397
8. A. H. Mueller and J. Qiu, Nucl. Phys. **B 256** (1986) 427
9. NMC Collaboration, P. Amaudruz et al. CERN-PPE/91-52 (1991)
10. E665 Collaboration, M. A. Adams et al. contributed to the 25th Conference on High Energy Physics, Singapore (1990)
11. D. E. Jaffe, Proceedings of the Rencontres de Moriond, High Energy Hadronic Interactions, 17-23 March 1991, to be published.
12. D. O. Caldwell et al. Phys. Rev. Lett. **42** (1979) 553
13. M. Arneodo et al., Phys. Lett. B**211** (1989) 493
14. P. N. Harriman et al., FAL-90-007 (1990)
15. J. G. Morfin, *Proceedings of the Workshop on Hadron Structure Functions and Parton Distributions*, ed. by D. F. Geesaman et al. World Scientific (Singapore) 1990
16. L. W. Whitlow: Ph. D. Thesis, Stanford University, 1990, SLAC-Report-357 (1990)
17. A. Milsztajn et al., Z. Phys. C. **49** (1991) 527
18. EMC-NA2, J. J. Aubert et al., Nucl. Phys. **B259**, 189 (1985); J. J. Aubert et al., Nucl. Phys. **B293**, 740 (1987).
19. BCDMS, A. C. Benvenuti et al., Phys. Lett. **B223** (1989) 485; A. C. Benvenuti et al., Phys. Lett. **B237** (1990) 592
20. A. Milsztajn, *Proceedings of the Workshop on Hadron Structure Functions and Parton Distributions*, ed. by D. F. Geesaman et al. World Scientific (Singapore) 1990
21. P. Z. Quintas, *Proceedings of the Workshop on Hadron Structure Functions and Parton Distributions*, ed. by D.F. Geesaman et al. World Scientific (Singapore) 1990
22. NMC Collaboration, P. Amaudruz et al. Phys. Rev. Lett. **66** (1991) 2712
23. M. Burkardt and R. Busch, SLAC-PUB-5426 (1991)
24. NMC Collaboration, D. Allasia et al., CERN PPE/90-178, (1990) unpublished.
25. R. Baier and R. Rückl, Nucl. Phys. B **218** (1983) 289 ; A. D. Martin, C.-K. Ng and W. J. Stirling, Phys. Lett. B **191** (1987) 200
26. D. M. Jansen, Ph. D. Thesis, University of Washington, unpublished, (1991)
27 T. Sjostrand, Lund Preprint LU TP 82-3 (1982) and Comp. Phys. Comm. **39** (1986) 347
28. M. Virchaux, *Proceedings of the Workshop on Hadron Structure Functions and Parton Distributions*, ed. by D.F. Geesaman et al. World Scientific (Singapore) 1990

This article was processed using Springer-Verlag TEX Z.Physik C macro package 1991
and the AMS fonts, developed by the American Mathematical Society.

Z. Phys. C - Particles and Fields 56, S169-S178 (1992)

Zeitschrift für Physik C Particles and Fields

Deep inelastic scattering of polarized muons

Rüdiger Voss

CERN, PPE Division, CH-1211 Geneva 23, Switzerland

20 February 1992

Abstract. An introduction is given to the physics of deep inelastic scattering with polarized muon beams on polarized nuclear targets. Earlier results on this subject are reviewed and an outlook is given on current and future experiments in this field.

1 Introduction

The physics of deep inelastic scattering with polarized high energy electron and muon beams is presently experiencing a Renaissance, following the seminal results obtained by the European Muon Collaboration (EMC) at CERN on the spin structure function of the proton. The EMC result has raised a number of questions about the understanding of the dynamics of the nucleon spin at the parton level which have not been answered so far and will require more experimental information to be resolved theoretically, including data on the spin structure of the neutron which is totally unknown today. This paper tries to give an introduction to the phenomenology and the experimental techniques of spin physics with high energy polarized muon beams, with emphasis on the SMC experiment currently underway at CERN.

This discussion is limited to the experimental study of spin-dependent structure functions. The equally rich topic of electroweak physics with polarized muon beams is not touched upon here [1]. The subject of unpolarized deep inelastic muon scattering has been reviewed at this conference by D. Geesaman [2].

2 Polarized high energy muon beams

Two high energy muon beams are presently in operation at the CERN SPS and at the Fermilab Tevatron. Muons are produced by the decay of high energy pions and kaons in flight and the designs of both beams follow similar principles. So far, only the SPS muon beam [3,4] has been used in the past – and will be used in the future – for experiments which exploit the polarization properties of such beams.

2.1 Layout of the CERN SPS muon beam

A schematic layout of this beam is shown in Fig. 1. A beam of secondary hadrons – mostly pions and kaons – is produced by the 450 GeV proton beam extracted from the SPS and impinging on a primary production target. Downstream of this target, the beam can be subdivided into four functional sections:

1. *Momentum selection of parent hadrons.* A series of 6 quadrupoles collects charged kaons and pions and focuses them onto a vertical bending magnet which selects the central value of the parent hadron momentum in a momentum band $\Delta p/p$ up to 10%. Five more quadrupoles provide for an optimal matching to the second section of the beam.
2. *The decay channel.* This channel is designed to transport both the hadron beam over its entire phase space, and the decay muons over the entire kinematically allowed momentum band ($p_\pi \gtrsim p_\mu \gtrsim 0.57 p_\pi$). This design allows to freely choose the μ/π momentum ratio and thus the polarization of the beam (see below). It is achieved with a regular lattice of alternately focussing and defocussing quadrupoles (FODO array) of 60° phase advance per cell, a structure more commonly found in storage rings. With a decay length of about 500 m, 10% of the parent hadrons decay into muons at 100 GeV beam energy.
3. *The hadron absorber.* The muon beam is focused by a set of quadrupoles onto a beryllium absorber of 9.9 m maximum length to remove the remaining hadrons. This absorber is subdivided into 9 elements which can be moved individually into the beam to obtain the desired compromise between multiple scattering and hadron contamination. The effect of multiple scattering on the beam size is minimized by locating the absorber in a

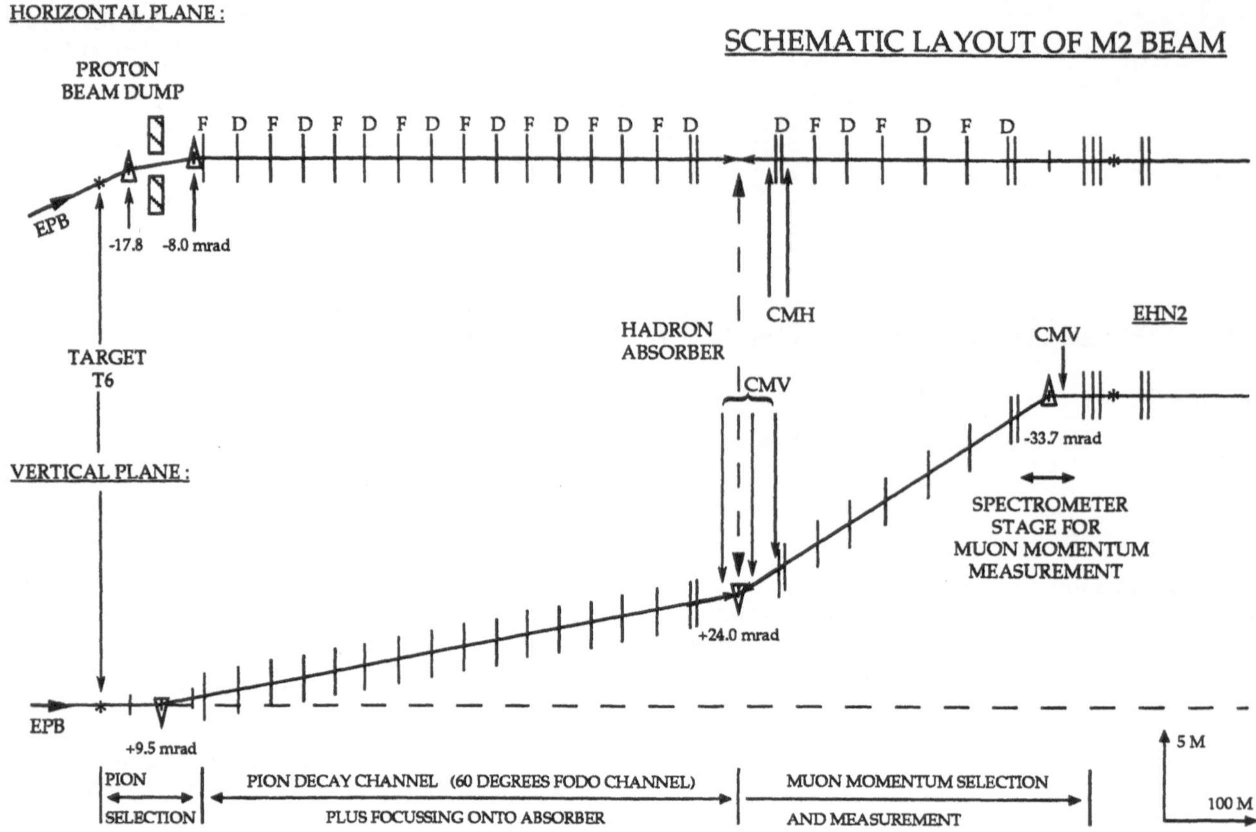

Fig. 1. Layout of the CERN SPS muon beam M2. F and D indicate focussing and defocussing quadrupoles, respectively; EPB is the extracted proton beam, and the CMH (CMV) are horizontal (vertical) magnetic collimators (scrapers). Note that the horizontal and vertical scales are strongly different.

focus of the beam. With all absorber elements in place, the π contamination is less than 10^{-6}.

4. *The muon transport system.* The absorber elements are installed inside the gap of a vertical bending magnet which selects the final μ momentum band of about $\pm 3\%$. The muon beam is transported from the underground absorber section to the experimental area by a FODO structure similar to the decay section. This section provides the necessary drift space to clean the beam from halo muons which originate mainly from π decays not captured in the acceptance of the decay channel. This halo is eliminated with a system of 2 horizontal and 5 vertical magnetic collimators (scrapers). A last vertical bending magnet serves to deflect the beam back into the horizontal direction before it is focussed onto the experiment. This magnet is instrumented with a system of fast scintillator hodoscopes and serves as a spectrometer magnet to measure the momentum of individual muons triggering the experiment.

The SPS muon beam is capable of producing intensities of several $10^8\,\mu$ per accelerator pulse of about 2 sec length, with a repetition rate of 14.4 sec. In practice, the beam intensity is limited by the proton intensity available at the primary production target and by the rate capabilities of the experiments. Recent experiments have operated at typical intensities of a few $10^7\,\mu$/pulse.

The beam was first commissioned in 1978 and operated until 1989 without significant modifications, mostly at beam energies between 100 and 280 GeV. The muon transport system has recently been modified to make the beam dispersion-free in the experimental area. This beam is better adapted to the needs of the NA47 experiment discussed below, allowing for small beam spot sizes of about 2 cm FWHM at all critical locations in the experiment. The price to pay for this improvement was a reduction of the maximum muon energy to 225 GeV/c.

2.2 Kinematics of beam polarization

High energy muon beams originating from $\pi(K)$ decays are naturally polarized due to parity violation. In the decay $\pi(K) \to \mu\nu_\mu$, the muon and the neutrino are fully polarized in the center-of-mass system, giving positive (negative) helicity to the positive (negative) muon. Lorentz boosting changes the longitudinal polarization of the massive muons which in the laboratory system depends on the ratio of muon to hadron energy. If both hadron and muon beams were monochromatic, the longitudinal polarization of the muons would be given by [5,6]

$$P_L = \pm \frac{u - (m_\mu/m_{\pi,K})^2(1-u)}{u + (m_\mu/m_{\pi,K})^2(1-u)} \tag{1}$$

where the + (−) sign refers to negative (positive) muons, and where

$$u = \frac{E_\mu/E_{\pi,K} - (m_\mu/m_{\pi,K})^2}{1 - (m_\mu/m_{\pi,K})^2} = \cos^2\frac{\theta^*}{2}, \tag{2}$$

where θ^* is the angle between muon and pion direction in the center-of-mass system.

In practice, hadron and muons are not monochromatic but the selection of both π and μ beams of limited phase space preserves a resulting net polarization of the muons arriving at the experimental target.

2.3 Beam polarization from Monte Carlo simulations

Monte Carlo programs exist [7] which simulate the phase space of parent hadrons and decay muons and model in detail their propagation through the beam transport system. These programs can be used to calculate the beam polarization in a straightforward way and the most reliable figures on the polarization of the CERN muon beam which are presently available come from Monte Carlo simulations. In practice, the accuracy of such calculations is limited to a few percent by effects which are poorly known and are difficult to simulate exactly, such as the exact shape of the secondary hadron spectrum, the kaon background to the pion spectrum, radiative energy losses of muons in the hadron absorber, and possible depolarizing effects in the beam optics. A reliable determination of the beam polarization to better than $\approx 5\%$ therefore calls for a good experimental measurement.

2.4 Polarization measurements from $\mu-e$ decay in flight

The standard technique to measure the polarization of a high energy muon beam also exploits parity violation and makes use of the fact that in the decay $\mu^+ \to e^+\nu_e\bar{\nu}_\mu$ the positron is preferentially emitted in the muon spin direction in the center-of-mass system, and in the opposite direction in the case of μ^- decay. The Lorentz boost will therefore yield a positron spectrum in the laboratory system which depends on the polarization of the muon beam [8,9],

$$\frac{dN}{dv} = N_0\left[\frac{5}{3} - 3v^2 + \frac{4}{3}v^3 - P_L\left(\frac{1}{3} - 3v^2 + \frac{8}{3}v^3\right)\right] \tag{3}$$

where $v = E_e/E_\mu$, $N_0 = 1.6 \cdot 10^{-4}\, l(m)/E(GeV)$, and where l is the length of the decay space available to the muons (Fig. 2). The polarization is best measured around $v = 0.75$ where the electron energies are high and the spectrum is most sensitive to the polarization.

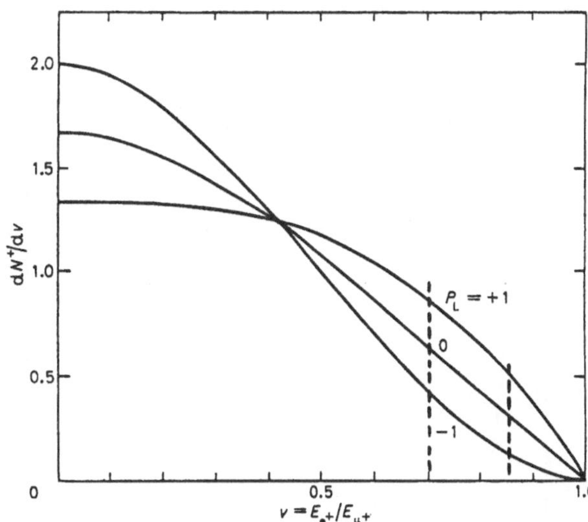

Fig. 2. Energy spectrum of positrons from muon decay, for longitudinal muon polarizations $P_L = -1, 0, +1$.

Polarization measurements based on this technique are experimentally demanding. They require a two-stage spectrometer to measure both the muon and the electron energy; the muon spectrometer has to operate at high rates to compensate for the small decay rate, and the electron spectrometer usually has to withstand a strong background of electrons and positrons from other sources, mostly from radiative energy losses of muons. In an earlier measurement in the SPS beam, a leadglass calorimeter to measure the positron energy was placed close to the beam right downstream of the last vertical bending magnet which serves as spectrometer magnet to measure the muon momentum (see above) [10]. Within errors of $\approx 10\%$, these measurements were in agreement with the Monte Carlo simulations (Fig. 3).

An improved implementation of such a decay polarimeter has been designed for the SMC experiment which is described in Sect. 5.1 below.

2.5 Polarization measurements by Møller scattering

Measuring the cross section asymmetry of elastic scattering off polarized electrons is the standard technique to measure the polarization of high energy electron beams, where the polarized electron target is usually realized by a thin magnetized iron foil [11]. In order to apply the same technique to a muon beam, one has to use much thicker targets to compensate for the much lower beam intensity, and the dilution of the cross section asymmetry by background from radiative energy losses has to be carefully controlled. A practical implementation of this

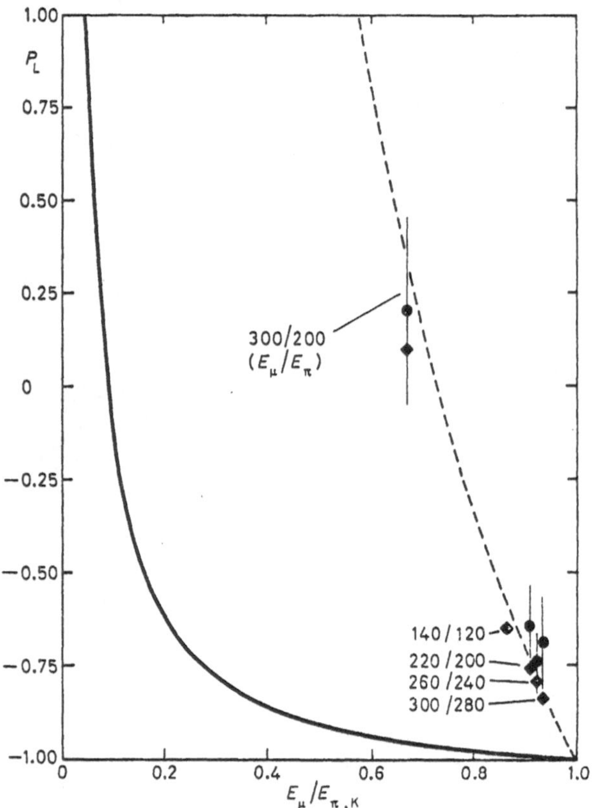

Fig. 3. Polarization vs. muon energy for different combinations of pion/muon energy in the SPS muon beam. Circles show results of a Monte Carlo calculation and rhombs show the measurements of ref. [10]. The kinematic relation of eq. (1) is also shown for pions (dashed line) and kaons (solid line).

technique has been proposed in ref. [12] and will be tried in the SMC polarimeter from 1992 onwards.

3 Polarized lepton-nucleon scattering

A detailed account of the cross sections relevant for deep inelastic scattering of polarized leptons can be found in refs. [1,13-15]. The following brief review is limited to the phenomenology of the scattering of longitudinally polarized electrons and muons. The scattering of transversely polarized beams is not discussed here.

3.1 The deep inelastic cross section

In the laboratory system, the scattering process is conveniently visualized in the two kinematic planes depicted in Fig. 4. The scattering plane is defined, as in the unpolarized case, by the momentum 3-vectors **k** and **k**′ of the incoming and scattered lepton, respectively; θ is the scattering angle. The polarization plane is defined

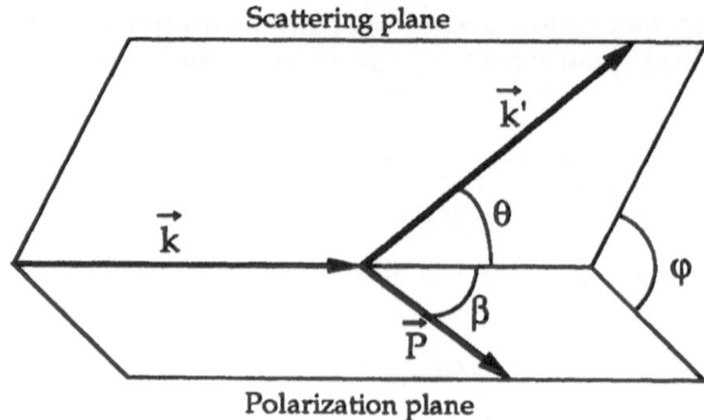

Fig. 4. Kinematic planes of scattering of longitudinally polarized leptons.

by **k** and by the polarization vector **P** of the nucleon. The angle between **k** and **P** is often referred to as β ($0 \leq \beta \leq \pi$) and ϕ is the angle between the scattering and the polarization planes. The kinematics of inclusive scattering are completely described by two independent variables. One often uses the squared four-momentum transfer Q^2 and the energy transfer ν. Both are Lorentz invariants and are given in the laboratory system and in the high energy limit by

$$Q^2 = 4EE' \sin^2 \theta/2, \tag{4}$$

$$\nu = E - E', \tag{5}$$

where E and E' are the energies of the incident and the scattered muon in this system, respectively. Alternatively, the scaling variables x (Bjorken variable) and y may be used:

$$x = Q^2/2M\nu, \tag{6}$$

$$y = \nu/E, \tag{7}$$

where M is the nucleon mass.

The differential deep inelastic cross section for the process shown in Fig. 4 can be decomposed into an unpolarized piece σ_0 and a polarized piece $\Delta\sigma$,

$$\frac{d^3\sigma(\beta)}{dxdyd\phi} = \frac{d^3\sigma_0}{dxdyd\phi} - \frac{d^3\Delta\sigma(\beta)}{dxdyd\phi}, \tag{8}$$

where the unpolarized part has been discussed in ref. [2]. In the Born approximation, it is given by

$$\frac{d^3\sigma_0}{dxdyd\phi} = \frac{2\alpha^2}{Q^2xy}\Big[xy^2F_1(x,Q^2)+ \left(1-y-\frac{Mxy}{2E}\right)F_2(x,Q^2)\Big], \tag{9}$$

where F_1 and F_2 are the *structure functions* of the unpolarized nucleon. In the same approximation, the polarized piece is given by

$$\frac{d^3\Delta\sigma(\beta)}{dxdyd\phi} = \frac{2\alpha^2}{MExy}\Big\{\cos\beta\Big[\Big(1-\frac{y}{2}-\frac{Mxy}{2E}\Big)g_1(x,Q^2)- \frac{Mx}{2E}g_2(x,Q^2)\Big]- \cos\phi\sin\beta\frac{\sqrt{Q^2}}{\nu}\Big(1-y-\frac{Mxy}{2E}\Big)^{\frac{1}{2}}\cdot \Big[\frac{y}{2}g_1(x,Q^2)+g_2(x,Q^2)\Big]\Big\}, \tag{10}$$

where g_1 and g_2 are the so-called *spin structure functions* of the nucleon. They play a central role in the understanding of the spin structure of nucleons.

An inspection of eq. (10) reveals immediately how these two structure functions can be disentangled experimentally from the measured differential cross section. For $\sin\beta = 0$, i.e. target polarization (anti)parallel to the beam direction, one mainly measures g_1 since g_2 is strongly suppressed at high energies by the factor $Mx/2E$. For $\cos\beta = 0$, i.e. transverse target polarization, g_1 and g_2 contribute to the measured cross section with approximately equal weight. So far, only the case of longitudinal target polarization has been studied experimentally and no data exist on g_2.

3.2 Structure functions in the Quark-Parton Model

In the so-called naive Quark-Parton Model (QPM) [16], deep inelastic scattering is described as *elastic* scattering of leptons on quasifree pointlike partons (quarks) inside the nucleon. The deep inelastic cross section (8) can then be computed as incoherent sum over the elastic cross section of these partons, provided a prescription is given for their kinematical distribution inside the target. This prescription is provided by the structure functions.

Most structure functions take a simple intuitive meaning in the QPM. This is especially true for the well-known unpolarized structure functions F_1 and F_2 which appear in eq. (9) and which can be written as simple linear combinations of probability densities of quarks, $q(x,Q^2)$. In the so-called scaling limit of non-interacting quarks, these quark distributions depend on the Bjorken variable x only:

$$F_1(x) = \frac{1}{2}\sum_i e_i^2[q_i(x)+\bar{q}_i(x)] \tag{11}$$

$$F_2(x) = x\sum_i e_i^2[q_i(x)+\bar{q}_i(x)] \tag{12}$$

where the index i refers to all quark flavours participating in the scattering process, the $\bar{q}$ distribution describes the antiquarks in the nucleon, and the e_i are the quark charges in units of the elementary charge. The Bjorken variable also finds a simple, intuitive interpretation in the QPM: in the limit of very high energies, it is the fraction of the total nucleon momentum p carried by the individual quark i on which the lepton is scattered, i.e. $x = p_i/p$.

From equations (11) and (12) one finds immediately the Callan–Gross relation

$$F_2(x) = 2xF_1(x). \tag{13}$$

Again, this relation holds in the high energy limit only and is violated at finite energies. This violation is characterized by the quantity

$$R(x,Q^2) = \frac{\Big(1+\frac{2Mx}{Ey}\Big)F_2(x,Q^2)-2xF_1(x,Q^2)}{2xF_1(x,Q^2)} \tag{14}$$

which can also be interpreted as the ratio of cross sections for longitudinally and transversely polarized virtual photons,

$$R = \frac{\sigma_L}{\sigma_T}. \tag{15}$$

The spin structure function g_1 also has a straightforward interpretation in the QPM:

$$g_1(x) = \frac{1}{2}\sum_i e_i^2[q_i^+(x)-q_i^-(x)], \tag{16}$$

where $q_i^+(x)$ ($q_i^-(x)$) is the density of quarks with helicity parallel (antiparallel) to the nucleon spin. This interpretation of $g_1(x)$ can be understood from the fact that a virtual photon with spin projection +1 can only be absorbed by a quark with spin projection −1/2, and vice versa.

The interpretation of the "transverse" spin structure function g_2 in the QPM is much less obvious and is presently the subject of much theoretical debate [17]. Wandzura and Wilczek [18] have shown that in Quantum Chromodynamics (QCD) it can be decomposed as

$$g_2(x,Q^2) = g_2^{WW}(x,Q^2)+\bar{g}_2(x,Q^2), \tag{17}$$

where the "trivial" piece g_2^{WW} is a "leading twist" (twist-2) contribution in the jargon of QCD, and is completely determined by $g_1(x,Q^2)$:

$$g_2^{WW}(x,Q^2) = -g_1(x,Q^2)+\int_x^1\frac{dy}{y}g_1(y,Q^2). \tag{18}$$

The term $\bar{g}_2(x,Q^2)$ is a twist-3 contribution which seems to be best understood in an Operator Product Expansion (OPE) analysis in QCD, where it is sensitive to a quark-gluon correlation function in the nucleon and thus contains unique new physics. In Regge theory, g_2 is shown to fulfill, under certain conditions, the Burkhardt-Cottingham sum rule [19,17]

$$\int_0^1 dx\,g_2(x,Q^2) = 0. \tag{19}$$

3.3 Sum rules for spin structure functions

Just as for unpolarized structure functions, no theoretical predictions exist yet for the x dependence of their spin dependent counterparts, although such predictions are expected to emerge ultimately from non-perturbative QCD. Predictions do exist, however, in the form of sum rules related to polarized structure functions. The most general of these, and one of the most fundamental predictions of the QPM indeed, is the celebrated Bjorken sum rule [20]

$$\int_0^1 [g_1^p(x) - g_1^n(x)]dx = \frac{1}{6}\left|\frac{g_A}{g_V}\right|, \tag{20}$$

where p and n denote the proton and the neutron, respectively, and where g_A and g_V are the axial and vector weak coupling constants of nuclear beta decay. In this form, the sum rule was derived by Bjorken from light cone algebra and from very general assumptions on the partonic structure of the weak and electromagnetic hadronic currents. Nowadays, the sum rule (20) can be rigorously derived in QCD in the limit $Q^2 \to \infty$. At finite values of Q^2 [21],

$$\int_0^1 [g_1^p(x,Q^2) - g_1^n(x,Q^2)]dx = \frac{1}{6}\left|\frac{g_A}{g_V}\right|\left[1 - \frac{\alpha_s(Q^2)}{\pi}\right], \tag{21}$$

where α_s is the strong coupling constant.

Separate sum rules for the proton and the neutron were derived by Ellis and Jaffe for the proton and the neutron [22]. Ignoring QCD radiative corrections, they read

$$\int_0^1 g_1^p(x)dx = \frac{1}{12}\left|\frac{g_A}{g_V}\right|\left[+1 + \frac{5}{3}\frac{3F/D-1}{F/D+1}\right] \tag{22}$$

and

$$\int_0^1 g_1^n(x)dx = \frac{1}{12}\left|\frac{g_A}{g_V}\right|\left[-1 + \frac{5}{3}\frac{3F/D-1}{F/D+1}\right], \tag{23}$$

where F (D) are the antisymmetric (symmetric) weak $SU(3)$ couplings measurable in hyperon decays. These predictions are less fundamental than the Bjorken sum rule since they assume exact flavour $SU(3)$ symmetry of the baryon octet decays, and zero net polarization of the sea of strange quarks and heavier flavours.

No experimental data exist on g_1^n and the only sum rule which is tested experimentally until now is the Ellis-Jaffe prediction (22) for the proton. These data will be discussed in Sect. 4. below.

3.4 Cross section asymmetries

Since the polarized piece (10) gives, in general, only a small contribution to the cross section, it is customary to evaluate it from measurements of cross section asymmetries in which the unpolarized part (9) cancels. In the most simple case where both the beam and the target are longitudinally polarized (i.e. $\sin\beta = 0$), this asymmetry is

$$A = \frac{\sigma^{\uparrow\downarrow} - \sigma^{\uparrow\uparrow}}{\sigma^{\uparrow\downarrow} + \sigma^{\uparrow\uparrow}} \tag{24}$$

where $\sigma^{\uparrow\downarrow}$ and $\sigma^{\uparrow\uparrow}$ are the cross sections for opposite and equal spin directions, respectively. From equations (10)–(16), neglecting terms of order M/E, one finds

$$A = D[A_1 + \eta A_2], \tag{25}$$

where

$$A_1(x) = \frac{g_1(x)}{F_1(x)} = \frac{\sum_i e_i^2[q_i^+(x) - q_i^-(x)]}{\sum_i e_i^2[q_i^+(x) + q_i^-(x)]}, \tag{26}$$

$$A_2(x) = \sqrt{\frac{2Mx}{Ey}}\,\frac{g_1(x) + g_2(x)}{F_1(x)}. \tag{27}$$

D is sometimes called the depolarization factor of the virtual photon and is given by

$$D = \frac{2y - y^2}{2(1-y)(1+R) + y^2}; \tag{28}$$

the factor η depends only on kinematic variables:

$$\eta = \frac{\sqrt{Q^2}}{E}\,\frac{2(1-y)}{y(2-y)}. \tag{29}$$

A_1 and A_2 can also be interpreted as virtual photon-nucleon asymmetries

$$A_1 = \frac{\sigma_{1/2} - \sigma_{3/2}}{\sigma_{1/2} + \sigma_{3/2}}, \tag{30}$$

$$A_2 = \frac{2\sigma_{TL}}{\sigma_{1/2} + \sigma_{3/2}}, \tag{31}$$

where 1/2 and 3/2 are the total spin projections in the direction of the virtual photon, and σ_{TL} is a cross section arising from the interference of amplitudes for longitudinal and transverse polarized virtual photons. The following bounds can be derived for A_1 and A_2 [23]:

$$|A_1| \le 1, \qquad |A_2| \le R; \tag{32}$$

for this reason, A_2 is expected to give a small contribution to A and is usually ignored.

Finally, the experimentally measured counting rate asymmetry is related to the cross section asymmetry (24) by

$$A_{exp} = f_t P_t P_\mu A, \tag{33}$$

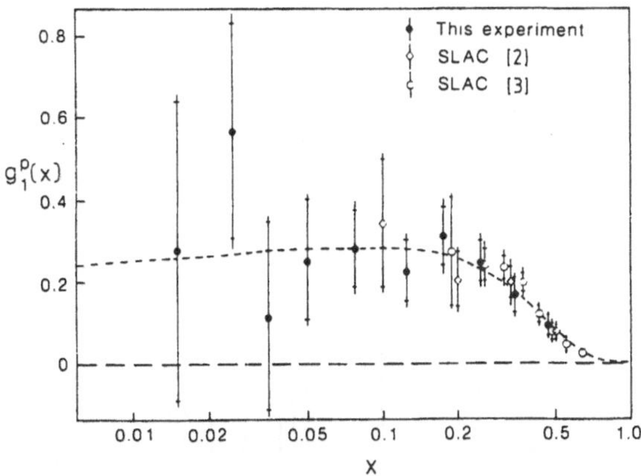

Fig. 5. The spin structure function $g_1^p(x)$ measured by the EMC Collaboration ("This experiment") and the SLAC-Yale Collaboration. The dashed curve is a phenomenological parametrisation.

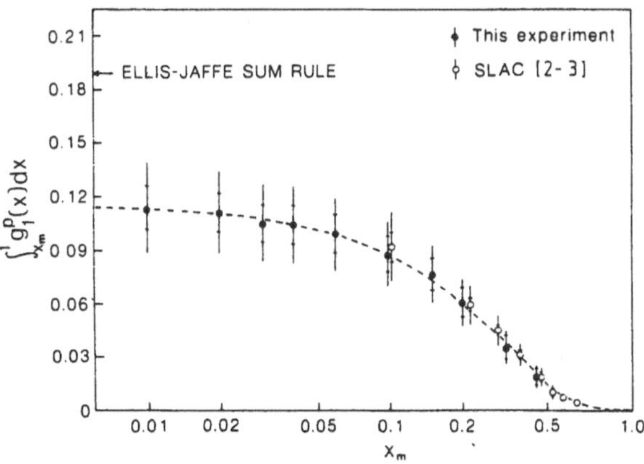

Fig. 6. The integral $\int_{x_m}^1 g_1^p(x)dx$ as a function of the lower integration limit x_m. The prediction of the Ellis-Jaffe sum rule is also shown.

where P_μ is the beam polarization, P_t the polarization of the target nucleons, and f_t the target dilution factor, i.e. the fraction of polarized nucleons in the target material.

4 The proton spin crisis

The most recent data on spin structure functions were presented in 1987 by the European Muon Collaboration (EMC) at CERN [24,25]. The EMC measurement of $g_1^p(x)$ was found to be in good agreement with earlier data from the SLAC-Yale collaboration [26-28] but covers a significantly larger kinematic range in the x variable (Fig. 5). These data therefore allowed the first significant test of the Ellis-Jaffe sum rule which is shown in Fig. 6.

Using the parametrisation of Fig. 5 to extrapolate the measured integral to $x = 0$, the result from the combined EMC and SLAC data is

$$\int_0^1 g_1^p(x)dx = 0.126 \pm 0.010\,(\text{stat.}) \pm 0.015\,(\text{syst.}).$$

The Ellis-Jaffe prediction, using the most recent data on F/D [29], is

$$\int_0^1 g_1^p(x)dx = 0.189 \pm 0.005,$$

i.e. there is a 3.5 standard deviation discrepancy between the Ellis-Jaffe prediction and the experimental data.

The quark contribution to the total spin of the proton is given by

$$\frac{1}{2}\Delta\Sigma = \frac{1}{2}(\Delta u + \Delta d + \Delta s) \qquad (34)$$

where heavier flavours have been neglected and where

$$\Delta q = \int_0^1 [q^+(x) + \bar{q}^+(x) + q^-(x) + \bar{q}^-(x)] \qquad (35)$$

is, apart from a factor 1/2, the contribution to the nucleon spin from an individual quark flavour. Using a generalization of the Ellis-Jaffe sum rule by Glück and Reya which includes QCD radiative corrections [30], assuming isospin invariance and the same experimental data on the $SU(3)$ couplings F and D, it can be shown that [25]

$$\frac{1}{2}\Delta\Sigma = 0.060 \pm 0.047\,(\text{stat.}) \pm 0.069\,(\text{syst.}),$$

i.e. the quark contribution to the total proton spin is compatible with zero within the experimental errors. The proton spin fulfills the sum rule

$$S_p = \frac{1}{2} = \frac{1}{2}\Delta\Sigma + \frac{1}{2}\Delta g + < L_z >, \qquad (36)$$

where Δg is the gluon equivalent of $\Delta\Sigma$ and $< L_z >$ is the mean z component of the orbital angular momentum of the partons. Most of the proton spin must therefore be carried by gluons and/or parton orbital momentum. This surprising result has triggered intense theoretical efforts to explain the spin composition of the proton. The recent literature on this subject is too vast to be reviewed here [31].

5 New experiments on spin dependent structure functions

Following the unexpected result of the EMC experiment, several new experiments to study spin dependent structure functions have also been proposed, most of which share the following main goals:

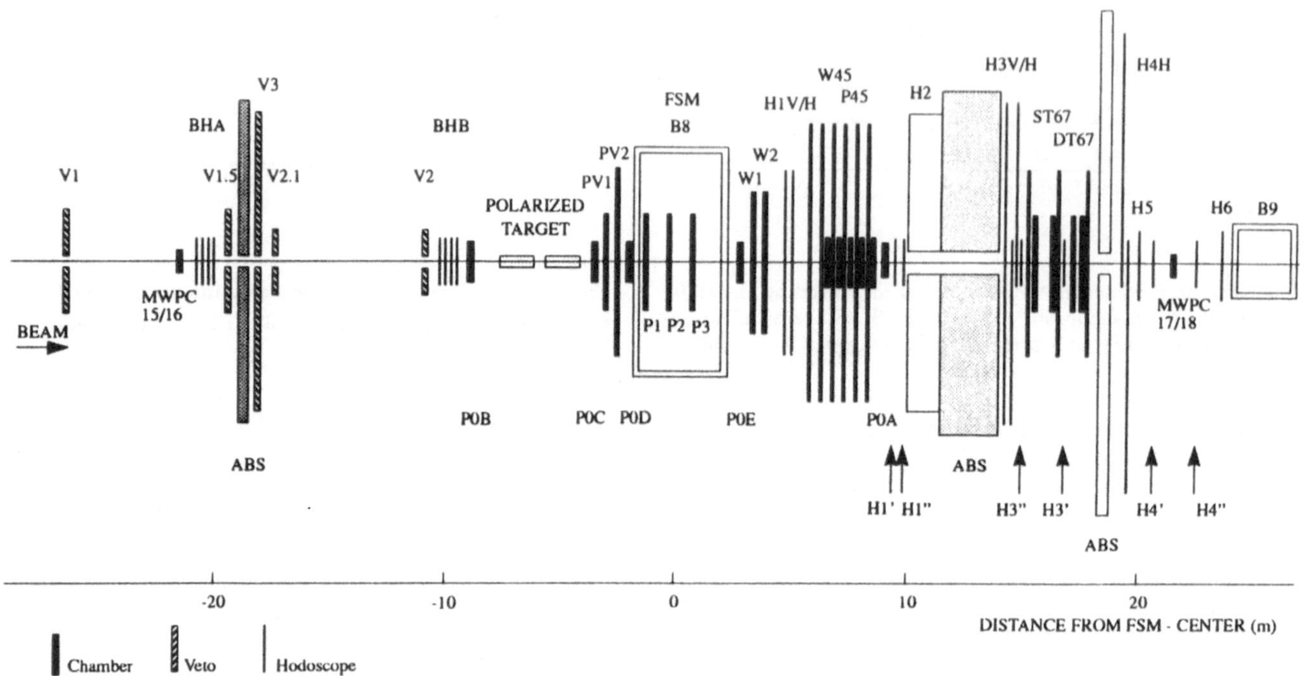

Fig. 7. Schematic layout of the SMC spectrometer. The muon beam arrives from the left and impinges on a twin solid state target. Downstream of the target, one distinguishes the momentum measurement stage with the Forward Spectrometer Magnet (FSM), and the muon identification system downstream of the hadron absorber (ABS). The detectors to the left of the target are beam defining counters to measure the track of the incident muon, and veto counters to shield the experiment from halo muons.

1. New measurements of $g_1^p(x)$ for an improved test of the Ellis-Jaffe sum rule for the proton,

2. measurement of the neutron distribution $g_1^n(x)$ and test of the corresponding Ellis-Jaffe sum rule,

3. test of the Bjorken sum rule,

4. measurement of the "transverse" spin structure function $g_2(x)$.

The study of other related physical questions has also been proposed by several of these next generation experiments [32].

The direct successor of the FMC is the SMC experiment which has recently started to take data at CERN and uses the muon beam described above in Section 2.1. All other experiments which are presently proposed or in preparation use polarized electron beams.

5.1 The SMC experiment

The SMC (Spin Muon Collaboration) experiment [33] uses an upgraded configuration of the apparatus built for the earlier EMC experiments [24,25,34].

In this experiment polarized high-energy muons are scattered off two solid targets polarized in opposite directions. The target materials are deuterated and normal hydrocarbon glasses, doped with a small amount of paramagnetic metallo-organic substance EHBA-Cr(V) or its deuterated version. The glass matrix consists mainly of 1-butanol C_4H_9OH(95%) and water (5%), or their deuterated forms. The dilution factors for free hydrogen (deuterium) nuclei in these materials are $f_p = 0.13$ and $f_d = 0.23$.

The Dynamic Nuclear Polarization (DNP) [35] technique is used to obtain high nuclear polarizations in the targets. In this technique the paramagnetic electron spin system in the material is saturated slightly off-resonance (70 GHz); this produces dynamic cooling of the spin-spin interactions among the electrons by a factor around $\pm 1/400$. The nuclear spins are cooled to a temperature very close to that of the electron spin-spin interactions, because no other thermal contact to the nuclear spin system is provided in the material. If the material is cooled to 500 mK by a $^3He - ^4He$ dilution refrigerator, nuclear spin temperatures around ± 2 mK can be obtained.

The target material is arranged inside a superconducting solenoid, with a 2.5 Tesla field of high homogeneity, in two cells so that each cell can be irradiated with microwave power from an independent source. The microwave frequencies are adjusted just below and just above the electron spin resonance line so that maximum positive and negative polarizations are obtained in the two target cells. In the present materials the proton polarizations of $\pm 85\%$ and deuteron polarizations of about $\pm 30\%$ have been obtained in the large target cells containing each about 500 g of solid target material.

The target is cooled with a powerful dilution refrigerator [36] so that the microwave losses of about 2 W in the material can be cooled at a helium temperature of 0.5 K. When the microwave power is turned off, the refrigerator cools the material to about 50 mK temperature, where the nuclear spin lattice relaxation becomes extremely slow, thus enabling the "freezing" of the target polarization. The polarization of such a frozen-spin target is insensitive to the magnetic field inhomogeneity, and reasonably slow relaxation is measured down to 0.5 Tesla field value. In this mode the target polarizations can be reversed by the rotation of the magnetic field, which is accomplished by exciting a dipole magnet superimposed on the solenoid, while ramping the solenoid current through zero value.

The target polarization is measured with $\pm 3\%$ relative accuracy by continuous wave NMR techniques using a series-resonant circuit and a Q-meter with real-part detector. The polarization can also be monitored continuously during frozen-spin operation in 0.5 T field.

During a first phase (1991/92), the SMC uses an improved version of the target set-up used which was originally built for the EMC experiment [37]. From 1993 onwards, a new target configuration will be used with longer target cells (60 cm instead of 36 cm each) in a bigger cryostat, a new solenoid with improved field homogeneity for higher polarization, and a more powerful refrigerator. In this configuration, a transverse dipole field can be superimposed to the solenoid field which is employed for fast polarization reversal by field rotation in frozen spin mode, or for tranverse target polarization to measure g_2. It is expected that a proton polarization above 80% and deuteron polarizations of up to 40% will be achieved with this target.

The muon spectrometer of the SMC experiment is an upgraded version of the well-known EMC apparatus (Fig. 7). A high precision measurement of the scattering angle and of the momentum of charged particles is provided by a large aperture dipole magnet ($\int B\, dl = 2.3\,\mathrm{Tm}$) instrumented with multiwire proportional chambers and drift chambers. The momentum measurement stage is followed by a muon identification stage which consists of an iron absorber to remove the hadrons produced in the deep inelastic interaction, an array of large-surface streamer tubes and drift tubes to measure the muon tracks behind the absorber, and two arrays of scintillator hodoscopes which provide the muon trigger of the experiment.

The muon spectrometer is followed by a beam polarimeter [33] which has been newly designed for the SMC experiment, following the principles outlined in Sect. 2.4 and 2.5 above. The muon energy is measured in the upstream magnetic spectrometer which is part of the beam transport system and was mentioned above in Sect. 2.1. The polarimeter provides a 30 m long decay space for the muons, the beginning of which is defined by an electromagnetic shower counter to suppress background from electromagnetic interactions in material exposed to the upstream beam. Decay electrons are identified and momentum analyzed by a simultaneous measurement of their momentum and energy in a magnetic spectrometer and in a lead-glass calorimeter. A magnetized iron target for polarization measurement with the Møller scattering method is presently under construction.

The SMC experiment will test the sum rules (21)-(23) to an accuracy of 10–20% which will be dominated by systematic errors. The main uncertainties are the measurement errors on the beam and target polarization and the uncertainty in the extrapolation of $g_1(x)$ to $x = 0$.

5.2 Electron beam experiments

Four experiments are presently under construction, or have been proposed, to study spin structure functions in polarized electron beams.

1. The SLAC E-142 experiment [38] will use an external polarized electron beam of 23 GeV from a GaAs source ($P_e \approx 40\%$) and a high-pressure (10 atm) polarized ^3He target of about 50% polarization. For a measurement of spin asymmetries, ^3He is effectively an almost pure neutron target since the proton spins are oriented antiparallel. The scattered electrons are detected in two point-focussing spectrometers under 4.5° and 7.5° with respect to the beam axis. The experiment has been approved and is scheduled to take data in the fall of 1992.

2. The SLAC E-143 experiment [39] uses the same spectrometer as E-142, but proposes a more advanced technology for the GaAs source and solid NH_3 and ND_3 targets polarized with the DNP method. The experiment has also been approved and is scheduled to take data in 1993.

3. The HERMES collaboration [40] proposes to use the high intensity internal electron beam of the HERA electron-proton storage ring at DESY, at an energy of 35 GeV. The target proposed for the HERMES experiment is a windowless storage cell fed with polarized hydrogen, deuterium, or ^3He gas. Scattered electrons are detected in a compact, large solid angle forward spectrometer. The experiment relies on sufficient longitudinal polarization of the HERA electron beam which is expected from the Sokolov-Ternov effect [41]. Transverse polarization arises naturally in electron storage

rings from synchrotron radiation and can become large when the machine is designed to minimize depolarizing effects. Transverse beam polarization can be transformed into longitudinal polarization with spin rotators. Final approval and scheduling of the HERMES experiment is subject to the experimental observation of sufficient transverse polarization ($\approx 50\%$) of the HERA electron beam; a polarization of $P_e = (8 \pm 1)\%$ has recently been observed [42].

4. The HELP collaboration [43] proposes to use the polarized internal 45 GeV beam of the LEP electron-positron storage ring at CERN. The proposal foresees a polarized jet gas target and a spectrometer which is very similar to the HERMES apparatus. The expected luminosity is, however, smaller than the one of HERMES by about two orders of magnitude due to the lower intensity of the LEP beams and the lower thickness of the jet target. The experiment is subject to similar uncertainties on the beam polarization as HERMES, and its prospects are unclear at present since its beam requirements are potentially incompatible with the physics program of the four large LEP detectors.

A more detailed comparison of most of the experiments discussed here can be found in ref. [32].

6 Conclusion

The EMC results on the spin structure function of the proton have resuscitated a major interest in the experimental and theoretical study of the internal spin structure of hadrons. A new generation of experiments, exploiting a large variety of different techniques for spectrometers and beam and target polarization, has recently started to collect data. In about five years from now our understanding of spin structure functions, which is scarce and incomplete today, should have substantially improved.

Acknowledgement. I wish to thank K. Jungmann and G. zu Putlitz for organizing a very stimulating meeting, and for their patience in waiting for a very late manuscript. L. Gatignon, T.O. Niinikoski and A. Staude have critically read the manuscript and contributed numerous clarifications. L. Gatignon also has kindly contributed Fig. 1 to this paper.

References

1. See e.g. T. Sloan, G. Smadja and R. Voss, Phys. Rep. 162 (1988) 45.
2. D. Geesaman, these proceedings.
3. R. Clifft and N. Doble, Proposed design of a high-energy, high intensity muon beam for the SPS North Experimental Area, CERN/SPSC/74-12 (CERN/Lab. II/EA/74-2).
4. L. Gatignon, M2 Handbook, CERN, March 1991 (unpublished).
5. L.M. Lederman and M.J. Tannenbaum, Advances in Particle Physics, Vol. I, New York 1968, p. 11.
6. F.-L. Navarria, Rivista Nuovo Cimento 6 (1983) 1.
7. Ch. Iselin, HALO, CERN 74-17 (1974) (CERN Yellow Report).
8. F. Combley and E. Picasso, Phys. Rep. C14 (1974) 20.
9. S.V. Golovkin et al., Nucl. Instr. Meth. 138 (1976) 235.
10. D. Bollini et al., Nuovo Cimento 63A (1981) 441.
11. P.S. Cooper et al., Phys. Rev. Lett. 34 (1975) 1589.
12. P. Schüler, Proc. 8th Int. Symp. on High Energy Spin Physics, Minneapolis 1988.
13. T. Kamae, Y. Shimiziu and M. Igarashi, eds., Report of the TRISTAN ep (ee) Working Group, University of Tokyo preprint UTPN-165 (UT-345) (1980).
14. N.S. Craigie et al., Phys. Rep. 99 (1983) 69.
15. V.W. Hughes and J. Kuti, Ann. Rev. Nucl. Part. Sci. 33 (1983) 611.
16. For an excellent introduction into the Quark-Parton Model, see: F. Halzen and A.D. Martin, Quarks and Leptons. New York, Chichester, Brisbane, Toronto, Singapore: John Wiley & Sons, 1984.
17. R.L. Jaffe, Comments Nucl. Part. Phys. 19 (1990) 239.
18. S. Wandzura and F. Wilczek, Phys. Lett. 72B (1977) 195.
19. H. Burkhardt and W.N. Cottingham, Ann. Phys. (N.Y.) 56 (1970) 453.
20. J.D. Bjorken, Phys. Rev. 148 (1966) 1467; Phys. Rev. D1 (1970) 465; ibid. D1 (1970) 1376.
21. J. Kodaira et al., Phys. Rev. D20 (1979) 627.
22. J. Ellis and R.L. Jaffe, Phys. Rev. D9 (1974) 1444; D10 (1974) 1669.
23. M.G. Doncel and E. de Rafael, Nuovo Cimento 4A (1971) 363; P. Gnädig and F. Niedermayer, Nucl. Phys. B55 (1973) 612.
24. EMC, J. Ashman et al., Phys. Lett. 206B (1988) 364.
25. EMC, J. Ashman et al., Nucl. Phys. B328 (1989) 1.
26. SLAC-Yale E80 Collaboration, M.J. Alguard et al., Phys. Rev. Lett. 37 (1976) 1261; ibid. 41 (1978) 70.
27. SLAC-Yale E180 Collaboration, G. Baum et al., Phys. Rev. Lett. 51 (1983) 1135.
28. SLAC-Yale Collaboration, G. Baum et al., Phys. Rev. Lett. 45 (1980) 2000.
29. M. Bourquin et al., Z. Phys. C21 (1983) 27.
30. M. Glück und E. Reya, Z. Phys. C39 (1988) 569.
31. For recent reviews see e.g.: H. Rollnik, Proc. 9th Int. Symp. on High Energy Spin Physics, Bonn 1990, K.-H. Althoff and W. Meyer, eds., p. 183; E. Reya, Dortmund preprint DO-TH 91/09.
32. K. Rith, Proc. 9th Int. Symp. on High Energy Spin Physics, Bonn 1990, K.-H. Althoff and W. Meyer, eds., p. 198.
33. The Spin Muon Collaboration (SMC), Measurement of the spin-dependent structure functions of the neutron and proton, CERN/SPSC 88-47 (SPSC/P242), 1988.
34. EMC, O.C. Allkofer et al., Nucl. Instr. Meth. 179 (1981) 445.
35. M. Borghini, Phys. Rev. Lett. 20 (1968) 419.
36. T.O. Niinikoski, Nucl. Instr. and Meth. 192 (1982) 151.
37. S.C. Brown et al., in Proc. Proc. 4th Workshop on Polarized Target Materials and Techniques, Bonn 1984, ed. W. Meyer, (Physikalisches Institut, Universität Bonn, Bonn 1984) p. 102.
38. R. Arnold et al., A proposal to measure the neutron spin dependent structure function, SLAC Proposal E-142, 1989.
39. E143 Collaboration, R. Arnold et al., Measurements of the Nucleon Spin structure at SLAC in End Station A.
40. The HERMES Collaboration, A Proposal to Measure the Spin-Dependent Structure Functions of the Neutron and the Proton at HERA, DESY-PRC 90/01, 1990.
41. A.A. Sokolov and I.M. Ternov, Sov. Phys. Dokl. 8 (1964) 1203.
42. R. Brinkmann, Status Report on HERA at the 50th Plenary ECFA Meeting, CERN, Geneva, December 1991.
43. G. Ballocchi et al., CERN/LEPC 89-10 (LEPC/M 88), 1989.

This article was processed using Springer-Verlag TEX Z.Physik C macro package 1991
and the AMS fonts, developed by the American Mathematical Society.

Z. Phys. C - Particles and Fields 56, S179–S185 (1992)

Zeitschrift für Physik C Particles and Fields

Attempts to understand $g_1(x)$ and $g_2(x)$

Andreas Schäfer

Institut für Theoretische Physik, Universität Frankfurt, Germany

7.-10.5.1991

Abstract. The present theoretical understanding of the polarized nucleon structure functions is reviewed. The results of the European Muon Collaboration for $g_1^p(x)$ have generated an enormous theoretical activity, resulting in a large number of possible explanations. The presently available data are not sufficient to decide between them, but much improved experiments are planned for the next years. The possibility of an anomalous gluon contribution is most interesting on theoretical grounds. Finally we discuss $g_2(x)$ which is extremely model dependent and thus could allow a clear decision between the various models proposed. Its measurement requires, however a much improved experimental precision.

1 The data and its interpretation

The EM collaboration measured the relative difference in the polarised muon-proton scattering cross sections between the case that the longitudinal polarisation of muon and proton are the same and the case that they are opposite. This quantity is called $A_p(x)$ and is measured as a function of the Bjorken variable x. The results are shown in Fig.1.

The asymmetry is approximately given by the ratio of the polarised and unpolarised structure functions $2xg_1^p(x)/F_2^p(x)$.

$$A_p(x) = \frac{d^2\sigma/d\Omega dE'(\uparrow\uparrow) - d^2\sigma/d\Omega dE'(\uparrow\downarrow)}{d^2\sigma/d\Omega dE'(\uparrow\uparrow) + d^2\sigma/d\Omega dE'(\uparrow\downarrow)}$$

$$= D\left(\frac{g_1 - \gamma^2 g_2}{F_1} + \eta\gamma\frac{g_1 + g_2}{F_1}\right) \tag{1}$$

D is the depolarization factor of the virtual photon and $\gamma = \sqrt{Q^2}/\nu$. The structure functions $g_{1,2}(x)$ are defined as parametrizations of the nucleon scattering tensor.

$$W_{\mu\nu} = \frac{1}{2\pi}\int d^4x\, e^{iq\cdot x} \;< N(p)|J_\mu(x)J_\nu(0)|N(p) >$$

$$= W_{\mu\nu}^S + iW_{\mu\nu}^A$$

$$W_{\mu\nu}^A = \epsilon_{\mu\nu\lambda\sigma}q^\lambda\left[\frac{s^\sigma}{p\cdot q}g_1(x) + \frac{q\cdot ps^\sigma - q\cdot sp^\sigma}{(q\cdot p)^2}g_2(x)\right] \tag{2}$$

From Equ.(2) one can see why longitudinal electron polarizations are needed to measure g_1 and g_2. Their contribution must be antisymmetric in μ and ν. On the other hand it is proportional to $\sum\epsilon_\mu^*\epsilon_\nu$, where ϵ_μ is the polarisation of the virtual photon. Obviously only transverse polarisation vectors like $\epsilon_\mu \approx (0,1,\pm i,0)$ give rise to such antisymmetric terms.

As $F_1^p(x)$, D, η and γ are all known one can extract $g_1^p(x)$ from A_p in the measured x region. From $g_1^p(x)$ in the measured range it is possible to extrapolate to all x values and thus to give an estimate for the integral

$$\int_0^1 g_1^p(x)\,dx \;=\; 0.123 \pm 0.013 \pm 0.019 \tag{3}$$

The validity of this extrapolation has been questioned [2] and it is indeed unclear whether non-perturbative effects could set in at very small values of x which are missed by any extrapolation. This is one of the reasons why it could be advantageous to concentrate on $g_1^p(x)$ than on its first moment. (The other is that it contains more information.) From the very solid Bjorken sumrule (which relies only on the fact that isospin symmetry holds on the nucleon level and that quarks have the charges $\pm\frac{1}{3}$, $\pm\frac{2}{3}$) and Equ.(?) one gets the first moment of the polarized neutron structure function. Finally adding some knowledge from weak decays one is able to derive values for the spin carried by the different quark flavours in a nucleon.

total spin carried by quarks and antiquarks

$$= 0.060 \pm 0.047 \pm 0.069 \tag{4}$$

spin carried by strange quarks and antiquarks

$$= -0.095 \pm 0.016 \pm 0.023 \tag{5}$$

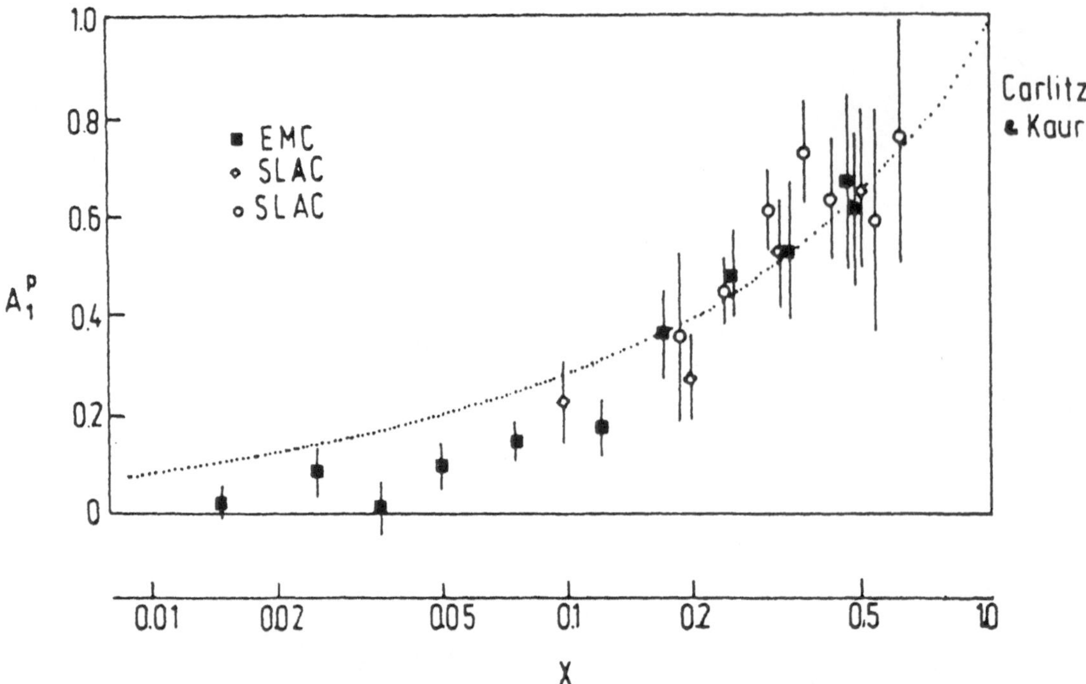

Fig. 1. The proton asymmetry measured by the EM collaboration

While the derivation of these results seems to be straight forward it was argued [3] that there is no compelling reason to assume that isospin symmetry is valid for polarisation phenomena at small values of x, i.e. to assume $(\Delta u(x))_p = (\Delta d(x))_n$ etc.. Keeping these caveats in mind Eqs.(4) and (5) are still most surprising as they imply that the spin-distribution inside of a nucleon differs drastically from what simple valence quark models suggest. In fact the whole theoretical discussion of these results suggests that the polarized structure functions are very sensitive to the precise internal structure of the nucleon, causing a lot of theoretical controversy but also opening most promissing possibilities to understand this structure better.

2 Models for $g_1^p(x)$

The line in Fig.1 gives the prediction of a simple phenomenological model by Carlitz and Kaur published in 1977 [4]. It obviously is already pretty close to the data. Several authors proposed extensions of this model which allow for a satisfactory fit of the new data [3,5]. All of these models use the information on the unpolarized structure functions and add specific assumptions about the spin distribution among the quarks. They are basically valence quark models.

Another class of phenomenological models starts from the observation that in any relativistic bag model the valence quarks have angular momentum [6,7]. In various bag models this was used to explain the small amount of spin carried by quarks.

A subclass of such models are the Skyrme model, non-topological soliton model or chiral bag-model [8,9]. In these models additional scalar and pseudoscalar fields enter, which can carry angular momentum and thus explain why the spin carried by the quarks should be small. The extent to which this happens depends on the details of the models: Topological and non-topological soliton models give different results, it is not clear how the pion field should be treated, and the boosting into the infinite momentum frame of solitons seems to be a problem. Thus these models may point towards the correct qualitative explanation but have difficulties to give reliable quantitative predictions. With all these caviats, such models can, however, give a good description not only of the first moment, but also of $g_1^p(x)$ as a function of x [9].

All these various nucleon bag models share the problem that very similar results can be obtained with completely different models. Also it should be noted that a spin-orbit force is usually needed in such models to get a good description of the hadron spectra. Such a force can, however, completely change the spin-distribution of the quarks.

Thus phenomenological models are capable of fitting the EMC spin data, but the physical interpretation of these data differes from model to model. Furthermore these models allow generally for a rather large number of modifications which alter their predictions for $g_1^p(x)$.

The most widely discussed possibility to interpret the EMC data is that $g_1^p(x)$ is modified by an anomalous contribution due to polarized gluons. It turned out that such a gluonic contribution is extremely sensitive to the precise understanding of QCD. Even slight conceptual

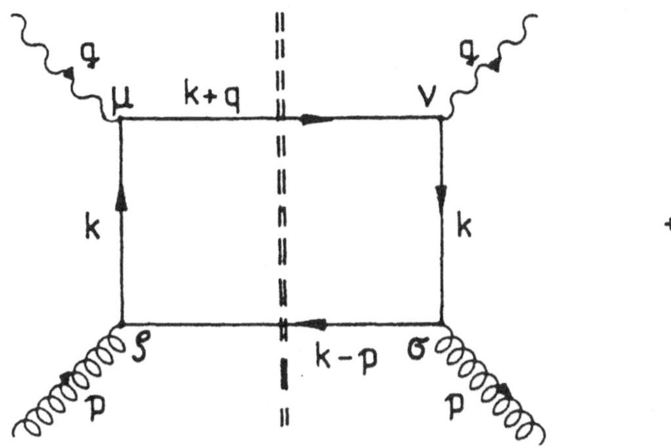

+

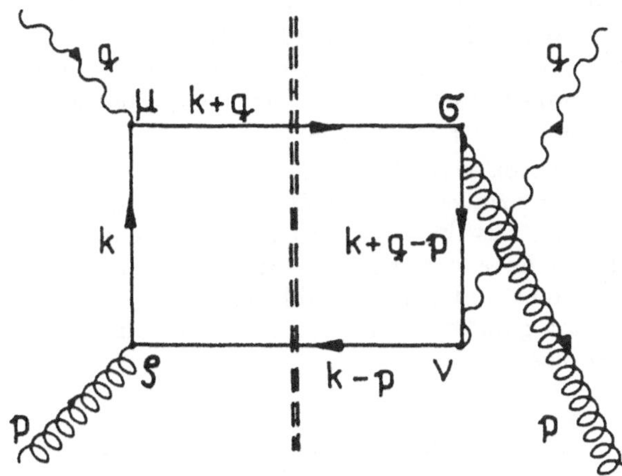

Fig. 2. The perturbative gluon contribution

differences can lead to completely different results. This fact generated rather fierce discussions and thus possibly helped a lot in understanding QCD better. Let me start the discussion with the perturbative hard contribution.

This contribution (Fig.2) turned out to be highly dependent on the kind and ratios of the infrared regulators (finite quark mass, minimal gluon virtuality, minimal transverse momentum etc.) one uses. As the use of infrared regulators is somewhat alien to the general concept of parton dynamics in the infinite momentum frame, several authors have put forward different ideas, leading to different results.

Adopting the usual procedure of the GLAP equations the anomalous contribution is written in the form

$$\Delta g_1^p(x) = < \frac{e^2}{2} > \int_x^1 A(\frac{x}{z}, \frac{Q^2}{\mu_{fac}^2})\, \Delta G(z, \mu_{fac}) \frac{dz}{z}. \quad (6)$$

Here ΔG ist the momentum distribution of polarised gluons and A is the splitting function to be calculated. The problems arising in its calculation are due to the fact that the dominating logarithmic contribution to A does not contribute to its first moment [10]. E.g. Carlitz. Collins and Mueller obtained

$$A(x) = -\frac{\alpha}{2\pi} N_f\, (1-2x) \left[\log \frac{Q^2}{-p^2} - \log x^2 - 2\right] \quad .(7)$$

While the logarithmic term is unique, the finite contributions depend on the regularization scheme adopted. Extreme care is needed when calculating $A(x)$ to keep all terms contributing to the next to leading order. It was shown by Bodwin and Qiu [11] that if the quark mass and gluon virtuality are used as infrared cut-offs the resulting splitting function is identical zero. We showed [12] that introducing an additional cut-off in the transverse momentum leads to a finite contribution for small quark masses, which vanishes if the quark mass becomes very large. (This is in agreement with the original observation by Carlitz, Collins and Mueller.) We also found strong scaling violating effects. Qualitatively similar but quantitatively different results were obtained by Bass, Nikolaev and Thomas [13]. The main problems of this interpretation are that the gluon spin must be as large as 10 $\hbar$, that strong scaling violation is predicted but not observed, and that it is hard to get an acceptable fit to $g_1^p(x)$ as a function of x [13,14].

It was argued by several authors [15,16] that a non-perturbative gluon contribution can exist. While this is possible in principle it is improbable that it could explain the data as it should occure only for very small x values. Its possible existence points, however, to the fact that the extrapolation of $g_1^p(x)$ to small x values might be problematic. So far nobody was able to actually calculate any non-perturbative gluon contribution to $g_1^p(x)$.

A very severe objection to the whole concept of an anomalous gluon contribution was raised early on by Jaffe and Manohar [6], namely that there is no corresponding gauge-invariant local operator in the operator product expansion. This observation suggests that there cannot be any genuine point-like photon-gluon coupling and thus the gluon contribution could be rewritten in terms of a coupling to quark currents. To circumvent this argument it was suggested that the gauge non-invariance might be unimportant [16] and non-local operators were tried [17]. My personal understanding of this point is the following. For finite Q^2 the contribution resulting from Fig.2 is not the anomaly. The loop moment is bound by Q^2 and thus the contribution is not truely point-like. This could be the reason for the strong scaling violation, and is in accordance with the observation by Jaffe. Only for $Q^2 \to \infty$ does the graph collapse to the anomaly. Therefore it is in principle possible to resolve the gluon coupling into a polarised sea quark contribution. However, as was pointed out by Al. Mueller, such a procedure is not reasonable. The quark loop containes quarks with very large momentum which cannot be included naturally into the sea quark distribution. It is much more sensible to speak of this contribution as due to gluons and to make model assumptions for the distribution function of polarized gluons, i.e. to interpret it as anomaly plus infra-red corrections than as due to a crazy sea-quark component.

Attempt were also made to tie the divergence of the isoscalar axialvector current to the η' in the same sense as PCAC ties the isovector current to the pion. Different authors have derived completely diffent results along these lines and it is probably fair to say that PCAC is just not good enough a symmetry in this case to be of much use.

3 The physical meaning of $g_2(x)$

The primary aim of future experiments will be to measure with high precission the polarized nucleon structure functions $g_1(x)$. To achieve this the second polarized structure function $g_2(x)$ has also to be known with some accuracy (see Equ.(1)). It can be determined from a comparison of the results for longitudionally polarized electrons scattering off either longitudionally or transversely polarized nuclei. This perspective of a rough experimental determination challenged theoreticians to analyse the properties of $g_2(x)$ and to make model-based predictions.

The most crucial y of $g_2(x)$ is that it contains important contributions from higher twist. This fact makes any predictions of the exact form of g_2 as a function of x very difficult. On the other hand g_2 offers the only chance known so far to measure isolated higher twist effects directly. Because of this complication $g_2(x)$ can be analysed only in a rather formal manner, using the language of operator product expansion. Before we review this analyses let us, however, give a intuitive, physical argument for the physical content of $g_2(x)$.

$g_2^p(x)$ parametrizes the difference in the cross-sections for a longitudionally respectively transversely polarized proton. If a probabilistic interpretation would be applicable (as it is for the leading twist contribution) this difference would be proportional to the difference between the probability to find a transversely polarized quark in a transversely polarized proton and the probability to find a longitudionally polarized quark in a longitudionally polarized proton. For free particles this difference should obviously vanish and indeed $g_2(x) = 0$ for free quarks. Thus $g_2(x)$ is exclusively determined by interaction effects. One such effect is the following. The structure functions are defined in the infinite momentum frame and describe e.g. the boosted proton. For an interacting field theory a boost changes the particle content. If this change depends on the polarization of the proton relative to the boost direction this leads to a contribution to $g_2(x)$. From this it should already be clear that $g_2(x)$ is much more sensitive to subtle effects than $g_1(x)$ or the unpolarized structure functions. It will even become clear that nuclear factorisation might not apply at all.

It should be kept in mind that the properties of $g_2(x)$ have some important experimental consequences. If one would conclude e.g. that the nuclear factorisation (i.e. the usual treatment of the nuclear binding effects on the distribution functions) does not apply for $g_2(x)$ one would have to measure $g_2(x)$ separately for the p,d and ^3He target. Otherwise a measurement for e.g. p and d would be sufficient.

To obtain the OPE one investigates the virtual photon forward Compton amplitude [19,20]

$$T_{\mu\nu} = i\int d^4x\, e^{iq\cdot x} \;< N(p)|T(J_\mu(x)J_\nu(0))\,|N(p)>$$

$$= T^S_{\mu\nu} \;+\; iT^A_{\mu\nu}$$

$$T^A_{\mu\nu} = i\epsilon_{\mu\nu\lambda\sigma}q^\lambda\left[\frac{s^\sigma}{p\cdot q}\alpha_1(x) + \frac{q\cdot ps^\sigma - q\cdot sp^\sigma}{(q\cdot p)^2}\alpha_2(x)\right] \quad (8)$$

The time ordered product of currents gives the propagator which is expanded in the usual manner.

$$T^A_{\mu\nu} = i\epsilon_{\mu\nu\lambda\sigma}\, q^\lambda \sum_{n=0,2,4,\ldots}\left(-\frac{2}{q^2}\right)^{2+1} q_{\mu_1}\ldots q_{\mu_n}$$

$$< PS|\Theta^{\sigma\{\mu_1\ldots\mu_n\}}|PS>$$

$$\Theta^{\sigma\{\mu_1\ldots\mu_n\}} = i^n\bar\psi\gamma^\sigma\gamma^5 D^{\{\mu_1}D^{\mu_2}\ldots D^{\mu_n\}}\psi$$

$$-\; traces \;+\; O(twist\ 4) \quad (9)$$

Θ has a mixed symmetry, it is symmetrized in all but the first indices, and contains twist 2 and twist 3 contributions. The twist 2 contribution is completely symmetric in all indices and can be extracted according to

$$\Theta^{\sigma\{\mu_1\ldots\mu_n\}} = \Theta^{\{\sigma\mu_1\ldots\mu_n\}} + \Theta^{[\sigma\{\mu_1]\ldots\mu_n\}} \quad (10)$$

$$< PS|\Theta^{[\sigma\{\mu_1]\ldots\mu_n\}}|PS>$$

$$= \frac{d_n}{n+1}\left[(s^\sigma p^{\mu_1} - s^{\mu_1}p^\sigma)p^{\mu_2}\ldots p^{\mu_n} + \ldots \;-\; traces\right]$$

$$< PS|\Theta^{\{\sigma\mu_1\ldots\mu_n\}}|PS>$$

$$= \frac{a_n}{n+1}\left[s^\sigma p^{\mu_1}p^{\mu_2}\ldots p^{\mu_n} + \ldots \;-\; traces\right] \quad (11)$$

Here $\Theta^{[\sigma\{\mu_1]\ldots\mu_n\}}$ is symmetric in all μ's and antisymmetric under exchange of σ with one μ index. As the total deep inelastic cross section is proportional to the imaginary part of the forward Compton scattering amplitude α_1, α_2 and g_1, g_2 are connected by a dispersion relation

$$\alpha_1(x,Q^2) = \frac{4}{x}\sum_{n=0,2,4,\ldots}\frac{1}{x^n}\int_0^1 dy y^n g_1(y,Q^2)$$

$$\alpha_2(x,Q^2) = \frac{4}{x^3}\sum_{n=0,2,4,\ldots}\frac{1}{x^n}\int_0^1 dy y^{n+2} g_2(y,Q^2) \quad (12)$$

$$\int_0^1 dy y^n g_1(y, Q^2) = \frac{a_n}{4} \qquad n = 0, 2, 4, \ldots \qquad (13)$$

$$\int_0^1 dy y^n g_2(y, Q^2) = \frac{1}{4}\frac{n}{n+1}(d_n - a_n)$$

$$n = 2, 4, \ldots \qquad (14)$$

Thus the moments of g_1 are determined exclusively by the twist 2 matrixelements a_n, while g_2 has a leading twist 3 contribution. This is the reason why it is so difficult to make reliable predictions for g_2. As the twist 3 contribution is hard to be calculated one might hope that it is negligable. Under this assumption all the moments of g_2 are determined by the moments of g_1 and both functions are related [21]

$$g_2^{WW}(x, Q^2) = -g_1(x, Q^2) + \int_0^1 \frac{dy}{y} g_1(y, Q^2) \qquad (15)$$

However, this approximation is not justified. Jaffe and Ji have shown that e.g. for the bag model g_2 calculated from

$$g_1(x) + g_2(x) = \frac{1}{8\pi\sqrt{2}} \int d\xi^- \; e^{iq^+\xi^-}$$

$$< PS^x|\bar{\psi}(\xi^-)\; Q_{ch}^2 \gamma^1\gamma^5\psi(0) + \ldots|PS^x > \qquad (16)$$

differs substantially from the Wandzura-Wilczek prediction, see Fig. 3. Other models give different predictions (Fig.4,5). However, none of these predictions is reliable. E.g. in the bag model all correlators of two quark and one gluon operator (see Eq.(19)) are substituted by the bag-boundary condition. This crude approximation works quite well for leading twist contributions but it cannot describe higher twist contributions correctly. On the other hand light-cone wavefunctions as used by Dziembowski [27] (Fig.4) contain in principle all higher twist terms correctly, but up to now their precise form is still not sufficiently well known.

As $g_2(x)$ is zero in the limit of infinitely heavy quark masses all higher twist corrections which are due to finite quark masses should cancel part of the twist two contribution, i.e. should have the oposite sign for most x values. For the models leading to Fig.4 and 5 this is obviously the case. In these models all higher twist effects depend on to the finite quark masses and would vanish for zero quark masses. For the MIT bag model a similar cancelation occures, however, in this case it is not clear whether the bag boundary condition can be related to any effetive mass term. Most people in the field think that twist 2 and twist 3 parts should allways have different signs, but we are not aware of any formal proof for this assumption.

Thus not much more is known about $g_2(x)$ than that its absolut value should be comparable and probably somewhat smaller than the pure twist 2 contribution from Eq(9). A precise measurement of g_2 would be very helpfull in determining the nucleon wave function at least in the long run but even a rough estimate which is all one can hope for with the limited precission of experiments feasible in the near future could rule out some model predictions. As the value of g_2 for very small x is very sensitive to the specific model, hotoproduction experiments ($x = 0$), if feasible, could give valuable additional information.

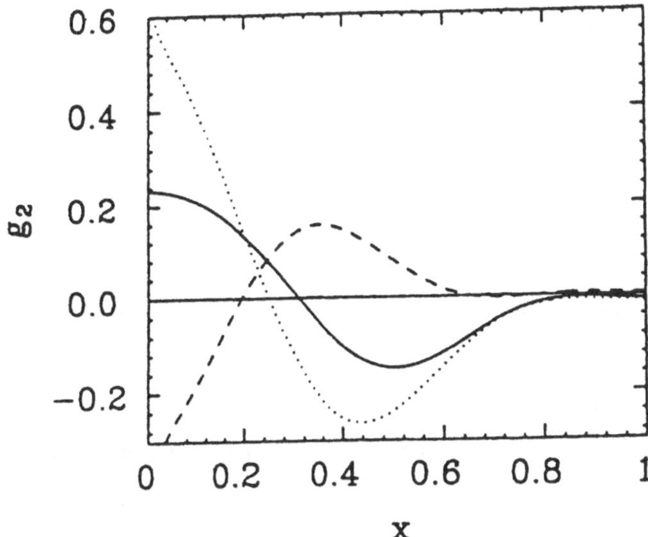

Fig. 3. $g_2(x)$ in the MIT bag model [22]. The dotted line is the Wandzura-Wilczek contribution of twist 2 (Eq.(9)). The solid line was calculated from Eq.(10). The dashed line is the difference to be identified with the twist 3 contribution.

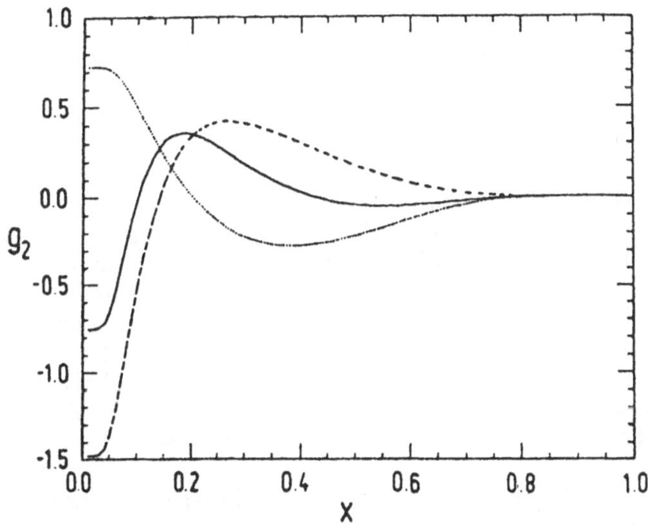

Fig. 4. Same as figure 1 for the light-cone quark model [27]. The predictions are obviously completely different for small x.

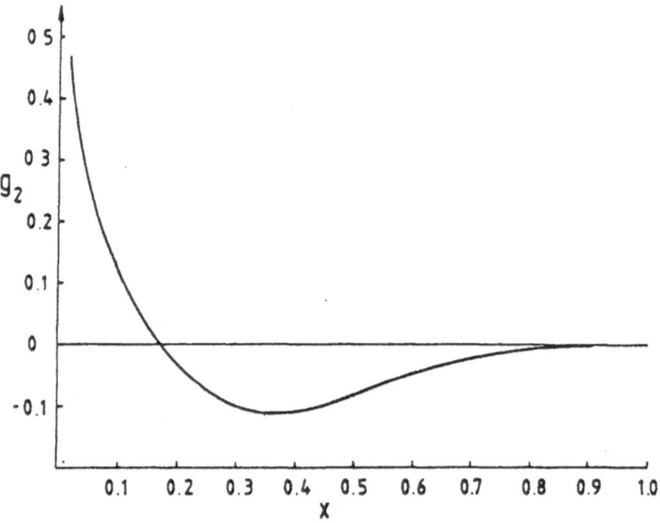

Fig. 5. $g_2(x)$ in a simple phenomenological model with only small higher twist contributions [28].

The predictions for $g_2(x)$ in Figures 3,4 and 5 fullfill the so called Burkhardt-Cottingham sumrule (BCSR) [22].

$$\int_0^1 dx\; g_2(x) \;=\; 0 \qquad (17)$$

This sumrule is also suggested by Eq.(8). While this equation is only valid for even values of n larger than zero one can think of continuing it analytically to $n = 0$. Although this continuation is not unique the ambiguity could be irrelevant. Two functions with the same higher moments can only differ by a contribution proportional to $\delta(x)$, which is unobservable in deep inelastic scattering anyway (though it would be seen in photoproduction). Higher twist contribution could give a small finite contribution to the BCSR for finite Q^2, corresponding to washing out this singular contribution to some extend, but still it seems somewhat unphysical and thus is assumed by most people to be absent. In the following we assume that the naive extrapolation to $n = 0$ is valid. Then the BCSR should hold if d_n stays finite for $n \to 0$ (a_0 is finite).

If the matrixelement d_0 is analysed in in the light-cone formalism [23] one finds not surprisingly that it can only diverge for $x \to \infty$, i.e. if the hit polarized quark has vanishing momentum fraction. More precisely the two relevant contributions can be written in the form

$$\lim_{n\to 0}\; n\; < PS_x|\frac{k_T^x}{2M_N x}\; x^n(\delta_{\uparrow\uparrow} - \delta_{\downarrow\downarrow})|PS_x > \qquad (18)$$

and

$$\lim_{n\to 0}\; n\; < PS_x|\frac{g}{M_N}\;\frac{\bar{Q}\gamma^+\gamma^5 G_x Q}{\sqrt{x_y}\,|y-x|}\;\frac{\log x/y}{y-x}|PS_x > \qquad (19)$$

The first term depends on the transverse momentum and is thus related to the angular momentum of the struck quark. The second term contains the coupling to a gluon. It can be shown, that soft gluons ($y - x \to 0$) nor soft outgoing quarks ($y \to 0$) can lead to divergences. Thus only the limit $x \to 0$ can lead to any divergence in any of the two relevant expressions. Thus whenever there is no substantial sea-polarization for $x \to 0$ BCSR must hold, as is the case for e.g the bag model. The fact that model calculations respect the BCSR is thus usually a direct consequence of the model assumptions and could be misleading.

The validity of the BCSR can also be analysed in terms of Regge theory. It turns out that in this language BCSR holds unless there is a J=0 fixed pole in the Compton amplitude. Whether this can be the case or not seems to be a question of debate among specialists. As said before the crucial question is whether the relevant matrix elements diverge as $1/x$ or not. One can show that in perturbation theory this does not happen for QCD. (It does e.g. for a scalar analog of QCD.) $1/x$ singularities can, however, also appear on the constituent quark level and it does not seem to be possible to decide unambiguously whether it does or not. A general problem of this approach is also that many arguments apply strictly only in the $Q^2 \to \infty$ limit such that it is not clear whether finite Q^2 corrections are important for realistic values.

This brings us to the more general question of Q^2 evolution for $g_2(x)$. Experiments at HERA like HERMES can measure the polarized structure functions only in a rather limited Q^2 range, depending on x. Thus for a very high precission comparison the precise Q^2 evolution of g_2 becomes important. In practice the experimental precission necessairy to observe this evolution will certainly not be reached in the near future. Still it is a very interesting theoretical problem. Ali, Braun and Hiller have recently presented a comprehensive analyses of it [24,25].

It was noted already some time ago that the Q^2 evolution of higher twist contributions can in general not be described by Gribov-Lipatov-Altarelli-Parisi type equations. Instead various contributions are mixed [24]. This holds also for the twist 3 part of g_2. Higher and higher moments get contributions from more and more different matrixelements.

The fact that g_2 contains a leading higher twist contribution, which has led to all the problems discussed so far, can still lead to another qualitative difference to the situation for g_1. For g_2 simple factorization of nuclear effects does not apply and thus it is not clear how large nuclear effects will be, i.e. it is possible that g_2^n determined for ^3He is quite different from that obtained from experiments with deuterium. Such nuclear effects would be analogous to the first EMC effect, namely the nuclear dependence of the unpolarised structure functions. However, as $g_2(x)$ is so strongly dependent on higher twist effects its nuclear dependence could be much more pronounced than that of $F_2(x)$. This fact implies also, that to constrain g_2 sufficiently to reduce the error on g_1 it should be measured for each nucleus separately. The size of these nuclear effects has not yet been determined theoretically but at least a formalism was developped [26]. Unless the nuclear effects are surprisingly large HERMES will not reach the precission to actually measure them. One reason why these effects could be large is the following: One part of the twist 3 contribution depends on the perpendicular momentum of the struck quark, which in turn is related to its angular momentum, see Eq.(12). As factorization does not apply g_2 could depend on the total hadronic angular momentum. For ^3He one has a strong D state admixture and thus one could hope for substantial nuclear effects.

4 Conclusions

Measurements of the spin structure functions play an important role in improving our understanding of the internal structure of the nucleons. The theoretical discussion of the last years has already lead to important new results. Still, without new, more precise experiments a decission between the various interpretations is not possible. Thus one of the major tasks at the moment is to find additional ways of extracting important information (e.g. Drell-Yan experiments, semi-inclusive experiments etc.). If such high precission data will become available they could become one of the most sensitive ways to investigate thge internal structure of the nucleons.

References

1. J. Ashman et al., EMC, Phys.Lett. 206B (1988) 364, Nucl.Phys. 328B (1989) 1
2. F.E. Close and R.G. Roberts, Phys.Rev.Lett. 60 (1988) 1471
3. A. Schäfer, Phys.Lett. 208B (1988) 175
4. R. Carlitz and J. Kaur, Phys.Rev.Lett. 37 (1976) 673
   J. Kaur, Nucl.Phys. 128B (1977) 219
5. D.J.E. Callaway and S.D. Ellis, Phys.Rev. 29D (1984) 567
   A. Giannelli, L. Nitti, G. Preparata, and P. Sforza, Phys.Lett. 150B (1985) 214
   R.M. Wolshyn, Nucl.Phys. 496A (1989) 749
6. R.L. Jaffe and A. Manohar, MIT preprint CTP no. 1706
7. P.Ratcliffe, Phys.Lett. 192B (1987) 180
   J. Stern and G. Clement, Phys.Lett. 231B (1989) 471
   A.Abbas, J.Phys. 15G (1989) L73
8. S.J. Brodsky, J.Ellis, M. Karliner, Phys.Lett. 206B (1988) 309
   Z.Ryzak, Phys. Lett. 217B (1989) 325
   H.Dreiner, J.Ellis, R.A. Flores, Phys.Lett. 221B (1989) 167
   T.D. Cohen and M.K. Banerjee, Phys.Lett. 230B (1989) 129
9. C.J. Benesh and G.A. Miller, Phys.Lett. 222B (1989)476
10. A.V. Efremov and O.V. Teryaev, Dubna rep. E2-88-287 (1988)
    G. Altarelli and G.G. Ross, Phys.Lett. 212B (1988) 391
    R.D. Carlitz, J.C. Collins, and A.H. Mueller, Phys.Lett. 214B (1988) 229
11. G.T. Bodwin and J. Qiu, Argonne preprint ANL-HEP-PR-89-83
12. L. Mankiewicz and A. Schäfer, Phys. Lett. 242B (1990) 455
13. S.D. Bass, N.N. Nikolaev, and A.W. Thomas, Adelaide Uni preprint ADP-133/180 (1990)
    J. Ellis, M. Karliner, and C. Sachrajda, Phys.Lett. 231B (1989) 497
    A. Schäfer, J.Phys.G16 (1990) L121
14. G. Altarelli and W.J. Stirling, Particle World 1 (1989) 40
15. S. Forte, Phys. Lett. 224B (1989) 189
16. Z.Ryzak, Harvard preprint HUTP-90/B003
17. A.S. Gorsky and B.L. Ioffe, Particle World 1 (1990) 114
18. T.P. Cheng and L.F. Li, Phys.Rev.Lett. 62 (1989) 1441
    H. Fritzsch, Phys.Lett. 229B (1989) 122
    A.V. Efremov, J. Soffer, N.A. Törnqvist, Phys.Rev.Lett. 64 (1990) 1495
    U. Ellwanger and B. Stech, Phys. Lett. 241B (1990) 409
    X. Ji, MIT preprint, CTP no. 1868 (1990)
19. E.V. Shuryak and A.I. Vainshtein, Nucl.Phys.B201(1982)141
    A.P. Bukhroster, E.A. Kureav and L.N. Lipatov, Pis'ma Zh. Exsp. Teor. Fiz. 37(1983)406
    A.V. Efremov and O.V. Teryaev, Phys.Lett. 200B(1988)363
20. R.L. Jaffe, Comm. Nucl. Part. Phys. 19(1990)239
21. W. Wandzura and F. Wilczek, Phys. Lett. 172B(1977)195
22. H. Burkhardt and W.N. Cottingham, Ann.Phys. (NY) 56 (1970)453
23. L.Mankiewicz and A. Schäfer, Phys. Lett. B
24. A. Ali, V.M. Braun, and G. Hiller, Phys.Lett. B266(1991)117
25. L. Mankiewicz and A. Schäfer, to be published
26. R.L. Jaffe and X. Ji, Phys.Rev. D43(1991)733
27. Z. Dziembowski and J. Franklin, Phys.Rev. D42(1990)905 and references therein. The calculations for $g_2(x)$ are unpublished.

This article was processed using Springer-Verlag TEX Z.Physik C macro package 1991
and the AMS fonts, developed by the American Mathematical Society.

Z. Phys. C - Particles and Fields 56, S186-S193 (1992)

Zeitschrift für Physik C Particles and Fields

Medium-energy neutrino physics

D. Hywel White

Medium Energy Physics Division, Los Alamos National Laboratory
Los Alamos, New Mexico 87545

Submitted 22 November 1991

Abstract. A selection of opportunities for neutrino physics from a source generated from pions that decay in flight and at rest is described. The present source at LAMPF has a duty factor of about 6%; improvements in opportunities that emerge from a source using a pulse 0.25 μsec long from a proton storage ring are also described.

In Table I is shown a list of the topics that we have considered in planning a program of low-energy neutrino physics at LAMPF. Neutrinos at LAMPF come from two basic sources, charged pions that decay at rest and those that decay in flight. There are no other particles that are expected to produce neutrinos. Pions are produced in a combination of water and copper targets and subsequently come to rest in materials surrounding the target where they decay to muons and muon neutrinos with a lifetime of 26 ns. In turn, muons decay to electron and muon neutrinos. In practice at LAMPF, positive pions are produced more copiously (~6x) and negative pions are mostly absorbed, so that neutrinos are produced dominantly from the reactions

$$\pi^+ \to \mu^+ + \nu_\mu \, , \; \mu^+ \to e^+ + \nu_e + \bar{\nu}_\mu \; . \qquad (1)$$

Notice that at this level of precision $\bar{\nu}_e$ are not produced. Pions can decay in flight before coming to rest and the rate depends in detail on the structure of the target area. In a solid target, rather less than 1% of the pions decay in flight, up to 30% of the pions decay in flight in an open region, which is optimized to produce neutrino flux at a detector. When the decay-in-flight component is significant, muons produced in the primary decay process also decay but at a lower rate, reflecting the muon lifetime. In the decay-in-flight component of the beam, π^- are also present so that $\bar{\nu}_\mu$ exist in this component of the beam although their energy is higher than the decay at rest spectrum. Energy spectra for both components in the beam are shown in Fig. 1.

a. Neutrino-electron scattering

- * i. $\sin^2\theta_W$ (and search for 17-keV neutrino)
- ii. ν electromagnetic properties
- iii. very low energy recoil, Magnetic Moment

* b. Neutrino-proton scattering

c. Inverse beta decay

- i. $\nu_e - {}^{12}C \to e^- + {}^{12}N$ (form factor and normalization)
- ii. $\nu_\mu - {}^{12}C \to \mu^- + {}^{12}N$

d. Threshold cross-section measurement, G_p

$$\nu_\mu - p \to \mu^+ + n$$

* e. Neutrino oscillations

- i. ν_μ disappearance
- ii. ν_e disappearance
- iii. $\nu_\mu \to \nu_e$
- iv. $\nu_\mu \to \bar{\nu}_e$
- v. Long baseline disappearance

f. ν - A coherent scattering (bolometric detectors?)

g. $\nu - {}^{12}C$ neutral-current inelastic scattering (15.1 MeV)

Table I. In this paper we consider only the topics that are marked with an asterisk.

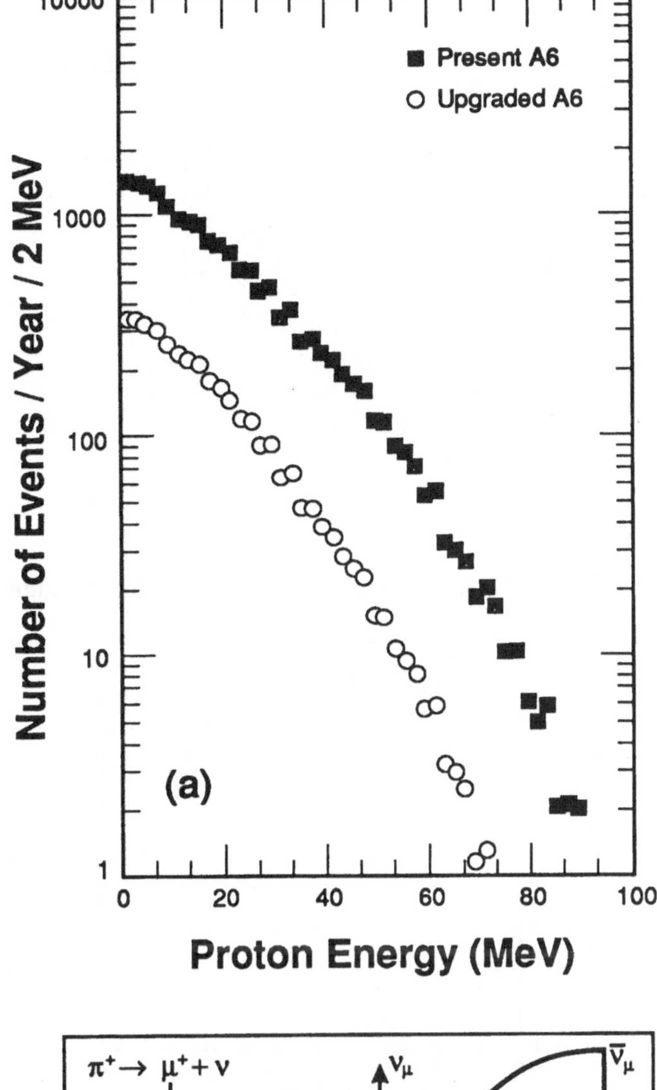

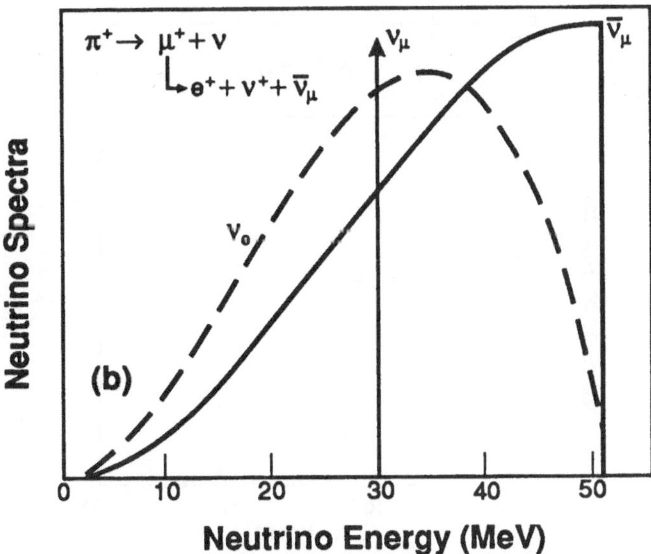

Figure 1. Energy spectra for decay-in-flight neutrinos (ν_μ), and decay at rest (ν_μ, ν_e, $\overline{\nu}_\mu$) are shown. The decay in flight neutrino spectra are for the source as configured at present and for the projected source with improved flux.

Neutrino events at low energy are characterized generally by simple topologies. At the same time, the cross sections are low, even by neutrino standards, so that massive detectors are very much at a premium. We have developed a technique that extends the successes of the Kamioka collaboration in detecting low-energy neutrino events. The essence of the method consists of a homogeneous transparent liquid viewed by a large number of phototubes mounted at the periphery of a containment vessel . A schematic of the detector tank and phototube disposition in the liquid scintillator neutrino detector (LSND) is shown in Fig. 2. These phototubes are capable of detecting single photoelectrons with high efficiency and good timing. Such phototubes have become available from a number of sources in the last few years. Water has been used previously as target medium and radiator for these low-energy neutrino detectors. Combining target mass and radiating medium yields advantages of cost and acceptance. The LSND collaboration is using very pure mineral oil, which has the advantage of lower density, Z, and higher refractive index, low water, reducing multiple scattering and increasing the Cherenkov photon yield. In addition, we have added a small quantity of a scintillating compound so that nonrelativistic particles may also be detected. For most events, electrons only contribute to detectable Cherenkov light and protons produce scintillation light. Pions and muons are mostly nonrelativistic except for cosmic radiation.

It is widely believed that Cherenkov light cannot be observed in liquid scintillator because of the short absorption length that arises when the scintillant and wavelength shifters are added in sufficient quantity to make an efficient scintillator. However, when sufficient phototube coverage is available and a very dilute scintillator is used, light from both sources may be seen simultaneously. In Fig. 3 is shown an angular distribution of light observed from 600-MeV/c electrons in the LAMPF test beam. When ~ 60 ppm of butyl PPD is added, the same angular distribution shows a Cherenkov peak of the same magnitude together with scintillator light giving a uniform angular distribution. This amount of additive gives a desirable ratio of scintillation to Cherenkov light. The absorption length at ~ 400 nm is also adequate.

It has been shown by the Kamioka experiment [1] that with adequate phototube coverage, low-energy electrons may be reconstructed in energy and direction to a precision limited only by multiple scattering and total light acceptance. Expected energy, angle, and position resolutions are shown in Fig. 4 for LSND. Electron-proton separation is also expected to be better than 10^{-3} at about 30 MeV. Some reactions of interest involve low-energy neutrons, which we expected to thermalize in about 200 μsec, giving a 2.2-MeV γ on capture by free protons. The detector is expected to be sensitive to these γs, including the capability to reconstruct the capture position.

The detector is located about 25 m from the beam stop at LAMPF in a shielded area shown in Fig. 5. An

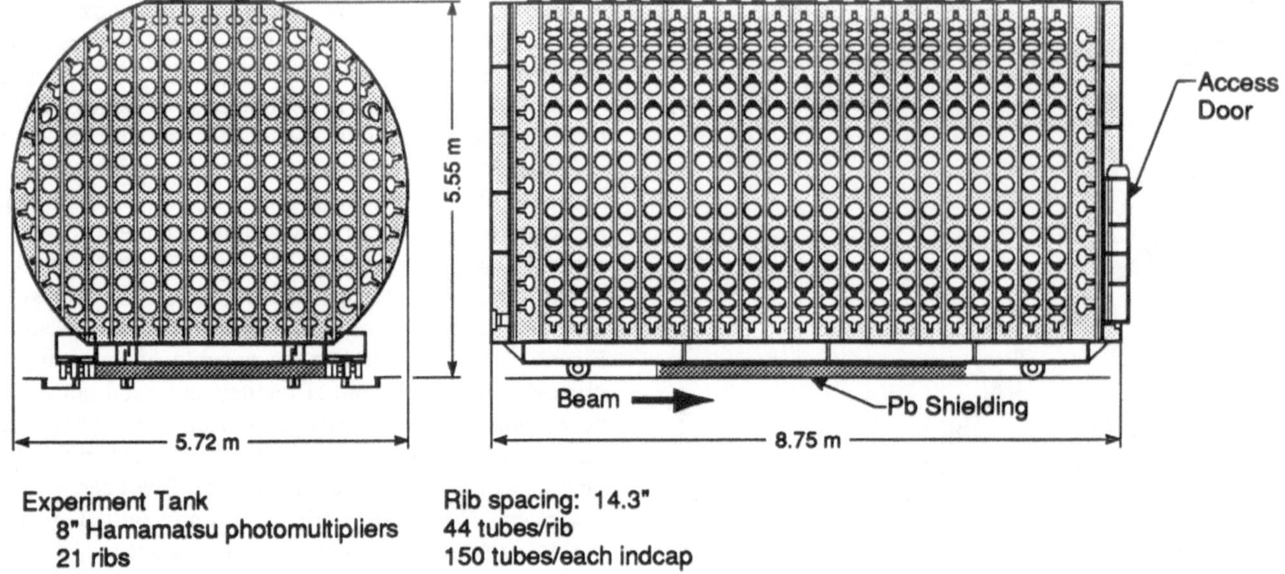

Figure 2. Schematic diagram of the Liquid Scintillator Neutrino Detector (LSND). The tank contains a dilute liquid scintillator viewed by 1250 phototubes.

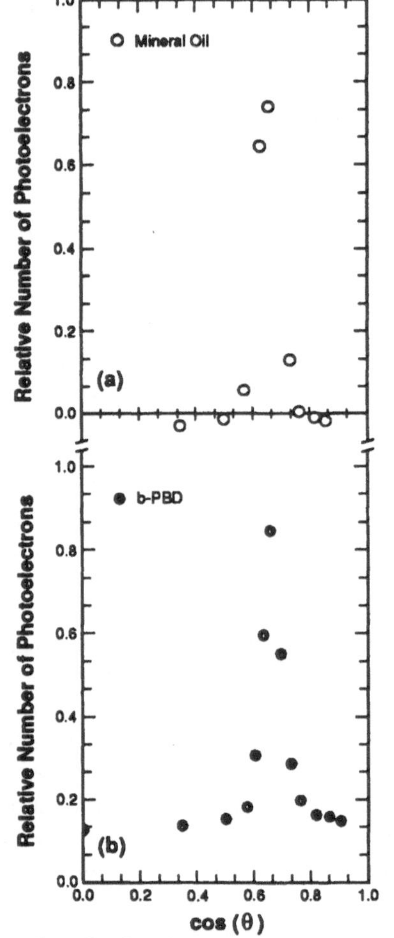

Figure 3. Angular distributin of light from a dilute mixture of mineral oil and butyl PBD. This mixture allows the observation of Cherenkov and Scintillator light simultaneously, a requirement for LSND.

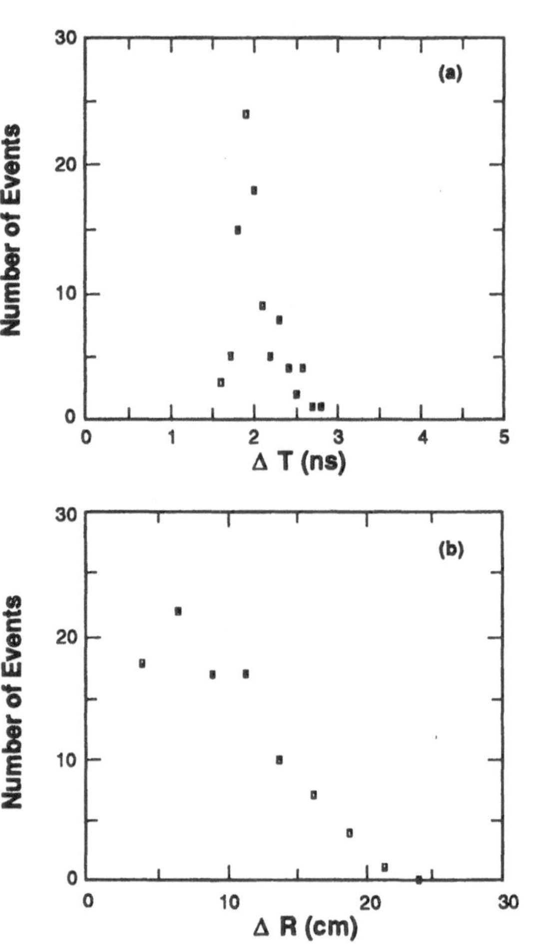

Figure 4. Resolution of the detector in time and position of the event vertex.

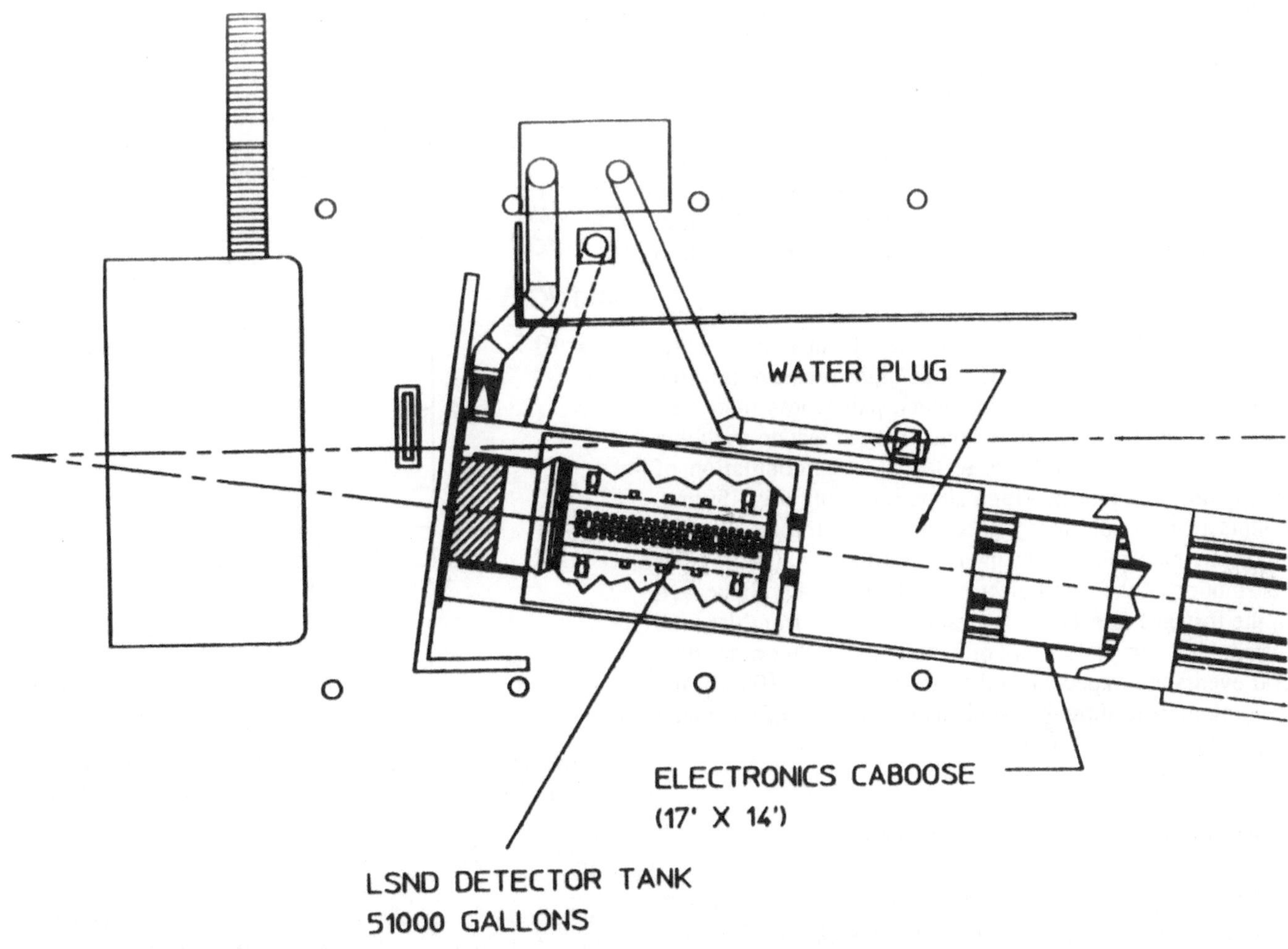

Figure 5. Plan of the experimental area in which the detector is located.

overburden of about 2.5 kg/cm^2 serves to filter out the hadronic component of cosmic rays. Cosmic rays that traverse the detector are excluded in a trigger, cosmic rays which stop in the target medium are likely to decay giving an electron which can mimic a neutrino interaction unless a correlation is made both in the position of the muon track end and the two microseconds characteristic of muon decay. A primary source of background for low-energy neutrino events comes from those muons that stop close to the detector in nearby material and either decay or are captured. Decay muons generate breamsstrahlung photons from electrons and captured μ^- generate neutrons , both of which may penetrate the shielding into the detector. The closest layer of material always becomes the primary source of these events, so that adding more shielding does not help. In order to remove this background, it is necessary to veto cosmic-ray muons in a counter outside a shielding layer. Such a system, in which liquid scintillator is used outside a thickness of lead to veto this secondary background, has been used successfully in a previous experiment at LAMPF and will be reused.

Neutrino-Electron Scattering

The ratio of neutrino electron scattering cross sections,

$$R = \frac{\sigma(\nu_\mu e)}{\sigma(\nu_e e) + \sigma(\bar{\nu}_\mu e)} ,$$

is sensitive to the weak mixing angle $\sin^2\theta_W = s$ through the relation,

$$R = \frac{3}{4}\frac{1 - 4s^2 + 16/3s^4}{1 + 2s^2 + 8s^4} = 0.144 ,$$

at the tree level in standard electroweak theory.

A measurement of R to 2% yields a measure of $\sin^2\theta_W$ to about 1% because of a favorable derivative between R and $\sin^2\theta_W$. Extraordinarily precise measurements of the Z mass at LEP, have concentrated interest on the effect of radiative corrections, both within and outside of the standard model. These radiative corrections depend on the overall structure of the standard model, for example the magnitude of the top quark mass has a major impact on radiative corrections to the Z mass. In the ratio R the effects on numerator and denominator largely cancel, leaving R rather insensitive to the top quark mass so that different measures of $\sin^2\theta_W$ through different processes can provide a critical test of the overall structure of the theory. Conversely, a Z' would affect the numerator, leaving the denominator, which is dominated by charged current amplitudes, unchanged. This ratio, along with other measurements like high-energy neutrino neutral current interactions and parity violation in atoms, forms an array of critical tests of the standard electroweak theory and its constituents.

Neutrinos that contribute to the numerator come from pion decay; those in the denominator from muon decay. Pion decay neutrinos that decay in flight preserve the fine time structure of the LAMPF beam, pulses less than 1 ns wide separated by 5 ns. Pions that decay at rest do so with a lifetime of 25 ns, which is reflected in a modulation of events between pulses with a slope reflecting this lifetime. Events from muon decay do so nearly uniformly in time. Neutrino electron scattering is identified by the angular distribution of the recoil electrons, which are sharply peaked in the forward direction. The ratio is derived from the time distribution. In three years of data taking, rather more than 500 events are expected in the numerator with 1000 in the denominator, leading to a statistical error on R of 6% and a precision of $\sin^2\theta_W$ of ± 0.006.

Neutrino-Proton Elastic Scattering

Neutrino-proton scattering is identified by a recoil particle in which no Cherenkov light is observed and no decay products correlated in time. The direction of this recoil cannot be determined, but multiple interactions such as might be observed from neutrons can be excluded. The neutron background in the experiment location has been measured and a signal to background in excess of three to one is expected after the time correlation of the neutrino events with the beam line structure has been used. The cross section for ν_μ elastic scattering on protons is given by

$$\frac{d\sigma}{dQ^2} \approx A \pm BW/m_p^2 + CW^2/m_p^4 , \qquad (2)$$

with $W = Q^2 + 2\, m_p E_\nu$.

The three terms

$$A = Q^2\, (\, G_A{}^2 - F_1{}^2 + F_2{}^2\, Q^2 / 4m_p{}^2\,)$$

$$B = Q^2\, G_A\, (\, F_1 + F_2\,)$$

$$C = 0.25\, (\, G_A{}^2 + F_1{}^2 + F_2{}^2\, Q^2 / 4m_p{}^2\,) \; .$$

The differential cross section for selected ν energies is shown in Fig. 6. This cross section becomes particularly simple in form at low Q^2 because $F_1(0) = 1 - 4\sin^2\theta_W$ is small and the F_2 terms involve a power of Q^2 and are also not important. The cross section is dominated by the axial

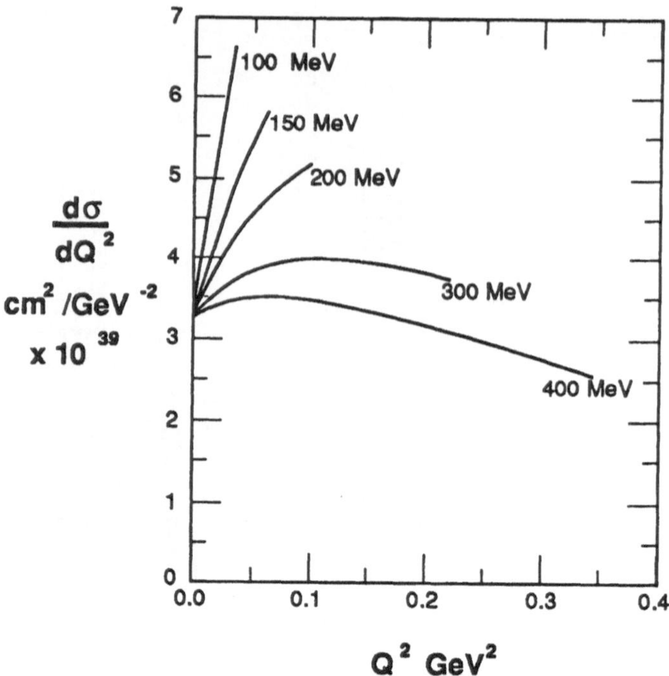

Figure 6. Cross section vs. Q^2 for ν_μ - p elastic scattering.

form factor G_A with small corrections from vector terms. The hadronic current may be written schematically as

$$\frac{1}{2}\,\bar{u}u - \frac{1}{2}\,\bar{d}d - \frac{1}{2}\,\bar{s}s \;,$$

with numerical coefficients reflecting the axial charge of the relevant quark. The first two terms are just those responsible for neutron decay so that any deviation from $G_A = 1.26$ is directly attributable to strange quark contributions in the proton. The cross section is measured absolutely by comparing with the reaction

$$\nu_\mu + {}^{12}C \rightarrow \mu^+ + {}^{12}N \;,$$

the magnitude of which is related to β decay. This cross section is known to a 7% uncertainty which comes from lack of knowledge of the Q^2 dependence of some form factors in the reaction.

The cross section is needed on free protons, of course, so the "background" from protons bound in ^{12}C is a problem. The extrapolation to $Q^2 = 0$ will help, together with the possibility of running the detector with a pseudocumene fill ($CH_{1.2}$), as well as the normal mineral oil fill so that a subtraction may be made for the contribution from bound protons.

Neutrino Oscillations

Neutrino oscillations and the possibility of a finite neutrino mass have received impetus from a number of observations in the last few years. The solar neutrino problem has a plausible solution through the Mikhail-Smirnov-Wolfenstein [2] effect producing matter oscillations (MSW). We do not wish to discuss this problem here, but only take note of the fact that the event rate seen in the Chlorine experiment [3], together with the rate of neutrino electron scattering in Kamioka [1], shows very interesting indications of finite neutrino mass and generation mixing. Initial indications from the Soviet-American Gallium Experiment (SAGE) are also supportive of this idea. If this hypothesis is correct, then the product of mass difference squared of electron neutrino and muon neutrino and the vacuum mixing between them is given by

$$\Delta m^2 \sin^2 2\theta \sim 10^{-8}\ \mathrm{eV}^2 \ .$$

The center of the allowed range for these two parameters from solar neutrino experiments gives $\sin^2 2\theta \sim 10^{-2}$ so that a continuation with this somewhat speculative reasoning gives a mass for the muon neutrino of 10^{-3} eV with a much smaller mass for the electron neutrino. Moreover, extending the seesaw mechanism leads us to believe that the tau neutrino mass might be in the range 0.1-10 eV. We do not pretend that this reasoning is other than a small contribution to the motives for searching for neutrino oscillations in this mass range.

The 17-keV neutrino observed first by Simpson and Hime [4] is still controversial but also influences the motives of experimenters with low-energy neutrinos. Langacker and Caldwell [5] have produced an analysis of the relevant data and have suggested two solutions, one that ν_μ and ν_τ form an almost degenerate doublet with both masses near 17 keV; a second possibility identifies the 17-keV neutrino with the tau and the muon neutrino has a mass between 180 and 270 keV. The lower limit is given by the limit on $\nu_\mu \rightarrow \nu_e$ oscillations from an experiment at BNL [6]. LSND [7] was designed to search for such oscillations and because of the low probability for finding ν_e in the LAMPF beam above 30 MeV, such a search has the potential for being extremely sensitive. In fact, as is shown in Fig. 7, the second solution of Langacker-Caldwell would either yield an observation of $\nu_\mu \rightarrow \nu_e$ or be eliminated in a year of data taking.

The MSW-motivated possibilities, as well as Langacker-Caldwell, lead us to consider ν_μ disappearance. The reaction that is most promising to sample ν_μ flux is

$$\nu_\mu + {}^{12}\mathrm{C} \rightarrow \mu^- + {}^{12}\mathrm{N} \ .$$

As we have remarked, the cross section is well known. If oscillations occur between two flavor states with mass difference squared $\Delta m^2 = m_1{}^2 - m_2{}^2$, then the probability of finding neutrinos of type 1 in a beam initially pure is

$$P = 1 - \sin^2 2\theta \sin^2[1.27\ \Delta m^2\ L(E_\nu)] \ ,$$

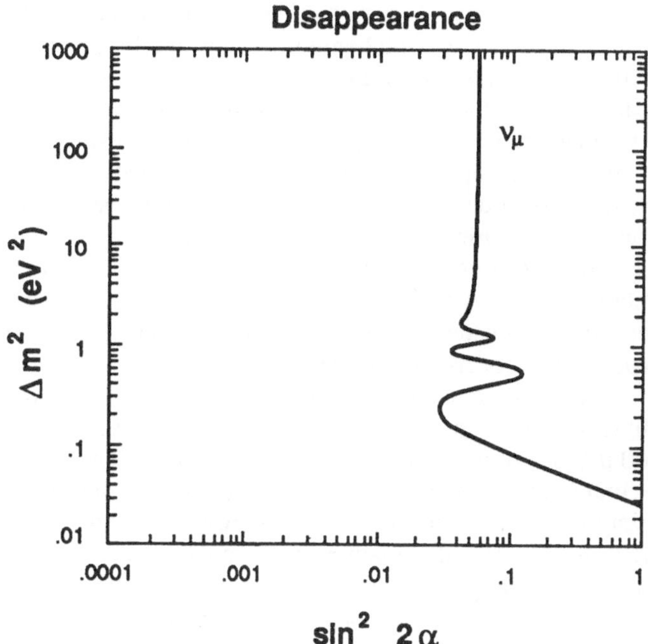

Figure 7. Expected neutrino oscillation limits for $\nu_\mu \rightarrow \nu_e$ using the LSND detector.

where L is the distance from the source in meters and E_ν is the neutrino energy in MeV. For a neutrino energy of 150 MeV and a $\Delta m^2 = 1$ eV^2, the oscillation wavelength is 370 m. At the higher end of the Δm^2 range the event distribution will show a modulation within the detector length of 9 m. For lower Δm^2 (~10) moving the detector to ~50 m will give good sensitivity. In Fig. 7 is shown the limits established by existing experiments.

Experiments that rely on measuring the ν flux as a function of distance from the target have systematic error problems stemming mostly from uncertainty in the characteristics of the source. For flavor oscillations, LSND has a major advantage, namely the simultaneous measurement of ν_μ - p elastic scattering, which is flavor independent. The traditional methods of simulating the source and the neutrino flux are subject to a very substantial set of constraints, namely the angular and distance distributions of elastic scattering events. It seems likely that a significant improvement in systematic error can be achieved. With LSND, a statistically limited exclusion area is shown in Fig. 8.

All this is in the relatively near future. LSND is under construction, and it is expected that data taking with beam will commence in 1993. Much will be learned about this new technique for neutrino physics at these low energies. In another paper, the pulsed lepton source for LAMPF (PLS) is described with emphasis on the source rather than the physics program.

An outline of the facility is shown in Fig. 9. The extracted proton beam will be shared continuously between

LANSCE, the neutron facility, and the pulsed lepton source. A single muon beam is shown schematically to indicate the general location in which muon beams might be built. With sufficient funds both sides of the target could be instrumented. Our present thinking is to use two targets close together for muons and neutrinos to provide sufficient flexibility. The upstream target would be relatively thin and provide a source tailored for a surface muon beam, for example. The primary neutrino target would be enclosed in the pion focusing structure. The shield is of the right thickness to provide adequate shielding for the neutrino detectors from neutrons from this target system and a typical neutrino detector size is indicated.

A question remains, what does the PLS do for neutrino physics? The answer is relatively simple. First with 200 μA of protons at the facility, the flux of neutrinos per second is about the same as at Area A when an optimized decay-in-flight source is used. Improvements in source efficiency offset the reduction in proton current. The big change is in the duty factor from about 10% to 10^{-4}. This makes it feasible to abandon the veto shield and allows the prospect of much larger detectors. A second advantage emerges from the use of pion focusing, which becomes feasible at this duty cycle. As the focusing is particle sign specific, the purity of the decay-in-flight component of each sign selected beam is enhanced, improving event identification.

A detector of 2000 tons mass would have about twenty times the counting rate of LSND at Area A. At the same

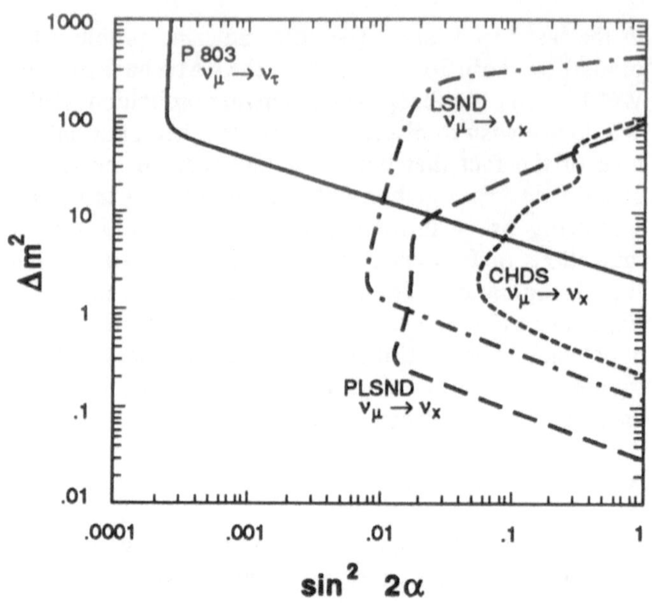

Figure 8. Limits for ν_μ disappearance using LSND using statistical errors only.

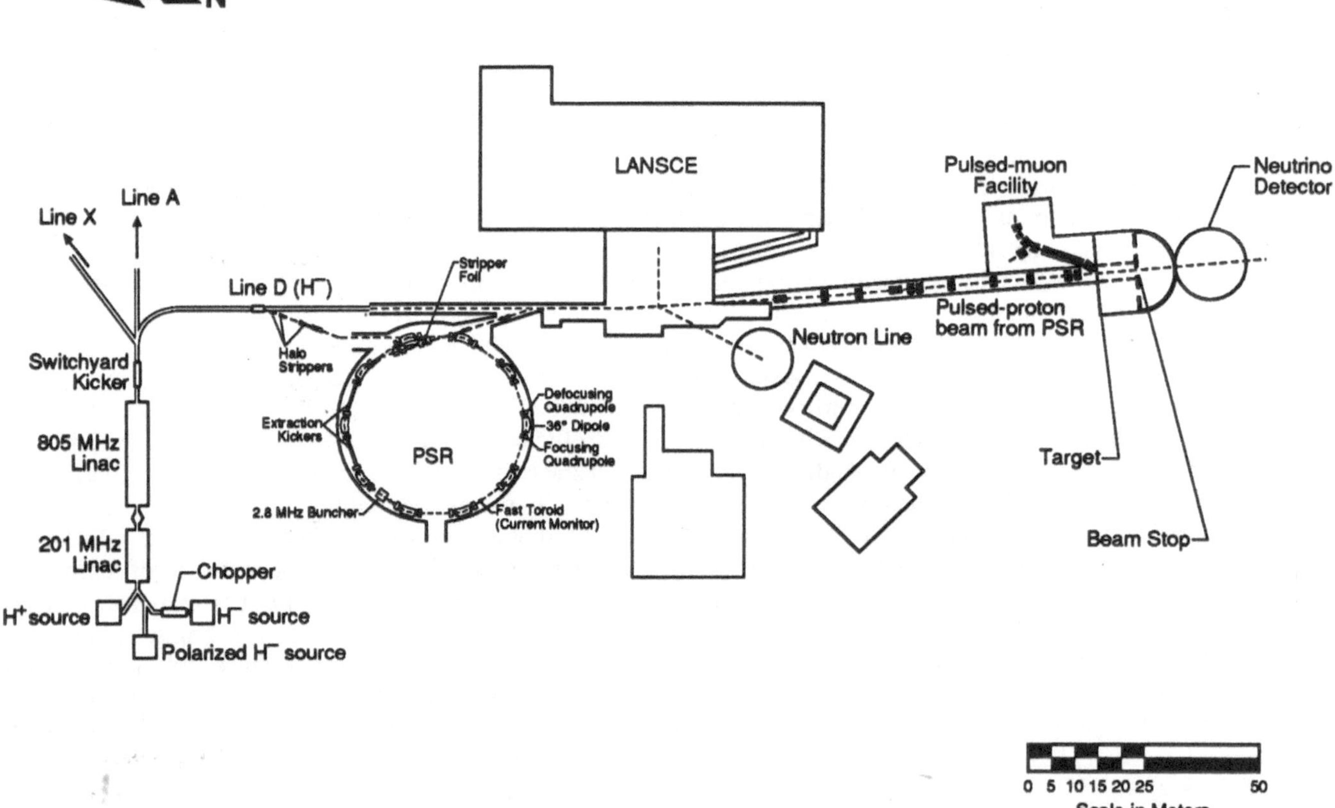

Figure 9. Plan of the pulsed lepton source area.

phototube coverage, this would require nearly 4000 phototubes at a cost of about 7M$ scaling from LSND costs. Of course, the possibilities of other neutrino detectors and experiments are considerable, given freedom from constructing an elaborate veto shield for each one. For example, a detector capable of detecting very low-energy electrons would extend the region in which small values of neutrino magnetic moment could be observed.

A major opportunity would exist at PLS for a long baseline ν_μ disappearance experiment. Assume that the 2000-ton detector alluded to above could be placed at an appropriate distance, and that a systematic limit of 1% can be achieved using both charged current

$$\nu_\mu + {}^{12}C \rightarrow \mu^- + {}^{12}N \quad ,$$

together with flavor-independent νp elastic scattering. Then the distance that the detector is placed should roughly balance systematic and statistical errors. For a four-year data-taking cycle, this puts the detector at 500 m. Putting the detector closer to the target will not help appreciably because the systematic limit will inhibit improvement in sensitivity. Putting the detector further away improves the sensitivity at low Δm^2 at the expense of sensitivity to low mixing angles. Of course, any hint of specific Δm^2 determines the optimum detector position.

The neutrino program at PLS gains from the flexibility that follows independence from elaborate veto requirements. Sign-selected beams are also advantageous. Such a facility shows remarkable promise in a field where developments are coming rapidly.

References

1. K. S. Hirata et al., Phys. Rev. Lett. 65(1990)1297
2. J. N. Bahcall, Neutrino Astrophysics, Chapter 7, University Press, Cambridge, 1989
3. R. Davis, Jr., in Proceedings of the 21st International Cosmic-Ray Conference, Adelaide, Australia
4. J. J. Simpson, Phys. Rev. Lett. 54(1985)1891
5. D. Caldwell and P. Langacker, preprints UCSB-HEP-91-02 and UPR-0469T
6. L. A. Ahrens et al., Phys. Rev. D31(1985)2732
7. Proposal to search for neutrino oscillations in the appareance channels $\nu_\mu \rightarrow \nu_e$ and $\bar{\nu}_\mu \rightarrow \bar{\nu}_e$, Los Alamos National Laboratory report LA-11842-P (1989)

Z. Phys. C - Particles and Fields 56, S194–S202 (1992)

Zeitschrift für Physik C Particles and Fields

The future of Neutrino Physics

Alfred K. Mann

Department of Physics, University of Pennsylvania, Philadelphia, PA 19104

1-September-1991

Abstract. A brief, selective overview is presented of recent progress in neutrino physics, with emphasis on several unanswered questions and possible future experiments which might address them.

1 Introduction

It is appropriate, perhaps, to devote a little time at this Workshop to the close relatives of the charged leptons, namely the chargeless, nominally massless neutrinos. As you know, the interactions of these leptons with one another and with quarks are successfully described by the present Electroweak Theory (EWT). This description is empirically validated at the few percent level of experimental precision, and represents one of the significant accomplishments of elementary particle physics in the last two decades.

It is this success that sets the goals of neutrino physics of the future, which fall into two broad categories:

(a) to test the EWT—particularly in a variety of weak neutral current measurements—at the $\lesssim 1\%$ level of experimental precision,and

(b) to determine the intrinsic properties of the three neutrinos: ν_e, ν_μ, and ν_τ, i.e., values of or improved limits on masses, electric charge, magnetic moments, and possible "structure"; for any of which a positive result would indicate new physics beyond the now standard EWT. Another property of interest that remains to be determined is the so-called electromagnetic charge radius of the neutrino, but a non-zero value of it would be accommodated naturally in the standard theory.

In what follows we discuss in Part I (chapter 2) past work as a guide to present and future experiments. In Part II (chapter 3) we turn to the on-going searches for neutrino mass and mixing with emphasis on the results of solar neutrino observations. Finally, in Part III (chapter 4)the future is addressed in a discussion of new measurements and improved interpretation of results, which to a large extent revisits the subject matter of Part I and Part II.

2 Part I. Mass, Magnetic Moment, Structure and Charge Radius

2.1 Extra-Terrestrial Neutrino Sources

It is worth emphasizing that attempts to uncover the intrinsic properties of neutrinos are often best made with neutrino sources other than those from accelerators. The sources available for this purpose are shown in Fig. 1 where the cross-hatched areas indicate the extra-terrestrial neutrino fluxes that have so far been explored in a preliminary way. Also shown is the huge flux of reactor antineutrinos which were used in the first detection of (anti) neutrino interactions with matter [1] and subsequently have been the antineutrino source for several fundamental experiments [2].

2.2 Mass

The current kinematic limits on neutrino masses are shown in Table I [3], where it is seen that the direct measurement of the mass of $\nu_e, m(\nu_e)$, obtained by studying the endpoint of the ^3H beta-spectrum, is temporarily stalled at about 10 eV. The limits on $m(\nu_\mu)$ and $m(\nu_\tau)$, despite much hard work, are orders of magnitude worse than the limit on $m(\nu_e)$. It is noteworthy that an astrophysical argument suggests an upper limit of roughly 100 eV for the sum of the masses of ν_e, ν_μ, and ν_τ, which is set by the requirement that the critical mass density of the universe, $\rho_c = 3/8\pi \times H^2/G_N$, not be exceeded by the product of the sum of ν_e, ν_μ, and ν_τ masses times their number density, ≈ 260 per cm^3, similar to the number density of the known $2.7^0 K$ electromagnetic background radiation. Clearly, the need for improvements of the limits on $m(\nu_\mu)$ and $m(\nu_\tau)$ in Table I presents a substantial challenge to experimentalists. It is necessary in this overview that the possibility of a neutrino of mass 17 keV be mentioned. This neutrino, if it exists at all, might be thought of as the ν_τ, but the experimental evidence is still being assembled, and no firm conclusions regarding its existence can be reached now.

2.3 Magnetic Moment

In Table II are presented the limits on a static neutrino magnetic moment in units of the Bohr magneton, μ_B [3]. One sees here an illustration of the earlier comment that astrophysical data play an important role in the determination of neutrino properties: the laboratory limits are significantly less restrictive than the limits from astrophysics. The latter are in the main based on general reasoning relating to stellar cooling by means of a sterile neutrino, e.g., ν_R, which is generated in the scattering by an electron of ν_L, the standard theory neutrino, via its magnetic moment. In the standard EWT the sterile ν_R is thought to interact only in higher order with matter and so would escape from and thereby more rapidly cool the interior of a stellar body, e.g., the collapsed core of a supernova, than is actually observed.

Incidentally, the limiting value of roughly $10^{-12}\mu_B$ at 90% c.l. in Table II, if taken in conjunction with the extent and conjectured magnetic field of the solar convection zone, appears to be too small to account for the observed nominal deficit of solar neutrinos, a result of interest in solar neutrino physics which remains still to be confirmed.

2.4 Charge Radius

Closely related to the possible existence of a neutrino magnetic moment is the property known as the electromagnetic charge radius. The Feynman diagrams are essentially similar, but the charge radius enters naturally as a higher order (loop) diagram in the standard theory, while a static magnetic moment of order $10^{-12}\mu_B$ probably would require non-zero neutrino mass and, as seen by the last entry in Table II, would not be explained as a loop correction in the standard theory.

The Feynman diagrams for the elastic scattering of ν_μ by electrons, $\nu_\mu e \to \nu_\mu e$, are shown in Fig. 2, in which the charge radius diagram and the first order diagram involving Z^0 exchange interfere to produce the common final state. The cross section for the interaction may be written as

$$\sigma(\nu_\mu e \to \nu_\mu e) = \sigma^{WNC}(\nu_\mu e)(\sin^2\theta_W + \delta) \tag{1}$$

where $\sigma^{WNC}(\nu_\mu e)$ is the cross section due to the WNC diagram in Fig. 2., $\sin^2\theta_W = 0.230 \pm 0.004 \pm 0.008$ (from deep inelastic neutrino scattering data) is the fundamental parameter of the EWT specifying the relative strength of the weak neutral current (WNC), and

$$\delta = (2\pi\alpha/G_F) < r^2 > = 3.37 \times 10^{30} cm^{-2} < r^2 > . \tag{2}$$

In Eq. 2 α is the fine structure constant, G_F the Fermi constant of the weak interaction, and $< r^2 >$ is the effective square of the charge radius. Experimentally [4], from measurements of $\sigma(\nu_\mu e)$ and $\sigma(\bar{\nu}_\mu e)$, the limits on $< r^2 >$ are

$$-2.1 \times 10^{-32} \leq < r^2 > \leq 0.24 \times 10^{-32} cm^2, \tag{3}$$

since there is no *a priori* specification of the sign of $< r^2 >$.

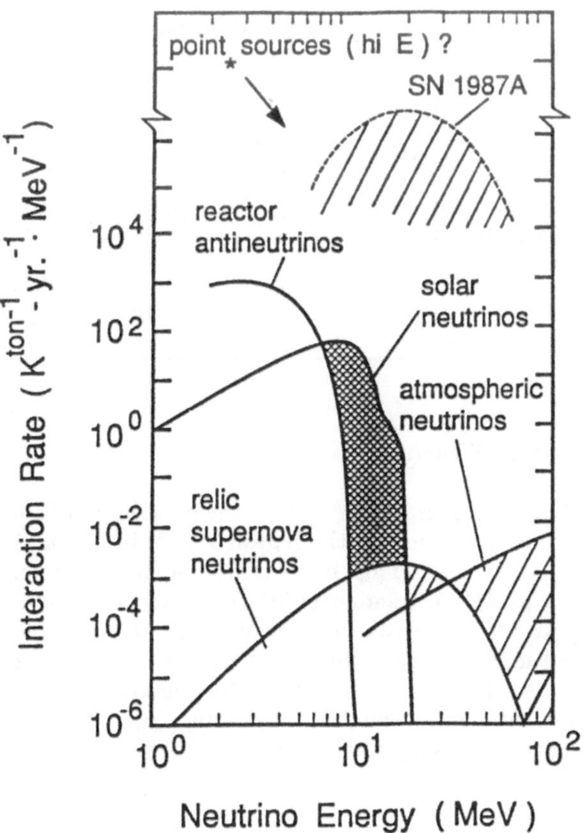

Fig. 1. Rough estimated interaction rates of extra-terrestrial neutrinos as a function of neutrino energy. Reactor neutrinos are shown for comparison.

Table 1. Mass limits for neutrinos from reference 3, which contains the references listed.

m_{ν_e}	$<$	$9.4eV$	LANL	[56]
m_{ν_e}	$<$	$18eV$	Zurich	[57]
m_{ν_e}	$<$	$29eV$	INS-Tokyo	[58]
$17eV$	$< m_{\nu_e} <$	$40eV$	ITEP	[59]
m_{ν_e}	$<$	$0(20eV)$	SN1987A	[61]
m_{ν_μ}	$<$	$0.27MeV$	SIN	[62]
m_{ν_τ}	$<$	$35MeV$	ARGUS	[63]

Table 2. Limits on neutrino magnetic moments from reference 3 which contains the references listed.

laboratory	[147]	$\mu_{\bar{\nu}_e}$	$< 4 \times 10^{-10}\mu_B$
		μ_{ν_μ}	$< 9.5 \times 10^{-10}\mu_B$
Stellar cooling ($\gamma \rightarrow \nu\bar{\nu}$)	[148]	μ_ν	$< 1.1 \times 10^{-11}\mu_B$
Red giants	[149]	μ_ν	$< (2-3) \times 10^{-12}\mu_B$
Nucleosynthesis ($\nu_L e \rightarrow \nu_R e$)	[150]	μ_ν	$< 0.5 \times 10^{-10}\mu_B$
SN1987A	[151] -[154]	μ_ν	$< (10^{-13} - 10^{-12})\mu_B$
Standard model (Dirac mass)	[155]	μ_ν	$\sim 3 \times 10^{-19}(\frac{m_\nu}{1\ eV})\ \mu_B$

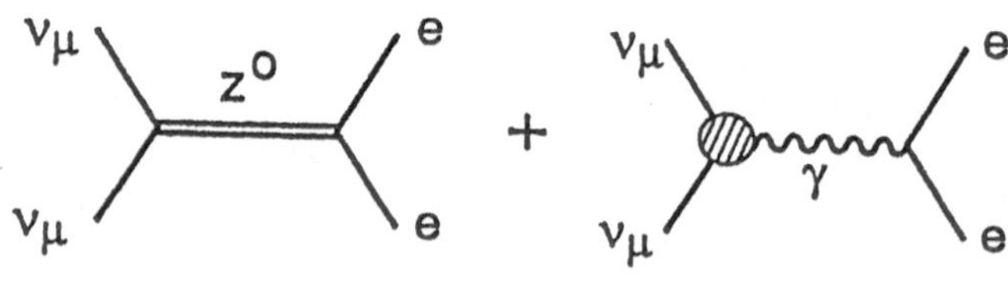

Fig. 2. Feynman diagrams for $\nu_\mu e$ elastic scattering showing the first order diagram on the left and the one-loop correction charge radius diagram on the right.

2.5 Intrinsic Structure

It is also of interest to consider the possibility of some non-specific structure at the lepton vertex in a neutrino scattering process. This has been done by comparing the x- and Q^2- dependences of the nucleon form factor, $F_2(x,Q^2)$, in weak charged current, deep inelastic neutrino-nucleon scattering with the x- and Q^2- dependences of the same form factor determined from deep inelastic muon-nucleon scattering [5]. In first order, $F_2^{\mu N}(x,Q^2)$ and $F_2^{\nu N}(x,Q^2)$ are related by

$$F_2^{\mu N}(x,Q^2) = \frac{5}{18}\ F_2^{\nu N}(x,Q^2)(1 \pm Q^2/\Lambda_{\nu\mu}^2), \tag{4}$$

where $x \simeq Q^2/2m_pE_h$, E_h/E_{inc} is the inelasticity of the interaction, m_p the proton mass, Q^2 the square of the momentum transferred from the lepton vertex, and $\Lambda_{\nu\mu}$ is an effective mass representing any residual quantum chromodynamic (QCD) effects at the quark vertex plus any intrinsic (non-specific) lepton structure at the lepton vertex. The situation is exhibited by the Feynman diagrams in Fig. 3. An early comparison of the two structure functions [5] is shown in Fig. 4 which yields after analysis

$$|\Lambda_{\nu\mu}| \gtrsim 30\ GeV \qquad (90\%\ c.l.), \tag{5}$$

which corresponds to point-like behavior at the lepton vertex down to a length of approximately 7×10^{-17} cm. Note that early studies of $e^+e^- \rightarrow \mu^+\mu^-$ indicated that $|\Lambda_{\mu\mu}| \gtrsim 114$ GeV (but see below). Hence the length limit, 7×10^{-17} cm, is applicable to any intrinsic structure of the ν_μ itself.

These attempts to probe for structure of the neutrino are at a level of the order of 10^{-16} cm. Electromagnetic radiative corrections, illustrated in general in Fig. 3 and more specifically by the charge radius diagram in Fig. 2 are expected to be of the order of $M_Z/\sqrt{\alpha} \approx 10^3$ GeV, which corresponds to $\approx 10^{-17}$ cm. Accordingly, there is roughly an order of magnitude below the present length limit in which to find any primitive structure over and above that due to radiative corrections. We return in Part III to this issue in mentioning a possible experiment at the $e+p$ collider, HERA, where much higher values of Q^2 will be accessible than are shown in Fig. 4.

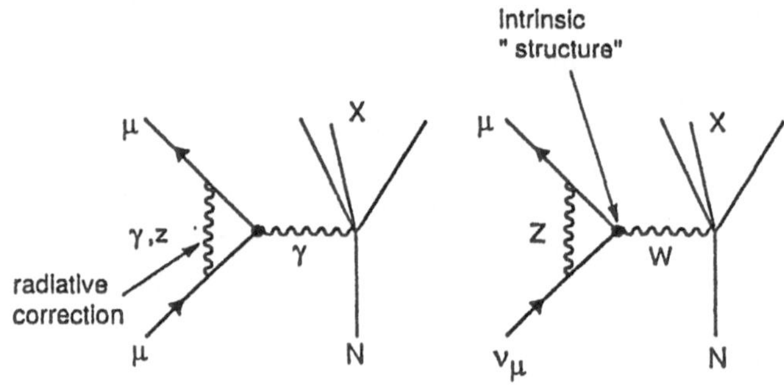

Fig. 3. Feynman diagrams to illustrate a known radiative correction and unknown intrinsic structure.

3 Part II. Mass and Mixing of Neutrinos: Solar Neutrinos

In recent years, the emphasis in neutrino physics has shifted from scattering experiments at accelerators to the observation and study of neutrinos from the extraterrestrial sources in Fig. 1. To restrict the length of this review, we pass over the observations of neutrinos from the supernova SN1987A, and the searches for neutrinos from past supernovae and other sources, as well as the measurements of atmospheric neutrinos generated by the primary cosmic ray component interacting in the earth's atmosphere. Instead, we concentrate on the recent substantial progress in the study of solar neutrinos. In many respects., the Sun is an "ideal" neutrino source because (a) the flux is large in magnitude, (b) the energy range is low (0 to 15 MeV) relative to accelerator fluxes, (c) the matter traversed (in the Sun) is large, and (d) the distance to the earth is long.

The spectrum of neutrinos (ν_e) thought to originate from the fusion reactions in the Sun is shown in Fig.

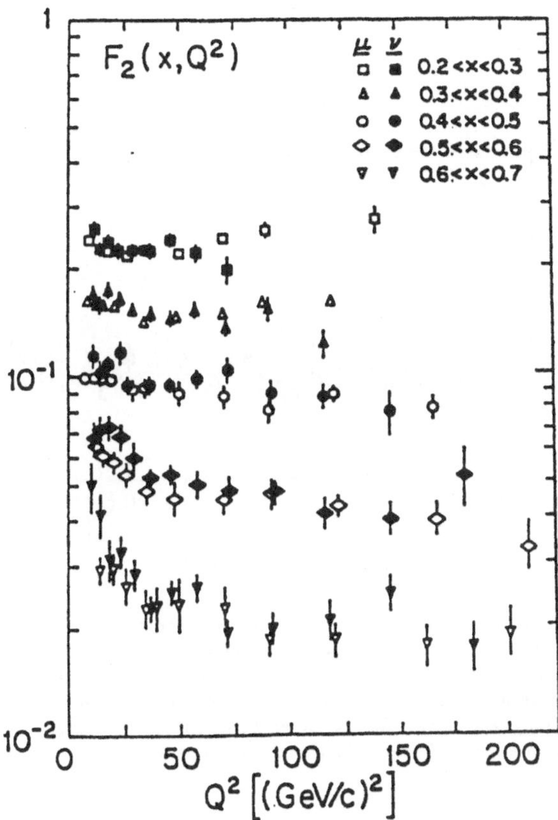

Fig. 4. Plot of $F_2(x, Q^2)$ *vs* Q^2 showing the similarity of results from $\nu_\mu N$ and μN scattering processes reproduced from reference 5.

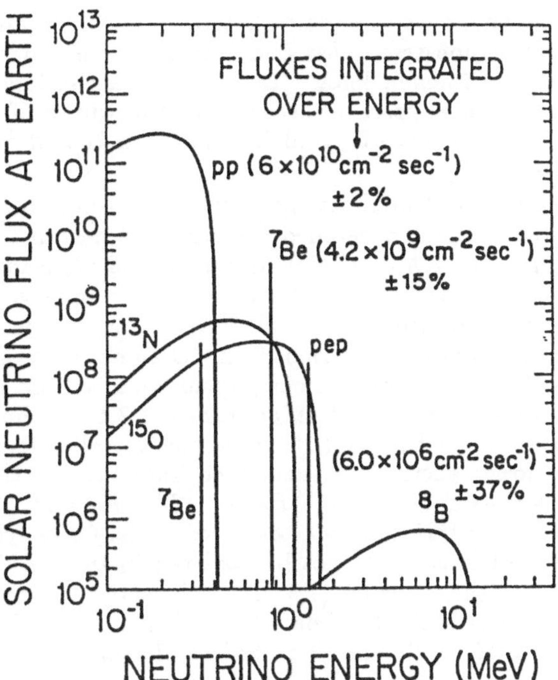

Fig. 5. Calculated solar neutrino fluxes *vs* energy. See references 8 and 9.

5. To date, the upper energy region of the ^8B decay spectrum is the only source of solar neutrinos detected with certainty. This has been accomplished in two detectors: a radiochemical detector utilizing the reaction $\nu_e + {}^{37}Cl \rightarrow {}^{37}Ar + e^-$ with subsequent isolation and detection of the beta decay of ^{37}Ar [6], and a real-time, directional, imaging, water Cerenkov detector [7] in which observation of the reaction $\nu_e + e^- \rightarrow \nu_e + e^-$ yields the magnitude of the ^8B neutrino flux and the shape of the final state electron energy spectrum between 7.5 and 14 MeV, and also correlates the direction of the incident ν_e with the position of the Sun at the instant of observation of the final state electron.

The directional correlation with respect to the Sun, observed in the water Cerenkov detector (KAM-II), is shown in Fig. 6, where a clear identification of the Sun as the source of the incident neutrinos is indicated. Note that the angle of the recoiling electron relative to the incident neutrino in the reaction $\nu_e+e \rightarrow \nu_e+e$ is limited kinematically by $\theta^2 \leq 2me/E_\nu$. The energy distribution of the final state electrons is shown in Fig. 7. The shape of the energy distribution depends only on the cross section for $\nu_e + e \rightarrow \nu_e + e$, on the shape of the energy spectrum of neutrinos from ^8B, measured in laboratory experiments, and on the detection efficiency of KAM-II; the shape is independent of the absolute value of the ^8B solar neutrino flux.

The results of the KAM-II detector are summarized as follows:

(i) for 1040 days of live detector time in the period January 1987 through April 1990, with $E_e \geq 7.5$ MeV, the ratio of the observed ^8B solar neutrino flux (Data) to the flux calculated from the standard solar model (SSM) is given by

$$\frac{Data}{SSM} = \begin{cases} 0.46 \pm 0.05(\text{stat}) \pm 0.06(\text{syst}) & (6) \\ 0.70 \pm 0.076(\text{stat}) \pm 0.09(\text{syst}) & (7) \end{cases}$$

where the results in (6) and (7) follow from two calculations [8] and [9], respectively, based on the SSM but using slightly different semi-empirical input data[1] .

(ii) the clear directional correlation and the electron energy distribution provide unequivocal evidence for the production of ^8B by fusion in the Sun.

(iii) no significant time variation is observed in the solar neutrino signal in the 1040 day observation period.

(iv) no significant day-night or seasonal variation is observed in the 1040 day observation period.

These results, taken in conjunction with the result of the ^{37}Cl radiochemical detector in the same time period, suggest, but do not compel one to believe, that the mass of either ν_e or ν_μ or both is non-zero, and that there

[1] *Note added in proof: The value of the ^8B flux claculated by reference [9] has been revised upward from 4.0 SNU to 5.5 SNU, so that equation (7) should now read* $= 0.51 \pm 0.06(stat.) \pm 0.07(syst.)$. *The remainder of the text in part II is unchanged.*

may be an as yet unknown interaction which mixes them. The non-zero mass and mixing are manifested in the phenomenon known as neutrino oscillations in matter which appears to account for the values of the solar neutrino flux observed by the KAM-II and ^{37}Cl detectors.

The essential implications of the results may be appreciated without a detailed discussion of neutrino oscillations. The KAM-II result in eqs.(6) and (7) leads to the prediction that the ^{37}Cl detector should observe 2.8 SNU (solar neutrino unit $= 10^{-36} Ar$ atoms per sec per target atom) due to neutrinos from ^8B alone. This value follows from both calculations [8] and [9]. The SSM calculations also agree that an additional 1.7 SNU should be observed by the ^{37}Cl detector due to solar neutrinos of lower energy, i.e., neutrinos from the reactions $p+e+p \to d+\nu_e$, $e+\ ^7Be \to\ ^7Li+\nu_e$, and from the decays of ^{13}N and ^{15}O, produced in the CNO cycle. These neutrinos would not be observed in the KAM-II detector with its higher (7.5 MeV) threshold compared with the 0.81 MeV threshold of the ^{37}Cl detector. Thus the KAM-II result and the SSM result (on which, as stated, the two calculations are in good agreement) indicate that the ^{37}Cl result should be 4.5 SNU and not the 2.3 ± 0.3 SNU actually observed. The observations of the two detectors and the SSM calculations can be brought into agreement by invoking neutrino oscillations in which the higher energy $^8B\ \nu_e$ would be partially oscillated into ν_μ while the lower energy ($\lesssim 1.5$ MeV) ν_e would be almost completely oscillated into ν_μ. In the ^{37}Cl detector the ν_μ are unobservable, and in the KAM-II detector observable only with a severely reduced (1/6) efficiency relative to ν_e.

Furthermore, the preliminary result from the Soviet-American gallium detector, searching for the very low ($\leq$ 0.44 MeV) energy ν_e from the reaction $p+p \to d+e+\nu_e$, which should be the most abundant neutrino source in the Sun, is a relatively low upper limit on that neutrino flux. This low limit, if confirmed, would be strong further evidence for the neutrino oscillation channel $\nu_e \to \nu_\mu$ or, possibly, $\nu_e \to \nu_{sterile}$, and consequently for neutrino mass and mixing.

The solar neutrino results may be the first indication of intrinsic properties of neutrinos other than zero electric charge, spin, and lepton number. Accordingly, these results hold the center of the stage in the present drama of neutrino physics. They remain to be substantiated and extended by experiments of the future, to which we turn now.

4 Part III. The Future: Physics Beyond the Standard Model

4.1 *Neutrino-proton elastic scattering*

In a more thorough review of the recent past, a discussion of measurements of the reaction $\nu_\mu + p \to \nu_\mu + p$ would have been included. Here, we look to the future in which the reaction $\nu_\mu+p \to \nu_\mu+p$, if carried out with relatively low $[O(10^2\text{ MeV})]$, energy neutrinos, will provide a direct measurement of the strange quark contribution to the proton spin. That contribution was expected to be zero, or at most very small, but an earlier $\nu_\mu+p \to \nu_\mu+p$ measurement [10] and recent measurements of polarized muon deep inelastic scattering by polarized nuclei [11] indicate that $\Delta s \neq 0$, where

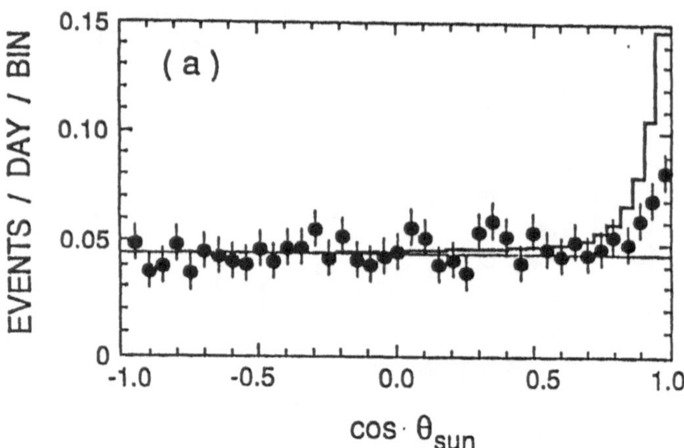

Fig. 6. Plot of ^8B solar neutrino event rate against the cosine of the angle, $\cos\theta_{Sun}$, between the direction of the recoil momentum of the target electron in $\nu_e e \to \nu_e e$ scattering and the direction of the incident neutrino from the Sun. Reproduced from reference 7. The histogram shows the shape and magnitude of the $\cos\theta_{Sun}$ distribution expected on the basis of reference 8.

$$\Delta s \equiv \int_0^1 dx \left[s^\uparrow(x) + \bar{s}^\uparrow(x) - s^\downarrow(x) - \bar{s}^\downarrow(x)\right] , \tag{8}$$

(Feynman) x is the relative quark momentum, s and $\bar{s}$ are the quark and antiquark constituents of the strange quark sea in the nucleon, and the arrows indicate the relative spin directions.

The differential cross section for $\nu_\mu + p$ scattering may be approximated at $Q^2/M_p^2 \approx 0$ by

$$\frac{d\sigma}{dQ^2} \approx \frac{G_F^2}{2\pi}\left[\left(-0.63 + \frac{G_1^S}{2}\right)^2 \left(1 + \frac{Q^2}{4E_\nu^2}\right) + (0.034)^2 \left(1 - \frac{Q^2}{4E_\nu^2}\right)\right] \tag{9}$$

where the only unknown term is $G_1^S \equiv \Delta s$, and, for example, at $E_\nu = 150$ MeV the difference between eq. (9) and the general cross section expression is only 5%. A proposed experiment at the Los Alamos Meson-Proton Facility (LAMPF), to be carried out at $50 \lesssim E_\nu \lesssim 200$ MeV, is likely to reduce substantially the uncertainty in the previous measurement [10] both statistically and

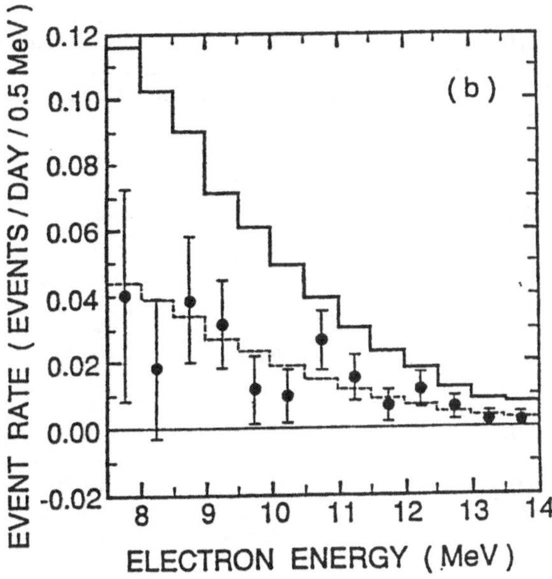

Fig. 7. Plot of the differential energy spectrum of the recoiling electron from the solar neutrino-induced $\nu_e e$ scattering process. The dashed histogram is the best fit to the data points based on the known cross section for $\nu_e e \rightarrow \nu_e e$, the known shape of the beta-decay spectrum of ^8B, and the detection efficiency of the Kamiokande II detector. The solid histogram has the same shape as the dashed histogram but the area predicted by the calculation in reference 8.

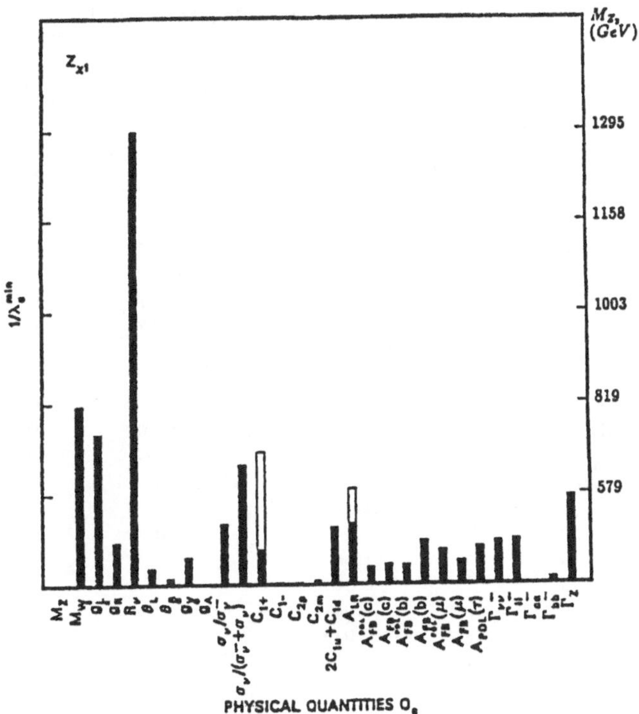

Fig. 8. Bar chart showing the relative sensitivity ($1/\lambda_a^{min}$) of the observables in Table III to an extra Z^0 boson of a given type (Z_X). The corresponding mass limit on the Z_X achievable in a measurement of a specified observable is given in the vertical right hand axis. A precise measurement of $R_\nu \equiv (\nu_\mu N \rightarrow \nu_\mu X)/(\nu_\mu N \rightarrow \mu^- X)$, i.e., deep inelastic neutrino-nucleus scattering, would lead to a mass limit on the Z_X of roughly 1300 GeV. The plot is reproduced from reference 13.

through elimination of the uncertainties due to the higher average Q^2 of that measurement.

In the future, polarized deep inelastic scattering experiments, such as the SMC experiment at CERN and the HERMES experiment at DESY, in conjunction with the $\nu + p$ scattering experiment at LAMPF, and nuclear beta-decay results will allow a model independent determination of $\Delta u + \Delta d + \Delta s$, the contribution to the total spin of the proton from its quark constituents.

4.2 Intrinsic Structure

In Part I a search for primitive structure of the neutrino was described in which a comparison of the form-factor $F_2(x, Q^2)$ extracted in deep inelastic scattering of muons and neutrinos was made at Q^2 values up to 200 $(GeV/c)^2$. There, one found the effective mass limit on $\Lambda_{\nu\mu}$ in the term $Q^2/\Lambda^2_{\nu\mu}$ to be greater than 30 GeV, corresponding to an upper limit on the "size" of the structure of $\sim 10^{-16}$ cm.

At HERA, the $e + p$ collider soon to be in operation, it will be possible to carry out a similar analysis of the reactions $e + p \rightarrow e + p$ and $e + p \rightarrow n + \nu_e$ (both of which might also involve pions in the two final states) at $Q^2 \gtrsim 10^4 (GeV/c)^2$. The experimental problems do not appear to be severe, and consequently an upper limit on intrinsic structure $\lesssim 10^{-17}$ cm might be expected.

It is useful to note in passing that a future (g-2) experiment at BNL is expected to yield a lower limit on $\Lambda_{\mu\mu}$ in the vicinity of 4-5 TeV. This is to be compared with the limiting value of 0.11 TeV available in 1981, and quoted above.

4.3 Mass and Mixing

Here, as noted above, emphasis will be on confirming or repudiating the indication of neutrino oscillations in the matter of the Sun. There are in operation or under construction a number of detectors which will provide important data for that purpose. The ^{71}Ga detectors (SAGE in the Soviet Union and GALLEX in Italy) will concentrate on measuring precisely the magnitude of the $p + p$ flux. Specialized detectors (SNO in Canada and Borexino in Italy) will attempt to observe the weak neutral current scattering of ν_μ and ν_τ arising from the neutrino oscillation channels $\nu_e \rightarrow \nu_\mu$ and/or $\nu_e \rightarrow \nu_\tau$, if they exist. Another specialized detector, still only in proposal form, would attempt to measure the neutrino flux from $e^- + {}^7Be \rightarrow {}^7Li + \nu_e$ which, when compared with the already observed flux from the decay of ^8B (created by $p + {}^7Be \rightarrow {}^8B + \gamma$) would lead to a precision measurement of the core temperature of the Sun. The most massive of the planned detectors – SUPERKAM in

Table 3. Reproduction of Table 1.1 in reference 13. The observables considered in this article, their SM predictions, and their present and future experimental uncertainties (including theoretical uncertainties where they are important). The SM predictions use the observed value of M_Z and assumes $m_t = 100$ GeV, and $M_H = 100$ GeV. $g_L^2, g_R^2, R_\nu, \theta_L$, and θ_R are quantities measured in νN scattering; $g_V^e, g_A^e, \sigma_\nu/\sigma_{\bar\nu}$, and $\sigma_\nu/(\sigma_{\bar\nu}+\sigma_\nu)$ are relevant to νe scattering; $C_{1\pm}, C_{2\pm}$ are measured in atomic parity violation, muonic atoms, and lN scattering; A_{LR}, A_{FB}, and A_{pol} are asymmetries at the Z-pole, and the Γ's are the partial and total Z widths. The e^+e^- asymmetries slightly off the Z- pole are briefly ...

Quantities		present			future	
O_a	O_a^{SM}	O_a^{exp}	ΔO_a^{exp}	$\Delta \sin^2\theta_W^{exp}$	ΔO_a^{exp}	$\Delta \sin^2\theta_W^{exp}$
$M_Z(GeV)$	—	91.177	0.031	0.0004	0.02	0.0003
$M_W(GeV)$	79.983	80.1	0.3	0.0018	0.105	0.0006
g_L^2	0.300	0.2977	0.0042	0.0057	—	—
g_R^2	0.030	0.0317	0.0034	0.013	—	—
R_ν	0.312	—	—	—	0.001	0.002
θ_L	2.46	2.50	0.03	—	—	—
θ_R	5.18	4.59	0.44	—	—	—
g_V^e	−0.036	−0.045	0.022	0.011	—	—
g_A^e	−0.504	−0.513	0.025	—	—	—
$\sigma_\nu/\sigma_{\bar\nu}$	1.152	1.083	0.10	0.012	0.046	0.005
$\sigma_\nu/(\sigma_{\bar\nu}+\sigma_\nu)$	0.146	—	—	—	0.0026	0.0025
C_{1+}	0.129	0.126	0.003	0.01	0.0013	0.003
$C_{1+}(iso)$	0.129	—	—	—	0.0003	0.0009
C_{1-}	−0.36	−0.45	0.1	0.07	—	—
C_{2p}	−0.014	—	—	—	0.046	—
$C_{2p}(1)$	−0.014	—	—	—	0.0046	—
C_{2m}	−0.054	—	—	—	0.11	0.03
$2C_{1u} + C_{1d}$	−0.033	—	—	—	0.004	0.002
$A_{LR}(SLC)$	0.131	—	—	—	0.0066	0.0008
$A_{LR}(LEP)$	0.131	—	—	—	0.0044	0.0005
$A_{FB}^{pol}(c)$	0.473	—	—	—	0.025	0.01
$A_{FB}(c)$	0.062	—	—	—	0.007	0.0017
$A_{FB}^{pol}(b)$	0.697	—	—	—	0.02	0.04
$A_{FB}(b)$	0.091	0.11	0.04	0.007	0.0054	0.001
$A_{FB}^{pol}(\mu)$	0.098	—	—	—	0.009	0.0015
$A_{FB}(\mu)$	0.013	0.024	0.007	0.004	0.0035	0.002
$A_{pol}(\tau)$	0.131	—	—	—	0.01	0.0014
$\Gamma_{inv}(GeV)$	0.499	0.482	0.016	0.007	—	—
$\Gamma_{l\bar{l}}(GeV)$	0.0835	0.0839	0.0007	0.0016	—	—
$\Gamma_{c\bar{c}}(GeV)$	0.296	0.235	0.038	0.02	0.03	0.016
$\Gamma_{b\bar{b}}(GeV)$	0.377	0.372	0.064	0.03	0.04	0.018
$\Gamma_Z(GeV)$	2.484	2.497	0.015	0.001	—	—

Japan at approximately 32 kilotons – will strive for high statistics measurements of the ^8B neutrino flux and the electron energy spectrum, both described above, and a search for possible time variations of small amplitude in the solar neutrino flux.

In laboratory experiments stimulated by the present solar neutrino data, there will be renewed efforts to search for the oscillation channels $\nu_\mu \to \nu_\tau$ and $\nu_e \to \nu_\tau$ at FERMILAB, CERN, and LAMPF. The new detectors of extra-terrestrial neutrinos, with primary motivation the study of solar neutrinos, will also be sensitive to a supernova in our Galaxy, which will give rise to hundreds, possibly thousands of interactions in several of these detectors. They will also continue and expand the observation of atmospheric neutrinos. Finally, the undersea detector, DUMAND, will search for very energetic neutrinos from stellar point sources.

4.4 Search for New Physics through High Precision Electroweak Experiments

The EWT has been successful in providing a quantitative description of a variety of phenomena which agrees without exception with all experimental data at the, roughly, 5% level [12]. To confront the theory with future experimental results requires a broader analytic framework for comparison of the results of experiments of different types with one another and with the theory. This is because the value of the fundamental parameter of the EWT, $\sin^2\theta_W$, is not specified by the theory and small departures from the world average value found in individual experiments are not unambiguously interpretable without such a framework. Furthermore, directly measured observables have different sensitivities to $\sin^2\theta_W$, so that direct comparison of the measured values of those observables with theory may be more revealing than proceeding indirectly through comparison

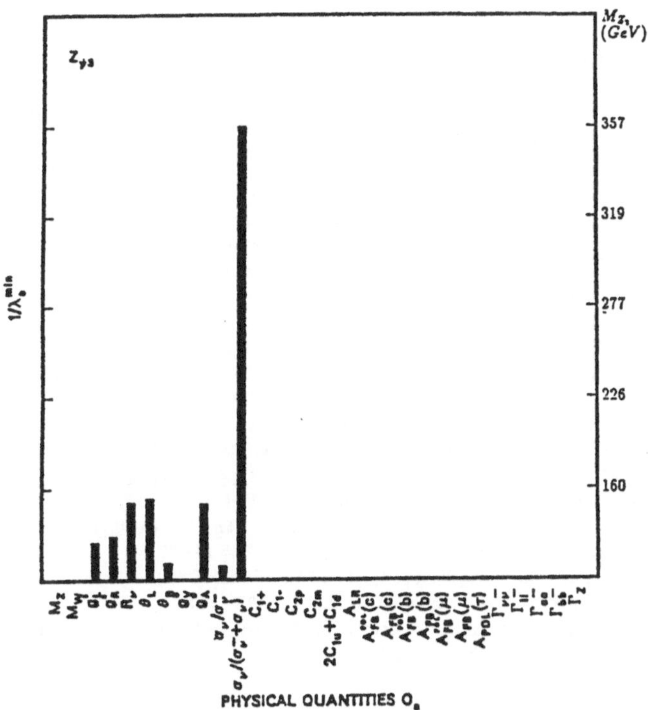

Fig. 9. Plot similar to that in Fig. 8, but for an extra Z^0 boson of different type, the Z_ψ. Here, the most sensitive observable is the ratio $\sigma(\nu_\mu e \to \nu_\mu e)/[\sigma(\nu_\mu e \to \bar{\nu}_\mu e) + \sigma(\nu_e e \to \nu_e e)]$ and the achievable mass limit on the Z_ψ is roughly 350 GeV.

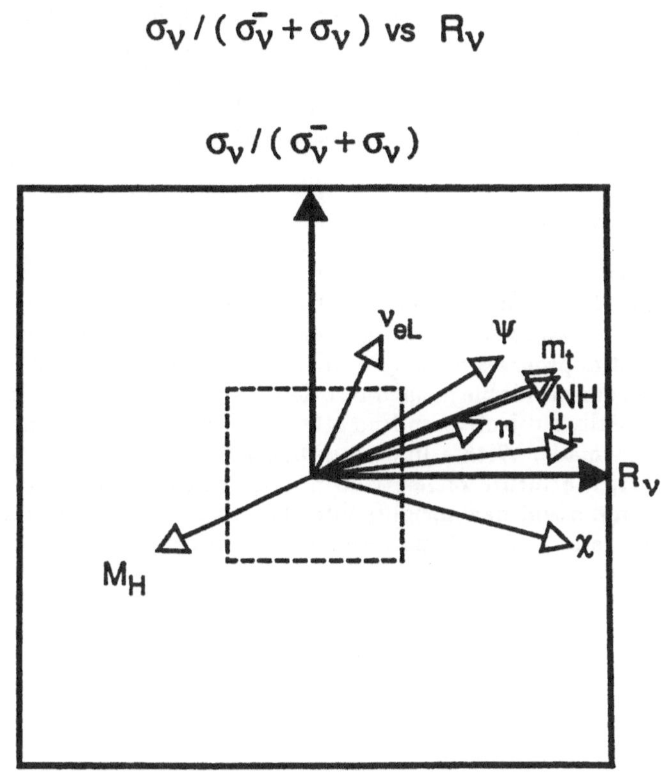

Fig. 10. Two dimensional plot of $\sigma(\nu_\mu e \to \nu_\mu e)/[\sigma(\nu_\mu e \to \bar{\nu}_\mu e) + \sigma(\nu_{\to}\nu_e e)]$ *vs* R_ν to illustrate the method of delineating new physics, in this plot a massive Higgs boson. Other possible new physics origins of departures from the standard model values of these two observables are also shown as vectors in the two dimensional space, but are not easily distinguishable one from the other. The dashed box indicates the relative size of the measurement errors and the expected values of the coupling strength of the new physics.

of the $\sin^2\theta_W$ values obtained from them. Lastly, the appearance of new physics, i.e., a departure from the EWT, if it exists at all, will be at a level less than the $\sim 5\%$ level at which the theory has so far been tested, and so will require a detailed, internally consistent analysis to elicit and delineate it. We briefly discuss this subject here because among possible future high precision electroweak experiments are neutrino scattering experiments.

One such framework has been described in a recent paper [13] in which it is assumed that many electroweak experiments in the coming decade will achieve a precision of $\lesssim 1\%$, and will therefore serve to tes the EWT at that level. We show in Table III from reference [13] a list of 32 observables that may be measured in deep inelastic neutrino-nucleus scattering, neutrino-electron scattering, experiments on parity violation in atoms, experiments on asymmetries in the final states of $e^+ + e^-$ scattering, and total and partial widths in Z^0-decay. The theoretical predictions of the values of the observables, present experimental values and errors, and estimates of future errors are also shown in Table III. In Table IV are listed the various types of new physics whose contributions to modifications of the standard model values of each of the observables was estimated.

Examples of the results of the analysis are shown in Figs. 8 and 9 where is plotted along the y-axis the reciprocal of that value of the single, undetermined coupling constant (λ) of the specified new physics for which the change in the observable is equal to the projected one

Table 4. Reproduction of Table 1.2 in reference 13. The types of new physics considered in this paper.

Tree-level physics:	extra Z: χ, ψ, η, Z_{LR} nonstandard Higgs representations leptoquarks: type I, type II extra fermions: $u'_{L,R}$, $d'_{L,R}$, $e'_{L,R}$, $\nu'_{L,R}$ compositeness: four-fermi contact operators
Loop-level physics:	m_t, M_H extra fermions S-T parameters: gauge boson self-energies two Higgs doublets supersymmetry

standard deviation experimental error. The observables are listed along the x-axis. One sees in Fig. 8 the relative sensitivity of deep inelastic neutrino-nucleus scattering to the new physics of an extra Z-boson of a specific type and mass indicated by the right hand scale. In Fig. 9 is shown the sensitivity of neutrino-electron scattering to the new physics of a different type of extra Z-boson. Finally, in Fig. 10 is shown an example of a general method of identifying new physics – in this case the mass of the Higgs boson at a value $\gtrsim$ 100 GeV – by the intersection of (neutrino scattering) observables with particular sensitivities to a given type of new (large M_H) physics.

5 Summary

In this brief review an attempt has been made to explore the future of neutrino physics by discussing briefly several issues from the past and present in which substantial progress is likely to be made in the next decade. As is unavoidable in such a review, the choice of issues is partially subjective and no attempt at completeness has been made. With a little luck and kindness on nature's part, the future of neutrino physics will be productive of important new insights into the subjects discussed in this talk, and perhaps into subjects not now perceived by us.

References

1. F. Reines and C.L. Cowan, Phys. Rev. *92*, 830 (1953); C.L. Cowan *et al.*, Science *124*, 103 (1956).
2. See, for example, G. Zacek *et al*, Phys. Rev. *D34*, 2621 (1986); F. Boehm, Proceedings of 13th Int'l Conference on Neutrino Physics and Astrophysics (World Scientific, 1989, p. 490).
3. P. Langacker, Proceedings of 13th Int'l Conference on Neutrino Physics and Astrophysics (World Scientific, 1989, P. 863).
4. K. Abe *et al.* Phys. Rev. Lett. *58*, 636 (1987); L.A. Ahrens *et al.*, Phys. Rev. *D41*, 3297 (1990).
5. A.K. Mann, Phys. Rev. *D23*, 1609 (1981).
6. R. Davis, Jr. *et al.*, Phys. Rev. Lett. *20*, 1205 (1968); K. Lande *et al.*, in Proceedings of the 25th Int'l Conference on High Energy Physics, Singapore, Aug. 1990, to be published.
7. K.S. Hirata *et al*, Phys. Rev. Lett. *63*, 16 (1989); *65*, 1297 (1990); *65*, 1301 (1990); Phys. Rev. *D44*, 2241(1991).
8. J.N Bahcall and R.K. Ulrich, Rev. Mod. Phys. *60*, 297 (1988).
9. S. Turck-Chièze *et al.*, Astrophys. J. *355*, 415 (1988).
10. L. A. Ahrens *et al.*, Phys. Rev. *D35* 785 (1987).
11. J. Ashman *et al*, Phys. Lett. *206B*, 364 (1988).
12. U. Amaldi *et al*, Phys. Rev. *D36*, 1385 (1987).
13. P. Langacker, M. Luo, and A.K. Mann, to be published in Rev. Mod. Phys., January, 1992.

Acknowledgement. It is a pleasure to thank Gisbert zu Putlitz, Klaus Jungmann and the staff of the Institute who organized this Workshop and extended such cordial hospitality to its participants. This work has been supported in part by the U.S. Department of Energy and the University of Pennsylvania Research Fund.

This article was processed using Springer-Verlag TEX Z.Physik C macro package 1991
and the AMS fonts, developed by the American Mathematical Society.

Z. Phys. C - Particles and Fields 56, S203-S209 (1992)

Zeitschrift für Physik C Particles and Fields

MUON-CATALYZED FUSION THEORY

M. Leon

Los Alamos National Laboratory, Los Alamos, NM 87545, USA

15 June 1991

Abstract. Some topics in muon-catalyzed fusion theory are discussed: Resonant formation of $dd\mu$ molecules appears to be well understood, with good agreement so far between theory and experiment. The situation for resonant $dt\mu$ formation is much less clear, because of the more complicated kinetics, the apparent three-body effect, and the evident need to treat thermalization and molecular formation together to compare theory and experiment. Recent theoretical progress in $pd\mu$ fusion by Friar et al. has resolved a serious discrepancy in the Wolfenstein-Gershtein effect, i.e., the increase in $pd\mu$ fusion yield with increased deuterium fraction.

1 Introduction

μCF theory encompasses the whole catalysis cycle, sketched in Fig. 1 for deuterium-tritium targets. Steps in the cycle include the slowing to very low (eV) energies through ionization of the target molecules, the transition from free to bound states, deexcitation of the initially formed highly excited muonic hydrogen atoms, transfer of the muon from lighter to heavier hydrogen isotopes (because of the reduced-mass effect), which can take place from excited or ground states, formation of the muonic molecular ion and its deexcitation, and finally nuclear fusion with the μ^- either stuck to a fusion product or free to go around the cycle again. Lack of time prevents me from discussing all of these steps, so I will concentrate on the molecular formation step, in particular on the intricate and fascinating resonant molecular formation mechanism, first in general and then the particulars of the two operative examples, $dd\mu$ and $dt\mu$ formation. Finally, I will discuss the new theoretical light shed on the oldest μCF reaction, that of $pd\mu$, and on the Wolfenstein-Gershtein effect, which was thought to be understood quite well nearly three decades ago but which became considerably more troublesome in recent years.

2 Resonant molecular fromation

In the authoritative 1960 review article of Zeldovitch and Gershtein [1] we find the following paragraph:

> "f) Formation of Mesomolecules
> In the collision of free mesonic atoms with nuclei of hydrogen molecules, formation of mesomolecules is possible. In such a process the binding energy of the mesonic molecule can, in general, be given off either as radiation or to the electron of the hydrogen molecule, or finally to a neighboring nucleus in the molecule. The last of these mechanisms might play an important role in the formation of mesomolecules in excited states with a binding energy close to the dissociation energy of the hydrogen molecule. Since, however, there are no such levels in mesomolecules (cf. Table 3), this mechanism need not be considered."

To explain the unexpectedly large and temperature-dependent $dd\mu$ molecular formation rate $\lambda_{dd\mu}$, Vesman in 1967 [2] postulated that there *must* exist such a state in the $dd\mu$ system, with binding energy less than the D_2 dissociation energy! L. I. Ponomarev and collaborators then set to work to determine theoretically whether such a state actually exists, developing for this purpose what they call the "adiabatic expansion" (which others call the "method of perturbed stationary states"). After a decade of effort, the "School of Ponomarev" had progressed sufficiently to conclude that the angular momentum $J = 1$, vibrational quantum number $v = 1$ state of $dd\mu$ is bound by about 2 eV, and furthermore the corresponding state in the $dt\mu$ muonic molecule is bound by only about 1 eV [3,4]. On the basis of these results, Gershtein and Ponomarev in 1977 published a Physics Letter [5] pointing out that (1) the Vesman idea appeared to be the correct explanation of the $dd\mu$ results, and (2) for $dt\mu$ $\sim 10^2$ fusions per muon are expected! This prediction of $\sim$100 fusions/μ^- did much to revive interest in μCF.

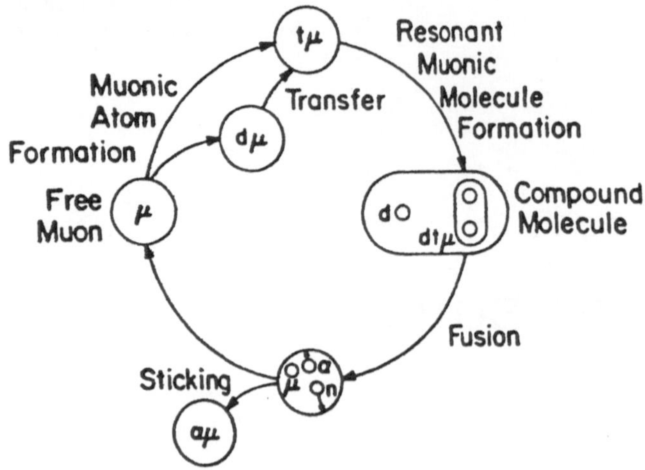

Fig. 1. Simplified μCF cycle for a deuterium-tritium target (from Ref. 6)

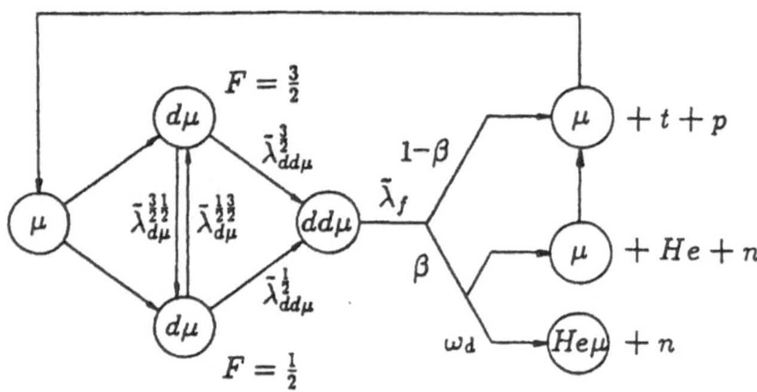

Fig. 2. Simplified μCF cycle in D_2 (from Ref. 6)

In recent years, variational calculations have overtaken in precision the adiabatic expansion in determining the nonrelativistic point Coulomb binding energies of the various muonic molecule systems; results quoted in a recent review [6] are displayed in Table 1. Corrections to these energies from the hyperfine interaction, relativity and QED, nuclear electromagnetic structure, etc., are important for the critical $(J,v)=(1,1)$ states of $dd\mu$ and $dt\mu$, and are shown in Table 2.

The energies of these loosely bound states, along with the excitation energies of the compound molecule formed in the resonant reaction

$$(t\mu)^F + (D_2)_{K_i} \rightarrow [(dt\mu)^S_{11}d2e]^*_{\nu,K_f} \quad , \tag{1}$$

determine the resonance energies ϵ_r. Here K_i and K_f are the rotational quantum numbers of the initial D_2 and final compound molecule (or *complex*) $[\ ]^*$, and ν the $[\ ]^*$ vibrational quantum number (initial D_2 $\nu=0$), whi le F and S denote the $t\mu$ and $dt\mu$ hyperfine states. The rate for this resonant molecular formation can be written

Table 1. Coulomb molecular binding energies in eV[a] (from Ref. 6).

Table 1. Coulomb molecular binding energies in eV[a] (from Ref. 6).

J, v	$pp\mu$	$pd\mu$	$pt\mu$	$dd\mu$	$dt\mu$	$tt\mu$
0,0	253.15	221.55	213.84	325.07	319.14	362.91
0,1	—	—	—	35.84	34.83	83.77
1,0	107.27	97.50	99.13	226.68	232.47	289.14
1,1	—	—	—	1.97[b]	0.66[c]	45.21
2,0	—	—	—	86.45	102.65	172.63
3,0	—	—	—	—	—	48.70

[a] See [6] for references.
[b] The accurate energy is 1.9749 eV.
[c] The accurate energy is 0.6603 eV.

Table 2. Corrections (in meV) to the energies of the $J=1$, $v=1$ states of $dt\mu$ and $dd\mu$ (from Ref. 6).

Table 2. Corrections (in meV) to the energies of the $J=1$, $v=1$ states of $dt\mu$ and $dd\mu$ (from Ref. 6).

	$dt\mu$[a]	$dt\mu$[b]	$dt\mu$[c]	$dd\mu$[a]	$dd\mu$[b]
Nuclear charge distribution	+13.3	+13.3	+10.4[d]	−1.5	−2.1
Darwin-type corrections	−2.5	−2.4	−1.8	−0.9	←[e]
Relativistic mass corrections	−0.5	←		+0.4	←
Recoil corrections	+3.8	←	+2.7	+1.9	←
Vacuum polarization	+16.6	+16.61	+17.1	+8.7	+8.66
Deuteron polarizability	−2.2	←	←	−0.1	←
Finite size of muonic molecule	+1.2	+0.29[f]	←	+1.0	+0.24[g]
Nuclear strong interaction	$\lesssim 10^{-4}$	←	$\sim 10^{-4}$	$\lesssim 10^{-4}$	←
Lower (para) hyperfine state	+35.9	+35.9	←	+16.2	←
Total $\Delta\epsilon$	+65.6	+64.8	+62.4	+25.7	+24.3

[a] Most recent values of Bakalov and colleagues (see [6] for references).
[b] G. Aissing and H. J. Monkhorst (unpublished).
[c] Kamimura and M. Kamimura et al. (unpublished).
[d] Using the triton charge form factor of Juster et al.; with the triton form factor of Collard et al. used in the other calculations, the calculated value is +13.3 (M. Kamimura, private communication).
[e] Arrows indicate which value is included in the sum if not calculated.
[f] Scrinzi & Szalewicz.
[g] Estimated.

$$\lambda_{mf} = N_0 \sum_{if} W_{K_i}(T) \int |M_{fi}|^2 2\pi\delta(\epsilon-\epsilon_r) f(\epsilon,T) d\epsilon \tag{2}$$

where $W_{K_i}(T)$ is the probability of initial rotational state K_i at temperature T, M_{fi} the transition matrix element, ϵ_r the resonance energy (which, of course, depends on K_i, K_f, ν, F, and S), and $f(\epsilon,T)$ the distribution of kinetic energy in the collisional center of mass (the Maxwell distribution for thermalized $t\mu$'s).

The pioneering calculation for the λ_{mf} was given by Vinitsky et al. in 1978 [7]. Subsequently, Lane [8] pointed

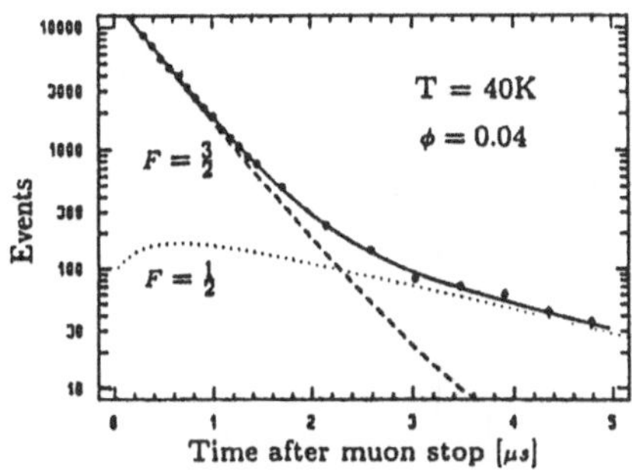

Fig. 3. Time spectrum of $dd\mu$ fusion neutrons (from Ref. 6)

out the importance of back decay, Leon [9] emphasized the need to include the plane-wave factor in M_{fi}, Cohen and Martin [10] showed how to include the effect of electron screening, and Menshikov and Faifman [11] stressed that undistorted wave functions need to be used in M_{fi}. Either the *post* or *prior* form of the interaction can be u sed, but except for Lane [12] the *post* form is the choice made; recently Faifman et al. [13] stressed its advantages. Here

$$H' = \overline{d} \cdot \overline{E} \quad , \tag{3}$$

where $\overline{d}$ is the dipole operator of the $dt\mu$ system and $\overline{E}$ the electric field at its c.m. from the spectator nucleus plus electrons. The dipole interaction, Eq. (3), takes the S-state $t\mu + d$ system to the $J = 1$ $dt\mu$ state. However, the $t\mu$+ D_2 system has an orbital angular momentum L, so that

$$\overline{L} + \overline{K}_i = \overline{J} + \overline{K}_f \quad , \tag{4}$$

with $J = 1$. Normally $L = 0$ is dominant, so that $K_f = K_i \pm 1$.

Once the $[(dt\mu)d2e]^*$ complex is formed, back decay competes with deexcitation and fusion [8], so that for the *effective* molecular formation rate we have

$$\widetilde{\lambda}_{mf} = \lambda_{mf} \cdot \frac{\widetilde{\lambda}_f}{\widetilde{\lambda}_f + \Gamma} \quad , \tag{5}$$

where $\widetilde{\lambda}_f$ is the sum of deexcitation and fusion rates and Γ is the back-decay rate (which in general is affected by collisions between the complex and the target molecules).

3 $dd\mu$ formation

μCF in pure D_2 targets has provided a valuable verification of resonant muonic molecule formation processes and allowed detailed comparison between theory and experiment.

The $F = \frac{3}{2}$ hyperfine state of $d\mu$ lies 48.5 meV above the $F = \frac{1}{2}$ state. The transition rate $F \to F'$ is an important quantity, denoted by $\lambda_{d\mu}^{FF'}$; $\lambda_{d\mu}^{\frac{3}{2}\frac{1}{2}}$ is directly measured experimentally, as described below [14].

Table 3. Results of the calculation of Friar et al. [43] compared to recent experiments or recent analyses of older experiments (from Ref. 43).

Table 3. Results of the calculation of Friar et al. [43] compared to recent experiments or recent analyses of older experiments (from Ref. 43).

Process	Theory	Experiment
$\lambda^{\mu}_{1/2}$ (10^6 sec^{-1})	0.062(2)	0.056(6) [45]
$\lambda^{\gamma}_{3/2}$ (10^6 sec^{-1})	0.107(6)	0.11(1) [42,46]
$\lambda^{\gamma}_{1/2}$ (10^6 sec^{-1})	0.37(1)	0.35(2) [42,46]
S_s (keV mb)	0.108(4)	0.12(3) [47]

The resonant molecular formation rate $\lambda_{dd\mu}^{FS}$ involves vibration excitation $\nu = 7$ for the complex. The much smaller nonresonant (Auger) rate is denoted by $\lambda_{dd\mu}^{nr}$. The $(dd\mu)_{11}$ state is very long-lived, because the identity of the deuterons implies that $\Delta J = 1$ is accompanied by $\Delta S_{dd} = 1$, which is forbidden. Thus, the $dd\mu$ is *stuck* in the $J = 1$ state, where fusion is rather slow: $\lambda_f \simeq 0.5 \times 10^9$ s^{-1} according to Bogdanova et al. [15]. As a result, back decay is actually dominant, and the effective molecular formation rate is [16]

$$\widetilde{\lambda}_{dd\mu}^{F} = \lambda_{dd\mu}^{nr} + \sum_S \lambda_{dd\mu}^{FS} \frac{\widetilde{\lambda}_f}{\widetilde{\lambda}_f + \sum_{F'} \Gamma_{SF'}} \quad . \tag{6}$$

Furthermore, resonant molecular formation followed by back decay contributes to hyperfine transitions, so for the effective HF transition rate we have [16,17]

$$\widetilde{\lambda}_{d\mu}^{FF'} = \lambda_{d\mu}^{FF'} + \sum_S \lambda_{dd\mu}^{FS} \frac{\Gamma_{SF'}}{\widetilde{\lambda}_f + \sum_{F''} \Gamma_{SF''}} \quad . \tag{7}$$

The muon kinetics in a pure D_2 target is shown in Fig. 2. Because the steady-state populations differ from the initial ones, *transients* appear in the detection of fusion neutrons as a function of time [14]. This is seen clearly in the AAS–PSI data shown in Fig. 3. The steep initial slope represents the emptying of the $F = \frac{3}{2}$ state, where the effective molecular formation rate is large (~4×10^6 s^{-1}), into the $F = \frac{1}{2}$ state where it is very small at low T. From this kind of data, the AAS–PSI group has been able to extract the individual molecular formation rates $\widetilde{\lambda}_{dd\mu}^{F}$, shown in Fig. 4(a), and the hyperfine transition rate $\widetilde{\lambda}_{d\mu}^{\frac{3}{2}\frac{1}{2}}$, shown in Fig. 4(b) [6]. The four hyperfine transitions that contribute are shown in Fig. 5.

The AAS–PSI data on $\widetilde{\lambda}_{dd\mu}^{F}$ can be fit using the beautiful *ab initio* calculations of Menshikov et al. [16], with only the resonance energy ϵ_r and $\widetilde{\lambda}_f$ adjustable. However, the theoretical values of $\widetilde{\lambda}_{d\mu}^{\frac{3}{2}\frac{1}{2}}$ clearly exceed the experimental ones by a significant amount, as seen in

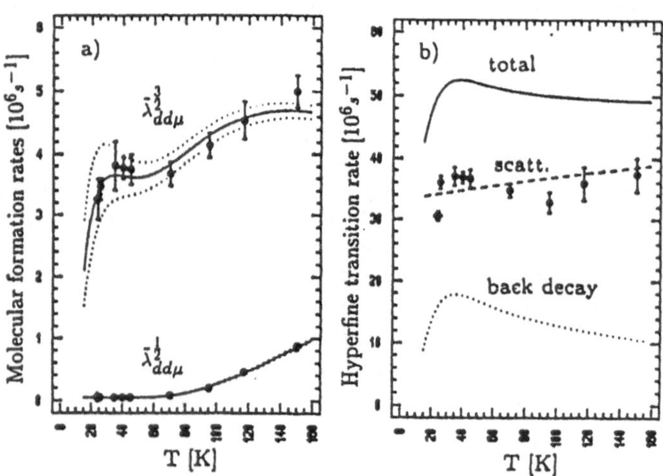

Fig. 4. Results of PSI experiments on hyperfine effects. (a) Molecular formation rates, (b) hyperfine transition rates (from Ref. 6)

Fig. 4(b); presumably the *nonresonant* contribution [18] is being overestimated. It should also be noted that $\widetilde{\lambda}^F_{dd\mu}$ is relatively insensitive to the magnitude of the matrix element $|M_{fi}|$, since this enters in both the numerator and the denominator of the resonant term of Eq. (6).

It should be possible to extract even more information from the $dd\mu$ system, by going to low target density ϕ (measured relative to liquid H_2 density). According to Menshikov et al. [16], for the complex equilibrated at the target temperature (20 K),

$$\Gamma \equiv \sum_{F'} \Gamma_{SF'} \simeq 1.5 \times 10^9 \text{ s}^{-1} \quad , \tag{8}$$

while Padial et al. [19] find for the rate for rotation relaxation of the $K_f = 1$ complex,

$$\lambda_{1\to 0} = 1.6 \times 10^{11} \phi \text{ s}^{-1} \quad . \tag{9}$$

Thus, for $\phi \lesssim 1\%$, the effect of the *initial* (time $t = 0$) back-decay rate should become visible. Since we expect

$$\Gamma(t=0) \simeq 2\Gamma(t=\infty) \equiv 2\Gamma \tag{10}$$

at this T, then for $\phi \ll 1\%$ $\widetilde{\lambda}^{\frac{3}{2}}_{dd\mu}$ should be a factor of two smaller, and the part of $\widetilde{\lambda}^{\frac{3}{2}\frac{1}{2}}_{d\mu}$ due to molecular formation a factor of two bigger, than the $\phi \gg 1\%$ values. Observation of this variation with ϕ would allow direct comparison of Γ and $\lambda_{1\to 0}$.

4 $dt\mu$ formation

For D/T targets, the kinetics, shown in Fig. 6, is, of course, much more complicated than for pure D_2. The steady-state cycling time can be written as a sum of the time the muon spends in the $d\mu$ ground state, plus the times in the $t\mu$ singlet and triplet ground states, so that for the steady-state cycling rate [9]

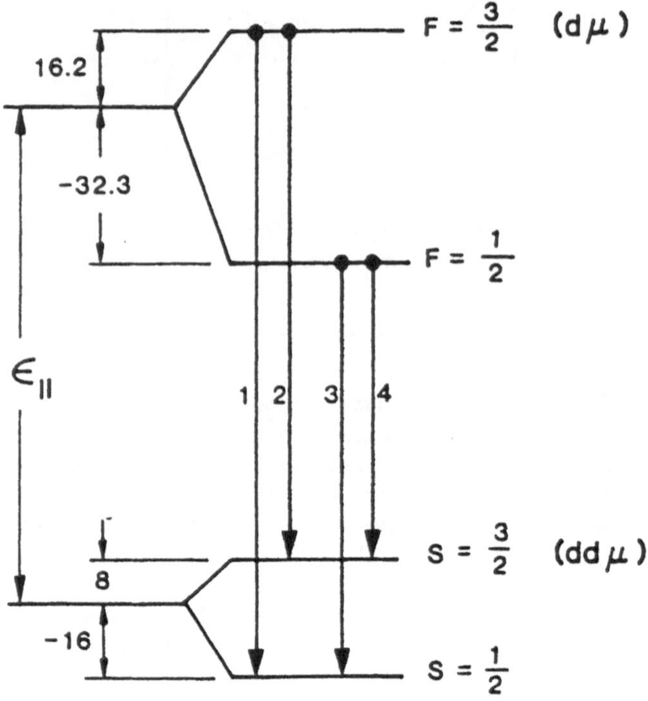

Fig. 5. The hyperfine transitions contributing to $dd\mu$ formation (from Ref. 17)

$$\lambda_c^{-1} = \frac{q_{1S} c_d}{\lambda_{dt} c_t} + \frac{\frac{3}{4}}{\lambda^{10}_{t\mu} c_t + \lambda^{1}_{dt\mu} c_d} + \frac{\frac{1}{4} + \frac{3}{4}\chi}{\lambda^{0}_{dt\mu} c_d} \quad ; \tag{11}$$

here the branching ratio

$$\chi \equiv \frac{\lambda^{10}_{t\mu} c_t}{\lambda^{10}_{t\mu} c_t + \lambda^{1}_{dt\mu} c_d} \quad , \tag{12}$$

and the molecular formation rates $\lambda^F_{dt\mu}$ have contributions from both D_2 and DT molecules.

At low temperature ($\lesssim$200 K), the only accessible resonances for thermalized $t\mu$'s are for $F = 0$ on D_2 [9]. Furthermore, for $c_t \gtrsim 0.7$ the first two terms on the RHS of Eq. (11) should be negligible, so that

$$\lambda_c \propto c_{D_2} \quad . \tag{13}$$

A recent LAMPF experiment to test this relation found that it does *not* seem to hold [20,21]! The apparent contribution of DT molecules to molecular formation is thought to be due to the contribution of epithermal $t\mu$'s to steady-state molecular formation. (The role of epithermal molecular formation in giving rise to the transients seen for low density ($\phi \simeq 1\%$) at PSI is well-established by now [6].) The $dt\mu$ molecular formation rates are evidently so large as to compete with thermalization [22,23]! (Examples of rates calculated by Faifman et al.

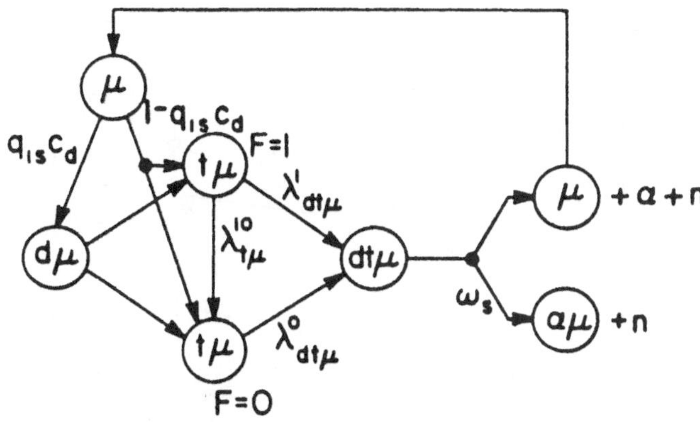

Fig. 6. Simplified $dt\mu$ cycle (from Ref. 6)

are shown in Fig. 7 [13].) Thus, all the elastic scatteri ng cross sections and an intricate kinetics calculation are needed just to compare the theoretical molecular formation rates to the experimental cycling rates.

As if that were not enough complication, the (normalized) cycling rates show a density dependence evidently due to a three-body contribution to molecular formation; LAMPF and PSI data exhibiting this effect are shown in Fig. 8. The three-body effect is thought to be due to the unique property of singlet $t\mu$ on D_2 having its strongest transitions lying just below threshold (for $\nu = 2$): $0 \to 1$ has $\epsilon_r \simeq -12$ meV, $0 \to 2$ has $\epsilon_r \simeq -2$ meV [6]. Menshikov and Ponomarev [24] have suggested that a spectator molecule picks up enough energy in a three-body collision,

$$t\mu + D_2 + X \to [(dt\mu)d2e]^* + X' \quad , \tag{14}$$

to move the $0 \to 1$ transition into the physically accessible region. Starting with Petrov [25], several workers [26] have used the idea of *collisional broadening* and the replacement

$$\delta(\epsilon - \epsilon_r) \to \frac{1}{2\pi} \cdot \frac{\Gamma_c}{(\epsilon - \epsilon_r)^2 + (\Gamma_c/2)^2} \tag{15}$$

to calculate three-body molecular formation.

This replacement is equivalent to the *impact approximation* in the theory of spectral line broadening. However, as pointed out by Cohen and Leon [27], the slowness of the molecular collisions implies that the impact approximation is valid only for $|\Delta\epsilon| \ll 1$ meV, and is therefore uninteresting!

Thus, we are left with the problem of how to calculate three-body molecular formation. Petrov and Petrov [28] have applied many-body perturbation theory to the problem, but had to replace the intermolecular potential by a hard-sphere interaction. Lane [29] attempted to extend the Baranger line-broadening theory to include the $t\mu$ momentum, but again had to assume hard-sphere interactions. Leon [30] used a quasistatic treatment to show that the torque exerted on the complex by a neighboring molecule implied a significant configuration mixing of the K_f states of the complex, and hence a significant three-body effect. The lack of a complete and convincing method of calculation for three-body molecular formation remains a glaring deficiency of μCF theory.

In a somewhat different vein, Fukushima [31] has recently calculated resonant formation in *solid* D_2; the three- or multi-body effect can be termed *phonon-assisted* molecular formation [32].

5 $pd\mu$ fusion

$pd\mu$ fusion was first predicted long ago in 1947 by F. C. Frank [33], and observed a decade later by Alvarez et al. in a hydrogen bubble chamber [34]. The fusion reactions are

$$d\mu + p \to {}^3\mathrm{He}\mu + \gamma \tag{16a}$$

$$\to {}^3\mathrm{He} + \mu \quad , \tag{16b}$$

with (16a) predominant. The Wolfenstein-Gershtein (W-G) effect says that the HF quenching, $(d\mu)^{3/2} \to (d\mu)^{1/2}$ (and therefore the deuterium fraction), affects the fusion yield [35]. That is, the statistical ratio of $d\mu$ hyperfine populations leads to a statistical distribution among the $S = 2,1,1',0$ HF states of the $pd\mu$ molecule , while complete quenching of the $(d\mu)^{3/2}$ state implies that only $S = 1,1'$, and 0 are populated; the former have a larger fraction in the nuclear quartet state than the latter combination. Since the nuclear reaction rate from the quartet is expected to be smaller than from the doublet, quenching of the $(d\mu)^{3/2}$ state increases the fusion rate and hence the fusion yield.

Cohen et al. in 1960 [36] estimated that the fusion rates satisfy

$$\lambda^\gamma_{3/2} \ll \lambda^\gamma_{1/2} \quad ; \tag{17}$$

thereafter it became traditional to neglect $\lambda^\gamma_{3/2}$ completely [4], although Carter in 1966 [37] warned that relation (17) was not at all justified. Thus, when Bleser et al. [38] in 1963 measured the W-G effect, they assumed $\lambda^\gamma_{3/2} = 0$ and the then current theoretical value of the quenching rate $\lambda^{\frac{3}{2}\frac{1}{2}}_{d\mu} \simeq 7\ \mu s^{-1}$ [35]. These values lead to a predicted ratio of γ-yields at the different deuterium fractions of

$$\frac{Y_\gamma\ (25\%)}{Y_\gamma\ (0.7\%)} = 1.18 \quad . \tag{18}$$

(Complete quenching of the $F = 3/2$ state would give 1.8 for this ratio.) The measured value was 1.17(1), in essentially perfect agreement [38].

When Bertl et al. [39] remeasured the W-G effect twenty years later, they found

$$\frac{Y_\gamma\ (22\%)}{Y_\gamma\ (0.6\%)} = 1.172(5) \tag{19}$$

in excellent agreement with the Bleser et al. [38] experiment. However, in the meantime the value of the quenching rate $\lambda^{\frac{3}{2}\frac{1}{2}}_{d\mu}$ had changed drastically: Matveenko

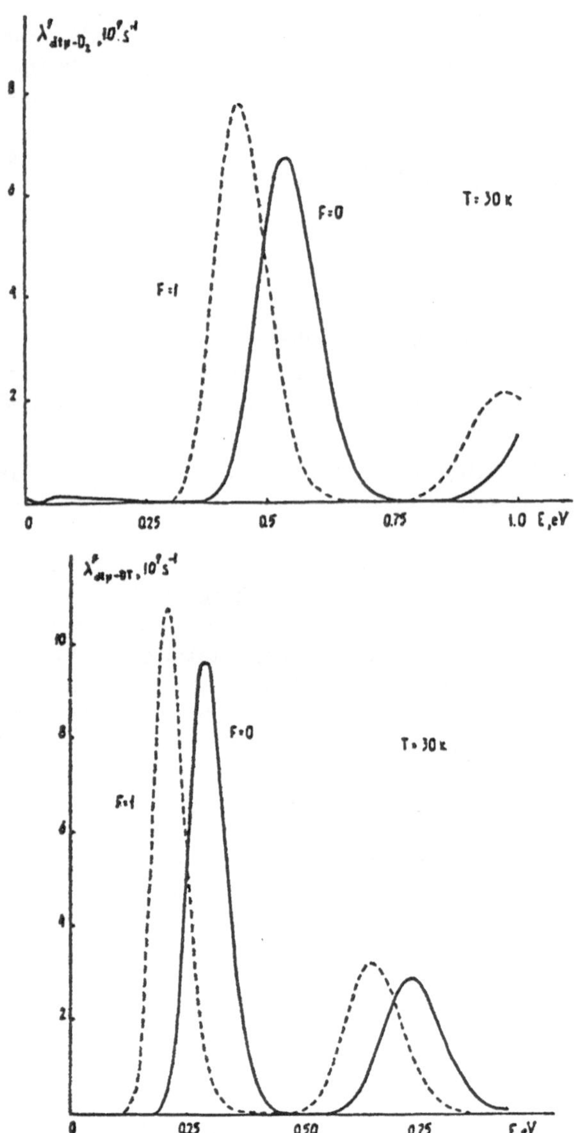

Fig. 7. $\lambda^F_{dt\mu}$ as functions of laboratory energy for T = 30 K for D_2 (upper) and DT (lower) targets (from Ref. 13)

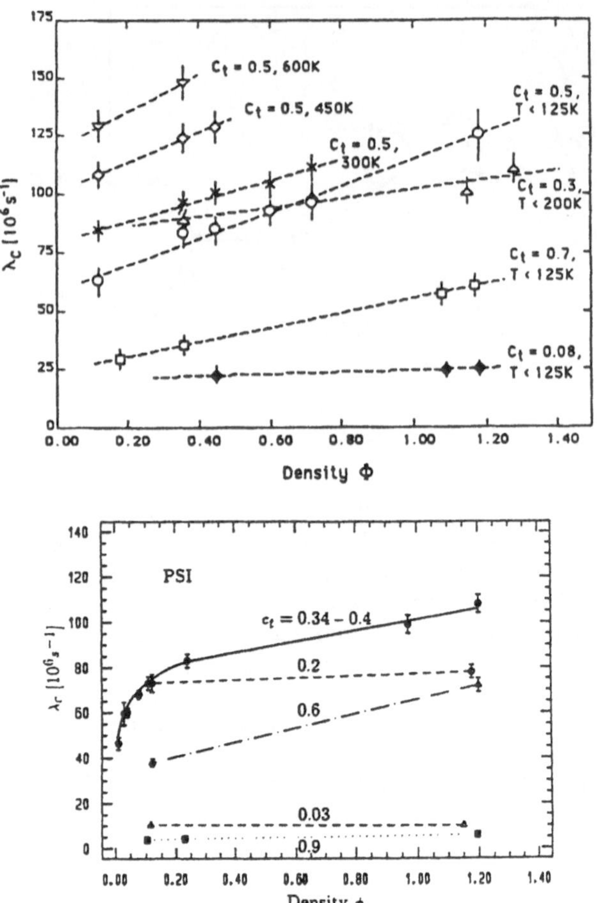

Fig. 8. Density dependence of normalized cycling rates λ_c; (top) LAMPF, (bottom) PSI, T between 20 K and 45 K (from Ref. 6)

and Ponomarev in 1971 [40] calculated 46 μs^{-1} for this value, while even more significantly Kammel et al. [14] measured this rate as 37(2) μs^{-1} using $dd\mu$ fusion (see Fig. 3). With this latter value, the predicted W-G ratio becomes 1.50! To patch up this discrepancy, Bertl et al. [39] postulated that in addition to the quenching from collisions with deuterons (from exchange scattering), there is a contribution from collisions with protons, and adjusted its value to give the observed W-G ratio. However, there is no mechanism known that can account for this process [41], so this solution was not very convincing.

In a more recent $pd\mu$ experiment at PSI, Petitjean et al. [42] instead fitted the data by allowing $\lambda^{\gamma}_{3/2}$ to differ from zero. Finally, Friar et al. [43] have very recently calculated the various $pd\mu$ fusion rates using accurate three-nucleon wave functions derived with realistic potentials; not only do they find a significant value for $\lambda^{\gamma}_{3/2}$, but all the calculated fusion rates agree remarkably well with the experimental values—as shown in Table 3. Thus, after 30 years we can now say that the W-G effect is finally understood! (As it was thought to be in 1963.)

Lest I leave you with the impression that the $pd\mu$ system is now completely understood, I mention that unexpected molecular effects, H_2+ D_2 vs. HD, have been reported in the yields of the fusion γ's [44]—these remain completely unexplained, and will doubtlessly be the subject of vigorous future investigation.

6 Summary

The agreement between theory and experiment for resonant $dd\mu$ formation is very encouraging, especially that for the magnitude of the resonance energies, which are determined to better than a meV, and the temperature dependence of the formation rates. For $dt\mu$ formation the situation is much less clear, because of the more complicated kinetics and the need to treat thermalization and molecular formation together to compare theory with experiment, and the difficulty in calculating three-body molecular formation. In contrast, $pd\mu$ fusion and the Wolfenstein-Gershtein effect now appear to be very well understood, while the reported dependence of $pd\mu$ formation on the molecular structure of the target remains mysterious.

Acknowledgement. This work was performed under the auspices of the U.S. Department of Energy.

References

1. Zeldovitch, Ya. B., Gershtein, S.S.: *Usp. Fiz. Nauk* **71**, 581 (1960) [*Sov. Phys. Uspekhi* **3**, 593 (1961)].
2. Vesman, E. A.: *Pisma Zh. Eksp. Teor. Fiz.* **5**, 113 (1967) [*JETP Lett.* **5**, 91 (1967)].
3. Vinitsky, S. I. et al.: Preprint JINR P4-10336, Dubna (1976).
4. Gershtein, S. S., Ponomarev, L. I., in *Muon Physics*, ed. V. W. Hughes, C. S. Wu, New York: Academic (1975), **3**, 141.
5. Gershtein, S. S., Ponomarev, L. I.: *Phys. Lett.* **72B**, 80 (1977).
6. Breunlich, W. H. et al.: *Ann. Rev. Nucl. Part. Sci.* **39**, 311 (1989).
7. Vinitsky, S. I. et al.: *Zh. Eksp. Teor. Fiz.* **74**, 849 (1978) [*Sov. Phys. JETP* **47**, 444 (1978)].
8. Lane, A. M.: *Phys. Lett.* **98A**, 337 (1983).
9. Leon, M.: *Phys. Rev. Lett.* **52**, 605 (1984).
10. Cohen, J. S., Martin, R. L.: *Phys. Rev. Lett.* **53**, 738 (1984).
11. Menshikov, L. I. Faifman, M. P.: *Yad. Fiz.* **43**, 650 (1986) [*Sov. J. Nucl. Phys.* **43**, 414 (1986)].
12. Lane, A. M.: *J. Phys. B* **20**, 2911 (1987).
13. Faifman, M. P. et al.: *Muon Catal. Fusion* **4**, 1 (1989).
14. Kammel, P. et al.: *Phys. Rev. A* **28**, 2611 (1983).
15. Bogdanova, L. N. et al.: *Phys. Lett.* **115B**, 171 (1982); errata **167B**, 485 (1986).
16. Menshikov, L. I. et al.: *Zh. Eksp. Teor. Fiz.* **92**, 1173 (1987) [*Sov. Phys. JETP* **65**, 656 (1987)].
17. Leon, M.: *Phys. Rev. A* **33**, 4434 (1986).
18. Bracci, L. et al.: *Phys. Lett. A* **134**, 435 (1989).
19. Padial, N. T. et al.: *Phys. Rev. A* **37**, 329 (1988).
20. Jones, S. E., Anderson, A. N.: *Proc. Int. Symp. Muon Catal. Fusion μCF-89*, ed. J. D. Davis, RAL-90-022, 13 (1990).
21. Leon, M.: see Ref. 20, p. 23.
22. Cohen, J. S.: *Phys. Rev. A* **34**, 2719 (1976).
23. Jeitler, M. et al.: *Muon Catal. Fusion*, 5/6, 217 (1991)
24. Menshikov, L. I., Ponomarev, L. I.: *Phys. Lett.* **167B**, 141 (1986).
25. Petrov, Yu. V.: *Phys. Lett.* **163B**, 28 (1985).
26. Leon, M.: *Muon Catal. Fusion* **1**, 163 (1987); Petrov, Yu. V. et al.: *ibid.* **2**, 261 (1988); Menshikov, L. I.: *ibid.* **2**, 273 (1988); Menshikov, L. I.: Preprint IAE-4606-2, Moscow (1988); Lane, A. M.: *J. Phys. B* **21**, 2159 (1988).
27. Cohen, J. S., Leon, M.: *Phys. Rev. A* **39**, 946 (1989).
28. Petrov, V. Yu., Petrov, Yu. V.: *Muon Catal. Fusion* **4**, 73 (1989).
29. Lane, A. M.: *J. Phys. B* **22**, 2817 (1989).
30. Leon, M.: *Phys. Rev. A* **39**, 5554 (1989).
31. Fukushima, K.: *Muon Catal. Fusion*, see Ref. 20, p. 62.
32. Leon, M.: *Workshop on Fund. Muon Physics*, Los Alamos, New Mexico (1986), LA-10714-C.
33. Frank, F. C.: *Nature* **160**, 525 (1947).
34. Alvarez, L. W. et al.: *Phys. Rev.* **105**, 1127 (1957).
35. Gershtein, S. S.: *Zh. Eksp. Teor. Fiz.* **40**, 698 (1961) [*Sov. Phys. JETP* **13**, 488 (1961)].
36. Cohen, S. et al.: *Phys. Rev.* **119**, 384 (1960).
37. Carter, B. P.: *Phys. Rev.* **141**, 863 (1966).
38. Bleser, E. J. et al.: *Phys. Rev.* **132**, 2679 (1963).
39. Bertl, W. H. et al.: *Atomkernenerg. Kerntech.* **43**, 184 (1983).
40. Matveenko, A. V., Ponomarev, L. I.: *Zh. Eksp. Teor. Fiz.* **59**, 1593 (1970) [*Sov. Phys. JETP* **32**, 871 (1971)].
41. Cohen, J. S.: *Phys. Rev. A*, to be published.
42. Petitjean, C. et al.: See Ref. 20, p. 42.
43. Friar, J. L. et al.: *Phys. Rev. Lett.* **66**, 1827 (1991).
44. Aniol, K. A. et al.: *Muon Catal. Fusion* **2**, 63 (1988).
45. Bogdanova, L. N. et al.: *Muon Catal. Fusion* **3**, 377 (1988); Bogdanova, L. N., Markushin, V. E. (unpublished).
46. Petitjean, C. et al. (unpublished).
47. Griffiths, G. M. et al.: *Can. J. Phys.* **41**, 724 (1963).

This article was processed using Springer-Verlag TEX Z.Physik C macro package 1991
and the AMS fonts, developed by the American Mathematical Society.

Z. Phys. C - Particles and Fields 56, S210–S214 (1992)

Zeitschrift für Physik C Particles and Fields

Resonant formation of $dt\mu$ mesic molecules in the triple $H_2 + D_2 + T_2$ mixture

M.P. Faifman and L.I. Ponomarev

I.V. Kurchatov Institute of Atomic Energy, Moscow, 123182, USSR

15-January-1992

Abstract. The processes of resonant $dt\mu$ molecule formation $t\mu + XD \to [(dt\mu)xee]$ are considered and their rates $\lambda_{dt\mu-x}$ are calculated as a function of the $t\mu$ mesic atom energy and the temperature of $H_2 + D_2 + T_2$ mixture. It is shown that at kinetic energies of $t\mu$-atoms of $E \sim 0.1 - 0.3$ eV the resonant rates $\lambda_{dt\mu-x}$ reach values of $10^9 - 10^{10} s^{-1}$. At $T = 900$ K the resonant formation rate $\lambda_{dt\mu-p}$ averaged over the Maxwellian distribution on $t\mu$ atom energies is $\approx 2.3\ 10^9 s^{-1}$. The consequences of these results are discussed.

1. In recent years interest has grown in the problem of the muon catalysis in deuterium-tritium mixture [1]

$$\mu^- + (D_2, T_2) \xrightarrow{\lambda_a} d\mu \xrightarrow{\lambda_{dt}} t\mu \xrightarrow{\lambda_{dt\mu}}$$

$$dt\mu \; - \begin{cases} \xrightarrow{\lambda_f} {}^4He + n + \mu^- & (1a) \\ \xrightarrow{\omega_s} \mu\,{}^4He + n & (1b) \end{cases}$$

From the point of view of possible muon catalysis practical applications the "bottle neck" in the sequence of reactions (1a) is the stage of $dt\mu$ mesic molecule formation.

It was shown in 1977 [2] that the $dt\mu$ mesic molecule resonant formation rate $\lambda_{dt\mu}$ in the reaction

$$t\mu + D_2 \to [(dt\mu)dee] \tag{2}$$

can reach the value of $10^8 s^{-1}$ in a $D_2 + T_2$ mixture and, as a consequence, one muon can catalyze about 100 fusions (1a). The experiments and calculations which followed confirmed this estimate (see reviews [1]).

In the triple mixture $H_2 + D_2 + T_2$ other resonant reactions are also possible

$$t\mu + DT \to [(dt\mu)tee] \tag{3}$$

$$t\mu + HD \to [(dt\mu)pee] \tag{4}$$

The rates of the reactions (2) and (3) as functions of the $t\mu$ mesic atom kinetic energy E at some temperatures T of $D_2 + T_2$ mixture were calculated in paper [3].

Here we discuss the results of our calculations of the rates of reaction (4) (they will be presented in detail in a separate paper [4]), and their significance from the point of view of the experiment and possible practical applications.

2. It is known [1] that the number of the muon catalysis cycles X_c can be approximately estimated by the formula

$$X_c = \frac{\lambda_c \phi}{\lambda_0 + \omega_s \lambda_c \phi} \tag{5}$$

where $\lambda_0 = 0.46\ 10^6 s^{-1}$ is the rate of muon decay $\mu^- \to e^- + \nu_\mu + \nu_e$, $\phi = N/N_0$ is the relative density of the mixture ($N_0 = 4.25\ 10^{22} cm^{-3}$ is the liquid hydrogen density), ω_s is the probability of muon sticking to helium in the fusion reaction (1b), and λ_c is the muon catalysis cycling rate. In the binary ($D_2 + T_2$) mixture the rate λ_c is mainly determined by the rate λ_{dt} of the isotope exchange $d\mu \to t\mu$, and the rate $\lambda_{dt\mu}$ which is determined by the rates $\lambda^F_{dt\mu-x}$, (X=d,t) of the $dt\mu$ molecule resonant formation [1] in reactions (2) and (3) in collisions with $(t\mu)_F$ - atoms with the total spin F (C_d and C_t - relative concentrations of nuclei d and t in $D_2 + T_2$ mixture, $C_d + C_t = 1$):

$$\lambda_c^{-1} \approx (\lambda_{dt\mu} \cdot C_d)^{-1} + C_d \cdot (\lambda_{dt} \cdot C_t)^{-1} \tag{6}$$

It is seen from (5) that in order to increase the number of catalysis cycles one has either to diminish ω_s, or to increase the cycling rate λ_c. The theoretical estimates and experimental results show that in the

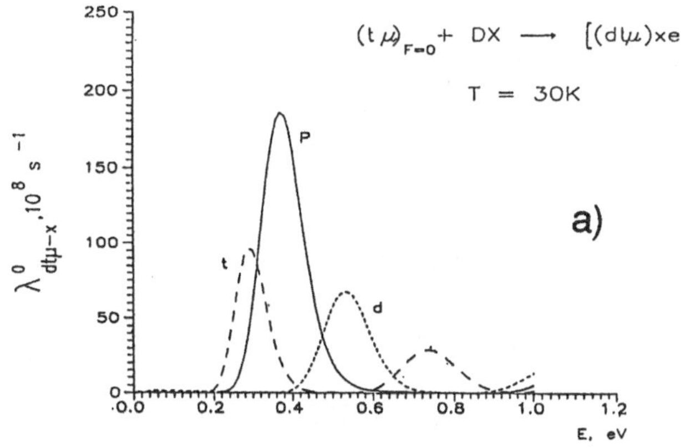

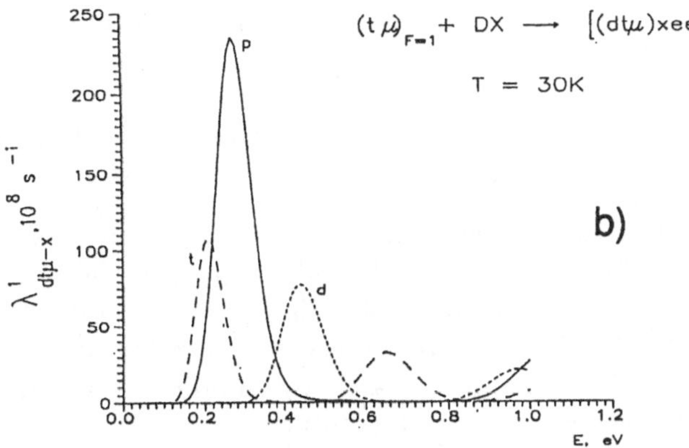

Fig. 1. The dependences of the resonant rates of reaction (2)-(4), $\lambda^F_{dt\mu-x}$, $X=(p,d,t)$, on the $t\mu$-atom kinetic energy E at temperature $T=30$ K for $F=0$ (a) and $F=1$ (b).

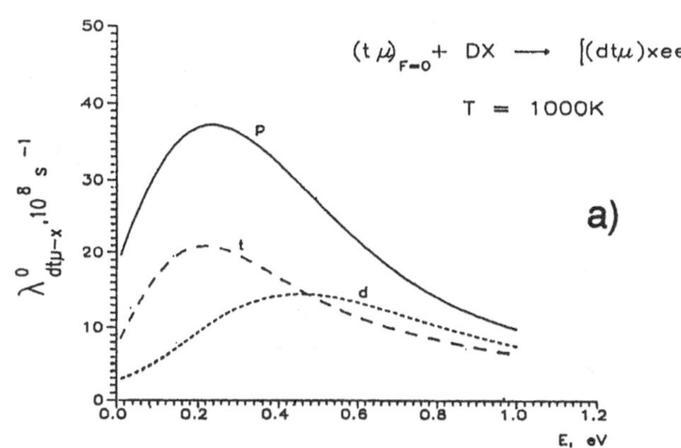

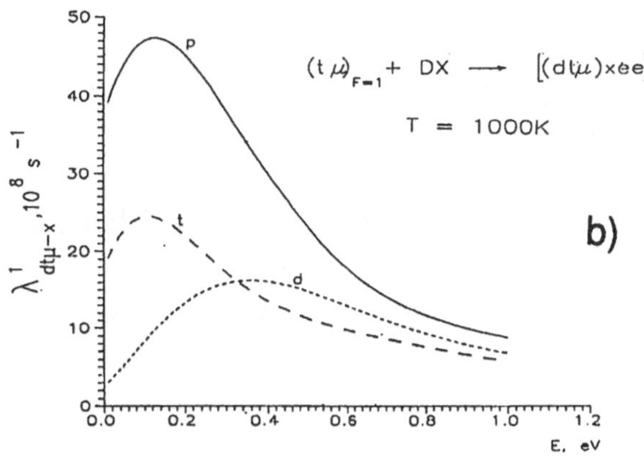

Fig. 2. The dependences of the resonant rates of reaction (2)-(4), $\lambda^F_{dt\mu-x}(E), X=(p,d,t)$, on the $T\mu$-atom kinetic energy E at temperature $T=1000$ K.

reaction chain (1) the rates $\lambda_a \approx 3\ 10^{12}s^{-1}$ and $\lambda_f \approx 1.2\ 10^{12}s^{-1}$, but the rates λ_{dt} and $\lambda_{dt\mu}$ (and, hence, λ_c) do not exceed $\sim 3\cdot 10^8 s^{-1}$ at $T \leqq 1000$ K of D_2+T_2 mixture. For "hot" $t\mu$ atoms with a kinetic energy of $E=0.3-0.5$ eV the rates $\lambda_{dt\mu-d}$ and $\lambda_{dt\mu-t}$ reach values of $\sim(0.3-1.0)\cdot 10^{10}s^{-1}$ [3,5], however, this does not result in a noticeable increase in X_c due to a fast thermalization of $t\mu$ atoms in elastic collisions $t\mu+d$ and $t+t$, and, the low value of the isotope exchange rate $\lambda_{dt}\approx 2.7\ 10^8 s^{-1}$ for termalized $t\mu$-atoms limits the cycling rate λ_c.

The isotope exchange rates $p\mu\to t\mu$ and $p\mu\to d\mu$ are $\lambda_{pt}=0.7\ 10^{10}s^{-1}$ and $\lambda_{pd}=1.7\ 10^{10}s^{-1}$ correspondingly (5). In these transitions "hot" $t\mu-$ and $d\mu-$ atoms are formed with an initial kinetic energies $E_{d\mu}\approx 20$ eV and $E_{d\mu}\approx 46$ eV correspondingly. For the "hot" $t\mu$-atoms with a kinetic energy $E_{d\mu}\approx 20$ eV the charge exchange rate $d\mu\to t\mu$ is $\lambda_{dt}\approx 10^{10}s^{-1}$ [5]. Therefore at moderate concentration C_t the main part of $t\mu$-atoms forms with an initial kinetic energy $E_{t\mu}\geqq 20$ eV due to the processes $p\mu\to t\mu$ and $p\mu\to d\mu\to t\mu$. At the same time the cross sections of the elastic processes $t\mu+p$ and $d\mu+p$ are smaller than the cross section of the process $t\mu+d$ and, besides, they have a Ramsauer minimum at collision energies $E\approx 1.2$ eV and $E\approx 1.1$ eV respectively [5]. All these features should increase the time of the $t\mu$ atom thermalization and essentially influence the physics of the muon catalysis phenomenon in the triple $H_2+D_2+T_2$ mixture. The latest experimental results [6] and the recent kinetic calculations [7] support this statement.

3. Fig. 1 and 2 display the results of calculations [4] of the rates of the reactions (2)-(4) $\lambda^F_{dt\mu-x}$, $X=(p,d,t)$, at temperatures $T=30$ K and 1000 K as functions of the kinetic energy E of the $(t\mu)$ mesic atom in the state with spin F. It is seen that at $T=30K$ the rates $\lambda^F_{dt\mu-p}$ reach values $\lambda^1_{dt\mu-p}=2.3\ 10^{10}s^{-1}$ (E = 0.27 eV) and $\lambda^0_{dt\mu-p}=1.9\cdot 10^{10}s^{-1}$(E = 0.37 eV), at $T=1000$ K $\lambda^1_{dt\mu-p}=0.5\ 10^{10}s^{-1}$ (E = 0.13 eV), $\lambda^0_{dt\mu-p}=0.4\ 10^{10}s^{-1}$ (E = 0.24 eV).

Fig. 3 shows the temperature dependences of the rates $\lambda^F_{dt\mu-x}(T)$ of the resonance reactions (2)-(4) which were obtained by averaging the func-

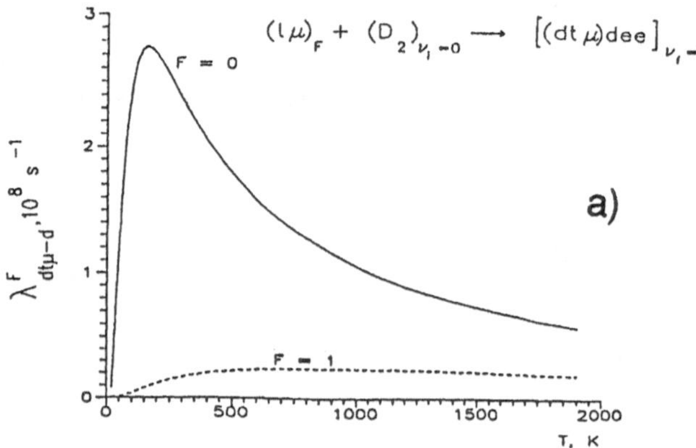

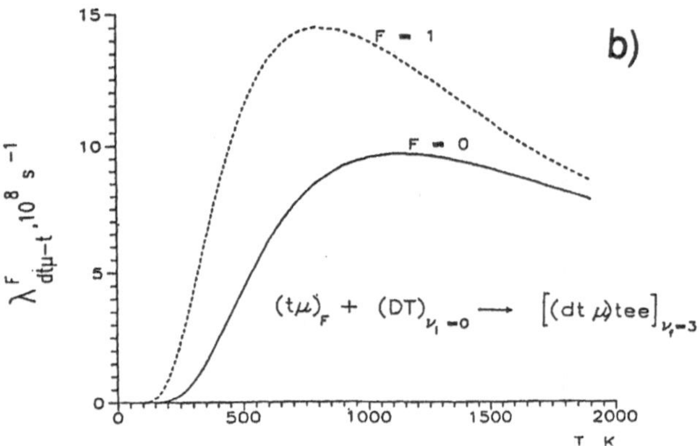

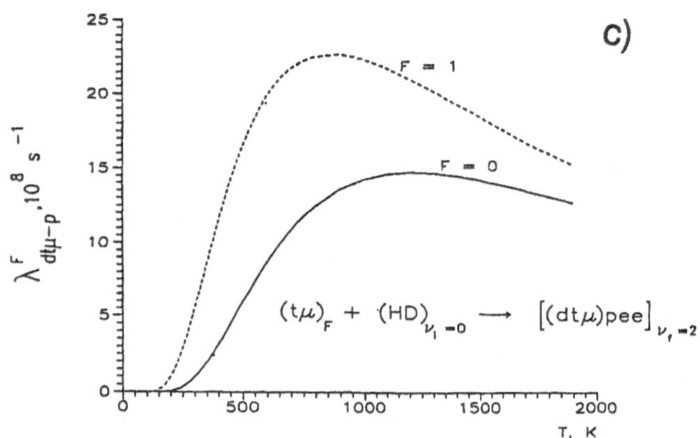

Fig. 3. The dependences of the resonant rates $\lambda^F_{dt\mu-x}(T), x = (p,d,t)$, on temperature T: a) x = d, reaction (2), transition $(\nu_i = 0) \rightarrow (\nu_f = 2)$; b) x = t, reaction (3), transition $(\nu_i = 0) \rightarrow (\nu_f = 3)$; c) x = p, reaction (4), transition $(\nu_i = 0) \rightarrow (\nu_f = 2)$.

tions $\lambda^F_{dt\mu-x}(E,T)$ over a Maxwellian distribution $f(E,T) = 2(E/\pi)^{1/2}T^{-3/2}$ exp(-E/T) in the $t\mu$ atom energy E.

$$\lambda^F_{dt\mu-x}(T) = \int_0^\infty \lambda^F_{dt\mu-x}(E,T)f(E,T)dE \qquad (7)$$

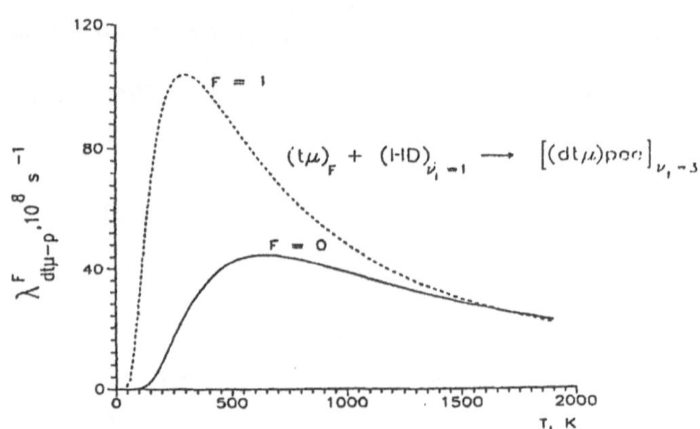

Fig. 4. The dependences of reaction (4) rate $\lambda^F_{dt\mu-p}(T)$ on temperature T for the transition $(\nu_i = 1) \rightarrow (\nu_f = 3)$ (to be compared with Fig. 3c).

(the populations of the rotational states K_i of the molecules XD at a given temperature also taken into account).

The rate of reaction (4) from state F = 1 of the $t\mu$ mesic atom reaches the value $\lambda^1_{dt\mu-p} \cong 2.3\ 10^9 s^{-1}$ at $T \cong 900$ K and rate $\lambda^0_{dt\mu-p} \cong 1.5\ 10^9 s^{-1}$ at $T \cong 1200$ K.These values exceed the rates $\lambda^F_{dt\mu-d}(T)$ of reaction (2) approximately by an order of magnitude in the same range of temperatures.

The largest contribution to the quantity $\lambda^F_{dt\mu-p}$ gives the transition from the triplet state F = 1 of the mesic atom $t\mu$ and initial vibrational state $\nu_1 = 0$ of HD molecule to the final vibrational state $\nu_f = 2$ of the mesic molecular complex $[(dt\mu)pee]$. (The resonant rate of just this transition has a maximum at the temperature $T \cong 900K$)[1]

4. The method peculiarities of the $dt\mu$ mesic molecule resonant formation in triple mixtures must manifest themselves in two ways: The Fig. 5-7 show the two dimensional plots $\lambda^F_{dt\mu-x}(E,T)$. It is clear today that the result $X_c \sim 100$ is the consequence of the samll bump at $E \approx 0.1$ eV, $T \approx 100$ K in the plot $\lambda_{dt\mu-d}(E,T)$ (see Fig.5a). It is clear also from these figures that we have several possibilities to enhance the cycling rate λ_c (and also X_c) by varying the temperature T, the density ϕ and the concentration C_p, C_t of the triple mixture.

Firstly, the essential part of the "hot" $t\mu$ mesic atoms realize reactions (2) - (4) with high rate $10^{10}s^{-1}$ at $t\mu$-atom energies $E = 0.2 - 0.5$ eV i.e., during the mesic atom deceleration before they are thermalized. As it has

[1] It is interesting to mention, that partial rates $\lambda^F_{dt\mu-p}(T)$ of reaction (4) for the transitions $(\nu_i = 1 \rightarrow (\nu_f = 3)$ essentially exceeds analogous rates for the transitions $(\nu_i = 0) \rightarrow (\nu_f = 2)$: $\lambda^1_{dt\mu-p} \cong 10^{10}s^{-1}$ $(T \cong 300$ K) and $\lambda^0_{dt\mu-p} \cong 4\ 10^9 s^{-1}$ $(T \cong 600$ K) (see Fig. 4). However, the contribution from this transition to the total rate of reaction (4) is insignificant, since the population of the vibrational state $\nu = 1$ of HD molecule is small at these temperatures.

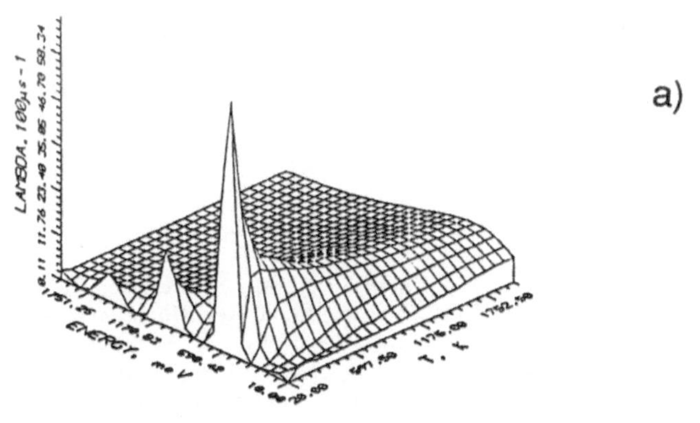

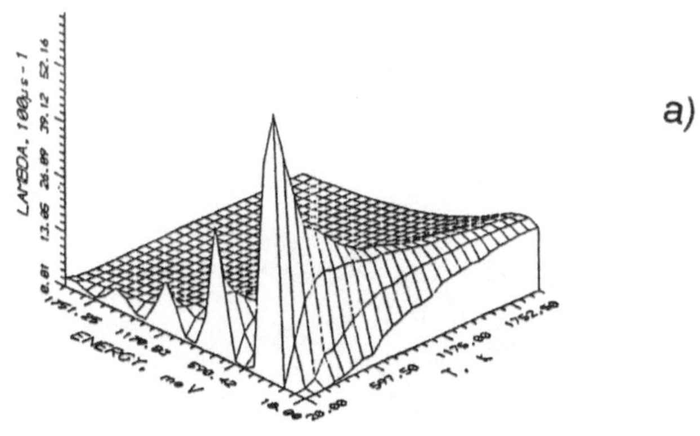

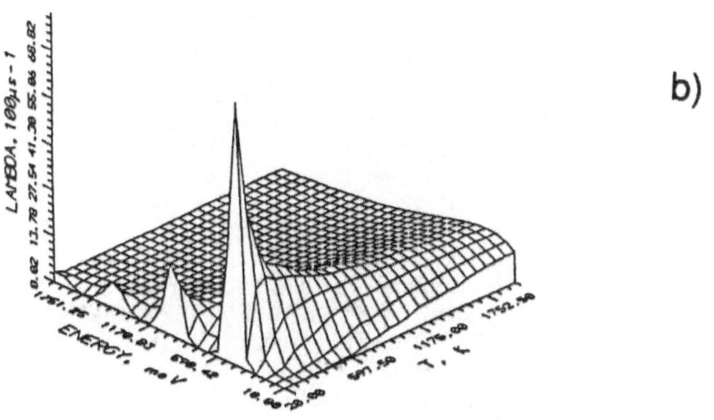

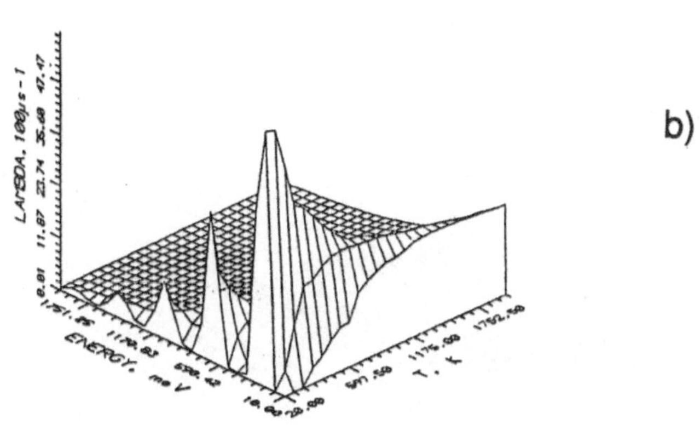

Fig. 5. The plots of the resonant rate $\lambda^F_{dt\mu-d}(E,T)$ as a function of the collision energy E and temperature T from the state F of $t\mu$-atom for $F=0$ (a) and $F=1$ (b).

Fig. 6. $\lambda^F_{dt\mu-t}(E,T)$ as a function of energy E and temperature T for $F=0$ (a) and $F=1$ (b).

been already noted, this effect shows up in the character of the neutron time distribution [6] (see also [8]).

Secondly, high rates $\lambda^F_{dt\mu-p}(T)$ of reaction (4) in a quasisteady regime, i.e., with thermalized $t\mu$-atoms, open the principal possibility to increase the rate λ_c of muon catalysis cycle in the triple mixture compared to the binary one, and, as a consequence, the number X_c of catalysis cycles, which is measured now in the binary mixture D_2+T_2 as $X_c \approx 120 \div 150$ [9]. Besides this, the high rates $\lambda^F_{dt\mu-p}$ and λ_{pt}, λ_{pd} and λ_{dt} in the triple mixture enable one to diminish the amount of tritium in the target, not diminishing the number of catalysis cycles. Of course, in order to make quantitative conclusions, detailed calculations of the muon catalysis kinetics in a triple mixture should be performed taking into account the deceleration of mesic atoms. These calculations will enable us to optimize the mixture $H_2 + D_2 + T_2$, its temperature and the density in order to increase the number of cycles X_c.[2]

5. The presented theoretical predictions are based on the calculation scheme described in detail in ref. [3], which has been successfully applied for the description of $dd\mu$ molecule resonant formation [10]. In recent years some modifications of this scheme have been developed [11]. The quasiresonant process of the $dt\mu$ molecule formations [12] in triple collisions and other many-body effects are also not considered here. These effects, which are essential at high mixture densities, can only increase the observed rates $\lambda^F_{dt\mu-x}(T)$ and does not change our conclusions, too.

[2] Surprisingly the kinetics of muon catalysis looks very similar to the kinetics of neutron deceleration in the natural uranium: first the resonant capture in ^{238}U at neutron energies $E \geq 5$ eV with the formation of ^{238}Pu and then the effective fission of ^{235}U at $E \leq 1$ eV.

tμ(F=0) + HD ---> [(dtμ)pee]

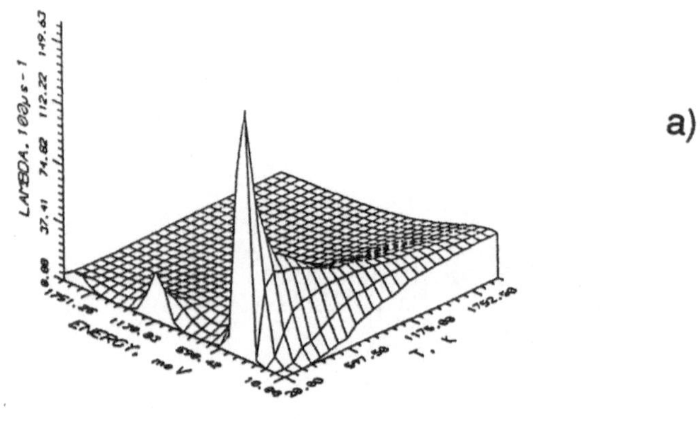

tμ(F=1) + HD ---> [(dtμ)pee]

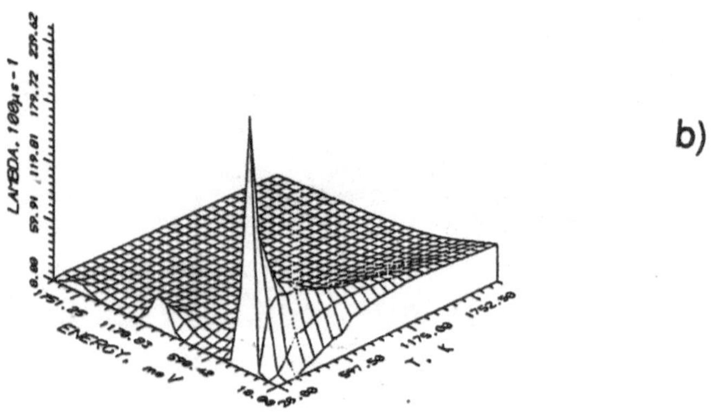

Fig. 7. $\lambda^F_{dt\mu-p}(E,T)$ as a function of enery E and temperature T for $F=0$ (a) and $F=1$ (b).

For the experimental test of the conclusions of this paper it is extremely desirable to study the muon catalysis phenomenon in the triple mixture $H_2 + D_2 + T_2$ at temperature $T = 500 - 1000$ K and for various concentrations of its components.

Acknowledgement. One of the authors (L.I.P.) is very glad for the possibility to present these new results in Muon Catalyzed Fusion at the Symposium "The Future of Muon Physics" and expresses many thanks to Professor G. zu Putlitz for the outstanding hospitality.

References

1 Ponomarev L.I. and Fiorentini G., Muon Catalyzed Fusion 1 (1987) 3; Breunlich W.H. et al., Annu. Rev. Nucl. Part. Sci. 39 (1989) 311; Ponomarev L.I., Cont. Phys. 31 (1990) 219.

2 Vinitsky S.I. et al., Zh. Eksp. Teor. Fiz. 74 (1978) 849 [Sov. Phys. JETP 47 (1978) 444]; Gerstein S.S. and Ponomarev L.I., Phys. Lett. 72B (1977) 80.

3 Faifman M.P. et al., Muon Catalyzed Fusion 4 (1989) 1.

4 Faifman M.P. et al., to be published

5 Bubak M. and Faifman M.P., Preprint JINR E4-87-464, Dubna, 1987. Ciccoli

6 Petitjean C. et al., Preprint PSI-PR-90-39, Villigen, 1990.

7 Markushin V.E., Afanasieva E.I. and Petitjean C., Preprint PSI-PR-91-21, Villigen, 1991

8 Kammel P., Nuov. Cim. Lett. 43 (1985) 349.

9 Jones S.E., in Muon Catalyzed Fusion and Fusion with Polarized Nuclei, Eds. Brunelli B. and Leotta G.G., Plenum Press, N.Y., (1987), p. 73; Petitjean C. et al., ibid., p. 99.

10 Menshikov L.I. et al., Zh. Eksp. Teor. Fiz. 92 (1987) 1173; [Engl. transl. - Sov. Phys. JETP 65 (1987) 656].

11 Lane A.M., J. Phys. (At. Mol. Phys.) B21 (1988) 2159; Yu. v. Petrov et al., Muon Catalyzed Fusion 2 (1988) 261; Scrinzi A., Thesis Ph. D., Vienna, (1989), The talk at the Conference on Muon Catalyzed Fusion, Vienna, (1990).

12 Menshikov L.I., Elem. Chast. Atomn. Yadra, 19 (1988) 1349 [Sov. Journ. Part. Nucl. 19 (1988)]; Petrov V.Yu. and Petrov Yu.V., Muon Catalyzed Fusion 4 (1989) 73; Cohen J. and Leon M., Phys. Rev. A39 (1989) 946.

This article was processed using Springer-Verlag TEX Z.Physik C macro package 1991
and the AMS fonts, developed by the American Mathematical Society.

Z. Phys. C – Particles and Fields 56, S215–S222 (1992)

Zeitschrift für Physik C Particles and Fields

Muon Science Research with Pulsed Muons at UT-MSL/KEK

K. Nagamine[1]

Meson Science Laboratory, Faculty of Science, Unversity of Tokyo (UT-MSL), Hongo, Bunkyo-ku, Tokyo, Japan

[1] Permanent joint address: Institute of Physical and Chemical Research (RIKEN), Wako, Saitama, Japan

Received 15 February 1992

Abstract. The facility construction as well as scientific achievements during the first 10 years operation of the Meson Science Laboratory, University of Tokyo (UT-MSL) located at KEK are summarized with special emphasis on the advantage of the use of sharply pulsed muon beam for new types of muon science experiments. Also, a short description is given on the nearest future project of ultra-slow μ^+ facility.

1 Introduction

In 1980, contrary to the widely-accepted concepts of intense d.c. muon beam realized at meson factories, the importance of the use of sharply pulsed beam has been recognized and the first pulsed muon facility of UT-MSL BOOM (Booster meson science facility) has been constructed at KEK [1]. There, instantaneously intense muons, without being identified the arrival of each muon, are stopped inside the target material of the research object and the muon associated observables such as decay e^+/e^-, muonic X-ray, etc. are detected with reference to the arrivals of beam pulse by somewhat ingeneous detection methods.

The concept of pulsed muons is highly assisted by the developement of modern accelerator technology, namely, the achievement of rapidly cycling intense proton synchrotron with the single-turn extraction of the accelerated protons; when the width of the beam is enough short, the pulsed protons can be directly used as a source of pulsed muons. It was really fortunate that, before we noticed the idea of pulsed muons, the 500 MeV proton synchrotron of the KEK booster associated with 12 GeV main ring had been constructed as an ideal source of such a pulsed proton beam with 50 ns pulse width and 50 ms pulse distance (20 Hz of a repetition frequency).

Before we started pulsed muon experiments in 1980, some muon experiments at muon facility of Saclay electron liniac had been done by adopting a similar concept of the pulsed muons [2]. Also, after we had started an extensive experimental program, some experiments at BNL were conducted by using a fast extracted beam to generate pulsed muons [3]. More extended program of pulsed muon experiments was started in 1987 at the intense proton synchrotron of ISIS at Rutherford Appleton Laboratory [4].

In this short review article, the authors would like to summarize the developments of muon science research program by using pulsed muons at UT-MSL/KEK. Annual progresses as well as progresses in some specific periods have been summarized elsewhere [5,6,7,8].

2 Facility developments at UT-MSL/KEK

2.1 Accelerator and BSF facility

As it is summarized elsewhere [1], the 500 MeV KEK-Booster proton synchrotron can provide a sharply pulsed proton beam which has a 50 ns pulse width and 20 Hz repetition frequency with an average current of 6 μA. The proton beam is delivered to the Booster Synchrotron Utilization Facility consisted of BOOM of UT-MSL, KENS (spallation neutron scattering facility) of KEK and PARMS (medical application facility of Tsukuba University). Currently, at every 4.0 sec, 72 pulses are delivered to the BSF facility with 8 pulses for 12 GeV main ring. The BSF beam is shared bi-weekly between UT-MSL and KENS (each 45 %) with some fractions of the day-time to PARMS for medical treatment (10 %).

2.2 Superconducting muon channel

Schematic view of the existing experimental area is shown in Fig. 1. The superconducting muon channel was constructed with the 3Q + B pion injector at zero degree take-off angle and the 2Q-B-Q-B-2Q achromatic muon

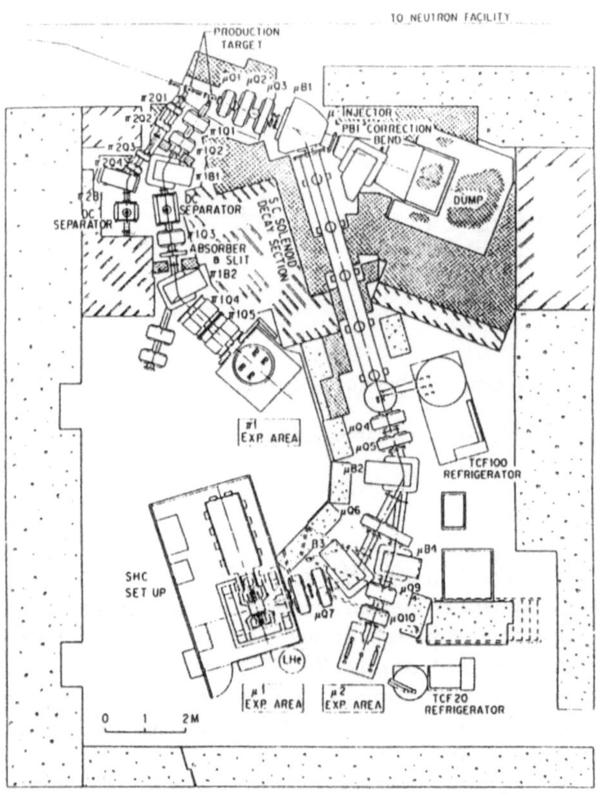

Fig. 1. Layout of Experimental Hall No.1 of UT-MSL/KEK

extraction. For the intense decay μ^+/μ^- production, 6m long, 12 cm bore and 5T superconducting solenoid and its supercritical He cooling system was constructed for a pion-to-muon decay section [9] and has been operational since 1980. Almost maintenance free operation was achieved over 30,000 h. At the main port ($\mu1$), the high quality backward-decay μ^+/μ^- can be obtained with a typical intensity of 10^4 μ^-/pulse in (6cm x 6cm) area with a range width of 1 g/cm^2 at 55 MeV/c.

The 3.5 T superconducting Helmholz coil set-up (SHC) was installed for a dual purpose; a 3.5 T field source for 500 MHz muon spin resonance experiment and a confinement field source of various types of radiation backgrounds in order to gain a signal to noise ratio for a weak muon-associated signal detection.

The muon extraction has two branches ($\mu1$ and $\mu2$), to which a backward μ^+/μ^- can be alternatively supplied. At the $\mu2$ a conventional μSR set-up is placed. Also, the $\mu1$ branch has an extention capability towards the new experimental areas (Experimental Hall No.2) with two 3Q configurations.

2.3 Surface μ^+ channel

The suface muon channel is constructed in the form of 2Q-B-Q-B-2Q made of conventional magnets with a DC separator. The final part has two branches to feed the surface μ^+ alternatively to either πA port or πB one. At the πA port a conventional μSR set-up is located, while the ultra high vacuum target set-up is located at the πB port for thermal muonium experiment. At either one of these two ports, more than 10^4/pulse surface μ^+ can be obtained in 5 cm x 5 cm with less than 0.1 % e$^+$ contamination.

2.4 Basic instrumentations for pulsed muon experiments

In order to handle the multiple events of *e.g.* decay e$^+$/e$^-$ from the stoping muons with an intensity of more than 10^3/pulse, various types of detection systems have been constructed. In order to reduce a load on the single detector, segmented decay e$^+$/e$^-$ counter system was employed. There, the most important instrumentation was a multi-hit (or Non-Stop) TDC, where timings of multiple stopping signals are recorded with a reference to the muon pulse without stopping its operation by each stopping signal. The shift-register based TDC with a time resolution of 16 ns has been constructed by Shimokoshi et.al. [10].

When the stopping events are far larger than the time resolution of TDC, the event can be detected by the amplitude profile in an accumulated pulse shape. Along this line, so-called analogue method has been developped [11].

3 Novel experimental results with pulsed muons

A number of excellent features of the pulsed muon experiments over the conventional dc muon have been appreciated throughout our 10 years experiences and they can be summarized as follows;

a. Long time range meausurement can be realized for muon-associated events like μe decay or μSR with a rate-unlimited manner.

b. Coupling with the extreme experimental conditions which are only realized in pulsed modes can be effectively realized, enabling muon spin rf resonance, muon state laser resonance etc.

c. Phase sensitive detection of the weak muon-associated signals can be achieved under large white-noise backgrounds.

In the followings, several novel muon science experiments at UT-MSL/KEK are summarized with an emphasis of these excellent features.

3.1 Long time range measurements

Among so many μSR experiments as well as μe decay experiments assisted by the long time-range measurement capability, we will point out only several high-light examples.

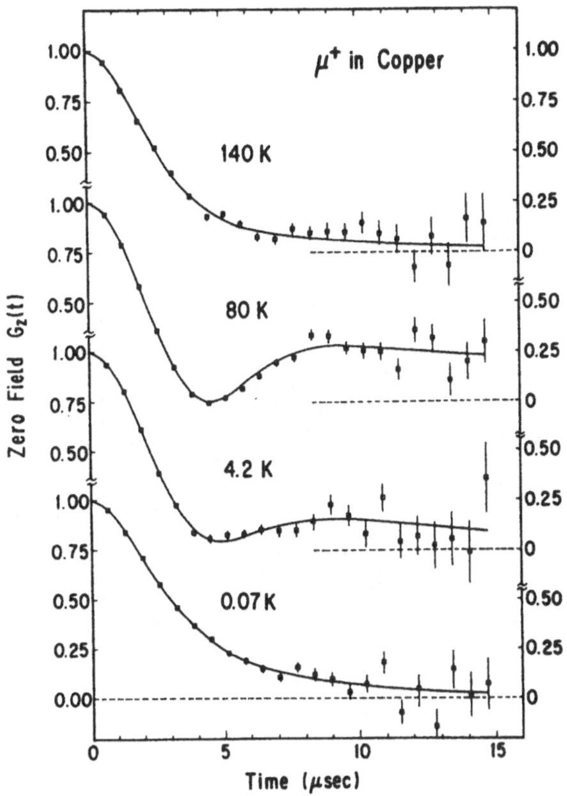

Fig. 2. ZF relaxation function of μ^+ in Cu at typical temperatures

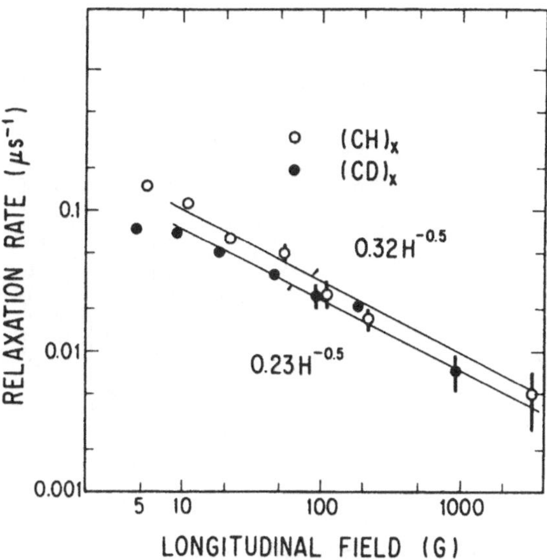

Fig. 3. The observed longitudinal relaxation rates vs external fields for μ^+ in trans-$(CH)_x$ and trans-$(CD)_x$ at 293 K.

3.1.1 Quantum diffusion of the μ^+ in Cu revealed by zero-field μSR method [12]

At low temperature, the light interstitial impurity of the μ^+ could diffuse rapidly throughout condensed matter. The first convincing experiments has been produced by observing zero-field μ^+ spin relaxation spectrum for the μ^+ in fcc Cu, where μ^+ microscopic diffusion is studied by the narrowing of μ^+ spin relaxation under static random fields from the surrounding Cu nuclear moments. With the help of pulsed muons, the characteristic decay of the tail region of Kubo-Toyabe function which is appearing at 10 μs after μ^+ arrival has been detected for the first time (Fig. 2), which is one of the most clear demonstrations of the excellent feature of pulsed muons. In this experiment, a basic understandings on the μ^+ quantum diffusion in metal have been obtained, demonstrating an importance of the effect of μ^+ to conduction electron interaction.

3.1.2 Solitons produced and probed by muons in polyacetylene [13]

When the μ^+ is injected into some semiconductor chemical solids like polyacetylene, the muonium (Mu) is formed at relatively high energy. It becomes chemically bonded to the host in the form of Mu radical at thermal energies. It is really so in the case of cis-polyacetylene, while the electron in Mu radical in trans-polyacetylene was found to take a rapid soliton-like motion throughout one-dimensional $(CH)_x$ chain, due to an existence of degenerate states formed by bond-alternations. This picture has only been evidenced by the field dependent decrease of the long relaxation time (Fig. 3). Thus, the μ^+ as an its own radiation-effect, produces soliton in trans-$(CH)_x$ which was sensitively detected by the μ^+ itself.

Recently, we found that this kind of μ^+SR studies could reveal the most fundamental processes in the rapid electron transfer phenomena through macro-molecules like nucleic acid, DNA, etc.

3.1.3 μ^+ location and probing magnetism in high T_c superconductors [14,15]

The μ^+ location in high T_c superconductor LaSrCuO was precisely determined by observing the μ^+ long-time range relaxation due to random nuclear dipolar fields from Cu and La and its crystal orientation dependence (Fig.4). Through these measurements, the μ^+ location as well as μ^+ diffusion properties through LaSrCuO was determined [14]. At the same time, the magetic properties of the superconducting phase of LaSrCuO were studied with the μ^+ probe producing a magnetic phase diagram [14].

Although experiment was conducted jointly at TRIUMF and at UT-MSL/KEK, the earliest studies on $YBa_2Cu_3O_x$ has revealed , for the first time, an existence of antiferromagnetic phase by reducing oxygen concentration [15]. This experiment has demonstrated an

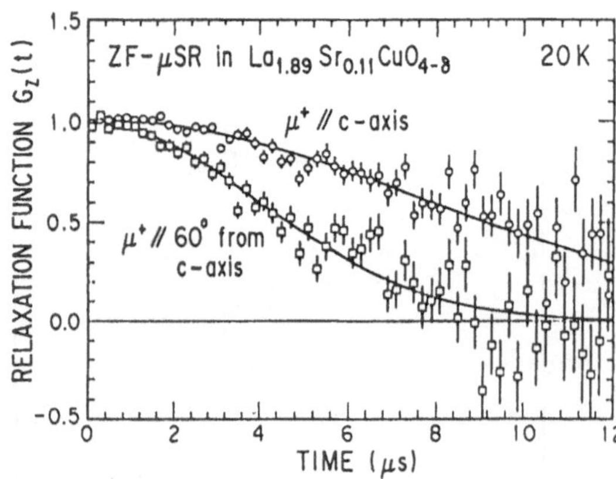

Fig. 4. The ZF relaxation function due to random nuclear dipolar fields from Cu and La and its crystal orientation dependence for μ^+ in $La_{1.89}Sr_{0.11}CuO_4$

important role of the μSR for the studies of microscopic magnetic properties in high T_c superconductors.

3.1.4 Detection of thermal muonium generation in vacuum [16]

The first successful experiment on thermal Mu production in vacuum was carried out for the μ^+ re-emission from hot W surface. The convincing detection of thermal Mu production was performed by the position dependent T.O.F. profile of the Mu moving with thermal energy by detecting the decay e^+ in a long time range (Fig. 5). The measurement was highly assisted by the long-time-rang-meausrement capability of the pulsed muons.

3.2 Coupling with extreme experimental conditions

The most significant examples of the pulsed muon experiments under low-duty factor experimental conditions are muon spin rf resonance and muon laser resonance.

3.2.1 Mu/chemical reaction in solids revealed by spin resonance [17]

In order to accomplish rf resonance of the diamagnetic muon spin, it is required to rotate the μ^+ magnetic moment around the applied rotating field H_1 within a time comparable to the muon life time; corresponding rf power becomes easily over a few 10 kW. Thus, resonance experiment with pulsed rf field which is operational in a pulsed muon arrival is the most effective.

The important feature of the muon spin rf resonance is a capability of detecting the final state after the state change of μ^+ or Mu which is formed at t=0; contrary

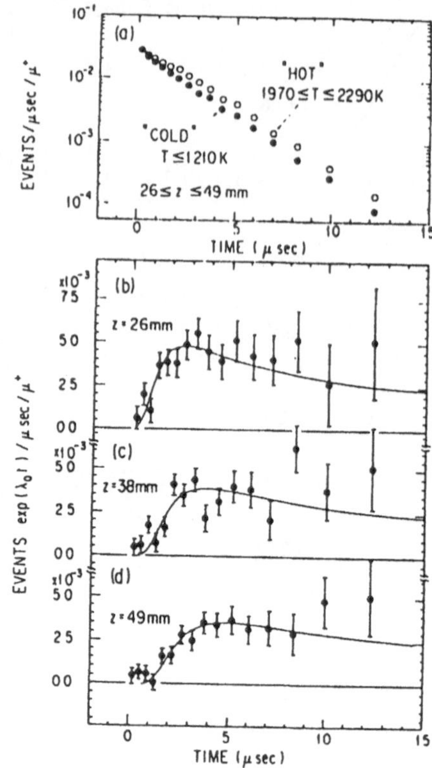

Fig. 5. T.O.F. profile of Mu from hot W surface: (a) Raw spectra obtained with the W target hot and cold; (b)-(d) "hot"-"cold" difference for three Mu decay positions from the W target.

to spin rotation or spin relaxation methods, the rf resonance method does not need the phase-coherency started from t=0.

With the help of this excellent feature, existence of chemical reaction in a μs time range of the Mu to diamagnetic μ^+ in alkali-halides has been revealed for the first time (Fig. 6).

Recently, this type of muon chemical reaction in solids has also been found in Mu/μ^+ in semiconductor Si.

3.2.2 Laser resonance of thermal muonium [18]

Many quantum states in either muonium or muonic atom can be subject to excitations by intense multi-photon laser transitions. Since such a laser excitation can only be done by pulse operation, it is inevitable to have an intense muon pulse in coincidence with the laser pulse.

The first successful laser resonance experiment on the muonium states was carried out in our laboratory in collaboration with AT&T Bell laboratories. There, the pulsed 244 nm laser with 15 mJ/pulse was irradiated

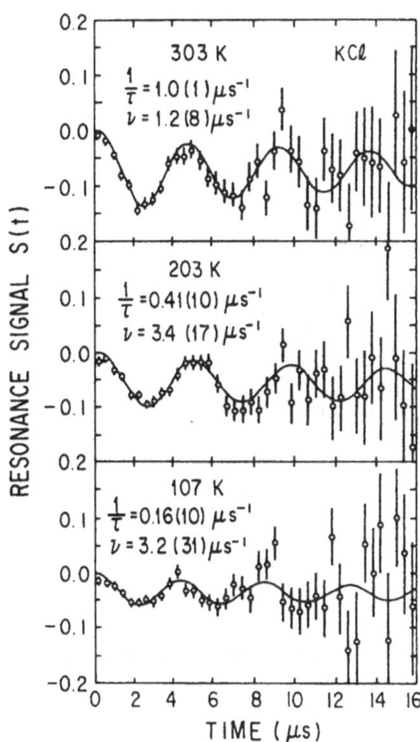

Fig. 6. Typical time-differential resonance signal for diamagnetic μ^+ in KCl.

pulsed 244 nm laser with 15 mJ/pulse was irradiated onto the thermal Mu produced from SiO_2 powder to make a transition of 1s $\rightarrow$ 2s by two photon absorption followed by the transtion of 2s $\rightarrow$ unbound by the third photon absorption. The detection of the resonance signal was made by measuring thus ionized μ^+ with an ion optics of 4 kV accelerator lens system. The obtained 1s(F=1) $\rightarrow$ 2s(F=1) transition frequency was in agreement within 300 MHz with the QED prediction.

As a result of this experiment, the method of laser resonance ionization was noticed to be one of the most efficient ways to produce slow μ^+ from thermal muonium. The idea was extended to the project of the Ultra-Slow Muon Facility to be mentioned later.

3.3 Phase-sensitive detection of weak signal

When muon-associated signal is very weak compared to the white-noise type backgrounds which are not associated with accelerator beam, the pulsed muon beam does help to increase significantly the signal to noise ratio for the weak signal. This phase-sensitive detection method can also be effective even when the muon-associated signal with a time constant of μs range exists under an existence of long-lived accelerator-beam-associated backgrounds with a time constant of say ms range. Typical and significant example is shown here.

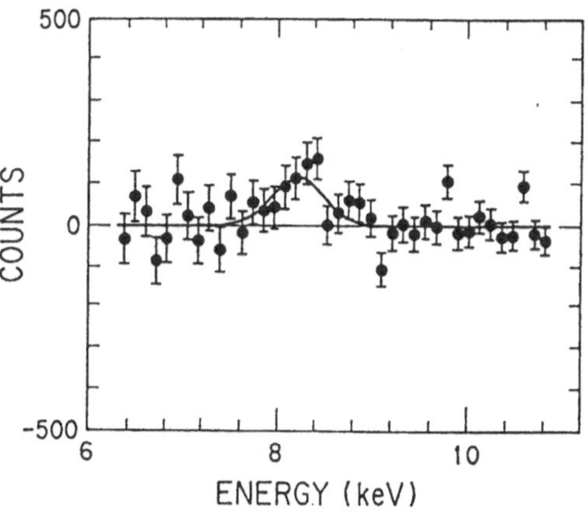

Fig. 7. The observed 8.2 keV characteristic muonic X-ray from sticked ($\alpha\mu$) atom in liquid D-T mixture. The data were obtained after removal of bremsstrahlung and beam associated background.

3.3.1 X-ray detection of α-sticking probability in muon catalyzed fusion of D-T mixture [19]

The direct knowledge of α-sticking probability (ω_s) in muon catalyzed fusion (μCF) of high density D-T mixture with high T-concentration is one of the most important observables in entire μCF studies, because the ω_s does place a stringent upper limit on the energy production capability from μCF. The straight answer to the ω_s can be obtained by measuring a charateristic muonic X-ray from the sticked ($\alpha\mu$) atoms (central energy of 8.2 keV with Doppler broadening of 0.5 keV). The significant difficulty for this X-ray method for the ω_s is an existence of huge radiation background of bremstrahlung from the t-decay (up to 17 keV) in D-T mixture which does kill the weak X-ray associated with the ω_s; easily over 10^5 photons/s background exist from 1cc liquid D-T target. The help of pulsed muon is really significant here, because the X-ray detection is only needed to be operational during the existence of the muons, increasing the signal/noise ratio drastically.

Experiments have been carried out by the UT MSL-RIKEN-JAERI collaboration. Successful observation of the 8.2 keV X-ray with a correct Doppler broadening and a correct disappearence rate was made for liquid $D_{0.70}T_{0.30}$ mixture as seen in Fig. 7. From this measurement, with the help of atomic physics theory for connecting X-ray intensity with the α-sticking probabiolity, the ω_s is determined to be 0.34(8) % after correcting the μ^- stripping probability of ($\alpha\mu$) during the slowing-down.

4 Ultra-slow Muon Facility project

So far, at any accelerator laboratories, the muons are obtained by either pion decay-in-flight (decay μ^+/μ^-) or π^+ decay at the surface skin of the pion prduction target (surface μ^+). There, even after careful design of the beam channel, the range width of muon stopping is

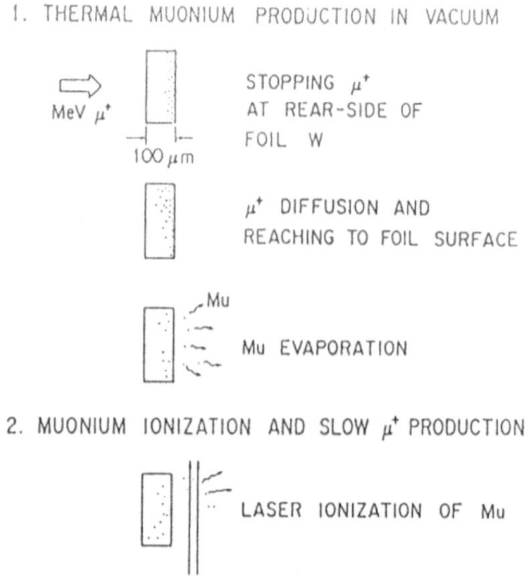

1. THERMAL MUONIUM PRODUCTION IN VACUUM

MeV μ^+

$100\,\mu m$

STOPPING μ^+ AT REAR-SIDE OF FOIL W

μ^+ DIFFUSION AND REACHING TO FOIL SURFACE

Mu

Mu EVAPORATION

2. MUONIUM IONIZATION AND SLOW μ^+ PRODUCTION

LASER IONIZATION OF Mu

Fig. 8. Schematic view of ultra-slow μ^+ production by laser ionization of thermal Mu

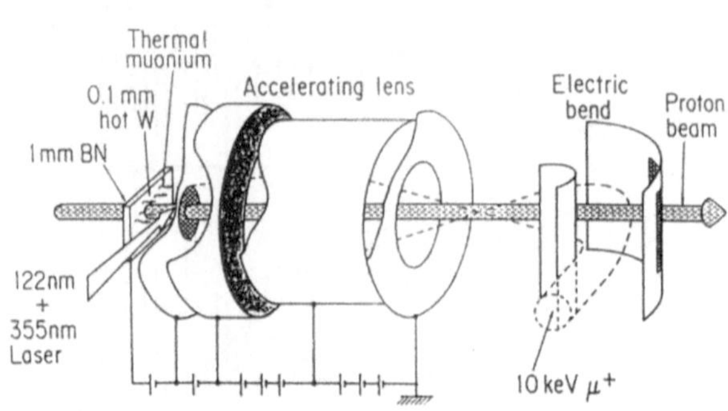

Fig. 9. Schematic view of direct production of ultra-slow μ^+ at the primary proton line

larger than 1 mm (10 mg/cm^2) and the size of the muon beam spot is larger than a few cm^2. Signicant intensity decrease is inevitable when we are forced to obtain more thinner stopping range or more smaller spot size. If we have a way to increase the slow μ^+ intensity, so many new types of muon science experiments become realized.

4.1 *Ultra-Slow μ^+ Facility at UT-MSL/KEK*

The project of Ultra-Slow μ^+ Facility has been proposed as the third steps following the two important achievements at UT-MSL/KEK, namely a) production of thermal muonium in vacuum (see 3.1.4) and b) laser ionization of thermal munium (see 3.2.2) as schematically summarized in Fig. 8. Once we can accomodate these two steps right at the primary proton beam line as seen in Fig. 9, a significant increase of the ultra-slow μ^+ intensity can be expected [20].

The project has been approved as the major part of upgrading program of the UT-MSL during 1989-1991 and a budget was given to this project including a new laboratory building dedicated to this project. Most of the facility components have been completed before the fall of 1991 (see Fig. 10). The commissioning experiments have been started before December of 1991. A summary of the completed facility is given in the followings.

4.1.1 Dedicated laboratory space and proton beam line

In order to obtain a freedom for the future development of slow μ^+ production technique, laboratory space with the dedicated proton beam line was constructed. Furthermore, the proton beam line has 4 m straight section at around the thermal Mu production target in order to accomodate ultra high vacuum target and ion optics systems. By using pulse magnet of PHB3, each of the proton pulses (50 ns width and 20 Hz repetition) can be selected for the new laboratory.

4.1.2 Thermal Mu producing target

As the first trial, we adopted our original hot tungsten as the thermal Mu producing target. Since "vapor pressure" of thermal Mu from hot W is less than 10^{-8} Torr, all the target area as well as the following ion optics of slow μ^+ transport is maintained below 10^{-9} Torr under the conditions of proton beam delivery and target heating.

Actual target is composed of 1 mm BN (boron nitrate) and 50 μm W. The use of BN which is practically the best light material to be placed beside 2000 K hot W, is for the efficient π^+ production with the minimum allowable divergence of proton beam. Thus, the intended scinario is the following; the low energy π^+ produced in BN stops in hot W where π^+ to μ^+ conversion, μ^+ stopping and thermal Mu production are taking place. By the Monte-Carlo calculation, the optimum thickness of BN was determined to be 1 mm [21].

4.1.3 Mu ionizing laser system

In order to ionize thermal Mu by lasers, we have adopted the transition of 1s $\rightarrow$ 2p $\rightarrow$ unbound due to the fact that reasonably high ionization is expected by employing recently developed laser technology. For this purpose, intense 122 nm VUV light has been produced

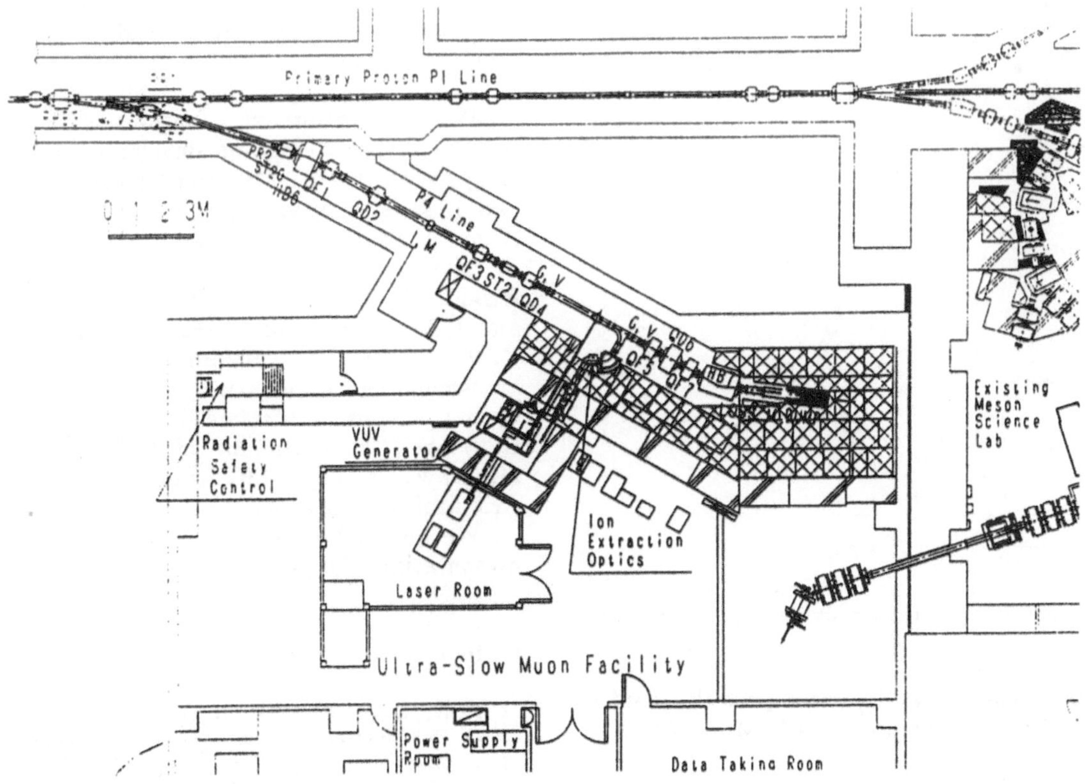

Fig. 10. Layout of the Experimental Hall No. 2 of the UT-MSL/KEK; the Ultra-slow μ^+ Facility completed at UT-MSL

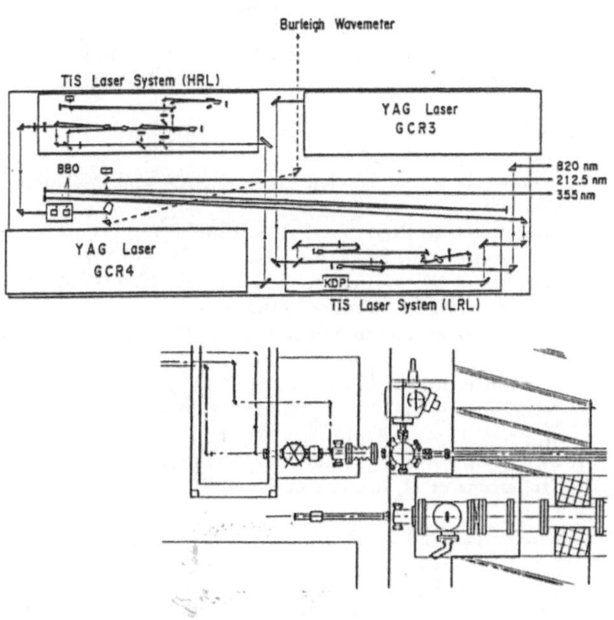

Fig. 11. The diagram of the laser system of thermal Mu ionization for ultra-slow μ^+ production and detailed arrangement around mixing cell

by so-called 4 wave-mixing method of two 212 nm photons and 820 nm photon in optimized Kr/Ar gas mixture [22]. It has 200 GHz(FWHM) width in order to match to the Doppler broadening of thermal Mu. All these lights plus ionizing (2p $\rightarrow$ unbound) 355 nm light are produced by Ti Sapphire solid lasers pumped by YAG lasers. Diagram of the laser system is shown in Fig. 11a and detailed arrangement around the mixing cell is shown in Fig. 11b. Already at the level of a few 10 μJ/5 ns of 122 nm and 10 mJ/5 ns of 355 nm were delivered from the mixing cell to the free space (5 mm gap) between hot W target and the front end of ion extraction optics located at 4 m distance from the cell.

The use of all solid-state lasers are essential for us to achieve a maintenance-free and long-term operation, which is a great improvement over Dye-based laser system.

4.1.4 Ion extraction optics and monitor system

The basic structure of ion extraction optics for ionized product of μ^+ is schematically shown in Fig. 12, where the SOA lens of 10 keV acceleration and focussing is followed by an electric bend (E) and a magnetic bend (B) with relevant focussing electric Q lens. Thus, focussing of the μ^+ is expected at the MCP monitor. Thus, two-fold monitoring, namely, mass analysis in E-B plane and T.O.F. spectrum, can be used for identification of the ionized μ^+.

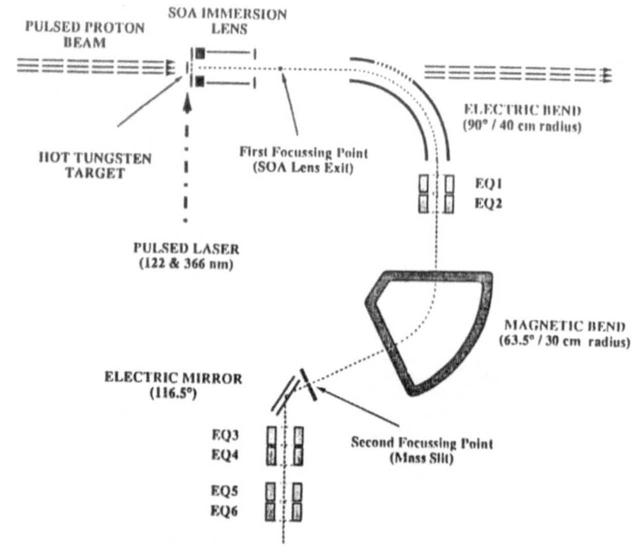

Fig. 12. The ion extraction optics for ultra-slow μ^+ production

As described separately [21], we expect the yield of slow μ^+ as 2 x 10^4/s at the present UT-MSL/KEK.

4.2 Possible new muon science experiments with slow μ^+.

Once slow μ^+ is produced, a tremendous amount of new muon sciences will be realized which may cover from the fundamental physics to the application fields. Some of the novel subjects are listed in the followings.

(1) surface science studies with slow μ^+

The intence, high quality polarized slow μ^+ will be used for exploring new aspects of materials surface such as the electronic structure around H-like single impurity on surface, probing surface magnetism, in particular, its dynamical behaviour, studying catalytic chemical reactions of Mu on surface, etc.

(2) intense production of thermal Mu and QED

Once we have the keV μ^+ source, thermal Mu can be produced in a controlled way; complete conversion can be expected from keV μ^+ to thermal Mu in vacuum on either hot W or SiO_2 surface. Intense and localized thermal Mu can be used for the improved measurements of QED e.g. 1S - 2S resonance, providing the most accurate Rydberg constant , etc.

5 Conclusion and perspectives

As summarized here, pulsed muons have quite excellent features in various muon science experiments. Higher flux pulsed beam is really needed for the next step. Some significant progresses might be expected at the forthcoming RIKEN muon facility at RAL.

Further improvements such as much sharper pulse (50 ns $\rightarrow$ a few ns, hopefully) and higher repetition rates (20 Hz $\rightarrow$ 1 kHz, etc) are called for. For this purpose, a realization of the intense high energy accelerator project like JHP (Japanese Hadron Project) is very much wanted.

We acknowledge contributions of the following persons to the creations and development of experimental program at the UT-MSL/KEK; Professor T. Yamazaki and Drs. H. Nakayama, J. Imazato, K. Nishiyama, Y. Kuno, T. Matsuzaki, R. Kadono, Y. Miyake, K. Ishida, M. Iwasaki, S. Sakamoto, R.S. Hayano, T. Kuga, N. Nishida, E. Torikai, A.P. Mills, Jr., S. Chu, S.E. Jones and K.M. Crowe and Messers K. Fukuchi, P. Strasser, Y. Watanabe and F. Shimokoshi. We also sincerely acknowledge encouragements and supports from the following persons; Professors T. Nishikawa, H. Sugawara, H. Sasaki, N. Watanabe of KEK and Professors H. Miyazawa, H. Fujita, H. Kamimura, K. Yazaki and A. Arima of the University of Tokyo.

We also acknowledge Dr. A. Schenck for his kind presentation of a part of this paper at the conference.

References

1. K. Nagamine, Hyperfine Interactions 8 (1981) 787
2. G. Bardin et al, Nuclear Physics A352 (1981) 365
3. A. Blaer et al, Phys. Rev. A40 (1989) 158
4. G. H. Eaton et al, Nuclear Instruments and Methods A269 (1988) 483
5. UT-MSL Newsletter 1-10, eds K. Nagamine et al. (1981-1990)
6. Collected Papers on Muon Science Research, eds K. Nagamine and T. Yamazaki (1987)
7. Meson Science vol. 1, eds T. Yamazaki, K. Nagamine and K. Nakai (1990)
8. K. Nagamine, AIP Conf. Proceedings, to be published (1992).
9. K. Nagamine et al, IEEE Transactions on Magnetics MAG17 (1981) 1892.
10. F. Shimokoshi et al, Nuclear Instruments and Methods, A297 (1990) 103
11. T. Yamazaki et al, Nuclear Instruments and Methods 196 (1982) 299
12. R. Kadono et al, Phys. Rev. B39 (1989) 23
13. K. Nagamine et al, Phys. Rev. Lett. 53 (1984) 1763
14. E. Torikai et al, Hyperfine Interactions 63 (1990) 271; E. Torikai et al, to be published (1992)
15. N. Nishida et al, Japan Journal of Applied Physics 26 (1987) L1856
16. A. P. Mills et al, Phys. Rev. Lett. 60 (1988) 101
17. K. Nishiyama et al, Phys. Lett. 111A (1985) 369; Y. Morozumi et al, Phys. Lett. 118A (1986) 93
18. S. Chu et al, Phys. Rev. Lett. 60 (1989) 101
19. K. Nagamine et al, Muon Catalyzed Fusion, 5/6 (1990/1991) 289
20. K. Nagamine and A. P. Mills, Jr., Los Alamos Report LA-1071c (1986) 216
21. K. Ishida and K. Nagamine, in this proceedings
22. Y. Miyake et al, to be published (1992)

This article was processed using Springer-Verlag TEX Z.Physik C macro package 1991
and the AMS fonts, developed by the American Mathematical Society.

Z. Phys. C - Particles and Fields 56, S223-S225 (1992)

Zeitschrift für Physik C Particles and Fields

Existing muon beams at LAMPF

Chandra Pillai*

Medium Energy Physics Division, Los Alamos National Laboratory
Los Alamos, New Mexico 87545

Submitted 6 December 1991

The Stopped Muon Channel (SMC) at LAMPF is heavily used for a variety of muon experiments: (1) Muon Catalyzed Fusion (μCF), Muon Spin Rotation (μSR), and Muonic X-rays using decay muon beams; (2) Muon Level-Crossing Resonance (μLCR), Rare Decays, and Muonium Hyperfine Structure & Muon Magnetic Moment using surface muon beams; (3) Laser Polarized Muonic Atoms using cloud beams.

There are two dedicated experimental caves for muon experiments. When one crew is taking the beam, the other one can set up and check out their equipment in the other cave. We also have crossed-field separators, beam profile monitors, and a fast high-voltage chopper, which can provide chopped muon beam at a rate of 60-70 kHz for experiments that are limited by pileup problems.

The SMC channel at LAMPF is of conventional design. It is a beam transport system designed to accept as large a fraction of pions from the primary production target as possible, to capture muons from pion decays, and finally to select from the resulting beams muons with the desired momenta and focus them onto the target. The details of the channel are described elsewhere [1]. In addition to the muons from pion decay in flight, the channel is also being used successfully for surface muons and cloud beams. For decays in flight, pions of a given momentum decaying into muons produce a muon momentum (laboratory) spectrum extending from about one-half the pion momentum (backward decays) to slightly greater than the pion momentum (forward decays). The decay section can capture muons with decay angles up to ±5 degrees (lab); because the maximum laboratory decay angles are much larger than 5 degrees for practical range of pion momentum, only extreme forward and backward decays are accepted, and the muon momentum spectrum for a given pion momentum is narrow (less than 2% for pion polarization is high (80-90%) and the muon momentum spectrum is determined mainly by the pion momentum spectrum. Table I gives the calculated rates for μ^- and pions at the end of the SMC channel.

Table I. Calculated rates for pions and decay muons.

Particle	Momentum (MeV/c)	Rate (/s/mA)	Contamination (%) π	e
Backward (μ^-)	40-150	1-30×10^6	0.2	10
Forward (μ^-)	100-250	30-60×10^6	10	10-100
π^-	100-250	30-300×10^6	-	50-100

1 Backward decay beam

Because of the large difference between the pion and muon momentum for the backward decay beam, it is easy to generate a very high-purity beam in this mode of operation. For this reason, many experiments that require good quality muon beams for stopped muons in the momentum range of 40-150 MeV/c use the channel in this mode. Table II gives some measured rates for μ^- for the backward decay beam; the rates for μ^+ are approximately four times higher.

2 Forward decay beam

Due to the kinematics of the decay, the forward decay beam should be significantly more intense than the backward decay beam. The channel can deliver muons of momenta from 100 to 250 MeV/c in this mode of operation. The main drawback of this mode is the high pion and electron backgrounds - up to 5 times the muon flux depending on the momentum selected. The channel is very seldom used in this mode.

*Presented by M. Leon.

Table II. Backward decay rates.	
Momentum (MeV/c)	Rate (μ^-/s/mA)
50	3×10^5
60	1.3×10^6
70	2.0×10^6
85	8.0×10^6
105	2.0×10^7
115	2.2×10^7
130	2.5×10^7
140	2.5×10^7

3 Surface beam

Many experiments in muon physics require stopping many muons in small, thin targets. This requires a high flux of low-energy muons with small momentum spread. Surface muons are produced from pions stopped near the surface of the production target; the maximum momentum for these muons is 29.8 MeV/c. Because of the high current of the proton beam at LAMPF (1 mA), the yield for the surface muon beam is also high; one can get a rate of 5×10^7 μ^+/s. Crossed-field separators in the channel reduce the high positron background (10 times μ^+ flux) to less than 5%. The beam size at the end of the channel is rather large, about 8 × cm^2 experiments normally use collimators to make the beam size fit their requirements. Furthermore, large superconducting solenoids are available for experiments that require large solid angle. At the same time, the solenoidal field focuses the beam to a smaller area. We have obtained beams of 3×3 cm^2 (FWHM) by using this technique. Table III gives some measured surface muon rates for a tune that is used to a get small beam size at the expense of rate. Figure 1 gives the rates for decay, clous, and surface beams as a function of momentum.

Table III. Surface muon rates.		
Momentum (MeV/c)	Rate (μ^+/s/mA)	e^+/μ^+
3.0	2.5×10^3	
5.0	7.5×10^4	1000
7.5	5.0×10^5	
10.0	7.0×10^5	
15.0	3.2×10^6	
20.0	1.0×10^7	
28.5	3.0×10^7	10

4 Cloud muon beam

Experiments which require low-energy negative muons for higher stopping densities in thin targets can use the cloud beam. These are muons from pion decay near the production target. Since both the backward and forward decay beams are accepted in this mode, the polarization of the beam is very low (about 10%). A crossed-field separator is used to reduce the electron background in the cloud beam below 5%. Table IV gives some measured rates for the cloud beams.

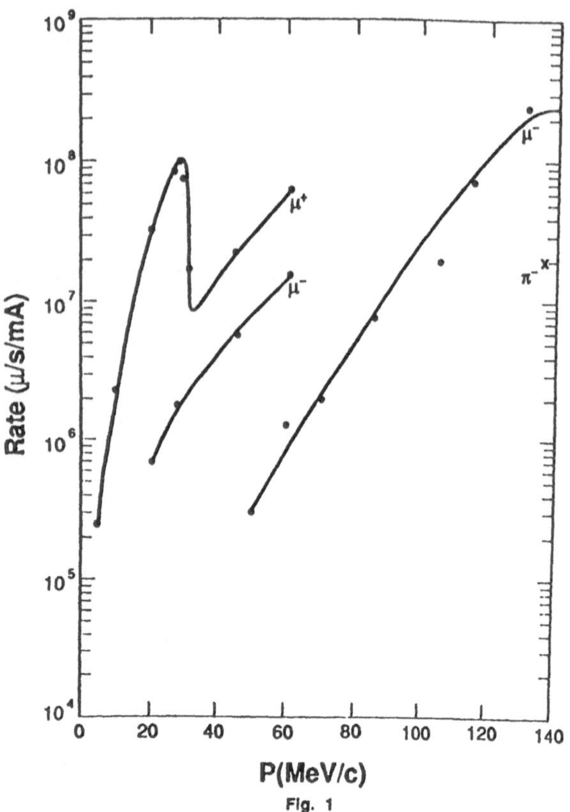

Figure 1. Measured rates of surface beam, cloud and backward decay μ^+ and μ^- beams as a function of momentum for the SMC at LAMPF.

Table IV. Measured cloud μ^- rates.		
Momentum (MeV/c)	Rate (μ^+/s/mA)	e^-
20.0	0.7×10^6	2.0×10^8
28.0	2.0×10^6	2.5×10^8
45.0	6.0×10^6	3.2×10^8
60.0	1.5×10^7	3.5×10^8

The SMC channel was originally designed for decay muons and has a long decay section having alternating quadrupole magnets. Therefore the channel is not very efficient in transporting mono-energetic beams like the surface and cloud beams because of the strong focusing in the decay section. We found theoretically as well as experimentally that one can transport monochromatic muons efficiently with most of the quads in the decay section turned off. This narrows the momentum spread and the chromatic aberration is also reduced, resulting in a cleanly-dispersed beam. This technique has been successfully used in stopping a large fraction of negative muons in gas targets [2]. But in this mode of operation, the intensity of the beam is reduced by a factor of approximately two compared to the tune designed to obtain the maximum flux. One can use the same technique to get a good quality surface beam at the expense of rate.

5 Pion beam

The channel can also deliver a good-quality pion beam. Normally, π^- beam is used for calibration purposes. One can tune the channel for the pion beam of momenta 40 to 250 MeV/c. The typical rate for π^- beam at 150 MeV/c is 2×10^7/s with 1 mA proton on a 4-cm-long carbon target.

References

1. P. A. Thomson et al., Nucl. Instrum. and Methods 161(1979)391
2. R. Holmes et al., Nucl. Instrum. and Methods A303(1991)226

Z. Phys. C - Particles and Fields 56, S226-S231 (1992)

Zeitschrift für Physik C Particles and Fields

Muon beams and facilities at TRIUMF

G.M. Marshall

TRIUMF, 4004 Wesbrook Mall, Vancouver, B.C., Canada V6T 2A3

Received 20 September 1991

Abstract. The facilities are described for performing muon experiments at TRIUMF. The properties of the muon beam lines are discussed and compared, emphasizing the advantages and disadvantages of each.

1 Introduction

Muon research has played a strong role in the science program at TRIUMF[1] since beams first became available in the mid 1970's. A combination of several factors, a timely scientific initiative in several fields, the realization of significant technical opportunities, and the right people to confront a new set of problems, allowed TRIUMF to remain competitive with the meson factories of Switzerland and the USA. In the intervening period a system of muon beams and facilities has developed to meet the scientific challenges. A detailed description of all facilities and operations at TRIUMF can be found in the TRIUMF Users Handbook,[1] which is a valuable reference for experimenters. Some aspects pertinent to muon science are summarized here.

2 Accelerator operations

TRIUMF is based on a cyclotron capable of accelerating negative hydrogen ions to an energy of 500 MeV. Protons are extracted by stripping the electrons to form an external beam. For the purposes of muon science, the highest possible intensity is usually desirable, and in this case it is 140 μA. The cyclotron operates at a radiofrequency of 23 MHz, emitting protons in bunches of about 3 ns width every 43 ns, with a macroscopic duty factor of 100%. This time structure can be useful in some modes of operation for diagnostics, particle identification, and particle separation. The cyclotron delivers high intensity beam 5.5 or 6.5 days per week, with short maintenance and beam development making up the remainder. This schedule is maintained for two periods of typically 13 weeks per year. The remaining weeks are used either for low intensity operation, delivering polarized proton beams of variable energy and intensity, or for scheduled shutdown periods for large scale maintenance, upgrading, and installation.

Beam allocation for the entire experimental program, including muon science, is controlled by an Experimental Evaluation Committee, which determines the priority for each project that requests time. The beam is then scheduled several months in advance by a TRIUMF committee according to the established priorities. The muon beam lines are generally fully or oversubscribed, but it has usually been possible to meet the demand for those experiments of medium priority and above. The system of allocation allows quick response to new proposals and to new and important ideas which occur from time to time. For example, it was possible to schedule a muon beam to investigate the properties of the new high temperature superconductors as soon as the samples were available.

3 Primary beam and targets

The high intensity beam line, known as BL1A, is used for all meson production. It passes through two production targets before dissipating its remaining energy in a beam dump (Fig. 1).

Both target stations use a ladder arrangement for the actual targets, so that any one of several may be chosen with ease. The first target, known as 1AT1, is limited in thickness to less than 4 g cm^{-2} by the requirement to efficiently transport beam to the second target and to the beam dump. The most frequently used target at 1AT1 is a piece of graphite of 10 mm thickness, cooled on one edge. The proton beam profile at this target is 7 mm

[1] TRIUMF is the meson facility of the University of Alberta, University of Victoria, Simon Fraser University and the University of British Columbia, and is operated with funding from the National Research Council of Canada

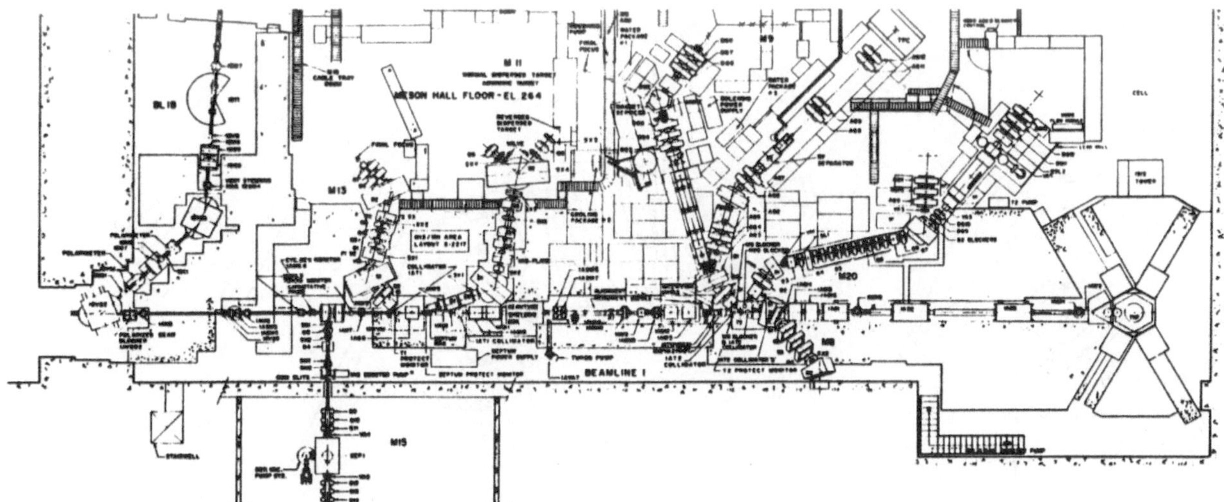

Fig. 1. Secondary beam placement on the primary proton beam, BL1A

high by 2 mm wide. The secondary channels M11, M13, and M15 collect particles produced; the latter two are used extensively for muon research. The second target, known as 1AT2, is thicker and can be up to 20 g cm^{-2}. Normally it consists of a piece of beryllium of 100 mm in the beam direction, encased in a stainless steel jacket through which cooling water flows. Secondary channels known as M8, M9, and M20 view 1AT2. The latter two again are used for muon production.

4 Beam lines

Three methods are generally used for the production of muon beams:

1.The conventional decay mode in which muons are produced in a long straight decay section composed of either a superconducting solenoid or a series of quadrupole magnets. Normally pions from the production target are momentum analyzed by an injection section before the decay section. The muons arising from decays within it, either forward or backward with respect to the pion direction, are further analyzed by an extraction section which transports them to the experiment.
2. The cloud muon mode in which pions decay in flight in a region close to the production target. The resulting muons are momentum analyzed and transported to the experiment.
3.The surface muon mode where positive muons arise directly from the decay of positive pions at rest in the surface of the production target.

Different modes of operation are optimized with different channel designs, although there is some compatibility among modes. All three have been used for muon production at TRIUMF. Because of the high demand for high stopping density positive beams, especially for μSR (muon spin rotation) experiments, surface muons have been most utilized. TRIUMF was able to take an early advantage with surface muons because the beam lines were constructed without thick windows between the target surface and the experimental position. Surface muons cannot penetrate more than about 150 mg cm^{-2} of carbon, so only thin windows and counters can be tolerated. The development of surface muon techniques at TRIUMF was aided by a previous collaboration with a group from the University of Arizona working at Berkeley. They were the first to build a beam to exploit surface muons.[2]

All of the secondary channel elements are controlled by a user-friendly microcomputer and control software package. Functions such as element scaling, automatic intensity optimization, and trip warning are standard in the system. We first describe the channels which use the thinner 1AT1 target as a source, M13 and M15.

The M13 beam line (Fig. 2) was designed to provide both low energy pions and surface muons.[3] It is a very short channel with good beam quality, although the muon intensity and positron contamination are poorer than the other surface muon channels. Particles are collected in the horizontal plane at an angle of 135° to the primary proton beam. M13 contains two bends of 60° in opposite directions, and has two dispersed foci between the bending elements where slits can be used to adjust the momentum width.

A short DC separator and an extra quadrupole triplet can be added to the beam line, resulting in a loss of about 90% of the muons, but it substantially reduces the positron contamination. Experiments which need only modest muon intensities but are sensitive to positrons can take advantage of the extension.

The M15 beam line (Fig. 3) was planned solely for the exploitation of a surface muon beam. It has two orthogonal bend planes, both vertical, with four dipole bending magnets. The 1AT1 production target is viewed from 150° in the vertical plane, with the final leg of the

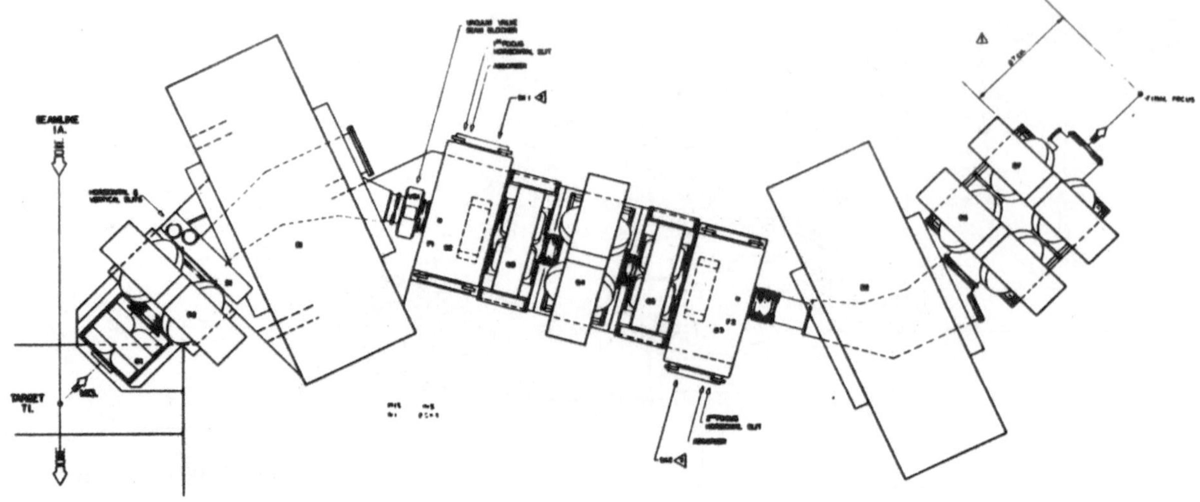

Fig. 2. Plan view of M13

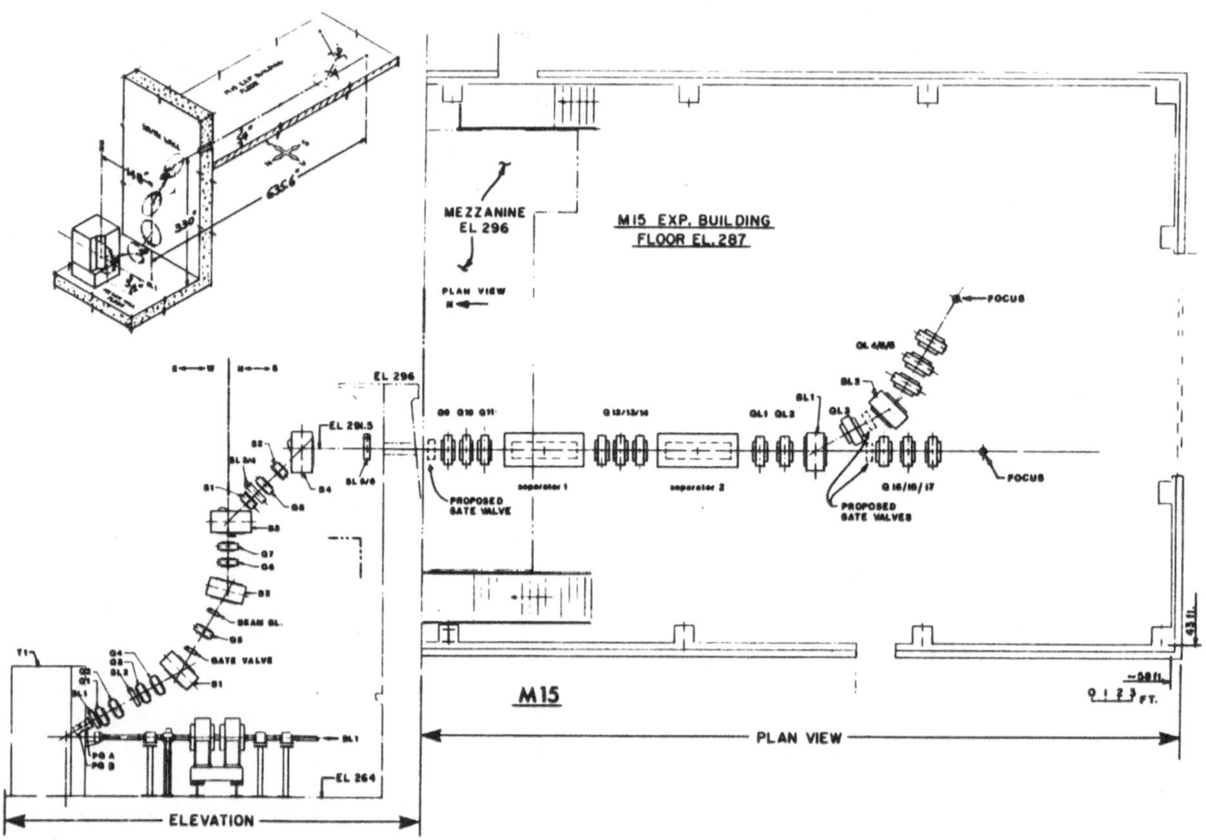

Fig. 3. Diagram of M15, showing two vertical bend planes and the planned split extension

beam horizontal at 7 m above the level of the target. The complex arrangement was chosen as effective in terms of cost and space.

A unique feature of M15 is the use of permanent magnets for the first two quadrupoles. The modest magnetic field needed for surface muons makes this approach practical. The advantage is that the elimination of electrical and water connections allows placement of the quadrupole magnets closer to the production target, and the acceptance is improved to about 35 msr. The design

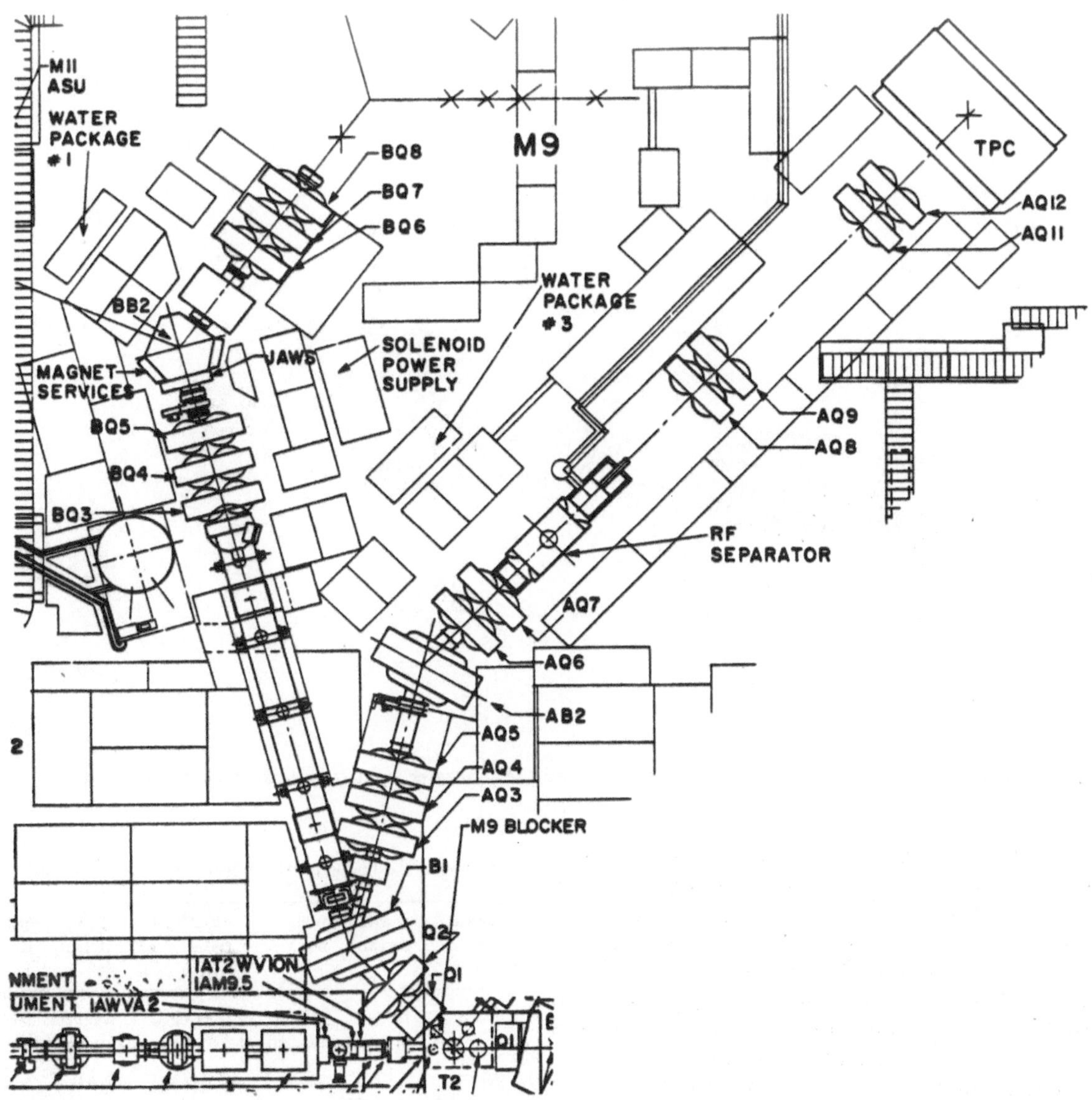

Fig. 4. Plan view of M9, showing both the RF separator leg (A) and the superconducting solenoid leg (B)

of the channel also allows very high luminosity beams. Since the fields of the two quads cannot be adjusted, there is a limitation in momentum for which an advantage can be gained, and of course performance with negative particles is compromised.

M15 has six sets of slits which can be used to limit angular acceptance, attenuate the beam, and control the beam spot and momentum width. An important feature, especially for the μSR program, is a pair of crossed field DC separators which not only reduce positron contamination but can rotate the polarization direction of the surface muon beam by 90°. The majority of beam time on M15 is allocated to the μSR program, and the transverse muon polarization is necessary to permit axial injection into a solenoid spectrometer for high transverse field μSR experiments.

The possibility exists with M15 to install a beam kicker after the final bend and to install a septum magnet and several extensions at the end of the channel (Fig. 3). This option is currently being studied.[4]

The thicker production target, 1AT2, is the source for the other muon channels, M9 and M20.

The M9 beam line (Fig. 4) has two legs, of which only one can be used at any time. They share the first two quadrupoles and dipole, taking particles at 135° to the incident primary beam.

The M9A leg, fed by a 60° bend, is normally dedicated to a large particle tracking spectrometer, such as the one used by the radiative muon capture (RMC) experiment in progress at the moment.[5] Its normal mode of operation is with cloud muons at a specific momentum (usually in the range of 60 MeV/c) compatible with the use of an RF separator after the second bend. The separator reduces pion and electron contamination below the level of about 1%.

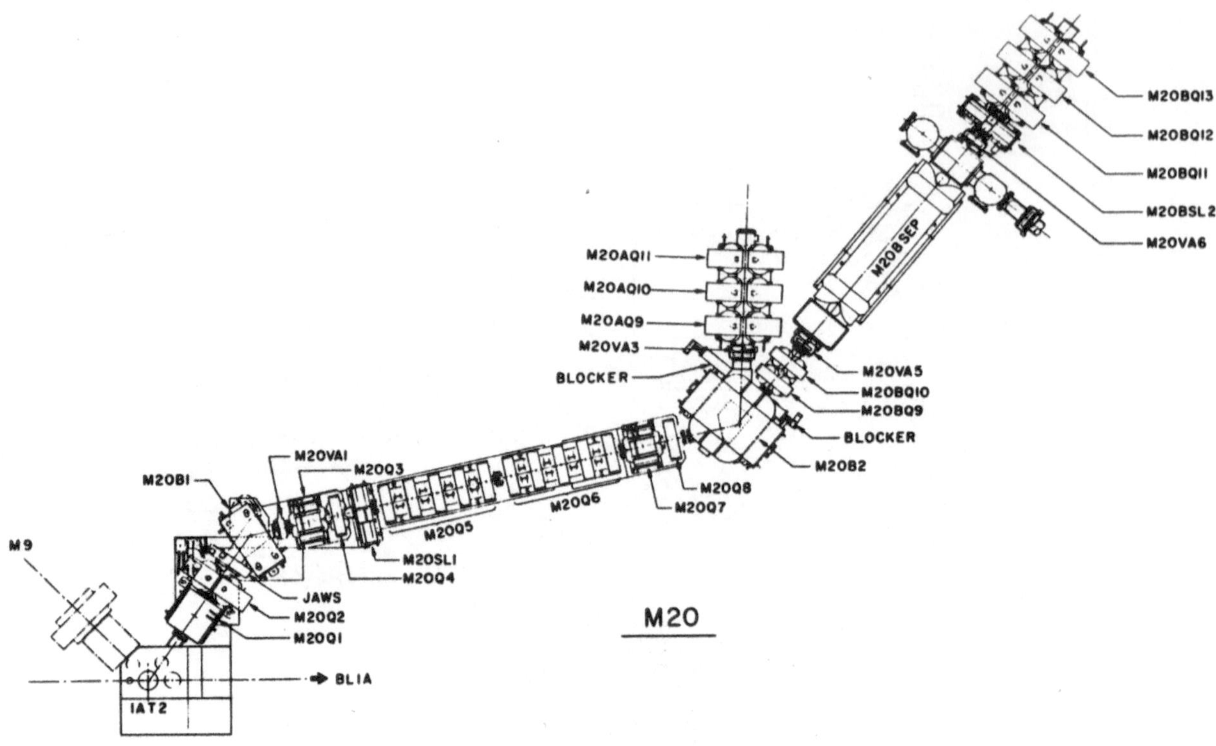

Fig. 5. Diagram of M20, showing the backward decay leg (A) and the separated spin-rotated surface muon leg (B)

The M9B leg, also known as the superconducting muon channel (SMC), is fed by a 30° bend from the first bending magnet, which injects pions into a superconducting solenoid decay section. This leg was designed primarily as a decay channel for negative muons. Funding for its construction came mostly from Japan, and the University of Tokyo and associated institutions are major users. The short length of the injection section allows lower energy pions to survive long enough to decay in the solenoid, and the result is that useful numbers of muons at momenta below 30 MeV/c are available to experiments. The muons are analyzed after the solenoid by two quadrupole triplets separated by a bending dipole. A design exists to extend M9B with one more analyzing bend and another quad triplet (or possibly two triplets, feeding two separate experimental areas). A second bend in the analyzing section would be very useful in reducing backgrounds, especially of neutral particles, and providing control of the beam size and momentum spread at the experimental target. The extension awaits the availability of the resources required to construct it.

The M20 beamline[6] also has two legs, M20A and M20B (Fig. 5), but in this case the split comes after the second bending magnet. After a short injection section ending at the first bending magnet, particles are transported through a long straight section which can be used to form a decay muon beam. The M20A leg is generally operated with a decay beam, but the M20B leg is often preferred because it has a crossed field separator and spin rotator for surface muons. The intensity of the surface beam is as great as for M15, but a longer production target and the concomitant larger beam size mean that the luminosity is lower. M20 is in high demand especially for μSR experiments.

Table I lists the muon beams at TRIUMF and summarizes their important characteristics.

Table 1. Existing muon beam lines at TRIUMF

beam line	main mode	p MeV/c	flux*(p) 10^6s^{-1} (MeV/c)	$\Delta p/p$ fwhm %	pol^n %	size h×v cm
M13	μ^+	20-30	1.8†(29)	1-10	>90	3×2
Excellent optics, short DC separator available (10% flux)						
M15	μ^+	20-30	2.0†(29)	2-10	>90	1.2×1.6
Mostly μSR, DC spin rotator/separator, high luminosity						
M9A	μ^-	30-100	1.4‡(62)	4-8	50	8×9
RF separator, dedicated to RMC spectrometer						
M9B	μ^-	20-80	1.4‡(40)	11	>90	10×10
Superconducting decay channel, funded from U Tokyo						
M20A	μ^+	25-85	3.5‡(85)	10	85	7×10
Backward conventional decay beam, seldom used						
M20B	μ^+	20-200	2.1‡(29)	3-8	>90	4×3
Mostly μSR, DC spin rotator/separator, high flux						

* For proton current of 140 μA at full $\Delta p/p$
† For production target of 1 cm graphite
‡ For production target of 10 cm Be

5 User facilities

TRIUMF provides certain general purpose equipment for its muon users, ranging from simple things such as cables and electronics to state-of-the-art spectrometers and cryostats.

The μSR facility services the majority of muon users at TRIUMF. It supports several general purpose μSR spectrometers, which include collimators, beam line extensions and windows, magnets, counters, electronics, and specialized data acquisition hardware and software. One of the spectrometers is a superconducting solenoid with high field capability. There is a selection of cryostats which can be installed in the spectrometers to control the temperature of most typical samples. Of particular note is a dilution refrigerator which allows samples to be cooled to the millikelvin range. Technical support for μSR experiments is also very good.

The RMC spectrometer[5] is another example of a facility supported by TRIUMF.

Various general purpose detectors are available, such as many sizes of plastic scintillator and several germanium diodes. An active detector group can loan a certain number of wire chambers to experimenters. Data acquisition hardware, based on the NIM and CAMAC standards, can be reserved for beam periods, and a versatile VAX-based data acquistion and monitoring system is supported for resident and visiting scientists.

6 Conclusion

TRIUMF is open to researchers regardless of nationality, subject to the approval of any proposal by the Experimental Evaluation Committee. Experiments may benefit from high quality muon beams and world class facilities. The list of measurements made at TRIUMF with muons is impressive, and the muon program is expected to dominate the list of active experiments in any conceivable future of the laboratory.

References

1. TRIUMF User's Handbook, ed. E.L. Mathie, unpublished
2. A.E. Pifer, T. Bowen, and K.R. Kendall: Nucl. Instr. and Meth. 135 (1976) 39
3. C.J. Oram, J.B. Warren, G.M. Marshall, J. Doornbos, and D. Ottewell, Nucl. Instr. and Meth. 179 (1981) 95
4. J.L. Beveridge, contribution to this workshop
5. W. Bertl et al., contribution to this workshop
6. J.L. Beveridge, J. Doornbos, D.M. Garner, D.J. Arseneau, I.D. Reid, and M. Senba, Nucl. Instr. and Meth. A240 (1985) 316

This article was processed using Springer-Verlag TEX Z.Physik C macro package 1991
and the AMS fonts, developed by the American Mathematical Society.

Z. Phys. C - Particles and Fields 56, S232–S239 (1992)

Zeitschrift für Physik C Particles and Fields

The ISIS Pulsed Muon Facility

G.H.Eaton

Rutherford Appleton Laboratory, Chilton, Oxfordshire, U. K.

4-June-1991

Abstract. The pulsed muon facility at the ISIS spallation neutron source of the Rutherford Appleton Laboratory is described. Work currently in-hand to increase the performance of the proton synchrotron from its present 100 microamperes at 750 MeV to its design goal of 200 microamperes at 800 MeV are described in conjunction with plans to expand the muon facility itself. Within a period of three years it will be possible to offer five simultaneous experimental ports with single muon pulses at 50Hz from a surface/cloud facility on the south side of the extracted proton beam, and a superconducting decay channel on the opposite side. Such an overall facility will offer unique opportunities for pulsed muon science to the end of the decade and beyond.

1 Introduction

The overall lay-out of the ISIS facility is shown in Figure 1. The accelerator system consists of a negative H linac A delivering 72 MeV protons after stripping into a fast-cycling 800 MeV Synchrotron B. Multi-turn injection is used to generate at present 100 microamperes time-averaged proton current at 50 Hz repetition rate. Use of twice-harmonic accelerating frequencies in the Synchrotron RF system results in two accelerated proton pulses per cycle. The proton pulse width is 100 ns(base width), 70ns (FWHM) with a peak to peak separation of 340 ns.

The presence of a double pulse has been an inhibiting feature of the facility to-date for many muon spin rotation experiments. The upgrade of this surface channel will utilise this double pulse structure to give simultaneous multi-port provision of single muon pulses at 50 Hz.

The extracted proton beam is principally used to generate pulsed neutrons from a depleted uranium target C which serves a large number of neutron beam lines D. Each neutron beam line views one of four moderators arranged around the spallation target. This pulsed neutron facility is the most intense in the world and is used by a European-wide neutron community and research workers from Japan.

Table 1 shows the ISIS performance from 1987 to 1990. The present operating current is close to 100 microamperes, with proton energy of 750 MeV up to the end of 1990. ISIS is now running at 800 MeV. The design goal is 200 microamperes and this will be realized in 1993 in stages funded by KFK Germany.

2 Restrictions imposed by ISIS Environment

The ISIS Pulsed Muon Facility, both in it's present form and in future upgraded version, uses an intermediate target station 20 m upstream of the spallation target. The environment of the muon facility with respect to the overall ISIS complex is shown in Figure 2. This environment is very different to those at the Meson Factories at PSI and TRIUMF and imposes severe restrictions on the design and lay-out of the muon facilities. In particular the muon facility which operates simultaneously with the ISIS neutron and neutrino facilities must satisfy the following conditions.

a) The thickness of the muon production target must be limited to those which result in acceptable reductions in neutron intensity from the spallation target.The neutron loss is 3.2 % for a target thickness of 10 mm and this is the present and future limit for simultaneous operation of these two facilities.

b) The induced activation of the proton channel downstream of the muon production target must be acceptable from the operational point of view for maintenance of the ISIS facility. This has been achieved by the introduction of water-cooled collimation after the muon target. Additional collimation may be introduced to reduce the activation in the extreme downstream section of the extracted proton channel in front of the spallation target.

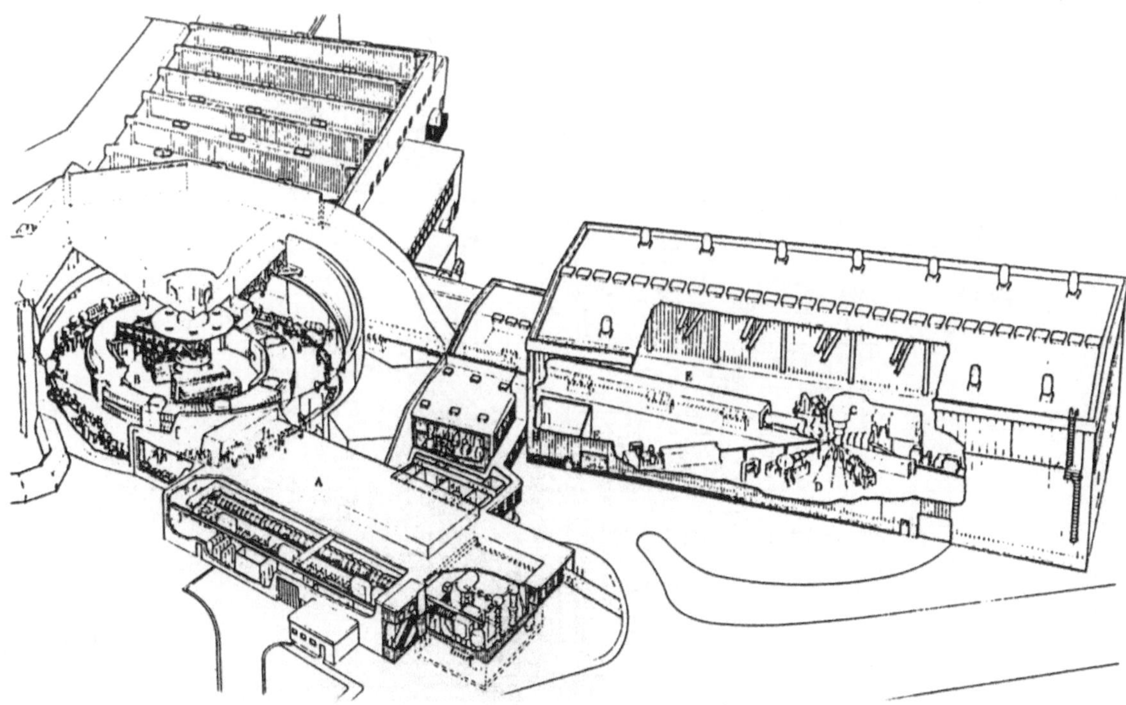

Fig. 1. Overall View of the ISIS Facility of the Rutherford Appleton Laboratory. The Pulsed Muon Facility is located upstream of the Spallation Target C.

c) Neutron backgrounds at the neutron scattering instruments and in the neutrino facility arising from the muon production target must be acceptable to both of these facilities. In the early operational phase of the muon facility, the induced backgrounds to both types of instruments were high, requiring improvements in the shielding local to the muon target.

3 The ISIS pulsed muon Facility 1991

The ISIS Pulsed Muon Facility as constructed and in its present more complex form has been described in detail elsewhere [1,2], so only a brief summary will be given here. The facility, which consists of a surface/cloud muon beam taken off at 90 degrees to the direction of the proton beam, is shown in Figure 3. A conventional double-bend achromatic beam-line delivers 400,000 surface muons/sec from the 5mm production target at 100 microamps proton current, in a final spot of 20 mm(H) * 10mm(V) [FWHM] at a repetition rate of 50 Hz. It was commissioned in March 1987 and now supports a European wide research program in materials science using the muon spin rotation technique, and in muonium and muon-catalysed fusion research.

Table 1. ISIS performance summary 1987 - 1990

		1987	1988	1989	1990
Total scheduled user time	(days)	176	182	178	167
Total lost time	(days)	48	57	47	36
Total time on target	(days)	128	132	131	131
Percentage of time on target	(%)	73	72	73	78
Total integrated current	(mAhrs)	120	230	288	302
Average current during scheduled time	(μA)	28	53	67	75
Average current with beam on	(μA)	39	72	91	96
Peak current averaged over 24hrs	(μA)	70	97	107	101
μAhrs per beam trip		6	18	31	61
Number of cycles for science		7	6	7	6

The double pulse structure of the ISIS extracted proton beam, which produces a similar time structure for the muons, imposes restrictions on the observable frequencies in muon-spin-rotation experiments.This problem was overcome in 1989 by the installation of a pulsed electrostatic kicker [2] into the beam- line; with the pur-

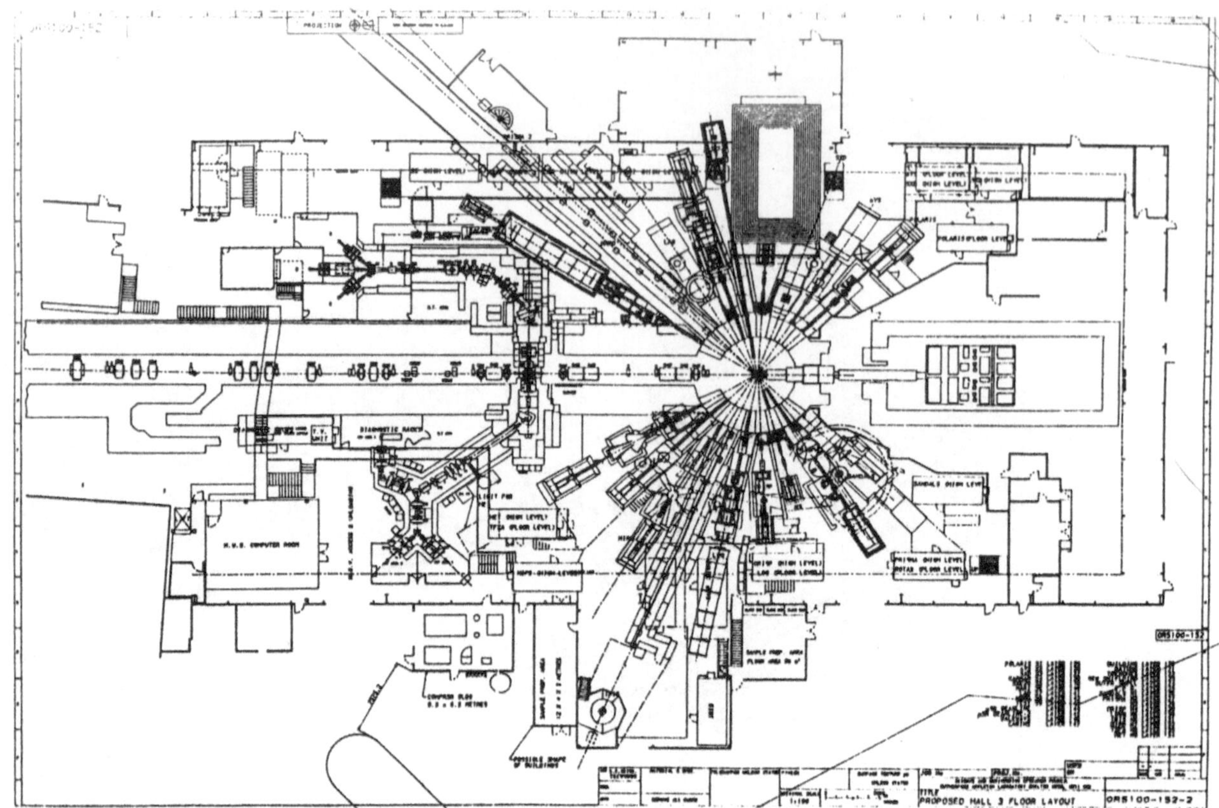

Fig. 2. The Environment of the Pulsed Muon Facility with respect to the other Facilities. The complex of neutron instruments radiating from the spallation target are shown together with the KARMEN neutrino detector on the south side of the target bulk shield.

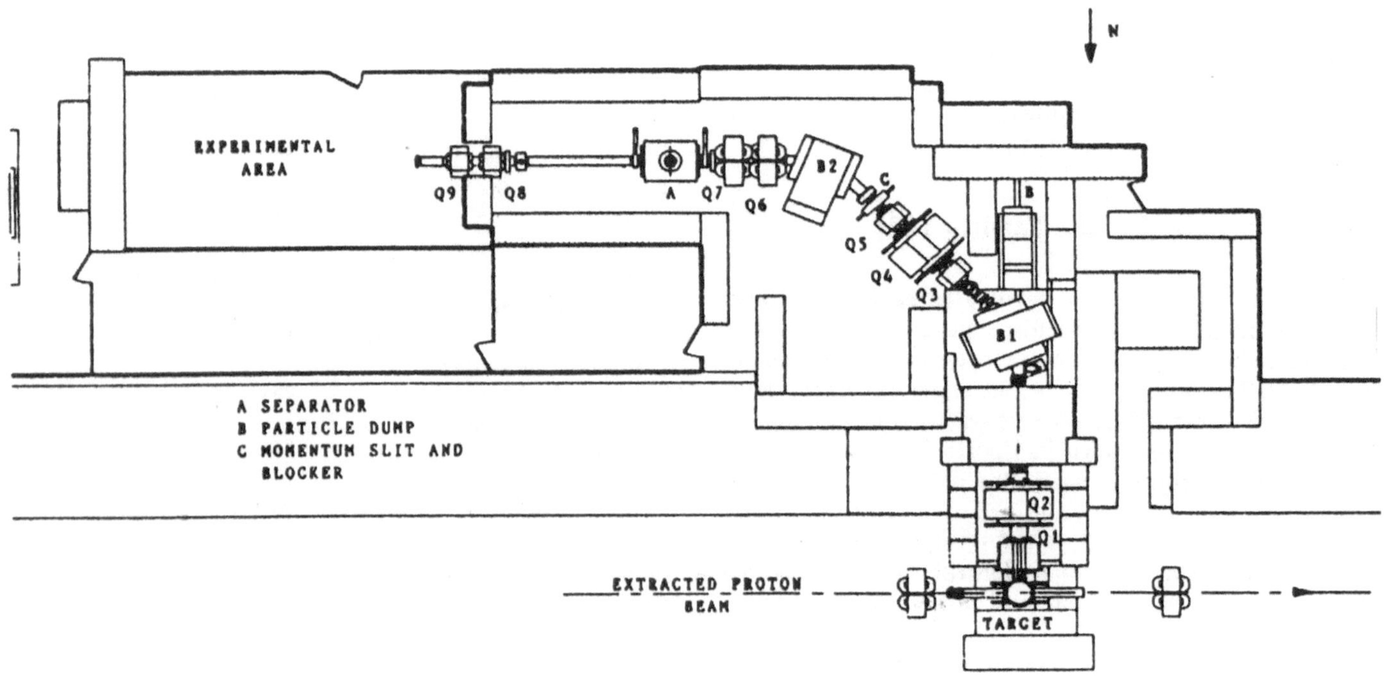

Fig. 3. The ISIS pulsed muon facility as built in 1987.

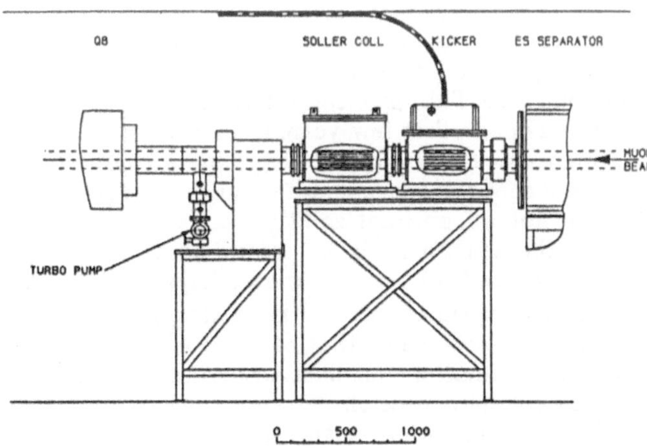

Fig. 4. The UPPSET electrostatic kicker and Soller collimator shown downstream of the cross-field separator in the muon beam. The kicker consists of closely spaced horizontal foils to which high voltage is applied during the passage of the muon pulses. The deviated muons in the second pulse are captured in the vertical divergence-limiting Soller collimator. Figure 5 shows the MuSR frequency response measured under double and single pulse operation, illustrating the efficacity of the UPPSET device in giving a smooth response function compared with the regions of reduced sensitivity with double pulses.

pose of eliminating the second muon pulse. This device called UPPSET (Uppsala Pulse Eliminator and Trimmer) is shown in Figure 4. It allows the operation of the facility with either single or double pulses reaching the sample position at the end of the beam-line. Transfer from each mode can be made in about 10 minutes. The halving of muon intensity so incurred from single pulse operation can be fully compensated by doubling the thickness of the production target from 2.5 mm to 5.0 mm, an operation which can be done in parallel with switching on the UPPSET device over the same time scale.

4 Experimental advantages of pulsed muon sources

The experimental advantages of pulsed sources for many experiments have been extensively discussed by the KEK group[3], but a few illustrative examples from the ISIS experience will be given here.

For MuSR experiments the principal advantage arises from the fact that after the muon pulse the backgrounds in the experimental hall are extremely low, thereby allowing the muon decays in the sample to be traced over many muon lifetimes. This is well illustrated by Figure 6, which shows a typical transverse MuSR spectra obtained at ISIS, where the behavoir of the muon in the sample is clearly observable above background for 12 lifetimes, a unique property of pulsed sources.

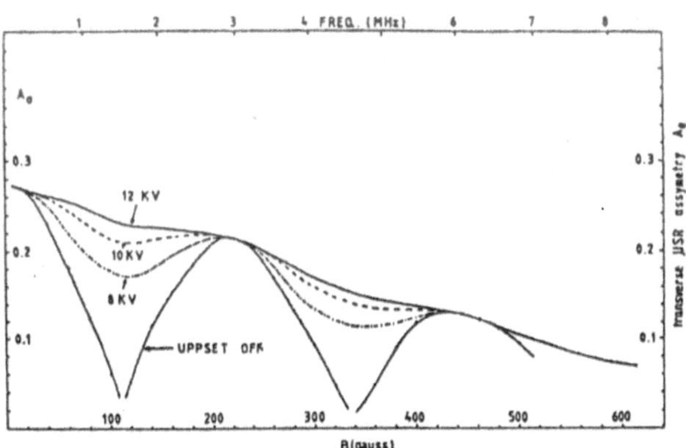

Fig. 5. Transverse MuSR frequency response plotted against external magnetic field for various operating voltages on the UPPSET kicker. It can clearly be seen that using the design voltage of 12 kV removes the 'holes' in the response function caused by the double-pulse structure.

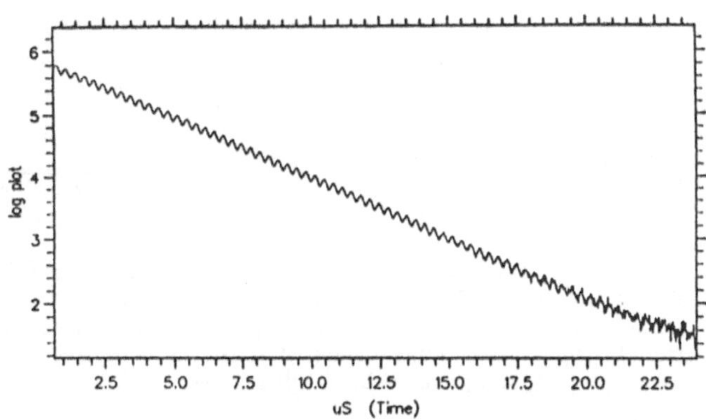

Fig. 6. Log-linear plot of the tranverse MuSR event rate against decay time for a Silver sample at ISIS. It can be seen that the muon behavoir can be traced clearly above background to more than 12 life-times. This is a unique advantage of a Pulse Muon source such as ISIS.

For MuCF experiments, this very low background level after the arrival of the muon pulses is also very advantageous for some experiments. An example of this is shown in Figure 7, which shows fusion neutron spectra observed at ISIS with low pressure DT mixtures. With such a pulsed muon source it is possible to clearly observe single fusion events above the noise at much lower pressures than possible at the continuous sources.

Finally a clear advantage of pulsed sources lies in the use of a pulsed environment in coincidence with the arrival of an intense bunch of muons. An example of this is the currently-running 1S to 2S Muonium transition experiment which utilizes an intense pulsed laser.Such experiments are only possible at sources such as the KEK and ISIS facilities.

Many other examples can be quoted of experiments which are more suited to pulsed muon sources. This is the principal justification for the developing world-wide interest in such facilities and new ones being proposed at the PSR at LAMPF and the Japanese Hadron Facility.

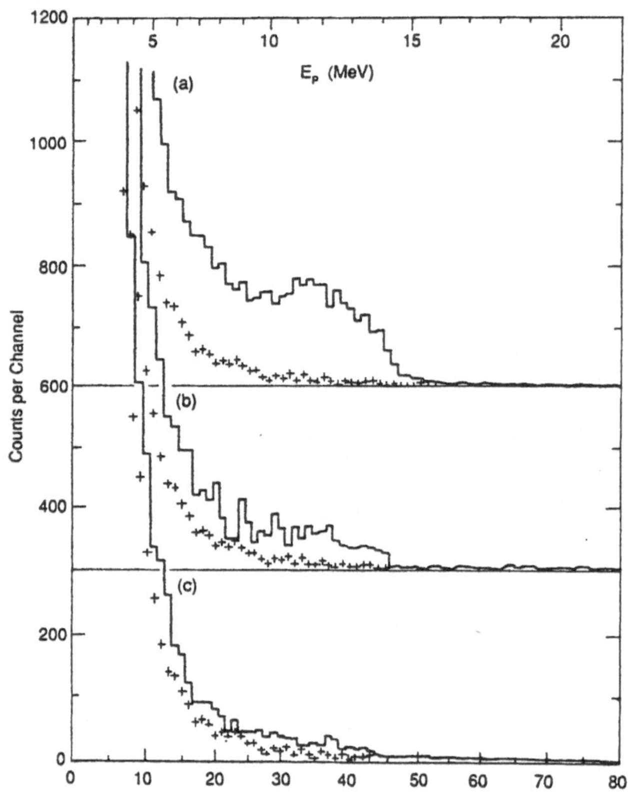

Fig. 7. Demonstration of muon catalysed nuclear fusion in DT mixtures at the ISIS pulsed muon beam. The recoil proton spectrum in neutron detectors show the single fusion neutrons emitted from DT mixtures at 2.0, 1.0 and 0.66 atmospheres compared with the measured backgrounds shown with crosses.

5 Operational Experience with Pulsed Muons

Various important lessons have been learnt during the short lifetime of the ISIS pulsed source, which are illustrative of the differences in continuous and pulsed facilities. In the design of pulsed sources several significant features must be incorporated to optimise the utilization.

5.1 Contamination in a Pulsed Beam

In a continuous muon beam, it is relatively easy to discriminate electronically against contaminant particles such as positrons at surface momentum. In a pulsed source this is not possible, hence the inclusion of a cross-field electro-static separator as a velocity selector is an absolute requirement. Since a large fraction of the contaminant positrons in a surface beam arise from muon decays in the production target, and hence have the characteristic 2.2 microsec life- time, it is self evident that these must be reduced to negligible proportions in a pulsed beam.

5.2 Control of Muon Spot Size.

In continuous muon sources, the muon spot size at the sample can easily be controlled using scintillation anticounters. This is clearly not feasible at a pulsed source. For the first few years operation at ISIS, a lead collimator at the end of the beam line was used to tailor the final spot to experimental requirements [4]. However this is not an optimal solution as considerable space inside the apparatus is necessary to accomodate the thicknesses of lead needed for an adequate collimator.

A more optimal solution is to arrange the optics of the muon beam-line so that at some intermediate point along it, a simultaneous horizontal and vertical waist is imaged to the end of the beam. Provision of remotely controlled collimation at this intermediate point allows one to tailor the final beam spot size over a wide range without flux loss.

5.3 Timing of the Muon Pulses.

All experiments which use pulsed muon sources are critically dependent on accurate timing of the arrival of the muon pulses at the apparatus, or at least a stable reference point. Our early experiments used timing signals from the accelerator, however these were found to be insufficiently stable for the requirements. A more stable and reliable timing signal is now derived from lead-glass Cerenkov counters which are positioned after the first bending magnet in the muon beam, viewing the direct spill from the muon production target.

5.4 Muon target induced backgrounds at Pulsed Sources.

In contrast to the situation at the continuous muon sources where pion and muon users are the principal or sole users of the facility, pulsed muon users have to co-exit with very different user facilities. As a result of this situation, it is especially important to design the shielding around the muon facility to the more stringent background requirements compared with the usual biological limits common to all accelerator complexes.

The initial shielding at ISIS was designed from health and safety requirements and early dose measurements indicated that these were being fully satisfied. However Figure 8 shows the fast neutron background measured in the KARMEN neutrino detector with this shielding, as a function of the thickness of the muon target being used. The linear increase in the background with thickness hardly constitutes an ideal situation for independent simultaneous operation of both facilities. In addition the absolute level of background so generated precluded a considerable fraction of the proposed experimental program of the neutrino program.

Careful examination of the shielding local to the muon facility identified the weaknesses responsible for the backgrounds at the KARMEN detector. The principal weakness was found to be the labyrinth into the

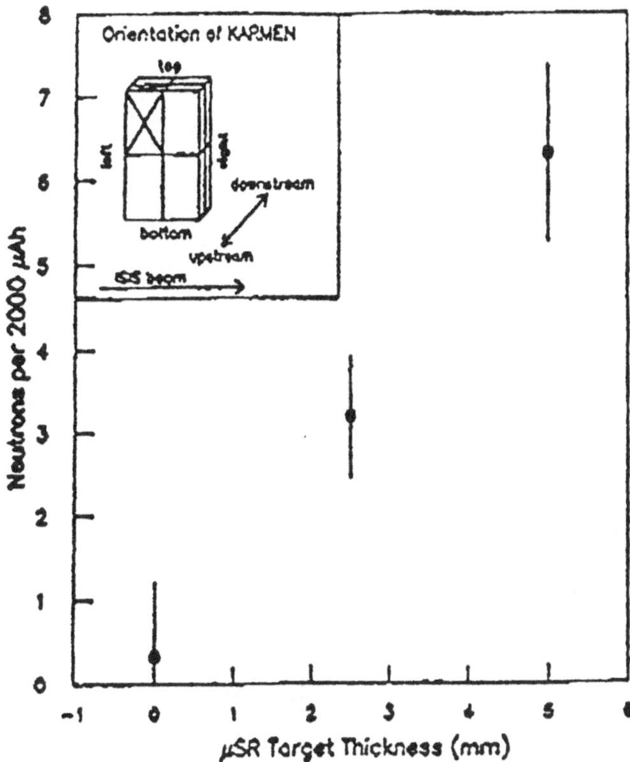

Fig. 8. Measured event rate in the KARMEN neutrino detector in 1990 against the thickness of the muon production target. This rate due to fast neutrons was measured in the top-left hand corner of the detector , which was most sensitive to the muon target. This sensitivity to the muon target has been removed by improvements to the shielding in early 1991.

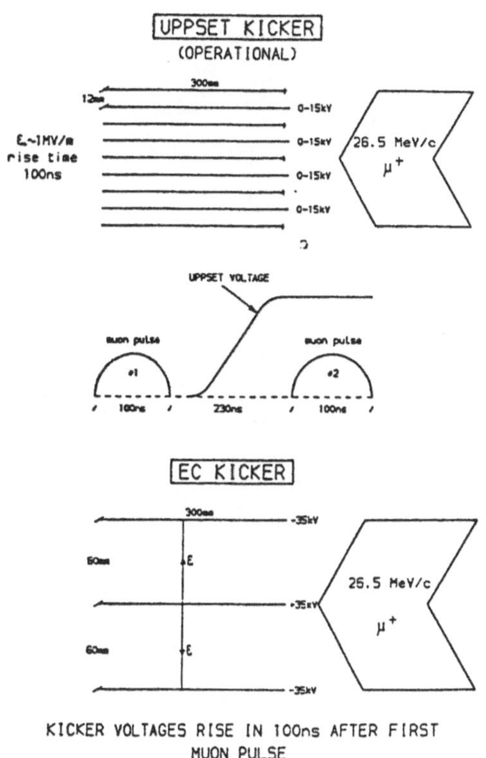

Fig. 10. Comparison between the electrostatic kickers used in the UPPSET device and for the EC upgrade.

proton channel, and the neutral particle dump in the muon beam itself. In early 1991 these weaknesses were removed by filling in the labyrinth with steel shielding and also enhancing the shielding around the muon beam dump. Data taken with the KARMEN detector since then indicate within the limited statistics obtained that the muon induced backgrounds have now become of a similar level to the general level of background in the experimental hall from all other sources.

6 Future Upgrades of the ISIS Muon Facility

At present, the principal restriction of the ISIS pulsed muon facility is the fact that only a single experimental port is available at any one time. Beam-time applications exceed time available on this port by a factor of three. In addition, the surface/cloud beam cannot provide polarized negative muons or polarized positive muons above surface momentum (26.5 MeV/c in the case of the ISIS beam).

In order to overcome all of these restrictions, major expansion plans have been approved by the EC Large Facilities Program (for the South beam) and by the STA Agency Japan through the RIKEN Laboratory (for the north side facility). In addition a short-term improvement to the south beam has been initiated. These plans are described below in chronological order of completion.

6.1 Remote Tailoring of South-side Muon Spot 1991

The beam-line optics of the present south-side beam have been re-optimised to provide a horizontal waist immediately downstream of the cross-field separator shown in Figure 3. This horizontal waist is then imaged to the final spot at the end of the beam-line. A horizontal slit will be installed after the separator to allow remote horizontal tailoring of the final spot over a factor of five from the normal width (20mm FWHM) without flux loss. This will obviate the need for the imperfect lead collimation currently in use for experiments with small samples.

6.2 EC Upgrade of South-side Surface/Cloud Facility 1991-Spring 1993

The approved EC upgrade of the south side muon facility shown in Figure 9, consists essentially of the inclusion in the beam-line of a pulsed electro- static septum, the purpose of which is to deliver single muon pulses at surface or sub-surface momentum to three experimental ports simultaneously. The septum kicker shown in Figure 9 has a wide-gap three electrode structure, whose voltages are raised to plus and minus 35 kv between

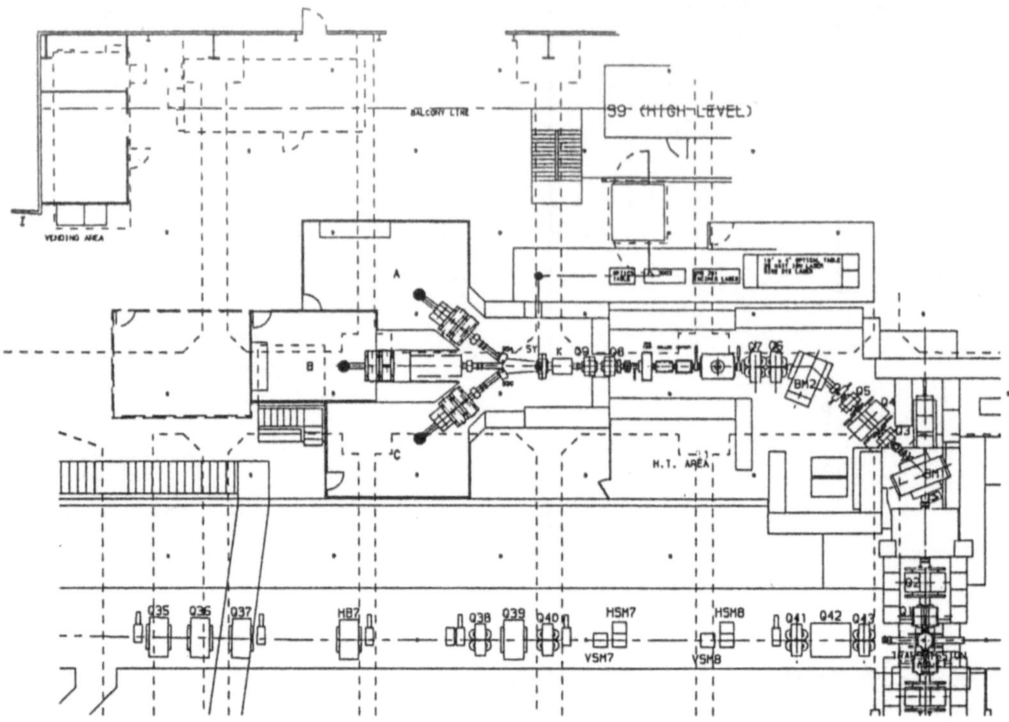

Fig. 9. Layout of the EC upgrade to the present ISIS south-side surface/cloud muon facility. A fast electrostatic septum magnet ramps up in electric field between muon pulses, so that each of the three experimental ports A,B and C receive single muon pulses at 50 Hz. Both muon pulses can be directed to an area of choice using small switchyard magnets SY with the electrostatic kicker K switched off.

the muon pulses. This layout is contrasted with that of the UPPSET kicker in Figure 10. The first muon pulse proceeds through the kicker undeviated to the central experimental port B, where the present MuSR spectrometer will be located. The second muon pulse arrives after the high voltages have been established on the electrodes (rise-time approx. 200 ns), so that 50 % of this pulse will be deviated to the left by 4.5 degrees, while the remaining 50% is deviated to the right. Septum magnets operated continuously will then direct these muons to experimental ports A and C.

The intensity reductions so incurred compared with the present facility will be made up by the higher proton currents expected to be available at this time, and also by using slightly thicker muon production targets. Area C will be equipped with a new MuSR spectrometer for new European users. Area A will be reserved for Muonium experiments with surface or sub-surface positive muons, MuCF experiments as sole user or MuSR experiments using analogue detection techniques.

Two small switchyard magnets will be included in the beam-line extension immediately after the septum kicker. With the kicker switched off, these will allow both muon pulses at surface momentum to be directed to an experimental port of choice for a sole-user experiment requiring the maximum muon intensity available. Cloud muons up to 40 MeV/c can be directed to area A for sole-user MuCF experiments. All the three beam lines will be equipped with horizontal collimation for remote tailoring of the final beam spot.

Such an upgrade will allow up to a three-fold increase in the number of MuSR experiments, thereby adequately catering to a European demand for low -energy pulsed muons. This upgraded facility is expected to be operational by the spring of 1993.

6.3 RIKEN Superconducting Muon Channel 1993

This facility which is shown in Figure 11, will ideally complement the south side surface/cloud complex, in being able to provide polarized negative muons and higher momenta for polarized positive muons. A coventional high acceptance pion injection section incorporating a doublet of large aperture quadrupoles and a 'C' type bending magnet (55 degree bend) will be followed by a 5.5 m long 5T superconducting solenoid in which the pions decay. The maximum momentum for pion injection will be 220 MeV/c.

The muon extraction section after the exit of the solenoid consists of a double bend achromatic system including a pulsed magnetic kicker,septum magnets and a short electrostatic separator.The magnetic kicker will allow simultaneous single pulse operation in the two extraction arms shown in Figure 11, for muon momenta below 45 MeV/c.For muons in the momentum range between 45 and 120 MeV/c, a single switchyard magnet will be used instead of the magnetic kicker to direct both muon pulses to an experimental port of choice.

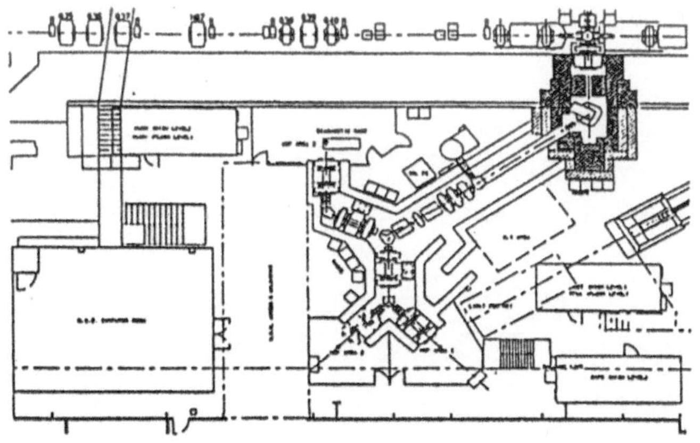

Fig. 11. Schematic layout of the RIKEN north side superconducting decay muon channel. This facility will incorporate a fast magnetic kicker to give single muon pulses in the north and south facing arms at 50 Hz up to a maximum muon momentum of 45 MeV/c. Above this momentum switchyard magnets will distribute both muon pulses to an area of choice up to a maximum momentum of 150 MeV/c.

It is expected to commission this facility in Autumn 1993. Half of the available beam-time will be used by Japanese scientists principally in MuCF research, while the remaining half will be used for SERC sponsored research.

References

1 Eaton GH,Carne A,Cox SJF,Davies JD,De Renzi R,Hartmann O,Kratzer A,Ristori C, Scott CA,Stirling GC and Sundqvist: Commissioning of the Rutherford Appleton Laboratory Pulsed Muon Facility. Nucl.Inst.and Methods A 269 (1988) 483
2 Borden AI,Carne A,Clarke-Gayther M,Eaton GH,Jones HJ and Thomas G.: UPPSET:A Pulsed Electrostatic Kicker To Improve The MuSR Frequency Response in the ISIS Pulsed Muon Beam. Nucl. Inst and Methods. A292 (1980) 21
3 Nagamine K and Yamazaki T.: Meson Science Laboratory, University of Tokyo. Collected Papers on Muon Science Research at MSL Tokyo 1987
4 Dalmas de Reotier P,Yaouanc A,Eaton GH and Scott CA: Positive Muon Data Analysis at ISIS. Hyperfine Interactions. 65 (1990) 1113.

This article was processed using Springer-Verlag TEX Z.Physik C macro package 1991
and the AMS fonts, developed by the American Mathematical Society.

Z. Phys. C - Particles and Fields 56, S240-S242 (1992)

Zeitschrift für Physik C Particles and Fields

Muon beams at PSI

R. Abela, F. Foroughi, D. Renker

Paul Scherrer Institute, CH-5232 Villigen PSI, Switzerland

5 December 1991

Abstract. The present upgrading of accelerators and beams at PSI will greatly improve the offered muon beams. A short description of the beam properties and a summary of the expected muon rates are presented.

1 Introduction

At PSI we almost finished a long improvement program. It started 6 years ago with the reconstruction of one of our two target stations (target M). In a long shutdown from the beginning of 1990 to June 1991 we rebuilt the second station (target E) together with the secondary beam lines and a proton beam transport system to the new spallation neutron source [1].

This long and expensive work was mainly done to be able to increase the current of the proton beam up to at least 1.5 mA [2,3]. This intense beam will carry almost one Megawatt and therefore calls for targets and a beam dump that can stand this power. In addition all elements, including dipoles and quadrupoles, all diagnostic devices and so on had to be prepared for remote handling because of the very high secondary activation.

In parallel we reconstructed all our pion and muon beams. There are only minor changes in the μE4 channel, but essential improvements in μE1, πE1 and πE3 and we built a completely new beam line, the πE5 (Fig. 1). The channels πM1 and πM3 were changed already 6 years ago and were not touched in the last shutdown, but there are plans to modify the πM3 such that it serves several μSR facilities by simple switching from one to the next with dipoles or even kickers.

All our beam lines with the exception of πM1, which is a high resolution pion beam and the μ-channels with superconducting solenoids, are now built without any windows between the vacuum system of the proton channel and the system of the secondary beams. By this they all can be used as low energy - cloud, surface and subsurface - muon beams.

2 Muon beam μE1

Up to now this beam line shared the first quadrupoles and a bending magnet with the πE1 channel and therefore the polarity and the momentum of the muons could only be changed in agreement with the users of the πE1. We decoupled these channels by installing half quadrupoles, focussing elements made of two poles - half of a quadrupole - and an iron mirror plate, which produce a perfect quadrupole field in half of the aperture [4] (Fig. 2).

We reinstalled the first coil of the superconducting solenoid, which tended to quench in earlier times because it was heated by the nearby beam dump, and built in new vacuum chambers and collimators. This will give higher muon rates and less contamination with pions and electrons. See Table 1 for rates and other beam properties. A flexible muon extraction system will allow to swith from SINDRUM II, which is permanently installed, to an other experiment within short time.

3 Pion beam πE1

As described in the previous section this channel became now independent from the μE1 beam line by a similar triplett of half quadrupoles for the first focussing. By greater bending angles in the dipoles, which forced us to introduce a third dipole in order to keep the geometry of the area, the momentum resolution of πE1 was improved. Since there is no more any window in the vacuum system between the production target and the experiment, this channel can now be used as a surface or low energy cloud muon beam, but the expected high contamination with electrons or positrons will always call for an electrostatic separator.

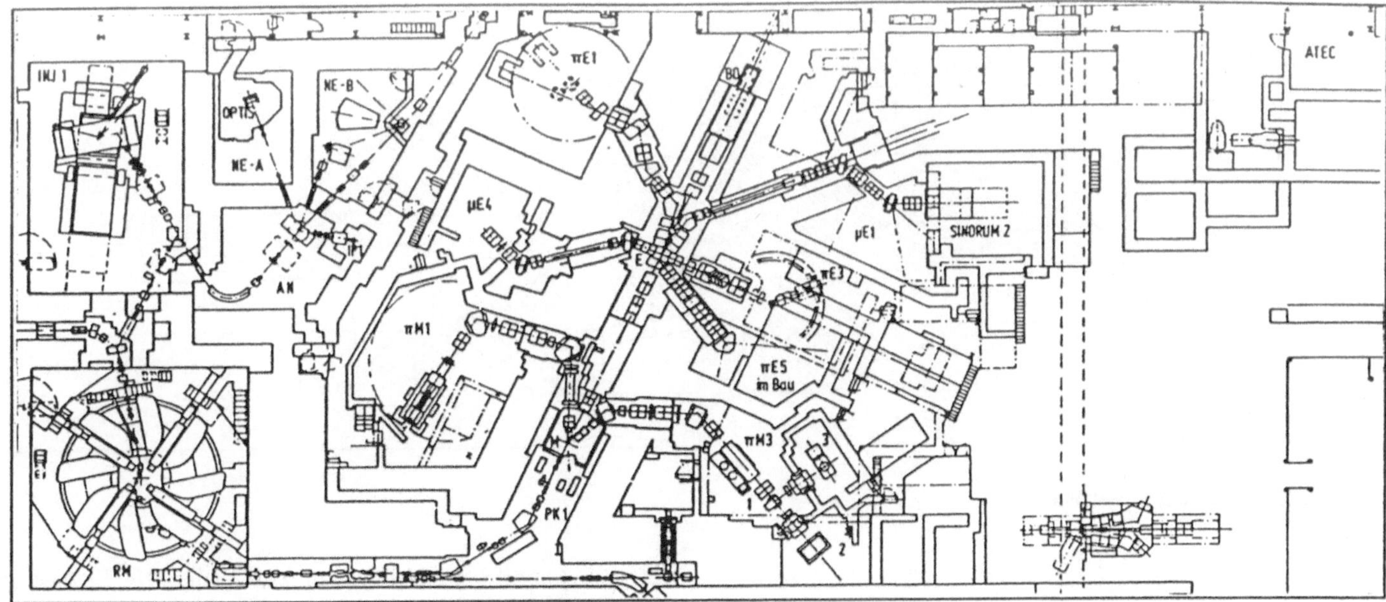

Fig. 1. Layout of the experimental hall

Table 1. Summary of expected muon fluxes and beam properties after the shut-down 1990-1991 from the various beam lines at 1 mA primary beam current

μ^+	mom. MeV/c	flux at 1 mA	$\delta p/p$ FWHM	spot size cm^2	pol.
μE1	40-125	2×10^8 at 125 MeV/c	3%	3 x 2	75 %
μE4	30-100	4×10^6 at 50 MeV/c	3%	6 x 4	75 %
πE1	5-30	2×10^6 at 28 MeV/c	0.8%	1.5 x 2	> 95 %
πE3	5-30	3×10^7 at 28 MeV/c	1 %	2 x 3	> 95 %
πE5	5-30	2×10^8 at 28 MeV/c	2 %	6 x 4	> 95 %
πM3	5-30	4×10^6 at 28 MeV/c	0.2 %	2 x 2	> 95 %

μ^-	mom. MeV/c	flux at 1 mA	$\delta p/p$ FWHM	spot size cm^2	pol.
μE1	40-125	6×10^7 at 125 MeV/c	3 %	3 x 2	75 %
μE4	30-100	1×10^6 at 50 MeV/c	3 %	6 x 4	75 %
πE1	5-280	4×10^7 at 100 MeV/c	0.8 %	1.5 x 2	(var.)
πE3	5-150	3×10^7 at 100 MeV/c	1 %	2 x 3	(var.)
πE5	5-120	2×10^8 at 100 MeV/c	2 %	6 x 4	(var.)
πM3	5-300	4×10^6 at 100 MeV/c	0.2 %	2 x 2	(var.)

4 Muon beam μE4

There is only one minor change in the μE4 channel: we installed a removable degrader made of 2.5cm Beryllium at the entrance of the superconducting solenoid. This, by reducing the pion decays in the injection system, will increase the rate of very low energy muons and makes the channel suitable for feeding the phase space compression device (PSC) proposed by D. Taqqu [5].

5 Pion Beam πE3

This channel was an excellent surface muon beam in the past and will be so in future. The rates are high because they are proportional to the projection of the production target seen by the channel and the πE3 looks to it under an angle of 90 degrees and images more than half of the 6 cm long target.

The main design criterium was a chromatic high resolution final beam spot matching the Low Energy Pion Spectrometer (LEPS), but an achromatic tune is possible for muon physics with bigger acceptance than in the high resolution mode [6].

6 Pion beam πE5

The πE5 is a new channel looking to the target from the back under an angle of 175 degree. The first focussing element of the πE5 is a bending magnet, which deflects the proton beam too. This limits the beam momentum to 120 MeV/c because at higher values it is no more possible to correct the proton deflections, but anyhow the pion production cross section at backward angles will fall steeply at higher momenta.

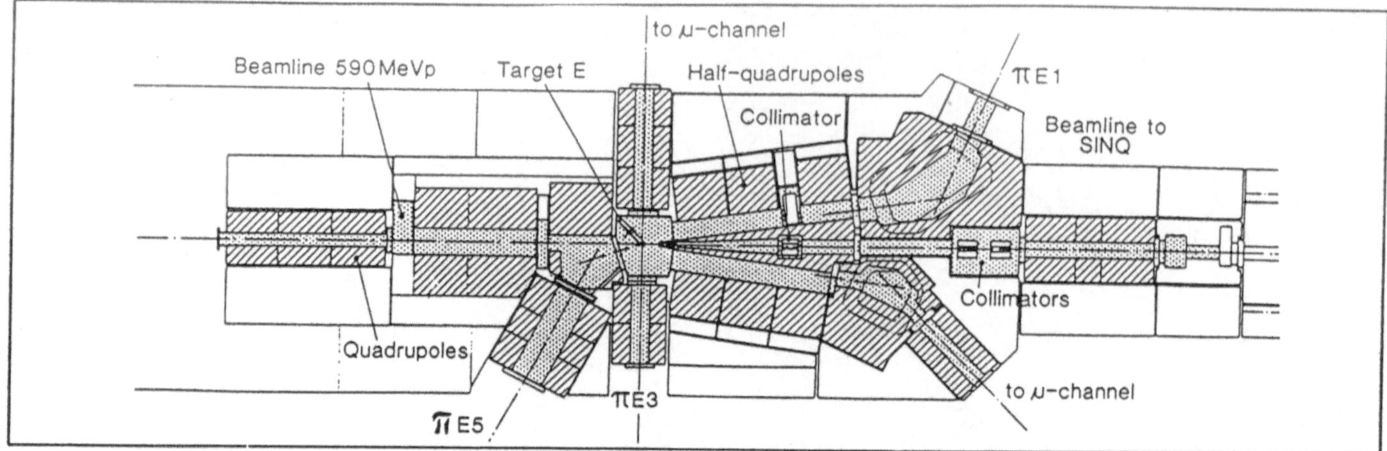

Fig. 2. Layout of the target E region

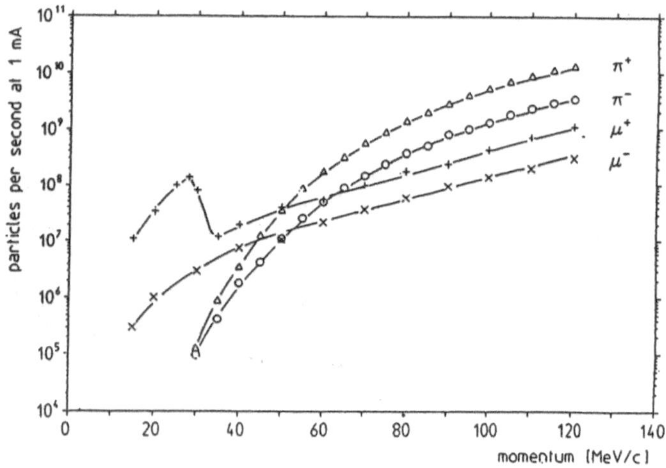

Fig. 3. Caculated particle fluxes from the new πE5 channel at 1 mA primary proton beam current.

This channel has an angular acceptance of 150 msr, which is five times bigger than in 'standard' beam lines, and transports a momentum band of 10% FWHM. To achieve this we had to develope special quadrupoles with an aperture radius of 20 cm and with built in octupole components for the correction of third order aberations. Together with the short channel length of 11 meter the big acceptance will give outstanding high pion rates (Fig. 3) [7]. To give high flexibility the beam line can serve two experimental ports just by switching the polarity of the last bending magnet.

This beam will be most suitable

1. for stopped pion experiments at highest,
2. for adaption to the planned "pion-muon-converter", a special, 6.5 m long superconducting solenoid to produce a high purity μ-minus beam with a pion contamination of 10 E-8 and with rates of the order 10 E8/sec in SINDRUM II, and
3. for injection into a future "low energy muon facility" which will produce muons at keV energies [8,9].

7 Outlook

The reconstruction of the target station E and of the secondary beams, except the new πE5, was finished by summer 1991. At that time only part of the enlarged accelerating RF-power is available, allowing an average proton current of approx. 500 μA until the end of 1991. Two short shutdowns at the beginning and at the end of 1992 are planned for the installation of the last parts of the acceleration system.

References

1. W. Fischer, Proc. 10th Meeting of the Int. Coll. on Advanced Neutron Sources, Los Alamos, 1988.
2. W. Joho, Proc. 11th Int. Conf. on Cycl. and their Appl., Tokyo, 1986, p. 31.
3. U. Schryber et al., Proc. 12th Int. Conf. on Cycl. and their Appl., Berlin, 1989.
4. D. George et al., Proc. 9th Int. Conf. on Mag. Tech., Zürich, 1985, p. 185.
5. D. Taqqu, Nucl. Instr. & Meth., A247 (1986) 288.
6. F. Foroughi, The new πE3: Achromatic Beam Optic, PSI-Technical Note (1990), unpublished.
7. D. Renker et al., PSI Annual Report 1989, Annex I, p. 12.
8. See contributions to these proceedings by D. Taqqu and E. Morenzoni.
9. R. Abela et al., PSI Annual Report 1990, Annex I, p. 19

This article was processed using Springer-Verlag TEX Z.Physik C macro package 1991
and the AMS fonts, developed by the American Mathematical Society.

Z. Phys. C - Particles and Fields 56, S243-S249 (1992)

Zeitschrift für Physik C Particles and Fields

Very slow positive muons

Morenzoni Elvezio

Paul Scherrer Institute, CH-5232 Villigen PSI, Switzerland

11 October 1991

Abstract. The present status and future possibilities for the production of slow (keV) and very slow (eV) positive muons are discussed. Particular emphasis is put on prospects for and open questions presented by the slowing down of MeV muons to eV energies by means of a moderation technique. The use of this technique to produce tertiary, tunable, low energy muon and muonium beams at PSI is outlined in view of the ongoing machine and beam lines upgrading program.

1 Introduction

Positive and negative muons beams in the energy range between a few eV and a few tens of keV and with good phase space properties have been on the wish list of scientists involved in muon physics for many years. All fields of muon science, which, as shown by the large spectrum of contributions to this volume, range from chemistry, condensed matter-, molecular-, atomic- to nuclear- and particle physics, would profit from such a development. Very slow polarized muons would also offer a new tool for surface physics, thin-film and interface studies.

In this paper we will concentrate on the most promising developments for the production of slow and very slow polarized positive muons. Negatively charged muons and their deceleration to few tens of keV are discussed in another contribution to this volume [1]. The paper is organized as follows. In the next section a short overview of the various efforts toward low energy μ^+'s is given. In Section 3 the moderation of MeV μ^+'s to eV energies is reviewed. Open questions and prospects offered by this technique in relation with the current improvement program and plans at PSI are discussed. Characteristics of a possible tertiary beam of very slow μ^+'s are presented in section 4. In the spirit of the title of this volume, some possible future fields of muon physics research which can be addressed with very low energetic μ^+'s are outlined in the last section.

2 Present status

In principle low energetic μ^+'s can be obtained by tuning a surface μ^+ beam line to lower momenta where the intensity has the well known $p^{3.5}$ momentum dependence [2]:

$$N(p) = N_S \frac{p^{3.5}}{30^{3.5}} \exp(-t/\tau_\mu)$$

(p in MeV/c, t: flight time, τ_μ: muon lifetime). This dependence has been tested down to 3 MeV/c [3]. Assuming an intensity at the surface edge N_S of 10^7 μ^+/sec the formula predicts only ~ 20 μ^+'s/sec at 5 keV. The intensity will be even lower in fact, since deflecting and focussing elements of present surface muons beam lines are not designed for such low momenta. The background level from positron contamination would be a disturbing factor for most experiments also.

These limitations led to various proposals of deceleration techniques during the last decade. The production of a low energy muon beam is intimately related to the increase of phase space density. Electronic and stochastic cooling which have been applied successfully to beams of protons , antiprotons and heavy ions to improve the beam quality, are too slow for unstable particles like the muons, where rapid action is required within their lifetime. For muons, the application of friction forces in the form of electric fields, for instance, or deceleration in thin foils combined with rf acceleration has been proposed [4,5]. A method to handle the time limitation and to achieve phase space compression has been proposed in [6] and prototype studies have been performed at PSI [7,8]. The rapid cooling action (within few μs) is obtained by measuring the phase space coordinates of each individual particle and by correcting the values with variable pulsed electromagnetic fields. The application of such a complex procedure to compress a beam from 40 MeV/c to few eV represents a tremendous challenge for the present state of the technological art, but the method could be used as an improvement stage of a precooled beam.

A very simple way to act with non-conservative forces on charged particles is to decelerate them in matter or to use the stop-reemission probability from matter for particles. A broad spectrum of slow muons is obtained by slowing down energetic μ^+ beams in degraders. If the degrader thickness is $\sim 1/2$ of the range, about 0.1 % of an incoming 28 MeV/c beam and few percents of a 10 MeV/c will emerge with energies below 20 keV and a roughly flat spectrum [3]. Through the process of electron capture in the last layers of the target foil, a fraction of these muons can form muonium (Mu) [9] or negative muonium ions (Mu^-) [10] with typically a few keV of kinetic energy.

Compression of part of the initial spectrum into a narrow band can also be realized.

For instance emission of polarized muonium at thermal energy from a hot W foil [11] or from SiO_2 powders or aerogel [12,13] has been observed after stopping μ^+'s inside the solid. For SiO_2 yields of thermal muonium up to 12%[14] with respect to the incoming μ^+ intensity have been reported, whereas for hot foils the muon production efficiency is $\sim 10^{-2}$. The thermalized muonium emitted from SiO_2 grains shows a rest polarization of $25.9 \pm 10.1\%$ [13,14]. From a hot foil source one expects a 50% polarization loss due to the muonium hyperfine oscillations. Thermal muonium can be ionized by lasers [15] and the resultant μ^+'s can be used as a source. This approach is being pursued at the Meson Science Laboratory (University of Tokyo) [16]. The beam intensity can be improved by orders of magnitude by placing the set-up for thermal muonium production at the primary proton beam so that protons can interact directly with the foil. This method needs major technical developments but it promises intensities of several 10^4 very slow muons at the MSL or several 10^5 if the system is installed at the facility proposed by the Japanese Hadron Project.

Recently in a pioneering experiment performed at TRIUMF by a Bell Lab-Triumf-UWW collaboration, it has been shown that very slow μ^+'s in the eV energy range are directly emitted from appropriate degrader materials. With a solid Ar layer deposited on a foil, $\sim 2 \cdot 10^{-5}$ of the incoming surface μ^+'s are emitted with energies around 10 ± 15 eV [17,18]. This work has been guided by the experience with positrons where moderation of high energy positrons to eV energies is widely used as a source; the development of low energy positron beams has been crucially dependent upon the discovery and design of efficient eV positron moderation schemes. Whereas μ^+ moderation is at its beginning, experience with positrons gives us confidence that this technique could also be used to produce a source of very slow μ^+'s. Combining the present known Ar yields with the intensity of $\sim 3 \cdot 10^8$ surface μ^+'s/sec, which is predicted to be available after completion of the PSI upgrading program, a tertiary low-energy μ^+ beam line, delivering several thousand μ^+'s/sec in the energy range between eV and some tens of keV can be envisaged. For this reason a project study has been started at PSI to systematically study and develop this technique.

3 Muon moderation

3.1 Results

Moderation in solids is routinely used to produce sources of eV e^+'s. Hot single-crystal tungsten of high purity, where the e^+ are not trapped in defects and the diffusion length is high, has proven to be a very efficient and practical moderator with an efficiency $(N_{out}(eV)/N_{in})$ of up to few part per thousand [19]. The reemission of the implanted slow positrons that have diffused to the moderator surface is due to the negative work function of the positrons in the metal. The energy distribution is narrow and the highest emission energy correlates with the work function. This does not seem to be the only mechanism for eV e^+ emission. Positrons injected into ionic crystals have been found to be reemitted with kinetic energies peaked at a few eV; the maximum energy is approximately equal to the electron band gap energy [20]. Positrons are reemitted from rare gas solids with kinetic energies less than the inelastic threshold. Solid Ne gives the best moderation efficiency of $7 \cdot 10^{-3}$ [21].

Some appropriate analogies can be drawn between e^+'s and μ^+'s, but their mass differences implies important differences in the physical properties of e^+'s and μ^+'s in condensed matter [22]. Even at thermal energies the μ^+ behaves like a well-localized classical particle, whereas for positrons the wave character is dominant. The neutral fractions (Ps and Mu) have different binding energies and sizes, positron diffusion rates are much greater than those of μ^+.

Drawing guidance from the positron results, two experiments have investigated the emission of slow muons from a few selected solids [17,23] by measuring the energy distribution of surface muons transmitted through thin moderating targets with a magnetic spectrometer and a time-of-flight system. With Cu , LiF and SiO_2 crystals, muons degraded to keV energies are observed with an efficiency of about 10^{-7} (normalized to the incoming 30 MeV/c particles). In the case of LiF (and possibly SiO_2), the time distribution of the degraded muons indicates an additional slower, separated component at the level of $\sim 2 \cdot 10^{-7}$ originating from muons with fractions of keV energy. The emission of very slow Mu^- is also observed with a rate of a few ions per 10^7 incident μ^+'s. Thin films of rare gas solids deposited on a substrate have been found to yield very slow μ^+'s with an intensity ($\sim 2 \cdot 10^{-5}$) of up to two order of magnitude higher than for LiF. The energy distribution of the moderated μ^+'s is peaked with a maximum (for Ar and Xe) at very low energy ($\sim 10\ eV$), a width of $\sim 15eV$ and a high energy tail falling off monotonically (Fig.1, [18]).

For Xe, Kr and Ar the moderation efficiency is correlated to the band gap energy, which is large for noble elements. With increasing band gap energy, an increase of the μ^+'s yields in the eV region is observed. To explain this behavior it has been suggested that the observed muons are epithermal, thus retaining part of their initial kinetic energy.

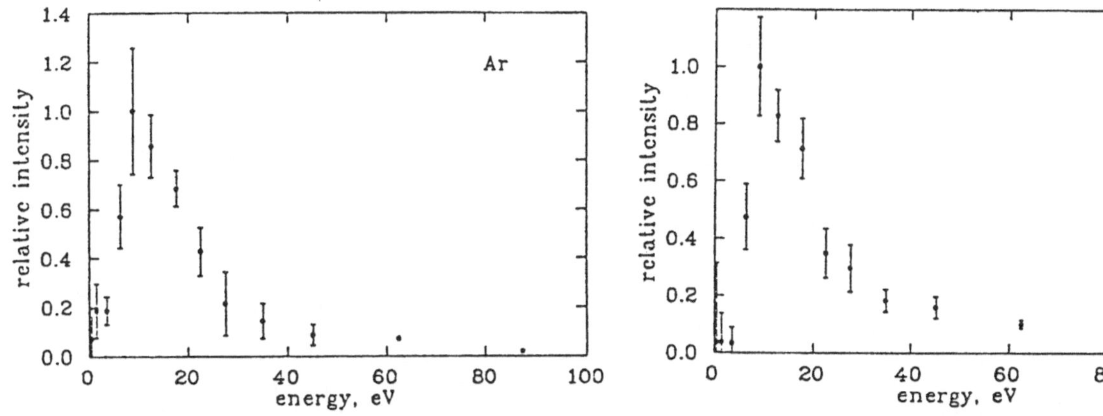

Fig. 1. Energy spectrum of muons moderated in solid Argon and Krypton (from [18]).

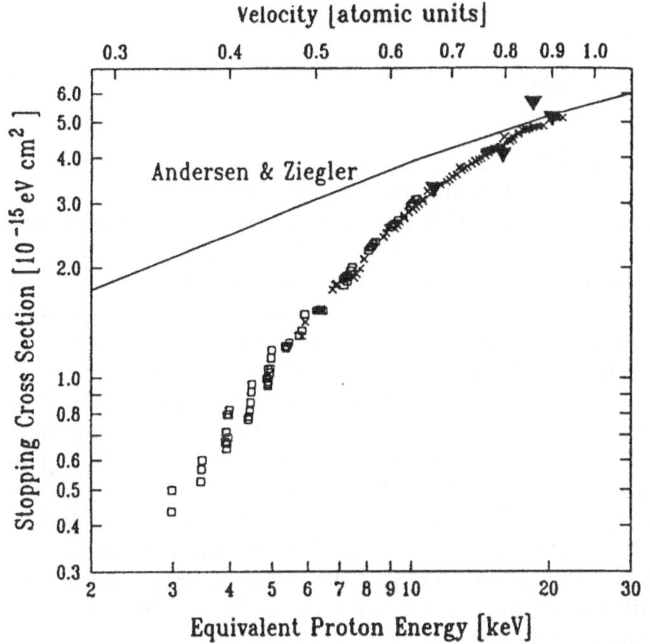

Fig. 2. Stopping cross section of He for protons (cross, triangle) and deuterons (square) [27]. The solid line from [28] shows a linear proportionality below 10 keV.

The emission is attributed to the reduction or suppression near and below the threshold of efficient electronic energy loss mechanisms such as electron-hole pair formation [17,24]. The decrease in energy loss rate leads to a relatively long escape depth for epithermal muons and therefore to an increased reemission probability. The same mechanism of reduced stopping power was first suggested to explain the positron emission data from alkali halides and rare gas solids [25]. Fermi and Teller [26] were the first to point out that the case of energy loss in insulators differs from that in metals: the amount of energy that may be delivered to a metal is arbitrary small, whereas in insulators it must be at least as large as the gap between two Brillouin zones, typically several eV. Therefore the loss of energy to electrons will be reduced when energy is transferred in small individual amounts. This effect has been shown very recently in stopping cross section measurements of low energy protons and deuterons in He gas [27] (Fig.2), where the presence of an appreciable minimum excitation energy leads to a significant reduction and to a deviation from the generally assumed linear v_p proportionality. A dependence on roughly the third power of v_p is found. The threshold effect is appreciable at energies well above the threshold, for protons up to 20 keV. Assuming velocity scaling, such an effect would affect muons with less than 2 keV energy.

3.2 Open questions

A better understanding of the processes governing the emission of eV μ^+'s from a solid would not only be intrinsically interesting but could be of great relevance for an efficient very slow μ^+ source. Although the reduced energy-loss mechanism in the few 10 eV region seems plausible, present data do not provide an unambiguous signature about the processes responsible for the production of eV muons. Very little experimental information exists concerning the energy loss of particles with few keV or less kinetic energy. Recently a formalism was developed to describe the slowing down of μ^+'s in gases in terms of the two-body atomic cross sections for target ionization, electron capture, electron loss in muonium and elastic collisions [24,29,30]. It predicts the charged and neutral fractions as a function of pressure at various stages of the slowing down process, as well as the polarization.

No similar calculations exist for the solid state. In condensed matter, the mean free path of a muon can be of the same order of magnitude as the atomic dimensions. This means that a muon cannot interact with one atom before coming under the influence of the next. A calculation of the slowing down based on atomic cross sections, allowed at high velocities where the cross sections are small, becomes therefore doubtful at low velocities. Col-

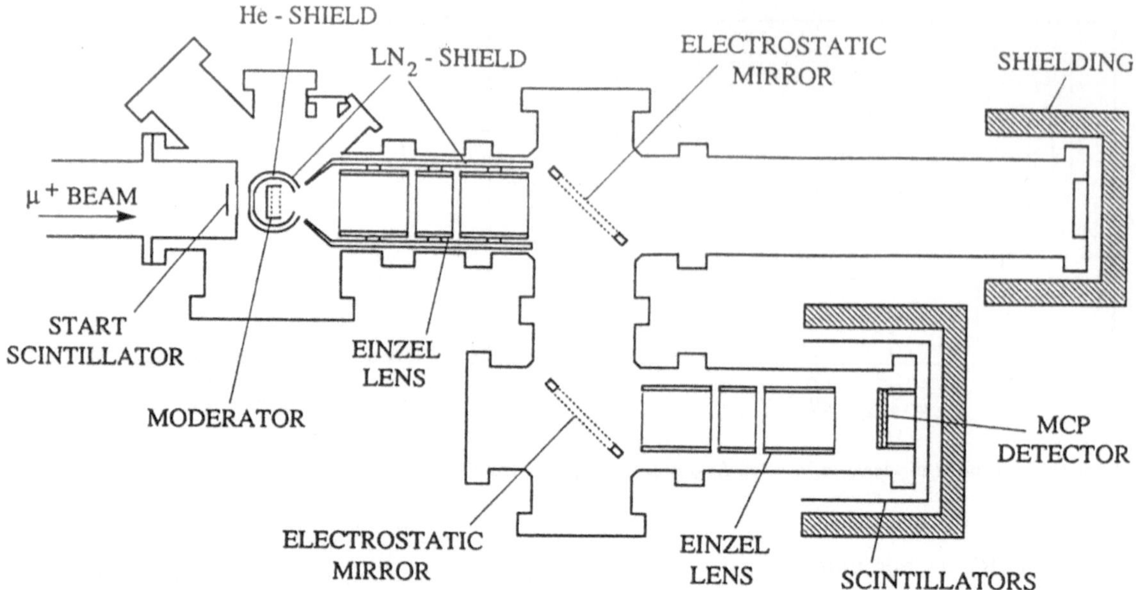

Fig. 3. Setup for the study of slow and very slow muons at PSI. For details see text.

lective effects associated with the solid state and possibly phonon interaction must be taken into account, also.

For the practical purpose of developing a source of very slow muons, the relevant parameters governing the yields and characteristics of the emitted particles must be determined.

The large difference in yields from LiF and Ar (about two orders of magnitude) in spite of almost equal band gap energies (13.7 eV for LiF and 14 eV for Ar) may indicate that this quantity is not the only one of relevance. A similar question arises when comparing the smaller moderation efficiency ($\sim 10^{-6}$) of Ne ($E_g = 22$ eV) with that of Ar. Most probably the Ne measurement has been influenced by nonideal experimental conditions.

The experience with low energy e^+ sources shows that only systematic and intense development studies together with technological improvements can exploit the potential of the moderation technique. The first measurement of slow e^+ emission reported an efficiency of $\sim 10^{-7}$ [31]. In the meantime the best value is in the percent region [21]. For various physical reasons, an improvement of many orders of magnitude cannot be expected for muons, but it is reasonable to assume that the ultimate value in efficiency has not yet been reached.

3.3 PSI Plans

The apparatus shown in Figure 3 is under construction at PSI. Its purpose is to study and optimize the emission of slow and very slow muons from solids and to perform first experiments. It will also serve as a basis for a tertiary beam of very slow muons.

The setup consists of a Ultra High Vacuum chamber with a design vacuum of 10^{-10} Torr. In a built-in UHV He-Cryostat, a thin layer of rare gas solids will be formed by vapor deposition on appropriate substrates. To yield a stable Ne film and low vapor pressure, the target will be cooled at liquid helium temperature. The thickness of the layer will be measured with a cooled quartz crystal. UHV conditions are needed to reduce the influence of target contamination on the moderation efficiency. With a vacuum between $5 \cdot 10^{-8}$ and $5 \cdot 10^{-9}$ Torr, a reduction of the Ar yields with time after the deposition was reported in ref. [17]. Recent investigations of slow proton reemission from rare gas solids indicate that the quality of the solified layer plays an important role [32]. This implies that the yields from carefully prepared rare gas solids could be significant better than the reported $2 \cdot 10^{-5}$. With a basis vacuum of 10^{-10} Torr, the formation time for a monolayer is of the order of hours, which is a sufficiently long time to provide stable conditions during a measurement.

The yields may depend not only on target composition but also on its structure. Thin films of rare gas solids prepared on a low temperature substrate are polycrystalline with a grain size of a few times the film thickness. Grain size and imperfections may influence the yields especially if diffusion processes are involved [33].

Another aspect to be investigated is possible channeling effects of epithermal muons in crystals of rare gas solids, which could result in a higher efficiency. Thin crystals can be epitaxially grown on single-crystal substrates and their structure has been studied by using low energy electron diffraction [34]. Spare ports in the UHV chamber are foreseen for future implementation of surface and structure analysis instrumentation. Instead of the cryostat, a target ladder could be mounted to study different crystal moderators.

Yields and energy spectra of the eV μ^+'s will be measured with a time-of-flight (TOF) system. The incident polarized surface muons traverse a thin scintillator starting the TOF measurement. Part of the muons are moderated in the target, which is electrically isolated to accelerate and extract the slow and very slow muons from the production region.

They are then focussed an a microchannelplate detector (giving the stop signal of the TOF measurement) with a system of einzel lenses and electrostatic mirrors. A second electrostatic mirror is foreseen to reduce the background due to the scattering of higher energy particles off the grid in the first one. In a short section of the TOF, the muons will be slightly accelerated in order to broaden the TOF spectrum and to determine accurately the energy distribution. Alternatively, an integral measurement of the energy spectrum can be performed by putting a grid to a higher voltage than the target. Formation of Mu^- can be determined by reversing the voltage polarities. The measurement of the fraction of slow muonium and Mu^- relative to μ^+ as a function of the energy should also help to elucidate the moderation mechanisms.

A very important characteristics is the polarization of the slow muons. In a solid, the total time for the μ^+ to decelerate to few eV is typically 10 ps. This is sufficiently rapid to avoid any significant loss of polarization during the stopping process even if there is depolarizing muonium formation. In the absence of a delayed emission mechanism, as in the case for epithermal muons, no depolarization is expected. For many applications the muon beam has to be polarized and the polarization must be known. It can be measured by measuring the μ^+ spin precession signal of muons implanted in a target for which the μSR signal is well known.

A complementary approach to the use of solid target moderators is the use of superfluid helium [35]. Here two questions are of relevance. First, is the mechanism of reduced stopping power near the threshold energy effective in superfluid helium too?. If yes, a superfluid helium target (eventually in the form of a thin film) could be used as a moderator. If the ionization energy is the relevant parameter ($\sim$ 25 eV in He), the efficiency is expected to be higher than with solid rare gases. The second question is whether it is possible to extract stopped μ^+'s from superfluid helium. In a μSR experiment with liquid helium, it was found that an applied external electric field strongly reduces the relaxation of stopped muons [36]. Moreover in electric fields, with $E > 600V/cm$ muons move with velocities of about 50 m/s. Krasnoperov [35] proposed to stop a surface μ^+ beam in superfluid helium at a temperature of 0.5 K. Within 100 ps the muons are thermalized. By applying an electric field perpendicular to the surface, muons drift to the surface. They can be extracted from the surface with a short electric pulse. This method may be efficient since the escape depth may be of the order of $1\mu m$ due to the high mobility in the electric field. Positive muons attract clusters of helium atoms around them (snowballs) [37]. The question is whether the μ^+'s can tunnel through the snowball and be extracted as free μ^+'s or whether they leave the surface as μ^+He_n clusters. For an assessment of the feasibility if this proposal, the kinetics and the behaviour of μ^+ in superfluid Helium will be investigated at PSI [38]. The motion of the particles through the liquid vapor interface and the existence of an energy barrier for leaving the liquid helium surface will be also studied.

4 Tertiary muon beam

The development program at PSI has been prompted by the upgrading program presently under way. A gradual increase of the proton current to its maximum value of 1.5 mA by 1994 is being pursued. The new large acceptance $\pi E5$ beam will be installed beginning of 1992 [39]. At the maximum current, $\pi E5$ is designed to deliver $3 \cdot 10^8$ surface μ^+/s (Fig. 4). Taking the present (not optimized) moderation efficiency for Ar, a source of $\sim$ 6000 μ^+/s with energies of 10 $\pm$ 8 eV can be provided by installing a moderator setup in the $\pi E5$ beam line. At other beam lines, several hundreds μ^+/s can be delivered.

The moderator represents the source of a tertiary beam line of polarized muons in the energy range between $\sim$ 10 eV and several tens of keV. Besides an optimized moderator geometry, such a tertiary beam consists of the following elements:

1) A beam handling system to extract the moderated μ^+'s from the production target and to separate them from background particles and high energy μ^+.

2) A section for tuning the energy. For fast time-differential measurements, a thin foil detector providing a clean start signal could be inserted in this section. The energy spread introduced by the measurement can be actively compensated with a pulsed electrode.

Since the initial energy spread of the source is small, good energy resolution, determined essentially by the transport and acceleration-deceleration system, can be achieved. The source dimension is determined by the size of the secondary beam. The size of the $\pi E5$ beam is 4 cm diameter; a smaller beam spot down to few mm squared is available at other beam lines at the expense of intensity. Since the very slow muons are accelerated at the target, they emerge with practically zero divergence. According to the Liouville theorem, the μ^+'s can be focussed on a small spot whose size is determined only by the transport system.

3) The last part of the facility is a UHV chamber where the various experiments can be accommodated. The moderator chamber, which is also a UHV chamber, is easily compatible with UHV conditions, the preferred medium for applications of very slow muons to surface studies.

A tertiary beam of this type could also be installed in a beam line of a pulsed machine or behind a surface μ^+ storage ring of the type proposed in [40].

A drastic increase of intensity of this tertiary beam line can be achieved by decelerating a surface μ^+ beam in the cyclotron trap and subsequently moderating the μ^+'s extracted with $\sim$ 20 keV energy [1]. The moderator efficiency is essentially determined by the ratio $\lambda/\Delta R$

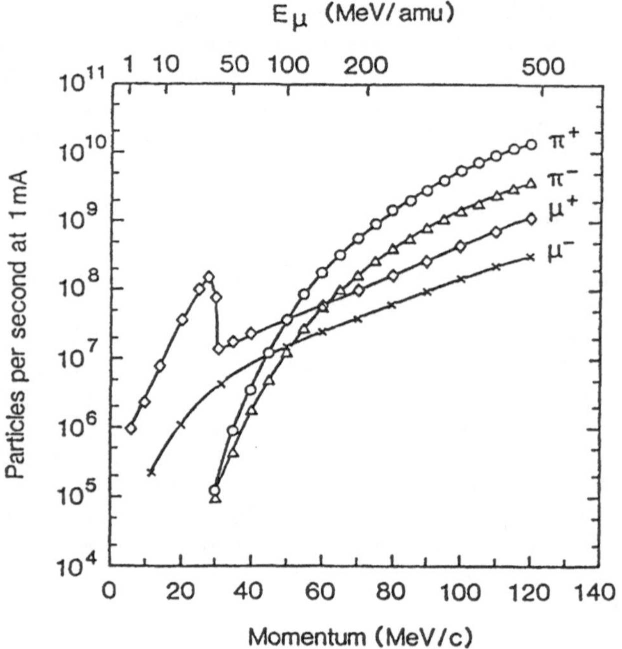

Fig. 4. Intensity of secondary beams at the $\pi E5$ beam line at PSI with 1 mA primary proton beam.

(λ, escape depth of eV muons; ΔR, stop width of the implanted muons). ΔR is proportional to $p^{3.5}$; therefore a reduction of the muon momentum prior to their implantation increases the moderation efficiency and the absolute number of epithermal muons. With a combined deceleration and extraction efficiency of $\sim 3 \cdot 10^{-3}$, a factor 40 improvement in eV yields with respect to the surface μ^+ intensity is expected. Possible μ^+ depolarization effects during the deceleration process in the cyclotron trap have to be studied in detail.

5 Future in muon physics

A tunable beam of polarized μ^+ with energies between some eV and some tens of keV with intensities of a few thousand per second offers a new powerful instrument for studies in fundamental muon physics as well as for new applications in surface and solid state physics. Without being exhaustive, we list some examples.

With a fine tuning of the μ^+ energy, the cross sections of basic atomic processes (ionization, electron capture, muonium ionization, scattering, energy loss in matter) can be obtained as a function of the energy. By comparing these results with those obtained with proton or hydrogen beams, mass effects and the velocity scaling of the cross sections can be tested. Related cross sections in μ^+-molecule collisions and molecular ion formation can be investigated as well. A detailed understanding of the fundamental interactions of very low energy μ^+'s with matter is also of importance for an optimization of the moderation method. Slow and very slow μ^+'s have an important impact in muonium production and muonium physics. A muonium beam with tunable energy can be formed by passing low energy μ^+'s through a thin foil or gas target, where the μ^+'s are neutralized with large probability [3]. The energy straggling can be kept small. With an appropriate target, the neutralization can result preferentially in excited states. For instance, from low energy collisions (below a few keV) of proton on Cs, it is known that practically all electron pick-up populates n=2 states. Up to 40% of those atoms are in the metastable 2s state [41]. Muonium in the 2s state , a preferred system for precision spectroscopy [42], could be formed this way.

With a low energy beam of μ^+'s, a detailed study of thermal Mu production in hot foils or Silica grains can be performed. On the other hand, by implanting keV or sub keV μ^+'s into such surfaces an efficient and clean thermal Mu source is obtained.

A beam of tunable low energy polarized μ^+'s provides a new tool for surface physics also. A polarized μ^+ is a natural candidate to study surface, thin-film and interface magnetism, subjects of considerable relevance to both fundamental research and to applications. New techniques such as scattering of μ^+'s on surfaces or the application of μSR type measurements to surfaces, thin films and interface studies could be developed. The μSR method is now limited to the study of bulk characteristics of the solid state, because the surface muons generally used have a typical penetration depth of some tenths of millimeter with a stopping distribution of comparable width. With a tunable beam of a few keV, the surface and thin film regions become accessible. The accessible range can be estimated by scaling the scarce proton results [43]: one expects for a 100 eV μ^+ a range in the order of 10Å and for a 10 keV a range of the order of 1000Å.

Detailed studies of the interaction of the very slow muons when the particle reaches the surface of a solid from outside (stopping profile, neutralization, reemission probability..) are needed in order to fully exploit the potential. At relatively high velocities (a few keV), the dominant process is probably deep implantation and full thermalization. At lower energies, the mean implantation depth becomes so small that a non negligible fraction comes back to the surface before being thermalized. Moreover, passing through the surface, the μ^+ can bind one or two electrons and emerge as Mu or Mu^-.

Magnetic order in the topmost layer could be studied by scattering slow μ^+'s at grazing incidence also. Spin polarized electrons at the surface are captured by polarized μ^+'s. If the electron spins are parallel to the muon spin, only muonium in the triplet state is formed and the muons maintain their polarization. If the electron spins are antiparallel, the Mu is in a 50% triplet and a 50% singlet combination and the μ^+ depolarizes due to the hyperfine interaction. Therefore, by measuring the μ^+ polarization, the electron spin polarization can be determined. Analogue experiments have been proven feasible with deuterons [44] and polarized positrons [45]. Surface studies can also be performed with muonium such as investigations of the electronic structure around H-like impurities on surfaces, of catalytic chemical reactions of Mu on metal surfaces, or of Muonium hyperfine field on surface.

6 Conclusions

Due to its own production mechanism, the muon has a large kinetic energy initially and a large spread in energy. Despite their energetic appearance, muons have found many applications down to thermal energies.

A new dimension in muon physics would be opened with the availability of tunable beams of polarized muons in the energy range between some eV and few tens of keV.

Much work remains to be done, but the continuous progress of experimental techniques and machines providing us with muons make us confident that this goal can be reached.

To paraphrase V. Hughes and G. zuPutlitz, to whom this volume is dedicated, muons and muonium have not yet decayed [46].

Acknowledgement. Stimulating discussions with J. Beveridge, K. Jungmann, M. Meyberg and L. Simons are gratefully acknowledged.

References

1. L. Simons, this volume
2. A. E. Pifer et al. Nucl. Instr. and Methods 135, 39 (1976)
3. A. Badertscher et al., Nucl. Instr. and Meth. A238 200 (1985)
4. H. Daniel, Z. Phys. A313, 249 (1983)
5. H. Daniel, Muon Cat. Fusion 4, 109 (1989)
6. D. Taqqu, Nucl. Instr. and Methods A247, 288 (1986)
7. A. Fuchs, E. Morenzoni, E. Pedroni, D. Taqqu, to be published
8. L. Simons, E. Morenzoni, F. Kottmann in Electromagnetic cascades and chemistry of exotic atoms, Ettore Majorana Institute (1989), Erice, Science Series, Phys. Sciences, Vol 52.
9. P.R. Bolton et al, Phys. Rev. Lett. 47, 1441 (1981)
10. Y. Kuang et al., Phys. Rev. A60, 3172 (1987).
11. A. P. Mills Jr. et al., Phys. Rev. Lett. 56, 1463 (1986)
12. G. A. Beer et al., Phys. Rev. Lett. 57, 671 (1986)
13. W. Schäfer, Ph. D. Thesis, University of Heidelberg, 1988
14. K.A. Woodle et al., Z. Phys. D9, 59 (1988).
15. S. Chu et al., Phys. Rev. Lett. 60, 101 (1988).
16. K. Nagamine, Hyperfine Interactions 65, 1149 (1990).
17. D.R. Harshman et al., Phys. Rev. B36, 8850 (1987)
18. G. D. Morris, Master of Science Thesis, University of British Columbia, 1990.
19. P.J. Schultz and K.J. Lynn, Rev. of Mod. Phys. 60, 701 (1988) and ref. therein.
20. A. P. Mills Jr. and W. S. Crane, Phys. Rev. Lett. 53, 2165 (1984).
21. A. P. Mills Jr. and E. M. Gullikson, Appl. Phys. Lett. 49, 1121 (1986).
22. F. M. Jacobsen, Hyperfine Int. 32, 501 (1986).
23. D.R. Harshman et al., Phys. Rev. Lett. 56, 2850 (1986)
24. M. Senba, J. Phys. B 21, 3093 (1988).
25. K.G. Lynn and B. Nielsen, Phys. Rev. Lett. 58, 81 (1987)
26. E. Fermi and E. Teller, Phys. Rev. 72, 399 (1947)
27. R. Golser, D. Semrad, Phys. Rev. Lett. 66, 1831, (1991).
28. J.F. Ziegler, J.P. Biersack, U. Littmark, The Stopping and Range of Ions in solids, Vol. 1, Pergamon Press 1985.
29. M. Senba, J. Phys. B 22, 2027 (1989).
30. M. Senba, J. Phys. B 21, 1545 (1990).
31. W. Cherry, Ph. D. Thesis Princeton University 1958.
32. A.P. Mills Jr. et al., Phys. Rev. B42, 5973 (1990).
33. J. A. Venables and B. L. Smith, in Rare gas Solids, Vol. II, M. L. Klein and J. A. Venables editors, Academic Press, 1977.
34. H.H. Farrel et al., Phys. Rev. B6, 4703 (1972).
35. E.P. Krasnoperov, Hyperfine Interactions, 65, 1049 (1990)
36. D. G. Eschenko et al., JETP Lett., 48, 616 (1988).
37. K.R. Atkins, Phys. Rev. 116, 1339 (1959).
38. E.P. Krasnoperov et al., PSI Proposal, Nov. 1990.
39. D. Renker, this volume.
40. J. Brewer, Hyperfine Interactions, 65, 1137 (1990).
41. V.N. Tuan, G. Gautherin, A.S. Schlachter, Phys. Rev. A49, 1242 (1974).
42. K. Jungmann, this volume.
43. G. G. Ross and B. Terreault, Nucl. Instr. and Methods B15, 61 (1986).
44. C. Rau, R. Sizmann, Phys. Lett. 43A, 317 (1973).
45. D. W. Gidley et al., Phys. Rev. Lett. 24, 1779 (1982).
46. V. Hughes and G. zuPutlitz, Comm. in Nucl. and Part. Physics, 2, 259 (1984).

This article was processed using Springer-Verlag TEX Z.Physik C macro package 1991
and the AMS fonts, developed by the American Mathematical Society.

Z. Phys. C - Particles and Fields 56, S250–S254 (1992)

Phase space compression of low energy muon beams

D. Taqqu

Paul Scherrer Institute, CH-5232 Villigen PSI, Switzerland

16 October 1991

Abstract. The phase space compression method (PSC) is applied to low energy muons produced in a high magnetic field. By making use of the existing apparatus, transverse and longitudinal PSC operations will lead to a standard low energy muon beam of energy variable between 2 keV and 20 keV. At 5 keV the beam has $\Delta p/p < 5\%$ and can be focused on a target of 2 cm diameter.

1 Introduction

The novel application possibilities of low energy muon beams have lately attracted many researchers at muon factories to look for methods to produce such beams with sufficient quality and intensity. In order to achieve the quality required, active phase space compression (PSC) has been proposed [1] and investigated at PSI in the last few years. In a specific test arrangement, many of the basic requirements for the PSC operation could be demonstrated. The detailed experimental results will be presented in a later publication. At the present time the experimental program is being continued in the direction of a scheme where only very low energy muons ($E \leq 20$ keV) are used. This leads to a PSC operation of the kind of the first stage of Ref. 1 with the advantage that it can already be implemented in the existing test device with minimal requirements in the development of new components. The principle of this scheme has already been presented at LEMS-90 [2]. The overall beam production line begins with a low energy muon production device. It is followed successively by the low energy muon selector, the PSC apparatus and a well adapted extraction stage for transformation into a standard muon beam of variable energy between 5 keV and 20 keV. In the present paper we concentrate on the two last components: the PSC apparatus and the extraction stage. We only refer to the various low energy muon source concepts that have been lately considered: the muon channel source [2], the magnetic mirror source [3], the cyclotron trap source [4], the degraded muon source [5], or the intense μCF source [6]. The simplest concept (the degraded muon source) has been recently successfully tested and another one (the cyclotron trap source) is presently the subject of an extended experimental program so that together with the development of the PSC stage it is expected that the first user-friendly negative muon beam will soon be available at PSI.

2 The PSC operation

We refer to Ref. [1] for the basic principles of the phase space compression method applied to muon beams. One essential feature of the PSC operation is the measurement of the incoming muon phase space with minimal disturbance of the muon trajectory. The proposed detector concept has been the subject of various test runs whose results have demonstrated its perfect suitability to the PSC operation. In a very recent run (July 1991) associated with a specific experimental proposal [5], muons of lowest energies (below 10 keV) have been subjected to the measurement of both their transverse and longitudinal phase space. The detection principle makes use of the secondary electron emission from thin foils. The crossing point of the muon has been measured by detecting the electrons with a position sensitive microchannel plate (MCP). The existing r-ϕ fast readout information allows one of the two transverse coordinate information (r) to be obtained with a precision of 200 μm and the crossing time information (longitudinal phase space) with a precision of 200 ps. The detection efficiency have been measured and, when scaled to the muons considered here, amounts to an average value exceeding 70% with MgO-coated carbon foils.

After detection of the muon crossing, the PSC principle makes use of the collected trajectory information to produce electric fields that displaces the muon trajectory to a fix point in both transverse and longitudinal phase space. In practice, 100–200 ns are required to build up the required fields and care has to be taken that the

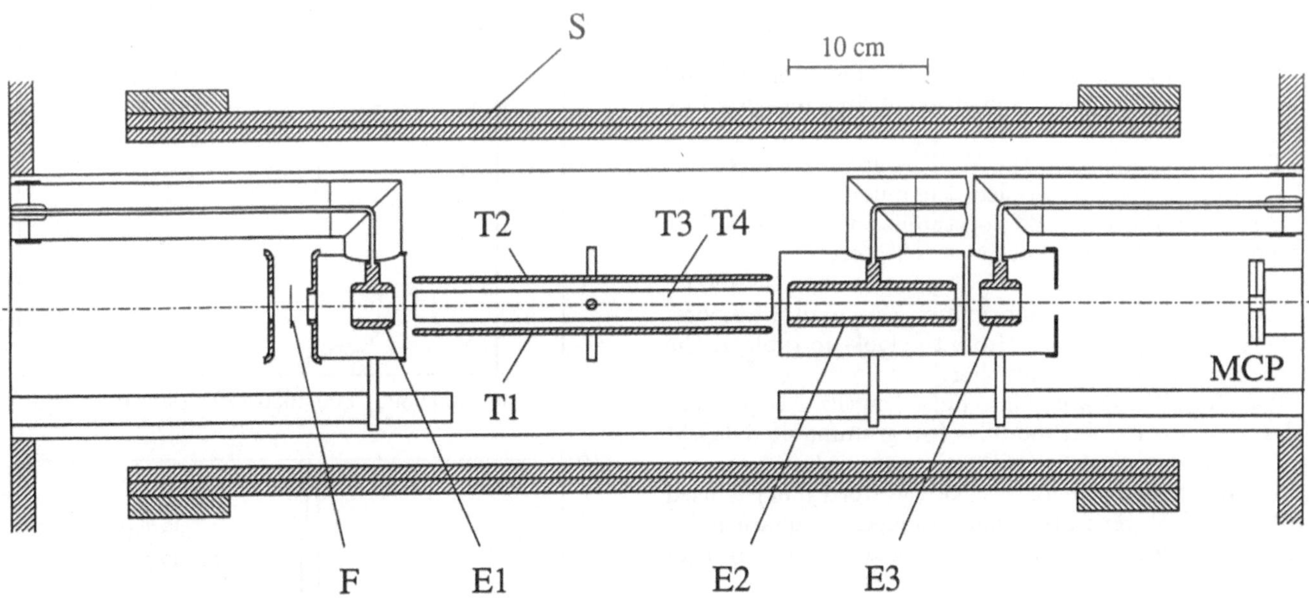

Fig. 1. Main components of the PSC apparatus; S: solenoid, F: detector foil, E1-E3: cylindrical electrodes; T1-T4: longitudinal electrodes.

particle conserves its phase space properties during this relatively long time. For the low energy muons obtained from any of the production schemes considered, the incoming beam divergence is so high that the only way to confine them is to use adiabatic transport in a high magnetic field. The particles spiral around the magnetic field lines according to the well known adiabatic laws and the collected transverse position information remain valid at any later time.

As the secondary electrons have an even lower momentum than that of the muons, they will also follow the field lines allowing the position information to be transported to a far away downstream MCP. On the other hand they can be made to travel much faster than the muons so that the PSC operation can be triggered much before the muon has advanced very far.

The practical configuration that has been used for various tests is shown in Fig. 1. The central components (detector foil and pulsed electrodes) are all placed in an homogeneous solenoidal magnetic field B_o of 50 cm length. The MCP is placed further downstream in a field B_m which is 4 times smaller. For $B_o = 5$ T, B_m is 1.25 T allows optimal operation of the MCP. With the detector foil at -4 keV, the secondary electrons require $\sim$ 20 ns to reach the MCP while muons with longitudinal kinetic energy $E_{\|}$=10 keV require about 100 ns to reach the cylindrical electrode E3. The 80 ns available is sufficient to allow the electron signal to trigger a high voltage negative pulse (for negative muons) that will be applied to E3 to reflect the muons of $E_{\|} < 10$ keV. Simultaneously, the electron signal triggers on E2 a pulse of special shape that will modify the $E_{\|}$ of the reflected muons to stabilize it near a fixed value (longitudinal PSC). Tests with a simple pulse have succeeded in reducing significantly the $E_{\|}$ spread. The pulse shape can be selected to achieved stabilization to $E_{\|}$= 10 keV and extend the acceptance in longitudinal energy to $E_{\|\ max}$= 12 keV. It is clear that the use of a longer solenoid or of faster HV pulsing will allow to increase $E_{\|\ max}$ but the existing system should be sufficient to achieve a first useful PSC beam. The reflected muon return toward E1 which in the maintain has been pulsed to -12 keV so that it reflects all μ^- with stabilized $E_{\|}$. By pulsing E2 to -12 keV also, the muon remain trapped between E2 and E3 for any required time.

This longitudinal PSC operation is independent of the transverse phase space so that it can be (and has been) used to confine muons with quite high transverse momentum $p_{\perp}$. However, as $p_{\perp}$ gets much higher than $p_{\|}$ the muon energy increases like $p_{\perp}^2$ and we are dealing less and less with low energy muons. Although this is not a fundamental obstacle [1], the highly simplified PSC operation associated with a relatively low maximum accepted muon energy have led to the present trend toward an increased effort in the production of low energy muons, so that it is, for the time being useful to limit $p_{\perp}$ to the same value as $p_{\|}$ or about 1.5 MeV/c ($E_{\perp\ max} \simeq 12$ keV) and reduce the transverse PSC operation to a centering of the muon spirals. This is achieved via the 4 transverse electrodes T1 to T4 which will be pulsed to the adequate HV according to the information stored in a fast look-up table and fetched by the encoded position information from the MCP (with x-y readout). The transverse electric field produced is selected in such a way that the $\mathbf{E} \times \mathbf{B}$ drift of the muon over the pulse length will bring the muon spiral to have one point of its circonference (the one with the same transverse coordinates as the foil crossing point) coincident with the solenoid axis.

The overall PSC operation will therefore involve the following steps:

1) Low energy muons that cross the detector foil will have their crossing point and crossing time measured by the MCP detector.
2) The time signal from the MCP is used to trigger:
a) a HV pulse on E3 to reflect the muon,
b) a shaped pulse on E2 to stabilize the longitudinal momentum and then to act as a reflector,
c) a pulse on E1 with $V > E_{\parallel\ max}/e$ to reflect the muon.
3) Encode the position information as a digital number.
4) Fetch two digital words from the look-up table at the location of the encoded number.
5) Transform these words into pulse lengths via digital to pulse length converter modules (programmable delays).
6) Each of these pulses (with adequate polarity as controlled by the sign bit of each of the word pair) is used to drive a HV generator that produces a square pulse with an amplitude nearly proportional to the length of the drive pulse.
7) After the centering operation is achieved, open the trap (voltage of E2 and E3 returned to zero) and let the muon escape via the hole in the MCP. (Extraction from the magnetic field will be described later in more detail).

All these steps [except for *4)* and *5)* which are standard readily available electronic operations] have been successfully tested in the PSC developmental phase. As an illustration, an on-line result of the trapping operation is shown in Fig. 2. It was detected by opening E2 and then E1 so that the muons are reflected by E3 toward the foil and the secondary electrons from the second foil impingement are measured in the MCP.

The basic $\mathbf{E} \times \mathbf{B}$ drift in the trapped configuration has also been investigated with electrons from a pulsed electron source. We could observe the displacement of the electron along the $\mathbf{E} \times \mathbf{B}$ drift direction after a definite trapping time between E1 and E2, by detecting the electron in the position sensitive MCP after the opening of E2.

The precision of the centering operation is determined, first by the accuracy in the position measurement and electric field action and second from the size of the muons spirals themselves. The first can be estimated to be less than 0.3 mm while the second is given by the maximal spiral radius of the accepted muons which amount to $\rho_m = 1$ mm for $p_{\perp\ max} = 1.5$ MeV/c in a field $B_o = 5$ T. The smallness of of the first term allows us to neglect it and as each of the 4 transverse phase space components (Δx, Δy, Δp_x, Δp_y) of the outgoing beam is proportional to $p_{\perp\ max}$ the overall final transverse phase space will be proportional to $p^4_{\perp\ max}$.

The phase space compression factor achieved is

$$f \simeq \frac{1}{\pi} \alpha \left(\frac{r}{\rho_m}\right)^2$$

where α is the intensity ratio with and without PSC and r is the radius of the accepted incoming beam. The value of α reflects the various loss processes (detection efficiency, muon decay) and the effect of the energy loss in the entrance foil. This depends on the momentum distribution of the incoming muons and may result in some cases to an increase in α. The present acceptance radius of the apparatus is $r = 10$ mm. With an estimated $\alpha = 0.3$ one gets $f = 10$. Ultimately the acceptance radius can be doubled to accept the wide incoming beam that is expected to be obtained in the most efficient production schemes. This increases f to 40 leaving the quality of the outcoming beam practically unchanged.

The final beam intensity depends mainly on the source intensity. For negative muons which are the focus of the present developments, the PSC apparatus was tested to show no limitation at trigger rates as high at 20 kHz, in excess the muon intensities expected to be available in the near future.

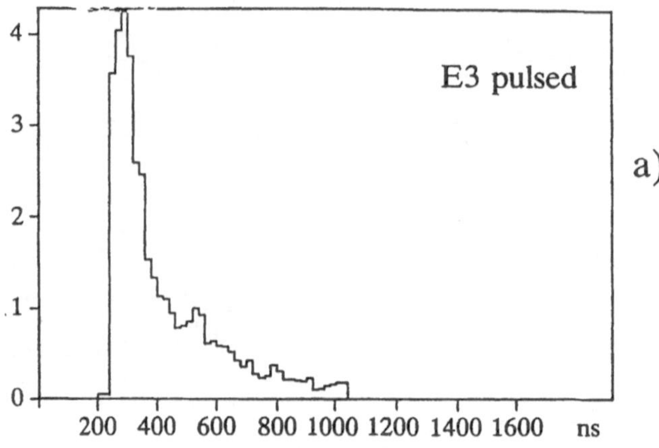

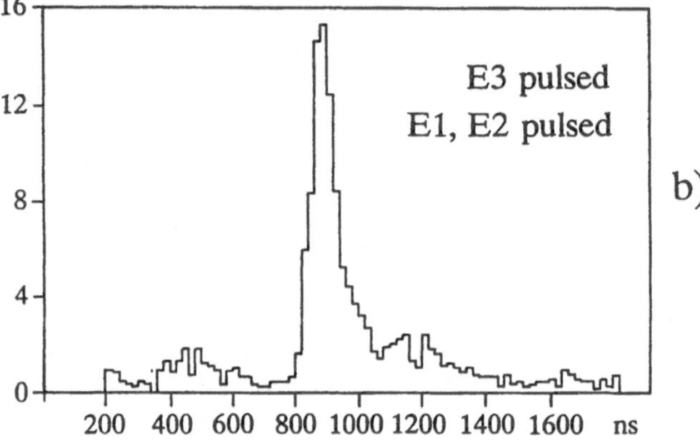

Fig. 2. Histograms of time delay between first foil crossing (incoming muon) and second foil crossing (reflected muon). a) Only E3 is pulsed; the peak after 250 ns is the time distribution (dependent on $E_{\parallel}$) of the muons that were slow enough to be reflected by E3. b) The muons are trapped between E1 and E2 for about 400 ns and then released; they reach the foil at a much later time.

3 Extraction

While the extraction of the centered muons from the apparatus is a straightforward operation (opening of E2 and E3 and letting the muon pass through the hole in the MCP detector), an optimal extraction from the axial magnetic to a zero field region requires special attention. Particle transport in a variable magnetic field depends on the parameter $\varepsilon = l(B)dB/Bdz$ where $l = p_{\|}/eB$ is the path length along the field lines for a one radian helical turn around them.

For $\varepsilon \ll 1$, we have adiabatic transport and both spiral radius ρ and distance to the axis vary like $1/\sqrt{B}$. With $p_{\perp}$ varying like $\sqrt{B}$, the transverse phase space of the beam remains fully conserved. On the other hand as B decreases the decrease in $p_{\perp}$ results in an increased $p_{\|}$ and the initial transverse momentum spread $\Delta p_{\perp o}$ adds partially to the initial longitudinal momentum spread $\Delta p_{\|o}$. As in our case $\Delta p_{\perp o} \simeq p_{\|o}$, the small relative spread $\Delta p_{\|o}/p_{\|o}$ initially achieved via the E2 pulsing will blow-up with the field reduction and result at low field to $\Delta p_{\|o} \simeq p_{\|o}$.

For $\varepsilon \gg 1$ the field varies rapidly compared to l and we have the well-known magnetic lens action which allows in principle to jump to $B = 0$ without any loss in phase space. Nevertheless stringent conditions have to be fulfilled in order to insure minimal aberration effects. First the beam has to have a small divergence ($p_{\perp} \ll p_{\|}$), which in our situation will only be achieved after a preliminary adiabatic reduction of B to quite a low value. With $B_o = 5$ T and $p_{\perp max} \cong p_{\|o}$ the divergence will be of the order of $\sqrt{B/2Bo}$ which falls to an acceptable 100 mrad at a field as low as 0.1 T. Second $\Delta p_{\|}/p_{\|}$ have to be sufficiently small in order to minimize the chromatic aberration. But as discussed before the longitudinal spread $\Delta p_{\|}$ at the low field preceeding the lens action have increased to almost equal the average $p_{\|}$ so that efficient extraction requires a preliminary reduction of $\Delta p_{\|}$.

Before describing how this will be done, it is appropriate to discuss here another fundamental limitation of the extraction from a magnetic field into a field free region. It is a well documented feature in electron (and positron) beam physics [7] but has not yet been sufficiently emphasized in connection with low energy muon beams. A charged particle in a magnetic field has a generalized momentum

$$\mathbf{P} = \mathbf{p} + e\mathbf{A}$$

(with $\mathbf{A}$ being the magnetic vector potential) that replaces the particle momentum $\mathbf{p}$ in the equations of motion. In a solenoidal field $|\mathbf{A}| = A_{\perp} = \frac{rB}{2}$ and $p_{\perp} = \rho eB$ so that when $r \gg \rho$ the generalized momentum $\mathbf{P}$ is much greater than the transverse particle momentum. In this case, as B jumps to zero (and thereby $\mathbf{A}$ too), angular momentum conservation imparts to the particle a transverse momentum as high as $(r/2\rho)p_{\perp}$ so that (if $p_{\|}$ is sufficiently high) it comes out along a skew trajectory of divergence $r/2\rho$ times higher than the value of $p_{\perp}/p_{\|}$ it had in the field. Since all efficient production schemes of low energy negative muons make use of high solenoidal magnetic fields, the muons produced far from the axis have necessarily $r \gg \rho$. Thus a direct extraction of such a beam into the field free region requires significant longitudinal acceleration and leads to such a high transverse phase space [increased by about $(r/2\rho)^2$] that, unless only the central portion of the beam is used, no useful low energy muon beam can be obtained. A method proposed recently to correct for this effect [7] in an intense source of slow positrons [8] requires $p_{\|} \gg p_{\perp}$ and $\Delta p_{\|} \ll p_{\|}$, conditions which are not reasonably achievable for the source muons without a PSC stage. This fundamental limitation excludes practically the direct conversion of a low energy muon beam produced in an efficient high field source into a standard low energy muon beam in a field free space. On the other hand, application of the PSC centering operation as previously described reduces the generalized angular momentum ($\mathbf{P} \times \mathbf{r}$) of the muons to practically zero making them identical to muons that would have been produced in a field free region. Under these conditions the extraction can take place without any loss in phase space. This consideration is of utmost importance in evaluating the usefulness of the PSC operation. It has the practical consequence that a paradoxically much higher field free low energy muon beam intensity can be achieved with PSC than without it and that the effective final phase space compression factor exceed significantly the estimate given in the previous section.

We now present a dedicated extraction scheme that will achieve the small $\Delta p_{\|}/p_{\|}$ required for optimal extraction and result in a low energy muon beam of standard quality. This is a significantly improved version of the extraction scheme presented at LEMS-90 for which the final energy spread had a lower limit of about 6 keV. It is essentially a longitudinal PSC stage which operates on the same principle as the stabilization of the longitudinal momentum described above. As shown in Fig. 3 a coated foil of about 15 μg/cm^2 and 5 mm diameter is placed on the extraction line after E3 at the beginning of the fall off of the 5 T field. The centered muons coming out from the trap cross the foil and emit secondary electrons that pass together with the muon through the hole of MCP. With an energy loss of about 7 keV for a zero divergence angle Θ, it increases with Θ together with the muon energy. This results in muons having $E_{\|} < 5$ keV and $E_{\perp} < 10$ keV. and $\Theta \lesssim 55^o$. As they reach lower B fields Θ decreases rapidly leading to a time of flight from the foil strongly correlated with the muon velocity (with a correlation error smaller than 10 ns). The foil exit time is obtained by detecting the secondary electrons with a second MCP (MCP2 in Fig. 3) placed the end of a 0.1 T magnetic channel within the region where B falls of to zero. The time information triggers a high voltage pulse of the shape required to stabilize the muon energy to a preselected value between 2 keV and 20 keV as it exits the long cylindrical electrode E placed in the low field region. For a 500 ns average time of flight (that requires

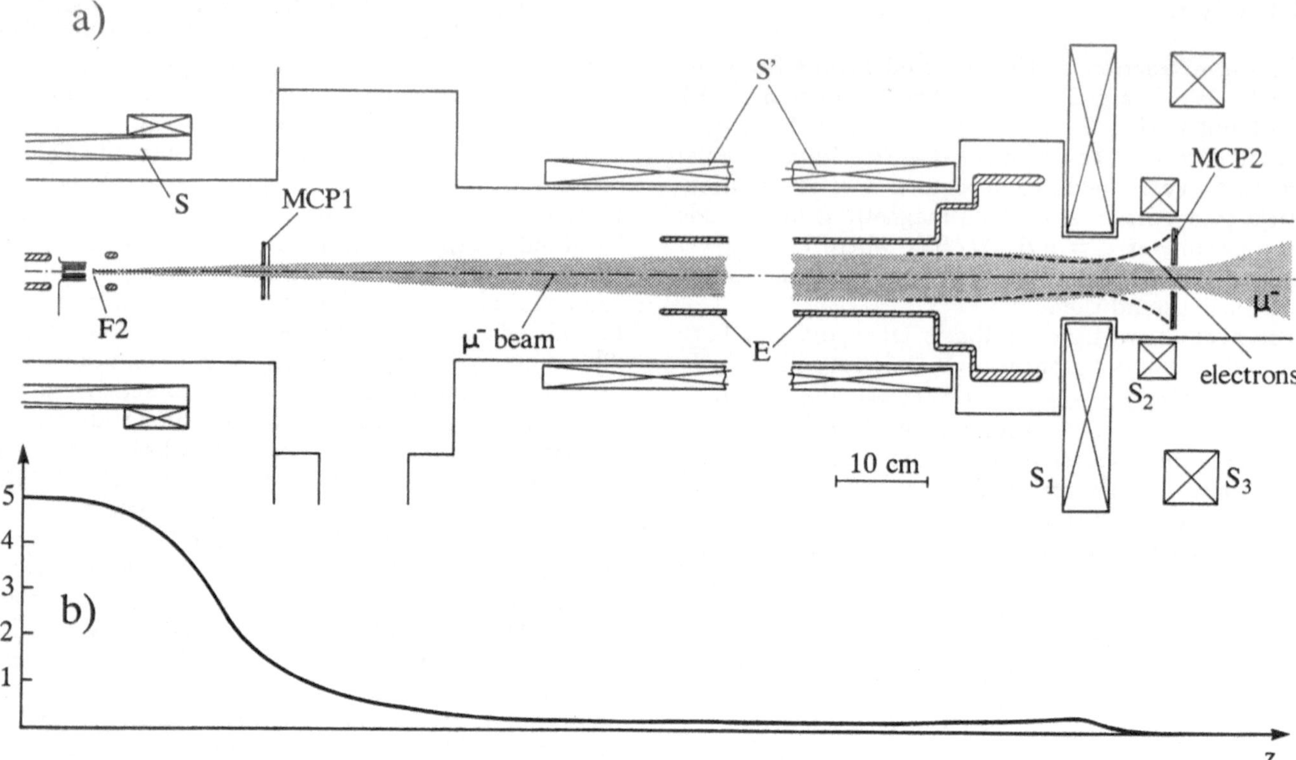

Fig. 3. The extraction scheme a) S: high field solenoid of fig. 1; S': low field solenoid; S_1, S_2, S_3: exit coils; S_2, and S_3 have a reverse current to obtain a sharp field fall-off, E: round electrode, MCP1: MCP of Fig 1. MCP; 2: MCP used for the extraction; F2: detector foil for extraction. b) Axial magnetic field along the extraction line.

a 2 m long 0.1 T solenoid) the small correlation error allows a final average energy spread of less than 500 eV. At $E = 10$ keV it corresponds to $\Delta p_{\|}/p_{\|} < 2.5\%$. This results in an optimal focusing of the muons into the hole of the MCP. The essential feature of this extraction scheme is the separation of the keV electrons from the 10 keV muons in the field fall-off region: while the muons have p = 1.4 MeV/c and are focused by the fast field variation ($\varepsilon \gg 1$) into the MCP hole, the electrons with 2 order of magnitude less momentum ($\varepsilon \gg 1$) follow adiabatically the field lines to impinge on the MCP.

The efficiency of the operation can be estimated to be about 50%. It achieves a longitudinal phase space compression of the order of 10.

The outcoming beam can be further transported with magnetic or electrostatic lenses. It can be accelerated, decelerated or sent to a next PSC stage operating with lenses and a deflector [1]. The energy spread is sufficiently low to allow the stopping in a short 1 torr hydrogen target or within the first 100 Å of the surface of materials. It allows therefore unprecedented experimental possibilities that are already being seriously considered by various experimental groups.

References

1. D. Taqqu, Nucl. Instr. & Meth. A247 (1986) 288.
2. D. Taqqu, talk presented at the Workshop on Low Energy Muon Science in the 90 (LEMS-90) at PSI, Villigen, Switzerland, April 2-4, 1990.
3. Reports of the Low Energy Muon development group, F. Kottmann, L. Simons, D. Taqqu, PSI, 1990.
4. L. Simons, these proceedings.
5. Described in PSI Proposal R-91-08.
6. D. Taqqu, in Muon Catalyzed Fusion 5/6 (1992) 543.
7. L. Brillouin, Phys. Rev. 67 (1945) 260.
8. Proposal for an intense slow positron beam facility at PSI, W.B. Waeber, D. Taqqu, U. Zimmermann and G. Solt, Appendix 2.
9. D. Taqqu, Helvetica Phys. Acta 63 (1990) 442.

This article was processed using Springer-Verlag TEX Z.Physik C macro package 1991
and the AMS fonts, developed by the American Mathematical Society.

Z. Phys. C - Particles and Fields 56, S255–S257 (1992)

A pulsed lepton source at LAMPF

D. Hywel White

Medium Energy Physics Division, Los Alamos National Laboratory
Los Alamos, New Mexico 87545

Submitted 22 November 1991

Abstract. A Pulsed Lepton Source is being considered at the LAMPF facility at Los Alamos National Laboratory. The source plan is described together with a description of the components and performance as they exist at present.

1 Introduction

In Fig. 1 is shown a schematic of the LAMPF accelerator complex. For the purposes of description it is convenient to divide the complex into three sections, injection, linear accelerator and the proton storage ring (PSR). At the injection area beams of protons, H^- ions and polarized H^- ions are produced. Beams of protons and H^- can be injected simultaneously although acceleration takes place at different phases of the RF. The accelerator produces pulses up to 1 ms in duration at 120 Hz. H^- beam is used at the PSR, but in discussing beam economics in the complex it is necessary to include the effect of other than H^- beam. The linear accelerator is capable of accelerating a total instantaneous beam current of about 30 mA, which may be divided between any of the components specified above. The linear accelerator is divided into two parts; an Alvarez section that

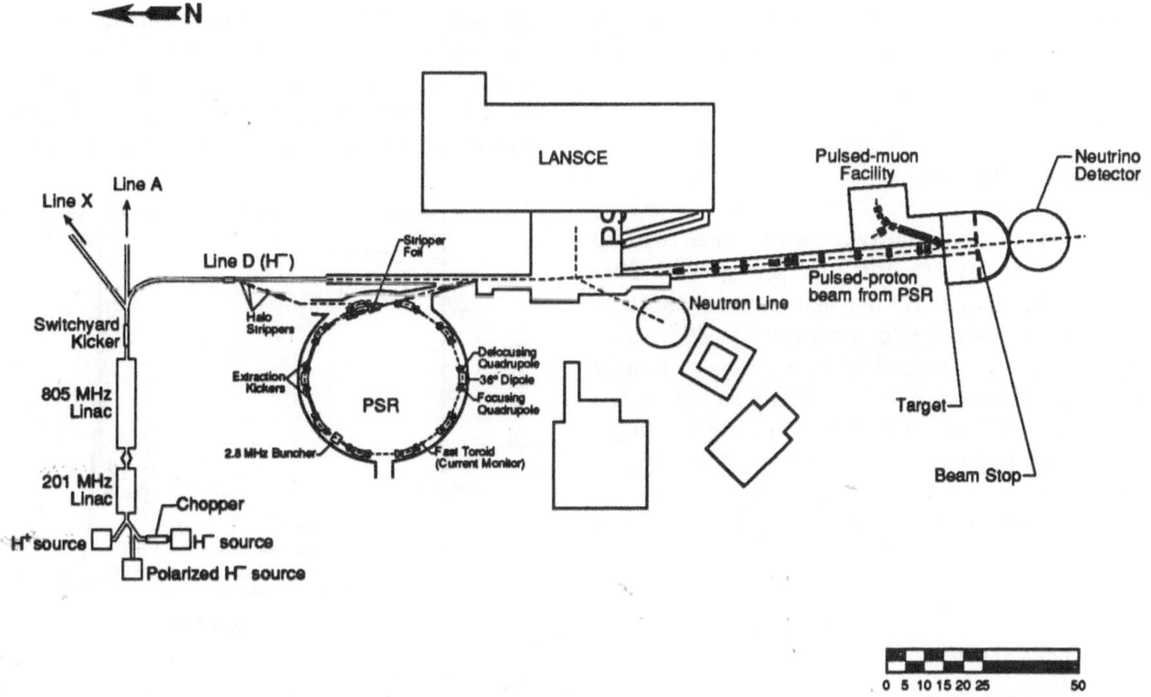

Figure 1. The LAMPF Accelerator Complex.

accelerates at 201 MHz and a side coupled cavity section that completes acceleration to 800 MeV at 805 MHz. The H^- beam is split at 20 Hz at present in a switchyard to the beam transport labelled line D in Fig. 1. The PSR accepts beam throughout the linac macropulse and this accumulation is ejected in a single turn in a pulse of 0.25-μs duration. It is this time compressed output that is relevant for PLS.

2 Injection

The injection area consists of three ion sources. The first produces protons at a peak current of 30 mA. After some scraping is done to produce an appropriate emittance for the linear accelerator, and the pulse length is matched to the accelerating RF acceptance, 14 mA is typically accelerated. Under normal conditions the linac has a macropulse of about 750 μs duration at 120 Hz, which gives an average current of 1 mA. The second source produces a relatively low current of H^- beam (30 μA) with protons polarized at ~ 50%. This beam is stripped after acceleration simultaneously with proton acceleration. An alternate H^- mode is of interest to us here, presently a H^- source produces an unpolarized current of 16 mA, of which 10 mA can also be accelerated at the same time as protons provided the peak beam loading of 30 mA is not exceeded. In fact this beam is chopped near the ion source at the revolution frequency of the PSR so that the ring circumference is not fully filled and only 7 mA is available for PSR injection. Normal operation has unpolarized H^- accelerated at 20 Hz, the PLS proposal envisages this repetition rate being increased to 60 Hz. At the same time it is expected that H^- source brightness can be increased so that a full charge of the proton storage ring can be accomplished in a shorter injection period.

3 The linear accelerator

Acceleration is accomplished in two sections, the first for low ion velocity is an Alvarez drift tube section with 201-MHz accelerating frequency. A transition region connects to a side coupled cavity accelerator operating at 805 MHz, which completes acceleration to 800 MeV. One in four of the high frequency RF buckets is filled giving an output 200 ps wide at intervals of 5 ns. This time structure is diluted by the energy spread in the beam when particles are allowed to drift for significant distances. Beam from the linac is switched to a beam line marked D in Fig. 1, and transported to the PSR area.

4 Proton storage ring

In Fig. 1 the proton storage ring is also shown. After scraping of the halo the beam is transported through a high magnetic field gradient where the H^- beam is stripped to H^0 with 90% efficiency. This beam is then injected into the ring where it is fully stripped by passing through a Carbon foil. The proton storage ring is operated at fixed B field; because of the chopping at the injector a portion of the circumference is particle free for extraction. This particle free region is maintained by a 2.8-MHz RF field. The stored beam passes through the stripping foil on a fraction of the turns determined by betatron oscillations. In Fig. 2 is shown loss rate as a function of time through the injection period. In the first turn, beam injected in the storage ring losses are relatively substantial on the limiting aperture of the ring, after which losses are proportional to the total charge accumulated from multiple scattering in the stripping foil. At the end of injection first turn losses cease and a drop in loss rate occurs. At extraction a further loss component exists mostly due to the finite aperture of the extraction magnet. The limiting aperture during fill is also believed to be the extraction system. With a peak injected current of 7 mA and a total fill of 3.8×10^{13} protons, the injection period must be 975 μs, rather longer than is shown in Fig. 2. The limitation on the amount of beam that can be stored are the losses described above that produce radioactive contamination of the ring making maintenance difficult. At present with a stored beam of 2.5×10^{13} at 20 Hz, the losses during accumulation are 0.4%, followed by 0.15% at extraction. In order to improve this situation a number of steps may be taken which are described below. The limitation on the charge that can be accumulated comes from an instability that is present in the ring at high currents. In Fig. 3 is shown accumulated charge as a function of time on the upper trace and the losses from the ring below. The linear rise of accumulated charge may be seen followed by a flat portion where current is circulating in the ring. During this period the losses gradually increase followed by a substantial increase of these losses during which the accumulated charge falls to one third. It is

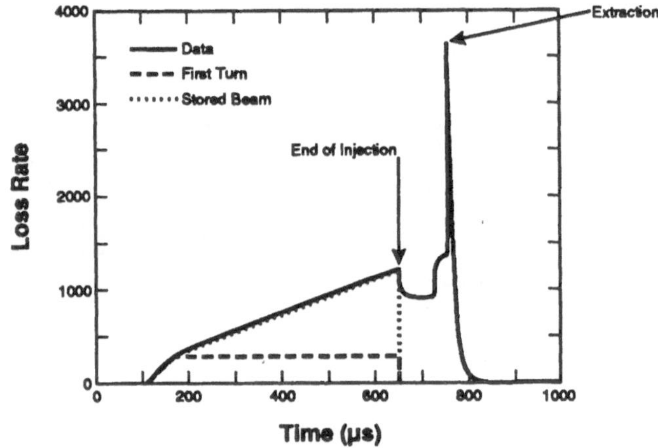

Figure 2. Loss rate as a function of time in the accumulation period. The precursor to the pulse at extraction is due to the finite rise time of the extraction kicker.

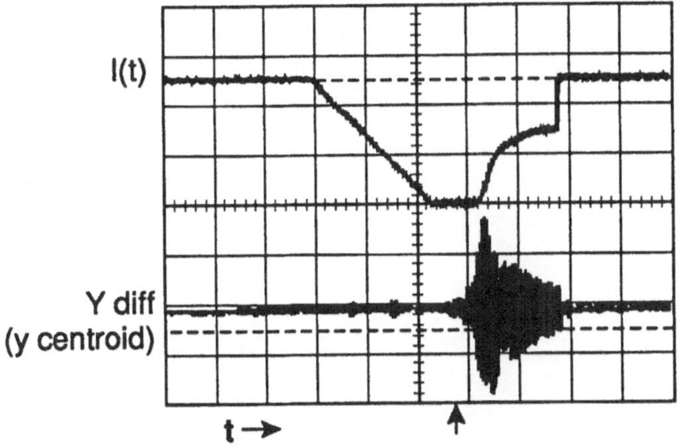

Figure 3. The upper trace is the charge accumulated in the ring. The lower trace shows loss rate, the instability after a period of coasting beam is clear accompanied by loss of beam in the ring.

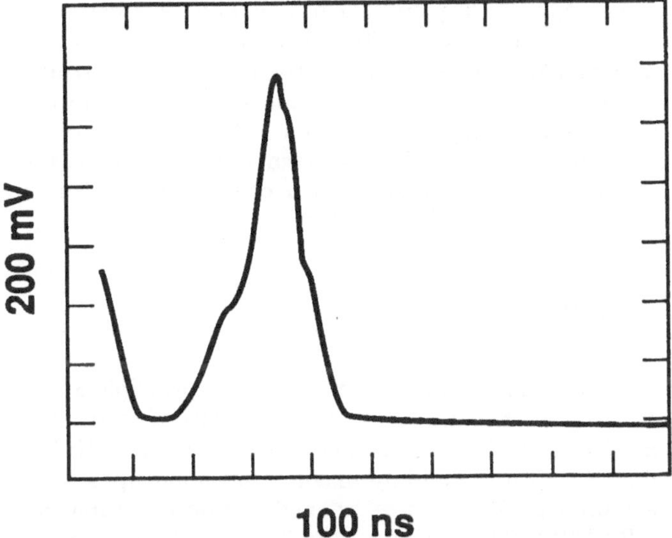

Figure 4. The last pulse in the ring at extraction is shown. This nearly triangular shape is close to that delivered to the experimental area.

believed that this instability is due to an interaction of the proton beam with residual electrons in the ring although this hypothesis is far from certain. What is certain is that this threshold is reasonably stable and the PLS proposal does not involve running the ring above this current.

Low loss at extraction is accomplished in part by the gap in the charge distribution around the circumference of the ring. This gap is maintained by the RF that is maintained during storage. For minimum loss the resulting charge distribution around the ring is not square as it is at injection but becomes almost triangular and this is reflected in the extracted pulse shape as shown in Fig. 4. This pulse shape is advantageous for most applications giving man apparently shorter duty factor than nominal.

5 External beam transport

The extraction as described is that used to feed the neutron spallation source that is presently in operation. The only modification that is envisaged is a kicker to transfer the beam into a PLS channel as is shown schematically in Fig. 3. It is assumed that 20 Hz will continue to be sent to the spallation source, the limitation on repetition rate is given by wraparound problems with neutron time of flight. Then if 40 Hz is sent to PLS this will still leave 60 Hz for polarized protons and other uses of H- from the linac. A beam channel similar to that shown in Fig. 1 has been designed in some detail for a previous proposal, and an appropriately low loss system seems to be straightforward.

6 Target and beam channels

At the target end of the transport it is envisaged that two targets would be mounted close together, one designed for muon channels, and a second designed to make the largest number of pions for neutrino generation. With a proton beam spill less than one microsecond it is straightforward to extend "old" technology and design a horn which would enhance neutrino flux as well as producing an enhancement of the flux from the pion sign that is selected. It is believed that the enhancement that can be achieved will approximately compensate for the difference between 1 mA available at the end of the linac and 200 μA available at PLS.

Z. Phys. C – Particles and Fields 56, S258–S260 (1992)

Zeitschrift für Physik C Particles and Fields

Kicked surface muon beams

J.L. Beveridge

TRIUMF, 4004 Wesbrook Mall, Vancouver, B.C., Canada V6T 2A3

1-August-1991

Abstract. The possibility of splitting a surface muon beam into three separate components using an electrostatic kicker is examined. Such a beam seems technically feasible and could be very effectively used to provide muons to time differential μSR experiments.

1 Introduction

The concept of a kicked surface muon beam or "magic beam line" was first discussed at TRIUMF by Spencer and Garner in the early 1980's [1]. Their idea was to address the muon pileup problem encountered in time differential (TD) μSR experiments while making more use of the $> 10^6$ muons available in modern surface muon beams. The principle is illustrated in Fig. 1. A kicker element is inserted in a beam line which is capable of directing the beam via a septum to either the upper or lower leg of the line. When a muon is detected entering the apparatus on either leg the kicker is turned off and the beam is allowed to continue in the straight through direction until another muon is required by the experiment. Since a typical TD μSR experiment can accept only one muon every 10-20 μsec (i.e. 10^5 μ^+/sec) such an arrangement can, in principle, provide beam simultaneously to two such experiments while the majority of the beam remains available for a third experiment. In addition, the TD μSR experiments can operate at up to 4 times the rate possible on a conventional beam line due to the lack of pileup effects.

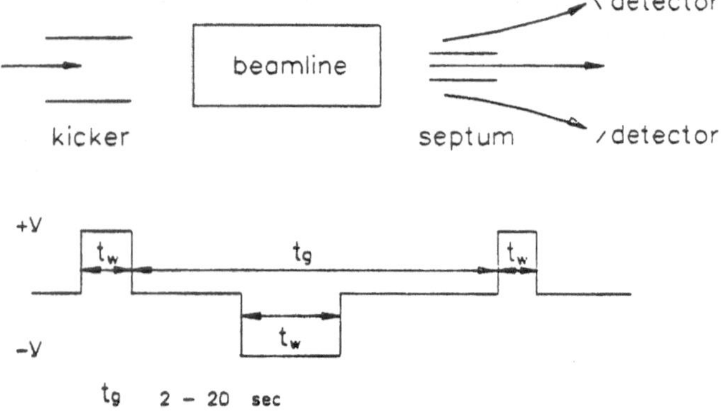

Fig. 1. Schematic diagram of a "kicked" surface muon beam and the kicker timing sequence.

2 Beam Line

A possible realization of the above principle at TRIUMF is shown in Fig. 2. Here, an electrostatic kicker is placed in the last section of the existing M15 beam line. A septum, bending magnet and quadrupole triplet are used to form a new "kicked" muon leg. First order TRANSPORT calculations indicate that an acceptable focus can be obtained with this arrangement and that good separation at the septum is attained with the kicker parameters listed below:

- length 50 cm
- gap 20 cm
- plate voltage 15-20 kV

Unfortunately, more realistic calculations using the program REVMOC indicate (Fig. 3) that higher order aberrations in the beam line create tails on the beam spot at the septum and that there is substantial overlap between the kicked and straight through beams. Higher kicker voltages give better separation but lead to beam loss in the elements prior to the septum. This implies that some improvement will have to be made in the beam optics before this kicker scheme will work practically in this beam line.

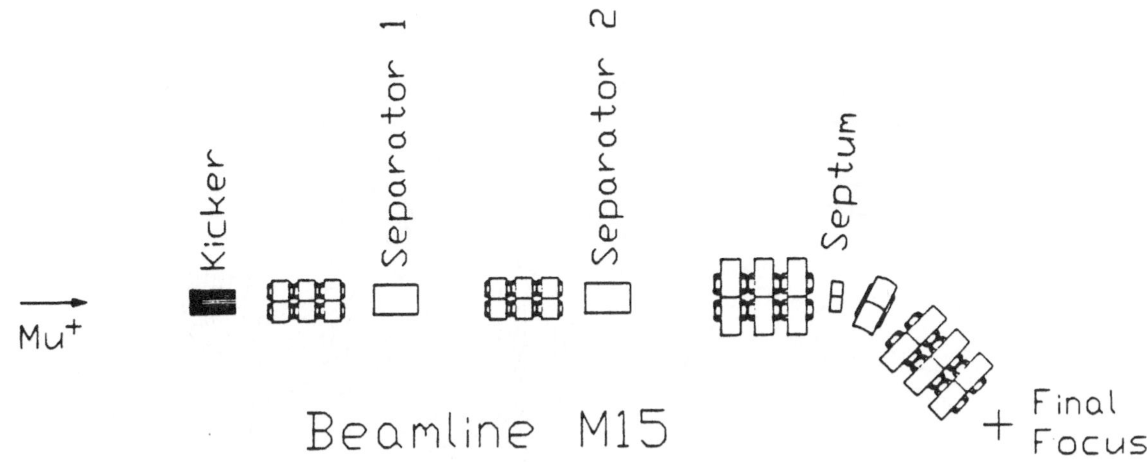

Fig. 2. Proposed insertion of a kicker in the existing TRIUMF M15 beam line.

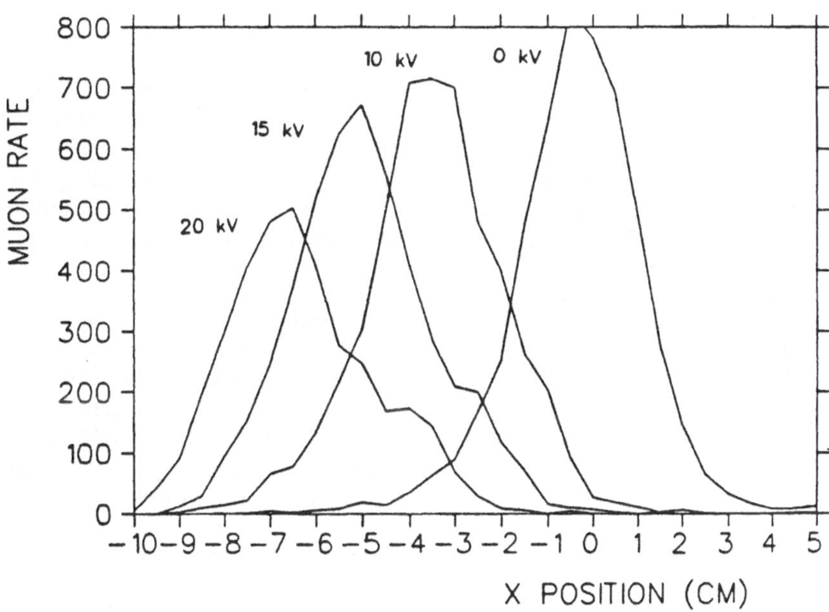

Fig. 3. Muon beam spot at the septum location for various kicker voltages.

3 Kicker

The kicker for the above arrangement must produce high voltage pulses (15-20 kV) with the shortest possible rise times (<100 nsec) and with repetition rates of up to 500 kHz. These specifications lead to very high instantaneous and average power requirements for the high voltage power supply. One possibility to circumvent such problems is to use a synchronous kicker similar to the 1 MHz Chopper developed for the TRIUMF KAON Factory [2]. This device (Fig. 4) is capable of producing pulses of the required voltages with very short (~10 nsec) rise times and has the advantage that power requirements are relatively low due to the storage of energy in the delay lines. Modification of this device required to lengthen the pulses and reduce their repetition rate by increasing the length of the delay lines seems feasible. However, such a chopper does not take full advantage of the "kicked" muon scenario as the muons are not delivered to the experiment on demand and the fixed frequency and pulse width can only be optimized for one muon beam rate. A more attractive option is an asynchronous pulser (Fig. 5) which could more closely reproduce the timing sequence of Fig. 1. This device appears to be within the capabilities of existing tetrodes provided rise times of the order of 100 nsec are accepted and would provide experiment driven pulse width and frequency. However, multiple high power, high voltage power supplies are required which make its realization expensive.

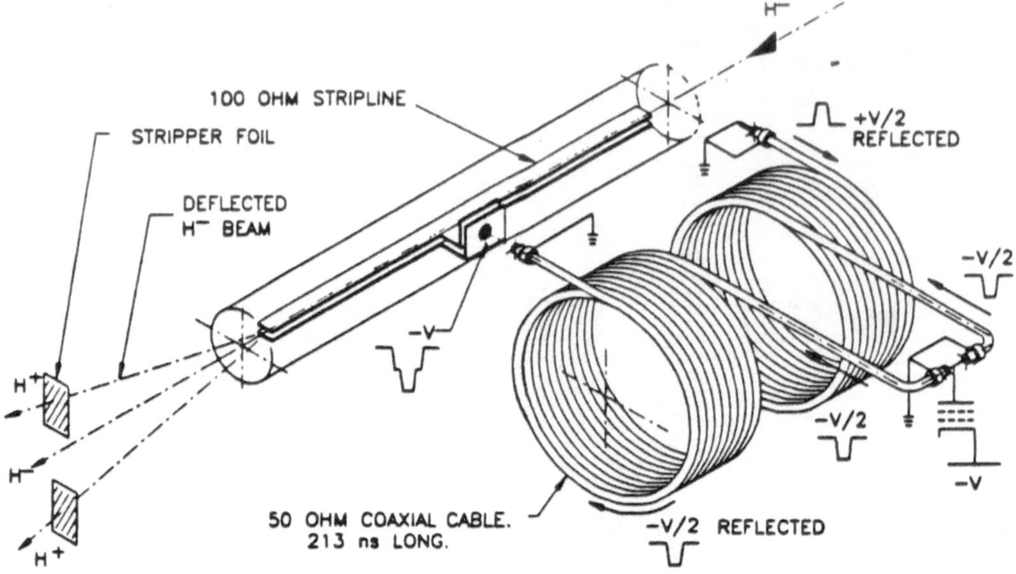

Fig. 4. The KAON prototype 1 MHz chopper.

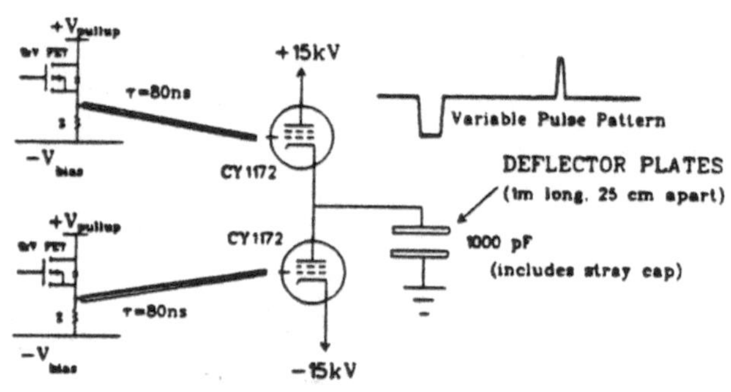

Fig. 5. A possible pulser for an asynchronous kicker.

4 Conclusions

A kicked surface muon beam has many attractive features for efficient use of available muons in certain experimental situations. It seems to be possible to produce the required kicker with existing high voltage pulse technology. Some care will be required in producing good beam optics to provide adequate beam separation at the septum, however, this should be possible and a kicker option should be seriously considered in the design of future surface muon beams.

5 References

1. D. Spenser and D. Garner, private communication.
2. G.D. Wait, M.J. Barnes, D. Bishop, G. Waters and C.B. Figley, "Prototype Studies of a 1 MHz Chopper for the KAON Factory", 1991 Particle Accelerator Conf., San Francisco (in press).

This article was processed using Springer-Verlag TEX Z.Physik C macro package 1991
and the AMS fonts, developed by the American Mathematical Society.

Z. Phys. C - Particles and Fields 56, S261-S268 (1992)

Recent applications of μ^+SR in magnetism: Novel magnetic features in *heavy electron* compounds

A. Schenck

Institut für Mittelenergiephysik der Eidgenössischen Technischen Hochschule Zürich, CH-5232 Villigen PSI, Switzerland

15-August-1991

This paper is dedicated to Prof. G. zu Putlitz on the occasion of his 60th birthday

Abstract. As an example for the power of μ^+SR-spectroscopy in solid state physics three applications to heavy electron systems, i.e. $CeA\ell_3$, $U_{1-x}Th_xBe_{13}$ and UCu_5 will be discussed. Each of these systems reveals very specific magnetic features of unusual characteristics which involve very small to extremely small magnetic moments and random or very short range magnetic order. This kind of small moment magnetic order can be studied relatively easily by μSR but will be hardly accessible by other methods such as neutron scattering.

1 Introduction

The fact that the μ^+ is a spin-1/2 particle and possesses a relatively large magnetic moment renders it a particularly well adapted probe for the investigation of magnetic phenomena in solids, especially weak ones, in view of the absence of any electrostatic quadrupolar interaction. In metals, due to its positive charge, the μ^+ is expected to reside in general at the center of an interstitial site (exhibiting, of course, a relatively large zero point vibration amplitude), where it interacts predominantly with conduction electrons and other less localized electron states, and therefore probes magnetism from a different perspective as compared with a substitutional probe as in nuclear magnetic resonance (NMR) and Mössbauer spectroscopy. Magnetic interactions of the implanted spin-polarized μ^+ are communicated to the outside world by means of the parity violating asymmetric emission of positrons from their decay (30% effect!) which allows to monitor the time evolution of the muon's spin auto correlation function, $< \mathbf{S}(t) \cdot \mathbf{S}(o) >$, quite easily, provided the involved time scale is compatible with the μ^+ lifetime of 2.2 μs [1]. Fortunately this latter requirement is in a majority of cases fullfilled.

Experiments can be performed with or without an external magnetic field $\mathbf{H}_{ext}$. Application of a transverse field ($\mathbf{H}_{ext} \perp \mathbf{S}(o)$) allows to observe the μ^+ Larmor precession and associated relaxation phenomena, a very important one being dephasing by a static inhomogeneous magnetic field distribution inside the specimen. Application of a longitudinal field ($\mathbf{H}_{ext} \parallel \mathbf{S}(o)$) allows to decouple the μ^+ spin from internal static field components and μ^+-spin relaxation is then only induced by fluctuating internal fields arising from the dynamical features of the host lattice magnetism (spin lattice relaxation). Of particular interest in certain cases is the possibility to study both static and dynamic aspects of internal fields in the absence of an external field which otherwise might interfere with and disturb the intrinsic magnetism. This proved to be of great importance in the study of spin glasses [2]. In ferro- or antiferromagnetically ordered systems one often finds a well defined internal field (e.g. the contact hyperfine field and or dipolar fields originating from the ordered moments) at a given type of interstitial site which then assumes the role of an external field and a well resolvable μ^+ Larmor-precession may become observable [3].

It is therefore no surprise that muon spin rotation spectroscopy (μSR) has found its widest application in the field of magnetism or in fields in which magnetic properties are an important feature, as e.g. in high temperature superconductivity [4]. In this contribution we would like to review applications to a particularly interesting family of compounds which are called **heavy electron** or **heavy fermion** systems and which, as it turned out, possess a wide variety of complex magnetic properties.

Heavy electron compounds are intermetallic compounds containing rare earth (Ce, Yb) or actinide (U, Np) elements and simple or transition metal elements [5]. Their most conspicuous feature is the occurance of a huge electronic specific heat at low temperatures (the Sommerfeld constant γ may assume values up to the order of severel J/mol K^2) and certain similarities to liquid ^{3}He suggesting that these systems may be described in terms of Fermi liquid theory, involving very large effec-

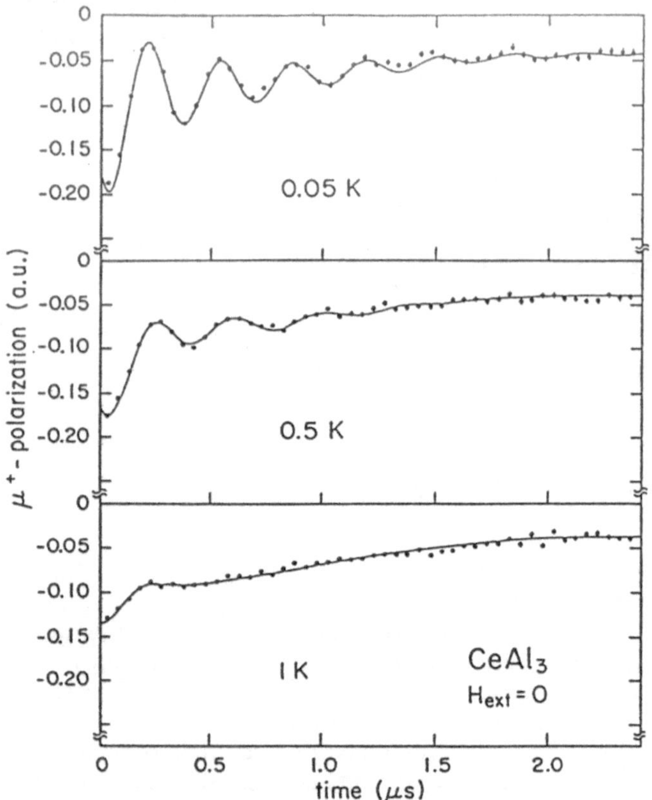

Fig. 1. ZF-μSR signal in $CeAl_3$ at three different temperatures. The solid lines represent three component fits. The low temperature frequency corresponds to an internal field of ~220 G (from ref. 13).

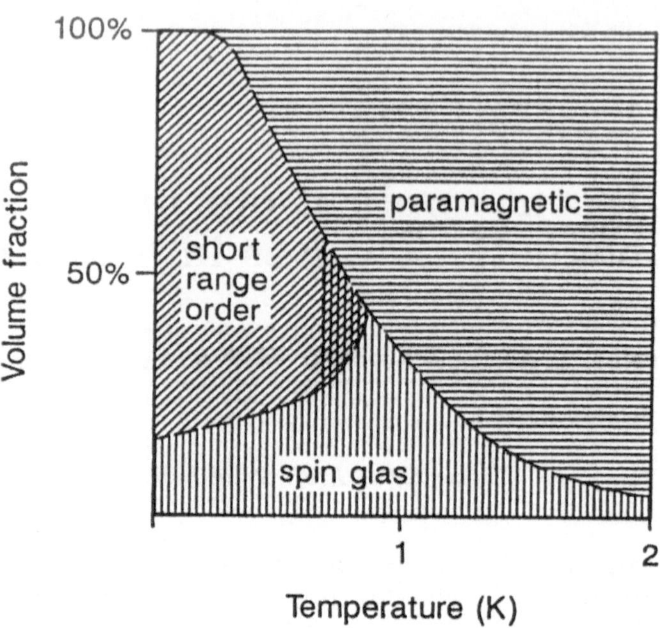

Fig. 2. Magnetic phase diagram of $CeAl_3$ as deduced from ZF- and TF-μSR data. Note the coexistence of different magnetic phases at a given temperature. No genuine phase transition can be identified.

tive quasi particle masses of the order of $200 \cdot m_e$ (i.e. of the order of the muon mass!) [6]. In other words these systems are characterized by strongly interacting and highly correlated electrons (f-electrons and conduction electrons) which — at low temperature — develop a ground state with unusual coherent features.

The groundstate may display other properties such as (perhaps unconventional) superconductivity ($CeCu_2Si_2$, $U_{1-x}Th_xBe_{13}$, UPt_3, URu_2Si_2) and antiferromagnetic (AF) order (U_2Zn_{17}, UCu_5, etc.) [5]. The magnetic properties seem to be determined by a competition between a Kondo type mechanism, screening the local f-moments, and the RKKY-interaction, which couples the f-moments via the conduction electrons by an indirect exchange mechanism. The Kondo type mechanism manifests itself by a change of the magnetic susceptibility from a Curie behavior at high temperatures to an essentially temperature independent strongly enhanced Pauli like behavior at low temperatures. Also the ordered moments are generally much smaller than the high temperature moments extracted from the Curie law [5].

The research on these, so far little understood systems, entered into a new phase when it was discovered first by μSR and subsequently also by neutron scattering that all those compounds exhibiting superconductivity also displayed — in coexistence with superconductivity! — some sort of magnetic ordering involving moments as small as possibly $10^{-3} \cdot \mu_B$ [7-12]. While the magnetic order in UPt_3 and URu_2Si_2 appears to be antiferromagnetic and long range in nature [10,11] the small moment magnetism in $U_{1-x}Th_xBe_{13}$ ($0.019 \leq x \leq 0.043$) [8,9] and in $CeCu_{2.1}Si_2$ [7] seems to be of a different type which may have more in common with spin glasses or very short range ordering. The $U_{1-x}Th_xBe_{13}$-results will be discussed in more detail in sect. 3.

Another interesting μSR result was obtained in $CeAl_3$ [13], which compound for a long time was considered to be an archetypical representative of a heavy electron system which remains paramagnetic down to the lowest temperatures. That this is not the case will be discussed in sect. 2 in which we will be confronted with a very strange magnetic phase diagram. Among the only remaining heavy electron systems not showing any sign of magnetic order are $CeCu_6$, $CeRu_2Si_2$, UPt_4Au and $YbCu_{4,.5}$. Future μSR studies on these compounds will show whether these compounds are truly paramagnetic at the lowest temperatures or not. We would not be surprised, if also these compounds would show some sort of small moment magnetism at low temperatures.

Finally also those heavy electron systems showing "conventional" AF order are perhaps more complicated than what is believed up to now. A good example in case is U_2Zn_{17} which has displayed in μSR studies an extremely complex behavior [14]. An even more interesting case is UCu_5 which seems to develop some small moment magnetism in coexistence with conventional AF-order [15]. This will be reviewed in sect. 4.

Finally, some future perspectives will be discussed in sect. 5.

2 Magnetic phase diagram of $CeA\ell_3$

The temperature dependence of the specific heat $c_p(T)$ of $CeA\ell_3$ does not show any anomaly which could be associated with a magnetic phase transition [16,17]. Usually, if there is a phase transition, one cannot fail to recognize it in form of a sharp cusp-like feature in $c_p(T)$. However, $c_p(T)/T$ displays a shallow maximum at around 0.5 K which was interpreted in terms of the onset of the heavy mass Fermi liquid state [17]. Also the susceptibility displays a shallow maximum at around 0.7 K [17] which in principle could be associated with an AF-phase transition. However, neutron diffraction results down to 60 mK provided no hint for the formation of static long range magnetic correlations [18]. The more it was a surprise when zero field (ZF) μSR measurements in polycrystalline $CeA\ell_3$ revealed the presence of static nonzero magnetic fields at the μ^+ below 2 K [13]. This is clearly visible in Fig. 1 which for temperatures below ~0.7 K, displays a nice Larmor precession signal associated with a local field of ~220 G. In fact the signal shown in Fig. 1, is composed of three components, two of which originate from the $CeA\ell_3$ sample and the third one from the Cu-target holder (the cold finger of a He^3-He^4 dilution refrigerator). This third component is of the well known Kubo-Toyabe type appropriate for Cu [1]. The second component arising from the $CeA\ell_3$ sample is also well described by a Kubo-Toyabe function, but describing a much faster relaxation than in the Cu-case. Hence μ^+ implanted in $CeA\ell_3$ experience two different magnetic environments: one associated with a nonzero magnetic field (220 G) and the other one associated with zero average field but a nonzero field spread around this value leading to the Kubo-Toyabe signal.

The signal showing the precession pattern must originate from domains inside the sample in which some coherent but short range magnetic order is established. The order cannot be long range otherwise it should have been seen in the neutron scattering study. The Kubo-Toyabe signal, on the other hand, must originate from domains exhibiting some static, random or extremely short range order resembling perhaps a spin glas phase.

The amplitudes of the two $CeA\ell_3$ related signals display a strong temperature dependence, their sum decreasing monotonously with rising temperature until the ZF-μSR signal becomes invisibly small on approaching 2 K. This behavior implies that there must be a third type of magnetic domain in which the μ^+ polarization is not affected suggesting that these domains are in a paramagnetic state. This was confirmed by transverse field (TF) μSR measurements which allowed to identify this component directly [19]. The amplitudes of the three components originating from the $CeA\ell_3$ sample provide a measure of the volume fractions occupied by the various types of domains. Their temperature dependence allows, therefore, to construct a phase diagram, which is shown somewhat schematically in Fig. 2.

The most surprising feature of this phase diagram is the obvious absence of any clear cut cooperative magnetic phase transition which involves the whole sample volume. Rather some random static order develops below ~2 K first in a vanishingly small fraction of the volume which fraction then increases as the temperature is decreased. In parallel the paramagnetic volume shrinks by the same proportion. Below (0.7–0.9) K, i.e. approximately where the susceptibility displays a shallow maximum, some new type of magnetic domain develops which is associated with coherent short range order. The corresponding volume grows with further decreasing temperature at the expense of both the paramagnetic and the "spinglas" like volume. The paramagnetic volume has essentially disappeared below 0.25 K. This phase diagram is very unusual and we do not know of any other example [20]. It is certainly consistent with the absence of any singularity in $c_p(T)/T$. A possible explanation may be given within the framework of magnetic frustration. Magnetic frustration is indeed not entirely impossible in view of the planar triangular arrangement of the Ce-moments in the hexagonal Ni_3Sn-type structure of $CeA\ell_3$ and intraplane AF-interactions between nearest neighbor Ce-moments [19].

More recently the μSR-findings where partially confirmed by NMR measurements which showed a drastic increase of the linewidth below ~1.2 K and a peak of T_1^{-1} at 1.2 K [21]. These results seem to point to the presence of a cooperative phase transition at ~1.2 K, which was suggested to lead to spin density wave-type order. Such a conclusion is at variance both with the c_p/T- and the μSR-results. More studies are clearly needed.

Finally, we wish to mention that a consistent analysis of all the TF-μSR data also allowed to determine the magnitude of the ordered moment (~0.5 μ_B in both types of ordered domains) and the μ^+ site [19].

3 Superconductivity and magnetism in $U_{1-x}Th_xBe_{13}$

The system UBe_{13} has attracted particular attention because it was the second heavy electron compound which displayed superconductivity [22] (after $CeCu_2Si_2$ [23]) associated with the heavy electron state and the first compound in which superconductivity seems to be of an unconventional nature [24]. Additional excitement arouse when it was discovered that Th doped UBe_{13} ($U_{1-x}Th_xB_{13}$) for a certain range of Th-concentrations ($1.9\% \leq x \leq 4.2\%$) possesses another second order phase transition at T_{c2} below the onset of superconductivity at T_{c1} [25]. The additional phase transition manifests itself by a typical singularity in the specific heat but the nature of this phase transition remained obscure. Moreover the x-dependence of T_{c1} shows a highly non-monotonic behavior. The x-dependence of T_{c1} and T_{c2} is displayed in Fig. 3. In order to learn more about the nature of the phase transition at T_{c2} ZF-μSR measure-

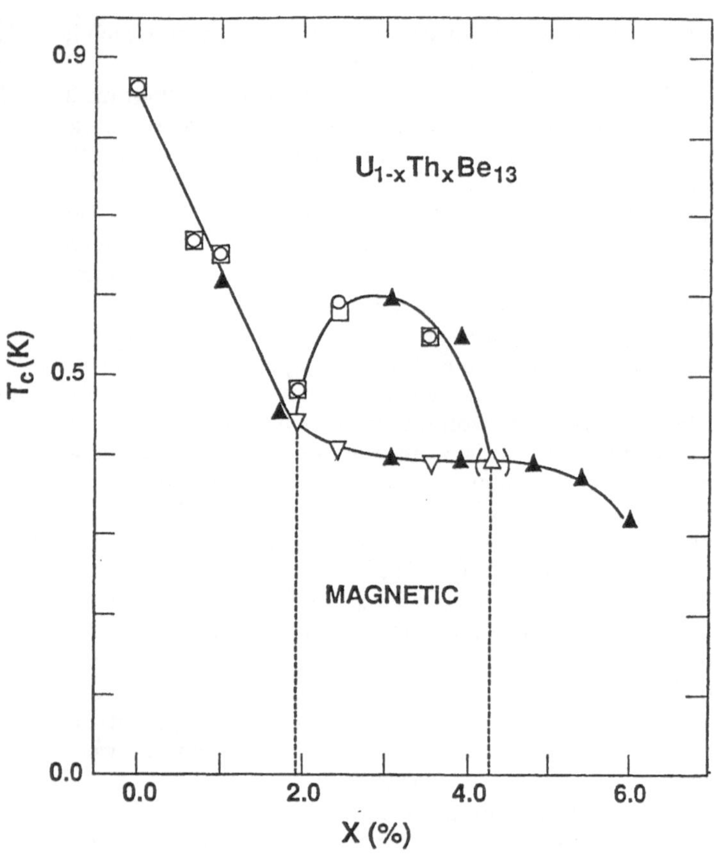

Fig. 3. Phase diagram for $U_{1-x}Th_xBe_{13}$. The data points stem from various techniques. Note the appearance of a second phase transition at T_{c2} between $x = 1.9\%$ and $x = 4.2\%$ (from ref. 28).

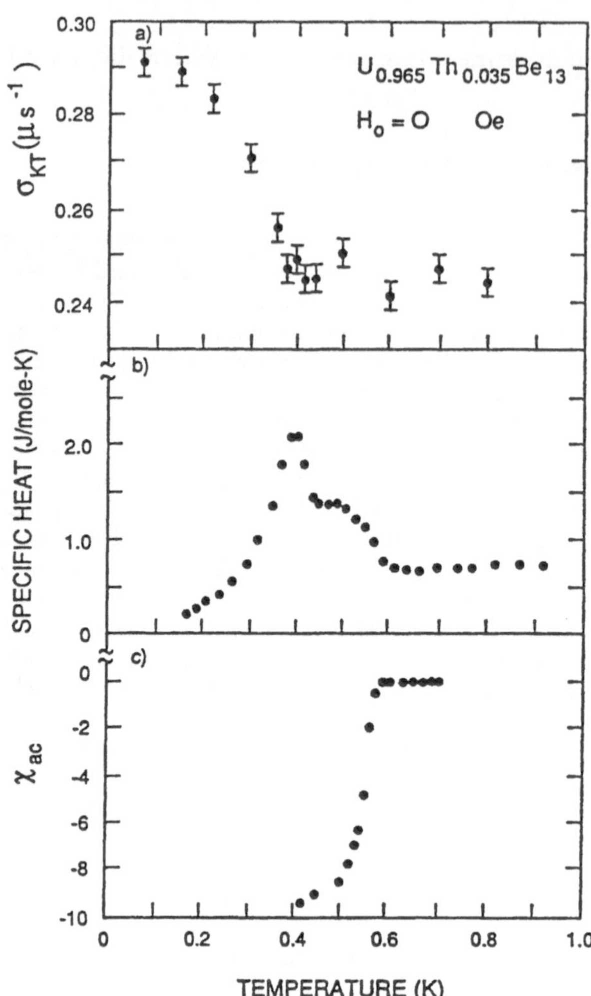

Fig. 4. Comparison of specific heat and ac-susceptibility data with ZF-μSR-relaxation rates in $U_{0.965}Th_{0.035}Be_{13}$. Note the onset of superconductivity at T_{c1} by the diamagnetic response in χ_{ac}, the appearance of the second phase transition at T_{c2} from the cusp in the specific heat and the increase of σ_{ZF} starting right at T_{c2} (from ref. 27).

ments were undertaken first at LAMPF [9,26] and — much improved — at PSI [27,28]. A typical result is displayed in Fig. 4. It shows the onset of superconductivity in ac-susceptibility data at T_{c1} and the position of the second phase transition in specific heat data at T_{c2}. At the latter temperature the ZF-μ^+ relaxation rate σ_{ZF} starts to increase significantly as the temperature is lowered below T_{c2}. The relaxation of the μ^+ polarization below and above T_{c2} is well described by a Kubo-Toyabe function. Above T_{c2} the observed temperature independent relaxation rate can be traced back to the spread in dipolar fields arising from the Be nuclear magnetic moments at a certain interstitial site. This site is shown in Fig. 5. For this site one calculates under certain additional assumptions $\sigma_{ZF} = 0.246\ \mu s^{-1}$ [24] which compares very well with the experimental value of $\sigma_{ZF}^{ex} = 0.245(2)\ \mu s^{-1}$.

Below T_{c2} an additional source for internal fields must become available which, since it cannot be of nuclear origin has to be of electronic origin. The preserved gaussian form of the internal field distribution below T_{c2}, as evidenced from the persisting Kubo-Toyabe type relaxation function, implies further that the involved electronic moments must be static, randomly oriented and fairly regularly positioned, e.g. at each U-site. The magnitude of these moments must be of the order of $(10^{-3}$–$10^{-2})\ \mu_B$, i.e. extremely reduced compared to the high temperature U-moment of $\mu_{5f} = 3.15\ \mu_B$ which results from the Curie behavior of the susceptibility. μSR measurements on samples with other x revealed that the enhanced relaxation rate shows only up in samples with $1.9\% \leq x \leq 4.2\%$ and the onset of the increase of σ_{ZF} is always correlated with T_{c2}. No anomalous features could be observed for $x < 1.9\%$ and $x > 4.2\%$.

The occurance of some sort of static magnetic order in coexistence with superconductivity has found so far two interpretations. The first one is more conventional and assumes the formation of a perhaps rather complex antiferromagnetic phase below T_{c2}. This view is not inconsistent with ultra sound attenuation data [27]. In view of the termination of the phase boundary associated with $T_{c2}(x)$ on the phase boundary $T_{c1}(x)$, separating the normal state from the superconducting state (see Fig. 3), the order parameters of the two phases must be strongly coupled.

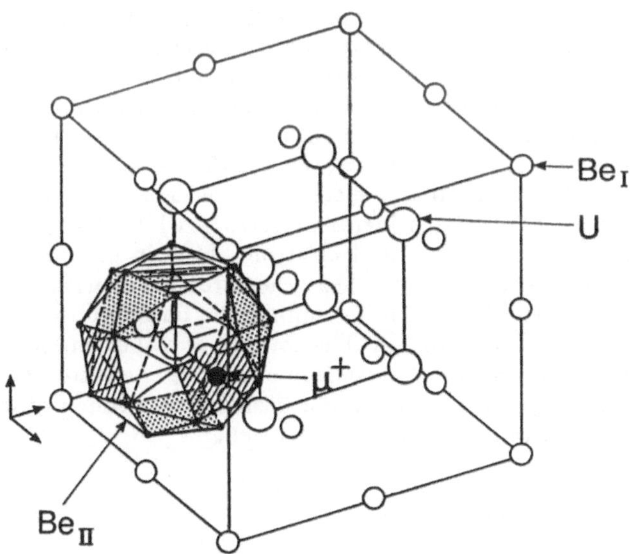

Fig. 5. Crystal structure of UBe_{13} and possible μ^+-position with two nearest U- and 4 nearest Be neighbors. Each U-atom is surrounded by a 'snub cube' of 32 Be-sites (Be_{II}). Not shown are additional Be_{II} positions which form icosahedrons around the Be_I positions at (0,0,0) and (0,0,1/2).

The second more exotic explanation postulates the existence of an intrinsically magnetic superconducting state below T_{c2} which would have to be described by a complex multicomponent order parameter and would violate time reversal invariance [30]. The phase transition at T_{c2} would be a phase transition between two different superconducting states which differ by their symmetry properties. Several other observations [28] add credibility to such an explanation, although the first interpretation cannot be ruled out on the basis of the present status of knowledge.

It should be emphasized that only the μSR technique up to now has the sensitivity to detect the kind of weak magnetic features associated with the phase-transition at T_{c2} in $U_{1-x}Th_xBe_{13}$.

4 'Strong' and 'weak' magnetism in UCu_5

UCu_5 shows a transition to an antiferromagnetic state close to 15 K [31]. The magnetic structure could be determined from neutron scattering measurements [32]. Interestingly the heavy electron state develops below 4 K, i.e. well inside the magnetically ordered state [31]. Specific heat measurements revealed a second continuous phase transition close to 1 K which displayed hysteretic features (see Fig. 6) but no latent heat [31]. In addition the electrical resistivity increased by an order of magnitude below ~1 K. This behavior is exceptional among all heavy electron systems which generally show a trend towards smaller resistivity values when the co-

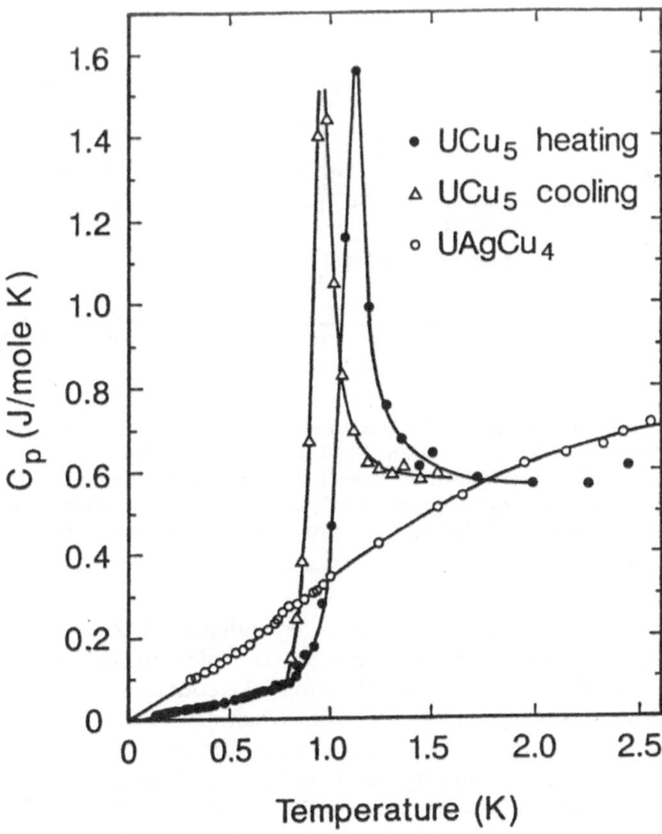

Fig. 6. Specific heat of UCu_5 and $UAgCu_5$ showing evidence for a phase transition in UCu_5 at ~1 K (from ref. 31).

herent heavy electron state develops. The nature of this additional phase transition could not be determined. Interestingly the 1 K anomaly in the specific heat and the rise in resistivity below this temperature are only observed in high quality UCu_5 samples.

In order to determine the nature of the 1 K phase transition we applied again the μSR-technique complemented this time by a neutron diffraction study at the reactor saphir of PSI [15]. The results were the following [15]. The neutron study showed no change in intensity and position of the nuclear and the magnetic Bragg peaks when changing the temperature from 1.3 K to 10 mK (see Fig. 7). Within the limits given by the accuracy of these measurements one has to conclude that at ~1 K no change in the magnetic structure nor in the crystalline structure takes place. The zero field μSR-signal below T_N proved to be quite complicated in that four components were contained in it. Three components showed a relaxing precession pattern, corresponding to average internal fields between ~1 kG and 1.46 kG, while the fourth component did not display any oscillation (corresponding to zero average field) but a significant

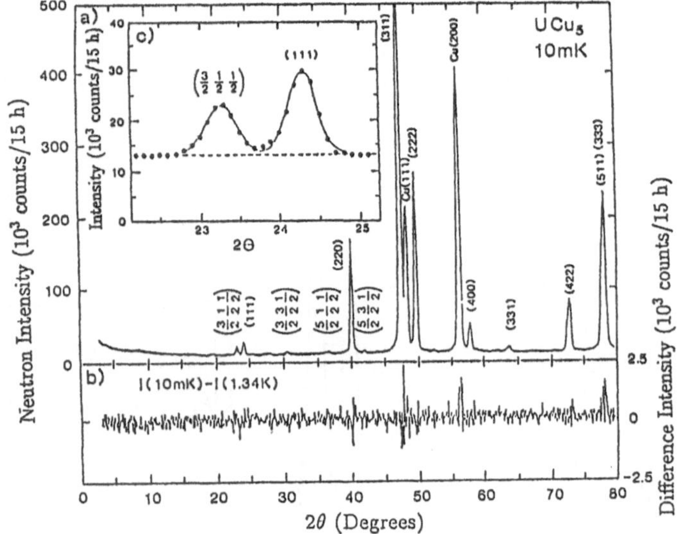

Fig. 7. (a) Neutron-diffraction pattern at 10 mK in UCu_5. (b) Difference neutron diagram I (10 mK) - I (1.39 K) showing the absence of any influence of the 1 K transition on the crystallographic and magnetic structure (from ref. 15). (c) A blow up of the magnetic (3/2,1/2,1/2) r eflection next to a nuclear peak.

relaxation. The temperature dependence of the precession frequencies did not show any conspicuous features at around 1 K, confirming therefore the conclusions drawn from the neutron results. Very drastic effects, however, were seen in the relaxation rates of three of the four components including the nonprecessing one (the other precessing component possesses an order of magnitude larger relaxation rate both above and below 1 K. It's origin is not clear to us). As can be seen from Fig. 8 the relaxation rates start to rise steeply as the temperature is lowered through ~1.1 K. This behavior implies that although the average fields remain unchanged the field spread around each average field rises dramatically below 1.1 K. The relaxation rates above 1.1 K can be explained in terms of the dipolar fields arising from the Cu nuclear moments. The question now is what causes this increase in field width sampled by the μ^+. (Another question concerns the origin of the four components and their different average fields. This question has not been answered yet.) Since the neutron diffraction results exclude more conventional explanations, including the possibility of a charge density wave [31], we interpret the increase in relaxation rate below 1.1 K along the same line as in $U_{1-x}Th_xBe_{13}$, URu_2Si_2 and UPt_3, namely as a result of the formation of small moments which order below 1.1 K in some random fashion or in a pattern which is incommensurable with the lattice structure. These small moments (~0.01 μ_B if we place them at the Cu sites) exist in parallel to the ordinary U-5f moments which are responsible for the antiferromagnetic order below 15 K.

If this interpretation is right, it has some far reaching consequences. First, the fact that the 1 K phase transition is seen both in the specific heat, which arises from the heavy electron ground state, and in the μSR linewidth data could imply that the small moment magnetism is carried by the heavy quasi particles. Secondly, the fact that the normal antiferromagnetic order is unaffected by the 1 K anomaly could imply that the U-5f moments are unrelated to the heavy electron state. In effect we are proposing that the ground state of UCu_5 is made up of two rather independent electron substates the first one involving those 5f-electrons which settle into an antiferromagnetic order below 15 K and a second one which forms the heavy electron state. The latter could be associated with itinerant electrons while the former one arises from localized electrons. Within such a picture the formation of a heavy electron state within an already established antiferromagnetic phase appears to pose no special problem. The concept of rather unrelated electron substates, if present in other heavy electron systems, may perhaps provide a basis for explaining other peculiarities in such systems as well.

5 Future perspectives

The three presented examples prove without doubt the power of μSR-spectroscopy in the field of magnetism. In particular weak effects are now susceptible to thorough investigations. The field of magnetism itself, although quite venerable in age, seems to be an inexhaustible reservoir of new and unexpected phenomena. The last 15 years have witnessed the emergence of such important topics as spin glasses (to which μSR has contributed prominently (see e.g. [2])), magnetism in low dimensional materials, magnetism in mixed valence systems, magnetism in the family of the cuprate oxides, famous for their ability to develop high temperature superconductivity, and last but not least magnetism in heavy electron compounds. This development was accompanied by corresponding efforts in solid state theory. Magnetism, of course, is just but one side of a more basic aspect of solids which concerns their electronic structure. And since solids are made up of practically an infinite number of constituents one can imagine that the phenomenology of solid state systems (or more generally of condensed matter systems), including their magnetic properties may be as varied as there are degrees of freedom. This perspective promises lots of work in the future and μSR spectroscopy is bound to remain an indispensible tool as many other present day techniques such as NMR and neutron scattering.

One import aspect in this respect is the complementary character of neutron scattering (measures in k-space) and μSR (measures in r-space) which should be more fully and systematically exploited in the future. It is therefore of much advantage that at several laboratories μSR and neutron scattering can be practised next door to each other (e.g. at RAL, PSI, Dubna, KEK).

In this contribution we have only discussed applications of μSR spectroscopy in magnetism. To fully assess the future of μSR spectroscopy one would have to consider all the other applications in physical chemistry and solid state physics as well. New and exciting possibili-

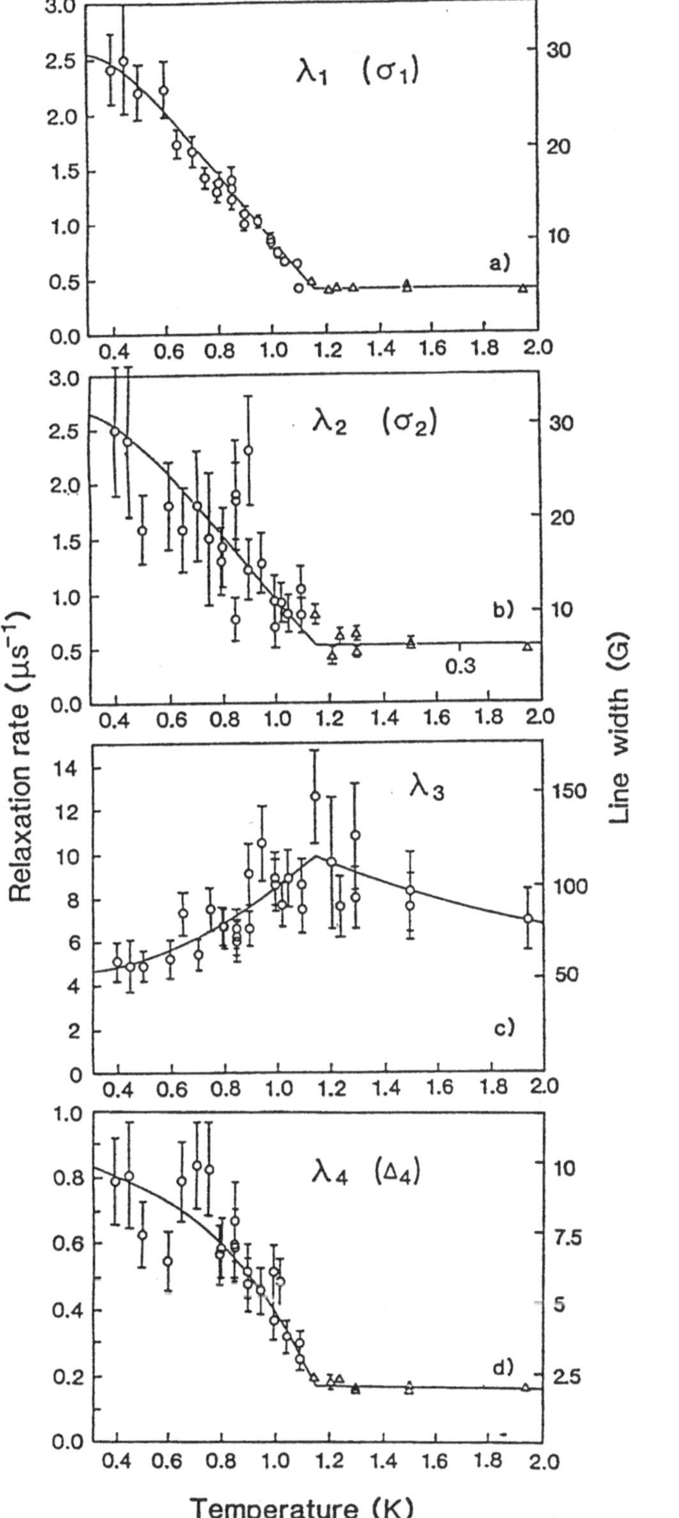

Fig. 8. Temperature dependence of the μ^+ relaxation rates associated with the four components. (a)-(c) are from the precessing components, (d) is from the non precessing one. (a), (b) and (d) clearly reflect the phase transition at ~1.1. K (from ref. 15).

ties would also arise with availability of ultra slow muon beams. It may then become possible to apply μSR to the study of magnetism in surfaces and thin layers and other artificial structures. In conclusion μSR spectroscopy will be a prime force in the demand for muon sources also in the future.

Acknowledgement. The reported work is the result of very fruitful collaborations with H.R. Ott, R.H. Heffner and D.E. MacLaughlin. The experiments were performed and analyzed with the help of S. Barth, P. Birrer, F.N. Gygax, B. Hitti, E. Lippelt and M. Weber. The neutron diffraction measurements in UCu_5 were performed by P. Böni and P. Fischer. My sincere gratitude extends to all these colleagues and to A. Amato for many helpful discussions.

References

1. See e.g. A. Schenck, Muon Spin Rotation Spectroscopy (Adam Hilger, Bristol, 1985).
2. H. Pinkvos, A. Kalk, and Ch. Schwink, Phys. Rev. **B41** (1990) 590.
3. See e.g. A.B. Denison, H. Graf, W. Kündig, and P.F. Meier, Helv. Phys. Acta **52** (1979) 460.
4. See e.g. Proceed. 5th Int. Conf. Muon Spin Rotation (Oxford, 1990) in Hyperfine Interact. **63** (1990).
5. Z. Fisk, D.W. Hess, C.J. Pethich, D. Pines, J.L. Smith, J.D. Thompson, and J.O. Willis, Science **239** (1988) 33.
6. P.A. Lee, T.M. Rice, J.W. Serene, L.J. Sham, and J.W. Wilkins, Comments Condensed Matter Phys. **12** (1986) 99.
7. Y.J. Uemura, W.J. Kossler, X.H. Yu. H.E. Schone, J.R. Kempton, C.E. Stronach, S. Barth, F.N. Gygax, B. Hitti, A. Schenck, C. Baines, W.F. Lankford, Y. Ōnuki, and T. Komatsubara, Phys. Rev. **B39** (1989) 4726.
8. R.H. Heffner, J.O. Willis, J.L. Smith, P. Birrer, C. Baines, F.N. Gygax, B. Hitti, E. Lippelt, H.R. Ott, A. Schenck, and D.E. MacLaughlin, Phys. Rev. **B40** (1989) 806.
9. R.H. Heffner, D.W. Cooke, and D.E. MacLaughlin, Theoretical and Experimental Aspects of Valence Fluctuations and Heavy Fermions, eds. L.C. Gupta and S.K. Malik (Plenum, New York, 1987) p. 319.
10. G. Aeppli, E. Bucher, C. Broholm, J.K. Kjems, J. Baumann, and J. Hufnagle, Phys. Rev. Lett. **60** (1988) 615.
11. C. Broholm, J.K. Kjems, W.J.L. Buyers, P. Matthews, T.T.M. Palstra, A.A. Menovsky, J.A. Mydosh, Phys. Rev. Lett. **58** (1987) 1467.
12. D.E. MacLaughlin, D.W. Cooke, R.H. Heffner, R.L. Hutson, M.W. McElfresh, M.E. Schillaci, H.D. Rempp, J.L. Smith, J.O. Willis, E. Zierngiebl, C. Boekema, R.L. Lichti, and J. Oostens, Phys. Rev. **B37** (1988) 3153.
13. S. Barth, H.R. Ott, F.N. Gygax, B. Hitti, E. Lippelt, A. Schenck, C. Baines, B. van den Brandt, T. Konter, and S. Mango, Phys. Rev. Lett. **59** (1987) 2991.
14. A. Schenck, A. Amato, P. Birrer, F.N. Gygax, B. Hitti, E. Lippelt, S. Barth, H.R. Ott and Z. Fisk, submitted to ICM 1991 (Edinburgh).
15. A. Schenck, P. Birrer, F.N. Gygax, B. Hitti, E. Lippelt, M. Weber, P. Böni, P. Fischer, H.R. Ott, and Z. Fisk, Phys. Rev. Lett. **65** (1990) 2454.
16. D. Jaccard and J. Flouquet, J. Magn. Magn. Mat. **47+48** (1985) 45.
17. H.R. Ott in: Progress in Low Temperature Physics, Vol. XI, ed. D.F. Brewer (North Holland, Amsterdam, 1987).
18. A.P. Murani, K. Knorr, K.H.J. Buschow, A. Bonoit, and J. Flouquet, Solid State Communic. **36** (1980) 523.
19. S. Barth, H.R. Ott, F.N. Gygax, B. Hitti, E. Lippelt, A. Schenck, and C. Baines, Phys. Rev. **B39** (1989) 11695.
20. Recently a similar phenomenon may have been observed in the high temperature superconductor $HoBa_2Cu_3O_7$ concern-

ing the Ho-moment ordering (P. Birrer et al., Phys. Rev. **B39** (1989) 11449).
21. H. Nakamura, Y. Kitaoka, Y. Asayama, and J. Flouquet, J. Phys. Soc. Jap. **57** (1988) 2644.
22. H.R. Ott, H. Rudigier, Z. Fisk, and J.L. Smith, Phys. Rev. Lett. **50** (1983) 1595.
23. F. Steglich, J. Aarts, C.D. Bredl, W. Lieke, D. Meschede, W. Franz, and H. Schäfer, Phys. Rev. Lett. **43** (1979) 1892.
24. H.R. Ott, H. Rudigier, T.M. Rice, K. Ueda, Z. Fisk, and J.L. Smith, Phys. Rev. Lett. **52** (1984) 1915.
25. H.R. Ott, H. Rudigier, Z. Fisk, and J.L. Smith, Phys. Rev. **B31** (1985) 1651.
26. R.H. Heffner, D.W. Cooke, A.L. Giorgi, R.L. Hutson, M.E. Schillaci, H.D. Rempp, J.L. Smith, J.O. Willis, D.E. Mac Laughlin, C. Boekema, R. Lichti, J. Oostens, and A.B. Denison, Phys. Rev. **B36** (1989) 11345.
27. R.H. Heffner, J.O. Willis, J.L. Smith, P. Birrer, C. Baines, F.N. Gygax, B. Hitti, H.R. Ott, A. Schenck, and D.E. Mac Laughlin, Phys. Rev. **B40** (1989) 806.
28. R.H. Heffner, J.L. Smith, J.O. Willis, P. Birrer, C. Baines, F.N. Gygax, B. Hitti, E. Lippelt, H.R. Ott, A. Schenck, E.A. Knetsch, J.A. Mydosh, and D.E. MacLaughlin, Phys. Rev. Lett. **65** (1990) 2816.
29. A Amato et al., (1990) unpublished work.
30. M. Sigrist and T.M. Rice, Phys. Rev. **B39** (1989) 2200.
31. H.R. Ott, H. Rudigier, E. Felder, Z. Fisk, and B. Batlogg, Phys. Rev. Lett. **55** (1985) 1595.
32. A. Murasik, S. Ligenza, and A. Zygmunt, Phys. Status Solidi (a) 23 (1979) K 163.

This article was processed using Springer-Verlag TEX Z.Physik C macro package 1991
and the AMS fonts, developed by the American Mathematical Society.

Z. Phys. C - Particles and Fields 56, S269-S279 (1992)

Muons and Pions as Probes in Condensed Matter

J. Major [1], A. Seeger [1,2], and Th. Stammler [2]

[1] Max-Planck-Institut für Metallforschung, Institut für Physik, Postfach 800665, D-7000 Stuttgart 80, Germany
[2] Universität Stuttgart, Institut für Theoretische und Angewandte Physik, Pfaffenwaldring 57, D-7000 Stuttgart 80, Germany

07-05-1992

Abstract. The basic concepts and requirements of condensed-matter studies employing positive and negative muons (μ^+, μ^-) or positive pions (π^+) are briefly reviewed. As examples for the various experimental techniques and for their application to solid-state problems the paper discusses the determination of the sites of thermalized π^+ or μ^+ in crystals by lattice steering (channelling or blocking), time-differential muon spin rotation (μ^+SR) in transverse magnetic fields, radio-frequency μ^+SR, and muon spin rotation of negative muons (μ^-SR) in semiconductors.

1 Introduction and general background

In the so-called *meson factories*, high-energy - high-current proton accelerators allow us to produce intense pion and muon beams of both charges ($\pi^+, \pi^-, \mu^+, \mu^-$). The muon beams are obtained from the decay of pions. Although the threshold proton energy for pion production is only about 260 MeV, for practical reasons the production of pion/muon beams for use in materials research requires strong proton beams of at least 500 MeV.

The main characteristics of the most important meson factories are as follows:

-Paul Scherrer-Institut (PSI) at Villigen/Switzerland: 500 μA/590 MeV proton beam, quasi-continuous on the time-scale of the muon lifetime ($\tau_\mu = 2.2$ μs) and pulsed (1 ns bursts every 20 ns) on that of the pion lifetime ($\tau_\pi = 26$ ns).

-ISIS facility at the Rutherford Appleton Laboratory (RAL) at Chilton/UK.: 150 μA (mean current)/790 MeV proton beam, pulsed (two 100 ns wide bursts with 230 ns separation every 20 ms), only surface μ^+ beam (see below) available.

-Canada's National Meson Facility TRIUMF in Vancouver, B. C./Canada: 200 μA/500 MeV proton beam, quasi-continuous.

From the viewpoint of *materials science*, the fact that muons and pions belong to different classes of elementary particles (pions are mesons and muons are leptons) is of secondary importance. The positive particles (π^+, μ^+) may be considered as light isotopes of the hydrogen nuclides. In experiments on the physics and chemistry of condensed matter they can be employed as short-lived radioactive probes. The negative particles (π^-, μ^-) may play the rôle of "heavy electrons". E. g., μ^- may be used to locally probe electronic interactions in semiconductors (cf. Sect. 5). The experimental techniques that are available for studying their behaviour in condensed matter, however, are determined by the basic differences between muons and pions. Whereas the former possess spin 1/2 and magnetic moments and are hence capable of responding to magnetic fields even when at rest (quite a number of techniques are based on this), this is not so for pions, which are spin-zero particles.

The information on the fate of positive muons and positive pions injected into matter is entirely transmitted through their weak-interaction decays (1a, 2a) with mean lifetimes τ_μ and τ_π. In the case of the "heavy electrons" π^- and μ^- we may obtain information from the X-rays emitted during the transitions between "pionic" or "muonic" orbits, in the case of μ^- in materials with not too high atomic numbers Z also from the weak-interaction decay (2b).

In *weak interactions* parity is not conserved. This has the important consequence that the muons resulting from the decays

$$\pi^+ \rightarrow \mu^+ + \nu_\mu \tag{1a}$$

or

$$\pi^- \rightarrow \mu^- + \bar{\nu}_\mu \tag{1b}$$

are fully longitudinally polarized in the rest system of the pions (ν_μ and $\bar{\nu}_\mu$ denote the muon neutrino or anti-neutrino). The muons decay according to

$$\mu^+ \rightarrow e^+ + \nu_e + \bar{\nu}_\mu \tag{2a}$$

or

$$\mu^- \to e^- + \bar{\nu}_e + \nu_\mu \tag{2b}$$

into positrons (e^+) or electrons (e^-), electron neutrinos (ν_e) or antineutrinos ($\bar{\nu}_e$), and $\bar{\nu}_\mu$ or ν_μ. In these decays the momentum directions of e^+ and e^- are correlated with the muon spin directions at the time of the decay. The experimental technique for detecting the magnetic interactions of spin-polarized μ^+ and μ^- are based on this. The fact that the e^+ and e^- emitted in the decays (2a,b) are spin-polarized, too, does not play a rôle in condensed-matter studies, since, in contrast to the muon spin polarization, their polarization cannot be determined easily.

For pion/muon production a target consisting of low-Z elements (e.g., carbon) is bombarded with high-energy protons. During the proton–nucleus collisions, π^+ and π^- are created among other particles. By properly setting the ion-optical parameters of the beamlines, the desired particles (π^+ or π^-) with the desired momentum can be extracted. The pion beams may be used either directly in experiments or in so-called muon channels (involving long straight beamline sections subject to high longitudinal magnetic fields), in which virtually all the pions decay into muons. The adjustable momenta of the π/μ beams can be as high as 250 MeV/c (corresponding to mean ranges in matter of about 100 g/cm^2). The beam intensities may reach up to 10^9 π/s or 10^8 μ/s.

An alternative approach to the extraction of *positive* muons uses the fact that a fraction of the π^+ are thermalized in the production target. For these pions the rest system and the laboratory system coincide, hence the decay muons possess full longitudinal spin polarization in the laboratory system. Their momentum depends on the depth of the pion decay site in the target and reaches a maximum of 29.8 MeV/c, corresponding to a kinetic energy of 4.12 MeV or a speed of 0.27 c. They may be used to generate a so-called surface muon beam, which has a limited range (300 mg/cm^2) but may nevertheless be very useful for many purposes, e.g. if the samples are so thin that high-momentum μ^+ will not be thermalized in them or if a *transverse* spin polarization of the implanted muons is required. (The polarization of non-relativistic μ^+ beams can be rotated by standard $\mathbf{E} \times \mathbf{B}$ equipment.) For μ^- an analogous procedure is not possible since most thermalized π^- will be captured before the decay by a target nucleus. An unpolarized slow μ^- beam, though with substantially smaller intensity, may be formed from so-called cloud muons.

The *time structure* of the beams may play an important rôle in the experimental concepts employed. In μ^+SR *time-differential* measurements (cf. Sect. 3) only one muon should be in the sample at any time. For this *quasi-continuous* beams (= continuous on the time-scale of the muon lifetime) are optimal. In a *pulsed* facility such as that at RAL, the pulse width ($\sim 10^{-7}$ s) limits the observable μ^+ rotation frequencies seriously. This virtually precludes, e.g., μ^+SR experiments on strongly ferromagnetic materials (cf. Sect. 3). By contrast, pulsed beams offer highly significant advantages in those *time-integral* measurements that involve large energy deposition in the samples. Examples are experiments in which the samples are exposed to strong light beams (as in some semiconductor measurements) and/or to high-intensity radio-frequency (RF) fields (so-called RFμ^+SR, cf. Sect. 4). The low duty cycle at RAL is particularly advantageous if such experiments are to be carried out at low temperatures.

Since the application of muon/pion research to materials science is a very wide field, a full treatment of this branch of "nuclear condensed-matter physics" is impossible within the scope of the present article. Only a few selected topics will be described. For more detailed accounts of the commonly used methods in the fields of pion/muon condensed-matter physics the reader may consult ref. [1] or [2].

2 Lattice steering (channelling)

2.1 π^+/μ^+ lattice steering

After implantation of fast π^+ into a crystal, scattering by phonons and electrons will reduce the π^+ kinetic energy to thermal energies within a time short compared to their mean lifetime $\tau_\pi = 26$ ns. The rest system of the thermalized pions coincides with the laboratory system; hence the decay μ^+ (Eq. 1a) are emitted *isotropically* with a fixed kinetic energy of 4.12 MeV. Because of the crystalline order the intensity distribution of the μ^+ leaving the sample is *not* isotropic, however, but shows typical patterns of *channelling* and *blocking*. These depend on the π^+ sites at the time of decay and can therefore provide information on these sites [3]. These lattice-steering patterns may be observed by means of a position-sensitive detector but only for muons with energy losses of less than 300 keV, which have undergone only moderate straggling inside the crystals.(In heavy metals this corresponds to a π^+ implantation depth of about 30 μm.) This means that the kinetic energy of the pions in the beam channels (typically 60 MeV to 150 MeV) must be reduced strongly by moderators before the π^+ enter the samples and that the μ^+ detectors must be energy-sensitive.

For π^+/μ^+ lattice-steering measurements monocrystalline samples of very good quality are necessary. Since in these experiments the Lindhard critical angle (cf. Sec. 2.2) is typically $< 1°$, the mosaic spread of the samples must not exceed a few minutes of arc [3]. As a rule, the demands on the angular resolution of the detection apparatus are high, since the desirable distance (6-10 m) from the samples can only rarely be realized in the meson factories.

The low-energy pions (E_{kin} = 5 - 8 MeV) obtained by slowing down the original π^+ beam may be implanted into any material. Under most conditions the radiation damage is negligible. Although the π^+ mass is only about 15 % of the proton mass m_p, with regard to site occupation the behaviour of positive pions in a crystal is similar to that of protons. In several respects, the π^+/μ^+ lattice-steering technique is superior to the various methods developed for studying protons and deuterons in crystals. Since the technique detects single decay muons, there is no solubility limitation. E.g., the decay of pions in W, which has a low hydrogen solubility, was observed from He temperatures up to 2200 K and showed that the decay sites are the tetrahedral interstices of the bcc structure [4]. In the semiconductors Ge and GaAs, whose hydrogen solubility is also low, low-temperature π^+/μ^+ lattice-steering experiments were accomplished over a wide range of temperatures [5,6].

A further general advantage of the use of π^+ or μ^+ as "light hydrogen isotopes" is that these radioactive probes are implanted into the samples *during* the measurements and that they decay without leaving any chemical residue behind. Since the concentrations of the π or μ in the sample are always negligibly small, clustering effects or interactions analogous to proton-proton pairing in hydrogen experiments are excluded.

An essential feature of the μ^+ detectors used in the μ^+/π^+ technique (usually a plastic scintillator disk coupled to several photomultipliers [3,7]) is that they are energy-sensitive and, therefore, allow us to differentiate muons ejected from different depths of the sample. This permits to achieve not only optimal lattice-steering conditions by eliminating the particles with high energy losses but also, by means of a careful energy-dependent data analysis, a separation of channelling and blocking effects that is supplementary to that based on the μ^+ intensity pattern. The separation based on energy discrimination makes use of the fact that the straggling effects in channelling and blocking possess different ranges. This is particularly important for bcc metals, in which the lattice-steering profiles for the tetrahedral and octahedral interstitial sites are very similar and, hence, a clear-cut distinction is possible only though the comparison of the blocking patterns. In α-iron the lattice-steering measurements showed that at very low temperatures the π^+ are mainly in octahedral interstices, whereas at elevated temperatures tetrahedral interstitial sites are occupied [8].

As long as *time-integral* measurements are performed, it is the average intensity but not the time structure of the π^+ beam that matters. *Time-differential* π^+/μ^+ lattice-steering investigations are possible with the pulsed π^+ beam available at PSI in the so-called 17 MHz accelerator mode. In this mode the proton burst separation is about 60 ns, hence on the time scale of the pion lifetime $\tau_\pi = 26$ ns the π^+ pulses are narrow but widely separated. By separating the muons emitted by π^+ that have decayed at different "ages" $t\,(=$ times spent in the crystal after implantation) we may obtain a series of lattice-steering patterns which reflect the evolution of the π^+ localization in the crystal as a function of time. The simplest procedure in such time-differential studies is to distinguish between "young" (individual pion ages $t < 20$ ns, say) and "old" (20 ns $< t <$ 60 ns) pions [9].

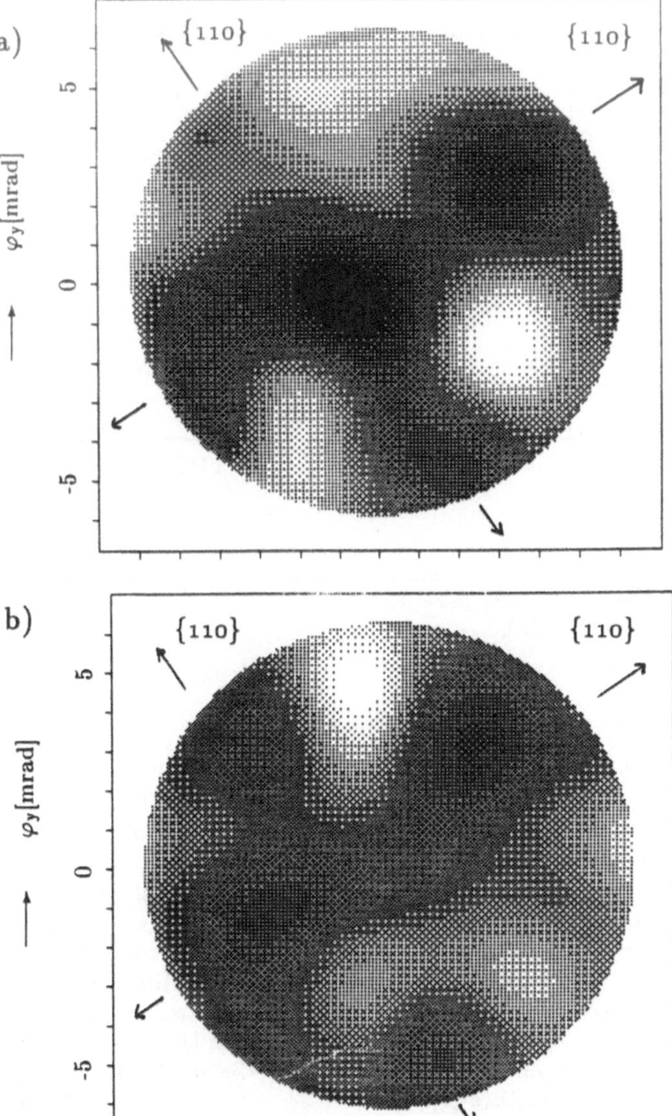

Fig. 1. Muon-flux distribution patterns in the neighbourhood of <100> in time-differential π^+/μ^+ lattice-steering experiments on α-Fe at 25 K. a) "young" π^+ (decay 0-20 ns after implantation), b) "old" π^+ (decay 20 - 60 ns after implantation) [10].

The time window of the π^+/μ^+ lattice-steering measurements is limited on the one end by the achievable time-resolution and on the other end by the lifetime of the π^+, which sets a practical limit of about 100 ns. At the present time, the time resolution of the *time-differential* experiments is about 20 ns. It is mainly determined by the electronics and the need for adequate measuring statistics. In the time-differential measurements based on the 17 MHz mode the time window closes at 60 ns; the extension to about 100 ns by operating in the 10 MHz mode would be very useful.

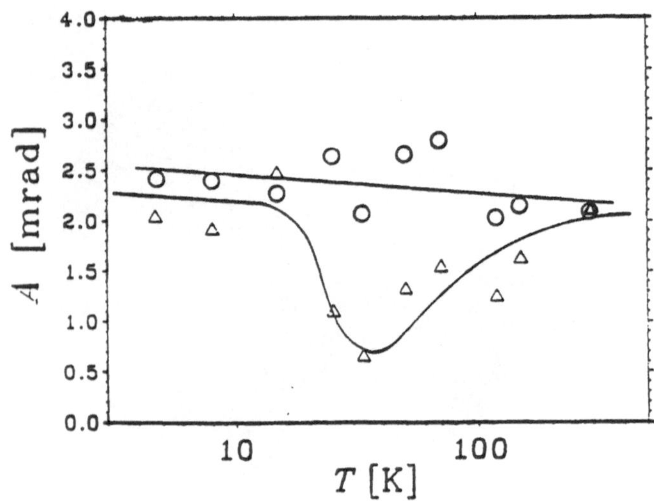

Fig. 2. Area under the normalized muon channelling peak in <100>-orientated α-Fe for "young" π^+ (○) and "old" π^+ (△) (see Fig. 1) [10].

A temperature- and time-dependent π^+/μ^+ lattice-steering study has been carried out on α-Fe [10], which possesses the bcc lattice structure. Fig. 1 shows the lattice-steering patterns for "young" and for "old" π^+ measured on a $\langle 100 \rangle$ orientated sample, Fig. 2 the area under the normalized muon channelling peak as a function of the measuring temperature. The vanishing of the muon channelling peak between 20 K and 70 K for the "old" pions is believed to indicate that the well-localized π^+ states are metastable with regard to states with more extended π^+ wavefunctions [10]. For another example, an investigation of Cu, the reader is referred to the literature [7].

2.2 μ^+/e^+ lattice steering

The positrons emitted by implanted and thermalized μ^+ according to the decay scheme (2a) are *not* monoenergetic, in contrast to the positive muons resulting from the decay (1a). They possess a rather wide continuous energy distribution with a maximum of the kinetic energy of $E^{\mathrm{kin}}_{\mathrm{e}^+,\mathrm{max}} = 52.8$ MeV, corresponding to $pv = 53$ MeV, where p denotes the momentum and v the speed of the emitted charged particles [$pv = 26$ MeV at $E^{\mathrm{kin}}_{\mathrm{e}^+,\mathrm{max}}/2$, to be compared with $E^{\mathrm{kin}}_{\mu^+} = 4.12$ MeV and $pv = 8.0$ MeV in the case of the pion decay (1a)]. Nevertheless, the relativistic positrons resulting from (2a) can be used for lattice-steering experiments [11], although compared with π^+/e^+ experiments a number of difficulties have to be overcome:

(a) Lindhard's critical angle for lattice steering, which is proportional to $(pv)^{-1}$ [12], is by a factor of 3 to 4 smaller than in the π^+/μ^+ case.

(b) As a consequence of (a), the demands on crystal quality are much higher than in π^+/μ^+ studies. The total mosaic spread of the samples to be used in μ^+/e^+ lattice steering should be smaller by at least a factor three, which makes it difficult to employ large metal crystals.

(c) In μ^+/e^+ experiments the depths of the decay events cannot be monitored by measurements of the e^+ energy. Hence, very careful ranging is mandatory in order to limit the μ^+ implantation depth to a few μm. At present this is feasible only with surface muon beams (cf. Sect. 1).

Items (a) and (b) taken together mean that well-developed μ^+/e^+ lattice-steering pattern necessitate much better statistics than π^+/μ^+ experiments. Taking into account that the μ^+ beams are necessarily less intense than the π^+ beams from which they are derived, we see that rather long measuring times are required even under favourable conditions. Such conditions prevail for Si and Ge, of which large highly perfect crystals are available. It is therefore not surprising that, with the exception of the early work on Au [11], μ^+/e^+ lattice steering has so far concentrated on semiconductors [13,14].

In summary, we conclude that at the present time μ^+/e^+ lattice-steering investigations can be recommended only on semiconductors (or, more generally, on valence-bond materials of which large and highly perfect crystals can be grown). The situation might change radically, however, when new high-current meson factories become operational. An increase of the intensity of surface muon beams by a factor ten to hundred will permit routine μ^+/e^+ measurements on high-quality metal crystals. Techniques for growing and testing such samples have recently been developed at the Max-Planck-Institut für Metallforschung for several refractory bcc metals [15].

Future high-current surface μ^+ beam should possess time-structures that are adequate for time-differential μ^+/e^+ work, i.e., consist of pulses separated by not less than about 10^{-5} s that are as narrow as possible compared with the muon lifetime of 2.2 μs. If available with a sufficiently high flux density, such a beam would allow us to obtain information on the site occupancy of implanted μ^+ that completes the information available from the various μSR techniques (see Sect. 3) in an ideal way.

An alternative interesting approach would be to develop a low-energy high-intensity μ^+ beam that allows the implantation of μ^+ in just the right depths (of the order of magnitude of a few μm, depending on the material) to give optimal lattice-steering conditions of the decay positrons.

3 Time-differential muon spin rotation (μ^+SR)

Among the techniques discussed in the present paper, *time-differential muon spin rotation in magnetic fields transverse to the μ^+ polarization* using muons from quasi-continuous beams is the most often used one. A so-called start counter, consisting of a thin plastic scintillator mounted upstream in front of the sample, records the arrival of every single μ^+ entering the sample and starts an "electronic clock". For surface μ^+, whose range is only ~ 1 mm in plastic scintillator materials, the scintillator thickness should not exceed a few tenth of a millimeter; nevertheless, by careful design the generated light suffices for proper timing. If, as is frequently the case, the measurements are done in high applied magnetic fields, the start scintillator has to be coupled to the photomultiplier by a long light guide. This reduces the time resolution somewhat, but resolutions better than 1 ns can nevertheless be achieved without difficulty. So-called e^+ telescopes, also made from plastic scintillators of a few mm thickness and possessing acceptance angles close to $\pi/2$ transverse to the applied magnetic field, serve as stop counters. From the time intervals between the start and stop signals and the information of the decay e^+ of individual muons histograms giving the number of detected events *vs.* μ^+ "ages" and representing typically 10^6 - 10^7 events are constructed. In the first place, they show the exponential decay of the μ^+ with $\tau_\mu = 2.2\mu$s; beyond that they allow us to derive the spin precession frequencies as well as information on the spin-phase coherence of a μ^+ ensemble.

For the time-differential μ^+SR technique it is essential that the relationships between arrival time, decay time, and flight direction can be established unambiguously. Ideally, at any one time at most one μ^+ should be in the sample. If a second μ^+ arrives before the decay e^+ of its predecessor has been detected, the measurement is vetoed. It will be continued only when with a high probability both μ^+ have decayed (say, after 8 μs). This means that for the rate of μ^+ arrivals at the sample an optimum rate of about $(1-2)\cdot 10^4 \mu^+/\mathrm{s}$ exists. The rate of data accumulation in transverse μ^+SR can be substantially increased beyond this rate only by using sufficiently intense pulsed beams and deriving the start signal from the arrival of an entire pulse. (In the present set-up at RAL/ISIS, each pulse contains about 10^4 μ^+.) As mentioned in Sect. 1, in this case the time resolution and hence the highest μ^+ spin precession frequencies that can be detected may seriously be limited by the pulse widths, however.

The muons implanted into the samples are thermalized within times that are very short compared with the μ^+ spin precession periods even in the highest magnetic fields that may act on μ^+ in condensed matter. As a consequence, the μ^+ maintain their spin polarization during the thermalization process. If in the thermalized state all μ^+ experience the same local magnetic field $\mathbf{B}_\mu$, their spins precess with the angular frequency

$$\omega_\mu = \gamma_\mu B_\mu, \tag{3}$$

where $\gamma_\mu = 8.516 \cdot 10^8\ \mathrm{T^{-1}s^{-1}}$ is the gyromagnetic ratio of the muons.

If the "magnetic history" of the μ^+ ensemble is complicated, e.g., if several interstitial sites with different $\mathbf{B}_\mu$ are occupied and/or if μ^+ diffusion or trapping – detrapping effects play a rôle, the extraction of the physical parameters involved may be quite complicated or even be virtually impossible without the help of a good model. As a simple example, let us suppose that the μ^+ can be located in two different interstitial sites with local μ^+ spin precession angular frequencies ω_μ^1 and ω_μ^2. Provided the exchange between these sites is slow enough, Fourier analysis of the events-*vs.*-age histogram gives us two sharp lines at the frequencies ω_μ^1 and ω_μ^2 (static or quasi-static case). If the μ^+ jump very rapidly between the two types of sites, a single relatively sharp Fourier line will be observed at a mean frequency (motional-averaging case). However, at intermediate rates of site changes the Fourier analysis may give a broad distribution of frequencies which may be difficult to discern from the noise.

The ultimate aims of the time-differential transverse μ^+SR are either the determination of the fields $\mathbf{B}_\mu$ acting on the μ^+ at known sites (e.g., in magnetically ordered materials) or the study of μ^+ location and motion in the matrix into which they have been implanted. The determination of magnetic fields in ordered magnetic materials by means of μ^+SR is complimentary to that by neutron diffraction in two respects: The time spent in the samples by μ^+ is several orders of magnitude larger than that spent by thermal neutrons. This has the consequences that μ^+SR gives information on a much longer time scale than neutrons and that, generally speaking, the information may be obtained from fewer μ^+. The second difference is that the information obtained by the μ^+ pertains to certain crystallographic sites only but is, under otherwise comparable circumstances, much more precise than that furnished by neutron diffraction. When comparing the results from neutron diffraction and μ^+SR, we have to keep in mind that from the point of view of magnetism neutrons are *soft* probes whereas the electronic charge of the μ^+ may significantly affect the μ^+ environment.

If we succeed in deducing the "magnetic history" of the μ^+ from μ^+SR measurements, knowledge of the B_μ values at possible μ^+ sites allows us, usually with the help of specific models, to derive information on the μ^+ diffusion and on the sites occupied, say, as a function of temperature [2,16,17]. Beyond that, in non-metals it is also of interest to find out whether the muons are in so-called diamagnetic states [μ^+ or $\mathrm{Mu}^- = (\mu^+e^-e^-)$] or whether they form muonium atoms $\mathrm{Mu} = (\mu^+e^-)$.

The four main groups of systems that are well suited for time-differential μ^+SR measurements are the following: (*i.*) Magnetically ordered materials (ferromagnets, ferrimagnets, antiferromagnets). (*ii.*) Para- or diamagnets with non-zero nuclear magnetic moments. (*iii.*) Type-II superconductors. (*iv.*) Materials in which muonium formation may take place.

(*i.*) If we neglect, for the time being, the contributions from nuclear magnetic moments, the general relationship for the local magnetic field felt by the μ^+ is [16]

$$\mathbf{B}_\mu = \mathbf{B}_{\text{appl}} + \mathbf{B}_{\text{demag}} + \mathbf{B}_{\text{Lorentz}} + \mathbf{B}_{\text{Fermi}} + \mathbf{B}_{\text{dip}}. \quad (4)$$

$\mathbf{B}_{\text{appl}}$ denotes the applied field, $\mathbf{B}_{\text{demag}}$ the demagnetizing field of the sample, $\mathbf{B}_{\text{Lorentz}}$ the Lorentz field, $\mathbf{B}_{\text{Fermi}}$ the contact magnetic field acting on the μ^+ due to the magnetic moments of the conduction or valence electrons at the μ^+ sites, and $\mathbf{B}_{\text{dip}}$ the dipolar magnetic field coming from localized electronic magnetic moments. The last two terms, $\mathbf{B}_{\text{Fermi}}$ and $\mathbf{B}_{\text{dip}}$, may be site-dependent; the remaining terms on the right-hand side of (4) are constant on an atomic scale. If the μ^+ occupy only one type of site, $\mathbf{B}_{\text{Fermi}}$ may be taken as constant, too. $\mathbf{B}_{\text{dip}}$, however, may show a strong site dependence if the μ^+ occupy interstices with reduced point symmetry [18]. We illustrate this by two examples. In ferromagnets with fcc crystal structure (e.g. Ni) both types of interstices (octahedral and tetrahedral) possess cubic point symmetry. This has the consequence that at both sites $\mathbf{B}_{\text{dip}} = 0$. It is therefore not easy to obtain reliable information on the behaviour of μ^+ in Ni from transverse μ^+SR [19]. In bcc crystals, e.g. in ferromagnetic α-Fe or in antiferromagnetic Cr, the interstices possess tetragonal symmetry, however. Here $\mathbf{B}_{\text{dip}}$ can be as high as several tenths of a Tesla. The directions of $\mathbf{B}_\mu$ at neighbouring interstitial sites will in general be different. This may lead to strong variations of $\mathbf{B}_\mu$ on an atomic scale that enables us to obtain detailed information on μ^+ diffusion and location.

Fig. 3 shows a transverse-μ^+SR histogram measured on a Cr single crystal sample at $T = 2.5$ K in the field $\mathbf{B}_{\text{appl}} = 0.26$ T in a $\langle 100 \rangle$ direction. The solid lines are the results of fits to three precession frequencies, *viz.* the frequency $\omega_a/2\pi = \gamma_\mu B_{\text{appl}}/2\pi$ coming from the μ^+ decaying in the sample holder (Cu) and feeling only the applied field, and two frequencies, placed symmetrically with respect to $\omega_a/2\pi$, from μ^+ decaying in well-defined interstices. For further details of the measurements on Cr see ref. [20,21].

A very interesting recent development is that at low temperatures the μ^+SR results on monocrystalline α-Fe are strongly affected by homogeneous elastic strains [22,23]. This is interpreted as strong evidence for large effects of elastic strains on the μ^+ hopping process and therefore of general importance.

(*ii.*) In metals such as Cu, Al, Nb, Ta and many others the dipole-dipole interaction between the magnetic moments of implanted μ^+ and those of the nuclei is strong enough to give rise to spin-phase relaxation that can be measured by the time-differential technique and that is reduced by μ^+ diffusion due to motional averaging. For quasi-continuous beams the finite μ^+ lifetime sets a lower limit to the measurable spin-phase (or transverse) relaxation rate of $(1-2) \cdot 10^4\ \text{s}^{-1}$. This limit can be lowered by using a pulsed μ^+ beam (e.g., in the ISIS facility); however, for reasons discussed above (finite widths of the pulses) the admissible spin precession

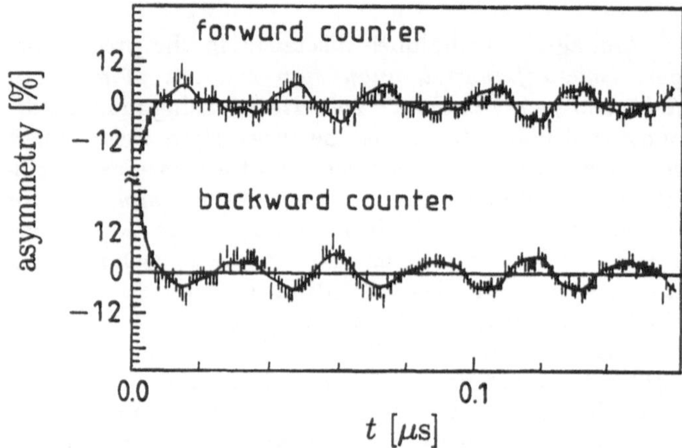

Fig. 3. Time-dependent μ^+ asymmetries measured on monocrystalline $\langle 100 \rangle$-orientated Cr ($B_{\text{appl}} = 0.26$ T, $T = 2.5$ K) after correction for the μ^+ lifetime. The solid lines are the results of a three-frequency fit ($\omega_a/2\pi = 35.0$ MHz, $\omega_1/2\pi = 51.4$ MHz, $\omega_2/2\pi = 118.3$ MHz).

frequencies and hence the applied magnetic fields are limited. This makes it difficult to separate the so-called quadrupolar effects [24] from those of the dipole-dipole interaction.

(*iii.*) The strongest spatial variation of magnetic fields on a scale that is large compared with the interatomic distances but still smaller than or comparable with the diffusion length $(2D_{\mu^+}\tau_\mu)^{1/2}$ of positive muons [$D_{\mu^+} = D_{\mu^+}(T) = \mu^+$ diffusivity] occur in the mixed state (the Shubnikov phase) of Type-II superconductors. Over distances of the order of magnitude of 10^2 nm the magnetic field may vary between values comparable with the thermodynamic critical field B_c of the superconductor and very small values. If the μ^+ diffusivity is low, so that the muons are immobile on that scale, time-differential transverse μ^+SR gives us the distribution function $n(B_\mu)$ of the magnetic field B_μ acting on the muon.

As an example Fig. 4 shows the real Fourier transform $\tilde{P}(\omega)$ of the μ^+ spin polarization *P(t)* (for details see [25]) as determined on plates of high-purity monocrystalline Nb at various temperatures and in various magnetic field B_{appl} in the direction of the plate normals ($\langle 111 \rangle$). Under these conditions the sample is in the mixed state, i.e., ideally it should contain a *periodic* arrangement of flux lines running perpendicular to the plane of the plates, the so-called *Abrikosov lattice.* B_μ possesses the periodicity of the flux-line lattice, hence $n(B_\mu)$ should show at least three van Hove singularities, *viz.* a logarithmic singularity at an intermediate B_μ value corresponding to a saddle point in the spatial B_μ variation (this gives rise to the sharp peaks in the $n(B_\mu)$ distributions of Fig. 4) and two step singularities at the highest B_μ (occurring in the flux line centres) and at the lowest B_μ (occurring somewhere in between the flux lines) [26]. The heights of these steps tend to zero as 0 K

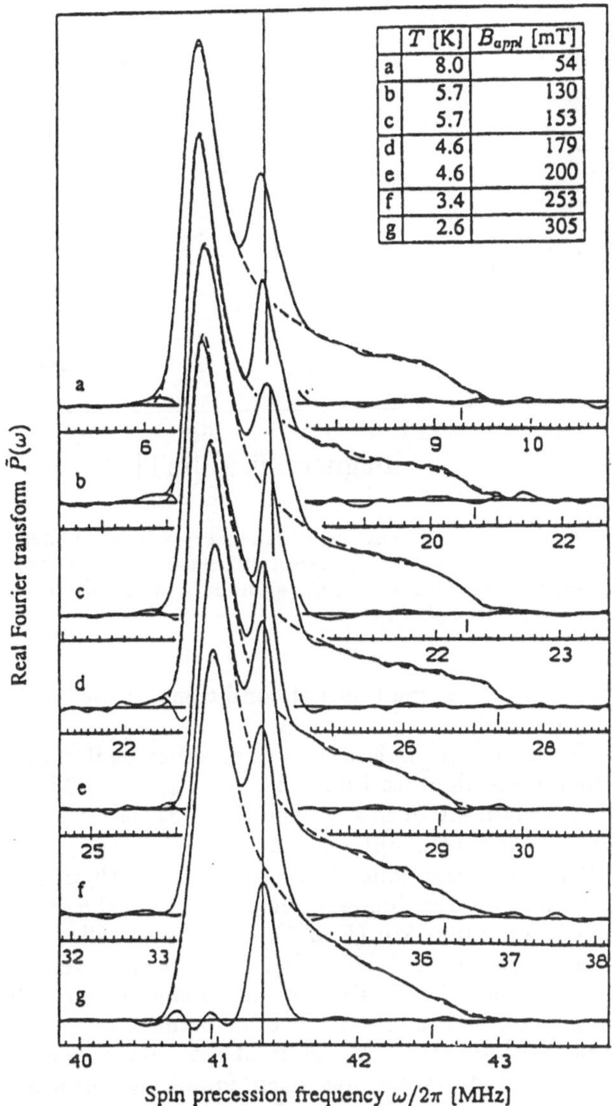

Fig. 4. Least-squares fits of theoretical field distributions in superconducting Nb calculated for Gaussian broadening of the van Hove singularities (dashed curves) to the real Fourier transforms of the experimental μ^+ polarization (solid curves). The vertical lines indicate the average precession frequencies $\bar{\omega}$. The peaks near $\bar{\omega}$ originating from μ^+ stopped in the sample holder were excluded from the fits. The positions of the van Hove singularities are indicated by vertical dashes above the frequency scales. The solid curves wiggling around $\tilde{P} = 0$ show the deviation of the fits [25].

is approached. Moreover, the low-B_μ step is close to the saddle-point singularity.

Under ideal conditions (perfect flux-line lattice, immobile μ^+) $\tilde{P}(\omega)$ should be proportional to $n(\omega/\gamma_\mu)$. If the artifact at $\omega_a = \gamma_\mu B_{appl}$ (due to μ^+ decaying in the sample holder) is disregarded, the agreement between the calculated $n(B_\mu)$ curves and the experimental data is excellent except for the neighbourhoods of the van Hove singularities. In these, however, the finite D_μ as well as imperfections in the flux-line lattice make themselves felt [25,26,27]. Separation of the two effects is not straight-forward but indicates nevertheless that the μ^+ diffusivity in the cores of the flux-lines in high-purity Nb at 8 K is of the order of magnitude 10^{-10} m^2s^{-1}, with an upper limit of $4{\cdot}10^{-10}$ m^2s^{-1}.

High-T_c superconductors are Type-II superconductors. Thus the technique just described is applicable in principle. However, since the samples of a similar quality as that of the Nb sample mentioned above are not available, a similarly detailed analysis has not yet been performed. A widely used procedure is to deduce the London penetration depth of high-T_c superconductors, which is difficult to measure by other techniques, from the width of the $n(B_\mu)$ distribution [29].

(*iv.*) Muonium, the bound state of μ^+ and e^- resembling the hydrogen atom, may be formed in organic materials, semiconductors, or ionic crystals but not in metals. Although the spatial part of its wavefunction can be strongly different from that of vacuum muonium, the spin wavefunction remains essentially the same. Apart from situations in which the muonium in matter is much more extended than in vacuum, the frequency of the "hyperfine splitting" between the triplet and the singlet states is too high to be detected directly in transverse measurements. The frequencies corresponding to the Mu intratriplet transitions in intermediate magnetic fields, however, are in the range of time-differential μ^+SR experiments [30].

4 Radio-frequency muon spin resonance (RFμ^+SR)

The time-differential transverse μ^+SR technique briefly described in Sect.3 loses its power if the magnetic fields $\mathbf{B}_\mu$ acting on the μ^+ vary strongly on a time scale comparable with the μ^+ spin-precession period $2\pi/\omega_\mu = 2\pi/\gamma_\mu B_\mu$. Then the μ^+ spin-phase coherence may be partly or wholly lost. This has the consequences that a substantial fraction of the implanted μ^+ will contribute to a time-independent background rather than to analysable spin precession signals. Such a loss of phase coherence may result from μ^+ trapping by lattice vacancies or other defects or from muonium (Mu) formation and may prevent the detection of well defined B_μ acting on the μ^+ magnetic moments.

This drawback can be overcome by radio-frequency (RF) μ^+SR. In contrast to time-differential μ^+SR, this technique does not require μ^+ spin-phase coherence but only that the μ^+ should stay in definite magnetic environments for several RF periods. The measuring set-up is longitudinal, i. e. the axis of the constant applied magnetic field $\mathbf{B}_{appl}$, the polarization axis of the μ^+ beam, and the detection axis of the decay asymmetry should coincide. The RF magnetic field $\mathbf{B}_1$ is transverse to this axis. If its frequency $\omega_{RF}/2\pi$ is properly chosen, it will induce transitions between the two μ^+ Zeeman levels in the $\mathbf{B}_\mu$ field. Since the μ^+ spin distribution of freshly implanted polarized μ^+ is far from thermal equilibrium (which corresponds to an almost unpolarized μ^+ ensemble), the RF-induced transitions are easily detectable by comparing the forward and backward e^+ counters. [The forward (f) = downstream, backward (b) = upstream positions are defined with respect to the beam momentum. The μ^+ spins are aligned opposite to the μ^+ momentum.] Since the RF field may be periodically switched on and off, difference measurements allow us to study rather small effects.

In terms of the time-integral count rates f and b in the two counters with RF on (subscript "on") or off (subscript "off") the RF asymmetry may be defined as

$$A_{\mathrm{RF}}^{(\mathrm{a})} = \frac{f_{\mathrm{on}} - f_{\mathrm{off}}}{f_{\mathrm{on}} + f_{\mathrm{off}}} - \frac{b_{\mathrm{on}} - b_{\mathrm{off}}}{b_{\mathrm{on}} + b_{\mathrm{off}}} \tag{5a}$$

or

$$A_{\mathrm{RF}}^{(\mathrm{a})} = \frac{f_{\mathrm{on}} - b_{\mathrm{on}}}{f_{\mathrm{on}} + b_{\mathrm{on}}} - \frac{f_{\mathrm{off}} - b_{\mathrm{off}}}{f_{\mathrm{off}} + b_{\mathrm{off}}}. \tag{5b}$$

The difference between these two RF asymmetries are of second order and thus usually very small. Should some instability influence the result, however, the asymmetry $A_{\mathrm{RF}}^{(\mathrm{b})}$ is the better choice if it is the μ^+ beam intensity that is unstable on the time scale of the on-off switching period, whereas the use of $A_{\mathrm{RF}}^{(\mathrm{a})}$ is better if the instability is related to the focusing of the μ^+ beam or the performance of the e^+ counters.

The condition for RF-induced transitions between the Zeeman levels reads

$$\omega_{\mathrm{RF}} = \gamma_\mu B_\mu^{\|}, \tag{6}$$

where $B_\mu^{\|}$ is the longitudinal component of $\mathbf{B}_\mu$. In a ferromagnetic sample $B_\mu^{\|}$ is usually not directly known, but in non-magnetic materials B_μ and $B_\mu^{\|}$ are equal in the set-up described above. If an RF resonance has been found, the linewidth may be obtained from a B_{appl} or an ω_{RF} scan. Which scanning method is preferable depends on how the B_1 field is generated. If the sample is inside the coil of a resonance circuit, the amplitude of B_1 is magnified by $Q^{1/2}$, where Q is the quality factor of the resonance. Such an arrangement gives access to high B_1 fields with relatively inexpensive RF amplifiers. The drawback of the resonance arrangement is that it is unsuitable for frequency scans, which in most cases is not a serious limitation. Off-resonance set-ups permit both frequency and magnetic-field scans.

Although RFμ^+SR experiments have been performed on a quasi-continuous μ^+ beam [31,32,33], a pulsed μ^+ beam has several advantages. The low duty cycle of such a beam makes it possible to operate the RF set-up in a low-duty-cycle mode, too. This facilitates the construction and use of cryostats as well as low-temperature measurements with large B_1 amplitudes. Moreover, low-duty-cycle amplifiers are much less expensive. Since the thermal characteristic time of the samples are usually much longer than the beam period, a pulsed beam allows us to perform these "RF-off" measurements under

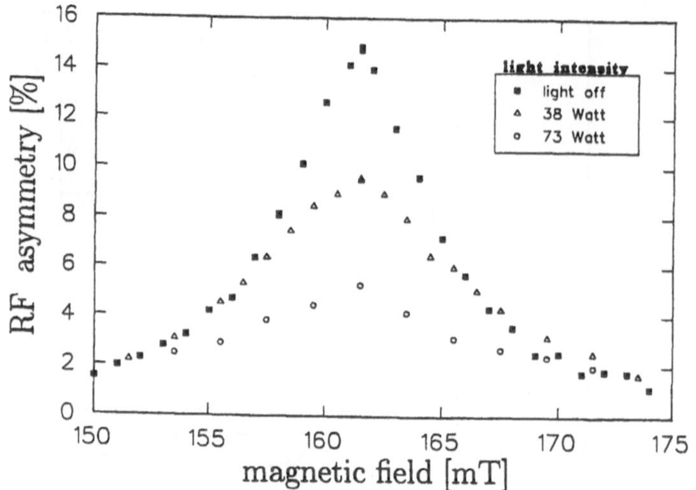

Fig. 5. RFμ^+SR resonance lines in monocrystalline Si at 300 K as detected using a magnetic-field scan with fixed $\omega_{\mathrm{RF}}/2\pi = 22$ MHz. Illumination with infrared light results in a strong decrease and broadening of the μ^+ line.

essentially the same thermal conditions as the "RF-on" measurements.

During the μ^+ pulses the samples may be illuminated with intense light as illustrated in Figs. 5-7. This was taken advantage of in a recent RFμ^+SR study of high-purity Si crystals employing the pulsed surface μ^+ beam at RAL [34]. Resonance lines associated with all known μ^+ states, *viz.* the diamagnetic μ^+ (Fig. 5), the so-called normal muonium Mu (Fig. 6), and the so-called anomalous muonium Mu* [30] (Fig. 7), were observed. Figs. 5-7 show that irradiation by an intense infrared light source reduces the intensities of all signals. Surprisingly, so far no indication of a shift of stability between the different muon states has been found. At present the nature of the illumination effect is not understood. It is not even clear whether we are dealing with a charge-density or photonic effect. Additional experiments are in progress.

RFμ^+SR in *metals* is beset by the difficulty that the RF skin depth can be smaller than the implantation distance even if a surface μ^+ beam is employed [35]. By careful ranging and/or by employing samples consisting of sets of parallel foils this difficulty can be overcome [31,36].

5 Muon spin rotation experiments with negative muons (μ^-SR)

The behaviour of negative muons implanted into condensed matter is radically different from that of μ^+. In contrast to positive muons, which are repelled by the positively charged ion cores and are therefore located in interstices, the negative muons are attracted by the nuclei. Within a very short time after implantation they will be captured in an atomic orbit. Estimates indicate that the μ^- are first bound in an orbit with main

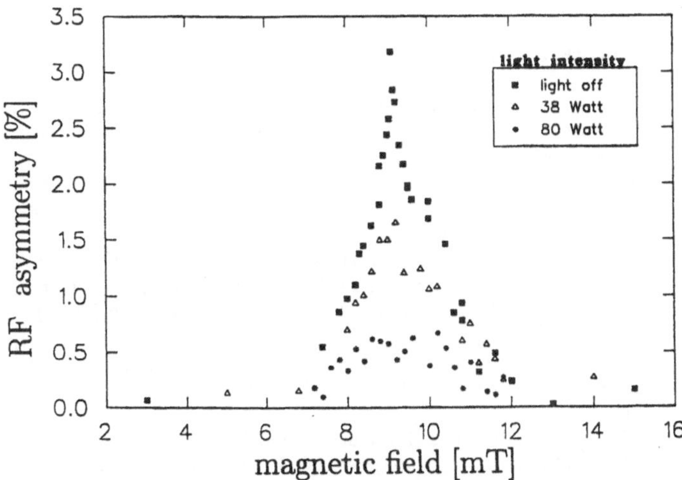

Fig. 6. $RF\mu^+SR$ resonance lines of triplet isotropic muonium (Mu) in monocrystalline Si at 300 K as detected using a magnetic field scan with fixed $\omega_{RF}/2\pi = 128$ MHz. Illumination with infrared light results in a decrease of the μ^+ line. Note that the spectrum is a superposition of three lines, which cannot be resolved at that low frequency.

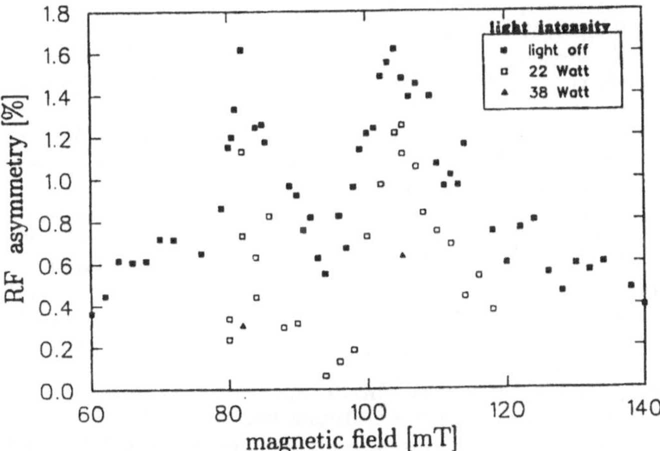

Fig. 7. $RF\mu^+SR$ resonance lines of Mu* (anomalous muonium) in monocrystalline Si at 22 K as detected using a magnetic field scan with fixed $\omega_{RF}/2\pi = 57$ MHz. Illumination with infrared light reduces the Mu* RF amplitude.

quantum number $n \simeq 14$. From such an orbit the μ^- gradually cascade down to the 1s orbit under emission of X-ray or gamma quanta, i.e. by radiative transitions, or by ejection of e^- in Auger processes. During the cascading, which is complete in about $\sim 10^{-14}$ s, electronic shells are heavily disturbed, but the electrons will very quickly return to the orbits corresponding to an atom with charge number $Z - 1$ and the neutrality of the atom will be restored. During the μ^- cascading and the refilling of the e^- orbits the muons lose most of their spin polarization. The residue, typically 15 - 20% of the original polarization, may still be sufficient to perform spin-rotation (μ^-SR) experiments.

In the case of high-Z atoms the radii of the 1s μ^- orbits are comparable with or even smaller than the nuclear radii. This means that the μ^- can be captured by the nucleus and that they may undergo the reaction

$$\mu^- + \mathrm{p} = \mathrm{n} + \nu_\mu \qquad (7)$$

(p = proton, n = neutron) instead of the decay (2b). For nuclear charge numbers $Z > 10$ the reaction (7) leads to a reduction of the μ^- lifetime, which levels off at 80 ns for $Z > 40$ [37,38]. In low-Z materials the majority of the μ^- do decay according to (2b). In this case we may obtain information on the μ^- spin polarization at the time of decay (which may be quite different from the residual polarization at the end of the capture process) from the distribution of the flight directions of the decay e^-, which are easily detected because of their high kinetic energy. It should be mentioned that the nuclear reaction (7) can lead to a slight activation of the sample, the sample holder, the cryostat etc. Although usually no long-lived radioactive isotopes are generated in low-Z samples, care is required in the disassembly of the measuring set-up.

As an example for the application of μ^-SR to solid-state physics we describe a recent experiment on the semiconducting element Si ($Z = 14$). A Si atom that has captured a μ^- in an 1s orbit acts as a quasi-aluminium (q-Al) atom owing to the screening of one nuclear charge by the μ^-. The orbiting μ^- is subject to the hyperfine interaction with the electrons of the q-Al atom (or ion). The polarization of the μ^- at the time of its decay thus contains information on the charge state of the q-Al.

According to the μ^-SR signal detected in an applied field $B_{appl} = 0.04$ T (Fig.8), at room temperature there is no depolarization effect that might be attributed to the above-mentioned hyperfine interaction. Fig. 9 shows that the depolarization increases with decreasing temperature. This cannot be due to processes occurring during the cascading, since these should be temperature-independent. On the other hand, the observed effective activation energy of about 1 meV is much too small to be attributed to electronic processes, e.g. to the filling of acceptor levels. The explanation presumably is that the above-mentioned disturbance of the electronic shells during the cascading is so strong that the small γ recoil energies (= 3.11 eV for the 2p-1s transition of the muonic Si) suffice to cause displacements of the q-Al "quasi nuclei" from their original lattice sites that at low temperatures last at least for times of the order of magnitude of τ_μ. According to this interpretation the positive temperature variation of the spin precession signal shown in Fig. 9. reflects the temperature dependence of the rate of return of the q-Al nuclei to their lattice sites, which is accompanied by changes in the electronic structure.

An attractive feature of the μ^-SR technique is that the detection of the γ quanta emitted during the cascading in coincidence with the start and stop signals allows us to distinguish between muons captured by different chemical constituents of compounds or alloys. The required energy-selective detection of the characteristic

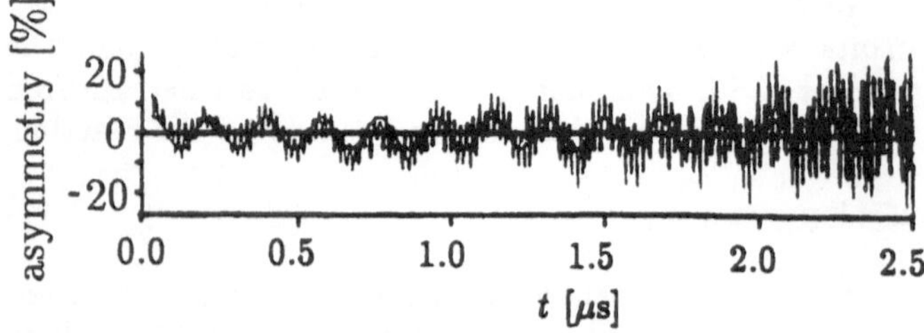

Fig. 8. μ^-SR signal of n-doped Si ($2.5 \cdot 10^{14}$ cm^{-3} phosphorous) in an applied magnetic field $B_{appl} = 40$ mT at room temperature after background subtraction and correction for μ^- lifetime [39].

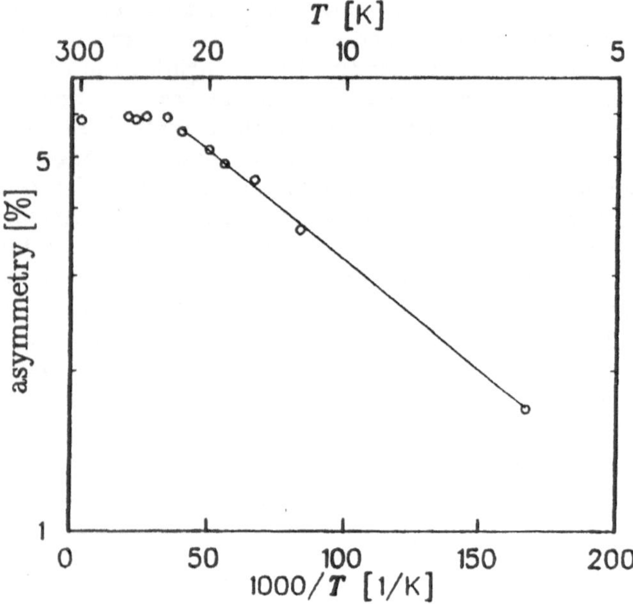

Fig. 9. Arrhenius plot of the measured μ^- asymmetries of n-doped Si ($2.5 \cdot 10^{14}$ cm^{-3} phosphorous) in an applied magnetic field $B_{appl} = 40$ mT. The straight line fitted to the data below 30 K corresponds to an activation enthalpy of ≈ 1 meV [39].

γ photons (with energies between about 50 keV and 500 keV) can be done by, e.g., NaI scintillator counters [39]. This coincidence technique should permit element-specific μ^-SR measurements in alloys and compounds consisting of low-Z elements, such as SiC.

Acknowledgement. The present paper reflects mainly ideas, experiences, failures, successes, and results of the Stuttgart "nuclear condensed-matter physics" group. In the seventeen years since the activities described in the paper were started, so many members of this group have contributed in essential ways that we cannot list names. We do wish, however, to mention the name of Karl Maier, without whose physical insight and experimental skills the field could not have developed so widely and so rapidly. We should also like to acknowledge the help of the personnel at the two accelerator facilities at which most of the experiments were carried out, *viz.* PSI (formerly SIN) at Villigen, Switzerland, and RAL/ISIS at Chilton, England, as well as the financial support of the Bundesministerium für Forschung und Technologie (BMFT) in Bonn throughout the years.

References

1 Schenck, A., Muon Spin Rotation Spectroscopy, Bristol: Adam Hilger 1985

2 Seeger, A., in: Muons and Pions in Materials Research, eds.: Chappert, J., Grynszpan, R. I., Amsterdam: North-Holland 1984, p. 251

3 Maier, K., in: Nuclear Physics Methods in Materials Research, eds.: Bethge, K., Baumann, H., Jex, H., Rauch, F., Braunschweig: Vieweg 1980, p. 264

4 Fabritius, G., Flik, G., Golczewski, J., Herlach, D., Jünemann, G., Krenke, M., Maier, K., Pathak, A. P., Rempp, H., Seeger, A., Sigle, W., Widmann, E.: Hyperfine Inter. **31** 229 (1986)

5 Flik, G., Bradbury, J. N., Cooke, D. W., Heffner, R. H., Leon, M., Paciotti, M. A., Schillaci, M. E., Maier, K., Rempp, H., Reidy, J. J., Boekema, C., Daniel, H.: Phys. Rev. Lett. **57** 563 (1986)

6 Flik, G., Bradbury, J. N., Cooke, D. W., Heffner, R. H., Leon, M., Paciotti, M. A., Schillaci, M. E., Maier, K., Rempp, H., Boekema, C., Reidy, J. J., Daniel, H.: Hyperfine Inter. **32** 595 (1986)

7 Fabritius, G., Feldmann, R., Flik, G., Gaul, T., Heinemann, V., Herlach, D., Krenke, M., Maier, K., Seeger, A., Staiger, W., Widmann, E.: Nuclear Instr. and Meth. **B33** 49 (1988)

8 Staiger, W., Widmann, E., Herlach, D., Krenke, M., Maier, K., Major, J., Seeger, A.: PSI Annual Report 1990, Annex III, Villigen/Switzerland 1991, 12

9 Jünemann, G., Carstanjen, H. D., Flik, G., Herlach, D., Maier, K., Rempp, H., Seeger, A., Sigle, W.: Hyperfine Inter. **17-19** 959 (1984)

10 Staiger, W., Widmann, E., Connell, S., Fabritius, G., Maier, K., Seeger, A.: Hyperfine Inter. **64** 701 (1990)

11 Maier, K., Flik, G., Herlach, D., Jünemann, G., Seeger, A., Carstanjen, H. D.: Physics Lett. **86A** 126 (1981)

12 Carstanjen, H.-D., Seeger, A., in: Muons and Pions in Materials Research, eds.: Chappert, J., Grynszpan, R. I., Amsterdam: North-Holland 1984, p. 293

13 Simmler, H., Eschle, P., Keller, H., Kündig, W., Odermatt, W., Patterson, B. D., Pümpin, B., Savic, I. M., Schneider, J. W., Straumann, U., Truöl, P.: Hyperfine Inter. **64** 535 (1990)

14 Simmler, H., Eschle, P., Keller, H., Kündig, W., Odermatt, W., Patterson, B. D., Savic, I. M., Schneider, J. W., Stäuble-Pümpin, B., Straumann, U., Truöl, P.: Mat. Sci. Forum **83-87** 1121 (1992)

15 Seeger, A., in: Relativistic Channeling, eds.: Carrigan, R. A., Ellison, J. A., New York: Plenum 1987, p. 423

16 Seeger, A., in: Hydrogen in Metals I, eds.: Alefeld, G., Völkl, J., Berlin: Springer 1978, p. 771

17 Seeger, A., Monachesi, P.: Phil. Mag. B **46** 283 (1982)

18 Kronmüller, H., Hilzinger, H. R., Monachesi, P., Seeger, A.: Appl. Phys. **18** 183 (1979)

19 Seeger, A., Schimmele, L., in: Perspectives of Meson Science, eds.: Yamazaki, T., Nakai, K., Nagamine, K., Amsterdam: Elsevier, in press

20 Major, J., Mundy, J., Schmolz, M., Seeger, A., Döring, K. P., Fürderer, K., Gladisch, M., Herlach, D., Majer, G.: Hyperfine Inter. **31** 259 (1988)

21 Templ, W., Hampele, M., Herlach, D., Major, J., Mundy, J., Seeger, A., Staiger, W.: Hyperfine Inter. **64** 679 (1990)

22 Fritzsche, A., Hampele, M., Herlach, D., Maier, K., Major, J., Schimmele, L., Seeger, A., Staiger, W., Templ, W., Baines, C.: Hyperfine Inter. **64** 691 (1990)

23 Schimmele, L., Seeger, A., Staiger, W., Templ, W., Baines, C., Fritzsche, A., Hampele, M., Herlach, D., Maier, K., Major, J.: Hyperfine Inter. **64** 671 (1990)

24 Hartmann, O.: Phys. Rev. Lett. **39** 832 (1977)

25 Herlach, D., Majer, G., Major, J., Rosenkranz, J., Schmolz, M., Schwarz, W., Seeger, A., Templ, W., Brandt, E. H., Essmann, U., Fürderer, K., Gladisch, M.: Hyperfine Inter. **63** 41 (1990)

26 Brandt, E. H., Seeger, A.: Adv. Phys. **35** 189 (1986)

27 Seeger, A.: Phys. Lett. **74A** 259 (1979)

28 Uemura, Y. J., Le, L. P., Luke, G. M., Sternlieb, B. J., Brewer, J. H., Kadono, R., Kiefl, R. F., Kreitzman, S. R., Riseman, T. M.: Physica C **162-164** 857 (1989)

29 Pümpin, B., Keller, H., Kündig, W., Odermatt, W., Savic, I. M., Schneider, J. W., Simmler, H., Kaldis, E., Rusiecki, S., Maeno, Y., and Rossel, C.: Phys. Rev. **B42** 8019 (1990)

30 Patterson, B. D.: Rev. Mod. Phys. **60** 69 (1988)

31 Majer, G., Messer, R., Seeger, A., Templ, W., Fürderer, K., Gladisch, M., Herlach, D.: Phil. Mag. Lett. **57** 57 (1988)

32 Kreitzman, S. R.: Hyperfine Inter. **65** 1055 (1990)

33 Lichti, R. L., Lamp, C. D., Kreitzman, S. R., Kiefl, R. F., Schneider, J. W., Niedermayer, Ch., Chow, K., Pfiz, T., Estle, T.L., Dodds, S. A., Hitti, B., DuVarney, R. C.: Mat. Sci. Forum **83-87** 1115 (1992)

34 Carne, A., Cox, S. F. J., Eaton, G. H., Scott, C. A.: Hyperfine Inter. **65** 1175 (1990)

35 Brandt, E. H., Messer, R.: phys. stat. sol. (b) **144** 343 (1987)

36 Hampele, M., Herlach, D., Kratzer, A., Majer, G., Major, J., Raich, H.-P., Roth, R., Scott, C. A., Seeger, A., Templ, W., Blanz, M. Cox, S. F. J., Fürderer, K.: Hyperfine Inter. **65** 1081 (1990)

37 Ford, K. W., Hughes, V. W., Wills, J. G.: Phys. Rev. **129** 194 (1963)

38 Yamazaki, T., in: Muons and Pions in Materials Research, eds. Chappert, J., Grynszpan, R. J., Elsevier Publishers B.V. 1984, p. 241

39 Koch, M., Maier, K., Major, J., Seeger, A., Sigle, W., Staiger, W., Templ, W., Widmann, E., Abela, R., Claus, V., Hampele, M., Herlach, D.: Hyperfine Inter. **65** 1039 (1990)

This article was processed using Springer-Verlag TEX Z.Physik C macro package 1991
and the AMS fonts, developed by the American Mathematical Society.

Z. Phys. C - Particles and Fields 56, S280–S284 (1992)

Zeitschrift für Physik C Particles and Fields

Ephemeral and/or coloured muonic hydrogen atoms

Hubert Schneuwly

Institut de Physique de l'Université, CH-1700 Fribourg, Switzerland

10-September-1991

Abstract. The unexpected and, in conventional terms, yet unexplained experimental results, obtained in systematic measurements of muon transfer from muonic hydrogen to sulphur and oxygen of sulphur dioxide, seem to violate the principle of reproducibility of muon transfer data. With the hypothesis of ephemeral muonic hydrogen atoms, the number of different hydrogen atoms can be reduced from four to two. This hypothesis does, however, not help to interpret the transfer data to helium, neon and argon, where the muonic hydrogen atoms seem to wear colours.

1 Introduction

A muonic hydrogen atom, which due to its small size and electric neutrality behaves in matter almost like a neutron, can irreversibly transfer its muon in collisions to nuclei of other elements. First experimental results [1,2] and the first theoretical paper [3] on this subject were published already thirty years ago. After the series of measurements performed at JINR [4] and at CERN [5-7] in the sixties, it seemed that the only interesting problem left open was a precise value of the transfer rate from muonic deuterium to helium, needed for the future of muon catalyzed fusion experiments [8,9]. In reality, the experimental muon transfer rates are a puzzle which is almost as old as the first muon transfer measurements to neon [10].

With the idea in mind to study later muon transfer to helium from excited states of muonic deuterium, we have measured muon transfer from the ground state of μp atoms to sulphur dioxide [11,12]. Indeed, muon transfer to the partners of a polar molecule should proceed differently from excited states than from the ground state, in particular the transfer rate to the positive ion should be greater than to the negative ion [13].

2 Principle of muon transfer measurements

Transfer rates can be determined by measuring the lifetime τ of the μp atom in hydrogen containing a small admixture of an element of charge number Z, e. g. a H_2+Z gas mixture

$$1/\tau = \lambda = \lambda_0 + \lambda_{pp} + \lambda_Z \tag{1}$$

where λ_0 is the decay rate of the muon bound to the hydrogen nucleus, λ_{pp} the formation rate of pμp molecules and λ_Z the transfer rate to element with charge number Z. In order to compare transfer rates measured under different experimental conditions, one defines a rate Λ_i, reduced to the atomic density of liquid hydrogen (ρ_0= $4.25\cdot10^{22}$ cm^{-3})

$$\Lambda_i = \frac{\rho_0}{N_i}\lambda_i \tag{2}$$

where N_i is the density of the i-th component in the target (λ_0=$0.455\cdot10^6$ s^{-1}, Λ_{pp}=$1.7\cdot10^6$ s^{-1}). The lifetime of the μp atom can be relatively easily measured. Indeed, the μp atom transfers its muon to an excited state of the element Z and the time distribution of the muonic X-rays of the deexcitation of μZ^* to the ground state has therefore the corresponding characteristic exponential structure (cf. Fig.1).

In a $H_2 + SO_2$ mixture, the lifetime of the μp atom can be measured through the time distributions of the muonic X-rays of either sulphur or oxygen:

$$1/\tau = \lambda = \lambda_0 + \lambda_{pp} + \lambda_S + \lambda_O \tag{3}$$

With the definition of the reduced transfer rate, Λ_{SO_2}, to the SO_2 molecule,

$$\Lambda_{SO_2} = \Lambda_S + 2\Lambda_O \tag{4}$$

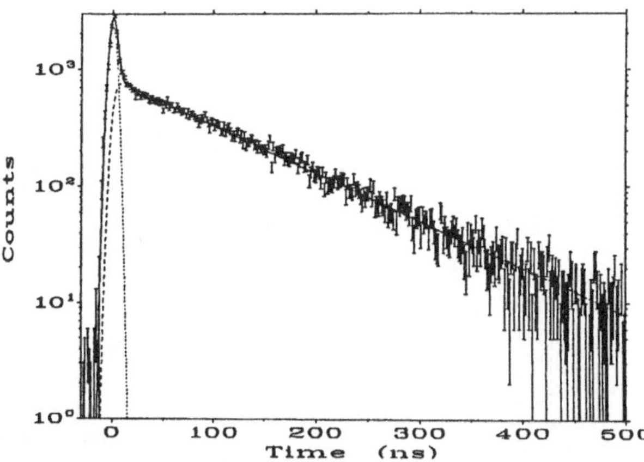

Fig. 1. Time distribution of the muonic sulphur μS(2-1) X-rays measured in a H_2+0.4%SO_2 gas mixture at 15 bar. The μp lifetime deduced from the exponential structure was τ_S = 107 ns. The prompt peak corresponds essentially to muons directly captured in sulphur of SO_2

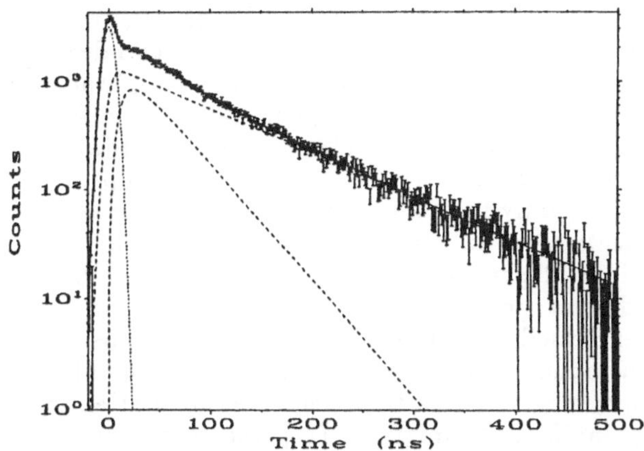

Fig. 2. Time distribution of the muonic oxygen $\mu O(2-1)$ X-rays measured in a $H_2 + 0.4\%$ SO_2 gas mixture at 15 bar. The flatter component has the same decay time $\tau_1 = \tau_S$ as the sulphur X-rays. The steeper component with a decay time τ_2 is attributed to transfer from μp atoms called "black" μp atoms.

Λ_O is the reduced per atom transfer rate to oxygen.

3 "Black" and "white" muonic hydrogen atoms

To our great surprise, the time distribution of the muonic oxygen X-rays had a completely unexpected structure (Fig.2), which could be fitted in its delayed part by using three exponential functions [12]. All four observed muonic X-ray transitions, namely $\mu O(2-1)$, $\mu O(3-1)$, $\mu O(4-1)$ and $\mu O(5-1)$, had the same structure, and the intensity pattern of the Lyman series correspondend to the theoretical prediction for muon transfer from the ground state of μp atoms [3,12,14]. The time distributions of the muonic sulphur X-rays of the Lyman series as well as of the Balmer series had all the same single-exponential structure with the same decay constant as the $\mu S(2-1)$ transition, and the intensity pattern of both series agreed with the prediction of muon transfer from the μp ground state [12].

Up to now, no convincing explanation of the observed phenomena has been proposed, though most of experts attribute the curious structure of the oxygen time spectra to a dependence of the transfer rate upon the energy of epithermal μp atoms. Under the particular assumptions that epithermal μp atoms transfer their muon exclusively to oxygen and get never thermalized, the measured time spectra can, indeed, be reproduced. In practice, this is however equivalent to assume the existence of two kinds of μp atoms: "black" ones which transfer their muon only to oxygen and "white" or normal ones which transfer their muon to both oxygen and sulphur [15].

The μp lifetime deduced from the time distribution of the muonic sulphur X-rays was always compatible with the longer component of the oxygen time spectra. From this lifetime, one determines the reduced transfer rate to sulphur dioxide, Λ_{SO_2}, given in Table 1.

Table 1. Time constants and reduced transfer rates, measured in $H_2 + SO_2$ gas mixtures with different SO_2 concentrations and at different pressures. The quoted values result partly from preliminary analyses.

Conc.	Press. [bar]	τ_S [ns]	Λ_{SO_2} [$\cdot 10^{11} s^{-1}$]	τ_2 [ns]	Λ_O^* [$\cdot 10^{11} s^{-1}$]	Ref.
0.2%	15	198.5	2.52	62.7	4.33	[16]
0.2%	10	257.7	2.54	91.1	4.27	[16]
0.2%	15	198.0	2.55	64.8	4.21	[16]
0.4%	15	107.0	2.51	40.9	3.42	[12]
0.4%	13	119.0	2.51	48.0	3.25	[12]
0.6%	15	69.8	2.60	31.0	2.91	[17]
0.6%	10	103.0	2.55	45.1	3.01	[17]

From the shorter of the two decay times observed in the oxygen spectra, and which is absent in the sulphur spectra, one can in analogy determine a reduced transfer rate to oxygen, Λ_O^* (Table 1). This transfer rate Λ_O^* from the "black" μp atoms to oxygen is much higher than the transfer rate Λ_O from the "white" or normal μp atoms, which is $\Lambda_O = 0.76 \cdot 10^{11} s^{-1}$ [12].

In a series of muon transfer measurements, which have been performed during the last five years at PSI, in gaseous $H_2 + SO_2$ mixtures at different pressures and different SO_2 concentrations (cf. Table 1), one observed that the reduced transfer rate to sulphur dioxide, Λ_{SO_2}, is a reproducible rate. The individual reduced rates, Λ_S and Λ_O, are also reproduced. The corresponding μp atoms are normal μp atoms in the sense that they behave as expected.

In all investigated mixtures, the time distributions of the muonic oxygen X-rays showed the same double-exponential decay structure, where the flatter component had the same decay time as the corresponding sulphur time distribution, $\tau_1 = \tau_S$. The steeeper component, interpreted as resulting from muons transferred from "black" μp atoms only to oxygen, with a decay time τ_2, was always present with comparable amplitudes. The reduced transfer rates to oxygen, Λ^*_O, deduced from this steeper component, are only partially reproduced. The reduced muon transfer rate from "black" μp atoms to oxygen is only the same in mixtures with the same SO_2 concentration. It is as if the reduced transfer rate Λ^*_O would be dependent on the SO_2 concentration but independent of the pressure. To try to save the basic principle of reproducibility of physical data, one may be tempted to introduce "coloured" μp atoms instead of "black" μp atoms, where unlike to other experimental situations, e. g. in $H_2 + Ar$ mixtures [15], the colour depends on the SO_2 concentration.

4 "Ephemeral" muonic hydrogen atoms

Another attempt to try to save the principle of reproducibility of the muon transfer data is to introduce "ephemeral" muonic hydrogen atoms [18]. Ephemeral μp atoms are black μp atoms which, in addition, disappear with a rate λ_ρ, which is proportional to the hydrogen density. The lifetime, τ_2, of ephemeral μp atoms is then defined by

$$1/\tau = \lambda_2 = \lambda_0 + \lambda_{pp} + \lambda'_O + \lambda_\rho \tag{5}$$

where λ'_O is again the transfer rate to oxygen. It is then possible to determine two reduced rates, Λ'_O and Λ_ρ, with which one reproduces all seven measured lifetimes τ_2 of Table 1 within better than 5%. The two reduced rates are:

$$\begin{aligned} \Lambda'_O &= 2.3 \cdot 10^{11} s^{-1} \\ \Lambda_\rho &= 4.5 \cdot 10^{8} s^{-1} \end{aligned} \tag{6}$$

The principle of reproducibility of experimental data is then saved with only two kinds of μp atoms, the normal or "white" ones and the "ephemeral" μp atoms.

The disappearance channel λ_ρ of "ephemeral" μp atoms remains obscure. The dependence upon the hydrogen density might suggest a thermalization process. The thermalization time of μp atoms in a hydrogen gas at 15 bar should then be of about 130 ns, which is reasonable and which coincides with the decay time of the low amplitude time component of the muonic neon X-rays measured in a $H_2 + Ne$ gas mixture at 15 bar [19]. However, if λ_ρ were really only dependent upon the hydrogen density, i. e. independent of the gas admixed to hydrogen, then the unexpected second component observed in the time spectra of the muonic fluorine X-rays in $H_2 + SF_6$ gas mixtures at 15 bar [20] should have a decay time of the order of $\tau_2 < 100$ ns, what is not the case (cf. Fig.3). In addition, from the single-exponential structure of the time distribution of the muonic sulphur X-rays, it follows that the λ_ρ channel does not convert "ephemeral" μp atoms into "white" or normal or thermalized ones.

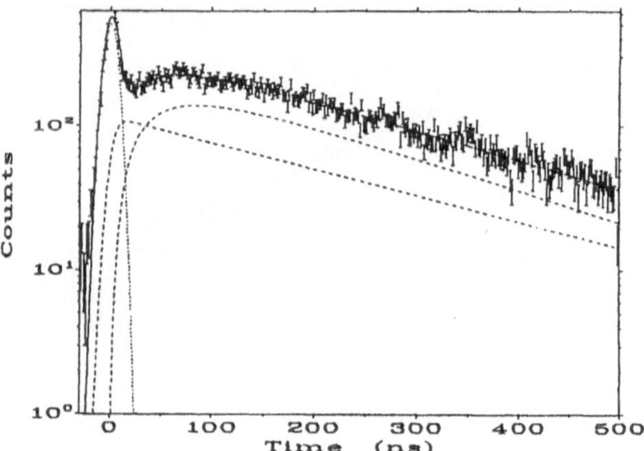

Fig. 3. Time distribution of the muonic fluorine μF(2-1) X-ray events measured in a binary gas mixture $H_2 + 0.1\%\ SF_6$ at 15 bar. The continuous line corresponds to a fit using in the delayed part three exponential functions [12]. The lower dashed line corresponds to delayed events with the same time structure as the muonic sulphur X-rays (μp lifetime, approx. 240 ns). The upper dashed line corresponds to the unexpected component of the time distribution with a decay time of approximately 200 ns and a build-up time of about 40 ns.

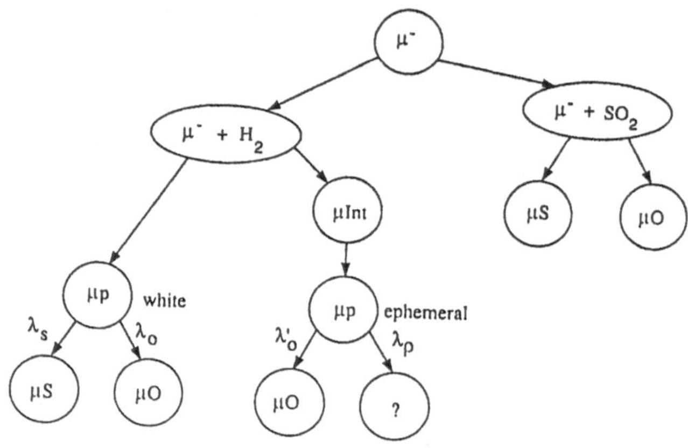

Fig. 4. Fate of a negative muon stopped in a $H_2 + SO_2$ gas mixture with a small molar SO_2 concentration, assuming the existence of ephemeral muonic hydrogen atoms [18]. Only the main channels are drawn [15].

Not only the disappearance channel λ_ρ has presently no plausible explanation, the build-up time observed in the oxygen and fluorine time spectra (Figs.2 and 3) is an additional open question. Contrary to the "white" μp atoms, the "ephemeral" ones are built-up from an unknown system (called μInt [15]) via an unknown mechanism and disappear partially through an unknown channel λ_ρ to an unkown final system (Fig.4). From the measured amplitudes of the two time components observed in the oxygen time spectra, about one third of the μp atoms formed in the investigated $H_2 + SO_2$ mixtures are "ephemeral" muonic hydrogen atoms [18].

Table 2. Experimental (reduced) muon transfer rates from muonic protium to neon.

Year qquad qquad	Ref.	pressure [bar]	Λ_{pNe} $[\cdot 10^{11} s^{-1}]$
1967 qquad qquad	[5]	26	1.16(28)
1990 qquad	[19]	15	1.15(20)
1990 qquad	[19]	15	0.062(4)
1990 qquad	[19]	15	0.058(48)
1990 qquad qquad	[19]	15	0.062(5)
1991 qquad qquad	[21]	15	0.079(4)
1991 qquad	[22]	15	0.081(4)
1991 qquad	[22]	15	0.084(8)

Table 3. Experimental (reduced) muon transfer rates from muonic deuterium to helium (^4He).

Year	Ref.	pressure [bar]	Λ_{dHe} $[\cdot 10^8 s^{-1}]$
1977	[23]	6	< 0.1
1984	[24]	92	3.2(3)
1985	[25]	88	3.68(18)
1988	[26]	liq.	13.1(12)
1988	[26]	liq.	6.7(13)
1990	[27]	1350	2.75(22)
1991	[28]	liq.	1.89(55)
1991	[28]	liq.	5.7(26)

5 "Coloured" muonic hydrogen atoms

The advantage of the muon transfer measurements performed in $H_2 + SO_2$ and $H_2 + SF_6$ gas mixtures is that one can distinguish between normal μp atoms and others which are not. In $H_2 + Z$ mixtures, the situation is completely different. As long as a muon transfer rate was measured only once, one had no reason to doubt about the normality of the studied muonic hydrogen atoms.

In a recent measurement of muon transfer to neon in a $H_2 + 0.7\%$ Ne gas mixture, the time distribution of the muonic neon X-rays showed a double-exponential structure [19]. The transfer rate deduced from the steeper slope agreed with the transfer rate measured twenty years ago. But the dominant transfer proceeded through a much slower channel with a rate about twenty times smaller. This small transfer rate has been confirmed in another measurement [19]. Such a situation is very uncomfortable. Three new measurements of muon transfer to neon, which yielded all three the same reduced transfer rate, were unable to confirm neither the first nor the second transfer rate measured [21,22] (cf. Table 2). Even the muon transfer rate from muonic deuterium to neon could not be reproduced [21].

Table 2 shows the present situation of the measured reduced muon transfer rates from μp atoms to neon. There are three categories of values, which are incompatible. Wheras in $H_2 + SO_2$ measurements one could distinguish between normal μp atoms and others (cf. Table 3), the three classes of reduced transfer rates to oxygen, Λ^*_O, were related to "black" μp atoms. With the postulate of an additional disappearance channel, the three classes could be reduced to only one class of μp atoms, called "ephemeral" μp atoms. In the case of the transfer rates to neon, an analogous reduction seems impossible because, among other reasons, the experimental results produced under identical initial conditions - at least to the best of our knowledge - are not reproduced ($H_2 + 0.7\%$ Ne).

The situation is not essentially different for the reduced muon transfer rates to argon [15] and to helium. In Table 3, all measured muon transfer rates from muonic deuterium to ^4He are listed. Four different experimental groups have measured these data at five different Meson Factories. The methods employed were also different (decay electrons, fusion neutrons, X-ray time distributions, X-ray intensity ratios). If such discrepancies among muon transfer rates would have appeared only in transfer to helium, one would have put in question the various methods used. The discrepancies are, however, not restricted to helium and the problem of the reproducibility of reduced muon transfer rates is, therefore, a more general one.

6 Conclusions

After the systematic transfer rate measurements from muonic protium and muonic deuterium to the heavier noble gases, performed at CERN by the italian groups [5-7], and which seemed to have shown that the rates are proportional to the charge numbers, the chapter of muon transfer was more or less closed during twenty years. The only interesting question, which remained, was a precise value of the transfer rate to helium. In 1977, it was found to be very low and in agreement with theory [23]. In 1981, N. P. Popov proposed a new mechanism for muon transfer to helium via an intermediate molecular state [29]. The predicted transfer rate turned out to be an order of magnitude greater and was promptly confirmed by new experimental results [30], which were in contradiction with the earlier ones [23]. Since then, the confusion among the measured muon transfer rates steadily increases. However, the first contradictions are already almost thirty years old [10]. The reproducibility of muon transfer rates should have been trivial. In reality, we are presently confronted to a chaos, where up to now explanations in conventional terms failed. Problems of this type exist not only in muon transfer but also in atomic capture of negative muons in noble gas mixtures [31] and in compounds [32]. Calculations will not help to solve the problems. In future, new experiments will have to be performed and a bright idea is wanted to escape the dead-end.

References

1. Schiff M., Nuovo Cimento 22, 65 (1961).
2. Dzhelepov V.P., Ermolov P.F., Kushnirenko E.E., Moskalev V.I. and Gershtein S.S, Zh. Eksp. Teor. Fiz. 42, 479 (1962) [JETP 15, 306 (1962)].
3. Gershtein S.S, Zh. Eksp. Teor. Fiz. 43, 706 (1962) [JETP 16, 501 (1963)].
4. Basiladze S.C., Ermolov P.F and Oganesyan K.O., Zh. Eksp. Teor.Fiz. 49, 1042 (1965) [JETP 22, 725 (1966)].
5. Quaranta A.A., Bertin A., Matone G., Palmonari F., Placci A., Dalpiaz P., Torelli G. and Zavattini E., Nuovo Cimento 47B, 92 (1967).
6. Placci A., Zavattini E., Bertin A. and Vitale A., Nuovo Cimento 52A, 1274 (1967).
7. Placci A., Zavattini E., Bertin A. and Vitale A., Nuovo Cimento 64A, 1053 (1969).
8. Leon M., contribution to this volume.
9. Ponomarev L.I., Contemp. Phys. 31, 219 (1990).
10. Conforto G., Rubbia C. and Zavattini E., Nuovo Cimento 33, 1001 (1964).
11. Schneuwly H., Jacot-Guillarmod R., Mulhauser F., Oberson P., Piller C. and Schellenberg L., Phys.Lett. A132, 335 (1988).
12. Mulhauser F., Schneuwly H., Jacot-Guillarmod R., Piller C., Schaller L.A. and Schellenberg L.; Muon Cat. Fusion 4, 365 (1991).
13. Bracci L. and Fiorentini G., Nuovo Cimento 50A, 373 (1973).
14. Holzwarth G. and Pfeiffer H.J., Z. Physik A272, 311 (1975).
15. Schneuwly H., Muon Cat. Fusion 4, 87 (1989).
16. Mulhauser F., Jacot-Guillarmod R., Piller C., Schaller L.A., Schellenberg L. and Schneuwly H., in *Electromagnetic Cascade and Chemistry of Exotic Atoms*, edited by L.M. Simons et al., Plenum Press, New York, 1990, p. 217.
17. Mulhauser F., Jacot-Guillarmod R., Piller C., Schaller L.A., Schellenberg L. and Schneuwly H., Helv. Phys. Acta 64 (1991) in press
18. Schneuwly H. and Mulhauser F., submitted to Phys. Lett. A.
19. Jacot-Guillarmod R., Mulhauser F., Piller C. and Schneuwly H., Phys. Rev. Lett. 65, 706 (1990).
20. Mulhauser F., Jacot-Guillarmod R., Piller C., Schaller L.A., Schellenberg L. and Schneuwly H., Muon Cat. Fusion 5 (1991) in press.
21. Jacot-Guillarmod R., Mulhauser F., Piller C., Schaller L.A., Schellenberg L. and Schneuwly H., Helv. Phys. Acta 64, 205 (1991).
22. Schneuwly H., Mulhauser F., Jacot-Guillarmod R., Mallinger M., Piller C., Schaller L.A. and Schellenberg L., Helv. Phys. Acta 64, (1991) in press.
23. Bertin A., Vitale A. and Zavattini E., Nuovo Cimento Lett. 18, 381 (1977)
24. Balin D.V., Maev E.M., Medvedev V.I., Semenchuk G.G., Smirenin Yu.V., Vorobyov A.A., Vorobyov An.A. and Zalite Yu.K., Phys. Lett. B141, 173 (1984).
25. Balin D.V., Vorobyov A.A., Vorobyov An.A., Zalite Yu.K., Markov A.A., Medvedev V.I., Maev E.M., Semenchuk G.G. and Smirenin Yu.V., Pisma Zh. Eksp. Teor. Fiz. 42, 236 (1985) [JETP Lett. 42, 725 (1985)].
26. Matsuzaki T., Ishida K., Nagamine K., Hirata Y. and Kadono R., Muon Cat. Fusion 2, 217 (1988).
27. Bystritsky V.M., Dzhelepov V.P., Zinov V.G., Ilieva-Sokolinova N., Konin A.D., Marcis L., Merkulov D.G., Rudenko A.I., Somov L.N., Stolupin V.A. and Filchenkov V.V., JINR Report P1-90-312 Dubna USSR, 1990.
28. Watanabe Y., Sakamoto S., Ishida K., Matsuzaki K., Strasser P., Iwasaki M. and Nagamine K., Muon Cat. Fusion 5 (1991) in press.
29. Aristov Yu.A., Kravtsov A.V., Popov N.P., Solyakin G.E., Truskova N.F. and Faifman M.P., Sov. J. Nucl. Phys. 33, 564 (1981).
30. Bystritsky V.M., Dzhelepov V.P., Petrukhin V.I., Rudenko A.I., Suvorov V.M., Filchenkov V.V., Khovanskii N.N. and Khomenko B.A., Zh. Eksp. Teor. Fiz. 84, 1257 (1983) [Sov. Phys.JETP 57, 728 (1983)].
31. Ehrhart P., Hartmann F.J., Köhler E.K. and Daniel H., Phys. Rev. Lett. 27, 575 (1983).
32. Schneuwly H., Proc. Int. Symp. on Muon and Pion Interactions with Matter, Dubna, June 30 - July 4, 1987, JINR Report D14-87-799 (Dubna, 1987), p. 347.

This article was processed using Springer-Verlag TEX Z.Physik C macro package 1991
and the AMS fonts, developed by the American Mathematical Society.

Z. Phys. C - Particles and Fields 56, S285-S288 (1992)

Zeitschrift für Physik C Particles and Fields

PILAC: A pion linac facility for 1-GeV pion physics at LAMPF

Henry A. Thiessen and D. Hywel White

Medium Energy Physics Division, Los Alamos National Laboratory
Los Alamos, New Mexico 87545

Submitted 22 November 1991

Abstract. A design study for a Pion Linac (PILAC) at LAMPF is under way at Los Alamos. We present here a reference design for a system of pion source, linac, and high-resolution beam line and spectrometer that will provide 10^9 pions per second on target and 200-keV resolution for the (π^+,K^+) reaction at 0.92 GeV. A general-purpose beam line that delivers both positive and negative pions in the energy range 0.4-1.1 GeV is included, thus opening up the possibility of a broad experimental program as is discussed in this report. A kicker-based beam sharing system allows delivery of beam to both beam lines simultaneously with independent sign and energy control. Because the pion linac acts like an rf particle separator, all beams produced by PILAC will be free of electron (or positron) and proton contamination.

1 The nuclear physics program of PILAC

There are five classes of experiments that require pions of energies up to 1.1 GeV [1]. These classes are:

(1) Λ-hypernuclear physics via the (π,K) reaction;
(2) Λ-nucleon scattering at threshold;
(3) rare decays of π and η;
(4) pion-nucleus elastic and inelastic scattering with 0.4-1.1-GeV pions; and
(5) baryon resonances.

2 PILAC concept

A concept for the reference facility is shown in Fig. 1. The reference PILAC facility requires the following items [2]:

(1) proton buncher;
(2) new target cell for zero-degree pion production;
(3) 0.38-0.53-GeV pion injection beam line;

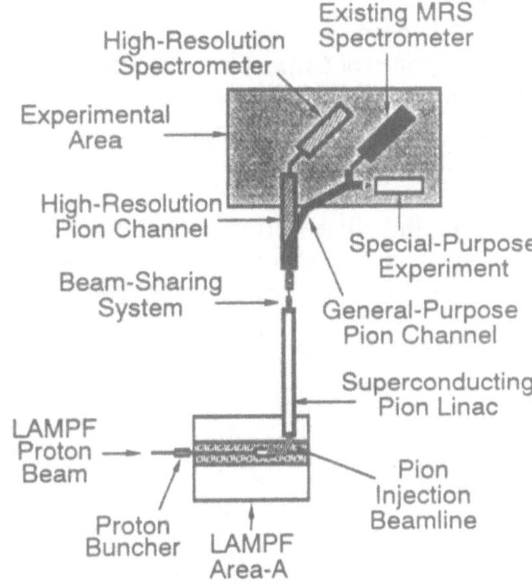

Figure 1. Concept for PILAC facility at LAMPF.

(4) 0.38-0.92-GeV, 12.5 MeV/m gradient superconducting pion linac;
(5) kicker-based beam-sharing system;
(6) 1.1-GeV high-resolution dispersed beam line and spectrometer;
(7) general-purpose beam line with dispersed mode for existing MRS spectrometer; and
(8) experimental area and related civil engineering.

3 Proposed development plan for superconducting cavities for PILAC

The reference design requires 12.5 MeV/m cavity gradient and has negligible beam loading. We believe that this large gradient can best be achieved by use of titanium heat-treated niobium cavities [3]. In order to establish the gradient and Q that will be achieved in the PILAC cavities, it is necessary to scale up the technology to the larger, lower-frequency cavities needed and to test several prototype cavities. This requires new, larger facilities at Los Alamos. To save time, we propose to develop the heat-treatment technology in a parallel effort using existing equipment at 3 GHz. We will also take an existing 805-MHz single-cell cavity and heat-treat it in the Cornell oven.

4 Injection beam line

The injection beam line proposed for PILAC consists of a matching section based on a strong quadrupole doublet placed as close as possible to the pion-production target followed by a second-order achromat. Sextupoles and octupoles correct all detrimental second- and third-order aberrations. This beam line has a solid angle of 10 msr, a momentum acceptance of 6.6%, and a transverse phase-space output of 225 π mm-mrad. In the reference version of this beam line design, 82% of the output beam is contained within the specified phase space.

We are also considering a possible design of the injection beam line that will deliver both π^+ and π^- simultaneously as is shown in Fig. 2. This version of the injection line has not yet been studied thoroughly.

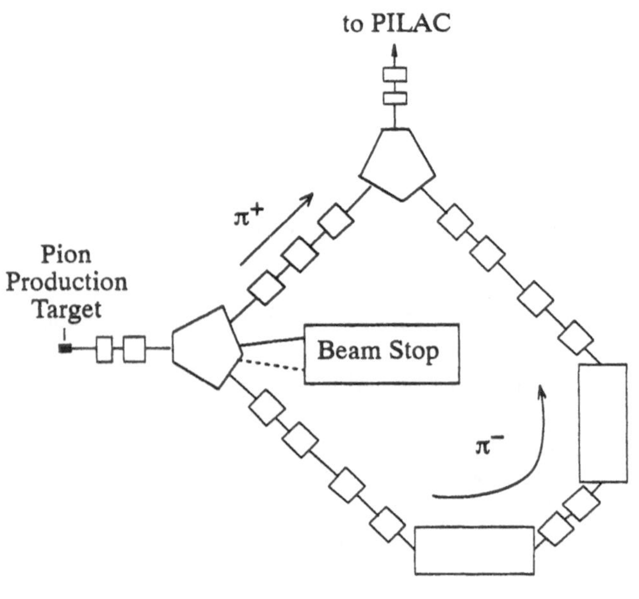

Figure 2. Possible design for simultaneous π^+ and π^- injection line for PILAC.

5 Choice of operating frequency for PILAC

A series of linac designs was studied to determine the optimum operating frequency for PILAC. The highest frequency that is a harmonic of the LAMPF injector and is also compatible with the required transverse phase space acceptance of 225 π mm-mrad is 805 MHz. This frequency has been chosen for the reference design.

6 Reference linac design

The reference linac design is based on a study maximizing the pion intensity in the desired output phase space. A conceptual layout of this design is shown in Fig. 3. The beam intensity is the product of longitudinal acceptance and pion survival in the linac. A total phase space rotation of $3\pi/4$ in the longitudinal plane maximizes the acceptance while minimizing the energy spread of the output beam. The layout of the linac minimizes pion decay. The number of cells per cavity was chosen to give the largest cavity that can be handled comfortably in the existing superconducting cavity lab. This results in a choice of seven cells per cavity. A total of 40 cavities is required to accelerate from 0.38-0.92 GeV. By raising the injection energy and rephasing the linac, energies up to 1.1 GeV can be provided, but at reduced intensity. Quadrupole doublets are required after each five cavities in order to contain the transverse phase space in

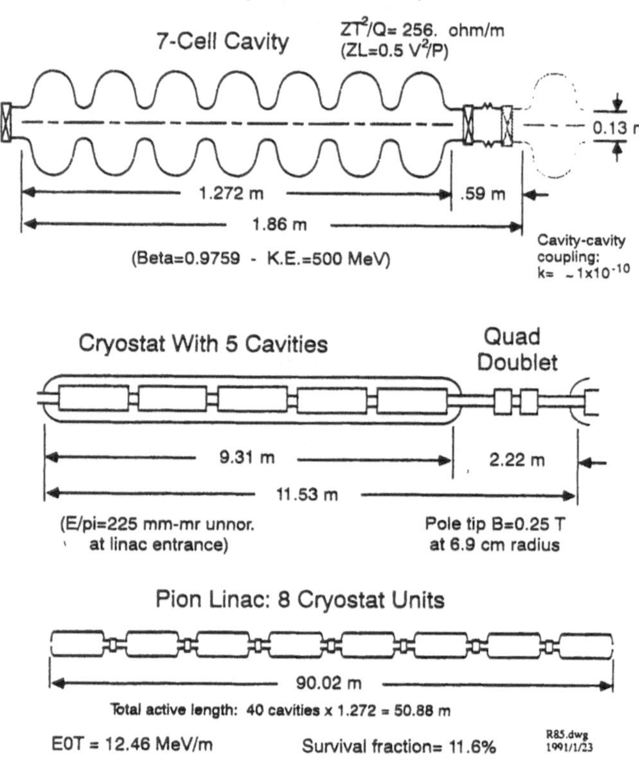

Figure 3. Reference design of pion linac.

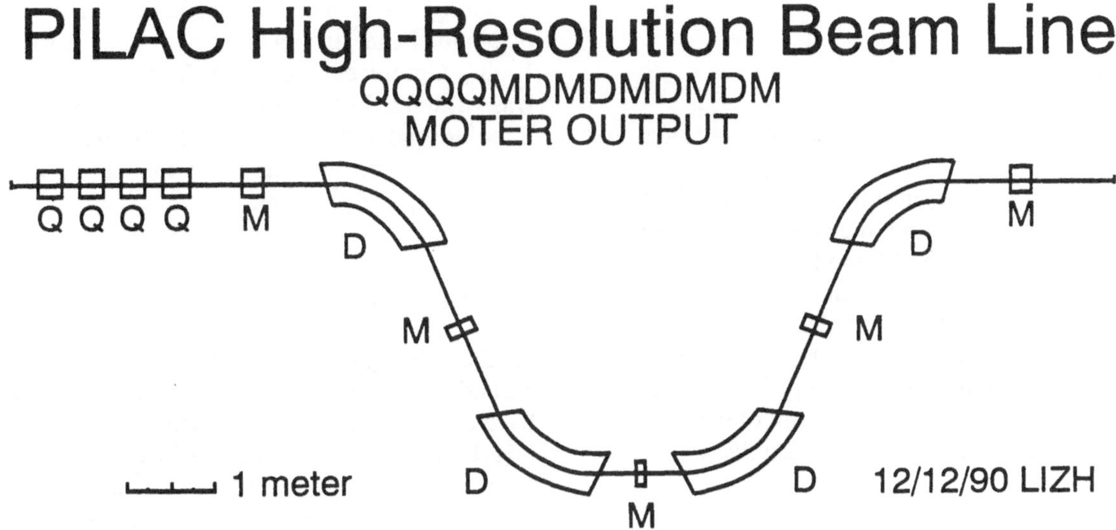

Figure 4. High-resolution beam line for PILAC.

the cavity bore. Five cavities will be mounted in a single cryostat as a module. Eight modules are combined to make the full linac. The total length of the linac is approximately 90 meters.

7 High-resolution beam line and spectrometer

A high-resolution beam line has been designed using program MOTER [4]. This beam line, a QQQQMDMDMDMDM design with vertical bends, has a momentum dispersion of 25 cm/% and a horizontal divergence of 5 mrad full width in the horizontal plane and is shown in Fig. 4. The full size of the beam spot will be 40-cm high by 10-cm wide. The momentum resolution of the high-resolution beam line calculated by MOTER is one part in 10^4 when second- and third-order optical corrections are introduced on the dipole entrance and exit faces.

A spectrometer with acceptance matched to this beam line has also been designed and is shown in Fig. 5. The design is very similar to that of the existing EPICS spectrometer at LAMPF except that the bending magnet has been made from a single unit in order to minimize the flight path of kaons in the spectrometer. By using iron-dominated superconducting quadrupole magnets similar to those planned for the Hall-C spectrometer at CEBAF, the acceptance of the spectrometer has been increased by almost a factor of six compared with that of the existing EPICS spectrometer. The acceptance of the proposed spectrometer is 27 msr for the 10-cm × 40-cm beam size of the high-resolution beam line.

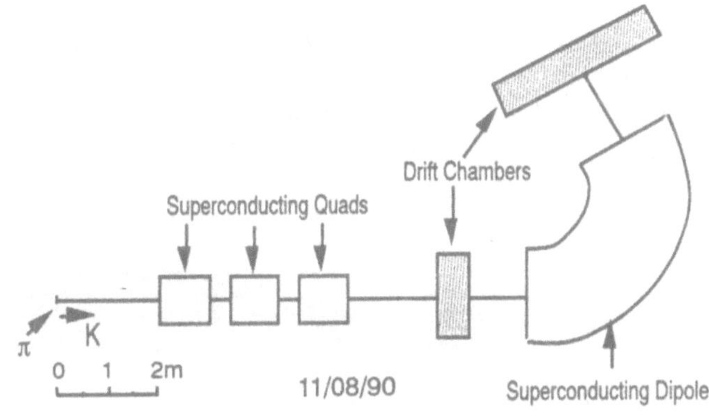

Figure 5. High Resolution Spectrometer for PILAC.

8 General-purpose beam line

A beam line capable of providing an achromatic beam spot and also 2-3 cm/% horizontal dispersion (for use with the existing MRS spectrometer) is being designed. The output beam spot and divergence in the achromatic mode will be 5-mm radius × 25 mrad (half widths). The output beam spot will be tunable over a wide range. We are also studying a second port for the output of this line. In addition to providing beam for experimenters, the new general-purpose line will serve as a pion injector for a future linac extending the energy range of PILAC to 1600 MeV.

9 Beam sharing in PILAC

A kicker-based beam-sharing system is envisaged for PILAC. With the kicker magnet off, the pion beam from the linac goes straight ahead into the high-resolution beam line. With the kicker magnet on, the beam is deflected into the general-purpose line. The basic idea is that the kicker rise time will be comparable to the time required to switch the phase program of the linac cavities. Then the phase program and the kicker can be switched simultaneously. With simultaneous injection of π^+ and p^- into the linac, it will be possible to have independent control of the pion sign and energy delivered to each of the two beam lines.

10 PILAC pion yield

The pion yield expected from PILAC is $1.2 \pm 0.6 \times 10^9$ π^+ per second at 0.92 GeV. This yield is comparable to that which can be achieved in the same phase space at the proposed KAON facility and is more than an order of magnitude larger than can be achieved at the Brookhaven AGS with the new booster operating, as is shown in Fig. 6. In this figure, the yield of PILAC is shown as a function of the cavity gradient. A design value of 12.5 MeV/m has been chosen for the Reference Design.

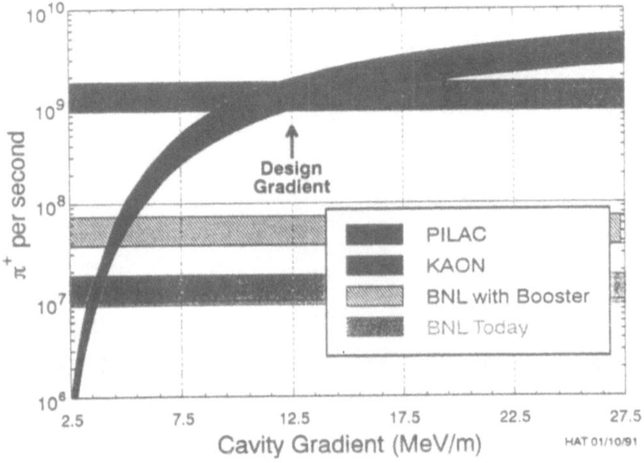

Figure 6. Pion yield of PILAC as a function of cavity gradient. For comparison, the expected yield of KAON, BNL with Booster, and BNL today are also shown.

11 Program MOTER

The program used for raytracing calculations of the high-resolution beam line and a spectrometer is MOTER (Morris Klein's Optimized Tracing of Enge's Rays). This program was written at Los Alamos in the early 1970s [4]. MOTOR is being updated to include rf cavities in order to make possible a precision study of the optics of PILAC in a single computer program.

12 Summary and conclusions

PILAC is being designed to provide a beam of 10^9 pions per second at 0.92 GeV, with a future upgrade to 1.6 GeV. We have demonstrated that the system resolution of 200 keV can be achieved in a high-resolution beam line and spectrometer. In order to provide the necessary flux, the linac requires superconducting capacities that achieve a gradient of 12.5 MeV/m. Although no linac presently operates at this high gradient, tests at laboratories around the world have shown that this gradient can be achieved by titanium heat treatment of the cavities. An R&D program is proposed to scale up the results from single-cell 1.5-3-GHz cavities to the necessary 7-cell 805-MHz cavities.

PILAC will provide an energy of 0.92 GeV with operation possible up to 1.1 GeV at reduced intensity. This energy is sufficient to optimize the yield of the (π^+,K^+) reaction and to access a broad range of interesting physics. The PILAC beam-sharing system will allow simultaneous operation of two or more lines with independent sign and energy control. The PILAC beams will be of unprecedented purity since the linac acts as a high-resolution rf separator. PILAC is cost-effective, since it is by far the least expensive upgrade to LAMPF that gives access to this physics. The new superconducting cavities represent a new technology that will open up applications in other fields. Finally, PILAC is feasible only at LAMPF, since only LAMPF has the necessary tightly bunched proton beam to produce pions that can then be accelerated in a superconducting linac.

References

1. "Physics with PILAC," Los Alamos Report (to be published)
2. PILAC Technical Notes 1-25, Los Alamos, 1990-91, and other contributions to this conference
3. H. Padamsee: IEEE Cat. No. 87CH2387-9(1987)1691
4. H. A. Thiessen and M. M. Klein: Proceedings of the 4th International Conference on Magnet Technology, Brookhaven, NY(1972)8-17

Z. Phys. C – Particles and Fields 56, S289–S291 (1992)

Accelerator plans at PSI

Hans-Christian Walter

Paul Scherrer Institute, CH-5232 Villigen PSI, Switzerland

10 December 1991

Abstract. A survey is given on the plans of the PSI accelerator division. It includes the completion of the Target E shut-down with preparation of the high current operation of the main ring, the future of injector I, the new experimental areas, the neutron spallation source and the new project of a Swiss synchrotron light source. Separate contributions to this conference will treat our muon beam lines and the developments towards very low energy muon beams.

1 The proton machines

In order to increase the proton current to 1.5 mA, which will be needed mainly for the injection into the neutron spallation source, the injector II was installed and the thin target station upgraded already years ago. In 1990 the thick target station was removed and replaced and a new RF system for the ring installed.

Simultaneously almost all secondary beams of this station had to be modified or changed; a totally new large acceptance beam (πE5) is being added. Also the first part of the proton beam for the spallation source has been implemented. A new liquid He plant will supply all main experimental hall users. The injector II has been successfully tested to deliver a 1.5 mA beam with a phase space acceptable by the ring, and the first of five new RF 500 kW amplifiers has been connected and works. The new thick target consists of a rotating radiation cooled carbon wheel with a thickness of 6 mm and a length in beam direction of 600 mm. The expected deposition of 100 kW necessitates good vacuum and careful shielding of the vacuum chamber by copper collimators, which in turn will unfortunately increase the electron and positron contaminations of the beams.

2 Secondary beams

Coupled to this new target station are five secondary beam lines πE1, μE1, μE4, πE3 and μE5. Some of the parameters of the corresponding pion and muon beams are listed in table 1.

Most of the beams have different operating modes, e.g. an achromatic mode yielding high fluxes, or a chromatic mode with smaller intensity but good momentum resolution. The new low energy πE3 channel with a length of only 12.8 m has been specially designed to match the optical properties of the existing Low Energy Pion Spectrometer (LEPS). During tuning of the channel it turned out, that higher order corrections and nonlinear effects are vital to optimize the channel and to compare its performances with the expectations. The new $\pi(\mu)$E5 channel is a low energy pion and muon channel at 165° to the proton beam with a solid angle acceptance of 150 msr. It will be the last channel to be implemented in the beginning of 1992.

Many channels are practically windowless so that surface and subsurface muon beams can be transmitted which are needed not only for particle physics experiments but also as starting points for muon cooling schemes and to satisfy the ever increasing demands of our μSR community. It is observed that the complexity of the accelerator installations increases drastically which on one hand is most satisfactory, which on the other hand concerns us because of the increasing possibilities of errors and the increasing load on our accelerator staff. The medical programs of pion therapy (PIOTRON), eye treatment (OPTIS) and the new proton therapy project (NA3), the irradiation facilities PIREX, PIF I and PIF II, the isotope production programs IPI and IP II and last not least the neutron spallation source (SINQ) require more and more F1 personal which is badly needed also to upgrade, maintain and repair the accelerator infrastructure itself.

The injector I cyclotron runs without major failures for an increasingly diversified community. Nuclear

Table 1. Secondary beams summary

π^+	mom. MeV/c	flux at 1 mA	$\delta p/p$ FWHM	spot size cm^2
πE1	120-600	2 x 10^9 at 300 MeV/c	0.26%	1.5 x 2
πE3	40-180	1 x 10^9 at 180 MeV/c	0.05%	10 x 4 disp.
πE5	30-120	1.5 x 10^{10} at 120 MeV/c	2%	6 x 4
πM1	120-500	2 x 10^8 at 300 MeV/c	0.1%	2 x 1.5

π^-	mom. MeV/c	flux at 1 mA	$\delta p/p$ FWHM	spot size cm^2
πE1	120-600	2 x 10^8 at 300 MeV/c	0.26%	1.5 x 2
πE3	40-180	1.7 x 10^8 at 180 MeV/c	0.05%	10 x 4 disp.
πE5	30-120	3.5 x 10^9 at 120 MeV/c	2%	6 x 4
πM1	120-500	2 x 10^7 at 300 MeV/c	0.1%	2 x 1.5

μ^+	mom. MeV/c	flux at 1 mA	$\delta p/p$ FWHM	spot size cm^2	pol.
μE1	40-125	2 x 10^8 at 125 MeV/c	3%	3 x 2	75 %
μE4	30-100	4 x 10^6 at 50 MeV/c	3%	6 x 4	75 %
πE1	5-30	2 x 10^6 at 28 MeV/c	0.8%	1.5 x 2	> 95 %
πE3	5-30	3 x 10^7 at 28 MeV/c	1 %	2 x 3	> 95 %
πE5	5-30	2 x 10^8 at 28 MeV/c	2 %	6 x 4	> 95 %
πM3	5-30	4 x 10^6 at 28 MeV/c	0.2 %	2 x 2	> 95 %

μ^-	mom. MeV/c	flux at 1 mA	$\delta p/p$ FWHM	spot size cm^2	pol.
μE1	40-125	6 x 10^7 at 125 MeV/c	3 %	3 x 2	75 %
μE4	30-100	1 x 10^6 at 50 MeV/c	3 %	6 x 4	75 %
πE1	5-280	4 x 10^7 at 100 MeV/c	0.8 %	1.5 x 2	(var.)
πE3	5-150	3 x 10^7 at 100 MeV/c	1 %	2 x 3	(var.)
πE5	5-120	2 x 10^8 at 100 MeV/c	2 %	6 x 4	(var.)
πM3	5-300	4 x 10^6 at 100 MeV/c	0.2 %	2 x 2	(var.)

Physics, the OPTIS program, irradiation of electronic components, isotope production is done in a round the clock schedule. New possibilities in radiology, radiochemistry and atomic and nuclear physics will be opened by the long awaited purchase of an ECR-source.

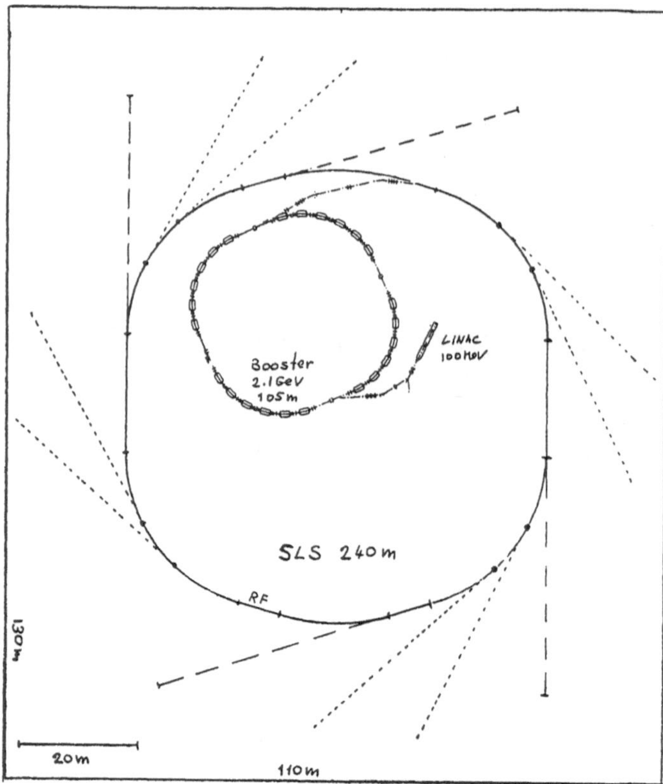

Fig. 1. Possible outline of the lattice for the Swiss Light Source

On 8-8-88 the digging of the 11 m deep proton transport tunnel to the neutron spallation source was started.

The two halls are covered meanwhile and design work on the target- and supplies systems is progressing well. A safety analysis report was prepared including investigations towards protection against airplane crashes and earthquakes and research to determine the retention factor of lead-bismuth for polonium and mercury. Apart from the liquid Pb-Bi, solid targets from Ta, W and a Pb-shot pebble-bed target have been studied. A detailed technical design of the cold D_2-moderator to be placed in the D_2O-moderator tank has been finished and provisions for a H_2-moderator as a long term option have been considered. Finally much work on the multilayers for neutron mirrors and polarizers has been done both theoretically and experimentally. Still the production of useful neutrons from SINQ, the first continuous high power spallation neutron source, is expected for 1995.

3 A Swiss Light Source (SLS)

A project definition study group has been established in spring 1991 to study a Swiss Light Source (Synchrotron Lichtquelle Schweiz, Source Lumière Suisse, Sorgente Luce Svizzera, SLS). Accelerator people from PSI and users organized by Prof. G. Margaritondo from Lausanne

Table 2. SLS general parameters

Energy	1.5 GeV		2.1 GeV
Emittance	$\geq$ 1 nm		$\geq$ 3 nm
Photon energy (s.c. dipoles)	3 keV		9 keV
Photon flux (ph/s/0.1%BW/ mrad/mA)			3×10^{10}
Circumference		240 m	
Straight sections		2×20 m, 4×7 m	
RF frequency		500 MHz	
Harmonic number		400	
Bunch length		few mm	
Beam lifetime	> 2 h possible on-axis injection		> 10 h

have sit together to specify a facility of the 4th generation which was presented at an international workshop at PSI on July 11-12, 1991. Starting from the wishes of "something special" and "something affordable " (< 150 MSF) the design of an intense VUV to X-ray synchrotron light source with very flexible parameters, allowing several modes of operation emerged. A very small emmittance (~1 nm) at lower energies (1.5 GeV electron energy) could be combined with a high flux, high photon energy operation at the highest electron energies of ~ 2.1 GeV. The source is based on insertion devices, i.e. it has two long (~20 m) straight sections, but can also deliver useful radiation from the bending magnets. It should be capable to run in a mode of delivering few, very short bunches with very good time definition.

Table 2 gives the main parameters for the two modes and Fig. 1 shows a possible outline of the lattice, where the incorporation of superconducting bending magnets is under study to increase the critical photon energy into the 10 keV range.

The scientific program ranges from atomic, molecular, solid state and surface physics over molecular and photochemistry including catalysis to crystallography of bio- and macromolecules for biological applications and angiography. The high energy mode will be used for applications in microstructure and -mechanics and in nanotechnology research.

A feasibility study is being prepared for mid 1992 to be submitted to the Schweizer Schulrat, after which a full project study is envisaged. The project could then be submitted for final approval late 1994 such that building can start beginning 1995 and first beam expected in 1999.

This article was processed using Springer-Verlag TEX Z.Physik C macro package 1991
and the AMS fonts, developed by the American Mathematical Society.

Z. Phys. C - Particles and Fields 56, S292-S295 (1992)

Zeitschrift für Physik C Particles and Fields

Accelerator plans at TRIUMF

J.L. Beveridge

TRIUMF, 4004 Wesbrook Mall, Vancouver, B.C., Canada V6T 2A3

12-August-1991

Abstract. A recently completed Project Definition Study has proposed a network of accelerators to take the existing 500 MeV 150 μA proton beam at TRIUMF to 30 GeV. This facility would be capable of providing beams of kaons, antiprotons and other hadrons of intensities 100 times greater than those presently available. In addition, large numbers of low energy muons should be available and this facility is potentially the most powerful muon source planned for the future. The proposed facilities are described and the potential for future muon beams reported.

1 Introduction

In 1989/90 an $11 million Project Definition Study (PDS) was jointly funded by the British Columbia Provincial Government and the Federal Government of Canada. During this study, the design and costing of a 100 μA, 30 GeV proton accelerator was carried out based on the present TRIUMF cyclotron as an injector. This TRIUMF KAON (Kaon-Antiproton-Otherhadron-Neutrino) Factory aims to increase the proton intensity at 30 GeV by roughly a factor 100 over present facilities. The intense secondary beams produced by such a Factory open a broad scope of physics opportunities which were extensively examined and delineated at a series of workshops conducted internationally. In addition to the science and accelerator designs, the PDS included environmental and economic impact studies, investigation of possible foreign participation in construction costs and the construction of prototypes of technically difficult components. The study was completed in 1990 and the results can be summarized in the following highlights:

- the cost of the project is forecast at $708 million (1989 dollars) over a six year construction period
- annual operating costs will be $98 million
- foreign participation is expected to be $200 million from USA, Japan, Europe, Korea etc.
- the province of British Columbia will provide $236 million (i.e., 1/3 of the construction costs)
- economic benefits include 17,000 person-years of employment
- the environmental assessment shows little risk
- international accelerator experts have pronounced the design to be technically mature
- the physics opportunities are broad and have extensive international support
- the study was presented to both levels of government on May 24, 1990.

The conclusions of the PDS could not be more positive and with the strong support of the Provincial government we are awaiting a favorable decision of commitment from the Federal government.

A	Accumulator	accumulates the 450 MeV cw beam from the TRIUMF cyclotron over 20 msec periods – H^- injection allows the painting of the beam into the accumulator phase space
B	Booster	a 50 Hz synchrotron; accelerates the beam to 3 GeV
C	Collector	collects 5 Booster pulses and manipulates the longitudinal emittance
D	Driver	the main 10 Hz synchrotron; accelerates the beam to 30 GeV
E	Extender	a 30 GeV stretcher ring which allows slow extraction of the beam

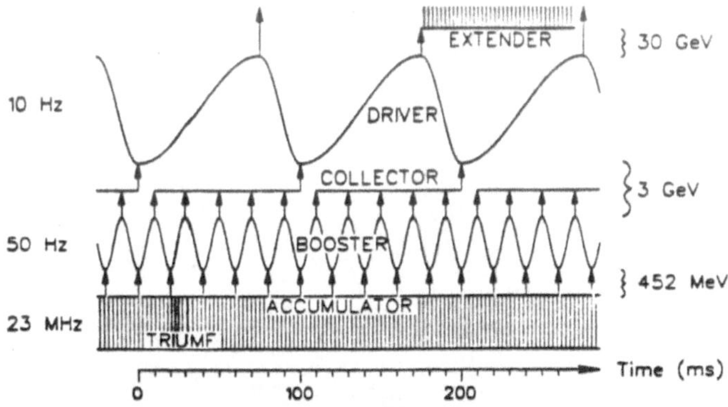

Fig. 1. Energy-time sequence for the five rings

2 Accelerators

The TRIUMF KAON accelerators have previously been described in detail[1-3]. The existing TRIUMF cyclotron which routinely accelerates 150 μA of H^- ions to 500 MeV will be used as an injector. For KAON, the H^- ions will be extracted from the cyclotron at 450 MeV (as opposed to the present extraction by stripping) and transported to a system of five rings consisting of two fast-cycling synchrotron accelerators interleaved with three storage rings. The function of each of the rings can be summarized as follows:

This system of accelerators and storage rings serves to transform the dc beam from the cyclotron to the appropriate pulse structure of the synchrotrons and allows them to run continuous acceleration cycles without flat tops or bottoms. The energy-time plot in Fig. 1 illustrates how this is accomplished.

A plan view of proposed rings is shown in Fig. 2. The Accumulator and Booster have basically circular lattices and are housed in a common tunnel. The Collector, Driver and Extender have racetrack lattices and are housed in a second tunnel with a circumference of about 1 km. The long straight sections in this design allow space for beam transfer between machines and for the slow extraction process. Calculations for the slow extraction system indicate beam losses below 0.2% which represent a great improvement over the usual 1% losses encountered in presently operating machines. However, even these losses are not insignificant and the Extender ring has been separated radially from the Collector and Driver rings to allow the insertion of shielding between the machines. Also, a long extraction hall has been provided over the slow extraction region to allow vertical access to activated components. This hall has been designed sufficiently long to cover not only the extraction components but also the components which allow the beam to be split continuously in up to four separate fractions. A tunnel, where beam losses are expected to be low, then connects this extraction hall to the experimental areas. The parameters of the extracted beams are given in Table 1. Included here are the Booster beam parameters although at present there is no design of an extracted beam from this machine.

3 Experimental Hall

The proposed layout of the primary and secondary beam lines is shown in Fig. 3. This layout is intended to address, at least in part, the expected scientific program as developed in the physics workshops conducted during the PDS. It envisages a full range of charged and neutral secondary beams, a neutrino facility and a polarized proton area. Six charged secondary channels are provided which view three production targets and taken together span the momentum range from 0.4 to 20 GeV/c. A neutral kaon line is taken from a fourth production target. Provision has been made for the extraction of low energy pion and muon beams in the backward direction from each of the production targets although no specific low energy channels have been designed. One such channel could be used for pion therapy. The four production targets are arranged with two in-line targets serviced by one of two slow extracted proton beams. These beams are sufficiently separated that the pairs of target stations can be operated independently. The fast extracted beam is transported to the neutrino facility and a third slow extracted beam services the polarized proton area. All production targets and beam dumps have been located in a single large experimental hall and can be accessed by a common overhead crane. This allows all highly radioactive components to be transported to a central hot cell facility. Long high-energy secondary channels are housed in separate buildings or tunnels attached to the main experimental hall. This proposed layout represents only one choice in a multitude of possibilities which satisfy the broad range of physics opportunities available at an intense hadron facility. As such, it should be regarded only as a useful tool in setting the scale and costs of the facilities required to exploit the proposed accelerators.

The low energy pion and muon channels are of particular interest to this workshop. It can be seen from the above layout that there will be a large amount of space available for such facilities as the production cross sections for other particles are strongly forward-peaked and therefore other secondary channels are positioned close to zero degrees. Some possibilities for channels and facilities which could be installed at KAON have been outlined by Brewer[4] and other possibilities will likely arise from experience with the new beam lines about to be commissioned at PSI. However, the performance of such facilities depends critically on the production of low energy pions by a 30 GeV beam.

Blackmore[5] has recently completed a survey of low energy pion production measurements for incident proton energies between 0.5 and 30 GeV. The results of this survey can be summarized as follows:

i The π^+/π^- cross section ratio changes from a factor 3–4 at 0.585 GeV to almost unity at 30 GeV.

ii There appears to be a significant (factors 2–4) angular dependence in the 200 MeV/c pion production cross sections between 45–168°.

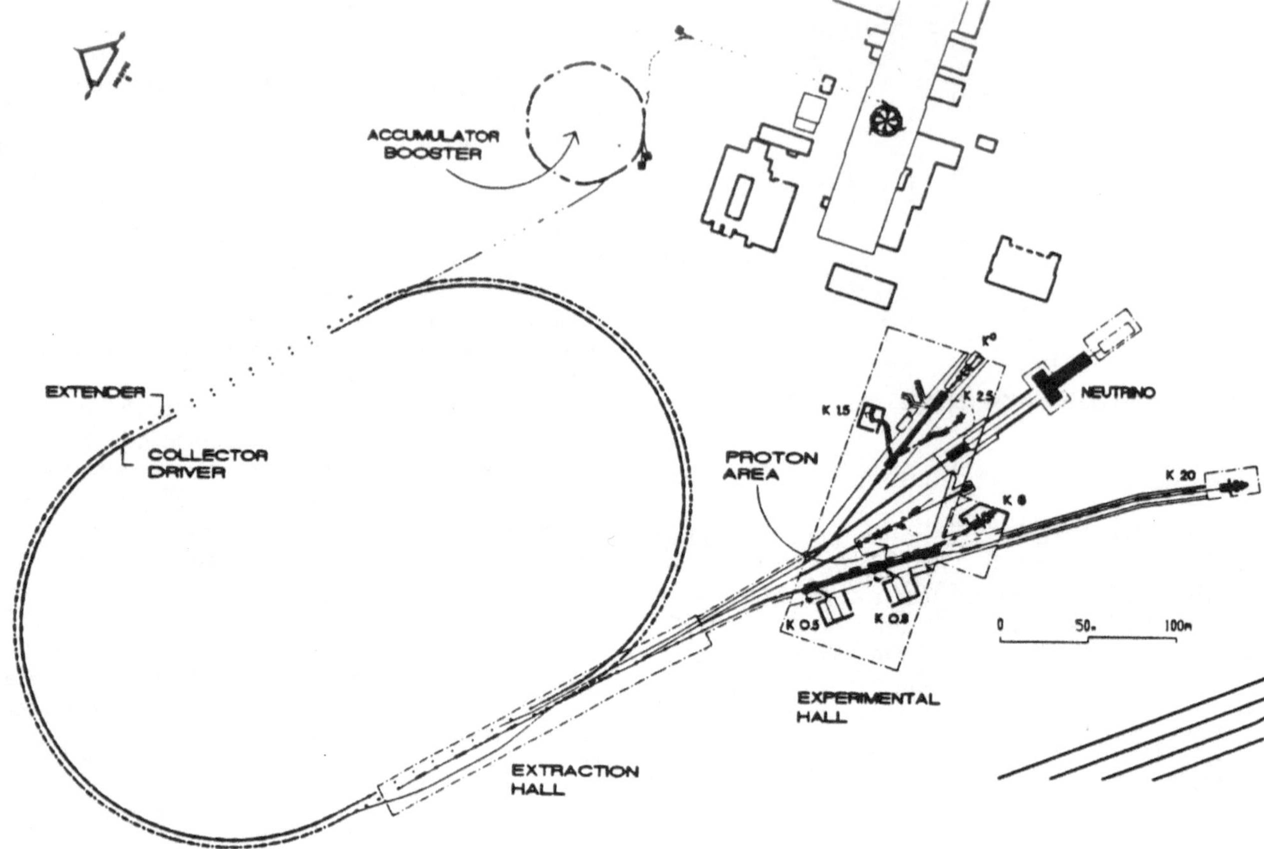

Fig. 2. Plan view of the proposed accelerator facilities

Table 1. Extracted beam parameters

Energy	30 GeV
Current	100 μA (6×10^{14} p/sec)
Fast Extraction	6×10^{13} ppp 3.5 μsec every 100 msec (10 Hz) rf microstructure 2 nsec every 16 nsec
Slow Extraction	80% - 90% duty cycle rf microstructure 2 nsec every 16 nsec can be dc can be split by 4 (continuously)
Polarized Beam	10 μA (6 $\times 10^{12}$ p/sec) 60% polarization
Booster	3 GeV 100 μA 1.2×10^{13} ppp 750 nsec every 20 msec (50 Hz)

iii For the same nuclear target, the 200 MeV/c production cross section at 90° is $\times 10$ for π^+ and $\times 30$ for π^- (30 GeV/0.5 GeV).

iv Data are not complete enough and not sufficiently consistent to have confidence that the optimum choice of secondary channel geometry can be made — a more complete study is therefore warranted before choosing channel geometry.

v Surface muon yields should be measured.

Blackmore comments that the present data indicate that a 100 μA, 30 GeV beam on a 7.5 cm tungsten target should be equivalent to 18 mA of a 585 MeV beam on a 6.0 cm carbon target (for π^- from the same channel at 90°). This therefore represents an order of magnitude increase in intensity over that expected from PSI operating at maximum anticipated proton currents. Kossler[6] has reported disappointing results from BNL, however, a number of technical problems could provide an explanation. Further cross section measurements must be done to verify the encouraging indications of the existing data.

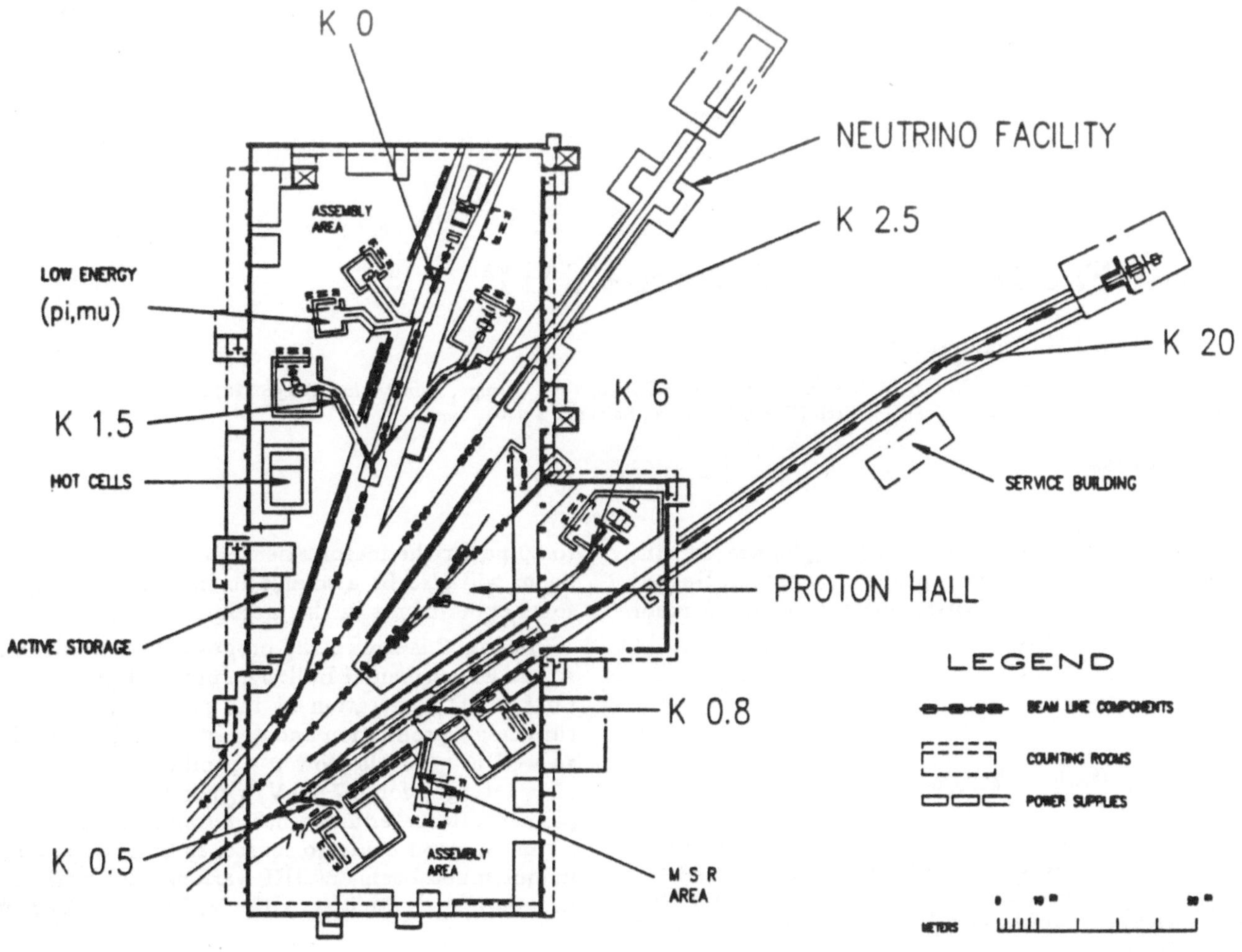

Fig. 3. Proposed layout of the experimental areas

4 Conclusions

The TRIUMF KAON facility has been designed, costed and is optimistically awaiting government approval for construction funds. This facility will open opportunities for a broad range of physics experiments and in particular promises to be a very intense source of low energy pions and muons. As such, it may well represent the muon physics machine for the start of the next century.

References

1. KAON Factory Study – Accelerator Design Report, TRIUMF (1990).
2. KAON Factory Study – Science and Experimental Facilities, TRIUMF (1990).
3. D. Frekers, D.R. Gill, J. Speth (Eds.): Proc. Int. Meeting Physics at KAON, Berlin: Springer-Verlag 1990.
4. J.H. Brewer: Proceedings of a Workshop on Science at the KAON Factory, vol. 2, sess. 9, D.R. Gill (Ed.), Vancouver: TRIUMF 1990.
5. E.W. Blackmore: TRIUMF report TRI-DN-91-K171 April 1991.
6. W.J. Kossler: Proc. Workshop on Science at the KAON Factory, vol. 2, sess. 9, D.R. Gill (Ed.), Vancouver: TRIUMF 1990.

This article was processed using Springer-Verlag TEX Z.Physik C macro package 1991
and the AMS fonts, developed by the American Mathematical Society.

Z. Phys. C - Particles and Fields 56, S296-S300 (1992)

Zeitschrift für Physik C Particles and Fields

Muon Facility Plans Towards JHP

K. Ishida[1,2], and K. Nagamine[1,2]

[1] Meson Science Laboratory, Faculty of Science, University of Tokyo, 7-3-1, Hongo, Bunkyo-ku, Tokyo 113, Japan
[2] Institute of Physical and Chemical Research (RIKEN), Wako, Saitama 351-01, Japan

31-December-1991

Abstract. The future muon facility planned for the future Japanese accelerator project (Japanese Hadron Project) and related activities towards advanced muon facilities are described.

1 Japanese Hadron Project (JHP)

The Japanese Hadron Project (JHP) aims to investigate many unexplored facets of matter utilizing various kinds of unstable particle beams such as mesons, neutrons, and short-lived nuclei. The JHP, formally proposed by the Institute for Nuclear Study, University of Tokyo, is now under discussion as the most promising Japanese accelerator project in the nearest future. In Fig. 1 is shown the proposed layout of the JHP. The accelerator complex consists of a 1 GeV proton linac and a compressor/stretcher ring. A high-intesity 1 GeV proton beam is supplied to three experimental arenas, namely Meson Arena, Neutron Arena and Exotic Nuclei Arena. It is planned to locate these facilities on available land to the south side of the National Laboratory for High Energy Physics (KEK) at Tsukuba. Also, the improvement of the existing 12 GeV proton synchrotron (Kaon Arena) at KEK is planned by the injection of a high-quality proton beam from the 1 GeV linac. The project is an extension of the existing project which were originally developed by the members composing the JHP. The pulsed muon project at the Meson Science Laboratory, University of Tokyo, located at KEK (UT-MSL/KEK) [1] is one of the typical examples.

The principal parameters of the proton linac are energy 1 GeV, average current 400 μA, repetition rate 50 Hz, pulse width 400 μs, and total length 500 m. The compressor/stretcher is a ring which shapes time structure of the beam. Receiving a beam of 400 μs in time length from the proton linac, the compressor/stretcher ring delivers a 200 ns pulse of proton beams for the neutron scattering facility and a short pulsed beam down to 20 ns for the meson science laboratory. A continuous beam will also be available to meet the requirement of some experiments in the meson laboratory.

In Fig. 2 is shown the proposed layout of the Meson Arena Experimental Hall. The proton beam will be used for keV μ^+ generation at the production target, MeV surface μ^+ production and decay μ^+ and μ^- production, as well as possible slow μ^- production. There will be also a target for 0.7 GeV/c pions for high resolution pion spectrometer and a beam dump used as a source of a time-sepated neutrino source. In order to realize these unique muon beams at JHP various pilot works are now in progress [2]. Such activities will be described below.

2 Activities towards JHP

2.1 Ultra slow μ^+ channel

It is a common dream of muon scientists to have an intense muon beam with advanced features such as ultra-low energy and a small spot size. There have been two proposals for slow muon beam production: a) the brightness-enhancement method, using reemission of the thermal muonium from the material surface after stopping conventional muons inside specially selected materials [3]; b) the beam cooling method, using electromagnetic confinements and acceleration/deceleration after detecting the phase space of the injected conventional muons. In the following, we would like to concentrate on the brightness-enhancement method proposed by the UT-MSL group. The beam cooling method proposed by Taqqu *et al* is now extensively under developement at PSI. We have been considering the following three steps towards the realization of the intense slow μ^+ source: 1) generation of thermal muonium after stopping μ^+ in special materials, which was initiated by the succesful original experiment done at UT-MSL/KEK on production of thermal Mu in vacuum from hot tungsten foil [3]; 2) production of slow μ^+ by ionization of thermal Mu, which was carried out in an experiment at UT-

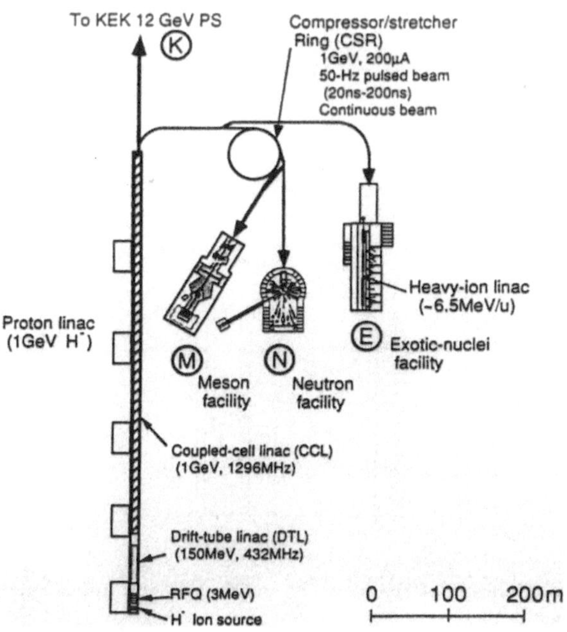

Fig. 1. Proposed layout of the Japanese Hadron Project

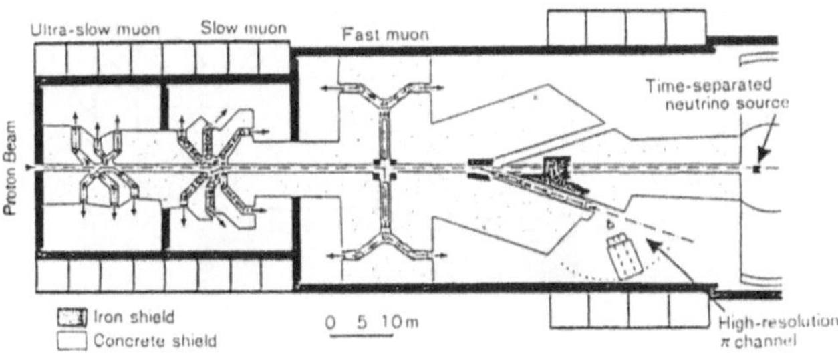

Fig. 2. Proposed basic structure of the Meson Arena Experimental Hall of JHP

MSL/KEK by using 3-photon ionization method (1S → 2S → unbound) [4]; and 3) further increase in slow μ^+ production by placing these steps at the primary proton beam [2,3]. Shematic picture of this idea is shown in Fig. 3. In this project, two photon ionization (1S → 2P → unbound) was adopted as the more effective ionization method. The project has been organized as the extension program of the UT-MSL/KEK. The new laboratory has already been completed and the dedicated proton beam was succesfully delivered to the production target. The details of the project will be seen in a separate contribution [5].

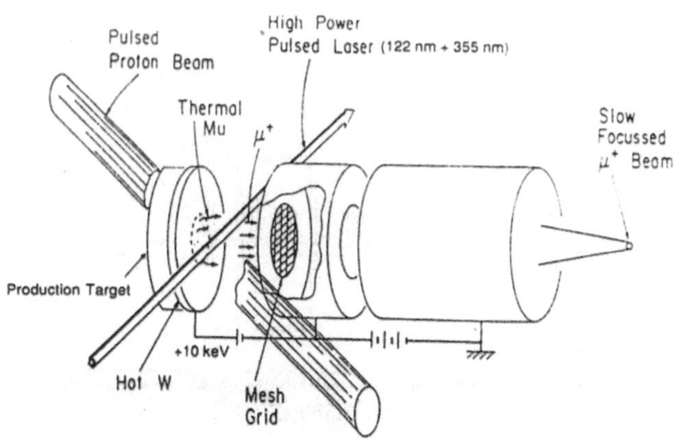

Fig. 3. Shematic picture of the intense thermal Mu production and its ionization at the production target of the primary proton line

The absolute yield of slow μ^+ (N_μ) can be expressed by yield of muons stopped in W foil ($N_{\mathrm{stop}\mu}$), thermal Mu production efficiency ($\varepsilon_{\mathrm{Mu}}$), ionization efficiency of Mu ($\varepsilon_{\mathrm{ion}}$) and collection efficiency ($\varepsilon_{\mathrm{coll}}$) as follows;

$$N_\mu = N_{\mathrm{stop}\mu}\varepsilon_{\mathrm{Mu}}\varepsilon_{\mathrm{ion}}\varepsilon_{\mathrm{coll}}. \quad (1)$$

where $N_{\mathrm{stop}\mu}$ can be represented by proton current (I_{p}), cross section of the production of π^+ (σ_π), number of target nuclei (N_{T}), and π^+ to μ^+ conversion inside the target ($\eta_{\pi\mu}$)

$$N_{\mathrm{stop}\mu} = I_{\mathrm{p}}\sigma_\pi N_{\mathrm{T}}\eta_{\pi\mu}. \quad (2)$$

The material and the thickness of the production target were carefully selected. Thick tungsten is not favorable as the pion production target, since the multiple scattering has to be kept low enough so that the proton beam can be transported to the downstream targets and to the beam dump. Also the yield of pions is low for high-Z nuclei target for a given target thickness, since the cross section increases only as $Z^{1/3}$ [6] while the number of target nuclei decreases as A^{-1}. On the other hand, we limited the thickness of tungsten to 100 μm for optimized thermal Mu production. So we placed a boron-nitride (BN) target for the production of pions just in front of the tungsten target (see Fig. 3). Boron-nitride was chosen because of its light element composition and of its high melting point, which is absolutely necessary in order to be placed close to the hot tungsten.

The expected number of muons stopped in W foil was calculated by a Monte-Carlo method taking both processes into account. The number is shown in Fig. 4 for various thicknesses of the BN target for the proton energy of 500 MeV and the proton current of 6 μA. Pions may be converted to surface muons in the BN target and then the muons are stopped in the W foil, or the pions stop in W foil and then the generated muons are stopped

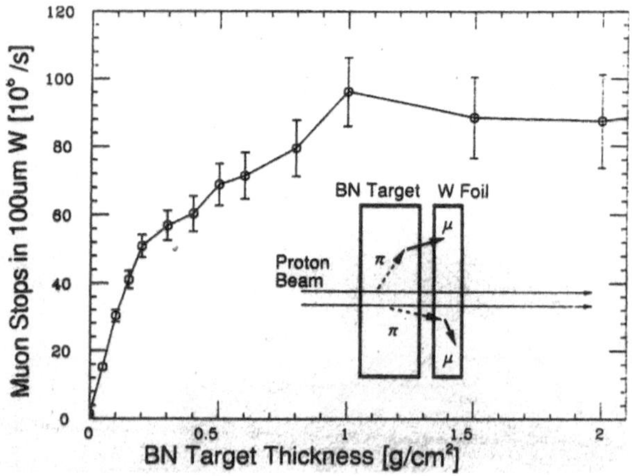

Fig. 4. Calculated number of muon stops in 100 μm W target from the pions produced in BN target of various thicknesses by 500 MeV proton beam with 6 μA intensity.

(see the inserted figure of Fig. 4). The latter process was found to contribute about 70 % of the total muon stops. The contribution from pions produced in tungsten was calculated to be 1.5 x 10^6 /s and is negligible. We have finally chosen 1 mm thickness for the BN target (0.225 g/cm^2) as a compromise between the maximum muon yield and the minimum proton scattering.

By using the above-mentioned number for muon stops and other reasonable numbers ($\varepsilon_{\mathrm{Mu}}$ = 0.04 [3], $\varepsilon_{\mathrm{ion}}$ = 0.05 [7], and $\varepsilon_{\mathrm{coll}}$ = 0.2), we expect the yield of slow μ^+ as 2 x 10^4 /s at the present UT-MSL/KEK. This can be extrapolated to 3 x 10^5 /s when the system is installed at JHP with I_p = 100 μA.

2.2 Generation of keV μ^- via reemission after muon catalyzed fusion

The production of slow μ^- beam is more difficult than slow μ^+, since all the injected μ^- in condensed matter are captured to the atomic nuclei to form the muonic atom. In order to overcome this difficulty, a new idea was proposed by K. Nagamine based upon the phenomena of muon catalyzed fusion [8]. Fig. 5 shows the concept of the slow μ^- generation. The principle is as follows: 1) the disappearance of the core nuclei at the instant of fusion reaction can produce μ^- of around 10 keV; 2) this production can be repeated up to 150 times in D/T mixture during μ^- lifetime, resulting in a significant fraction of μ^- to be reemitted from the target surface. One remarkable improvement idea was produced with the help of G.M. Marshall (TRIUMF). When a μ^- is stopped in a thick (~mm) H_2 layer containing small concentration of D_2 or T_2, rapid transfer of the muon occurs from μp to μd or μt. During slowing-down of the μd or μt, the Ramsauer effect helps μd or μt atoms to penetrate almost freely through thick H_2 layer once the energy reaches 3 eV. If the surface is coated with thin D_2(/T_2) layer, slowing down and successive formation

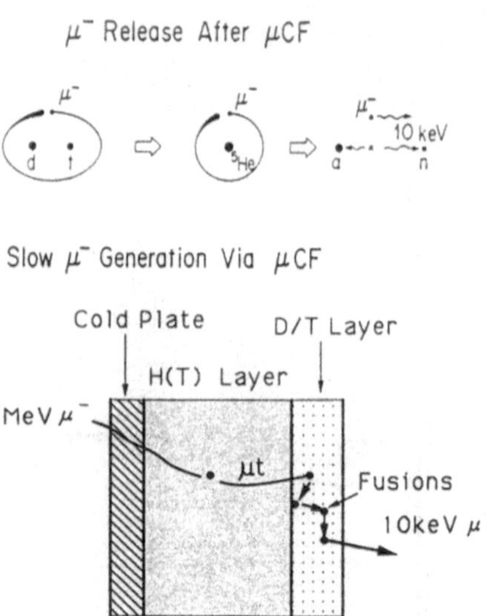

Fig. 5. Concept of the slow μ^- production via muon catalyzed fusion

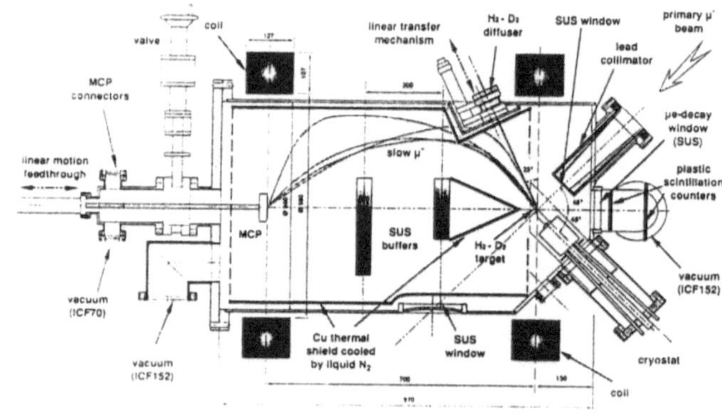

Fig. 6. Layout of the slow μ^- production experiment

of muonic molecules will occur in this thin layer followed by nuclear fusions and emission of slow μ^-.

Preliminary experiments using Ramsauer assisted ddμ process are being carried out by K. Nagamine, P. Strasser *et al* at UT-MSL and at TRIUMF. Fig. 6 shows the device used for the production and detection of slow μ^-. Conventional negative muons (40 MeV/c) are stopped in a layer of 2 mm thick solid hydrogen (H_2 with 1000 ppm D_2) at 3 K. The formation of hydrogen layer is monitored by detecting high-energy electrons from the decay of muons stopped in H_2. The keV μ^- is expected to be collected using an axial magnetic field spectrometer and to be detected by a micro channel

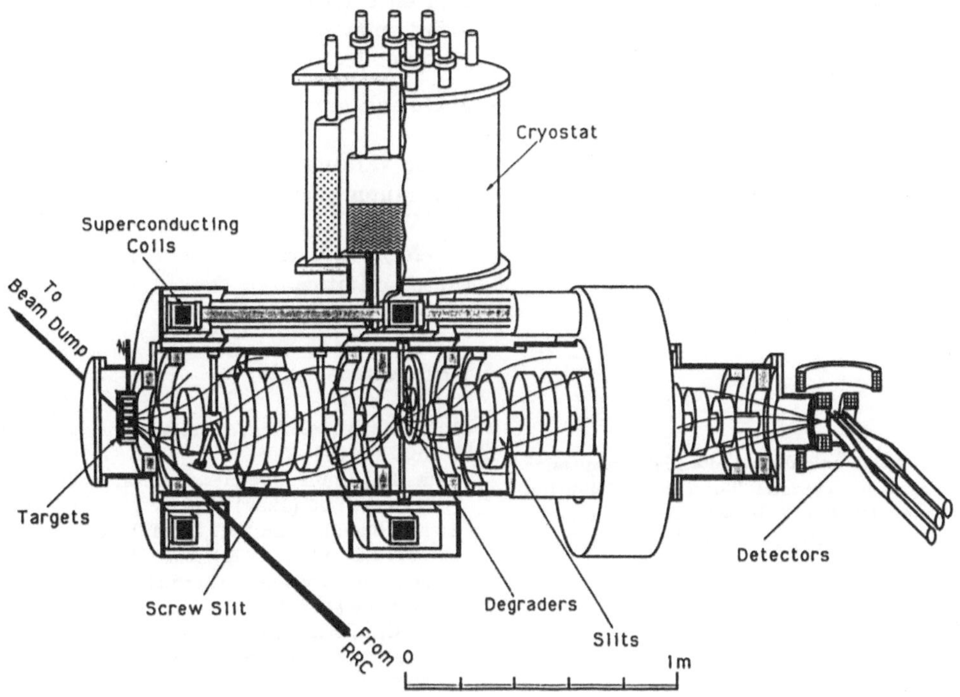

Fig. 7. Schematic view of the RIKEN surface muon beam channel

plate (MCP). This spectrometer has a wide collection efficiency over a wide momentum range. Although preliminary result has been obtained for various thicknesses of the thin coating D_2 layer, more statistics seems to be necessary before conclusive result can be obtained.

2.3 Large solid-angle injector

Intensity of the conventional decay muon beam may be increased by one order of magnitude with a high acceptance pion injector composed of axisymmetric superconducting coils. In an axisymmetric magnetic field, a particle emitted from a target on the axis makes helical motion and crosses the axis again. If a suitable field distribution is chosen, most of the particles are made to focus on a small point.

A beam course utilizing this principle for surface muons was constructed at RIKEN Ring Cyclotron (RRC) of the Institute of Physical and Chemical Research [9]. The schematic view of the beam course is shown in Fig. 7. The axial field of maximum 1.0 T is produced by three superconducting coils. The beam channel covers an emission angle between 35 to 47 degrees, the solid angle of which amounts to 600 msr. This channel has been used to investigate surface muon production from heavy-ion collisions at subthreshold energies for pion production (<280 MeV/nucleon), since the collective effect of the nuclei is considered to enhance the pion production. Even with a beam of ^{14}N at an energy of 135 MeV/nucleon, a substantial number of muons were detected (1000/μA) owing to the high collection efficiency of the beam course.

This focusing principle may be applied to a pion injection system of a decay muon beam channel with a superconducting solenoid $\pi\mu$ decay section. Main advantages are: 1) the large acceptance for pions compared with that of conventional systems, 2) simultaneous injection of π^+ and π^- into the solenoid, resulting in simultaneous generation of μ^+ and μ^-.

One example of the design calculation is shown in Fig. 8 [10]. The calculation was done by solving the equation of motion three dimensionally under the magnetic field calculated by TRIM code . The axial magnetic field is shown in the lower part of the figure, trajectories of the pions surviving as muons are shown in the upper part. The pion momentum was set to 150 MeV/c, the pion source was a square of 2 cm x 2 cm. The time and the emission angle of $\pi\mu$ decay was Monte-Carlo generated.

The calculated yield of transmitted muons is shown in Fig. 9. The vertical scale is normalized by the total number of 150 MeV/c pions generated isotropically from the target; the horizontal axis shows the muon emission angle at the solenoid exit. The number of transmitted muons originating from pions with a large emission angle in the range of 10-45 degrees ("large solid angle injection") is compared with that from pions with a small emission angle of 0-10 degrees ("small divergence injection"). The latter case, which covers 100 msr solid angle

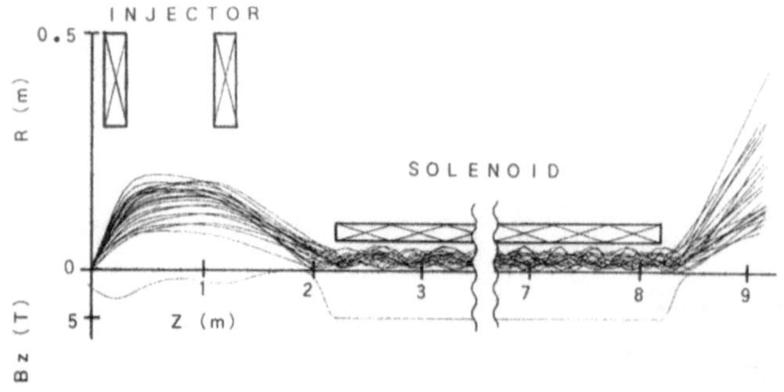

Fig. 8. An example of the design calculation of a high acceptance pion injector for the decay muon channel. The field distribution is shown in the lower part of the figure, trajectories of the pions surviving as muons are shown in the upper part of the figure.

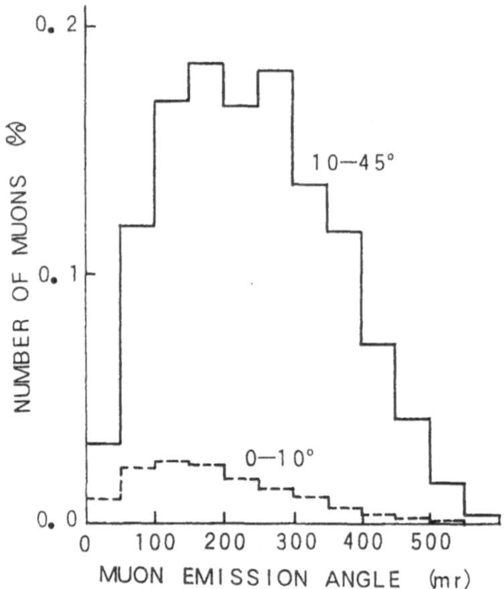

Fig. 9. The angular distribution of muons at the exit of the solenoid decay section. The solid line is for the muons originating from pions with emission angle in the range of 10-45 degrees. The broken line is for those from pions in the range of 0-10 degrees.

near axis, was intended to simulate the small divergence injection with conventional systems. It is clearly seen from Fig. 9 that the number of transmitted muons in "large solid angle injection" is about an order as large as in "conventional injection". It may be possible to design an even further intense muon channel by using the axisymmetric system for both the pion injector and the muon collector.

3 Conclusion

These various activities will help realizing next generation muon beams at JHP.

We would like to thank many people associated with UT-MSL and RIKEN who are joining various parts of these activities.

References

1. K. Nagamine: Pulsed μSR facility at the KEK Booster. Hyp. Int. 8 (1981) 787
2. K. Nagamine: Japanese hadron facility plans and future μSR. Hyp. Int. 65 (1990) 1149
3. K. Nagamine and A. P. Mills, Jr.: Los Alamos Report LA-10714-C, p.216 (1986); A. P. Mills, Jr., J. Imazato, K. Nagamine et al: Generation of thermal muonium in vacuum. Phys. Rev. Lett. 56 (1986) 1463
4. S. Chu, A.P. Mills, Jr., A. Yodo, K. Nagamine et al: Laser excitation of the muonium 1S-2S transition. Phys. Rev. Lett. 60 (1988) 101
5. K. Nagamine: Slow μ^+ and μ^- beam production and next generation of muon science. In: Proc. of INS Int. Symp. on Physics with High-Intensity Hadron Accelerators, p.79, Singapore: World Scientific 1990; K. Nagamine et al: this volume
6. J.F. Crawford et al: Measurement of cross sections and asymmetry parameters for the production of charged pions from various nuclei by 585-MeV protons. Phys. Rev. C 22 (1980) 1184
7. K. Kojima: private communication
8. K. Nagamine: Ultra-slow negative muon production via muon catalyzed fusion. Proc. Japan Academy 65 (1989) 225; K. Nagamine, P. Strasser and K. Ishida: Proposal for slow μ^- production via muon catalyzed fusion. Muon Catalyzed Fusion 5/6 (1990/1991) 371.
9. K. Ishida et al: Surface muon production in medium energy heavy ion reaction at RIKEN. Hyp. Int. 65 (1990) 1159; K. Ishida, T. Matsuzaki and K. Nagamine: Nucl. Instrum. Meth. in press
10. K. Ishida, T. Matsuzaki and K. Nagamine: Large solid angle superconducting coil and solenoid muon channel for future μCF experiment. Muon Catalyzed Fusion 5/6 (1990/1991) 467.

This article was processed using Springer-Verlag TEX Z.Physik C macro package 1991
and the AMS fonts, developed by the American Mathematical Society.

Z. Phys. C - Particles and Fields 56, S301-S305 (1992)

Conference summary

Vernon W. Hughes

Yale University, New Haven, Connecticut 06520, USA

9-May-1992

It's late and we have heard many talks today, so I'm sure you don't want some serious attempt by me to present a summary of our Workshop on The Future of Muon Physics.

The best talk I ever heard under the title of a conference summary was by Lev Okun at the International Symposium on Lepton and Photon Interactions at High Energies at Bonn in 1981. He didn't try to review or summarize the material presented at the conference but rather choose to discuss: What is problem No. 1 in particle physics?

I would like to read the first few paragraphs of his talk.

What is the Problem No.1?

"There are several competitors for this title. Of course, the problem of confinement is very fundamental. It is imperative to understand the structure of the QCD vacuum with its quark and gluon condensates. However, it seems that in spite of many unsolved intriguing puzzles we have already past the QCD crest: we know the Lagrangian.

Electroweak gauge bosons if not observed at CERN in the near future may become problem No. 1, but I hope they are at the right place.

Grand unification is exciting, but at the moment it has not so much connections with our everyday particle physics. After all, proton decay may be too slow to be detectable, and there is no reliable estimate of the abundance of relic magnetic monopoles on the earth, moon, planets and in cosmic rays.

Superunification (including gravity) is fascinating, but even in our dreams we cannot hope to run experiments on a Planck mass accelerator. The situation may change drastically as soon as very heavy magnetic monopoles are discovered. Their annihilation may bring us quite close to the experiments near the Planck threshold. But meanwhile supergravity does not look as a problem No. 1 for experimentalists.

I am aware that many physicists would place on the top of the list the so called preons, hypothetical structure elements of leptons and quarks, called by many other names. When and if discovered, preons will represent a major step on our way into the nature of matter. However, they do not look ripe enough.

It seems to me that the problem No.1 of high energy physics are scalar particles. The search for these particles is extremely important mainly because of their vital role in symmetry breaking. The whole picture of the physical world consists of two parts, which are complementary like yin and yang brought in another context into quantum physics by Niels Bohr: Here the yang comprises the principles of local symmetry which could be symbolized by the gauge derivative D. It represents the kinetic terms of the Lagrangian and interactions with the (and of the) gauge fields. The yin, which is not less important, comprises symmetry breaking which gives masses M to various particles including the gauge particles.

It seems now that the way to the understanding of the symmetry breaking inevitably goes through the land of scalars: scalarland. Fundamental scalars protect the renormalizability of the theory by moderating the cross-section growth and hence, the loop divergencies. They give masses to all particles. They violate CP- and may be, even P-invariance. There are theoretical models, which contain no fundamental scalars. In such models nevertheless tightly bound spinless particles inevitably appear.

At present the scalarland exists only in the dreams of theoreticians, who describe it in many ways, which are quite far from being self consistent. The aim of this talk is to urge experimentalists and accelerator builders to join their efforts in discovering this land, which lies below and not far above 1 TeV."

I would say that this analysis applies today for particle physics 10 years later. The SSC and LHC are aimed primarily at understanding the Higgs sector of scalarland. Our standard theory requires that a Higgs particle be found with a mass no greater than about 1 TeV; otherwise there is a crisis for the theory.

How might Okun's viewpoint have been applied in the mid-1940's? At that time the quantum field theory of electrodynamics was the leading advanced theory. Heitler's book the **Quantum Theory of Radiation** was the standard authoritative presentation. He discussed how the theory evaluated to lowest order in perturbation theory gave correct answers in agreement with experiment. But evaluated to next order the calculations diverged and the answer was ∞. Hence Heitler concluded that we didn't have any real theory at all. In other terminology the tree level diagram calculations gave finite answers but the loop diagram calculations diverged.

As you know the breakthrough came through the experimental discoveries of the Lamb shift in hydrogen and of the anomalous g-value of the electron in precision atomic beam spectroscopy experiments at Columbia. Before Lamb did his experiment there had been some experimental indication from precision optical spectroscopy of an energy splitting between $2^2S_{1/2} - 2^2P_{1/2}$ states and also there was the Uehling estimate of vacuum polarization. I think it is probably fair to say that Willis Lamb and Poly Kusch were intelligently addressing problem No. 1 of their day.

Actually there were three relevant precise atomic beam spectroscopy experiments done at Columbia at that time (1947). The first published result was $\Delta\nu$(H) (Nafe, Nelson, Rabi), not explicitly motivated by problem No.1. The new method of molecular beam magnetic resonance for precision radiofrequency and microwave spectroscopy was applied for the first time to measure $\Delta\nu$(H). The experiment achieved an accuracy of about 4 ppm (parts per million) giving $\Delta\nu$(H)=1420.410(6) MHz. This value disagreed with the theoretical value of 1416.97(54) MHz from the Fermi theory by 2 parts in 10^3 and was the first reported discrepancy, soon understood as due to a virtual radiative correction. Soon thereafter the Lamb shift in hydrogen was discovered. There was some initial theoretical confusion in attempts to understand the discrepancy between the experimental value for $\Delta\nu$(H) and the value predicted by the Fermi theory, until a correct calculation of the recoil correction m/M was done and Breit suggested that the discrepancy might be due to an anomalous spin magnetic moment for the electron. This suggestion stimulated the experiment of Kusch and Foley in which the anomalous spin magnetic moment of the electron was discovered through measurements of the Zeeman effect in medium Z atoms and reported several months later. There followed rapidly Schwinger's explanation of the anomalous spin magnetic moment of the electron as due to a virtual radiative correction whose value was a fractional change in g_s by $\alpha/2\pi$. These three complementary experiments thus provided the experimental basis for the solution of Problem No.1 in the late 1940's.

There is clearly a dramatic difference in scale of experiments to approach problem No.1 in 1945 when a small human scale atomic physics laboratory and some 50 K sufficed, and problem No.1 in 1991 when the inhuman scale of SSC or LHC and some 5 to 10 billion are believed necessary.

For our Workshop on The Future of Muon Physics it doesn't seem appropriate to try to identify a problem No. 1. One of the beauties of our Workshop is the breadth of the physics covered which extends from TeV muons at FNAL and high energy neutrinos to condensed matter and eV energy atomic physics. However I may perhaps try to identify several important problems in the different subfields.

Many of the topics discussed in our Workshop refer specifically to the standard model. Indeed from a general viewpoint, as far as we know, the current standard theory of particle physics adequately covers condensed matter physics as well, even though μSR studies in high temperature superconductors don't need to worry about the Higgs particle.

The standard theory itself is not completed until we find the top quark and study the Higgs sector. Also we need to see if CP violation is understandable in the standard theory. Thus far the theory is remarkably successful. A dominant current viewpoint is to search for a violation of the standard model indicating "new physics", and surely such a discovery would be most significant and exciting and indeed can be expected. However, Scheck(in this volume) advocated that we continue to deepen our theoretical understanding and experimental tests of the standard model and that we regard any fine experiment which is explained by the standard theory as a significant success for physics and not as a failure because we haven't found a violation of the standard model. I very much sympathize with this view.

A number of current and future experiments on the muon, muonium and simple muonic atoms aim at more precise determinations of the electromagnetic properties of the muon and at more sensitive tests of quantum electrodynamics and indeed also of the electroweak interaction. These include the muon g-2 value, muonium energy levels and laser spectroscopy of simple muonic atoms.

The measurement of the g-2 value of the muon has been an active topic since the discovery of parity nonconservation in the weak interactions in the late 1950's. The value of $g_\mu - 2$ has provided strong evidence that the muon is a lepton behaving as a heavy electron. F. Farley (in this volume) gave an illuminating talk about the early CERN $g_\mu - 2$ experiments and B.L. Roberts (in this volume) reported on the status of the ambitious new BNL experiment which aims to measure $g_\mu - 2$ to a precision of 0.35 ppm, an improvement in accuracy by a factor of 20 compared to the most recent CERN experiment. With this precision the experiment will determine the weak interaction contribution to $g_\mu - 2$ arising from loop diagrams with Z and W bosons and indeed will provide a most sensitive test of the standard model capable of observing contributions predicted by speculative theories beyond the standard model. (T. Kinoshita, in this volume)

Muonium as a two-lepton system is an ideal atom for the study of quantum electrodynamics. Because leptons are structureless as far as we know muonium is simpler than hydrogen in which the proton has a structure due to the strong interactions which limits significantly the precision with which the energy levels of hydrogen can be calculated. This relative simplicity of muonium compared to hydrogen is exemplified in the hyperfine structure interval $\Delta\nu$ in the ground state. (D. Yennie, in this volume) For hydrogen proton structure contributes at the level of 40 ppm to the theoretical value for $\Delta\nu$ and the accuracy of $\Delta\nu_H$ (theory) is limited to several ppm due to lack of knowledge of the spin-dependent proton polarizability. (Interestingly this contribution of proton polarizability can in principle be evaluated from the spin-dependent structure functions measured in deep inelastic polarized lepton-proton scattering. These measurements for muon-proton scattering are discussed in this volume by R. Voss.) On the other hand, for muonium the theoretical value for $\Delta\nu$ in the ground state is known to 0.4 ppm and is limited dominantly by uncertainty in our knowledge of the fundamental constant μ_μ/μ_p. The theoretical terms involving higher orders in α and m_e/m_μ arise from radiative and recoil contributions and are now known to 0.2 ppm; calculations in progress of additional higher order terms should reduce the uncertainty from the theoretical terms to about 10 ppb. For muonium $\Delta\nu(expt)$ is known to 4 parts in 10^8 whereas for hydrogen $\Delta\nu$ has been measured to better than 1 part in 10^{12} and indeed provides a basic frequency standard.

Several precision experiments on muonium were discussed. A new experiment is getting started at LAMPF to measure $\Delta\nu$ and the Zeeman effect or μ_μ/μ_p in the ground state to precisions of 10 ppb and 50 ppb respectively. The general method of this microwave magnetic resonance experiment is that employed in earlier measurements, but the use of line-narrowing with a chopped muon beam, of a higher intensity μ^+ beam from which e^+ have been removed by an $\vec{E}$ x $\vec{B}$ separator, and of a magnetic field of better homogeneity and larger value should make possible the improved precision. If the goals in precision are achieved, a value for the fine structure constant α accurate to about 25 ppb can be obtained. This accuracy is comparable to that presently obtained from condensed matter determinations and will allow a test of the theoretical value for the electron g-2 value without recourse to condensed matter theory. Alternatively the new muonium measurements can provide an unusually sensitive test of the QED theory of the two-body relativistic bound state problem.

A second important and on-going muonium experiment is the measurement of the 1S $\rightarrow$ 2S transition by a laser spectroscopy method discussed by K. Jungmann (in this volume). Using the quite intense pulsed μ^+ beam provided by the 800 MeV proton synchrotron at the Rutherford-Appleton Laboratory, a value for this energy interval accurate to 3 parts in 10^8 has been measured. At present a value for the Lamb shift in the 1S state of muonium accurate to about 1% and in agreement with the theoretical value of $S = 8091$ MHz is provided by this measurement. Further improvement in the experimental accuracy can be anticipated, particularly if a higher intensity source of pulsed muons becomes available, and then the 1S-2S interval in muonium would provide a precise value for the muon mass m_μ or of m_μ/m_e or, alternatively, a still better measurement of the Lamb shift in the 1S state. Since the precision of the Lamb shift measurement in H is about 10 ppm and since that is also the level at which theoretical uncertainty arises because of lack of knowledge about proton structure, the measurement of the Lamb shift in muonium must achieve an accuracy of 10 ppm to advance the test of Lamb shift theory.

A future field with great potential and importance but largely undeveloped as yet is laser spectroscopy of simple muonic atoms. The only experiment done as yet was the pioneering experiment of Zavattini et al. which measured the fine structure intervals in the n=2 state of the muonic helium ion, $(^4He\mu^-)^+$. The dominant contribution to the fine structure was due to electron vacuum polarization and the measurement confirmed the expected theoretical value to 0.2%. The full accuracy of the measurement could not be used to test the QED theory of electron vacuum polarization because of uncertainty about the radius of the ^4He nucleus, so the measurements have been used to determine the most accurate value for $< r^2_{^4He} >^{1/2}$.

This impressive, pioneering experiment in laser spectroscopy of muonic atoms has been clouded a bit by an interesting puzzle, or at least the development of the field has been slowed by this puzzle. In order that a resonance can be observed from a 2S state to a 2P state it is necessary that the 2S state of $(^4He\mu^-)^+$ have a minimum lifetime in the high pressure ($\sim$ 40 atm) He gas in which it is formed and studied. Subsequent to the experiment theoretical and experimental studies of the lifetime indicated that the 2S state would be quenched to the ground 1S state in a collision in the gas in a time much shorter than the required minimum value. A resolution of this problem may involve the formation of a multiatom structure at high pressures which shields the $2S(^4He\mu^-)^+$ from collision quenching.

It would be very important to measure precisely by laser spectroscopy a number of energy intervals in simple muonic atoms. These include for μ^-p the hyperfine structure interval in the n=1 state and the fine structure in the n=2 state. Proton structure contributes at the several percent level to these intervals and hence measurements could determine proton structure effects which would improve the precision of theoretical values for various energy intervals in H where very precise measurements have been made. Measurement of the fine structure intervals in the n=2 state of $(^3He\mu^-)^+$ will determine a very precise value for $< r^2_{^3He} >^{1/2}$. A measurement of the 3P to 3D transition in $(^4He\mu^-)^+$ would provide an important test of QED vacuum polarization theory substantially independent of ^4He nuclear structure effects.(L.. Braci, A. Vacchi, and E. Zavattini, in this volume)

These laser spectroscopy experiments on muonic atoms require high laser power because of the small size of muonic atoms and the small value of the muon magnetic moment. Hence pulsed lasers are required and correspondingly pulsed muon sources with high intensity are needed. At present the pulsed muon beam at Rutherford-Appleton Laboratory is the best such source; the proton storage ring at LAMPF could provide a still more intense source if a pulsed muon/neutrino beam is developed there.

Although no successful experiment has yet been done to study the electroweak interference in muonic atoms, despite the early recognition of the importance of the topic, J. Bernabeu (in this volume) emphasizes that the problem is still important, especially to study nuclear-spin dependent parity nonconserving effects. It is a worthy but difficult challenge to meet and we can note that it required about 12 years after the experimental possibility was recognized for studies of electronic atoms to make a contribution to the field of electroweak interactions, but the current results are of great importance.

It is clear of course that a deviation from QED or electroweak theory observed in any of these basic experiments which are extending the sensitivity of tests of the standard theory would be of unusual importance.

For the weak interactions the muon and the muon neutrino play a central role. The muon decay $\mu^+ \rightarrow e^+ + \nu_e + \bar{\nu}_\mu$ is a purely leptonic process and one of the simplest and probably the most extensively and carefully studied weak process. (W. Fetscher, in this volume.) The basic weak coupling constant, the Fermi constant G_F, is determined to a precision of 2 parts in 10^5 from the measurement of the μ^+ lifetime. Nine different μ decay experiments and measurement of the inverse μ decay, $\nu_\mu + e^- \rightarrow \mu^- + \nu_e$ determine coupling constants for a general weak interaction Hamiltonian and confirm thus far that the V-A form of the interaction of the standard theory applies. Future experiments are planned to determine more precisely the parameters ρ, ξ, δ and $\xi\prime$ and thus reduce the upper limits for exotic couplings beyond the standard model.

A most significant current problem is that of the mass of ν_μ. In the talk by R. Frosch (M. Daum et al. in this volume) we learned that a negative and hence unphysical value for $m^2_{\nu_\mu}$ is obtained from the combination of the measurement of p_μ from π^+ decay at rest and the reported mass values for m_μ and m_π. It is expected that the value for m_{π^-} obtained from a measurement of X rays from medium Z pionic atoms by crystal diffraction spectrometry is in error due to the effect of inner electrons in the pionic atoms. It should be possible to improve the precision of the measurement of m_π to the level of 1 ppm using a higher resolution crystal diffraction spectrometer-perhaps a flat rather than a curved crystal spectrometer - which could resolve the spectral lines associated with different configurations of inner electrons.

The basic muon capture process, whose generic reaction is $\mu^- p \rightarrow n\nu_\mu$, is being actively studied with some new experimental approaches to address several fundamental questions. Polarized gas targets, in particular ^3He, can be produced by laser-atomic techniques with polarizations of greater than 30% and at pressures up to 8 atm so that the reaction $\mu^- +^3 He_{pol} \rightarrow^3 H_1 + \nu_\mu$ can be measured to determine the important pseudoscalar coupling constant g_p.(P. Souder in this volume) Radiative muon capture on hydrogen is also being studied to determine g_p (W. Bertl et al., in this volume). Experiments have been done to search for heavy ν_μ neutrinos and are planned to test T-invariance in the μ capture reaction. (J. Deutsch, in this volume)

The search for as yet unobserved modes of muon decay in particular those that violate muon number conservation which is postulated in the present standard theory has been and continues as a very active field. (H. Walter and K. Jungmann, in this volume) The motivation is to discover new processes or set significant limits to speculative theories beyond the standard model. Processes being actively looked for include $\mu^+ \rightarrow e^+\gamma$, $\mu^+ \rightarrow e^+e^+e^-$, $\mu^-(Z) \rightarrow e^-(Z)$, and $\mu^+e^- \rightarrow \mu^-e^+$. Unfortunately, theoretical work although imaginative and important can offer very little guidance as to the level at which these decay modes might occur. (R. Mohapatra and P. Herczeg, in this volume).

Neutrino physics is of course a large and broad field extending over a large energy range up to the multi-GeV range. It has involved extensive experiments with multi-GeV neutrinos to study nucleon structure in deep inelastic scattering similar to those with muon-nucleon deep inelastic scattering discussed by R. Voss (in this volume). Now neutrino experiments emphasize the study of the weak interaction and questions beyond the standard model and also astrophysics aspects. At LAMPF with neutrinos from π and μ decay in the 100 to several hundred MeV range and with a large liquid scintillator neutrino detector H. White (in this volume) reported plans to study $\nu_\mu e$ elastic scattering to make a precision measurement of the Weinberg angle $\sin^2\theta_w$, to search for neutrino oscillations particularly of the type $\nu_\mu \rightarrow \nu_\tau$ and to study the strangeness content of the proton through $\nu_p p$ elastic scattering. Experiments which study neutrinos from the sun or other astrophysical source provide at present the most exciting suggestions of new physics such as neutrino oscillations (A.K. Mann, in this volume) and provide a strong stimulus for laboratory-based experiments.

High energy muons (100 GeV to 500 GeV) through deep inelastic muon-nucleon scattering experiments have provided our most detailed knowledge of the internal structure of the nucleon and important tests of perturbative QCD (D. Geesaman, in this volume). Relatively recently polarized muon-proton scattering provided data which violated an approximate QCD sum rule due to Ellis and Jaffe and gave the surprizing result that in the naive quark-parton model the quark spins contribute only a small fraction of the proton spin. As a consequence polarized deep inelastic muon-nucleon scattering will certainly be an active field in the future. (R. Voss, A. Schäfer in this volume)

Atomic and molecular collision processes involving muons are studied in connection with many of the experiments mentioned above. In particular, although not expected to have practical application for power production, the fascinating process of muon catalyzed fusion is most actively pursued by a world-wide group of physicists. (M. Leon, L. Ponomarev in this volume). The field involves a wide range of rather unusual atomic and molecular processes, and indeed using a deuterium-tritium mixture a high yield of about 150 neutrons per μ^- has been achieved.

The use of muons (both μ^+ and μ^-) for condensed matters studies by μSR is a well-established field. (A. Seeger and A. Schenck, in this volume) The range of problems is broad but recently work has emphasized heavy electron compounds such as are involved in high-temperature superconductors. Many different μSR techniques have been developed based on the various observables, and muon beams of different intensity and time structure can be useful.

The development and health of the field of muon physics has depended and will continue to depend heavily on the accelerator laboratories and the muon beams they provide. At this Workshop we heard of new ideas and plans for muon beams ranging from thermal energy to several hundred GeV.

At present in Europe, in the United States and Canada and in Japan there is a real danger of inadequate financial support to the intermediate energy laboratories where much of muon physics is done. Our field has been most significant and is clearly still rich, promising and very broad.

Indeed because of the breadth of the field spanning particle, nuclear, atomic and condensed matter physics there is not a well-defined cohesive constituency to support and argue for the facilities and support the field requires. The cost is relatively modest for the product produced and we must be alert whenever possible to make strong joint support for the needed resources.

Finally Gisbert zu Putlitz, Klaus Jungmann and I want to thank the participants very much for coming to the Workshop on rather short notice, for providing excellent talks and stimulating discussions and for working so hard.

This article was processed using Springer-Verlag TEX Z.Physik C macro package 1991
and the AMS fonts, developed by the American Mathematical Society.

Index of contributors